The Minerals, Metals & Materials Series

The Minerals, Metals & Materials Series publications connect the global minerals, metals, and materials communities. They provide an opportunity to learn about the latest developments in the field and engage researchers, professionals, and students in discussions leading to further discovery. The series covers a full range of topics from metals to photonics and from material properties and structures to potential applications.

More information about this series is available at http://www.springer.com/series/15240

Mohammad Shamsuddin

Physical Chemistry of Metallurgical Processes, Second Edition

 Springer

Mohammad Shamsuddin B.Sc. (Met. Engg.), M.Sc. (Met. Engg.), Ph.D. (Met. Engg.)
Former Professor and Head
Department of Metallurgical Engineering
Indian Institute of Technology
Banaras Hindu University
Varanasi, India

ISSN 2367-1181   ISSN 2367-1696   (electronic)
The Minerals, Metals & Materials Series
ISBN 978-3-030-58071-1   ISBN 978-3-030-58069-8   (eBook)
https://doi.org/10.1007/978-3-030-58069-8

This Springer imprint is published by the registered company Springer Nature Switzerland AG
The registered company address is: Gewerbestrasse 11, 6330 Cham, Switzerland

Foreword to the First Edition

The limited availability of textbooks in chemical and extractive metallurgy is evident from their paucity in different library websites. Hence, there is a need to enrich the extractive metallurgy library. In this field, hardly one or two books are published within a span of 5–10 years. Considering the requirement and circumstances, this book by Professor M. Shamsuddin, which discusses the physical chemistry of various steps involved in the extraction of different types of metals, is an important contribution in the field of chemical metallurgy. It is well known that the exploitation of many low-grade and complex ores/minerals has been possible in recent years by a thorough understanding of slag-metal reactions with the aid of thermodynamics and reaction kinetics. The fundamental principles of chemical metallurgy are based on physical chemistry that includes thermodynamics and kinetics. In general, textbooks on physical chemistry deal mainly with contents that are appreciated by the students of chemistry but are of less interest to metallurgists and chemical engineers. In this book, physical chemistry is presented for aspects concerning chemical metallurgy with appropriate examples drawn, as much as possible, from extractive metallurgical processes.

The physical chemistry principles that are key to extraction technologies play a decisive role in the development and improvement of processing methods. As a consequence, metallurgists and chemical engineers often face the problem in selecting the appropriate technique for treatment of the concentrate. In order to overcome such a challenging task, a sound knowledge of physical chemistry of different extraction methods is extremely useful. Since the chemistry of the extraction process varies according to the nature of the metal, which may fall under the categories of common, rare, reactive, or refractory, a comprehensive and collective treatment in one book is very much desired at the present time. Depending on the interest, one may further study the details in a book/monograph dealing with the particular metal. This is the main objective of this book.

This is a very special book for three reasons: Firstly, it includes discussions on physicochemical principles involved in different steps, namely, roasting of sulfide minerals, matte smelting/converting, reduction smelting, iron-and steelmaking, deoxidation, refining, degassing, leaching, purification of leach liquor, precipitation, cementation, etc., during extraction/production of not only common metals but also rare, reactive, and refractive metals by pyro- and hydrometallurgical methods. Secondly, it provides a number of worked-out examples in each chapter, which make understanding of the process easier. Thirdly, the author has systematically summarized and presented scattered information on physicochemical aspects of metal extraction from previously published books and journal articles.

The book will undoubtedly fulfill the need of students and teachers by providing information on the principles and methods of extraction of different metals—common, rare, reactive, and refractory—in one place. I am confident that the book will be in demand throughout the world by universities and institutes offering courses in metallurgy as well as chemical engineering and technology, and also by various metallurgical and chemical research laboratories. It will be more useful to students of metallurgical engineering specializing in chemical/extractive metallurgy, but the basic principles of various unit operations involved in extraction will also be appreciated by chemical engineering students.

In addition to his long tenure at Banaras Hindu University, Professor Shamsuddin has had diversified interactions with faculty members of two premier institutions, namely the Department of Metallurgical Engineering, University of Utah, Salt Lake City, and the Department of Materials Science and Engineering, Massachusetts Institute of Technology, Cambridge, on various aspects of metal extraction, thermodynamics, and kinetics. I have no reservation in stating most strongly that this book *Physical Chemistry of Metallurgical Processes* will achieve a high standard in the field of chemical/extractive metallurgy and be appreciated by metallurgists and chemical engineers.

Fellow, The Minerals, Metals &   H. Y. Sohn
Materials Society; Professor,
Departments of Metallurgical
Engineering and Chemical Engineering,
University of Utah,
Salt Lake City, UT, USA
http://www.metallurgy.utah.edu

Preface to the First Edition

I had been planning to write this book *Physical Chemistry of Metallurgical Processes* based on my experiences of teaching a graduate course with the same title in the Department of Materials Science and Engineering, Massachusetts Institute of Technology, and association/interaction with the faculty members in the Department of Metallurgy and Metallurgical Engineering, University of Utah, Salt Lake City, during my 3 years (1978–1981) of visit to the United States. But, while taking account of the rapid development of theoretical knowledge in recent years in gas–solid (roasting and reduction) and liquid–solid (leaching and precipitation) reactions, I was in a dilemma regarding the extent to which mathematical expressions should be incorporated in the book. After spending a lot of time on the mathematical contents while teaching the same course in the Department of Metallurgical Engineering, Banaras Hindu University, I concluded that it should be possible to discuss the new developments in a satisfactory manner without going into the use of advanced mathematics by giving more emphasis on thermodynamics, which brings out more convincing evidence as compared to kinetics involving complex expressions. This decision has helped me in the preparation of a book of reasonable size covering various process steps in production of different types of metals, namely common, reactive, rare, and refractory.

In the past, during 1950–1970, most textbooks on extractive metallurgy described processes for production of different metals emphasizing the technology rather than the basic principles involved. The physical chemistry of the processes has been restricted to mere listing of chemical reactions expected to be taking place. However, the book entitled *Physical Chemistry of Iron and SteelMaking* by Professor R. G. Ward published in 1962 has been an exception. With rapid increase in the number of extraction processes on the industrial scale, it became difficult to bring out the comprehensive idea of all metallurgical fundamentals for the development of future technology. During 1970–1995, Professors C. Bodsworth, W. G. Davenport, J. F. Elliott, F. Habashi, E. Jackson, J. J. Moore, R. H. Parker, R. D. Pehlke, T. Rosenqvist, H. Y. Sohn, M. E. Wadsworth, and Dr. E. T. Turkdogan paid due attention to the physicochemical aspects of metallurgical fundamentals in their books. I have benefited from their texts while preparing this manuscript and owe them most sincerely. In addition, I have tried to collect information on this subject from different journals and proceeding volumes. For the benefit of readers, important references have been listed in each chapter. This may also be considered as a tribute to various academicians, researchers, and investigators associated with the publication of various books and research articles in different journals.

By giving more emphasis on the physical chemistry of different metallurgical processes, I aim to solve some of the problems. Attention has not been paid to how different processes are carried out; instead, the emphasis has been on why the step has been adapted in a particular manner. These queries, with a clear understanding of the physical chemistry, may open ways and means for future developments. Lecturing on physical chemistry of metallurgical processes is associated with a number of challenging exercises and difficulties. It requires not only a thorough understanding of chemical reactions taking place in a process, but also a sound knowledge of chemical thermodynamics and reaction kinetics. In addition, technical principles of heat and mass transfer are also needed in designing a metallurgical reactor. Lastly, the chemical-extractive metallurgist must know about the existing processes and should be capable of employing his imagination in encouraging students/investigators in improving the existing techniques.

Currently, university courses provide inadequate background in chemical/metallurgical thermodynamics. At the majority of institutions, thermodynamics courses are formal. Often teachers feel satisfied by solving a few problems by plugging data in the thermodynamic expressions derived in the class. In this book, the thermodynamic interrelationships concerning the problems have been summarized in Chap. 1, and for clarity, the thermodynamics quantities have been defined together with an explanation of their physical significance. It has been presumed that readers are familiar with the undergraduate course in chemical/metallurgical thermodynamics. For details, readers are advised to consult the textbooks and necessary compilations listed in this book. Reaction kinetics of different processes has not been covered in detail, and topics on heat and mass transfer have not been included with the primary objective of publishing a book of reasonable volume. Depending upon the response from readers, it may be taken up in the second volume or edition.

The book deals with various metallurgical topics, namely, roasting of sulfide minerals, sulfide smelting, slag, reduction of oxides and reduction smelting, interfacial phenomena, steelmaking, secondary steelmaking, role of halides in extraction of metals, refining, hydrometallurgy, and electrometallurgy in different chapters. Each chapter is illustrated with appropriate examples of application of the technique in extraction of some common, reactive, rare, or refractory metal together with worked-out problems explaining the principle of the operation. The problems require imagination and critical analysis. At the same time, they also encourage readers for creative application of thermodynamic data. Exercises have not been given because I am confident that the worked-out examples provide ample platform for the framework of additional problems. In selecting these problems, I am grateful to the late Professor John F. Elliott of MIT and late Dr. Megury Nagamori of the Noranda Research Center, Canada. I am not consistent in using the SI unit throughout the book. Based on my long teaching experience of about four decades, I strongly feel that the use of different units will make students mature with regard to the conversion from one unit to another. The principal objective of the book is to enlighten graduate students of metallurgy and metallurgical engineering specializing in chemical-extractive metallurgy and chemical engineering with the basic principles of various unit operations involved in the extraction of different types of metals. It will also be useful to senior undergraduate students of metallurgy and chemical technology.

However, the success of the process is dictated by economic evaluation. The use of thermodynamic principles and reactor design becomes insignificant if the process is uneconomical and/or the product has poor demand. This aspect has not been considered in this book. I advise the industrial metallurgists and researchers to be careful about the economic consequences of their work.

I am grateful to a number of friends and colleagues for necessary help in the preparation of the manuscript. It is not possible to mention all, but I shall be failing in my duty if I do not thank Professor H. Y. Sohn of the University of Utah, Salt Lake City, USA, Professor Fathi Habashi of the Laval University, Canada, and Professor T. R. Mankhand of the Banaras Hindu University for their advice and constructive criticism and also Dr. C. K. Behera and Dr. S. Jha of the Banaras Hindu University for their assistance in the preparation of diagrams and the manuscript. I am also thankful to library staffs in the Department of Metallurgical Engineering for their active cooperation in locating the reference materials. Although due acknowledgments have been given to authors at appropriate places in the texts for adapting their tables and figures published in various books and journals, I take this opportunity to thank the publishers (authors as well) listed below for giving permission to reproduce certain figures and tables:

I. ASSOCIATION FOR IRON & STEEL TECHNOLOGY (AIST)

1. Pehlke, R. D., Porter, W. F., Urban, P. F. and Gaines, J. M. (Eds.) (1975) BOF Steelmaking, Vol. 2, Theory, Iron & Steel Society, AIME, New York for Table 6.2 and Figs. 4.12, 4.14, 7.7, 8.1, 8.3, and 8.4.
2. Taylor, C. R. (Ed.) (1985) Electric Furnace Steelmaking, Iron & Steel Society, AIME, New York.

    i. Elliott, J. F. (1985) Physical chemistry of liquid steel, In Electric Furnace Steelmaking, Taylor, C. R. (Ed.), Iron & Steel Society, AIME, New York (Chapter 21, pp 291–319) for Figs. 8.1, 8.2, 8.3, 8.4, 8.7, 8.8, 8.9, 8.10, and 8.11.

   ii. Hilty, D. C. and Kaveney, T. F. (1985) Stainless steelmaking, In Electric Furnace Steelmaking, Taylor, C. R. (Ed.), Iron & Steel Society, AIME, New York (Chapter 13, pp 143–160) for Figs. 8.5 and 8.6.

3. Chipman, J. (1964) Physical chemistry of liquid steel, In Basic Open Hearth SteelMaking, Derge, G. (Ed.), AIME, New York (Chapter 16, pp 640–724) for Figs. 7.4 and 8.7.

II. ADDISON-WESLEY PUBLISHING COMPANY, INC.

Muan, A. and Osborn, E. F. (1965) Phase Equilibria among Oxides in Steelmaking, Addison-Wesley Publishing Company, Inc. Reading, Massachusetts for Fig. 4.14.

III. AMERICAN CERAMIC SOCIETY

Osborn, E. F. and Muan, A. (1960) Phase Equilibrium Diagrams of Oxide Systems, American Ceramic Society for Figs. 4.10 and 4.11.

IV. ASM INTERNATIONAL

Marshall, S. and Chipman, J. (1942) The carbon-oxygen equilibrium in liquid iron, Trans. Am. Soc. Met. **30**, pp 695–741 for Fig. 7.6.

V. BUTTERWORTH–HEINEMANN

Moore, J. J. (1990) Chemical Metallurgy, 2nd Edition, Butterworth–Heinemann, Oxford for Figs. 4.1, 4.3, 4.7, 4.8, 5.1, 5.2, and 9.1.

VI. CRC PRESS

Bodsworth, C. (1990) The Extraction and Refining of Metals, CRC Press, Tokyo for Fig. 10.1.

VII. DISCUSSION FARADAY SOCIETY

Peretti, E. A. (1948) Analysis of the converting of copper matte, Discuss. Faraday Soc. **4**, pp 179–184 for Fig. 3.1(B).

VIII. EDWARD ARNOLD

Ward, R. G. (1962) An Introduction to Physical Chemistry of Iron and SteelMaking, Edward Arnold, London for Figs. 7.2, 7.3, and 7.5.

IX. ELLIS HORWOOD LIMITED

Jackson, E. (1986) Hydrometallurgical Extraction and Reclamation, Ellis Horwood Ltd. (a division of John Wiley), New York for Figs. 11.3, 11.4,

11.5, 11.6, 11.7, 11.8, 11.10, and 11.13 and Resin structures and chemical equations related to solvent extraction equilibria.

X. ELSEVIER

Davenport, W. G., King, M., Schlesinger, M. and Biswas, A. K. (2002) Extractive Metallurgy of Copper, 4th Edition, Elsevier Science Ltd., Oxford for Figs. 3.1(A), 3.1(B), and 11.12 and data concerning smelting, converting and leach liquors.

XI. GORDON & BREACH

1. Habashi, F. (1970) Principles of Extractive Metallurgy, Vol. 1 General Principles, Gordon & Breach Science Publishing Co., New York for Table 5.1
2. Habashi, F. (1970) Principles of Extractive Metallurgy, Vol. 2 Hydrometallurgy, Gordon & Breach Science Publishing Co., New York for Table 11.1, and Figs. 11.11 and 11.14.

XII. JOURNAL OF IRON & STEEL INSTITUTE

1. Turkdogan, E. T. and Pearson, J. (1953) Activity of constituents of iron and steelmaking slags, Part I, iron oxide, J. Iron Steel Inst. **173**, pp 217–223 for Fig. 4.12.
2. Balagiva, K., Quarrel, A. G. and Vajragupta, P. (1946) A laboratory investigation of the phosphorus reaction in the basic steelmaking process, J. Iron Steel Inst. **153**, pp 115–145 for Fig. 7.3.

XIII. McGRAW-HILL

Rosenqvist, T. (1974) Principles of Extractive Metallurgy, McGraw-Hill, New York for Figs. 2.1, 2.2, 4.15, 5.3, 5.4, 5.5, 11.2, and 11.9.

XIV. MESHAP SCIENCE PUBLISHERS, MUMBAI, INDIA

Shamsuddin, M., Ngoc, N. V. and Prasad, P. M. (1990) Sulphation roasting of an off-grade copper concentrate, Met. Mater. Process. **1**(4), pp 275–292 for Table 2.1, and Figs. 2.3, 2.4, 2.5, and 2.6.

XV. PERGAMON

1. Parker, R. H. (1978) An Introduction to Chemical Metallurgy, 2nd Edition, Pergamon, Oxford, Fig. 7.8.
2. Coudurier, L., Hopkins, D. W. and Wilkomirski, I. (1978) Fundamentals of Metallurgical Processes, Pergamon, Oxford for Tables 4.1, 4.2, 4.3 and 4.4 and Figs. 4.2, 4.4, 4.5, 4.6, 4.7, 4.8, 4.9, 4.10 and 4.11.

XVI. PRENTICE HALL INTERNATIONAL

Deo, B. and Boom, R. (1993) Fundamentals of Steelmaking Metallurgy, Prentice Hall International, London for Figs. 7.11 and 7.12.

XVII. THE INSTITUTE OF MATERIALS, MINERALS & MINING, LONDON

Turkdogan, E. T. (1974) Reflections on research in pyrometallurgy and metallurgical chemical engineering, Trans. Inst. Min. Metall., Sect. C, 83, pp 67–23 for Figs. 7.9 and 7.10.

XVIII. THE MINERALS, METALS & MATERIALS SOCIETY (TMS)

1. Rocca, R. N., Grant, J. and Chipman, J. (1951) Distribution of sulfur between liquid iron and slags of low iron concentrations, Trans. Am. Inst. Min. Metall. Eng. **191**, p 319, for Fig. 7.1.
2. Elliott, J. F. (1955) Activities in the iron oxide-silica-lime system, Trans. Am. Inst. Min. Metall. Eng. **203**, p 485, for Fig. 4.13.
3. Queneau, P. E., O'Neill, C. E., Illis, A. and Warner, J. S. (1969) Some novel aspects of the pyrometallurgy and vapometallurgy of nickel, Part I: Rotary top-blown converter, JOM **21**(7), p 35, for Table 3.1.
4. Pehlke, R. D. and Elliott, J. F. (1960) Solubility of nitrogen in liquid iron alloys, I Thermodynamics, Trans. Met. Soc., AIME. **218**, p 1088, for Figs. 8.8 and 8.9.
5. Weinstein, M. and Elliott, J. F. (1963) Solubility of hydrogen at one atmosphere in binary iron alloys, Trans. Met. Soc., AIME. **227**, p 382, for Figs. 8.10 and 8.11.

XIX. THE INDIAN INSTITUTE OF METALS

Sundaram, C. V., Garg, S. P. and Sehra, J. C. (1979) Refining of reactive metals, In Proceedings of International Conference on Metal Sciences – Emerging Frontiers, Varanasi, India, November, 23–26, 1977, Indian Institute of Metals, Calcutta, pp 361–381, for Table 10.2.

Last but not the least, I am extremely grateful to Mrs. Abida Khatoon for taking care of the family, and for offering moral support and patience for my long hours spent working on the book.

Varanasi, India   M. Shamsuddin

Preface to Second Edition

The first edition of "Physical Chemistry of Metallurgical Processes" was published in February 2016. I read the book again in 2018 after receiving comments from academicians and researchers working in the field of chemical/extractive metallurgy. In light of these comments, I realized the need of improvement. This has been done by incorporating additional information from recent publications, and deleting some tables, figures, and texts from a few chapters. In this process, the sequence of presentation in Chap. 3 on sulfide smelting and Chap. 7 on steelmaking has been modified along with the additional information. Chapters on slag, reduction smelting, secondary steelmaking, halides, refining, hydrometallurgy, and electrometallurgy have been enriched with recent researches and developments with the prime objective of developing new eco-friendly and low-cost processes. I take this opportunity in thanking Dr. H. Y. Sohn, Distinguished Professor, Departments of Metallurgical Engineering and Chemical Engineering, University of Utah, Salt Lake City, USA, for his help in providing necessary publications, advice, and constructive suggestions in a few chapters.

Varanasi, India

M. Shamsuddin
kausar.shams.72@gmail.com

Contents

Abbreviations

List of Symbols

A   Area of surface, pre-exponential factor, frequency factor

a Raoultian activity

atm   Atmosphere (pressure)

b   Bacisity

c   Concentration, constant

C   Heat capacity, component, constant

C_p Heat capacity at constant pressure

C_v Heat capacity at constant volume

d   Diameter

D   Diffusion coefficient

D_P Dephosphorization index

D_S Desulfurization index

e   Electron charge, interaction coefficient

E   Energy, electrode potential, emf, electron field mobility

E_A Activation energy

E_η Activation energy for viscous flow

f   Henrian activity coefficient, fraction reacted

F   Degree of freedom, Faraday constant

$f_{(\theta)}$ Shape factor

(g)   Gaseous phase

g   Acceleration due to gravity

G   Free energy

$G^{\ddagger}$ Free energy of activation

h   Planck constant, height (metal head)

H   Enthalpy

J   Flux, (mass transferred per unit area per unit time), Joule

k   Rate constant, partition/segregation/distribution coefficient, Sievert's constant, Boltzmann constant, kilo

k_m	Mass transfer coefficient
K	Equilibrium constant, distribution coefficient
K'	Equilibrium quotient of cations
L	Latent heat of transformation, liter
L^e	Latent heat of evaporation
L^f	Latent heat of fusion
(l)	Liquid state, liter
ln	Naperian logarithm (loge)
m	Mass
M	Atomic/molecular weight
n	Number of moles/atoms
N	Avogadro number
N′	Electrically equivalent ionic fraction
p	Partial pressure
P	Total pressure
p_i	Partial pressure of the component i
p^o	Partial pressure of the pure component
q	Quantity of heat
Q'	Total extensive thermodynamic quantity
Q	Molar extensive thermodynamic quantity
r	Radius, rate of evaporation
R	Gas constant, rate of reaction
(s)	Solid phase
S	Entropy
t	Time
T	Absolute temperature (Kelvin), temperature °C
T^e	Temperature of evaporation
T^f	Temperature of fusion
U	Internal energy
v	Rate of rise of the deoxidation product, volume
V	Volume
ω	Weight, work done
x	Atom/mole fraction, ionic fraction, distance in the direction of x
z	Valency, a factor in reduction of an oxide, electrochemical equivalent

Greek Symbols

α	Stoichiometry factor, separation factor
γ	Raoultian activity coefficient
γ_i^o	Raoultian activity coefficient at infinite dilution
δ	Thickness of the stagnant boundary layer
ε	Interaction parameter
η	Viscosity, over voltage due to polarization
θ	Contact angle made by nucleus with the substrate
μ	Chemical potential (partial molar free energy), mobility

ρ Density
σ Interfacial surface tension
$\varnothing$ Fugacity

Prefixes

Δ Change in any extensive thermodynamic property (between product and reactant), e.g.,
$$\Delta G, \Delta H, \Delta S, \Delta U, \Delta V$$
d Very small change in any thermodynamic variable, e.g., dG, dT, dP

Suffixes[1]

aq   Dissolved in water at infinitely dilute concentration, e.g., M^{2+} (aq)
g   Gaseous, e.g., Cl_2 (g)
l   Liquid, e.g., H_2O (l)
s   Solid, e.g., SiO_2 (s)

Subscripts

a Anode potential E_a, anode polarization η_a
c Cathode potential E_c, cathode polarization η_c
cell E_{cell}, ΔG_{cell}
f Formation, e.g., ΔG_f
M   Metal, e.g., E_M
sol   Solution, e.g., ΔG_{sol}
vol   Volume, e.g., ΔG_{vol}, ΔH_{vol}

Superscripts

$\ddagger$ Transition of activated state
$-$ Partial molar functions, e.g., $\overline{G_i}$
o   Thermodynamic standard state, e.g., ΔG^o
id   Ideal thermodynamic functions, e.g., $\Delta G^{M, id}$
M   Mixing, e.g., ΔG^M, ΔH^M
xs   Excess thermodynamic functions, e.g., G^{xs}

Additional Symbols

( )   Solute in slag phase
{ }   Gaseous phase
[ ]   Solute in metallic phase, e.g., [S], concentration of a species in solution, e.g., [CN$^-$]
c_i^m Concentration of the metal at the interface
c_b^m Concentration of the metal in the bulk of the metal phase

---

[1]The physical state of a substance is indicated by the following symbols, placed in bracket ( ) after the chemical formulae of the substance.

c_i^s Concentration of the slag at the interface

c_b^s Concentration of the slag in the bulk of the slag phase

$\frac{dc}{dx}$ Concentration gradient

$\frac{dm}{dt}$ Rate of mass transfer

k_m^m Mass transfer coefficient in the metal phase

k_m^s Mass transfer coefficient in the slag phase

List of Figures

List of Tables

Chapter 1
Introduction

Metals generally occur in combined states in the form of ores and minerals as oxides, for example, cassiterite (SnO_2), cuprite (Cu_2O), chromite (Feo·Cr_2O_3), hematite (Fe_2O_3), pyrolusite (MnO_2), rutile (TiO_2), wolframite [Fe(Mn)WO_4]; sulfides, for example, chalcopyrite ($CuFeS_2$), cinnabar (HgS), galena (PbS), molybdenite (MoS_2), pentlandite [(NiFe)$_9S_8$], sphalerite (ZnS), stibnite (Sb_2S_3); silicates, for example, beryl (3BeO·Al_2O_3·6SiO_2), zircon [Zr(Hf)SiO_4]; titanate, for example, ilmenite (FeO·TiO_2); carbonates, for example, azurite [2$CuCO_3$·Cu(OH)$_2$], dolomite ($MgCO_3$·$CaCO_3$), malachite [$CuCO_3$·Cu(OH)$_2$], magnesite ($MgCO_3$); phosphate, for example, monazite [$Th_3(PO_4)_4$]; vanadate, for example, carnotite (K_2O·2UO_3·V_2O_5) and so on. A few precious metals like gold, silver, and platinum are found in the native or uncombined form because they are least reactive. As naturally occurring ores and minerals are associated with gangue such as silica, alumina, etc., the first step in the extraction of metals is the removal of gangue from the ore containing the metal value by mineral beneficiation methods incorporating comminution, preliminary thermal treatment and concentration by magnetic separation, heavy media separation, jigging, tabling, and flotation. The choice of the method depends upon the nature of the gangue and its distribution in the ore and the degree of concentration of the metal value required, which depends on the extraction technology to be adopted. The extraction methods incorporate various steps to obtain the metal from the concentrate, ore or some mixture, or from chemically purified minerals; occasionally, the mineral may be first converted to a more amenable form. The mineral beneficiation step lies between mining and extraction. The extraction processes are classified into three main groups, namely:

1. *Pyrometallurgical* methods including smelting, converting, and fire refining are carried out at elevated or high temperatures. A step called roasting or calcination may also be incorporated in the flow sheet in the treatment of sulfide or carbonate minerals.
2. *Hydrometallurgical* methods incorporate leaching of metal values from the ores/minerals into aqueous solution. The resultant solution is purified before

© The Minerals, Metals & Materials Society 2021
M. Shamsuddin, *Physical Chemistry of Metallurgical Processes, Second Edition*,
The Minerals, Metals & Materials Series, https://doi.org/10.1007/978-3-030-58069-8_1

precipitation of the metal by pH and p_{O_2} control, gaseous reduction, or cementation. Roasting or calcination also forms an important step in the treatment of sulfide and carbonate ores. In the production of rare metals like uranium, thorium, zirconium, and so on, the leach liquor may be purified by fractional crystallization, ion exchange, and/or solvent extraction techniques.

3. *Electrometallurgical* methods use electrical energy to decompose the pure mineral that is present in aqueous solutions or in a mixture of fused salts. If the metal is extracted from the electrolyte using an insoluble anode the method is called electrowinning. On the other hand, if the impure metal (in the form of the anode) is refined using a suitable electrolyte, the method is known as electrorefining.

The choice of the technique mainly depends on the cost of the metal produced, which is related to the type of ore, its availability, cost of fuel, rate of production, and the desired purity of the metal. The fuel or energy input in the process flow sheet may be in the form of coal, oil, natural gas, or electricity. Being an electrically based process electrothermic smelting is an expensive method. This process can only be adopted if cheap hydroelectric power is available. Highly reactive metals like aluminum and magnesium can be produced in relatively pure states by fused salt electrolysis. Electrowinning is often employed as a final refining technique in hydrometallurgical extraction. Hydrometallurgy seems to be a better technique for the extraction of metals from lean and complex ore although it is slower than pyrometallurgical methods. Major quantities of metals are obtained by the pyrometallurgical route as compared to the hydrometallurgical route because kinetics of the process is much faster at elevated temperatures. This is evident from the discussion in the following chapters on matte smelting, slag, reduction smelting, steelmaking, refining, and halides, which deal with the pyrometallurgical methods of extraction. Separate chapters have been included on hydrometallurgy and electrometallurgy.

In addition to the well-established tonnage scale production of the ferrous and six common nonferrous metals (aluminum, copper, lead, nickel, tin, and zinc), in recent years many other metals, namely, beryllium, uranium, thorium, plutonium, titanium, zirconium, hafnium, vanadium, niobium (columbium), tantalum, chromium, tungsten, molybdenum, and rare earths have gained prominence in nuclear power generation, electronics, aerospace engineering, and aeronautics due to their special combination of nuclear, chemical, and physicochemical properties. Many of these metals are categorized as rare despite their more abundant occurrence in nature compared to copper, zinc, or nickel. This is due to the diversified problems associated with their extraction and conversion to usable form. Production of some of these metals in highly pure form on tonnage scale has been possible recently by efficient improvement of the conventional extraction methods, as well as through the development of novel unit processes.

On account of the refractory nature of the minerals and stability of the oxides and carbides of many rare metals, direct smelting of the ores with carbon is not feasible for rare metal extraction. The refining methods like fire refining, liquation, distillation, and so on are also not applicable. Hence, the flow sheets for rare metal extraction and refining involve many steps, each with the specific objective of successfully removing a particular impurity. On account of the co-occurrence of chemically similar elements, for example, uranium/thorium, niobium/tantalum, zirconium/hafnium, and rare earths, there are often problems in rare metal extraction. For the separation of such elements, unconventional techniques like ion exchange and solvent extraction have to be incorporated in the process flow sheet for production of high-purity metals. Finally, during the reduction and consolidation stages, one has to be extremely careful because rare metals in general, and titanium, zirconium and hafnium in particular, are very sensitive to atmospheric gases that affect their physical, chemical, and mechanical properties.

It would be appropriate to outline here the general steps in the extraction of rare metals:

1. *Physical mineral beneficiation*: Beach sand, a source of many rare metals like titanium, zirconium, hafnium, and thorium, is processed by exploiting the characteristic differences in the size, shape, density, and electromagnetic and electrostatic behavior of mineral constituents, that is, rutile, ilmenite, zircon, monazite, and so on.
2. *Selective chemical ore breakdown*: In order to bring the metal values to an extractable state, hydrometallurgical unit processes like strong acid or alkali leaching or pyrometallurgical techniques like fusion with alkalis and alkali double fluorides are employed.
3. *Ion exchange*: The technique developed long back for purification and deionization of water is currently used extensively for concentration and purification of lean leach liquor and for separation of chemically similar elements.
4. *Solvent extraction*: An analytical technique once developed for selective transfer of specific metal ions from aqueous solution to an organic phase has presently come up to the stage of large-scale unit process for purification and separation of a number of rare and nuclear metals.
5. *Halogenation*: For the production of oxygen-free reactive metals like titanium, uranium, zirconium, and so on, it has become essential to adopt intermediate routes by converting oxides into chlorides or fluorides prior to reduction.
6. *Metallothermic reduction*: The traditional "thermit process" has been very successfully employed in rare metal extraction. For example, uranium tetrafluoride is reduced with calcium for tonnage production of uranium metal required in atomic reactors. Similarly, magnesium is used for the production of titanium and zirconium from their respective tetrachlorides.
7. *Consolidation and vacuum refining*: As most of the metals mentioned above are high melting and very corrosive in the molten state, they pose problems during melting and consolidation. Special consumable electrode arc melting with supercooled copper hearths have been developed for the production of titanium and zirconium alloys. Electron beam melting technique has been practiced for melting, and refining of niobium and tantalum. The high superheat at temperatures around 3000 °C under vacuum helps in removing all impurities including oxygen, nitrogen, and carbon.

8. *Ultra-purification*: The performance of rare and reactive metals during usage depends on purity. For proper assessment, it is important that metals are free from impurities. Similarly, high order of purity is specified for semiconducting elements like silicon and germanium, required in electronic industry. In recent years, a number of ultra-purification methods, for example, thermal decomposition, zone refining, and solid-state electrolysis have been developed for large-scale purification of these metals.

A number of textbooks dealing with ironmaking, steelmaking, extraction of nonferrous metals, and principles of extractive metallurgy are available. Each book has some edge over the other in certain aspects of presentation in terms of theory and practice. Some emphasize on technology and some on principles. Thermodynamics and kinetics have been discussed. In this book, an attempt has been made to discuss the physical chemistry of different steps, for example, roasting, sulfide smelting/converting, reduction smelting, steelmaking, deoxidation, degassing, refining, leaching, precipitation, cementation, involved in the extraction of metals. A chapter on slag which plays an important role in the extraction of metals from sulfide as well as oxide minerals has been included. Similarly, another chapter highlights the significance of interfacial phenomena in metallurgical operations. The physicochemical aspects of desulfurization, dephosphorization, decarburization, and silicon and manganese reactions in steelmaking have been discussed along with brief accounts on various steelmaking processes highlighting the differences in their chemistry of refining and pretreatment of hot metal. Role of halides, ion exchange, and solvent extraction in metal production and refining have been discussed in different chapters. Methods of construction of predominance area diagrams applicable in selective roasting and leaching have been explained with suitable and appropriate examples in Chapters 2 and 11. At the end of the book, flow sheets demonstrating various steps in the extraction of copper, lead, nickel, zinc, tungsten, beryllium, uranium, thorium, titanium, zirconium, aluminum, and magnesium from their respective ores have been presented.

Relevant worked-out examples have been included in each chapter to illustrate principles. While reading the topics on suspension smelting, bath smelting, and continuous smelting for treatment of chalcopyrite concentrate in different books and journals, one may feel that the chapter on roasting is outdated, but one must realize that these developments have been possible only after a sound understanding of the physical chemistry and thermodynamics of all the steps involved in the extraction of metals. Although, currently, almost the entire production of steel comes from top-blown (LD), bottom-blown (OBM), and combined-blown (Hybrid) converters and electric arc furnaces, a discussion on the obsolete Bessemer process has been included to highlight the contributions of Henry Bessemer whose invention laid the foundation for the modern steelmaking processes. Readers may also raise questions on SI units not being used uniformly throughout the book. I want to stress here that based on my 40 years of teaching experience, I strongly feel that solving problems in different units will make students mature with respect to conversion from one system to another. As some basic knowledge of under-graduate level chemical/metallurgical thermodynamics is necessary to understand the worked-out problems in different chapters, a brief account on thermodynamic quantities and their interrelationships has been included in this chapter.

1.1 Thermodynamic Quantities and Their Interrelationships

1.1.1 General Thermodynamics

First law of thermodynamics: Energy can neither be produced nor destroyed in a system of constant mass, although it can be converted from one form to another. According to the first law of thermodynamics, the total heat content of the system called enthalpy (H) is expressed as:

$$H = U + pv \tag{1.1}$$

that is, heat content (enthalpy) = internal energy (U) + energy term, dependent on the state of the system (pv), where p and v are, respectively, the pressure and volume of the system.

Heat capacity at constant volume and constant pressure: Heat capaeity, C, may be defined as the ratio of the heat, Q, absorbed by a system to the resulting increase in temperature $(T_2 - T_1)$, that is, ΔT. Since the heat capacity usually varies with temperature,

$$C = \lim_{T_1 \to T_2} \frac{Q}{\Delta T} = \frac{q}{dT} \tag{1.2}$$

where q = quantity of heat, dT = small rise in temperature.

At constant volume, $q_v = \Delta U_v$

$$\therefore C_v = \frac{dU_v}{dT} = \left[\frac{\partial U}{\partial T}\right]_v \tag{1.3}$$

Hence, the heat capacity of a system at constant volume is equal to the rate of increase of internal energy content with temperature at constant volume.

Similarly, at constant pressure,

$$C_p = \frac{q_p}{dT} = \left[\frac{\partial H}{\partial T}\right]_p \tag{1.4}$$

Thus, the heat capacity of a system at constant pressure is consequently equal to the rate of the increase of heat content with temperature at constant pressure.

The well-known expression $C_p - C_v = R$ can be established from the knowledge of differential calculus.

Effect of temperature on heat of reaction: The variation of the heat of reaction with temperature can be expressed as:

$$\Delta H_2 - \Delta H_1 = \int_{T_1}^{T_2} \Delta C_p.dT \tag{1.5}$$

where ΔH_1 and ΔH_2 are the heat of reaction at temperatures T_1 and T_2, respectively, and ΔC_p is the difference in the total heat capacities of the reactants and products taking part in the reaction. Equation 1.5 known as Kirchhoff's equation is often used to calculate the heat of reaction at one temperature, if that is known at another temperature. In case reactant(s) and/or product(s) undergo any transformation at T^t, the above equation is modified as:

$$\Delta H_2 - \Delta H_1 = \int_{T_1}^{T^t} \Delta C_p(T_1 \leftrightarrow T^t).dT \pm L^t$$

$$+ \int_{T^t}^{T_2} \Delta C_p(T^t \leftrightarrow T_2).dT\,(T_1 \leq T^t \leq T_2) \tag{1.6}$$

According to the first law of thermodynamics, the total energy of the system and surroundings remains constant. The thermodynamic variables like enthalpy, entropy, and free energy changes for the reaction depend only on the initial and final states, not on the path chosen. This is the basis of *Hess law of constant heat summation*. The law states that the overall heat change of a chemical reaction is the same whether it takes place in one or several stages, provided the temperature and either the pressure or the volume remain constant.

Second law of thermodynamics: The first law of thermodynamics is concerned with the quantitative aspects of inter-conversion of energies. This law neither allows us to predict the direction of conversion nor the efficiency of conversion when heat energy is converted into mechanical energy. In the last half of the nineteenth century many scientists put their concentrated efforts to apply the first law of thermodynamics to the calculation of maximum work obtained from a perfect engine and prediction of feasibility of a reaction in the desired direction. These considerations led to the development of the second law of thermodynamics, which has had a far-reaching influence on the subsequent development of science and technology. The Carnot cycle made it possible to assess the efficiency of engines and it also showed that under normal conditions all the heat supplied to the system cannot be converted into work even by perfect engines. A perfect engine would convert all the heat supplied into work if the lower temperature of the process could be made equal to zero. This proof of practical impossibility of complete conversion of heat energy into mechanical work/energy was the starting point of the second law of thermodynamics. Based on this law, the concept of entropy was introduced by Clausius.

Entropy: In Carnot cycle, the calculated heat absorbed or evolved during isothermal steps depends on the temperature at which these steps occur. However, the numerical values of the ratios q_{rev}/T were the same, so that, for example, $q_1/T_1 = -q_2/T_2$, and the sum of these terms for a complete cycle irrespective of the number of stages comes to zero. Thompson gave the standard notation of $\sum(dq_{rev}/T) = 0$, for a number of steps forming a cyclic process. In 1850, Clausius recognized the fact that $\sum dq_{rev}/T$ was characteristic (state property) of the system and not the value dq_{rev}, since this varies from temperature to temperature. The name entropy change was finally given to the ratio $\partial q/T$ by Clausius who denoted the entropy change by the symbol ΔS and considered it to be a state function. Entropy depends on the state of a substance or system and not on its previous history, irrespective of whether the path is thermodynamically reversible or not, and it is also irrespective of the substance involved. Entropy is a state property. It is an extensive property of the system as it depends on the mass of the system and is a thermodynamic variable. Since entropy = energy/temperature, its unit would be cal deg^{-1} mol^{-1} (e.u.) or J deg^{-1} mol^{-1}. As the system absorbs heat, its entropy increases, for example, during melting and boiling.

Calculation of entropy change from heat capacities: From Clausius' mathematical definition, entropy change $\Delta S = q_{rev}/T$, since heat capacity C_p has exactly the same units as entropy, that is, cal deg^{-1} mol^{-1}, for a limiting case of an infinitesimal change in the process, the entropy change can be expressed by the equation:

$$\frac{dq_{rev}}{dT} = C_p \tag{1.7}$$

and entropy change of an element

$$dS = \frac{dq_{rev}}{T} \tag{1.8}$$

from (1.7 and 1.8) we get:

$$dS = \frac{C_p.dT}{T} \tag{1.9}$$

Equation 1.9 is a general differential equation for change in entropy and if we assume that entropy is zero at absolute zero, then the entropy of a substance at temperature T can be calculated from the equation:

$$S_T = S_0 + \int_0^T \frac{C_p.dT}{T} \tag{1.10}$$

$$\Delta S = S_T - S_0 = \int_0^T \frac{C_p.dT}{T} \tag{1.11}$$

Equation 1.11 is applicable in cases where there is no transformation from 0 to T K. In case of solid–solid, solid–liquid, and liquid–gas transformations (between 0 and T K), at T^t, T^f, and T^e, respectively, with the corresponding latent heat of transformation (L^t), heat of fusion (L^f), and heat of evaporation (L^e), Eq. 1.11 is modified as:

$$\Delta S = \int_0^{T^t} \frac{C_{p1}}{T} .dT + \frac{L^t}{T^t} . + \int_{T^t}^{T^f} \frac{C_{p2}}{T} .dT + \frac{L^f}{T^f} + \int_{T^f}^{T^e} \frac{C_{p3}}{T} .dT + \frac{L^e}{T^e}$$
$$+ \int_{T^e}^{T} \frac{C_{p4}}{T} .dT \tag{1.12}$$

where $\frac{L^t}{T^t} = \Delta S^t$, $\frac{L^f}{T^f} = \Delta S^f$, $\frac{L^e}{T^e} = \Delta S^e$, and C_{p1}, C_{p2}, C_{p3}, and C_{p4}, are respectively, the heat capacity of the substance in the temperature ranges $(0-T^t)$, $(T^t - T^f)$, $(T^f - T^e)$, and $(T^e - T)$.

Driving force of a chemical reaction: The driving force of a reaction can be calculated as $(\Delta H - \Delta S)$; the more negative this factor, the greater the driving force, and if the factor is positive, the reaction will not proceed spontaneously.

Free energy: The factor $(\Delta H - \Delta S)$ has dimensions of energy because ΔH is an energy term and ΔS is the heat absorbed divided by the absolute temperature. It has been called the change in the "free energy" of the system. Free energy is a thermodynamic function of great importance. Consider a system undergoing a thermodynamically reversible change at constant temperature and constant volume. From the first law of thermodynamics, $\Delta U = q_r - w$, where q_r is the heat absorbed reversibly by the system at temperature T and w is the maximum work done by the system.

Since $\Delta S = \frac{q_r}{T}$

$$\Delta U = T\Delta S - w \tag{1.13}$$
$$\therefore - w = \Delta U - T\Delta S \tag{1.14}$$

where $-w$ is the maximum work (whether mechanical or electrical) that can be obtained from the system. A thermodynamic function, A, the "work function," or Helmholtz free energy (after H. von Helmholtz) can be defined as:

$$-w = \Delta U - T\Delta S = \Delta A \tag{1.15}$$

A is a thermodynamic variable, depending only on the state of the system, not on its history, because U, T, and S are all thermodynamic variables. At constant temperature and constant pressure, work may be done by the system as a result of a volume change. This work $(-P\Delta V)$ is not a "useful" work. The useful work will then be the maximum work, $-w$, less the energy lost due to volume change $(-P\Delta V)$.

We can then define the "useful work" as ΔG, the Gibbs free energy change of the system:

$$\Delta G = -w - (-P\Delta V) = \Delta A + P\Delta V \qquad (1.16)$$

From Eqs. 1.15 and 1.16, we get:

$$\Delta G = \Delta U - T\Delta S + P\Delta V = (\Delta U + P\Delta V) - T\Delta S = \Delta H - T\Delta S \qquad (1.17)$$

In Eq. 1.17: $\Delta G = \Delta H - T\Delta S$, G is the Gibbs free energy of the system (after J. Willard Gibbs), which depends only on the thermodynamic variables, H, T, and S. It is the maximum work available from a system at constant pressure other than that due to a volume change. Most metallurgical processes work at constant pressure rather than constant volume, so we are more concerned with G than with A. We had already seen the fundamental importance of the factor $(\Delta H - T\Delta S)$, which is equal to ΔG. Hence, ΔG is a measure of the "driving force" behind a chemical reaction. For a spontaneous change in the system, ΔG must be negative; the more negative the value of ΔG, the greater will be the driving force.

Some more thermodynamic relationships: By definition $G = H - TS = U + PV - TS$. On differentiation, we get:

$$dG = dU + PdV + VdP - TdS - SdT \qquad (1.18)$$

Assuming a reversible process involving work due only to expansion at constant pressure, and according to the first law, $dU = dq - PdV$, and from the second law, $dq = TdS$ (Eq. 1.8).

$$\therefore dU = TdS - PdV \qquad (1.19)$$

From (1.18) and (1.19) we get:

$$dG = VdP - SdT \qquad (1.20)$$

at constant pressure, $dP = 0$,

$$\therefore \left(\frac{\partial G}{\partial T}\right)_P = -S \qquad (1.21)$$

and at constant temperature, $dT = 0$, and $\therefore dG = VdP$, and since for 1 mol of an ideal gas, $PV = RT$

$$\therefore dG = \frac{RT}{P} \, dP \qquad (1.22)$$

If P_A and P_B, respectively, denote the initial and final pressures of the system with corresponding free energies as G_A and G_B, on integration of Eq. 1.22 between the limits P_A and P_B at constant temperature, we get:

$$\Delta G = G_B - G_A = RT \int_{P_A}^{P_B} \frac{dP}{P} = RT \ln \frac{P_B}{P_A}$$

This corresponds to the maximum work done by a gaseous system at constant temperature when pressure alters from P_A to P_B, $-w = RT \ln \frac{P_B}{P_A}$

Since $\Delta G = -w$,

$$\therefore \Delta G = RT \ln \frac{P_B}{P_A} = -RT \ln \frac{P_A}{P_B} \tag{1.23}$$

When system undergoes a change at constant pressure, using Eq. 1.21, we can write:
$$dG_A = - S_A. \, dT \text{ and } dG_B = - S_B. \, dT$$
Hence, $(G_B - G_A) = - (S_B - S_A)dT$
but $G_B - G_A = \Delta G$ and $S_B - S_A = \Delta S$, $\therefore (\Delta G) = - \Delta S. \, dT$ and $\left(\frac{\partial \Delta G}{\partial T}\right)_P = -\Delta S$; on substitution in Eq. 1.17, we get:

$$\Delta G = \Delta H - T\Delta S = \Delta H + T \left(\frac{\partial \Delta G}{\partial T}\right)_{P,T} \tag{1.24}$$

Equation 1.24 is known as the Gibbs–Helmholtz equation.

Since free energy is an extensive thermodynamic quantity, it can be added and subtracted in the same way as enthalpy changes. Values of ΔS are usually tabulated at 298 K, and thus, ΔG_{298} can be calculated. ΔH and ΔS vary with temperature, and this variation can be calculated from the following equations. Combining Eqs. 1.5, 1.10, and 1.17, we get:

$$\Delta G_T = \Delta H_{298} + \int_{298}^{T}\Delta C_P.dT - T\Delta S_{298} - T\int_{298}^{T} \frac{\Delta C_P}{T}.dT$$

Neglecting any heat of transformation this leads to a generalized formula of the type:

$$\Delta G_T = a + bT \log T + cT^2 + eT^{-1} + fT$$

where a, b, c, e, and f are constants in ΔG versus T relationship of a particular system. However, experimental errors involved in the determination of the data do not often justify such complex formulae and normally two or three term formulae of the following types suffice:

$$\Delta G_T = a + bT$$

$$\Delta G_T = a + bT \log T + cT$$

van't Hoff isotherm: The van't Hoff isotherm relates free energy change of a reaction with the equilibrium constant (K) and activities of the reactants and products (concentrations/partial pressures) taking part in the reaction:

$$\Delta G = \Delta G^o + RT \ln \frac{\Pi a_{\text{products}}}{\Pi a_{\text{reactants}}} \tag{1.25}$$

If all the reactants and products are maintained at unit activity/atmospheric pressure the relation between ΔG^o and K_P (equilibrium constant at constant pressure) is expressed as:

$$\Delta G^o = -RT \ln K_P$$

On differentiation with temperature at constant pressure we get:

$$\left[\frac{\partial (\Delta G^o)}{\partial T} \right]_P = -R \ln K_P - RT \left[\frac{\partial \ln K_P}{\partial T} \right]_P$$

multiplying by T we get:

$$T \left[\frac{\partial (\Delta G^o)}{\partial T} \right]_P = -RT \ln K_P - RT^2 \left[\frac{\partial \ln K_P}{\partial T} \right]_P = \Delta G^o - RT^2 \left[\frac{\partial \ln K_P}{\partial T} \right]_P$$

From the Gibbs–Helmholtz Eq. 1.24, we have:

$$\Delta G^o = \Delta H^o + T \left[\frac{\partial (\Delta G^o)}{\partial T} \right]$$

Hence, from the above equations, on comparison we get:

$$T \left[\frac{\partial (\Delta G^o)}{\partial T} \right] = \Delta G^o - \Delta H^o = \Delta G^o - RT^2 \left[\frac{\partial \ln K_P}{\partial T} \right]$$

$$\therefore \Delta H^o = RT^2 \left[\frac{\partial \ln K_P}{\partial T} \right]$$

$$\text{or } \frac{d \ln K_P}{dT} = \frac{\Delta H^o}{RT^2} \tag{1.26}$$

The above relation, known as van't Hoff's equation, shows the effect of temperature on the equilibrium constant of a reaction. If K_1 and K_2 are equilibrium constants at temperatures T_1 and T_2, respectively, assuming enthalpy to be independent of temperature in the close temperature interval $(T_1 - T_2)$, Eq. 1.26 can be integrated:

$$\int_{k_1}^{k_2} d \ln K_P = \frac{\Delta H^o}{R} \int_{T_1}^{T_2} \frac{dT}{T^2}$$

$$\ln\left[\frac{k_2}{k_1}\right] = \frac{\Delta H^o}{R}\left[\frac{T_2 - T_1}{T_1 T_2}\right] \tag{1.27}$$

From $G = H - TS$, we can write: $G/_T = H/_T - S$
on differentiation with respect to T at constant pressure:

$$\left[\frac{\partial(G/_T)}{\partial T}\right]_P = \left[\frac{\partial(H/_T)}{\partial T}\right]_P - \left[\frac{\partial S}{\partial T}\right]_P$$

$$= H\left[\frac{\partial(1/_T)}{\partial T}\right]_P + \frac{1}{T}\left[\frac{\partial H}{\partial T}\right]_P - \left[\frac{\partial S}{\partial T}\right]_P$$

Since $\left[\frac{\partial H}{\partial T}\right]_P = C_P$ (from Eq. 1.4) and $\left[\frac{\partial S}{\partial T}\right]_P = \frac{C_P}{T}$ (from Eq. 1.9)

$$\left[\frac{\partial(G/_T)}{\partial T}\right]_P = -\frac{H}{T^2} + \frac{C_P}{T} - \frac{C_P}{T} = -\frac{H}{T^2}$$

$$\text{Similarly}\left[\frac{\partial(\Delta G)}{\partial T}\right]_P = -\frac{\Delta H}{T^2} \tag{1.28}$$

This is known as the Gibbs–Helmholtz equation, useful in determining ΔH at a particular temperature if ΔG vs T equation is known.

Phase rule: Thermodynamics has been found useful in predicting the maximum number of possible phases in a given system and in establishing simple equilibrium phase diagrams from very limited thermodynamic data. The phase rule mathematically relates phase, component, and degree of freedom by the following relation:

$$F = C + P - 2 \tag{1.29}$$

where P, C, and F refer to the phase, component, and degree of freedom, respectively, which are defined as follows:

Any homogeneous and physically distinct part of a system, separated from other part(s) of the system by a bounding surface is known as *Phase*. For example, ice, water, and water vapor coexisting in equilibrium at 273.15 K constitute a three-phase system. When ice exists in more than one crystalline form, each form will represent a separate phase because it is clearly distinguishable from each other. By and large, every solid in a system constitutes a separate phase. But a homogeneous solid or liquid solution forms a single phase irrespective of the number of chemical components present in it. However, two immiscible liquids constitute two phases since they are separated by a boundary. Gases on the other hand, pure or as a mixture, always give rise to one phase due to intimate mixing of their molecules.

The number of *Components* at equilibrium is the smallest number of independently variable constituents by means of which the composition of each phase present in a system can be expressed directly or in the form of a chemical equation. As an example, let us consider the decomposition of calcium carbonate: $CaCO_3(s) =$

$CaO(s) + CO_2(g)$. According to the above definition, at equilibrium this system will consist of two components since the third one is fixed by the equilibrium conditions. Thus, we have three phases: two solids ($CaCO_3$ and CaO) and a gas (CO_2), and the system has only two components.

The number of *Degrees of freedom* is the number of variables, such as temperature, pressure, and concentration that need to be fixed so that the condition of a system at equilibrium is completely stated.

Clausius–Clapeyron equation: The Clausius–Clapeyron equation is extremely useful in calculating the effects of temperature and pressure changes on the melting point of solids, boiling point of liquids, and any solid–solid phase transformations. It also enables us to quantify the effect of the changes in the concentration of solutions on their freezing points, and boiling temperatures and is therefore very useful for calculation of phase boundaries of systems, which are either immiscible or partially miscible in the solid state. The equation is expressed in its simplest form as:

$$\frac{dP}{dT} = \frac{\Delta S}{\Delta v} \tag{1.30}$$

In case of a solid–liquid transformation, the entropy of fusion, $\Delta S^f = \frac{q_{rev}}{T} = \frac{L^f}{T^f}$

$$\therefore \frac{dP}{dT} = \frac{L^f}{T^f(v_l - v_s)} = \frac{L^f}{T^f \Delta v} \tag{1.31}$$

where q_{rev} is the quantity of heat absorbed during fusion of 1 g mol of the substance, and L^f refers to the corresponding latent heat of fusion, and v_s and v_l are the volumes in solid and liquid states, respectively.

The above equation is known as the Clausius–Clapeyron equation and has been derived for a single chemical substance undergoing a change from one phase to another. This equation will also be applicable to liquid–vapor, solid–vapor transformations and for transformations between two crystalline substances. In each case the appropriate latent heat and the volume change have to be substituted into this equation. In every case, L is the heat absorbed during the process and Δv is the accompanying volume change.

The Clausius–Clapeyron equation in the above form enables us to calculate the changes in either pressure with temperature or latent heat of solid–solid, solid–liquid, solid–gas, and liquid–gas transformation. The above Eq. 1.31 does not allow us to calculate the absolute value of pressure and temperature for a given system. The equation will be of much wider application when transformed into a form suitable for *integration* by making the following assumptions:

(a) In the liquid–gas transformation, the volume of 1 g atom of the gaseous phase is much larger compared to the volume of 1 g atom of the liquid phase, that is, the volume of the liquid is negligible. This assumption is reasonable because in case of $Fe(s) = Fe(1)$, v (liquid iron) $= 10$ cc g atom^{-1}, and v (iron vapor) $= 22,400$ cc g atom^{-1}.

(b) Since metallic vapors behave ideally, $PV = RT$.

(c) The latent heat of transformation is constant over the range of pressure and temperature under consideration.

Thus, for liquid–gas transformation, we have:

$$\frac{dP}{dT} = \frac{L^e}{T^e(v_g - v_l)} \quad (\text{since } v_g \gg v_l)$$

$$\therefore \frac{dP}{dT} = \frac{L}{Tv_g}$$

from the second assumption: $PV = RT$, $v_g = RT/P$

$$\therefore \frac{dP}{dT} = \frac{LP}{RT^2} \tag{1.32}$$

from the third assumption, L/R can be taken out of the integral,

$$\therefore \int_{p_1}^{p_2} \frac{dP}{P} = \frac{L}{R} \int_{T_1}^{T_2} \frac{dT}{T^2}$$

where p_1 and p_2 are the initial and final pressures at temperatures T_1 and T_2, respectively.

On integration, we get:

$$\ln\left[\frac{p_2}{p_1}\right] = \frac{L}{R}\left[\frac{T_2 - T_1}{T_1 T_2}\right] \tag{1.33}$$

This equation allows us to calculate:

1. The latent of transformation provided p_1 and p_2 and the corresponding T_1 and T_2 are known.
2. The variation in boiling point with change in pressure, provided p_1, p_2, and T_1 are known.
3. The value of p, provided T_1, T_2, and L are known.

Integration of Eq. 1.32 may also be expressed as:

$$\ln P = -\frac{L}{RT} + C \tag{1.34}$$

This form of equation is very useful because plot of $\ln P$ vs $1/T$ gives a straight line, the slope of which is $-\frac{L}{R}$.

1.1.2 Solution Thermodynamics

Solutions: A solution may be defined as a homogeneous phase composed of different chemical substances, whose concentration may be varied without the precipitation of a new phase. It differs from a mixture by its homogeneity and from a compound by being able to possess variable composition. The composition may be expressed in terms of either weight percent (wt% or simply %) or atom percent (at%) or mole percent (mol%). The atom/mole fraction (x), most widely used in thermodynamic equations, is defined as the number of atoms/moles of a substance divided by the total number of atoms/moles of all the substances present in the solution. If n_A, number of moles of A, and n_B, number of moles of B, form a solution: A–B, atom fractions of A and B are given as:

$$x_A = \frac{n_A}{n_A + n_B} \text{ and } x_B = \frac{n_B}{n_A + n_B} \text{ and } x_A + x_B = 1$$

The properties of solutions are discussed by Raoult's law. The law states that the relative lowering of the vapor pressure of a solvent due to the addition of a solute is equal to the mole fraction of the solute in the solution. Suppose x_A atom/mole fraction of A and x_B atom/mole fraction B form a solution in which p_A and p_B are the partial pressures exerted by vapors of A and B, respectively, and P is the total pressure of the solution. If p_A^o and p_B^o are the partial pressure of pure A and pure B, respectively, at the same temperature at which solution exists, then according to the Raoult's law, we have:

$$\frac{p_A^o - p_A}{p_A^o} = x_B \text{ and } \frac{p_B^o - p_B}{p_B^o} = x_A$$

$$\text{or } 1 - \frac{p_A}{p_A^o} = x_B \text{ and } 1 - \frac{p_B}{p_B^o} = x_A$$

$$\text{or } \frac{p_A}{p_A^o} = 1 - x_B = x_A \text{ and } \frac{p_B}{p_B^o} = x_B$$

$$\therefore p_A = p_A^o . x_A \text{ or } p_A \propto x_A$$

$$\text{and } p_B = p_B^o . x_B \text{ or } p_B \propto x_B$$

or in general, for the species i in a solution, we can write as:

$$p_i = p_i^o . x_i \tag{1.35}$$

This means if the solution obeys Raoult's law, the vapor pressure of the component is directly proportional to its atom/mole fraction in the solution. The constant of proportionality is the vapor pressure of the component in the pure state. A solution that obeys Raoult's law is called an ideal solution. In order to form an ideal solution: A–B, the moles of A and B must be of similar size and must attract one another with the same force the molecules of A attract other molecules of A or molecules of B attract other molecules of B, and also the vapor should behave as an ideal gas. Thus, the solution obeying Raoult's law satisfies the following condition:

$$(A \leftrightarrow A) = (B \leftrightarrow B) = \frac{1}{2}\{(A \leftrightarrow A) + (B \leftrightarrow B)\}$$

Ideal solutions: An ideal solution obeys Raoult's law, which may be represented by plotting vapor pressure against the mole fraction. This gives straight lines with p_A^o and p_B^o being the intersection of the line with the vapor pressure axes as shown in Figs. 1.1 and 1.2. In ideal solutions, gas pressures p_A and p_B obey the ideal gas equation: $PV = RT$, and their physical properties will be additive, that is, total pressure, $P = p_A + p_B$.

Nonideal or real solutions: Deviations from Raoult's law occur when the attractive forces between the molecules of components A and B of the solution are weaker or stronger than those existing between A and A or B and B in their pure states. For example, if there were an attractive force between components A and B in the solution, which is weaker than the mutual attraction between molecules of A in pure A and molecules of B in pure B, there would be a higher tendency for these components to leave the solution. In this case, the vapor pressure would be more than that predicted from Raoult's law. This is known as the positive deviation from Raoult's law (Fig. 1.1). In case of a stronger attraction between A and B as compared to A and A or B and B, vapor pressures of both A and B (separately) would be less

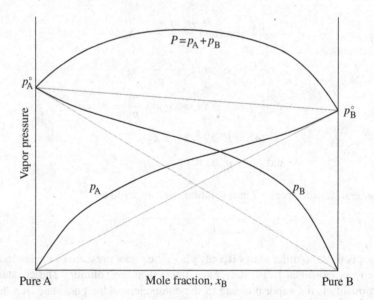

Fig. 1.1 Schematic representation of positive deviation from Raoult's law (broken lines represent ideal behavior)

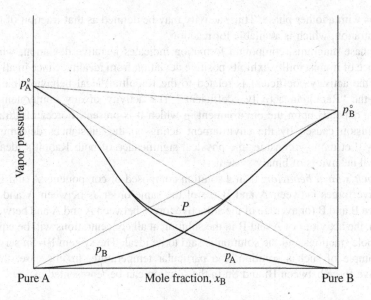

Fig. 1.2 Schematic representation of negative deviation from Raoult's law (broken lines represent ideal behavior)

than expected according to Raoult's law. This is known as the negative deviation from Raoultian behavior (Fig. 1.2). Systems with intermetallic compounds exhibit negative deviation since the attractive force between the components is large.

Activity: In actual solution the vapor pressure of a component is not directly proportional to the mole fraction of that component. It may be either greater or lesser than that expected from the solution if it obeyed Raoult's law. We can now define "activity," a_A of component A in the solution.

For ideal solutions, we have $p_A = p_A^o.x_A$
In case of nonideal solutions, $p_A \neq p_A^o.x_A$

So, in order to maintain equality in case of nonideal solutions, we introduce a new term activity, a_A, hence,

$$p_A = p_A^o.a_A \text{ and } p_B = p_B^o.a_B \tag{1.36}$$

Thus, for an ideal solution, $a_A = x_A$. In order to account for any deviation from ideality, we introduce a factor, γ_A, and write: $a_A = \gamma_A.\ x_A.\ \gamma_A$ is known as Raoultian activity coefficient. It may be greater or less than unity for a positive or negative deviation, respectively. For pure A: $x_A = 1$, $a_A = 1$, and hence, $\gamma_A = 1$. Thus, a pure substance having unit activity is said to be in its "standard state."

Since the vapor pressure of a substance is a measure of its attraction to the solution in which it exists, it is therefore a measure of its availability for reaction, perhaps with another phase. Thus, activity may be defined as that fraction of molar concentration, which is available for reaction.

In phase diagrams, compound formation indicates negative deviation, whereas existence of immiscibility exhibits positive deviation from ideality. Since in all these cases, the activity coefficient is related to the Raoultian ideal behavior line, it is called the "Raoultian activity coefficient." The activity of any component in a mixture depends upon the environment in which it is present. Forces of attraction or repulsion caused by the environment acting on the substances determine its activity. Let us now consider the physical significance of both Raoult's ideal and nonideal behaviors of binary systems.

Raoult's ideal behavior: If in a solution composed of components A and B, the attractive forces between A and B are of the same order as between A and A or between B and B (or average of the attractive forces between A and A and between B and B), the activities of A and B in the solution at all concentrations will be equal to their mole fractions and the solution is said to be ideal. The system Bi–Sn serves as an example of such a solution at a particular temperature. In this case, the net attractive force between Bi and Sn in the solution can be represented as:

$$(Bi \leftrightarrow Sn) = \frac{1}{2} \{(Bi \leftrightarrow Bi) + (Sn \leftrightarrow Sn)\}$$

In general, the ideal behavior of any solution A–B can be expressed as:

$$(A \leftrightarrow B) = (A \leftrightarrow A) = (B \leftrightarrow B) = \frac{1}{2} \{(A \leftrightarrow A) + (B \leftrightarrow B)\}$$

Positive deviation: When the net attractive force between components A and B is less than the average those between A–A and B–B, the solution A–B exhibits positive deviation from Raoult's law. In terms of the forces of attraction, this can be represented as:

$$(A \leftrightarrow B) < \frac{1}{2} \{(A \leftrightarrow A) + (B \leftrightarrow B)\}$$

In this case, the Raoultian activity coefficient is always greater than unity except when approaching the concentration of $x_A \rightarrow 1$. Pb–Zn liquid solutions show such a behavior at temperature above 1071 K. In general, the systems showing positive deviation are endothermic in nature.

Negative deviations occur when the attractive force between dissimilar components A and B is larger than the average those between A–A and B–B, that is, $(A \leftrightarrow B) > \frac{1}{2} \{(A \leftrightarrow A) + (B \leftrightarrow B)\}$. Negative deviations generally indicate a tendency for compound formation. For example, formation of Mg_3Bi_2 in a Mg–Bi system shows such a behavior. The systems exhibiting negative deviation are usually exothermic.

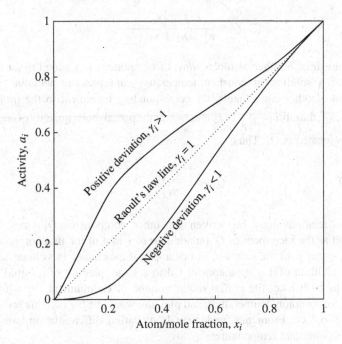

Fig. 1.3 Variation of activity with mole fraction of a component i in ideal and nonideal solutions showing positive and negative deviations from Raoult's ideal behavior

Occasionally, both negative and positive deviations from Raoult's law occur in the same system. Zn–Sb, Cd–Bi, and Cd–Sb systems are outstanding examples of such combined deviations. Ideal, positive, and negative behaviors are shown in Fig. 1.3.

Partial molar quantities: While discussing thermodynamics of solutions, a question is raised as how to express the molar quantities of a component in the solution. When two liquids are mixed, the total volume of the solution is often not equal to the sum of the individual volumes before mixing. This is due to the difference between the inter-atomic forces in the pure substance and in the solution. For example, on mixing liquid A and liquid B, if ΔV is the change in volume on formation of the solution, one is really confused in assessing the individual contribution made by A and B in the expansion or contraction that occurs on mixing. Similar problems may arise for other thermodynamic properties of the components in solutions. The problem can, however, be sorted out by the introduction of partial molar quantities. Since the same general treatment is applicable to any extensive thermodynamic quantity such as volume, energy, entropy, and free energy, we shall use the symbol Q to represent any one of these. Prime is used to indicate any arbitrary amount of solution rather than 1 mol, molar quantities are represented as unprimed.

Let us assume that Q' and Q represent the total quantity of solution and the molar solution, respectively. If n_1, n_2, n_3 ... are number of moles of components 1, 2, 3, ..., respectively, in the solution, we can write:

$$Q = \frac{Q'}{n_1 + n_2 + n_3 + \dots} \qquad (1.37)$$

If an infinitesimal number of moles, dn_1, of component 1 is added to an arbitrary quantity of a solution at constant temperature and pressure without changing the amount of other constituents, the corresponding increment in the property Q' is noted as dQ', the ratio $\left[\frac{\partial Q'}{\partial n_1}\right]_{P,T,n_2,n_3}$ is known as the partial molar quantity of component 1 and is designated as $\overline{Q}_1$. Thus,

$$\overline{Q}_1 = \left[\frac{\partial Q'}{\partial n_1}\right]_{P,T,n_2,n_3} \qquad (1.38)$$

Analogous relations may be written for other components. $\overline{Q}_1$ may also be represented as the increment of Q' on addition of 1 mol of the first component to a very large quantity of the solution. For example, if the change in volume accompanying the addition of 1 g mol/atom of Cd to a large quantity of a liquid alloy is observed to be 6.7 cc, the partial molar volume of cadmium in the alloy at the particular composition, temperature, and pressure would be 6.7 cc. This is expressed as $\overline{V}_{Cd} = 6.7$ cc. From the fundamentals of partial differentiation, we have at constant pressure and temperature:

$$dQ' = \left[\frac{\partial Q'}{\partial n_1}\right]_{n_2,n_3} dn_1 + \left[\frac{\partial Q'}{\partial n_2}\right]_{n_1,n_3} dn_2 + \left[\frac{\partial Q'}{\partial n_3}\right]_{n_1,n_2} dn_3 + \dots \qquad (1.39)$$

or $dQ' = \overline{Q}_1 dn_1 + \overline{Q}_2 dn_2 + \overline{Q}_3 dn_3 + \dots \qquad (1.40)$

If we add to a large quantity of solution n_1 moles of component 1, n_2 moles of component 2, and so on, the increment in Q' after mixing is given as $n_1\overline{Q}_1 + n_2\overline{Q}_2 + \dots$.

If we now mechanically remove a portion containing $n_1 + n_2 + n_3$ moles, the extensive quantity Q' for the main body of solution is now decreased by $(n_1 + n_2 + n_3)\, Q$. Since at the end of these addition and removal steps, the main body of the solution is the same in composition and amount as it was initially, Q' has the same value finally as it did initially, and the increment in Q' accompanying the individual addition is equal to the decrement accompanying their mass withdrawal,

$$(n_1 + n_2 + n_3 + \dots)\, Q = n_1\overline{Q}_1 + n_2\overline{Q}_2 + n_3\overline{Q}_3 + \dots$$

Dividing by $(n_1 + n_2 + n_3 + \dots)$ and noting that $\left[\frac{n_i}{n_1+n_2+n_3+\dots}\right] = x_i$, we get:

$$Q = x_1\overline{Q}_1 + x_2\overline{Q}_2 + x_3\overline{Q}_3 + \dots \qquad (1.41)$$

On multiplying by $(n_1 + n_2 + n_3 + \cdots)$, we can write as:

$$Q' = n_1\overline{Q}_1 + n_2\overline{Q}_2 + n_3\overline{Q}_3 + \cdots \tag{1.42}$$

For a binary solution, Eqs. 1.41 and 1.42 are expressed as:

$$Q = x_1\overline{Q}_1 + x_2\overline{Q}_2 \tag{1.41a}$$

$$Q' = n_1\overline{Q}_1 + n_2\overline{Q}_2 \tag{1.42a}$$

On differentiation of Eq. 1.42a, we get:

$$dQ' = n_1 d\overline{Q}_1 + n_2 d\overline{Q}_2 + \overline{Q}_1 dn_1 + \overline{Q}_2 dn_2 \tag{1.43}$$

From Eqs. 1.40 and 1.43, for a binary solution, we can write:

$$n_1 d\overline{Q}_1 + n_2 d\overline{Q}_2 = 0 \tag{1.44}$$

On dividing by $(n_1 + n_2)$, we get,

$$x_1 d\overline{Q}_1 + x_2 d\overline{Q}_2 = 0 \tag{1.45}$$

Equation 1.45 is one of the forms of the Gibbs–Duhem equation. This is the most important form of all the equations dealing with solutions, and a number of subsequent relations have been derived from this. From Eqs. 1.41a and 1.45, expressions for calculating the partial molar quantities of components 1 and 2 can be derived as:

$$\overline{Q}_1 = Q + (1 - x_1)\frac{dQ}{dx_1} \tag{1.46}$$

$$\text{and } \overline{Q}_2 = Q + (1 - x_2)\frac{dQ}{dx_2} \tag{1.47}$$

Chemical potential: The general equation for the free energy change of a system with temperature and pressure, $dG = VdP - SdT$, does not take into account any variation in free energy due to concentration changes. The total free energy of a system must vary when small amounts of constituents dn_A, dn_B, and so on are introduced into a homogeneous solution. Thus, the free energy of a system is dependent on four different variables or $G = f(T, P, V, n_i)$, where n_i denotes the addition of n_i moles of a constituent i. The change in free energy can be expressed as total differential using three of the four variables. Any infinitesimal free energy change in terms of a change in temperature, pressure, and composition can be expressed as:

$$dG = \left(\frac{\partial G}{\partial T}\right)_{P,n_i} dT + \left(\frac{\partial G}{\partial P}\right)_{T,n_i} dP + \sum \left(\frac{\partial G}{\partial n_i}\right)_{T,P} dn_i$$

$$= -SdT + VdP + \sum \left(\frac{\partial G}{\partial n_i}\right)_{T,P} dn_i \tag{1.48}$$

The coefficient $\left(\frac{\partial G}{\partial n_i}\right)_{T,P}$ is called the "chemical potential" and is denoted by μ_i.

$$\therefore dG = -SdT + VdP + \sum \mu_i dn_i \tag{1.49}$$

At constant pressure and temperature the first two terms are zero in Eq. 1.49 and if the system is at equilibrium, $dG = 0$, hence, $\sum \mu_i dn_i = 0$.

Physical meaning of chemical potential: Consider the change in free energy (dG) of a system produced by the addition of dn_A mole of component A at constant pressure and temperature. This change in free energy can be expressed as: $dG = \mu_A dn_A = \overline{G}_A dn_A$, where $\overline{G}_A$ is the partial molar free energy of component A in the solution. Chemical potential of either 1 g mol or 1 g atom of the substance dissolved in a solution of definite concentration is the partial molar free energy. Thus, $\overline{G}_A = \mu_A = \left(\frac{\partial G}{\partial n_A}\right)_{T,P}$.

Chemical potential in ideal solutions: In order to calculate the value of a chemical potential, it is necessary to obtain a relation between μ and some measurable property. Consider the possibility of calculating the chemical potential of a single gas in an ideal gaseous mixture, and then applying this result to other states of matter. It follows therefore that for a very small change of pressure at constant temperature, $dG = VdP$ (but since $\overline{G}_A = \mu_A$ for 1 mol of A).

Then $d\mu = VdP$, and for 1 mol of perfect gas, $(V = RT/P)$

$$\therefore d\mu = RT \left(dP/P\right)$$

on integration: $\mu = RT \ln P + C$

when $P = 1$ atm, $\mu = C = \mu^o$ (standard chemical potential)

$$\therefore \mu = \mu^o + RT \ln P$$

Thus, the chemical potential of component A in a gaseous mixture exerting partial pressure p_A can be calculated from the equation:

$$\mu_A = \mu_A^o + RT \ln p_A$$

$$\text{or } \Delta\mu_A = \mu_A - \mu_A^o = RT \ln p_A$$

$$\text{and } \Delta\mu_B = \mu_B - \mu_B^o = RT \ln p_B \tag{1.50}$$

Since $\Delta\mu$ (similar to the change in free energy of a system) is an extensive property, it depends on the quantity or mass of the system; therefore, the total free energy change associated with the formation of 1 g mol of the solution having x_A mole fraction of the constituent A and x_B mole fraction of the constituent B is the sum of the product of appropriate changes in chemical potential multiplied by the mole fractions of the components. Thus,

$$\Delta\mu_{sol} = \Delta G_{(1 \text{ g mol sol})} = x_A\Delta\mu_A + x_B\Delta\mu_B$$

$$\therefore \Delta G_{(sol)} = RT(x_A \ln p_A + x_B \ln p_B) \tag{1.51}$$

These equations applicable to gases can also be used for liquid solutions. If the solution at a given temperature T is at equilibrium with its own vapors, the chemical potential of each component in the gaseous phase and in the liquid phase must be equal.

The partial pressure p_A of the component A in a binary solution of A and B can be expressed in terms of its mole fraction x_A (from Raoult's law): $p_A = p_A^o \cdot x_A$

$$\therefore x_A = \frac{p_A}{p_A^o} \text{ (ideal) and } a_A = \frac{p_A}{p_A^o} \text{ (non ideal)}$$

Thus, from Eq. 1.50, we get:

$$\Delta\mu_A = \mu_A - \mu_A^o = \overline{G}_A - G_A^o = RT \ln \frac{p_A}{p_A^o} = RT \ln x_A \tag{1.52}$$

Where G_A^o is the free energy of 1 g mol of component A in its standard state. Hence, from Eq. 1.52, for 1 g mol of a binary solution consisting x_A mole fraction of A and x_B mole fraction of B, we have:

$$\Delta G_{(sol)} = \Delta G^{M,id} = RT(x_A \ln x_A + x_B \ln x_B) \text{ for an ideal solution} \tag{1.53}$$

$$\Delta G_{(sol)} = \Delta G^M = RT(x_A \ln a_A + x_B \ln a_B) \text{ for a non ideal solution} \tag{1.54}$$

The free energy of 1 mol of component i in the solution designated as $\overline{G}_i$ is known as the partial molar free energy of the component i. It represents simply the change in total free energy of the solution when 1 mol of component i is added to a large amount of the solution.

From the first and second laws of thermodynamics, we have derived: $dG = VdP - SdT$ (Eq. 1.20), and hence, at constant temperature: $dG = VdP$ and for 1 g mol of an ideal gas: $V = \left(\frac{RT}{P}\right)$.

$$\therefore dG = \frac{RT}{P} dP = RTd\ln P \tag{1.55}$$

The term activity of the component in a solution has been defined as $a_i = \frac{p_i}{p_i^o}$. In case of a nonideal gas, the deviation from ideal behavior may be accounted by introducing in a term fugacity Ø, for pressure. Hence, Eq. 1.55 can be modified as:

$$dG = RTd \ln \emptyset \qquad (1.56)$$

Thus, activity can be defined as the ratio of the fugacity of the substance in the state in which it happens to be to its fugacity in its standard state. That is, $a = (\emptyset/\emptyset^o)$. Substituting $\emptyset = a. \emptyset^o$ in Eq. 1.56, we get $dG = RTd \ln(a. \emptyset^o)$. As fugacity in its standard state $(\emptyset^o)$ is constant at a particular temperature and composition, we can write:

$$dG = RTd \ln a \qquad (1.57)$$

The standard state of a substance is commonly chosen as the pure liquid and solid at one atmospheric pressure and temperature under consideration. The activity of a pure substance in its standard state is unity.

Since under standard conditions the free energy of a component i expressed by G_i^o and its partial molar free energy in the solution by $\overline{G}_i$, integration of Eq. 1.57 gives:

$$\int_{G_i^o}^{\overline{G}_i} dG = \int_{a_i^o}^{a_i} RTd \ln a = \overline{G}_i - G_i^o = RT \ln a_i - RT \ln a_i^o \qquad (1.58)$$

since $a_i^o = 1, \overline{G}_i - G_i^o = RT \ln a_i = \Delta \overline{G}_i^M$

The difference $(\overline{G}_i - G_i^o)$ represents the partial molar free energy of mixing (on formation of the solution) of the component i and is denoted by $\Delta \overline{G}_i^M$. This is simply the free energy change accompanying the dissolution of 1 mol of i in the solution. In a binary solution, A–B, containing x_A and x_B mole fractions of A and B, respectively, the integral molar free energy of mixing can be estimated as:

$$\Delta G^M = x_A \Delta \overline{G}_A^M + x_B \Delta \overline{G}_B^M = RT(x_A \ln a_A + x_B \ln a_B) \qquad (1.59)$$

since for an ideal solution $a_i = x_i$

$$\Delta G^{Mid} = RT(x_A \ln x_A + x_B \ln x_B) \qquad (1.60)$$

If at constant T and P, n_A number of mole of A and n_B of moles of B are mixed to form a binary solution, the free energy before mixing $= \left(n_A G_A^o + n_B G_B^o\right)$ and after mixing $= (n_A \overline{G}_A + n_B \overline{G}_B)$. The free energy change for the entire body of solution due to mixing, $\Delta G^{\prime M}$, referred as the integral free energy of mixing, is the difference between the two quantities, that is,

$$\Delta G'^M = n_A \overline{G}_A + n_B \overline{G}_B - \left(n_A\, G_A^o + n_B G_B^o \right) = n_A \left(\overline{G}_A - G_A^o \right) + n_B \left(\overline{G}_B - G_B^o \right)$$
$$= n_A \Delta \overline{G}_A^M + n_B \Delta \overline{G}_B^M$$
$$= RT(n_A \ln a_A + n_B \ln a_B)$$

$$(1.61)$$

Dividing by $(n_A + n_B)$, we get integral molar free energy of mixing for nonideal and ideal solutions as expressed by Eqs. 1.59 and 1.60, respectively. The integral molar entropy of mixing for an ideal solution is given as:

$$\Delta S^{\mathrm{Mid}} = -R \left(x_A \ln x_A + x_B \ln x_B \right) \qquad (1.62)$$

Equilibrium constant: Consider a general chemical reaction taking place between different numbers of moles of reactants to form different numbers of moles of products at constant temperature and pressure.

$$lL + mM + \cdots = qQ + rR + \cdots$$

The change in free energy for the above reaction may be expressed as:

$$\Delta G = q\overline{G}_Q + r\overline{G}_R + \cdots - l\overline{G}_L - m\overline{G}_M - \cdots \qquad (1.63)$$

and in the special case when all products and reactants are in their standard state as:

$$\Delta G^o = qG_Q^o + rG_R^o + \cdots - lG_L^o - mG_M^o - \cdots \qquad (1.64)$$

On subtracting Eq. 1.63 from Eq. 1.64, we get:

$$\Delta G^o - \Delta G = q\left(G_Q^o - \overline{G}_Q \right) + r\left(G_R^o - \overline{G}_R \right) + \cdots - l\left(G_L^o - \overline{G}_L \right) - m\left(G_M^o - \overline{G}_M \right)$$
$$- \cdots$$

substituting $\overline{G}_i - G_i^o = RT \ln a_i$, we get:

$$\Delta G^o - \Delta G = -qRT \ln a_Q - rRT \ln a_R - \cdots + lRT \ln a_L + mRT \ln a_M + \cdots$$

$$\Delta G = \Delta G^o + RT \ln \frac{a_Q^q . a_R^r}{a_L^l . a_M^m}$$

Thermodynamic equilibrium constant for the above chemical reaction,

$$K = \frac{a_Q^q . a_R^r}{a_L^l . a_M^m}$$

Since, at equilibrium,

$$\Delta G = 0, \Delta G^o = -RT \ln K \tag{1.65}$$

Regular solutions: In the case of nonideal solutions, it is still possible to assume random mixing in certain cases, but the enthalpy of mixing will no longer be zero because there will be heat changes due to changes in binding energy. This assumption of random mixing can only be made where there is a small deviation from ideal behavior, so that the enthalpy of mixing is quite small. Solutions of this type are called regular solution. For regular solutions the entropy of mixing is the same as for the ideal solution, hence,

$$\Delta H^M = \Delta G^M + T\Delta S^M = RT(x_A \ln a_A + x_B \ln a_B) - RT(x_A \ln x_A + x_B \ln x_B)$$
$$= RT(x_A \ln \gamma_A + x_B \ln \gamma_B)$$

$$\tag{1.66}$$

where γ_A and γ_B are the actively coefficient of the components A and B, respectively. Since $\Delta H^M = x_A \Delta \overline{H}_A^M + x_B \Delta \overline{H}_B^M$,

$$\Delta \overline{H}_A^M = RT \ln \gamma_A \text{ and } \Delta \overline{H}_B^M = RT \ln \gamma_B \tag{1.67}$$

Gibbs–Duhem integration: Thermodynamic equations for calculation of excess free energy and integral molar free energy of a solution need both the activity coefficient and the activity of all the components of the solution. However, experimental techniques, namely, chemical equilibria, vapor pressure, and electrochemical, can measure activity of only one component. For example,

1. Activity of mercury in mercury–thallium alloy is obtained by measuring the partial pressure of mercury over the alloy at a temperature, where, $p_{Hg} \gg p_{Tl}$, that is, at the working temperature partial pressure of thallium is negligible as compared to that of mercury. The same experiment cannot be used for measuring the partial pressure (hence activity) of thallium accurately.
2. Activity of carbon in Fe–C (or steel) can be determined by equilibrating CO–CO_2 or H_2–CH_4 gaseous mixture with steel at a particular temperature. However, we cannot obtain the activity of iron by this method. Similarly, a_S in Fe–S can be measured by using H_2–H_2S gaseous mixture but not a_{Fe} .
3. Concentration galvanic cell like Cd(l)/LiCl – KCl + $CdCl_2$/Cd – Pb(l) determines the activity of cadmium only by measuring the open circuit emf of the cell, which is related to the partial molar free energy of mixing and activity as follows:

$$\Delta \overline{G}_{Cd}^M = RT \ln a_{Cd} = -n\text{FE} \tag{1.68}$$

We cannot use the cell Pb(l)/LiCl – KCl + $PbCl_2$/Cd – Pb(l) to get a_{Pb} because the formation of more stable $CdCl_2$ will cause exchange reaction of the type, $PbCl_2$(l) + Cd(l) = $CdCl_2$(l) + Pb(l). Thus, cell emf will not be a true representative of the cell reaction but will represent a mixed potential.

Now the question is how to get the activity of the second component in a binary system. In order to get the activity of the second component, we must couple the activity and atom/mole fractions of both the components with the aid of the Gibbs–Duhem equation:

$$\sum x_i d\overline{Q}_i = 0$$

where $\overline{Q}_i$ is any partial molar extensive property.

Since activity of a component is related to the partial molar free energy, we can write the Gibbs–Duhem equation as under:

$$x_A d\Delta \overline{G}_A^M + x_B d\Delta \overline{G}_B^M = 0, \text{ since } \Delta \overline{G}_i^M = RT \ln a_i$$

$$\therefore RT(x_A d \ln a_A + x_B d \ln a_B) = 0$$

$$\text{or } d \ln a_A = -\frac{x_B}{x_A} d \ln a_B$$

If the variation of a_B with composition is known, the value of $\ln a_A$ at the composition $x_A = x_A$ can be obtained by integration of the above equation. Since problems dealing with Gibbs–Duhem integration have not been included in this book, there is no need to derive expressions for the calculation of activities. Interested readers may consult books on thermodynamics.

Dilute solutions: According to Henry's law, the partial pressure of a solute in a dilute solution is proportional to mole/atom fraction. If B is a solute, we can write: $p_B \propto x_B$.

$$\text{that is, } p_B = k.x_B \tag{1.69a}$$

dividing by p_B^o, we get:

$$\frac{p_B}{p_B^o} = \frac{k}{p_B^o}.x_B, \text{ that is, } a_B = k'x_B \tag{1.69b}$$

Hence, $a_B \propto x_B$

The constant, known as the activity coefficient of the solute B at infinite dilution, is equal to the slope of the curve at zero concentration of B and is designated by γ_B^o (Fig. 1.4). Like Raoult's law, Henry's law is valid within a concentration range. The extent varies from one system to another, but it is valid only at low concentrations.

In concentrated solutions, the standard state has been defined as unit atmospheric pressure and unit activity, that is, pure substance at any temperature. In dilute solutions, relative standard states other than pure substances are being used. Henry's law offers two such standard states called alternative standard states.

1. Infinitely dilute, atom/mole fraction standard state
2. Infinitely dilute, wt% (w/o or %) standard state

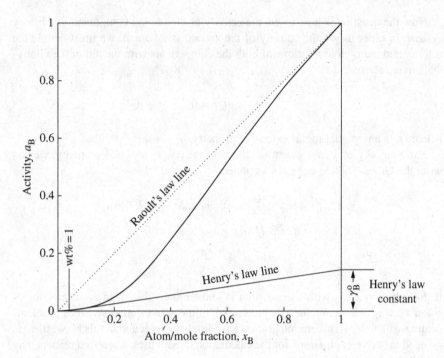

Fig. 1.4 Conversion of activities from one standard state to another

Solubility of gases: It is important to note that the validity of Henry's law depends upon the proper choice of solute species. For example, nitrogen dissolves as different species in water and in liquid iron.

(a) In water, nitrogen dissolves molecularly as N_2:

$N_2(g) \leftrightarrows N_2$ (dissolved in water)

$K = \frac{a_{N_2}}{p_{N_2}}$ (as solubility of N_2 in water is low, according to Henry's law: Eq. 1.69b), we can write:

$a_{N_2} = k' x_{N_2}$

$\therefore K = \frac{k' x_{N_2}}{p_{N_2}}$

$$\text{Hence,} \, x_{N_2} \, (\text{solubility}) = \frac{K}{k'} \cdot p_{N_2} = k'' \cdot p_{N_2} \tag{1.70a}$$

Thus, the solubility of nitrogen in water is proportional to the partial pressure of nitrogen gas in equilibrium with water. Solubility can be expressed as mole fraction, cc per 100 g of water, or any other unit.

(b) In the second case under consideration, nitrogen dissolves atomically in solid or liquid metals:

$N_2(g) \leftrightarrows 2[N] \, (\text{in Fe})$

$K = \dfrac{[a_N]^2}{p_{N_2}} = \dfrac{[k' x_N]^2}{p_{N_2}}$

$\therefore [x_N] \, (\text{solubility}) = \dfrac{\sqrt{K}}{k'} \sqrt{p_{N_2}} = k'' \sqrt{p_{N_2}}$

Both the cases are in accordance with the experiment. Since all the common diatomic gases N_2, O_2, H_2, and so on dissolve atomically in metals, the general expression for solubility is given as:

$$s = k \sqrt{p_{N_2}} \tag{1.70b}$$

This is known as Sievert's law and can be stated as—solubility of diatomic gases in metals is directly proportional to the square root of partial pressure of the gas in equilibrium with the metal.

Alternative standard states: In dilute solutions, two different standard states are used.

1. *Infinitely dilute, atom fraction standard state*: The standard state is so defined that the Henrian activity approaches the atom fraction at infinite dilution.

$$a_B^H = x_B \text{ as } x_B \to 0 \tag{1.71}$$

a_B^H denotes Henrian activity of B. In other words, in the concentration range where Henry's law is obeyed, $a_B^H = x_B$, that is, Henrian activity = atom fraction. Beyond this concentration range, the activity can be related to the atom fraction by the expression:

$$a_B^H = f_B.x_B \tag{1.72}$$

where f_B is the Henrian activity coefficient of B, relative to the infinitely dilute atom fraction standard state.

The relation between the activity of B relative to the pure substance standard state and the activity of B relative to the infinitely dilute atom fraction standard state is given by

$$\frac{\text{activity of B relative to pure substance standard state}}{\text{activity of B relative to the infinitely atom fraction standard state}}$$

$$= \gamma_B^o \, (\text{at constant } x_B)$$

$$\left[a_i = \gamma_i.x_i, \; \gamma_i = \frac{a_i}{x_i} = \frac{a_i^R}{a_i^H} = \gamma_i^o \right]$$

where γ_B^o is the Raoultian activity coefficient of B at infinite dilution.

The free energy change accompanying the transfer of 1 mol of B from pure B as standard state to the infinitely dilute, atom fraction standard state may be calculated as follows:

B (pure substance standard state) $\rightarrow$ B (dilute, atom fraction standard state)

$$\Delta G^o = \Delta G_B^o(\text{pure substance standard state})$$

$$-\Delta G_B^o(\text{infinitely dilute, atom fraction standard state})$$

$$= RT \ln \left[\frac{\text{activity of B relative to pure substance standard state, } a_B^R}{\text{activity of B relative to the infinitely dilute atom fraction standard state, } a_B^H} \right]$$

$$= RT \ln \gamma_B^o \tag{1.73}$$

2. *Infinitely dilute weight percent standard state*: This is the most widely used standard state in metallurgy and may be so defined that the Henrian activity approaches the wt% at infinite dilution, that is,

$$a_B^{wt\%} = \text{wt\%B as wt\%B} \rightarrow 0 \tag{1.74}$$

Assuming that the solution obeys Henry's law up to 1 wt% B, then Henrian activity $a_B^{wt\%}$ is unity at this concentration and the standard state is 1 wt% solution. Deviation from the equality is measured in terms of Henrian activity coefficient, relative to infinitely dilute wt% standard state, that is,

$$a_B^{wt\%} = f_B.\text{wt\%B} \tag{1.75}$$

In both the cases, f_B has been used for Henrian activity coefficient of B.

The relation between the activity of B, a_B relative to the pure substance standard state and the activity of B relative to the infinitely dilute wt% standard state is given by

$$\frac{\text{activity of B relative to pure substance standard state, } a_B^R}{\text{activity of B relative to the infinitely dilute wt\% standard state, } a_B^{wt\%}}$$

$$= \gamma_B^o \frac{x_B}{\text{wt\%B}} \text{ (at constant } x_B)$$

$$\left[\text{In Henrian activity } a_B^R = \gamma_B^o.x_B \right]$$

The free energy change accompanying the transfer of 1 mol of B from pure B as standard state to the infinitely dilute wt% standard state may be calculated as follows:

B(pure substance standard state) $\rightarrow$ B (dilute wt%standard state)

$\Delta G^o = \Delta G_B^o$(pure substance standard state) $- \Delta G_B^o$(infinitely dilute wt%standard state)

$$= RT \ln \left[\frac{\text{activity of B relative to pure substance standard state, } a_B^R}{\text{activity of B relative to the infinitely dilute wt\%standard state, } a_B^{wt\%}} \right]$$

$$= RT \ln \left[\gamma_B^o \frac{x_B}{\text{wt\%B}} \right]$$

$$(1.76)$$

$$= RT \ln \left[\gamma_B^o \frac{M_A}{100 M_B} \right]$$

$$(1.77)$$

$$= RT \ln \gamma_B^o + RT \ln \left[\frac{M_A}{100 M_B} \right]$$

where M_A and M_B refer to atomic weights of A and B, respectively.

The atom/mole fraction of B in solution of A – B can be expressed as:

$$x_B = \frac{\text{wt\%B}/M_B}{(\text{wt\%B}/M_B) + (100 - \text{wt\%B}/M_A)} = \frac{\text{wt\%B}/M_B}{100/M_A}$$

(at infinitely dilute concentration of the solute B, *wt %* B is negligible as compared to 100), hence,

$$\frac{x_B}{\text{wt\%B}} = \frac{M_A}{100 M_B}$$

$$(1.78a)$$

$$\therefore \text{wt\%B} = \frac{x_B . 100 . M_B}{M_A}$$

$$(1.78b)$$

Relation between different standard states: According to the Raoult's law for ideal solution: $a_B^R = x_B$. In Henrian solution, $a_B^R = \gamma_B^o . x_B$, where γ_B^o is the Raoultian activity coefficient at infinite dilution. This can also be expressed as:

$$\frac{a_B^R}{\gamma_B^o . x_B} = 1$$

$$(1.79)$$

Henry's law gives $a_B^H = f_B . x_B$ ($f_B = 1$ for Henrian solution)

$$\therefore a_B^H = x_B, \text{ that is, } \frac{a_B^H}{x_B} = 1$$

$$(1.80)$$

Similarly,

$$a_B^{1 \text{ wt\%}} = \text{wt\%B, that is, } \frac{a_B^{1 \text{ wt\%}}}{\text{wt\%B}} = 1 \tag{1.81}$$

from Eqs. 1.79, 1.80, and 1.81, we get:

$$\frac{a_B^R}{\gamma_B^o \cdot x_B} = \frac{a_B^H}{x_B} = \frac{a_B^{1 \text{ wt\%}}}{\text{wt\%B}} \tag{1.82}$$

$$\text{Hence, } \frac{a_B^R \text{ (pure)}}{a_B \text{ (dilute wt\%)}} = \frac{\gamma_B^o \cdot x_B}{\text{wt\%B}} \tag{1.83}$$

For example in a Fe–Si system, when $x_{Si} \rightarrow 0$

$$\frac{a_{Si} \text{ (pure)}}{a_{Si} \text{ (dilute wt\%)}} = \frac{\gamma_{Si}^o \cdot x_{Si}}{\text{wt\%Si}} \text{ since } \frac{x_{Si}}{\text{wt\%Si}} = \frac{M_{Fe}}{100 M_{Si}}$$

$$\therefore \frac{a_{Si} \text{ (pure)}}{a_{Si} \text{ (dilute wt\%)}} = \frac{\gamma_{Si}^o \cdot M_{Fe}}{100 M_{Si}}$$

Activity and interaction coefficients in multicomponent systems: Although we have considered only binary solutions so far, in metallurgical processes, we rarely deal with binary solutions. Hence, we must take into account the interaction between various solutes in multicomponent systems. If the components of binary solutions interact with one another, those of multicomponent solutions will certainly interact in a more complex manner.

As an example, take a dilute solution of carbon in liquid iron. The Henrian activity coefficient of carbon in the binary solution, f_C^C, is raised by the addition of small amounts of sulfur and lowered by the addition of small amounts of chromium in the solution. In liquid iron, several solutes are present in dilute concentrations, and each thermodynamic property of the system is influenced by the change in the interatomic forces. Interaction causes marked changes in solute activities and the data for binary solutions do not apply to more complex systems.

Probable effects of one solute on the thermodynamic behavior of another solute can be expressed by the interaction parameter (ε) and interaction coefficient (e), which are defined as:

$$\varepsilon_i^j = \left[\frac{\partial \ln \gamma_i}{\partial x_j} \right]_{x_1 \rightarrow 1} \text{ and } e_i^j = \left[\frac{\partial \log f_i}{\partial \%j} \right]_{\text{wt\%}1 \rightarrow 100} \tag{1.84}$$

Thus, (ε) and (e) are related to the alternative standard states in dilute solutions. If the solute i interacts more strongly with the solute j than with the solvent 1 (in the system $1 - i - j$), the activity coefficient of i, $\gamma_i/(f_i)$ decreases by the addition of j, so ε_i^j. (and also e_i^j) is a negative quantity. Conversely, if j interacts more strongly with the solvent than i, $\gamma_i/(f_i)$ will tend to increase and ε_i^j and also e_i^j will be positive (i.e., ε_i^j/e_i^j is positive, if j interacts more strongly with the solvent).

Thus, if carbon is the solute considered, f_C is lowered by the addition of solutes that tend to form more stable carbide, for example, Cr, V, or Nb; on the other hand, S and P raise the f_C. It is important to note that adequate representation of the equilibria encountered in steelmaking is possible only if the effects of the various solutes on the activity coefficient are taken into account. For instance, the presence of 2 wt% C in liquid iron increases the activity coefficient of silicon by a factor of 2.

The interaction coefficients result from the Taylor series expansion of the partial molar excess Gibbs free energy of mixing $\overline{G}_i^{xs}$ of the solute component i, in the solvent, iron denoted as component 1. In atom fraction as the standard state we have:

$$\overline{G}_i^{xs} = RT \ln \left[\gamma_i\right]_{T,P,x_1,x_2,...x_n} \tag{1.85}$$

Expansion of Eq. 1.85 in case of infinitely dilute solution as reference state:

$$RT \ln \gamma_i = RT \left[\ln \gamma_i^o\right]_{T, P, x_1 \to 1} + \sum_{j=2}^{m} RT \left[\frac{\partial \ln \gamma_i}{\partial x_j} x_j\right]_{T, P, x_1 \to 1}$$

$$+ \sum_{j=2}^{m} \frac{1}{2} RT \left[\frac{\partial^2 \ln \gamma_i}{\partial x_j^2} x_j^2\right]_{T, P, x_1 \to 1} + \sum_{j=2}^{m} \sum_{k=2}^{m} RT \left[\frac{\partial^2 \ln \gamma_i}{\partial x_j \partial x_k} x_j x_k\right]_{T, P, x_1 \to 1} + \cdots \tag{1.86}$$

Accuracy of data does not permit to account for the third-order terms, hence,

$$\ln \gamma_i = \ln \gamma_i^o + \sum_{j=2}^{m} \varepsilon_i^j x_j + \sum_{j=2}^{m} \rho_i^j x_j^2 + \sum_{j=2}^{m} \sum_{k=2}^{m} \rho_i^{j,k} x_j x_k \tag{1.87}$$

Equation 1.87 describes the thermodynamic properties of the ith component in an n-component system. By convention, γ_i is the activity coefficient based on atom fraction as the composition coordinate, γ_i^o is the value at infinite dilution.

Partial derivative of γ is named as interaction parameter.

$\varepsilon_i^j \equiv$ interaction parameter of j on i in $1-i-j$ system simply tells the effect of j on the activity coefficient of i.

$$\varepsilon_i^j = \left[\frac{\partial \ln \gamma_i}{\partial x_j}\right]_{x_1 \to 1} \text{ in } 1-i-j \text{ system (read as epsilon } j \text{ upon } i)$$

$$\varepsilon_i^i (\text{self interaction coefficient}) = \left[\frac{\partial \ln \gamma_i}{\partial x_i}\right]_{x_1 \to 1},$$

$$\varepsilon_2^3 = \left[\frac{\partial \ln \gamma_2}{\partial x_3}\right]_{x_1 \to 1} \quad (\text{epsilon } 3 \text{ upon } 2)$$

and ρ denotes second-order interaction parameter

$$\rho_i^j = \frac{1}{2}\left[\frac{\partial^2 \ln \gamma_i}{\partial x_j^2}\right]_{x_1 \to 1}$$

$$\rho_i^{j,k} = \left[\frac{\partial^2 \ln \gamma_i}{\partial x_j \partial x_k}\right]_{x_1 \to 1} \quad \text{(cross product)}$$

In a simplified manner for the system 1–2–3–4, we can write:

$$\ln \gamma_2 = \ln \gamma_2^o + \varepsilon_2^2 x_2 + \varepsilon_2^3 x_3 + \varepsilon_2^4 x_4 + \cdots + \rho_2^2 x_2^2 + \rho_2^3 x_3^2 + \cdots + \rho_2^{2,3} x_2 x_3$$
$$+ \cdots. \tag{1.88}$$

In a system of 1–2–3 (2 and 3 are solutes and 1 is a solvent) at constant temperature and pressure, it can be demonstrated that $\varepsilon_2^3 = \varepsilon_3^2$. This is known as the Wagner's Reciprocal Relationship.

In many instances it is convenient to use a weight percent (wt%) as the composition coordinate. Hence, the activity coefficient, f_i, in the zeroeth order form, $\log f_i^o$, disappears since the activity coefficient at infinite dilution is assigned the value of unity.

$$e_i^j = \left[\frac{\partial \log f_i}{\partial \%j}\right]_{\text{wt}\%1 \to 100} \quad \text{and} \quad e_2^3 = \left[\frac{\partial \log f_2}{\partial \%3}\right]_{\text{wt}\%1 \to 100}$$

$$f_i = 1, \quad f_i^o = 1, \quad \log f_i^o = 0$$

Taylor series expansion:

$$\log f_i = \left[\log f_i^o\right]_{\%1 \to 100} + \sum_{j=2}^{m}\left[\frac{\partial \log f_i}{\partial(\%j)}\right]_{\%1 \to 100}(\%j)$$

$$+ \sum_{j=2}^{m}\frac{1}{2}\left[\frac{\partial^2 \log f_i}{\partial(\%j)^2}\right]_{\%1 \to 100}(\%j)^2$$

$$+ \sum_{j=2}^{m}\sum_{k=2}^{m}RT\left[\frac{\partial^2 \log f_i}{\partial(\%j)\partial(\%k)}\right]_{\%1 \to 100}\partial(\%j)\partial(\%k) + \cdots \tag{1.89}$$

or

$$\ln f_i = \sum_{j=2}^{m} e_i^j(\%j) + \sum_{j=2}^{m} r_i^j(\%j)^2 + \sum_{j=2}^{m}\sum_{k=2}^{m} r_i^{(j,k)}\,\partial(\%j)\partial(\%k)$$

$$+\,...\tag{1.90}$$

where e and r denote first- and second-order interaction coefficients, respectively.

Relation between ε and e: (x, pure 2 standard state) = 2 (wt% hypothetical standard state)

$$K = \frac{a_2^{(wt\%)}}{a_2} = \frac{f_2^{(wt\%)}\cdot(wt\%2)}{x_2\gamma_2}\tag{1.91}$$

$$\ln\gamma_2 = \ln\left(\frac{wt\%2}{x_2}\right) + \ln f_2^{(wt\%)} - \ln K\tag{1.92}$$

or

$$\left[\frac{\ln\gamma_2}{\partial x_3}\right]_{x_1\to1} = \left[\frac{\partial\ln(\%2/x_2)}{\partial x_3}\right]_{x_1\to1} + \left[\frac{\partial\ln f_2^{(wt\%)}}{\partial(wt\%3)}\right]_{\%1\to100}\left[\frac{\partial(wt\%3)}{\partial x_3}\right]_{x_1\to1}\tag{1.93}$$

We know that:

$$\left[\frac{\partial\ln f_2^{(wt\%)}}{\partial(wt\%3)}\right]_{\%1\to100} = \left[\frac{2.303\,\partial\log f_2^{(wt\%)}}{\partial(wt\%3)}\right]_{\%1\to100} = 2.303 e_2^3\tag{1.94}$$

From Eqs. 1.93 and 1.94, we can write:

$$\therefore e_2^3 = \left[\frac{\partial\ln(\%2/x_2)}{\partial x_3}\right]_{x_1\to1} + 2.303 e_2^3\left[\frac{\partial(wt\%3)}{\partial x_3}\right]_{x_1\to1}\tag{1.95}$$

If M_1, M_2, M_3 are, respectively, the atomic weights of the elements 1, 2, and 3, %3 (wt% 3) can be expressed as:

$$\%3 = \frac{x_3 M_3 100}{x_1 M_1 + x_2 M_2 + x_3 M_3} = \frac{x_3 M_3 100}{x_3(M_3 - M_1) + x_2(M_2 - M_1) + M_1}\tag{1.96}$$

$$\left[\frac{\partial\%3}{\partial x_3}\right]_{x_1\to1} = 100\frac{M_3}{M_1}\tag{1.97}$$

$$\left[\frac{\partial\ln(\%2/x_2)}{\partial x_3}\right]_{x_1\to1} = \frac{M_1 - M_3}{M_1}\tag{1.98}$$

Substituting the above values in Eq. 1.95, we get:

$$\varepsilon_2^3 = 230.3 e_2^3 \frac{M_3}{M_1} + \frac{M_1 - M_3}{M_1}$$ (1.99)

or $\varepsilon_i^j = 230.3 \frac{M_j}{M_1} e_i^j + \frac{M_1 - M_j}{M_1}$

In 1–x–y–z system, considering only first-order interaction coefficients, the above relations can be summarized as:

$$\log f_x = \%x.e_x^x + \%y.e_x^y + \%z.e_x^z$$ (1.100)

$$\ln \gamma_x = \ln \gamma_x^o + \varepsilon_x^x x_x + \varepsilon_x^y x_y + \varepsilon_x^z x_z$$ (1.101)

where

$$\varepsilon_x^x = \left[\frac{\partial \ln \gamma_x}{\partial x_x} \right]_{x_1 \to 1} \text{ and } e_x^x = \left[\frac{\partial \log f_x}{\partial \%x} \right]_{wt\%1 \to 100}$$

$$\varepsilon_x^y = \left[\frac{\partial \ln \gamma_x}{\partial x_y} \right]_{x_1 \to 1} \text{ and } e_x^y = \left[\frac{\partial \log f_x}{\partial \%y} \right]_{wt\%1 \to 100}$$

Relation between e_2^3 and e_3^2 in 1–2–3 system: In the system 1–2–3, if 1 is the solvent and 2 and 3 are the solutes (at infinitely dilute concentrations), according to Wagner's reciprocal relationship:

$$\varepsilon_2^3 = \varepsilon_3^2$$

and according to Eq 1.99 ε and e are related as

$$\varepsilon_2^3 = 230.3 e_2^3 \frac{M_3}{M_1} + \frac{M_1 - M_3}{M_1}$$

If M_1 and M_3 are very close in atomic weights, we can approximate as

$$\varepsilon_2^3 = 230.3 e_2^3 \frac{M_3}{M_1}$$ (1.102)

and also

$$\varepsilon_3^2 = 230.3 e_3^2 \frac{M_2}{M_1}$$ (1.103)

$$\therefore e_2^3 = \frac{M_1}{230.3 M_3} \, \varepsilon_2^3 \qquad\qquad\qquad \text{(from Eq.1.102)}$$

$$= \frac{M_1}{230.3 M_3} \, \varepsilon_3^2 \qquad\qquad\qquad \text{since } \varepsilon_2^3 = \varepsilon_3^2$$

$$= \frac{M_1}{230.3 M_3} \, 230.3 \, e_3^2 \frac{M_2}{M_1} \qquad\qquad \text{(from Eq.1.103)}$$

$$\therefore e_2^3 = e_3^2 \frac{M_2}{M_3}$$

Hence, for Fe–C–N system, we can write: $\quad e_C^N = e_N^C \frac{M_C}{M_N}$

Further Reading

1. Darken, L. S., & Gurry, R. W. (1953). *Physical chemistry of metals*. London: McGraw-Hill Co. Ltd..
2. Lewis, G. N., & Randall, M. (1961). Revised by Pitzer, K. S. and Brewer, L. *Thermodynamics* (2nd ed.). London: McGraw-Hill Co. Ltd.
3. Mackowiak, J. (1966). *Physical chemistry for metallurgists*. New York: American Elsevier Publishing Co. Inc..
4. Gaskell, D. R. (2003). *Introduction to the thermodynamics of materials*. New York: Taylor & Francis.
5. Parker, R. H. (1978). *An introduction to chemical metallurgy* (2nd ed.). Oxford: Pergamon.
6. Lupis, C. H. P., & Elliott, J. F. (1966). Generalized interaction coefficients. Part I: Definitions, *Acta Metallurgica, 14*, 529–538.
7. Moore, J. J. (1966). *Chemical metallurgy* (2nd ed.). London: Butterworth Heinemann.
8. Bodsworth, C. (1990). *The extraction and refining of metals*. Boca Raton: CRC Press.
9. Kubaschewski, O., & Alcock, C. B. (1979). *Metallurgical thermochemistry* (5th ed.). Oxford: Pergamon.
10. Chase, M. W., Curnutt, J. L., Prophet, H., McDonald, R. A., & Syverud, A. N. (1975). *JANAF thermochemical tables, 1975 supplement*. Midland: The Dow Chemical Company.
11. Barin, I., Knacke, O., & Kubaschewski, O. (1977). *Thermochemical properties of inorganic substances: Supplement*. New York: Springer.
12. Dube, R. K., & Upadhyay, G. S. (1977). *Problems in metallurgical thermodynamics and kinetics*. Oxford: Pergamon.

Chapter 2
Roasting of Sulfide Minerals

In addition to common metals like copper, lead, zinc, and nickel a number of other metals such as antimony, bismuth, cadmium, cobalt, mercury, and molybdenum occur as sulfide minerals. Sulfides are not reduced with the most widely used reducing agents, carbon and hydrogen, because the free energy change for the reactions $2MS + C = 2M + CS_2$ and $MS + H_2 = M + H_2S$ is positive due to the lesser stability of CS_2 and H_2S, compared to most sulfides. Reduction of sulfides with metals is not economical. Furthermore, from most sulfide minerals the metal value is not brought into aqueous solution by leaching with common acids and alkalis. In the presence of an oxidant, chalcocite is leached quickly in dilute sulfuric acid whereas bornite and covellite are leached slowly. Leaching is speeded up in the presence of bacteria. But chalcopyrite, the major source of copper is not leached. Commercially, pentlandite, a nickel sulfide mineral, is treated with ammonia under pressure (8 atm) at 105 °C to dissolve nickel. Nickel is precipitated by blowing hydrogen at 30 atm in the purified leach liquor at 170 °C. Otherwise, by and large, the hydrometallurgical route for treatment of sulfides has failed. Under the circumstances, the only alternative seems to be the conversion of sulfide concentrates into oxides by dead roasting [1–3], which can be easily reduced with carbon (production of lead and zinc) or into mixed oxide and sulfate by partial roasting and sintering, which can be dissolved in dilute sulfuric acid, and the resultant solution is subjected to electrowinning (extraction of zinc).

Since the free energy of formation of SO_2 and SO_3 is lower than the free energy of formation of many sulfides (CuS, Cu_2S, NiS, FeS, PbS, ZnS, MnS, and so on under standard conditions), oxygen is conveniently used to remove sulfur from sulfides. Hence, the typical step, roasting (controlled oxidation) is employed to remove sulfur from sulfide minerals by forming SO_2 and SO_3. Until the very recent past, roasting happened to be a preliminary chemical treatment in the extraction of copper from chalcopyrite via the pyrometallurgical route incorporating steps, namely, concentration by froth flotation, reverberatory smelting, converting, fire refining, and electrorefining. Although 80% of the total world production of copper comes from the chalcopyrite concentrate even today, the roasting step has been eliminated from

M. Shamsuddin, *Physical Chemistry of Metallurgical Processes, Second Edition*, The Minerals, Metals & Materials Series, https://doi.org/10.1007/978-3-030-58069-8_2

the flow sheet on account of recently developed faster smelting and converting processes, for example, flash smelting by the International Nickel Company and Outokumpu, bath smelting by Noranda, Teniente Ausmelt/Isasmelt matte smelting, continuous converting by Outokumpu and Mitsubishi, and direct smelting/converting by Mitsubishi. However, roasting plays an important role in the extraction of nickel from pentlandite, zinc from sphalerite, lead from galena, and molybdenum from molybdenite.

In roasting, air in large amounts, sometimes enriched with oxygen, is brought into contact with the sulfide mineral concentrate. This is done at elevated temperatures when oxygen combines with sulfur to form sulfur dioxide and with the metal to form oxides, sulfates, and so on. The oxidation must be done without melting the charge in order to prevent reduction of particle surface–oxidizing gas contact area. Stirring of the charge in some manner also ensures exposure of all particle surfaces to the oxidizing gas. Exception to this procedure is blast roasting–sintering, where the particle surfaces are partially melted and there is no stirring of the charge. The degree of sulfur elimination is controlled by regulating the air supply to the roaster and by the degree of affinity of the mineral elements for sulfur or oxygen. For example, iron sulfide may all be oxidized because iron has more affinity for oxygen than for sulfur, while a copper mineral in the same roaster feed will emerge in the calcine still as a sulfide due to greater affinity of copper for sulfur than for oxygen.

Roasting is essentially a surface reaction where the oxide layer is formed first and continues to remain as a porous layer through which oxygen can pass into the still unreacted inner sulfide portion of the particle, and the SO_2 gas formed comes out. Roasting is an exothermic reaction. This heat helps to keep the roaster at the required roasting temperature so that the process can continue with little extra heat supplied by the burning fuel. Hence, sulfide roasting is an autogenous process, that is, where no extra fuel is supplied.

2.1 Methods of Roasting

There are several commercial roasters used on the industrial scale. The multiple hearth roaster that consists of a number of horizontal circular refractory hearths placed one above the other in a steel shell for the purpose of charging the feed on the top hearth as well as for discharging the roasted calcine from the bottom hearth, has been virtually replaced by the flash or suspension roaster that has only the top and bottom hearths. The capacity of such roasters is three times larger than that of multiple hearth roasters. In flash roasting, the preheated ore is injected through a burner. This process is most appropriate for the roasting of sulfides, which oxidize exothermally and require no additional fuel. In flash roasting, the benefits of a counter flow operation are partially lost.

The use of fluidized beds for roasting fine concentrate is obviously attractive. If a gas is passed upward through a bed of solid particles of small and preferably regular size in the range 2–0.02 mm diameter, the behavior of the bed will depend upon the velocity of the gas. At very low flow rates, the gas permeates through the bed without moving the particles at all and the pressure drop across the bed is proportional to the flow rate. An increase in gas velocity to a critical value causes the bed to expand as the effective weights of the particles become balanced by the drag forces of the gas stream upon them. Over a short range of velocities, the particles remain individually suspended, each with a downward velocity relative to the gas stream approximately equal to its terminal velocity.

Ore fines may be sintered due to incipient fusion in Dwight Lloyd Sintering Machine. It consists of a linked grate section, which forms an endless belt that moves on rollers. A suction box is located under the grate and the speed of the belt movement is adjusted.

2.2 Objectives

In the extraction of metals from sulfides, roasting is applied to achieve different objectives depending on the overall process flow sheet. Such objectives include one or more of the following:

Oxidizing roast is carried out to prepare a totally or a partly oxide product. For example, dead roasting of sphalerite concentrate to an all-oxide product is the prerequisite for the pyrometallurgical smelting processes. For copper and nickel extraction, only partial roasting is practiced. In order to produce a mixed oxide–sulfate product *sulfation roast* is preferred. The most outstanding example of this is found in the commercial production of zinc by the hydrometallurgical route. *Roast-reduction* is aimed at producing a metal directly by the interaction between the oxide formed in situ and the unroasted sulfide. This is employed in the old and now-obsolete "Newnam Ore Hearth Process" of making lead. Such a principle is also partially made use of in the sintering (agglomeration) of galena concentrates. The objective of *chloridizing roast* is to convert certain metals to their water-soluble chlorides. Notable among such applications is the treatment of pyrite cinders with $CaCl_2$ for the recovery of nonferrous values. *Volatilizing roast* is used to produce and recover metals as volatile oxides or for the elimination of unwanted metals as volatilized oxides. Roasting of speiss for As_2O_3 recovery and of bismuth ores for the removal of As, Sb, or Zn (as their oxides) are very interesting applications under this category.

Obviously, roasting is a very vast field with multifarious aspects covering the fundamental, applied, and industrial aspects of many metal sulfides. It would not be possible to survey the whole field with any justice. However, it would be interesting to discuss the relevant chemistry and thermodynamics of roasting with the primary aim of understanding the stability regions of different phases in M–S–O systems in the following sections.

2.3   Chemistry of Roasting

The chemistry of sulfide roasting is complex involving numerous reactions, important of which are as follows:

1. Decomposition of higher sulfides to lower sulfides, for example,

$$MS_2(s) = MS\ (s) + \frac{1}{2}S_2\ (g) \tag{2.1}$$

$$2MS_2(s) = M_2S_3\ (s) + \frac{1}{2}S_2\ (g) \tag{2.2}$$

2. Oxidation of sulfides to form oxides or sulfates

$$2MS\ (s) + 3O_2\ (g) = 2MO\ (s) + 2SO_2\ (g) \tag{2.3}$$

$$MS\ (s) + 2O_2\ (g) = MSO_4\ (s) \tag{2.4}$$

3. Burn-up of sulfur to form its oxides

$$S_2\ (g) + 2O_2\ (g) = \ 2SO_2\ (g) \tag{2.5}$$

$$2SO_2\ (g) + O_2\ (g) = \ 2SO_3\ (g) \tag{2.6}$$

4. Sulfation of metal oxides

$$MO\ (s) + SO_3\ (g) = MSO_4\ (s) \tag{2.7}$$

$$2MO\ (s) + 2SO_2\ (g) + O_2\ (g) = 2MSO_4\ (s) \tag{2.8}$$

5. Decomposition of sulfates to basic (oxy) sulfates

$$2MSO_4\ (s) = MO.MSO_4\ (s) + SO_3\ (g) \tag{2.9}$$

$$MO.MSO_4\ (s) = 2MO\ (s) + SO_3\ (g) \tag{2.10}$$

$$MSO_4\ (s) = MO.ySO_3\ (s) + (1-y)SO_3\ (g) \tag{2.11}$$

6. Sulfide–sulfate interaction

$$MS\ (s) + 3MSO_4\ (s) = 4MO\ (s) + 4SO_2\ (g) \tag{2.12}$$

7. Sulfide–oxide interaction

$$MS\ (s) + 2MO\ (s) = 3M\ (s) + SO_2\ (g) \tag{2.13}$$

8. Reaction between product oxides or product and impurity oxides to form complex compounds such as ferrites and silicates

$$2MO \ (s) + 2FeO \ (s) = 2MFeO_2 \ (s) \qquad (2.14)$$

$$2MO \ (s) + SiO_2 \ (s) = M_2SiO_4 \ (s) \qquad (2.15)$$

9. Formation of sub- or higher-oxides

$$2MO \ (s) = M_2O \ (s) + \frac{1}{2}O_2 \ (g) \qquad (2.16)$$

$$2MO \ (s) + \frac{1}{2}O_2 \ (g) = M_2O_3 \ (s) \qquad (2.17)$$

10. Other reduction or oxidation reactions such as

$$3M_3O_4 \ (s) + MS \ (s) = 10MO \ (s) + SO_2 \ (g) \qquad (2.18)$$

$$M \ (s,l) + SO_2 \ (g) = MS \ (s) + O_2 \ (g) \qquad (2.19)$$

Where more than one metal sulfide is present in the concentrate, the reactions are truly complex and very large in number. The composition of a product in sulfide roasting depends not only on the chemical and mineralogical composition of the concentrate, temperature, and partial pressures of O_2, SO_2, and SO_3 but also on the other process parameters such as particle size, mixing, time of reaction, and the technique of roasting.

2.4 Thermodynamics of Roasting

The necessary conditions for the formation of different products can be illustrated by the relationship between the equilibriums in any M–S–O system. In the simplest case, we have three components, and according to the phase rule ($P = C - F + 2$), there can be five phases, that is, four condensed phases and one gaseous phase ($P = 3 - 0 + 2 = 5$). P, C, and F, respectively, stand for the number of phases, components, and degree of freedom. If temperature is fixed, we have $P = 3 - 0 + 1 = 4$ (3 condensed phases and 1 gaseous phase). The gas phase normally contains SO_2 and O_2 but some SO_3 and even sulfur vapor (S_2) may be present. Among the gaseous compounds, the following equilibria exist:

$$S_2 \ (g) + 2O_2 \ (g) = 2SO_2 \ (g) \qquad (1)$$

$$2SO_2 \ (g) + O_2 \ (g) = 2SO_3 \ (g) \qquad (2)$$

At any selected temperature, the composition of the gas mixture is defined by the partial pressures of two of the gaseous components. Further, for any fixed gas composition, the composition of the condensed phase gets fixed. Thus, the phase relationship in the ternary system at constant temperature may be described in a two-dimensional diagram, where the two co-ordinates are the partial pressures of two of the gaseous components. Generally, p_{SO_2} and p_{O_2} are chosen because during roasting, sulfide reacts with oxygen to produce SO_2, which is a predominant gas species in the flue. Such isothermal plots are called Predominance Area Diagrams. They are also known as Kellogg Diagrams [4–6], after Kellogg, who conceived them. Commencing with the pioneering and innovative work of Kellogg and his coworkers, several researchers have constructed the predominance area diagrams not only for M–S–O systems but also for M_1–M_2–S–O systems. Many such diagrams are available in literature [7–16] and have been made use of to understand roasting, decomposition, smelting, refining, and so on.

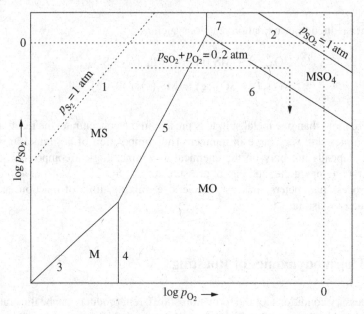

Fig. 2.1 Predominance area diagram of the M–S–O system. (From Principles of Extractive Metallurgy by T. Rosenqvist [3], © 1974, p 247, McGraw-Hill Inc. Reproduced with the permission of McGraw-Hill Book Co.)

Figure 2.1 shows equilibria and predominance areas at a constant temperature for the M–S–O system. Lines describing the equilibrium between any two condensed phases are given by equations:

$$M(s, l) + SO_2(g) = MS\ (s) + O_2\ (g) \tag{3}$$

$$2M(s, l) + O_2(g) = 2MO\ (s) \tag{4}$$

$$2MS\ (s) + 3O_2\ (g) = 2MO\ (s) + 2SO_2\ (g) \tag{5}$$

$$2MO\ (s) + 2SO_2\ (g) + O_2\ (g) = 2MSO_4\ (s) \tag{6}$$

$$MS\ (s) + 2O_2\ (g) = MSO_4\ (s) \tag{7}$$

If metal forms several sulfides and oxides, additional equilibria would have to be considered for the formation of MS_2, M_2O_3 and $M_2(SO_4)_3$ and so on. In addition, basic sulfates, $MO.MSO_4$, may exist. For the above reactions for all condensed phases in their standard states, the equilibria are given by the expressions:

$$K_3 = \frac{a_{MS} \cdot p_{O_2}}{a_{M_1} p_{SO_2}} = \frac{p_{O_2}}{p_{SO_2}} \quad \text{i.e., } \log p_{O_2} - \log p_{SO_2} = \log K_3$$

$$K_4 = \frac{1}{p_{O_2}} \quad \text{i.e., } \log p_{O_2} = - \log K_4$$

$$K_5 = \frac{p_{SO_2}^2}{p_{O_2}^3} \quad \text{i.e., } - 3 \log p_{O_2} + 2 \log p_{SO_2} = \log K_5$$

$$K_6 = \frac{1}{p_{SO_2}^2 \cdot p_{O_2}} \quad \text{i.e., } 2 \log p_{SO_2} + \log p_{O_2} = - \log K_6$$

$$K_7 = \frac{1}{p_{O_2}^2} \quad \text{i.e., } 2 \log p_{O_2} = - \log K_7$$

We notice that for a given reaction the form of the equilibrium expression is the same for all metals, that is, the slope of the corresponding lines in the figure is the same. Only the values of the equilibrium constant K_3, K_4, and so on, may differ from one metal to another. This means that the position of the equilibrium lines may change and, consequently, the size and the position of the areas between the lines. These areas are called the predominance area for that particular phase. From the figure we find that

1. As long as only one condensed phase exists, the partial pressure of SO_2 and O_2 may be changed independently of each other; that is, the system at constant temperature has two degrees of freedom.
2. Along the lines for equilibrium between two condensed phases, the system has one degree of freedom.
3. Finally, where three phases are in equilibrium, the system at constant temperature is nonvariant.

There are also lines in the figure to describe reactions (1) and (2) that is, for the formation of SO_2 and SO_3. These are given by the expressions:

$$K_1 = \frac{p_{SO_2}^2}{p_{S_2} \cdot p_{O_2}^2} \quad \text{i.e., } 2 \log p_{SO_2} - 2 \log p_{O_2} = \log K_1 + \log p_{S_2}$$

$$K_2 = \frac{p_{SO_3}^2}{p_{SO_2}^2 \cdot p_{O_2}} \quad \text{i.e., } 2 \log p_{SO_2} + \log p_{O_2} = - \log K_2 + 2 \log p_{SO_3}$$

This means that for fixed values of K_1 and K_2, the relationship between $\log p_{SO_2}$ and $\log p_{O_2}$ depends also on the partial pressures of S_2 and SO_3. In the figure, the lines are drawn for $p_{S_2} = 1$ atm and $p_{SO_3} = 1$ atm. For other pressures, the lines are to be shifted up and down in accordance with the above expressions. Thus, p_{S_2} becomes large when p_{O_2} is small and that p_{SO_2} is large, and p_{SO_3} is large for large values of p_{SO_2} and p_{O_2}. When roasting is carried out in air, $p_{SO_2} + p_{O_2} = 0.2$ atm, the conditions during roasting are as described by the dotted lines in the figure. First, the sulfide is roasted/converted into the oxide by reaction (5). Then the oxide may be converted into sulfate, which by prolonged heating in air at constant temperature again may be converted to give the oxide.

In roasting of a mixed sulfide concentrate, simultaneous reaction/conversion of different metals will not take place because the stability regions for different metals have different locations. Hence, during roasting of chalcopyrite (mixed Cu-Fe sulfide) iron sulfide will first oxidize to form Fe_3O_4, leaving copper as Cu_2S. Continued operation transforms Fe_3O_4 into Fe_2O_3 and Cu_2S into Cu_2O and then into CuO. However, the presence of iron oxide as a gangue in the concentrate complicates roasting by forming spinel, for example, in the roasting of sphalerite, zinc ferrite ($ZnO.Fe_2O_3$) is formed. On the other hand, depending on the temperature of roasting basic sulfates, namely, $ZnO.ZnSO_4$ and $PbO.PbSO_4$, $PbO.2PbSO_4$ and $PbO.4PbSO_4$ are formed in the roasting of sphalerite and galena, respectively. The formation of these complex phases alter shape and size of predominance areas because of lesser activity of the compound in the complex phase compared to the activity of the pure compound. This means, a lower partial pressure of oxygen than what is required for zinc oxide formation, would form zinc ferrite from sphalerite. Based on a similar argument, higher partial pressures of oxygen and sulfur dioxide are required for ZnO–$ZnSO_4$ equilibrium when ZnO is complexed as a ferrite phase ($ZnO.Fe_2O_3$) instead of its presence as pure ZnO. Thus, depending on the requirement of the process flow sheets, stability diagrams of different M–S–O systems at different temperatures become useful in carrying out selective roasting [17].

The effect of temperature on the roasting equilibria may be given in a two-dimensional diagram for constant values of p_{SO_2}. The Fig. 2.2 shows $\log p_{O_2}$ versus $1/T$ plots for $p_{SO_3} = 1$ atm and 0.1 atm. Slope of the curves giving enthalpy of the reaction is expressed in terms of 1 mol of oxygen:

$$M(s,l) + SO_2(g) = MS\ (s) + O_2\ (g), \quad K_3 = \frac{p_{O_2}}{p_{SO_2}}$$

$$\text{For a fixed value of } p_{SO_3} : \frac{\partial \log p_{O_2}}{\partial \left(\frac{1}{T}\right)} = \frac{\partial \log K_3}{\partial \left(\frac{1}{T}\right)} = \frac{\Delta H_3}{4.575} \quad (2.20)$$

Figure marked with two possible roasting paths is useful in discussing the effect of temperature on the roasting equilibria. Along the path (a) normally, as roasting starts, the sulfide gets roasted to oxide with an increase of temperature, which may partially get converted to sulfate on decrease of temperature toward the end of the roasting process. On the other hand, along the path (b), sulfate first formed at low temperature decomposes to oxide on rise of temperature due to additional heat.

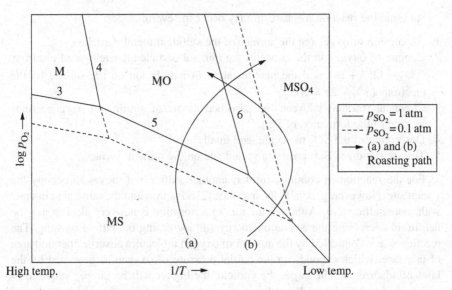

Fig. 2.2 Effect of temperature on equilibria and predominance areas in the M–S–O system at constant pressure of SO_2 gas. (From Principles of Extractive Metallurgy by T. Rosenqvist [3], © 1974, p 250, McGraw-Hill Inc. Reproduced with the permission of McGraw-Hill Book Co.)

2.5   Kinetics of Roasting

Roasting of sulfides, in general, is exothermic and hence the temperature of the ore lump remains high enough for the desired chemical reactions to occur fairly early during the roasting process. Every particle gets oxidized from the outside, which leads to the formation of a case of solid product and leaves a core of unchanged zone. The required reaction occurs at the interface between these zones if the gas ratio p_{O_2}/p_{SO_2} is locally higher than the equilibrium ratio for the reaction at the temperature under consideration. Take a generalized case of roasting of a metal sulfide (MS):

$$2MS\ (s) + 3O_2\ (g) = 2MO\ (s) + 2SO_2\ (g) K = \frac{p_{SO_2}^2}{p_{O_2}^3} \tag{2.21}$$

For the above reaction to proceed, it is essential that at least three molecules of oxygen should reach the interface to form two molecules of sulfur dioxide, that is, $J_{O_2} \nless \frac{3}{2} J_{SO_2}$. These two flux values (J_{O_2} and J_{SO_2}) are controlled by the respective diffusion coefficients and partial pressure of O_2 and SO_2. The diffusion coefficient of O_2 is larger than SO_2 molecules, but there is always a minimum ratio of p_{O_2}/p_{SO_2} in the atmosphere outside the particle. This minimum ratio has to be maintained, so as to keep the ratio at the reaction front always higher than the equilibrium value.

As usual the reaction mechanism may occur in several stages:

1. Adsorption of oxygen at the surface of the sulfide mineral particles.
2. Capture of oxygen in the lattice of the mineral particle after release of electrons ($\frac{1}{2}O_2 = O^{2-} + 2e$) and the neutralization of sulfide ion on the surface by the electrons ($S^{2-} + 2e = S$).
3. Chemical reaction between the adsorbed atoms of sulfur and oxygen atoms leading to the formation of SO_2.
4. Desorption of the SO_2 molecule, and finally
5. Transportation of SO_2 creates a vacant site on the mineral surface.

For the reaction to continue further, another sulfide ion moves to occupy the vacant site. However, generally the interface [2] advances into the mineral to interact with more sulfide ions. Although the role of adsorption is not very clear, it may be helpful in decreasing the activation energy and increasing the rate of roasting. The reaction rate is controlled by the number of oxygen molecules adsorbed per unit area of interface, which depends on the partial pressure of oxygen as governed by the laws of adsorption. The larger the particle, the higher will be the p_{O_2} required to maintain a particular flux, J_{O_2} at the reaction interface. In case of complete oxidation (i.e., dead roasting), attempt should be made to increase p_{O_2} in the gaseous mixture around the mineral particles in the last stages.

The following points must be kept in mind while carrying out roasting:

1. In case of nonuniform temperature throughout the ore lump, diffusion should be considered going down a "chemical potential gradient" rather than a partial pressure gradient.
2. As ore particles are not homogeneous lumps of sulfide, the presence of inert gangue may reduce or even stop reactions.
3. Exothermicity of roasting reactions may lead to fusion or agglomeration of particles, thereby altering the diffusion coefficients and diffusion paths and thus, ultimately, slowing down the process.
4. Depending on operating conditions, side reactions, for example, sulfate formation, may occur in the cooler part of the outer oxide shell.

Many investigators have studied the kinetics of roasting of metal sulfides. By and large, roasting of zinc sulfide is carried out at and above 800 °C to produce zinc oxide [2ZnS (s) + 3O$_2$ (g) = 2ZnO (s) + 2SO$_2$ (g)]. This is known as dead roasting. The rate of oxidation is very slow at 700 °C. At lower temperatures [18], ZnS reacts with SO_3 to form $ZnSO_4$ according to the reactions:

$$ZnS\ (s) + 4SO_3\ (g) \rightarrow ZnSO_4\ (s) + 4SO_2(g) \tag{2.22}$$

$$ZnS\ (s) + O_2\ (g) + 2SO_3\ (g) \rightarrow ZnSO_4\ (s) + 2SO_2(g) \tag{2.23}$$

In the absence of SO_3, sulfation does not take place when ZnS is treated with SO_2 and air. Sulfates are generally stable at temperatures below 800 °C at a lower p_{O_2}/p_{SO_2} ratio. Hence, sulfation of fine concentrate is restricted in a separate part of the kiln by maintaining a lower temperature and higher SO_2 in the gaseous mixture.

A bulk of fine concentrate behaves like a single large lump against the reacting gases. In such cases, the solid–gas contact area is increased by raking or fluidization in order to improve the rate of roasting because gases normally penetrate the solid particles through inter-granular routes rather than through the crystals. As the individual crystal gets converted topochemically to the product, the radial rate of conversion is the same all the way around. The rate of reaction decreases toward the center of the ore particle. Hence, the rate of reaction on the inner grains (deeper into the particles) will be determined by an entirely different mechanism from the particle as a whole. It is most likely that a solid-state diffusion mechanism may be operative within the grains. In the case of coarse grains, this mechanism can take control of the reaction rate for the particle as a whole.

Natesan and Philbrook [19, 20] studied the kinetics of roasting of zinc sulfide spherical pellets [19] and powder under suspension [20] in a fluidized bed reactor. In the temperature range of 740–1040 °C, roasting of pellets of 0.1.6 cm diameter according to the reaction (2.22) is controlled by gaseous transport through the product layer of zinc oxide formed during the course of the reaction. On the other hand, in the fluidized bed reactor for the suspended particles, the kinetics is governed by the surface reaction at the ZnS–ZnO interface in the same temperature range.

Based on their detailed studies on roasting of covellite (CuS) in oxygen, Shah and Khalafallah [21] have proposed the following sequence of conversion:

$$CuS \rightarrow Cu_{1.8}S \rightarrow Cu_2O \rightarrow CuO \rightarrow CuO.CuSO_4 \rightarrow CuSO_4 \qquad (2.24)$$

$Cu_{1.8}S$ is considered to be a defective form of copper sulfide, known as digenite. After establishing the sequence of roasting, the kinetics of the first [22, 23] and last steps [24] in Eq. 2.24 was systematically investigated. The conversion of covellite (CuS) to digenite ($Cu_{1.8}S$) was studied in nitrogen [22] and oxygen [23] atmosphere for the reactions:

In nitrogen $(340 - 400\,°C)$: $1.8\,CuS\,(s) \rightarrow Cu_{1.8}S\,(s) + 0.4\,S_2\,(g)$ \qquad (2.25)

In nitrogen $(260 - 400\,°C)$: $1.8\,CuS\,(s) + 0.8O_2\,(g) \rightarrow Cu_{1.8}S\,(s) + 0.8SO_2\,(g)$ \quad (2.26)

In nitrogen, the reaction was found to be topochemical with the activation energy of 24 ± 2 kcal mol^{-1}. The rate of conversion in oxygen was first order with respect to the partial pressure of oxygen with an average apparent activation energy of 23 ± 3 kcal mol^{-1}.

Rao and Abraham [7] have reported two stages in the kinetics of roasting of cuprous sulfide in the temperature range of 750–950 °C. The kinetics of the initial non-isothermal stage with an activation energy of 25 kcal mol^{-1} is controlled by heat and mass transport, whereas the second isothermal stage with an apparent activation energy of about 6 kcal mol^{-1} is controlled by mass transport. The mass transport involves diffusion through an outer boundary layer, diffusion through the layer formed during a non-isothermal period, and diffusion through a product layer formed during a continuing reaction.

Coudurier et al. [25] conducted studies on roasting of molybdenite (MoS_2) in an experimental multiple hearth roaster and reported that the rate of reaction was controlled by gaseous diffusion on the upper hearth and by surface reaction on the lower hearth. Different equations developed for the upper and lower hearths were useful in deriving optimum operating conditions by matching the heat balance with the reaction kinetics. In conclusion, they have proposed that the kinetics of roasting of molybdenite was controlled by gaseous diffusion in the early stages and by surface reaction in the later stages of the reaction. Amman and Loose [26] investigated the effects of temperature (525–635 °C), gas composition (5–20% oxygen), and particle size on the oxidation kinetics of molybdenite concentrate containing 90% MoS_2. They proposed a mathematical model to explain the reaction kinetics involving diffusion of oxygen from the bulk gas phase to the solid interface, diffusion through reaction product layer, and a first-order reaction kinetics at the molybdenite–oxide interface with an activation energy of 35.4 kcal mol^{-1}. This order of activation energy simply indicates that the oxidation kinetics of molybdenite in the temperature range of 525–635 °C is chemically controlled.

In the temperature range of 690–800 °C, lead sulfide is converted to lead sulfate:

$$PbS \text{ (s)} + 2O_2 \text{ (g)} = PbSO_4 \text{ (s)} \tag{2.27}$$

According to Khalafallah [27], the above roasting reaction is sensitive to particle size and obeys the following kinetic law:

$$1 + 2(1 - z.f) - 3(1 - z.f)^{\frac{2}{3}} = \frac{k}{4r_o^2} t \tag{2.28}$$

In Eq. 2.28, f is fraction of PbS converted into $PbSO_4$, and z the volume of $PbSO_4$ formed per unit volume of PbS and k, r_0 and t represent, respectively, the rate constant, initial radius of the sulfide particle, and time of reaction. From the diffusion measurement experiments, it has been established that PbS diffuses from the inside through the $PbSO_4$ layer to react on the surface. The kinetics of roasting of galena expressed by the rate Eq. 2.28 is very much similar to the well-known Gnistling-Brounshtein [28] product layer diffusion equation.

2.6   Predominance Area Diagrams as a Useful Guide in Feed Preparation

In the preceding paragraphs, it has been mentioned that roasting is one of the main steps in the extraction of nickel, zinc, lead, and molybdenum from pentlandite, sphalerite, galena, and molybdenite, respectively. Although roasting was one of the major steps in production of copper from chalcopyrite, this step has been eliminated in recently developed faster smelting and converting processes. In this section, an attempt has been made to demonstrate the usefulness of predominance area diagrams of M–S–O systems in the production of some nonferrous metals.

Though currently not practiced, it is interesting to note how Cu–S–O and Fe–S–O stability diagrams were used in the past for the preparation of feed for reverberatory smelting and leaching circuit by roasting of chalcopyrite at different temperatures. The composition of roaster products prepared for reverberatory smelting was not critical, but it was desirable to avoid excessive oxidation of iron. The formation of Fe_3O_4 or Fe_2O_3 led to high oxidizing conditions in the subsequent smelting slag, and these oxidizing conditions caused an increased loss of copper in the slag. The formation of high melting ferrites was also favored by over oxidation of the iron minerals. Figure 2.3 that exhibits the predominance area diagram of the Fe–S–O system at different temperatures directs that the best way to avoid over oxidation, that is, Fe_2O_3 formation, was to roast at relatively lower temperatures. This is demonstrated by the position of the roaster gas composition partial pressures square within the $Fe_2(SO_4)_3$ region at temperatures below about 600 °C.

The principal objectives of roasting prior to leaching were (a) to produce a controlled amount of $CuSO_4$ with the remainder of Cu being in oxide form (b) to produce iron oxides rather than soluble iron sulfates. The latter objective minimized contaminations of the leach liquor with iron. Figure 2.4 shows that $CuSO_4$ formation is favored below 677 °C, $CuSO_4$ and CuO are favored between 677 and 800 °C, and CuO is favored above 800 °C. Therefore, roasting to produce a mixed $CuSO_4$/CuO product should be carried out between 677 and 800 °C. The higher temperature favors higher proportion of CuO. Figure 2.3 also indicates that temperature in this range also favors the formation of insoluble iron oxides, which is the second requirement. Thus, roasting at a temperature of 675 °C fulfilled both the objectives.

The predominance area diagram of the Zn–S–O system [13] (Fig. 2.5) at different temperatures is useful in the preparation of feeds for smelting and leaching by roasting sphalerite concentrates at appropriate temperatures. For the Imperial Smelting furnace, the sphalerite concentrate is first partially roasted at 800 °C to obtain a product with 4% sulfur. Since the blast furnace requires lumpy charge (size varying from 2.5 to 10 cm), the resultant mass is then sintered at 1200 °C to produce lumpy zinc oxide sinter on Dwight-Lloyd sintering machine. Since the stability regions of both $ZnSO_4$ and $PbSO_4$ in the predominance area diagrams of Zn–S–O and Pb–S–O (Figs. 2.5 and 2.6) systems shrink with an increase of temperature, the product resulting from the roasting/sintering of sphalerite and galena concentrates at 800 °C will consist of mixed oxide (larger amount) and little sulfate (respectively, $ZnO + ZnSO_4$ and $PbO + PbSO_4$).

From the Mo–S–O diagram (Fig. 2.8), it is clear that MoS_2 will get converted to MoO_3 while roasting molybdenite at 700 °C. Therefore, in order to prevent melting due to local overheating and high volatility of MoO_3, the temperature is never allowed to exceed beyond 650° C.

Shamsuddin et al. [13] have effectively demonstrated the possibility of selective roasting of an off-grade chalcopyrite concentrate containing appreciable amounts of sphalerite and galena making use of the predominance area diagrams of Cu–S–O, Fe–S–O, Zn–S–O, and Pb–S–O systems.

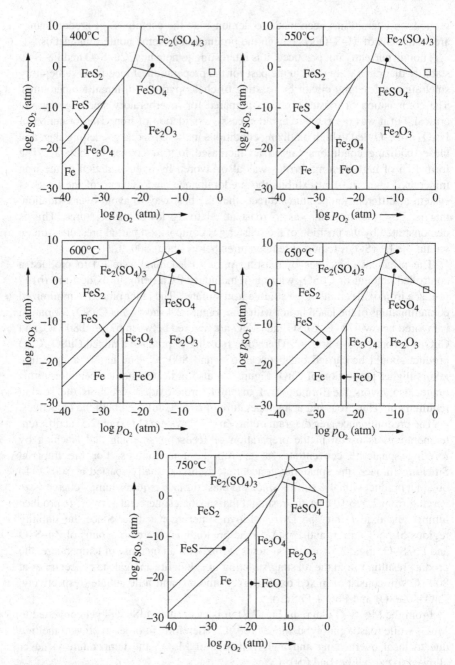

Fig. 2.3 Predominance area diagrams of the Fe–S–O system at 400, 550, 600, 650, and 750 °C. □ Usual roaster gas composition. (Reproduced from M. Shamsuddin et al. [13] with the permission of Meshap Science Publishers)

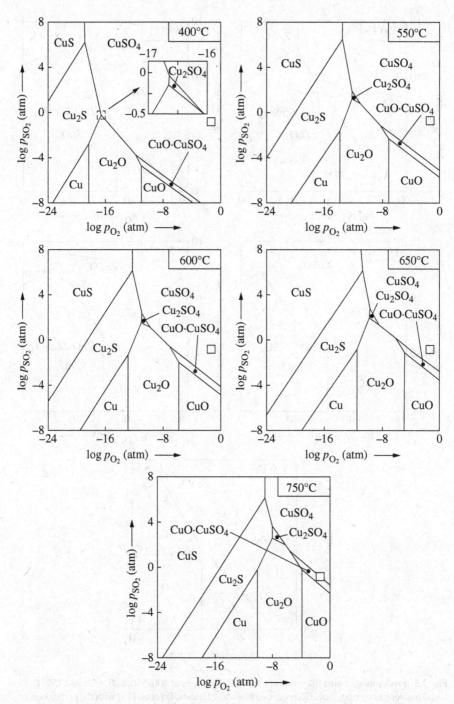

Fig. 2.4 Predominance area diagrams of the Cu–S–O system at 400, 550, 600, 650 and 750 °C. □ Usual roaster gas composition. (Reproduced from M. Shamsuddin et al. [13] with the permission of Meshap Science Publishers)

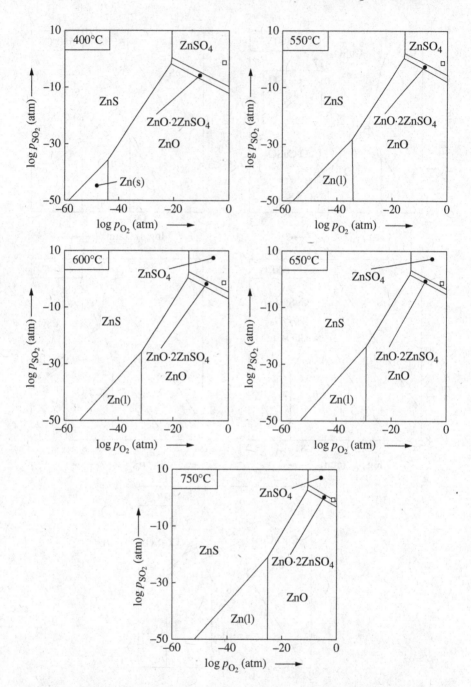

Fig. 2.5 Predominance area diagrams of the Zn–S–O system at 400, 550, 600, 650, and 750 °C. □ Usual roaster gas composition. (Reproduced from M. Shamsuddin et al. [13] with the permission of Meshap Science Publishers)

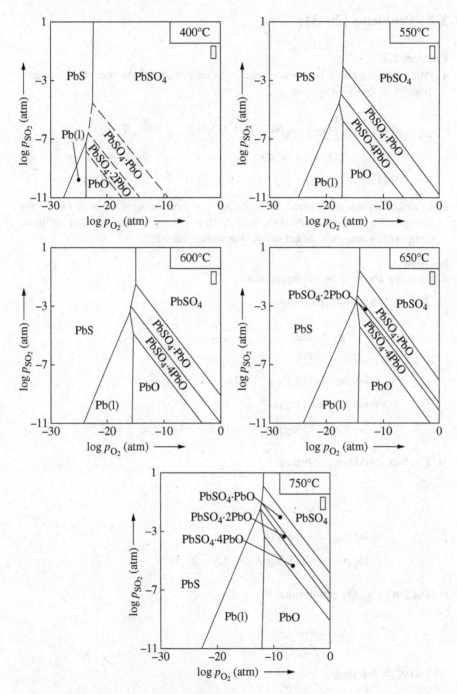

Fig. 2.6 Predominance area diagrams of the Pb–S–O system at 400, 550, 600, 650, and 750 °C. □ Usual roaster gas composition. (Reproduced from M. Shamsuddin et al. [13] with the permission of Meshap Science Publishers)

2.7 Problems [29–31]

Problem 2.1

a. From the data given below construct relevant portion of the predominance area diagram of Ni–S–O system:

$$NiS\ (s) + \frac{3}{2}O_2(g) = NiO\ (s) + SO_2(g), \qquad \log K_1 = 18.7$$

$$NiS\ (s) + 2O_2(g) = NiSO_4(s), \qquad\qquad \log K_2 = 21.59$$

$$NiO\ (s) + \frac{1}{2}O_2(g) + SO_2(g) = NiSO_4(s), \qquad \log K_3 = 2.72$$

b. If nickel sulfide is roasted in a roaster gas containing SO_2 and O_2 in the composition range 2–10% SO_2 and 2–10% O_2 over what portion of these composition ranges is nickel sulfate the stable phase?

Solution

a. From the above chemical equations:

i. For the NiS–NiO equilibrium,

$$K_1 = \frac{p_{SO_2}}{p_{O_2}^{3/2}}$$

or $3/2 \log p_{O_2} = \log p_{SO_2} - \log K_1$

or $\log p_{O_2} = 2/3 \log p_{SO_2} - 2/3 \log K_1$

$\therefore\ \log p_{O_2} = 2/3 \log p_{SO_2} - 2/3 18.87 = 2/3 \log p_{SO_2} - 12.58$

ii. For NiS–NiSO$_4$ equilibrium,

$$K_2 = \frac{1}{p_{O_2}^2}$$

$2 \log p_{O_2} = -\log K_2$

$\therefore \log p_{O_2} = -0.5 \log K_2 = 0.5 \times 21.59 = -10.795$

iii. For NiO–NiSO$_4$ equilibrium,

$$K_3 = \frac{1}{p_{O_2}^{1/2} \cdot p_{SO_2}}$$

$1/2 \log p_{O_2} + \log p_{SO_2} = -\log K_3$

or $\log p_{O_2} = -2 \log K_3 - 2 \log p_{SO_2}$

$\therefore \log p_{O_2} = -2 \times 2.72 - 2 \log p_{SO_2} = -5.44 - 2 \log p_{SO_2}$

Summary of the above calculations

Equilibria	log$K$	log p_{O_2}
NiS (s) + 3/2 O$_2$(g) = NiO (s) + SO$_2$(g)	18.87	2/3 $\log p_{SO_2} - 12.58$
NiS (s) + 2O$_2$(g) = NiSO$_4$(s)	21.59	−10.795
NiO (s) + 1/2O$_2$(g) + SO$_2$(g) = NiSO$_4$(s),	2.72	$-5.44 - 2 \log p_{SO_2}$

Figure 2.7 shows the predominance area diagram of the Ni–S–O system in terms of $\log p_{O_2}$ and $\log p_{SO_2}$.

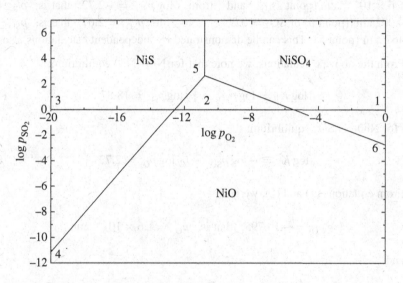

Fig. 2.7 Relevant portion of the predominance area diagram of the Ni–S–O system

Point 1: when $p_{O_2} = 1$ atm, $\log p_{O_2} = 0$
  A horizontal line is drawn up to $\log p_{O_2} = -20$
  Vertical lines are drawn at $\log p_{O_2} = 0$ and $\log p_{O_2} = -20$
Point 2: at $\log p_{O_2} = -10.795$, a vertical line is drawn
Point 3: $\log p_{O_2} = -20$
Point 4: NiS–NiO equilibrium

$$\log p_{SO_2} = \tfrac{3}{2} \log p_{O_2} + \log K_1 \text{ at point 3, } \log p_{O_2} = -20$$

$$= \tfrac{3}{2}(-20) + 18.87 = -11.13 (\text{point 4})$$

Point 5: NiS–NiO equilibrium, at point 2, $\log p_{O_2} = -10.795$

$$\log p_{SO_2} = \tfrac{3}{2} \log p_{O_2} + 18.87$$
$$= \tfrac{3}{2}(-10.795) + 18.87 = 2.67 (\text{point 5})$$

Point 6: NiO–NiSO$_4$ equilibrium, at point 1, $\log p_{O_2} = 0$

$$\log p_{SO_2} = -\log K_3 - 0.5 \log p_{O_2} = -2.72 (\text{point 6})$$

b. From Fig. 2.7, we find that NiSO$_4$ is a stable phase in the region: from $\log p_{O_2} = 0$, that is, $p_{O_2} = 1$ atm (point 1) to $\log p_{O_2} = -10.795$, that is, $p_{O_2} = 1.6 \times 10^{-11}$ atm (point 2)   and   from $\log p_{SO_2} = -2.72$,   that is, $p_{SO_2} = 0.002$ atm (point 6) or SO$_2 = 0.2\%$ to $\log p_{SO_2} = 2.67$,   that is, $p_{SO_2} = 467$ atm (point 5). This can be demonstrated by independent calculations also.

From the above calculations, we note that for NiS–NiO equilibrium:

$$\log K_1 = \log p_{SO_2} - \tfrac{3}{2} \log p_{O_2} = 18.87 \tag{1}$$

and for NiO–NiSO$_4$ equilibrium:

$$\log K_3 = -\log p_{SO_2} - \tfrac{1}{2} \log p_{O_2} = 2.72 \tag{2}$$

From equations (1) and (2), we get:

$$\log p_{O_2} = -10.795, \text{ that is, } p_{O_2} = 1.6 \times 10^{-11} \text{ atm,}$$

and

$$\log p_{SO_2} = 2.67, \text{ that is, } p_{SO_2} = 467 \text{ atm}$$

Thus,   NiSO$_4$ is   stable   from $p_{O_2} = 1.6 \times 10^{-11}$ to 1 atm $\left(\log p_{O_2} = 0, \text{ where } \log p_{SO_2} = -2.72, \text{ that is, } p_{SO_2} = 0.002 \text{ atm or } SO_2 = 0.2\% \right)$ and $p_{SO_2} = 0.002$ to 467 atm.

Problem 2.2
From the data given below, construct the predominance area diagram of the system Mo–S–O at 1000 K:

1. $2SO_2(g) = S_2(g) + 2O_2(g)$, $\Delta G_1^o = 173240 - 34.62T$ cal
2. $MoS_2(s) = Mo(s) + S_2(g)$, $\Delta G_2^o = 85870 - 37.33T$ cal
3. $MoO_2(s) = Mo(s) + O_2(g)$, $\Delta G_3^o = 140500 + 4.6T \log T - 56.8T$ cal
4. $MoO_3(s) = MoO_2(s) + 1/2 O_2(g)$, $\Delta G_4^o = 38700 - 19.5T$ cal

Solution

The following equilibria between compounds of Mo have to be considered:

Mo–MoO$_2$, MoO$_2$–MoO$_3$, Mo–MoS$_2$, MoS$_2$–MoO$_2$ and MoS$_2$–MoO$_3$. ΔG^o vs T equations for these equilibria may be obtained by making use of the given four ΔG^o vs T equations.

a. Mo(s) + O$_2$(g) = MoO$_2$(s), $\quad\quad \Delta G_a^o = -\Delta G_3^o = -140500 - 4.6T \log\ T$
$$+56.8T \text{ cal}$$

b. MoO$_2$(s) + 0.5 O$_2$(g) = MoO$_3$(s), $\ \Delta G_b^o = -\Delta G_4^o = -38700 + 19.5T$ cal

c. Mo(s) + 2SO$_2$(g) = MoS$_2$(s) $\quad\quad \Delta G_c^o = \Delta G_1^o - \Delta G_2^o = 87370 + 2.71T$ cal
$\quad$ + 2O$_2$(g),

d. MoS$_2$(s) + 3O$_2$(g) = MoO$_2$(s) $\quad\quad \Delta G_d^o = \Delta G_2^o - \Delta G_1^o - \Delta G_3^o$
$\quad$ + 2SO$_2$(g), $\quad\quad\quad\quad\quad\quad\quad\quad = -227870 - 4.6T \log\ T + 54.09T$

e. MoS$_2$(s) + 7/2 O$_2$(g) $\quad\quad\quad\quad \Delta G_e^o = \Delta G_2^o - \Delta G_1^o - \Delta G_3^o - \Delta G_4^o$
$\quad$ = MoO$_3$(s) + 2SO$_2$(g), $\quad\quad\quad\quad = -266570 - 4.6T \log\ T + 73.59T$

From the above ΔG vs T equations, calculated values of ΔG_{1000}^o, log K, and log p_{O2} for different equilibria are summarized below:

Equilibria	ΔG_{1000}^o	log$K$	logp_{O_2}
Mo(s) + O$_2$(g) = MoO$_2$(s)	−97500	21.3	−21.3
MoO$_2$(s) + ½ O$_2$(g) = MoO$_3$(s)	−19200	4.2	−8.4
Mo(s) + 2SO$_2$(g) = MoS$_2$(s) + 2O$_2$(g)	+90080	−19.7	log p_{SO_2} − 9.85
MoS$_2$(s) + 3O$_2$(g) = MoO$_2$(s) + 2SO$_2$(g)	−187580	41.0	2/3 log p_{SO_2} − 13.66
MoS$_2$(s) + 7/2 O$_2$(g) = MoO$_3$(s) + 2SO$_2$(g)	−206780	45.2	2/3.5 log p_{SO_2} − 12.91

a. Mo–MoO$_2$ equilibrium,

$$K_a = \frac{1}{p_{O_2}}$$

$$\therefore \ \log p_{O_2} = -\log K_a = -21.3$$

b. MoO$_2$–MoO$_3$ equilibrium,

$$K_b = \frac{1}{p_{O_2}^{1/2}}$$

$$\therefore \log p_{O_2} = -2 \log\log K_b = -2 \times 4.2 = -8.4$$

c. $MO–MoS_2$ equilibrium,

$$K_c = \frac{p_{O_2}^2}{p_{SO_2}^2}.$$

$$\therefore \log p_{O_2} = \frac{1}{2} \log K_c + \log p_{SO_2} = -\frac{1}{2} \times 19.7 + \log p_{SO_2} = -9.85 + \log p_{SO_2}$$

d. $MoS_2–MoO_2$ equilibrium,

$$K_d = \frac{p_{SO_2}^2}{p_{O_2}^3}$$

$$\therefore \log p_{O_2} = -\frac{1}{3} \log K_d + \frac{2}{3} \log p_{SO_2} = -\frac{41.0}{3} + \frac{2}{3} \log p_{SO_2}$$

$$= -13.66 + \frac{2}{3} \log p_{SO_2}$$

e. $MoS_2–MoO_3$ equilibrium,

$$K_e = \frac{p_{SO_2}^2}{p_{O_2}^{3.5}}.$$

$$\therefore \log p_{O_2} = -\frac{1}{3.5} \log K_e + \frac{2}{3.5} \log p_{SO_2} = -\frac{1}{3.5} \times 45.2 + \frac{2}{3.5} \log p_{SO_2}$$

$$= -12.91 + \frac{2}{3.5} \log p_{SO_2}$$

Figure 2.8 shows the predominance area diagram of the Mo–S–O system at 1000 K.

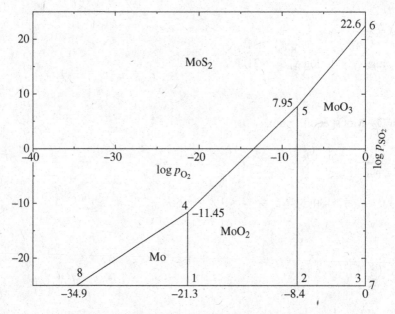

Fig. 2.8 Predominant area diagram of the Mo–S–O system

Construction of the Diagram

I. Draw a vertical line at $\log p_{O_2} = 0$, i.e., $p_{O_2} = 1$ atm (Y-axis) and a horizontal line at $\log p_{SO_2} = 0$, i.e., $p_{SO_2} = 1$ atm (X-axis).

Draw vertical lines at $\log p_{O_2} = -8.4$ and -21.3, respectively, demonstrating MoO_2–MoO_3 and Mo–MoO_2 equilibria. Mark $\log p_{O_2} = 0$ and -40 on X-axis and $\log p_{SO_2} = \pm 25$ on Y-axis.

Thus, at $\log p_{SO_2} = -25$, we have at point 1, $\log p_{O_2} = -21.3$, at point 2, $\log p_{O_2} = -8.4$ and at point 3, $\log p_{O_2} = 0$.

II. At point 4 on the vertical line exhibiting Mo–MoO_2 equilibrium, consider Mo–MoS_2

equilibrium ($\log p_{O_2} = -21.3$)

$$\log p_{O_2} = -9.85 + \log p_{SO_2}$$
$$\text{or } \log p_{SO_2} = \log p_{O_2} + 9.85 = -21.3 + 9.85 = -11.45$$

III. At point 5, on the vertical line exhibiting MoO_2–MoO_3 equilibrium, consider MoO_2–MoS_2 equilibrium ($\log p_{O_2} = -8.4$).

$$\log p_{O_2} = -13.66 + \frac{2}{3} \log p_{SO_2}$$
$$\text{or } 2/3 \log p_{SO_2} = \log p_{O_2} + 13.66 = -8.4 + 13.66 = 5.26$$
$$\text{or } \log p_{SO_2} = 3/2 \times 5.3 = 7.95$$

IV. At point 6, on the vertical line exhibiting MoO_3–O_2 equilibrium, consider MoS_2–MoO_3 equilibrium ($\log p_{O_2} = 0$)

$$\log p_{O_2} = -12.91 + 2/3.5 \log p_{SO_2} = 0$$
$$\text{or } \log p_{SO_2} = 22.6$$

V. At point 7, $\log p_{SO_2} = -25$ and $\log p_{O_2} = 0$

VI. At point 8, $\log p_{SO_2} = -25$, consider Mo–MoS_2 equilibrium,

$$\log p_{O_2} = -9.85 + \log p_{SO_2} = -9.85 - 25 = -34.85 \simeq 34.9$$

Problem 2.3

Given the following data, construct the predominance area diagram for copper compounds in the presence of oxygen and sulfur dioxide at 900 K.

Reactions	ΔG^o(cal)
$2Cu(s) + \frac{1}{2} O_2(g) = Cu_2O(s)$	$-40500 - 3.92\ T \log T + 29.5\ T$
$Cu_2O(s) + \frac{1}{2} O_2(g) = 2CuO(s)$	$-34950 - 6.10\ T \log T + 44.3\ T$
$2Cu(s) + \frac{1}{2} S_2(g) = Cu_2S(s)$	$-34150 - 6.20\ T \log T + 28.7\ T$
$2Cu_2S(s) + S_2(g) = 4CuS(s)$	$-45200 + 54.0\ T$
$S_2(g) + 2O_2(g) = 2SO_2(g)$	$-173240 + 34.6\ T$
$SO_2(g) + \frac{1}{2} O_2(g) = SO_3(g)$	$-22600 + 21.36\ T$
$Cu(s) + 2O_2(g) + \frac{1}{2} S_2(g) = CuSO_4(s)$	$-183000 + 88.4\ T$
$SO_3(g) + 2CuO(s) = CuO. CuSO_4(s)$	$-49910 - 3.32\ T \log T + 50\ 1\ T$
$Cu_2O(s) + SO_3(g) = Cu_2SO_4(s)$	$-44800 + 39.9\ T$

Solution

Based on the above ΔG^o vs. T equations given in the problem, ΔG^o_{900}, and $\log K_{900}$ for various reactions related to different compounds in the system, Cu–S–O, were calculated. Relations between p_{O_2} and p_{SO_2} for various reactions, which determine the predominant areas, are listed below:

Reaction		ΔG^0_{900}kcal	$\log K_{900}$	$\log p_{O_2} = f(\log p_{SO_2})$
1.	$2Cu + \frac{1}{2} O_2 = Cu_2O$	-24.37	5.92	$\log p_{O_2} = -11.84$
2.	$Cu_2O + \frac{1}{2} O_2 = 2CuO$	-11.30	2.74	$\log p_{O_2} = -5.48$
3.	$2CuS + O_2 = Cu_2S + SO_2$	-72.75	17.66	$\log p_{O_2} = -17.66 + \log p_{SO_2}$
4.	$Cu_2S + O_2 = 2Cu + SO_2$	-46.16	11.21	$\log p_{O_2} = -11.21 + \log p_{SO_2}$
5.	$Cu_2S + 3/2\ O_2 = Cu_2O + SO_2$	-70.62	17.15	$\log p_{O_2} = -11.43 + 2/3 \log p_{SO_2}$
6.	$CuS + 2O_2 = CuSO_4$	-91.89	22.31	$\log p_{O_2} = -11.16$
7.	$Cu_2S + 3O_2 + SO_2 = 2CuSO_4$	-111.03	26.96	$\log p_{O_2} = -8.99 - 1/3 \log p_{SO_2}$
8.	$Cu_2O + 3/2\ O_2 + 2SO_2 = 2CuSO_4$	-40.41	9.81	$\log p_{O_2} = -6.54 - 4/3 \log p_{SO_2}$
9.	$CuO. CuSO_4 + SO_2 + \frac{1}{2} O_2 = 2CuSO_4$	-12.09	2.93	$\log p_{O_2} = -5.87 - 2 \log p_{SO_2}$
10.	$Cu_2O + SO_2 + O_2 = CuO. CuSO_4$	-28.32	6.88	$\log p_{O_2} = -6.88 - \log p_{SO_2}$
11.	$2CuO + SO_2 + \frac{1}{2} O_2 = CuO. CuSO_4$	-17.02	4.13	$\log p_{O_2} = -8.26 - 2 \log p_{SO_2}$
12.	$Cu_2S + 2O_2 = Cu_2SO_4$	-82.88	20.13	$\log p_{O_2} = -10.07$
13.	$Cu_2O + SO_2 + \frac{1}{2} O_2 = Cu_2SO_4$	-12.27	2.98	$\log p_{O_2} = -5.96 - 2 \log p_{SO_2}$
14.	$Cu_2SO_4 + SO_2 + O_2 = 2CuSO_4$	-28.14	6.83	$\log p_{O_2} = -6.83 - \log p_{SO_2}$
15.	$Cu_2SO_4 + \frac{1}{2} O_2 = CuO. CuSO_4$	-16.06	3.93	$\log p_{O_2} = -7.80$

Based on the above table and following the procedure outlined in problems 2.1 and 2.2, the predominance area diagram of the Cu–S–O system constructed is shown in Fig. 2.9.

Problem 2.4

Galena is roasted at 750 °C and one atmosphere pressure in a roaster gas containing 10% SO_2 and 5% O_2. (a) Do you think $PbSO_4$ is a stable phase at this temperature in the roaster? (b) Calculate the partial pressure of SO_2 required to form $PbSO_4$ at 750 °C,

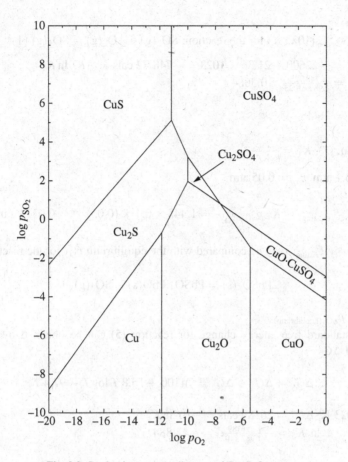

Fig. 2.9 Predominance area diagram of Cu–S–O system

if the gaseous mixture contains 1% O_2. (c) What is the maximum partial pressure of SO_2 at which Pb and PbO can coexist at 750 °C without forming $PbSO_4 \cdot 4PbO$?

Given that

$$SO_2(g) + \frac{1}{2}O_2(g) = SO_3(g), \Delta G_1^o = -22600 + 21.36\,T\ \text{cal} \qquad (1)$$

$$2PbSO_4(s) = PbSO_4 \cdot PbO(s) + SO_2(g) + \frac{1}{2}O_2(g),$$

$$\Delta G_2^o = 96700 + 16.81\,T \log\,T - 118.76\,T\ \text{cal} \qquad (2)$$

$$2PbO\ (s) = 2Pb\ (1) + O_2\ (g), \Delta G_3^o = 106600 - 51.40\,T\ \text{cal} \qquad (3)$$

$$PbSO_4 \cdot 4PbO(s) = 5PbO\ (s) + SO_2(g) + \frac{1}{2}O_2(g),$$

$$\Delta G_4^o = 95470 - 52.10\,T\ \text{cal} \qquad (4)$$

Solution

(a) At 750 °C (1023 K) for the reaction: $SO_2(g) + \frac{1}{2}O_2(g) = SO_3(g)$ (1).

$\Delta G_1^o = -22600 + 21.36 \times 1023 = -748.72$ cals $= -RT \ln K_1$

$\ln K_1 = \frac{748.72}{1.987 \times 1023} = 0.3684$

$\therefore K_1 = 1.445$

For Eq. (1), $K_1 = \frac{p_{SO_3}}{p_{SO_2} \cdot p_{O_2}^{1/2}}$

$p_{SO_2} = 0.1$ atm.$p_{O_2} = 0.05$ atm

$p_{SO_3(existing)} = K_1 \cdot p_{SO_2} \cdot p_{O_2}^{1/2} = 1.445 \times 0.1 \times (0.05)^{1/2} = 0.0323$ atm

This $p_{SO_3(existing)}$ may be compared with the equilibrium p_{SO_3} for the reaction:

$$2PbSO_4(s) = PbSO_4 \cdot PbO(s) + SO_3(g) \qquad (5)$$

$K_5 = p_{SO_3(equilibrium)}$

The standard free energy change for reaction (5) can be obtained by adding ΔG_1^o and ΔG_2^o

$$\therefore \Delta G_5^o = \Delta G_1^o + \Delta G_2^o = 74100 + 16.8\,T \log T - 97.4\,T$$

At 1023 K, $\Delta G_5^o = 26189$ cals $= -RT \ln K_5$

$\ln K_5 = -\frac{26189}{1.987 \times 1023} = -12.884$

$$\therefore K_5 = 2.54 \times 10^{-6} = p_{SO_3(equilibrium)}$$

Since this equilibrium $p_{SO_3} = 2.54 \times 10^{-6}$ atm is much lower than existing $p_{SO_3} = 0.0323$ atm in the roaster gas, $PbSO_4$ is a thermodynamically stable phase.

(b) If the roasting gas containing 1% O_2 is in equilibrium with $PbSO_4$ and $PbSO_4$. PbO, p_{SO_2} may be calculated as:

$$p_{SO_2} = \frac{p_{SO_3}}{K_1 \cdot p_{O_2}^{1/2}} = \frac{2.54 \times 10^{-6}}{1.445 \times (0.01)^{1/2}} = 1.8 \times 10^{-5} \text{ atm} \qquad \text{Ans.}$$

Since the partial pressure of SO_2 (= 1.8×10^{-5} atm) in the roaster gas is very low, $PbSO_4$ will get converted into $PbSO_4$. PbO in the presence of 1% O_2.

(c) At 750°C (1023 K) for the reaction: $2PbO\ (s) = 2Pb\ (l) + O_2\ (g)$ $K_3 = p_{O_2}$

$\Delta G_3^o = 106600 - 51.40 \times 1023 = 54018$ cal $= -RT \ln p_{O_2}$

$$\log p_{O_2} = -11.5422 \tag{6}$$

Similarly for reaction: $PbSO_4.4PbO(s) = 5PbO\ (s) + SO_2(g) + \frac{1}{2}O_2(g)$, we can express:

$$\log p_{O_2} = -18.022 - 2\log p_{SO_2} \tag{7}$$

On substituting the value of $\log p_{O_2} = -11.5422$, from Eq. 6 in Eq. 7, we get:

$$\log p_{SO_2} = -3.2399,$$

$$\therefore p_{SO_2} = 5.7 \times 10^{-4} \text{ atm} \qquad \text{Ans.}$$

Problem 2.5

Sphalerite is roasted at 750 °C and one atmosphere pressure in a roaster gas containing 10% SO_2 and 5% O_2. (a) Comment on the relative stability of $ZnSO_4$ and $ZnO.2ZnSO_4$ phases at this temperature. (b) Calculate the partial pressure of SO_2 required to form $ZnO.2ZnSO_4$ at 750 °C, if the gaseous mixture contains 1% O_2. (c) Calculate the maximum partial pressure of SO_2 at which 3 phases Zn (l), ZnS (s), and ZnO (s) will coexist at 750 °C.

Given that

$$SO_2(g) + \frac{1}{2}O_2(g) = SO_3(g), \Delta G_1^o = -22600 + 21.36\ T \text{ cal} \tag{1}$$

$$3ZnSO_4(s) = ZnO.2ZnSO_4(s) + SO_3(g), \Delta G_2^o = 39281 - 31.87\ T \text{ cal} \tag{2}$$

$$2Zn\ (l) + O_2(g) = 2ZnO\ (s), \Delta G_3^o = -171680 - 13.80\ T\log\ T + 93.30\ T \text{ cal} \tag{3}$$

$$ZnS\ (s) + \frac{3}{2}O_2(g) = ZnO(s) + SO_2(g), \Delta G_4^o = -107070 - 2.30T\log T + 24.77T\text{cal} \tag{4}$$

Solution

From the previous calculation in problem 2.4, for the reaction: $SO_2(g) + \frac{1}{2}O_2(g) = SO_3(g)$, at 1023 K, $K_1 = 1.445$ and $p_{SO_3(existing)} = 3.23 \times 10^{-2}$ atm

This $p_{SO_3(existing)}$ may be compared with the equilibrium p_{SO_3} in the reaction:

$$3ZnSO_4(s) = ZnO.2ZnSO_4\ (s) + SO_3(g) \tag{2}$$

$$\Delta G_2^o = 39281 - 31.87\ T \text{ cal}$$

at 1023 K, $\Delta G_2^o = 39281 - 31.87 \times 1023 = +7701$ cal

$$\ln K_2 = -\frac{7701}{1.987 \times 1023} = -3.7886$$

$$\therefore K_2 = 2.3 \times 10^{-2} \text{ atm} = p_{SO_3(\text{equilibrium})}$$

(a) Since this equilibrium, $p_{SO_3} (= 2.3 \times 10^{-2} \text{ atm})$ is very close to the existing $p_{SO_3} (= 3.23 \times 10^{-2} \text{ atm})$ of the roaster gas, $ZnSO_4$ as well as $ZnO.2ZnSO_4$ will coexist at 750 °C in the roaster at the given gas composition.

(b) If the roaster gas containing 1% O_2 is in equilibrium with $ZnSO_4$ and $ZnO.2ZnSO_4, p_{SO_2}$ may be calculated as:

$$p_{SO_2} = \frac{p_{SO_3}}{K \cdot p_{O_2}^{1/2}} = \frac{2.3 \times 10^{-2}}{1.445 \times (0.01)^{1/2}}$$

$$= 0.159 \text{ atm (about 16\% } SO_2 \text{ in the gas mixture).}$$

Hence, the roaster gas containing 16% SO_2 and 1% O_2 will covert (sulfatize) ZnO into $ZnO.2ZnSO_4$ Ans.

(c) At 750 °C (1023 K) for the reaction: $2Zn$ (l) $+ O_2$ (g) $= 2ZnO$ (s), $K_3 = \frac{1}{p_{O_2}}$

$$\Delta G_3^o = -118256 \text{ cals} = RT \ln p_{O_2}$$

$$\log p_{O_2} = -25.3675 \qquad\qquad\qquad (5)$$

Similarly for the reaction (4): ZnS (s) $+ \frac{3}{2} O_2(g) = ZnO(s) + SO_2(g)$, we can express:

$$\log p_{O_2} = -12.6507 + \frac{2}{3} \log p_{SO_2} \qquad\qquad\qquad (6)$$

On substituting the value of $\log p_{O_2} = -25.3675$, from Eq. 5 in Eq. 6, we get:

$$\log p_{SO_2} = -19.0773,$$

$$\therefore p_{SO_2} = 8.37 \times 10^{-20} \text{ atm} \quad \text{Ans.}$$

Problem 2.6

An off-grade copper concentrate contains chalcopyrite, pyrite, sphalerite, and galena. Referring to the predominance area diagrams of the systems: Fe–S–O (Fig. 2.3), Cu–S–O (Fig. 2.4), Zn–S–O (Fig. 2.5), and Pb–S–O (Fig. 2.6), comment whether selective sulfation roasting of copper and zinc is possible at about 600 °C. Explain by listing the stable phases at different temperatures. The usual roaster gas compositions are in the range: $p_{SO_2} = 10^{-1.5}$ to $10^{-0.5}$ and $p_{O_2} = 10^{-2}$ to 10^{-1} atm.

Solution

Since the roaster gas compositions of SO_2 (varying from $p_{SO_2} = 10^{-1.5}$ to $10^{-0.5}$ atm) and O_2 (varying from $p_{O_2} = 10^{-2}$ to 10^{-1} atm) are marked in figures, a summary of stable phases expected to form at different temperatures can be listed by referring to the predominance area diagrams of the systems: Fe–S–O (Fig. 2.3), Cu–S–O (Fig. 2.4), Zn–S–O (Fig. 2.5), and Pb–S–O (Fig. 2.6).

Stable Phases Expected from the M^a–S–O Diagrams at Different Temperatures and Usual Roaster Gas Compositions

Temperature, (°C)	Stable phases			
400	$CuSO_4$	$Fe_2(SO_4)_3$	$ZnSO_4$	$PbSO_4$
550	$CuSO_4$	$Fe_2(SO_4)_3$	$ZnSO_4$	$PbSO_4$
600	$CuSO_4$	$Fe_2(SO_4)_3$, Fe_2O_3	$ZnSO_4$	$PbSO_4$
650	$CuSO_4$	Fe_2O_3	$ZnSO_4$	$PbSO_4$
750	$CuO. CuSO_4$	Fe_2O_3	$ZnO.2 ZnSO_4$	$PbSO_4$

^{a}M stands for Fe, Cu, Zn, and Pb

On the basis of the information listed above, it may be concluded that during roasting at temperatures varying from 400 to 600 °C, all the four metals form only their sulfates except iron, which may form Fe_2O_3 at 600 °C. At 750 °C, copper and zinc may be converted to their basic sulfates, $CuO.CuSO_4$ and $ZnO.2 ZnSO_4$. Some of the copper may report as CuO and $CuSO_4$ at 650 °C. Hence, selective sulfation roasting of copper and zinc should be possible at about 600 °C. The product is expected to contain $CuSO_4$, $ZnSO_4$, $PbSO_4$, and Fe_2O_3, out of which the latter two are insoluble in water.

References

1. Moore, J. J. (1990). *Chemical metallurgy* (2nd ed.). Oxford: Butterworth-Heinemann Ltd.. (Chapter 7).
2. Gilchrist, J. D. (1980). *Extraction metallurgy* (2nd ed.). Oxford: Pergamon. (Chapter 12).
3. Rosenqvist, T. (1974). *Principles of extractive metallurgy*. New York: McGraw-Hill. (Chapter 8).
4. Coudurier, L., Hopkins, D. W., & Wilkomirski, I. (1978). *Fundamentals of metallurgical processes*. Oxford: Pergamon. (Chapter 5).
5. Pehlke, R. D. (1973). *Unit process of extractive metallurgy*. Co, New York: American Elsevier Pub. (Chapter 4).
6. Yazawa, A. (1979). Thermodynamic evaluation of extractive metallurgical processes. *Metallurgical Transactions B, 10*, 307–321.
7. Ramakrishna Rao, V. V. V. N. S., & Abraham, K. P. (1971). Kinetics of oxidation of copper sulfide. *Metallurgical Transactions, 2*, 2463–2470.
8. Nagamori, M., & Habashi, F. (1974). Thermodynamic stability of Cu_2SO_4. *Metallurgical Transactions, 5*, 523–524.
9. Ingraham, T. R., & Kellogg, H. H. (1963). Thermodynamic properties of zinc sulfate, zinc basic sulfate and the system Zn-S-O. *Transactions of the Metallurgical Society, AIME 227*, 1419–1426.
10. Kellogg, H. H., & Basu, S. K. (1960). Thermodynamic properties of the system Pb-S-O. *Transactions of the Metallurgical Society, AIME, 218*, 70–87.

11. Sohn, H. Y., & Goel, R. P. (1979). Principles of roasting. *Minerals Science and Engineering, 11* (3), 137–153.
12. Turkdogan, E. T. (1980). *Physical chemistry of high temperature reactions*. New York: Academic Press. (Chapter 8).
13. Shamsuddin, M., Ngoc, N. V., & Prasad, P. M. (1990). Sulphation roasting of an off-grade copper concentrate. *Metals Materials and Processes, 1*, 275–292.
14. Ahmadzai, H., Blairs, S., Harris, B., & Staffansson, L. I. (1983). Oxidation aspects of the lime-concentrate pellet roasting process. *Metallurgical Transactions B, 14*, 589–604.
15. Biswas, A. K., & Davenport, D. W. (1976). *Extractive metallurgy of copper*. New York: Elsevier Science Press. (Chapter 3).
16. Sohn, H. Y. (2014) Principles of copper production. In *Treatise on process metallurgy. Vol. 3, Industrial processes, Part A* (pp. 534–591). Oxford, UK/ Waltham: Elsevier (Section 2.1.1).
17. Shamsuddin, M., & Sohn, H. Y. (2019). Constitutive topics in physical chemistry of high-temperature nonferrous metallurgy – A review: Part 1. Sulfide roasting and smelting. *JOM, 71*, 3253–3265.
18. Sommer, A. W., & Kellogg, H. H. (1959). Oxidation of sphalerite by sulfur trioxide. *Transactions of the Metallurgical Society, AIME, 215*, 742–744.
19. Natesan, K., & Philbrook, W. O. (1969). Oxidation kinetic studies of zinc sulfide pellets. *Transactions of the Metallurgical Society, AIME, 245*, 2243–2250.
20. Natesan, K., & Philbrook, W. O. (1970). Oxidation kinetic studies of zinc sulfide in a fluidized bed reactor. *Metallurgical Transactions, 1*, 1353–1360.
21. Shah, I. D. & Khalafallah, S. E. (1970) *Chemical reactions in roasting of copper sulfides*, U.S. B.M., RI 7549, 21 pp.
22. Shah, I. D., & Khalafallah, S. E. (1971). Thermal decomposition of CuS to $Cu_{1.8}S$. *Metallurgical Transactions, 2*, 605–606.
23. Shah, I. D., & Khalafallah, S. E. (1971). Kinetics and mechanism of the conversion of covellite (CuS) to digenite ($Cu_{1.8}S$). *Metallurgical Transactions, 2*, 2637–2643.
24. Shah, I. D. and Khalafallah, S. E. (1972) *Kinetics of thermal decomposition of copper (II) sulfate and copper (II) oxysulfate*, U.S.B.M., RI 7638, 21 pp.
25. Coudurier, L., Wilkomirski, I., & Morizot, G. (1970). Molybdenite roasting and rhenium volatilization in a multiple hearth furnace. *Transactions of the Institution of Mining and Metallurgy, Section C 79*, C34–C40.
26. Amman, P. R., & Loose, T. A. (1971). Oxidation kinetics of molybdenite at 525–635 °C. *Metallurgical Transactions, 2*, 889–893.
27. Khalafallah, S. E. (1979). Roasting as a unit process. In H. Y. Sohn & M. E. Wadsworth (Eds.), *Rate processes in extractive metallurgy*. New York: Plenum Pres. (Chapter 4, Section 4.1).
28. Gnistling, A. M., & Brounshtein, B. I. J. (1950). The diffusion kinetics of reactions in spherical particles. *Journal of Applied Chemistry of the USSR, 23*, 1327–1338.
29. Kubaschewski, O., & Alcock, C. B. (1979). *Metallurgical thermochemistry* (5th ed.). Oxford: Pergamon.
30. Curnutt, J. L., Prophet, H., McDonald, R. A., & Syverud, A. N. (1975). *JANAF Thermochemical tables*. Midland: The Dow Chemical Company.
31. Barin, I., Knacke, O., & Kubaschewski, O. (1977). *Themochemical properties of inorganic substances (supplement)*. New York: Springer-Verlag.

Chapter 3
Sulfide Smelting

Platinum, iridium, and mercury can be obtained by thermal decomposition of the respective sulfides. Sulfides of cadmium, cobalt, molybdenum, and zinc can be reduced with metals like calcium, magnesium, manganese, sodium, and iron but economics does not permit large-scale production by metallothermic reduction. Nickel sulfide cast into anode can be directly electrolyzed to deposit nickel on cathode. The process has been successful on commercial scale for nickel but not for copper and lead. Conventional pyrometallurgical route based on the combination of roasting of sulfide concentrate in multiple hearth/flash/fluidized bed roasters and subsequent reduction smelting of oxides in a blast furnace of rectangular cross-section has been most widely adopted for production of zinc and lead. Until the very recent past, controlled roasting, matte smelting, and converting have been the most noteworthy method for production of blister copper from chalcopyrite concentrate. Roasting converts most of iron sulfide into iron oxide and copper sulfide into copper oxide/sulfate. But the roasting step has been eliminated from the flow sheet of copper extraction in recently developed faster smelting and converting processes [1, 2]. Smelting happens to be one of the main steps in extraction of metals from sulfide and oxide minerals via the pyrometallurgical route. It is essentially an operation of melting and slag/metal separation involving chemical reactions. Smelting may be defined in two different ways. In general, any process of metal production where metal is obtained in molten state is given the name "smelting." Thus, production of aluminum by electrothermic reduction of alumina dissolved in cryolite and those of iron, lead, and zinc by carbothermic reduction of hematite and calcined sinters (in case of lead and zinc) would fall in this category. The next level of definition is the overall process of producing primary metals from sulfide ores/minerals by going through a molten stage. The specific definition is the first step of the two-step oxidation of sulfur and iron from sulfide minerals, mainly for copper and nickel production, that is, matte smelting as against to "coverting" in which the matte is further oxidized to produce molten copper or a higher grade nickel matte. The two stages of the metal production operation, namely matte smelting and converting, are related to oxygen potential in two stages as well as heat generation

© The Minerals, Metals & Materials Society 2021

M. Shamsuddin, *Physical Chemistry of Metallurgical Processes, Second Edition*,

The Minerals, Metals & Materials Series, https://doi.org/10.1007/978-3-030-58069-8_3

[3]. Oxygen potential does not only affect the slag chemistry (e.g., magnetite formation) but also changes the behavior of impurities. In case, the sulfide mineral is oxidized all the way to metal in one step many more of the impurities would join the metal, instead of the slag and thus much larger amount of heat will be generated. Hence, in the first stage: the "matte smelting step," much iron, sulfur, and harmful impurities are removed into the slag formed in that stage. The resultant matte is separated and treated in a subsequent step, usually the converting step.

During smelting, constituents present in the charge in molten state separate into two immiscible liquid layers, namely liquid matte or metal and liquid slag. Matte, a mixture or solution of molten sulfides, is obtained by smelting of mixed sulfide concentrates. Some components of the charge may appear in the flue gases. Slag formation may be facilitated by addition of flux(es). Presence of arsenic and antimony in the concentrate may lead to the formation of a third immiscible liquid layer, known as speiss [4], which is a mixture of arsenides and antimonides of iron, cobalt, and nickel. On the other hand, smelting of oxide concentrate under highly reducing conditions with suitable reducing agent separates liquid metal and liquid slag into two layers.

In general, refining is not the primary objective of smelting but conditions may be developed by adjusting the operating variables to collect unwanted elements into slag, speiss, or flue gases [5]. For example, copper is collected into matte in lead smelting, cobalt into speiss in copper smelting, and sulfur into slag in iron smelting. High-temperature operation helps quick partitioning of selected elements in different phases but complete separation cannot be achieved. In treatment of both, the sulfide as well as oxide concentrates, smelting facilitates liquid–liquid separation, which may be carried out to transfer impurities into the slag phase to produce matte or metal or a layer richer in metal values for further processing. Thus, matte smelting is a concentrating step in the flow sheet of metal extraction from sulfide ores whereas reduction smelting (of oxide ores and calcines) happens to be the prime extraction step for production of primary metal, which is subjected to further refining. In both the cases, for smooth transfer of impurities from metal to the slag phase, basicity and fluidity (by maintaining proper viscosity and surface tension) of the slag are controlled by adjusting the slag composition. Thus, the knowledge about structure, properties, and constitution of slag becomes important in understanding the slag–metal reactions. The next chapter is devoted to these aspects.

3.1   Matte Smelting of Chalcopyrite

Since metal sulfides are low melting compared to metal oxides, lower temperatures are required for matte smelting. Chalcopyrite concentrate containing mainly sulfides of copper and iron as the major components and some gangue consisting of Al_2O_3, CaO, MgO, and SiO_2 is first smelted with the prime objective of obtaining a liquid matte (solution of Cu_2S and FeS) containing almost all the copper present in the charge and a slag with least copper [1, 3]. Silica is also added as a flux to produce a low melting (~1150 °C) and fluid slag but smelting is carried out at 1250 °C to achieve a better rate of smelting. Significant difference in densities of the matte and slag (respectively, 5.2 and 3.3 g cm^{-3}) brings out clear separation between different phases in two layers. An appropriate amount of FeS present in the matte prevents oxidation of more valuable copper sulfide. In order to facilitate separation and collection of droplets of matte, a slag with relatively low viscosity is desired. The physical chemistry of the matte–slag systems is controlled by Fe, Cu, S, O, and their sulfides and oxides. It is further influenced by the oxidation/reduction potential of the gases used for heating and melting the charge. The following reactions occur in the matte smelting of the chalcopyrite concentrate:

$$2CuFeS_2 + (4 - 1.5x)\,O_2 = [Cu_2S + xFeS]_{matte}$$
$$+(2 - x)\,FeO + (3 - x)SO_2; \tag{3.1}$$
$$x \text{ depends on the grade of matte produced.}$$

$$Cu_2S + 1.5O_2 = Cu_2O + SO_2 \tag{3.2}$$

$$FeS\,(1) + Cu_2O\,(1, slag) = FeO\,(1, slag) + Cu_2S\,(1) \tag{3.3}$$

The first and foremost purpose of matte smelting is to enrich the matte by sulfidizing the entire copper present in the charge. This is achieved by the presence of FeS in the matte (according to reaction 3.3). The above reactions take place because iron has higher affinity for oxygen than what copper has for oxygen. FeS gets oxidized to FeO. Higher oxides of iron, if formed, are reduced to FeO by FeS:

$$10Fe_2O_3(1) + FeS\,(1) = 7Fe_3O_4(s) + SO_2(g) \tag{3.4}$$

$$3Fe_3O_4(s) + FeS\,(1) = 10FeO\,(s) + SO_2(g) \tag{3.5}$$

A low-melting fayalite slag is formed according to the reaction:

$$2FeO\,(1) + SiO_2(s) = 2FeO.SiO_2(1) + 8.6\ kcal \tag{3.6}$$

Cu_2O is almost completely sulfidized to join the matte phase as Cu_2S because at the smelting temperature of 1200 °C the value of the equilibrium constant for reaction (3.3) is very high [6] $\left(K = \frac{a_{Cu_2S}.a_{FeO}}{a_{Cu_2O}.a_{FeS}} = \sim 10^4\right)$. Matte smelting may be carried out in blast or reverberatory furnaces using a considerable amount of hydrocarbon fuels or in electric arc furnaces using electrical energy. Though reverberatory smelting process is slow, it permits good slag/matte separation. The quality of matte is affected by the excess oxygen present in the furnace atmosphere. Iron in the slag will be oxidized to iron oxide, approximately to the composition of Fe_3O_4, even if the partial pressure of oxygen in the furnace atmosphere is 10^{-2} atm, that is, the furnace gas containing 1% oxygen. Thus, very little sulfide is found in the slag due to the oxidation of FeS to Fe_3O_4 in the presence of excess oxygen. Diffusion of oxygen from the magnetite-rich surface to the slag-rich interface oxidizes iron in the matte, thereby increasing the O/Fe ratio in that phase above unity. This occurs first in the cooler parts, on the hearth bottom where the solubility of Fe_3O_4 is minimum. In adverse cases it can take place anywhere, even at the slag–matte interface. Magnetite cannot be re-dissolved easily because of its high melting point of 1597 °C. Its formation makes the slag highly viscous and consequently hinders the separation of matte and slag and, therefore, continuous operation becomes very difficult. The formation of magnetite must be avoided by maintaining a high steady temperature and minimizing the excess oxygen in the combustion gases. It is important to note that a discardable slag with possibly minimum copper can be produced by (i) keeping the slag silica saturated, (ii) maintaining the furnace sufficiently hot to make the slag reasonably fluid, and (iii) avoiding oxidizing conditions. Although majority of plants [7] have closed down reverberatory furnaces, a brief discussion given above would be helpful in understanding the chemistry of matte smelting.

In recent years, copper production industry has made tremendous progress in developing more efficient and environment friendly technology incorporating increased oxygen enrichment, larger plants, continuous process with better control, more efficient acid plants to reduce sulfur dioxide emission, minor element recovery, lesser labor input employing more skilled personnel, and increased recycling of wastes [2]. These objectives led to the development of many new smelting processes such as Suspension smelting (INCO [8]), Outotec (formerly Outokumpu [9, 10] flash smelting units), Cyclone smelting (Contop [11] and Kivcet [12] processes), and Bath smelting (Noranda [8], Teniente [13], and Ausmelt/Isasmelt [8, 14] processes). The modern matte smelting processes produce high-grade mattes (defined by percent Cu or percent Cu_2S in the matte) containing nearly 70% Cu and slag with most of the gangue minerals.

3.1.1 Suspension Smelting

The objectives of roasting and smelting can be achieved by suspension smelting in a single unit.

3.1.1.1 Flash Smelting

Currently, flash smelting is more widely practiced for treatment of chalcopyrite concentrates with the prime objective of better utilization of heat generated during oxidation of sulfur present in the charge. The principal advantages of flash smelting include low energy cost compared to reverberatory and electric furnace smelting, high SO_2 in the effluent gases, and high rate of production of high-grade liquid matte containing 60–70% Cu. However, there is high loss of copper in the slag. Presently, more than 50% copper matte of the total world production is produced by flash smelting. In flash smelting, a mixture of dry fine particulate chalcopyrite concentrate (100 μ), quartzite flux, and recycle material is blown into a smelting unit through burners with oxygen or hot air or mixture of both. After entering, the ore fines get superheated in the furnace atmosphere above the slag. The fineness provides large surface area of contact between the ore particles and roasting gas and flux. The reaction kinetics is improved by mixing fine particles of the concentrate with the gas before injecting it into the smelting furnace. It is further improved by the use of oxygen-enriched air in place of air. Use of 95% O_2 makes flash smelting autogenous because lesser heat is taken away by the flue gas with nitrogen. Thus, a major amount of heat generated by oxidation of sulfides is utilized by the matte and slag. This amounts to using less or no hydrocarbon to generate the required temperature of approximately 1250 °C for matte smelting.

The sulfide particles react rapidly with oxygen with large evolution of heat resulting in controlled oxidation of iron and sulfur present in the concentrate and melting. The main reactions in flash smelting are summarized below:

$$2CuFeS_2(s) + 2.5O_2 = (Cu_2S.FeS)_{matte}(l) + FeO\ (s) + 2SO_2(g) \\ + 156.0\ kcal \tag{3.7}$$

$$2CuS\ (l) + O_2(g) = Cu_2S\ (l) + SO_2(g) + 65.2\ kcal. \tag{3.8}$$

$$FeS_2(l) + O_2(g) = FeS\ (l) + SO_2(g) + 54.1\ kcal. \tag{3.9}$$

$$2Cu_2S\ (l) + 3O_2(g) = 2Cu_2O\ (l) + 2SO_2(g) + 184.6\ kcal. \tag{3.10}$$

$$2FeS\ (l) + 3O_2(g) = 2FeO\ (s) + 2SO_2(g) + 225.0\ kcal \tag{3.11}$$

$$3FeO\ (s) + 0.5O_2(g) = Fe_3O_4(s) + 72.6\ kcal \tag{3.12}$$

Since the above reactions are highly exothermic, a large amount of thermal energy is provided for heating, melting, and superheating of the charge. The process is continuous and product gets separated into two layers consisting of liquid matte and liquid slag. The flue gas containing 30–70% SO_2 is loaded with dust. Any oxidized copper is reduced back to Cu_2S according to the reaction (3.3). The resulting copper matte contains copper, iron, and sulfur as its major components with up to 3% dissolved oxygen together with minor impurities such as arsenic, antimony, bismuth, lead, nickel, zinc, and precious metals. This matte is charged into a side-blown Pierce–Smith converter in molten state at 1100 °C for production of blister copper.

The two widely employed flash smelting furnaces were developed by the International Company of Canada (INCO [15, 16]) and the Outotec [17, 18] of Finland around the same time (1953–1955). The INCO unit based on commercial oxygen (95% O_2) is autogenous. Oxygen of the blast reacts with the chalcopyrite concentrate to produce molten matte (55–60% Cu) and molten slag (1–2% Cu) at the smelting temperature of 1250 °C. In both, the flash smelting unit dry concentrates from fluid bed driers are fed to regulate uninterrupted flow of particles through the burners. Flash smelting reactions in the INCO process produce flue gas containing 80% SO_2 that can be directly used to manufacture sulfuric acid. Matte is sent to the converter for production of blister copper and slag is recycled for copper recovery. The INCO flash furnace is also employed to smelt pentlandite (Ni–Cu–Co–Fe–S) concentrate [16] to produce approximately 45% Ni–Cu–Co matte and approximately 1% Ni–Cu–Co slag. In the INCO process, the matte grade is controlled by adjusting the ratio [19] of the oxygen input rate to the dried charge feeding rate whereas the slag composition by maneuvering the ratio of the rate of feeding of flux to the rate of feeding of the dried concentrate. By this process, copper content in slag can be lowered down below 1% which can be discarded to cut down copper recovery cost so as to economize the overall cost of production.

The Outotec process based on the use of preheated air or preheated oxygen-enriched air is not autogenous. Hydrocarbon or electrical energy is used to make up the thermal deficit. These flash smelting furnaces are more widely employed as compared to the INCO flash units because of better facilities and simplicity, such as recovery of heat from the off gas in waste heat boilers, water cooled reaction shaft capable of handling large amount of heat released, and single concentrate burner as compared to four burners in the INCO furnace. The Outotec furnaces can smelt up to 3000 tons of concentrate per day. Currently, they are operated with high oxygen blast and little hydrocarbon to produce off gas with high percentage of sulfur dioxide for efficient conversion into sulfuric acid. Automatic control of the process produces matte of uniform composition at high rate with low energy consumption at constant temperature. The matte and slag compositions [19] are controlled by maneuvering the ratios of the rate of O_2 input to the concentrate feed rate and the rate of flux input to the concentrate feed rate, respectively. Temperature is controlled by adjusting the ratio of oxygen to nitrogen in the blast and the rate of combustion of hydrocarbon.

3.1.1.2 Cyclone Smelting

In this smelting process, a mixture of fine chalcopyrite concentrate and quartzite flux particles with the gas stream is injected into a cyclone reactor. Since the interaction between the particles and gas in the cyclone is much faster as compared to the flash smelting unit, the process is more intense. The intensity of the large amount of heat causes severe erosion of refractories and other reactor materials. On account of the operational difficulties, cyclone smelting has not achieved commercial status.

3.1.2 Bath Smelting

In bath smelting processes, a mixture of the concentrate is either charged over a molten bath or injected into the melt together with the oxygen-enriched air.

3.1.2.1   Submerged Tuyere Smelting

In the submerged tuyere smelting processes developed by Noranda [20, 21] and Teniente [20–22], concentrate is brought into contact with oxygen into a mixture of molten matte and slag contained in a furnace. In both these processes, smelting is carried out in long horizontal cylindrical furnaces (5 m diameter and 20 m long) lined with magnesite chrome bricks and fitted with a horizontal row of submerged tuyeres. The furnace can be rotated to prevent molten matte and slag entering into tuyeres when blowing is interrupted. Oxygen-enriched air is blown through tuyeres into high-grade molten matte (72–75% Cu) layer. In these processes matte and slag layers are always maintained. Moist or dry concentrate is blown into the matte through tuyeres to get better distribution of heat and concentrate, high thermal efficiency, and lesser dust evolution. Vigorous stirring of the matte/slag bath caused by submerged blowing expedites melting and oxidation of the charge and also reduces deposition of solid magnetite in the furnace. About 15–20% of the total matte smelting of copper concentrate all over the world is executed by these two processes to produce high-grade matte (72–75% Cu), slag containing about 6% Cu, and flue gas with 15–25% SO_2. The matte is sent to the converter for production of blister copper and the slag to a copper recovery system and the off gas to cooling, dust recovery, and sulfuric acid plant. Almost entire heat for heating and melting the charge is derived from oxidation of iron and sulfur present in the concentrate. In case of additional requirement natural gas, coal, or coke may be burnt. In both the processes matte composition is controlled by maneuvering the ratio of the rate of oxygen input to the rate of feed of concentrate input and the slag composition is controlled by adjusting the ratio of rate of the flux input to the rate of feed of the solid charge input (aiming $SiO_2/Fe = 0.65$). The mechanism [19] of the process may be summarized in the following four steps:

1. Chalcopyrite concentrate and silica flux get quickly melted after falling into the violent matte/slag bath.
2. While entering into matte the heavier sulfide drops get oxidized by oxygen blown and oxides of copper and iron.
3. Iron silicate slag is formed by iron oxide and silica (reaction 3.6) above the matte.
4. SO_2 formed by oxidation of sulfur rises through the bath and leaves the furnace.

3.1.2.2   Ausmelt/Isasmelt—Top-Submerged Lancing (TSL) Technology

In recent years, conventional reverberatory furnaces for matte smelting of roasted chalcopyrite concentrate have been replaced to a major extent by the flash smelting furnaces developed by INCO and Outotec and also by the submerged tuyere smelting furnaces introduced by Noranda and Teniente due to high rate of production of high-grade matte and rich SO_2 flue gas at lower cost. But flash smelting suffers from the disadvantage of generating large amount of dust. In order to reduce dust a large settling area has to be provided into the flash furnace, which increases the size of the unit and, hence, the cost. This problem has been solved to a great extent by the researchers of the Commonwealth Scientific and Industrial Research Organization (CSIRO) of Australia by developing top-submerged lancing (TSL) technology [23, 24]. The idea of the TSL system introduced by Floyd [24] in 1970 has brought revolutionary changes in production of copper, lead, zinc, nickel, tin, and platinum group metals. This technology has set up a new trend in pyrometal-lurgical extraction of metals from sulfide, oxide, and complex minerals, as well as in the recovery of metals from a variety of metallurgical wastes [25] including slag and residues at lower operating and materials handling costs together with high rate of production. In addition, it also reduces the cost of fuel and expenses toward pollution control. Accumulation of all these benefits in a single unit has provided the TSL technology a unique status. The smelting technology developed at CSIRO in 1970s is commercially available in the name of Isasmelt/Ausmelt.

The TSL unit is based on a cylindrical furnace fitted with a central lance through which fuel, air, and oxygen are injected into a slag bath. The chemical reactions at the lance tip and in the region (zone) of slag–gas contact are controlled by generating oxidizing, neutral, or reducing atmospheric conditions through the lance. In order to remove impurities from the charge, the slag composition (normally containing CaO, FeO, SiO_2, and some Al_2O_3) is controlled by adding flux to attain proper viscosity at the working temperature. As TSL is capable of treating ore lump as well as fines (dry or wet), the feed preparation is not a serious requirement [23, 25]. The lance fitted with the Ausmelt furnace consists of two concentric pipes. The central stainless steel pipe (~0.5 m diameter) is used for blowing oxygen-enriched air through annulus space whereas the inner steel pipe is employed for inserting oil or natural gas. The outer pipe is immersed in the slag and the inner pipe is kept above the slag surface by about 1 m. Vanes are provided for swirling the blast down the lance tip. The swirling action created in the annulus space between the two pipes helps in cooling the lance tip by extracting heat from the outer pipe. As a result, slag freezes on the surface of the pipe and provides protection to the lance tip. Cooling by helical swirl vanes causes rapid extraction of heat from the outer pipe into the blast and formation of a protective slag coating after solidification on outer surface of the lance is the unique feature of the Ausmelt technique. The Ausmelt furnace produces high-grade matte and slag containing approximately 60% copper and 0.7% copper, respectively, and off gas with about 20% sulfur dioxide.

In the Ausmelt furnace mixture of moist concentrate, flux and recycle material is dropped (charged) into a molten slag/matte bath contained in a tall cylindrical furnace (3.5 m diameter and 12 m height). Simultaneously, oxygen-enriched air is blown through lance into the bath. The Ausmelt/Isasmelt smelting differs from flash smelting because, in the former, reactions take place mainly in the bath whereas, in the latter, it takes place above the bath. This difference causes a change in the sequence of reactions [19], which are summarized below:

$$2(CuFeS_2) \rightarrow (Cu_2S + 2FeS)_{matte} + \frac{1}{2} S_2(g) \tag{3.13}$$

$$2(FeO)_{slag} + \frac{1}{2} O_2(g) \rightarrow (Fe_2O_3)_{slag} \qquad\qquad (3.14)$$

$$(FeS)_{matte} + 3(Fe_2O_3)_{slag} \rightarrow 7(FeO)_{slag} + SO_2(g) \qquad\qquad (3.15)$$

$$\frac{1}{2}S_2(g) + O_2(g) \rightarrow SO_2(g) \qquad\qquad (3.16)$$

It is believed that the above process is catalyzed in the presence of the dissolved magnetite (~5 %) in the slag. The major requirement of energy for smelting is derived from oxidation of sulfur and iron (FeS) present in the concentrate.

The TSL technology has provided superior processes for smelting sulfide ores as compared to the conventional processes based on combination of roasters/sinter plants, reverberatory/blast furnaces, and electric/rotary furnaces. The conventional processes suffer from a number of disadvantages such as high capital investments and operating costs, low fuel efficiency, and environmental problems. On the other hand, owing to the high rates of reactions with sufficiently high rates of heat/mass transfer, Ausmelt technology lowers down the capital and operating costs [25]. Depending upon the quality of the concentrate, Ausmelt furnace produces matte containing 30–70% Cu, which is blown into blister copper in side-blown Pierce–Smith converters.

The Ausmelt/TSL technology can also be employed in smelting of complex ores and concentrates, which may not be successfully treated by the conventional methods. The process has been tested on pilot plant scale to demonstrate that lead, zinc, and copper can be recovered from a complex concentrate. In the first step of smelting of such complex concentrates, there is fuming of volatile PbS under low oxygen potential. There is high recovery of lead as oxide and sulfate in the fume. The furnace bath contains Cu/Zn-rich matte and slag. In the second step, matte is converted to blister copper whose purity depends on the extent of lead removal in the first step. In the third step, the slag is reduced with coal to recover zinc. In the fourth step, the lead fume produced in the first step is reduced with coal to obtain lead bullion. Gold and silver accumulated in copper and lead are recovered by subsequent treatments.

In recent years, a number of plants based on TSL technology have been installed throughout the world for processing concentrates, slag and residues of copper, lead, and zinc. More plants are under construction since TSL technology offers a number of advantages such as low capital and operating costs, flexibility of handling different types of feed with minimum preparation, strict environmental control, flexibility in operating continuous, semicontinuous or batch furnaces, efficient handling of small-scale (<100,000 tpy of feed) as well as large-scale plants (greater than half a million tons of feed) and solution to the waste problem in metallurgical industries such as zinc leach residue, Zn–Pb slag. The technology has better future because with passage of time pollution abatement laws will compel for stringent control toward emission of SO_2, CO, and CO_2 in atmosphere and dumping of leach liquor, slag, and other wastes in rivers or on the ground. Furthermore, in coming years, pressure on metallurgical industries to control dust and gaseous emission will increase; coke oven, blast furnaces, sinter plants, and rotary furnaces have to be replaced keeping in mind the protection of the city, town, and agricultural production. Toward this end, Ausmelt technology offers better alternatives.

3.2   Converting of Copper Matte

Pig iron was first refined into steel in the Bessemer converter. Impurities like carbon, silicon, manganese, and so on get oxidized when air blast penetrates through the molten pig iron. Oxidation being exothermic blowing raises the bath temperature from 1300 °C to 1600 °C, the temperature suitable for steelmaking. Such a bottom-blown converter is not suitable to convert copper matte into blister copper because air blast oxidizes the copper produced that settles at the bottom of the converter. For blowing copper matte, a horizontal side-blown Pierce–Smith converter is employed.

3.2.1   Pierce–Smith Converting

The copper converter is a cylindrical vessel (steel shell: ~10 m long and 4 m diameter) lined with magnesite bricks. Air for oxidation is introduced through the tuyeres located on side wall of the converter. To start the operation, the converter is turned so that the tuyeres are above the molten bath level. An appropriate amount of matte is introduced at about 1200 °C and the blast is turned on and the converter is turned to place the tuyeres in molten matte and mouth under the hood so that the gases are taken to the stack. The products of the converter are slag and blister copper, produced at different stages of the process. These are poured separately from the converter mouth by rotating it around its long axis. Large volumes of hot gases containing 5–15% SO_2 are also generated. The reactions are exothermic and the process is autogenous.

3.2.1.1   Stages of the Converting Process

Converting operation takes place in two chemically and physically distinct stages. Both of which involve blowing air into the molten sulfide phase.

First Stage of Blowing   During the first stage, the air blast stirs the bath vigorously and FeS is oxidized on the surface of the bubble to FeO as per the reaction:

$$2FeS\ (l) + 3O_2(g) = 2\ FeO\ (s) + 2SO_2(g) + 225\ kcal. \qquad (3.17)$$

Some copper sulfide is also oxidized as:

$$2Cu_2S\ (l) + 3O_2(g) = 2Cu_2O\ (l) + 2SO_2(g) + 184.6\ kcal. \tag{3.18}$$

But Cu_2O formed, gets re-sulfidized to Cu_2S in the bulk by reacting with FeS since Cu_2O is less stable (by virtue of possessing less negative, ΔG^o) as compared to FeO:

$$Cu_2O\ (l) + FeS\ (l) = Cu_2S\ (l) + FeO\ (s) + 19.6\ kcal. \tag{3.19}$$

The blowing causes turbulence that enhances both the rate of reaction and recovery of Cu_2S (reaction 3.19). FeO is slagged off by the quartz, which is added as soon as blast is turned on:

$$2FeO\ (l) + SiO_2(s) = 2FeO.SiO_2(l) + 8.6\ kcal. \tag{3.20}$$

The first stage of blowing is known as the slag-forming stage because FeO is slagged off according to the reaction (3.20). The oxidation of FeS in the matte in the presence of silica, which is the principal source of heat in the converter, may be described by a combined reaction (reactions: 3.17 and 3.20):

$$2FeS\ (l) + 3O_2(g) + SiO_2(s) = 2FeO.SiO_2(l) + 2\ SO_2(g) + 233.6\ kcal \tag{3.21}$$

The activity of FeO in slag decreases in the presence of acid silica, which helps in pushing reaction (3.21) in the forward direction to facilitate the removal of iron from the melt. This can be explained by van't Hoff isotherm for reaction (3.17) under nonstandard conditions, caused by the presence of silica:

$$\Delta G = \Delta G^o + RT \ln \left[\frac{(a_{FeO}^2) \cdot p_{SO_2}^2}{a_{FeS}^2 \left\{ p_{O_2}^3 \right\}} \right] \tag{3.22}$$

The second term in the above equation becomes negative because the quotient is less than unity as a_{FeO} is much less than unity in the presence of SiO_2. Thus, the resulting more negative value of ΔG increases the overall driving force for the transfer of iron from molten matte to the slag. After the slag-forming stage, the principal product is molten Cu_2S (white metal, $\sim$1200 °C) containing less than 1% Fe. The converter is turned down, the blast is turned off, and the slag is drained into a ladle.

During the slag-forming stage, the converter is charged with the matte several times. The slag is tapped after every oxidation step and thus the amount of Cu_2S gradually increases.

Second Stage of Blowing In the second stage, called the copper-making stage, the white metal (Cu_2S) is oxidized to Cu_2O. In this stage, there is no need for a slag because the white metal contains little iron to form iron oxide. The interaction between Cu_2S and Cu_2O leads to the production of metallic copper and SO_2 according to the reactions:

$$Cu_2S \ (l) + 1.5O_2(g) = Cu_2O \ (l) + SO_2(g) \qquad (3.23)$$

$$Cu_2S \ (l) + 2Cu_2O \ (l) = 6Cu \ (l) + SO_2(g) \qquad (3.24)$$

With continued blowing, the amount of blister copper goes on increasing due to additional removal of sulfur from the white metal. The metal obtained contains 98% Cu and is known as blister copper because bubbling of sulfur dioxide through the melt produces blister effect. The precious metals not being oxidized are retained in the blister. The small amount of iron left over from the first stage is oxidized at the beginning of the second stage, producing a small amount of slag high in copper. After the blister is tapped, this slag is left in the converter for the next operation. The blow to blister usually lasts for 2–3 h.

3.2.1.2 Physical Chemistry of Matte Converting

The constitution diagram of the system $Cu–Cu_2S$, shown in Fig. 3.1 indicates that at the converting temperature of 1200 °C the copper formed initially dissolves in Cu_2S, and thus the molten metal in the converter remains as a single phase. With the progress of oxidation, the bath separates into two phases: a Cu solution in Cu_2S and another Cu_2S solution in Cu. When the amount of Cu_2S in the converter lowers down to its limit of solubility in copper, the bath again becomes a single-phase solution of Cu_2S in copper. The equilibrium constant for the reaction (3.24) is given as:

$$K = \frac{a_{Cu}^6 . p_{SO_2}}{a_{Cu_2O}^2 . a_{Cu_2S}} \qquad (3.25)$$

If the mutual solubility of the reactants is neglected K may be expressed in terms of the equilibrium partial pressure of SO_2 (i.e., $K = p_{SO_2}$). The reaction will proceed as long as partial pressure of SO_2 in the gases above the molten bath in the converter is less than the equilibrium pressure. The partial pressure of SO_2 will not exceed 0.21 atm (air contains 21% O_2 and one molecule of O_2 produces one molecule of SO_2) even if the entire oxygen of the blast is fully used. The respective values of the equilibrium constant, $(K = p_{SO_2})$ for the reaction (3.24) 1.14, 10.99, and 55.12 atm at 900 °C, 1100 °C, and 1300 °C, indicate that the reaction producing metallic copper may begin at as low as 900 °C. However, it is definitely faster at the converting temperature of 1250 °C. Since toward the end of the converting operation a_{Cu_2O} and a_{Cu} reach a constant value (unity), p_{SO_2} is proportional to a_{Cu_2S} (Cu_2S remaining in blister copper).

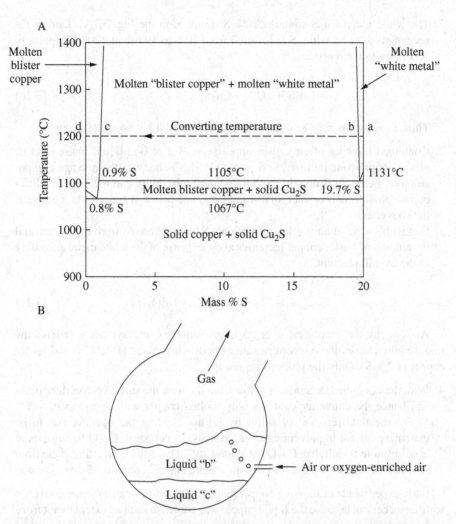

Fig. 3.1 (**A**) Copper–sulfur equilibrium phase diagram showing the converting reaction path (a, b, c, d, 1200 °C). (**B**) Sketch of the Peirce–Smith converter and its two immiscible liquids, (**b**) and (**c**), during the copper-making stage of converting (From Extractive Metallurgy of Copper by W. G. Davenport et al. [19], © 2002, p 135, Elsevier Science Ltd. Reproduced with the permission of Elsevier Science Ltd. (**B**) Originally from E. F. Peretti [26] with the permission of Discussion Faraday Society)

The oxidation stops when p_{SO_2} has reached a value slightly above 1 atm. If the reaction has to proceed further, the bubbling of SO_2 should overcome the atmospheric pressure and the hydrostatic pressure of the molten metal. The total pressure is greater than 1 atm even on the surface of the molten bath and some sulfur will therefore be always left in the resultant blister. The mechanism of converting operation may be discussed with the aid of Cu–S phase diagram in the following steps [19]:

1. The white metal Cu_2S contains 20% S (point a, in the Fig. 3.1A). During the second stage of blowing, S is lowered from 20% to 19.6% at 1200 °C (point b) according to the reaction:

$$Cu_2S + xO_2 \rightarrow Cu_2S_{1-x} + xSO_2 \qquad (3.26)$$

Thus, initially we have a sulfur deficient white metal but no metallic copper.

2. Continued blowing of air causes appearance of a second liquid phase, that is, blister copper containing 1.2% S (point c). This is because the average composition of the liquid is now in the liquid–liquid miscibility gap region. The blister copper phase is denser than S-deficient Cu_2S and hence, it sinks to the bottom of the converter.
3. Further blowing of air results in additional S being removed from the system and the amount of blister copper increases at the expense of the white metal according to the overall reaction:

$$Cu_2S \,(l) + O_2(g) = 2Cu \,(l) + SO_2(g) \qquad (3.27)$$

As long as the combined average composition of the system is within the immiscibility range, the converter contains both white metal (19.6% S) and blister copper (1.2% S). Only the proportions change.

4. With the continued decrease of sulfur from the melt the sulfide phase disappears and hence the converter contains only molten copper with approximately 1% S. For the final removal of sulfur by further blowing the operator has to be extremely careful in preventing over oxidation of copper to Cu_2O because there is no sulfur to re-sulfidize Cu_2O back into Cu_2S. The entire converting cycle from start-to-finish takes 6–12 h [19] and the metal obtained contains 98–99% copper.

Early experiments of blowing copper matte in bottom-blown Bessemer converter were unsuccessful because the liquid copper was cooled to such an extent that it froze in the tuyeres by the incoming air. The side-blown (Pierce–Smith) converter blows air into the sulfide (matte) phase rather than into the blister copper (Fig. 3.1B). Appendix: A.1 presents various steps in extraction of copper from chalcopyrite concentrate.

3.3   Continuous Converting

In Pierce–Smith converting operation, the vessel has to be rotated while charging and tapping to keep the tuyeres above the melt. Hence, it is a batch process, which has to be discontinued intermittently that causes difficulties in the collection of SO_2. These limitations led to the development of continuous converting processes by Noranda, Kennecott/Outotec, and Mitsubishi.

3.3.1 Noranda Continuous Converting Process

A rotary furnace is employed in Noranda continuous converting [27] process which is the further development of the Noranda submerged tuyere smelting. Molten matte and scrap are charged through a large mouth and the off gas is directed into a hood/ acid plant through another opening. Flux and coke are fed through a hole. The converter is operated continuously by maintaining layers of molten copper, matte (predominantly Cu_2S) and slag and blowing O_2-enriched air through tuyeres and collecting off gas containing nearly 20% SO_2. Blister copper and slag are tapped intermittently.

Molten matte containing 65–70% Cu, 5–10% Fe, and 20–24% S, on pouring into the converter, joins the matte lying between molten copper and the molten slag. O_2-enriched air oxidizes the matte. The slag and metal formation reactions may be represented as:

$$3(FeS)_{matte} + 5O_{2(blast)} \rightarrow (Fe_3O_4)_{slag} + 3SO_2(g) \tag{3.28}$$

$$(Fe_3O_4)_{slag} + (FeS)_{matte} \rightarrow (FeO)_{slag} + SO_2(g) \tag{3.29}$$

$$2FeO\ (l) + SiO_2(s) \rightarrow (2FeO.SiO_2)_{slag}(l) \tag{3.30}$$

Molten copper is produced by oxidation of Cu_2S:

$$(Cu_2S)_{matte} + O_2 \rightarrow 2Cu\ (l) + SO_2(g) \tag{3.31}$$

The resultant molten slag and molten copper rises and falls above and below the tuyeres to join the slag and molten copper layers, respectively. Blister copper containing 1.3% S and slag with a high percentage of Cu (about 10%) are tapped intermittently. The high oxygen potential required in this converting process for the production of copper is responsible for high residual copper in the slag.

3.3.2 Kennecott–Outotec Flash Converting Process

The Outotec continuous converting [28] process is the subsequent development of the Outotec flash smelting. Its combination with Kennecott's Solid Matte Oxygen Converting concept has popularized the Kennecott–Outotec Flash Converting Process [29]. Such a converting unit was first installed in 1995 in Utah, USA with the help of Kennecott Copper. In this process, the flash smelting furnace matte containing about 70% Cu is first granulated to about 0.5 mm size and then crushed to 50 microns and dried. The dried matte powder mixed with limestone is blown with 80% O_2-enriched air continuously into a flash furnace, which is similar in design to the Outotec flash smelting furnace. Copper is produced according to the following overall reaction by appropriate adjustment of oxygen supply:

$$(Cu_2S + FeS)_{matte} + O_{2(blast)} \rightarrow Cu(l) + (Fe_3O_4)_{calcium\ ferrite\ slag} + SO_2(g) \quad (3.32)$$

The flash converting unit produces (i) molten copper containing 0.2% S and 0.3% O, (ii) molten calcium ferrite slag containing approximately 16% CaO and nearly 20% Cu, and (iii) off gas loaded with 35–40% SO_2. The molten copper and slag are intermittently tapped and sent, respectively, to the refinery to electrodeposit copper and a water-granulation plant to produce slag-granules to be recycled to the matte smelting furnace. The off gas is subjected to de-dusting before being transported to the sulfuric acid plant. The calcium ferrite slag having lesser tendency to foam and being capable of absorbing magnetite/hematite while maintaining better fluidity than the fayalite slag is necessary because the oxygen potential required to produce the copper phase in the presence of a slag in a continuous process oxidizes iron to higher oxides.

Copper metal and SO_2 gas are formed below the slag layer according to the following reactions:

$$(Cu_2S)_{matte} + 2(Cu_2O)_{slag} \rightarrow 6Cu(l) + SO_2(g) \quad (3.33)$$

$$(Cu_2S)_{matte} + 2(CuO)_{slag} \rightarrow 4Cu\ (l) + SO_2(g) \quad (3.34)$$

Cu_2S may also interact with Fe_3O_4 present in the molten calcium ferrite slag to produce copper and FeO:

$$(Cu_2S)_{matte} + 2(Fe_3O_4)_{slag} \rightarrow 2Cu(l) + 6FeO + SO_2(g) \quad (3.35)$$

As SO_2 evolution causes slag foaming in Outotec flash converting accumulation of Cu_2S is avoided by maneuvering the ratio of O_2 input rate to the matte feeding rate toward oxidizing. This setting favors the formation of Cu_2O instead of Cu_2S. In the absence of Cu_2S, slag foaming is minimized. The productivity of the Kennecott–Outotec flash converter is two to three times larger than the Pierce–Smith converter. The process is simple and efficient for oxidation of matte and collection of dust and off gas, hence, useful in sorting out the environmental problems to a major extent. However, it is not suitable for melting copper scrap and it also requires solidification of the matte followed by granulation and crushing, which increase energy consumption.

3.3.3   Mitsubishi Continuous Converting Process

In the Mitsubishi continuous converting [30, 31] process, O_2-enriched air is blown through downward lances on the surface of slag–matte–molten copper bath contained in a 12.5 m diameter converter. Molten matte continuously fed into the converter through an opening immediately spreads out across the molten copper bath and pushes the slag toward the overflow notch. Five to ten vertical lances, each consisting of two concentric pipes, are inserted from the roof of the converter. The central pipe is used to blow air along with $CaCO_3$ flux and recycle materials whereas O_2-enriched air is blown through the annulus space between the two pipes. The outer pipe is kept above the liquid bath by 0.5–0.7 m and it is lowered when its tip is consumed. The central pipe is placed only up to the roof surface. The outer pipe is continuously rotated to prevent it being jammed by slag/metal splashing. Flux and recycled materials get mixed with O_2-enriched air at the end of the central pipe. The mixture impinging on the surface of the molten bath forms emulsion of molten copper–matte–slag–gas, which enhances solid–liquid–gas reactions to form new copper and new slag on account of matte consumption. In the Mitsubishi converter reactions between matte, flux, and oxygen may be represented as:

$$3(FeS)_{matte} + 5O_2(g) \rightarrow (Fe_3O_4)_{slag} + 3SO_2(g) \tag{3.36}$$

$$(CaO)_{flux} + (Fe_3O_4)_{slag} \rightarrow \text{molten calcium ferrite slag} \tag{3.37}$$

$$(Cu_2S)_{matte} + O_2(g) \rightarrow 2Cu(l) + SO_2(g) \tag{3.38}$$

Copper droplets descend through the copper layer and, thus, molten copper underflows through the siphon. On the other hand, slag droplets push the slag layer to overflow through the slag notch. Metallic copper oxidized as Cu_2O joins the fluid calcium ferrite slag. The off gas containing 25–30% SO_2 passes through a waste heat boiler, electrostatic precipitator and wet gas cleaning system and is finally treated in a sulfuric acid plant. Such a converter can produce 400–900 tons of copper per day. Viscosity of the Mitsubishi converter slag is low due to the formation of ternary low melting slag in the presence of CaO.

In all the continuous converting processes discussed above, a great deal of attention has to be paid for maintenance of the refractory lining and tap holes to prevent rapid erosion because of the constant presence of molten copper (dense and fluid) and calcium ferrite slag in the converter.

3.4   Direct Copper Extraction from Concentrate

From the above discussions, it is evident that production of copper from sulfide concentrates involves two major steps, namely, smelting and converting. These steps are chemically similar because both derive heat from the oxidation of sulfur and iron and discard iron in slag by fluxing FeO with silica. These common features encouraged extractive metallurgists to carry out smelting and converting in one furnace. Since 1974, copper-producing industries have been involved in developing continuous processes for production of copper and lead from sulfide ores in a single unit combining roasting/smelting and converting (and if possible refining also) with the prime objective of reducing the materials handling, energy consumption, capital and operating costs. In order to achieve these objectives, idealized requirements like unit separated into smelting, converting, and oxidizing zones; high reaction rates by increasing the area of contact between reactants by using fine ore particles; autogenous processing by use of oxygen; smooth separation of slag and metal by facilitating movement of the two liquids in opposite directions; recovery of valuable metals from fumes and slag cleaning; production of chemicals; and enhancement in the reaction kinetics by lancing to provide turbulence, have to be fulfilled. In this context several attempts were made on pilot plant scale for production of copper directly from chalcopyrite concentrate by processes known as WORCRA (Australia) and Noranda (Canada). However, these processes could not achieve commercial status due to the high loss of copper into slag. Cost of recovery of this copper is high because a large amount of slag is generated in treatment of chalcopyrite concentrate having relatively higher percentage of iron. The high copper slag in WORCRA [derived from the first three alphabets of the inventor, H. K. Worner [32] and CRA (Conzinc Riotinto of Australia Limited)] process is treated with low-grade matte by moving two liquids in opposite directions to reduce copper content from the slag to about 0.5–1%. In the Noranda [33] process slag is solidified and then subjected to ore dressing to recover most of the copper. Noranda submerged-tuyere process produced copper directly from chalcopyrite concentrate for several years but currently it produces only high-grade matte [34] (72–75% Cu). In order to reduce the cost of recovery of copper from the slag, chalcocite (Cu_2S) and bornite (Cu_5FeS_6) concentrates having lesser iron as compared to chalcopyrite are more suitable for direct conversion. Such a process has been developed by Outotec but Mitsubishi process is more versatile in the treatment of variety of copper concentrates including chalcopyrite.

3.4.1 Outotec Blister Flash Smelting Process

In this process smelting of chalcocite/bornite concentrates in the presence of flux and recycle materials by blowing O_2-enriched air produces molten copper (99% Cu, 0.04–0.9% S, 0.01% Fe and 0.4% O), slag (14–24% Cu), and off gas (15–25% SO_2). Use of highly O_2-enriched air blast generates enough heat to melt the entire charge. The process is autothermal as well as continuous. The temperature of the furnace is controlled by adjusting the extent of O_2-enrichment of the blast and the rate of combustion of fossil fuel. Direct production of copper from the chalcocite/bornite concentrates may be chemically represented as per the reaction:

$$Cu_2S \text{ (chalcocite)}/Cu_5FeS_6 \text{ (bornite)} + O_{2(\text{blast})} + (SiO_2)_{\text{flux}}$$
$$\rightarrow Cu(l) + (FeO.Fe_3O_4.SiO_2)_{\text{slag}} + SO_2(g) \tag{3.39}$$

Oxygen supply is regulated to produce metallic copper instead of Cu_2S or Cu_2O. Depending on the supply of oxygen, the smelting furnace produces a mixture of over oxidized Cu_2O (from the outside) and under oxidized Cu_2S (from the inside). The interaction of the two over oxidized and under oxidized fractions produce molten copper, molten slag, and gas according to the reactions:

$$(Cu_2S)_{matte} + 2(Cu_2O)_{slag} \rightarrow 6Cu\ (l) + SO_2(g) \qquad (3.40)$$

$$(Cu_2S)_{matte} + 2(CuO)_{slag} \rightarrow 4Cu\ (l) + SO_2(g) \qquad (3.41)$$

$$(Cu_2S)_{matte} + 2(Fe_3O_4)_{slag} \rightarrow 2Cu(l) + 6FeO + SO_2(g) \qquad (3.42)$$

The net reaction (3.41) is controlled by (i) analyzing the percent copper in the slag and percent sulfur in molten copper and (ii) adjusting the ratio of oxygen input in the blast to the concentrate feed rate. The copper content in the slag can be reduced by decreasing the O_2/concentrate ratio, whereas the ratio of the flux to the concentrate input rate controls the overall composition of the slag. Possibility of slag foaming is avoided by operating the smelting furnaces with high O_2/concentrate ratio, which also eliminates building up of Cu_2S layer. Molten matte layer formed between molten copper and molten slag gives rise to the formation of SO_2 gas according to the reactions (3.40), (3.41), and (3.42). However, in the absence of Cu_2S layer, high percent of copper is transferred into the slag (as Cu_2O) because reaction (3.40) responsible for elimination of Cu_2O formed cannot take place.

From the smelting furnace slag is transferred to an electric slag cleaning furnace where it is allowed to settle for about 10 h under a thick layer of coke. During the settling period, copper oxides present in the slag get reduced to metallic copper. Some magnetite also gets reduced first to FeO and then to Fe.

Iron joins the reduced copper melt. The above procedure lowers down copper in slag from 14% to about 0.6% at Glogow in Poland [35] whereas at Olympic Dam [36], it is reduced from 24% to 4%. For further reduction of copper content in the slag, methods based on solidification/comminution/flotation has to be adopted.

In the recently developed Outotec blister flash smelting unit, under operation at Zambia and China, a chalcopyrite concentrate blend with copper matte [37] is processed. In this way, the problem of lower Cu/Fe ratio in chalcopyrite ($CuFeS_2$) has been solved because mixing increases this ratio to 2, which is equivalent to what chalcocite, Cu_2S has. However, the Outotec process will probably be restricted to the treatment of chalcocite and bornite for direct conversion to metallic copper because of the high cost involved in the reduction and recovery of copper from the slag.

3.4.2   Mitsubishi Process

Chalcopyrite with higher iron content generates larger amount of slag on smelting and, hence, is not suitable for the Outotec process. Larger occurrence of chalcopyrite ($CuFeS_2$) as compared to chalcocite (Cu_2S) and bornite (Cu_4FeS_6), problem of copper recovery from slag and diversified benefits of direct conversion of concentrates into molten copper encouraged Mitsubishi Materials Corporation [30], Japan, to develop a continuous process for treatment of chalcopyrite to produce metallic copper directly. The Mitsubishi process involves a system of three furnaces for smelting, slag cleaning, and converting. These furnaces are connected in series, where molten matte and slag flow from one to another under gravity. This arrangement offers benefits like treatment of variety of copper sulfide concentrates including chalcopyrite, simpler recovery of copper from slag, collection of high SO_2 off gas, and lesser materials handling.

The smelting furnace has nine vertical lances made of two concentric pipes (of 5 and 10 cm diameter) for charging from top. The charge consisting of dried concentrate, silica flux, and recycle materials is fed through the central pipe of the lance and O_2-enriched air (55% O_2) is blown through the annulus space between the two pipes. The outer pipe is continuously rotated to prevent from jamming. The central pipe is terminated at the furnace roof whereas the outer pipe is extended downward and held above the molten bath by about 0.7 m. The charge interacts with the blast at the exit of the central pipe. The solid–gas mixture impinging on the surface of the molten bath forms a matte–slag–gas foam/emulsion [19]. A tremendous increase in the availability of surface area to the reacting species markedly increases the rate of reaction between the liquids, solids, and gas. The resulting products, matte (65–70% Cu), and iron silicate slag flow continuously through the tap hole to the electric slag cleaning furnace. It is held for about 1–2 h in an elliptical 3600 kW electric furnace with three or six electrodes to allow adequate separation of the matte and slag into two layers. The off gas loaded with 20–25% SO_2 is directed to the sulfuric acid plant after passing through a waste heat boiler, electrostatic precipitator, and a wet gas cleaning unit. The slag is electrically heated to 1250 °C to attain proper fluidity for efficient settling and separation of the matte droplets from the slag. In order to minimize the loss of copper in the slag, liquid matte/slag system is held for longer time in the slag cleaning furnace. From the electric furnace matte continuously underflows into the converting furnace and the slag overflows through a tap hole to the slag granulating plant.

The converting furnace that continuously receives matte from the electric slag cleaning furnace is also charged with copper scrap and anode. O_2-enriched air (30–35% O_2) along with limestone flux and granules of converter slag are blown on the surface of the matte to produce molten copper (with approximately 0.7% S), molten slag (14% Cu), and off gas (25–30% SO_2). The matte continuously flowing into the converter spreads over the molten copper layer present and reacts with oxygen supplied through lances to form FeO and molten copper according to the reactions:

$$2(FeS)_{matte} + 3O_2 \rightarrow 2(FeO)_{slag} + 2SO_2(g) \tag{3.43}$$

$$(Cu_2S)_{matte} + O_2 \rightarrow 2Cu(l) + SO_2(g) \tag{3.44}$$

FeO and Cu formed may undergo further oxidation:

$$3FeO + 0.5O_2 \rightarrow 2(Fe_3O_4)_{slag} \tag{3.45}$$

$$2Cu(l) + 0.5O_2 \rightarrow (Cu_2O)_{slag} \tag{3.46}$$

CaO formed by decomposition of limestone forms a low-melting slag with Cu_2O, FeO, and Fe_3O_4:

$$CaO + Cu_2O + FeO + Fe_3O_4 \rightarrow (CaO - Cu_2O - FeO - Fe_3O_4)_{slag} \tag{3.47}$$

The converting operation in the Mitsubishi process requires the use of CaO-based slag [38] because SiO_2 cannot absorb the magnetite and wustite formed under the high oxygen potential required to form metallic copper. In the absence of CaO, the solid magnetite crust stops further converting operation. In the Mitsubishi process, 15–20% CaO [19] in the slag [12–16% Cu (60% as Cu_2O and rest Cu), 40–55% Fe (70% Fe^{3+} and rest Fe^{2+}) and 15–20% CaO] has been found optimum for the converting operation. The higher residual sulfur in copper (about 0.7%) produced by this method as compared to that obtained from the Pierce–Smith converter (0.02% S) can be reduced by increasing the percentage of oxygen in the blast. But this will lead to the transfer of more copper into slag as Cu_2O. Hence, an optimum higher oxygen in the blast is used to eliminate building up of Cu_2S matte layer in order to avoid foaming (as practiced in Outotec blister smelting process). This is ensured by the presence of less than 0.7% S in the molten copper produced. Copper is continuously transferred to one of the two anode furnaces and slag is sent to a water-granulated plant. Off gas is directed to sulfuric acid plant via waste heat boiler, electrostatic precipitator, and a wet gas cleaning unit. The off gases from smelting and converting furnaces are mixed before entering into the electrostatic precipitator.

Although in the Mitsubishi process, continuous copper making is carried out in a unit consisting of smelting, electric slag cleaning, and converting furnaces, the process is simple and efficient for recovery of copper from the slag. The process is capable of treating different types of concentrates including low iron chalcocite and bornite as well as high iron chalcopyrite with less material handling, but it suffers from the disadvantage of collecting off gas from two furnaces and poor refractory life. However, with the aid of recent innovations in the Mitsubishi process, life of the furnace lining has enhanced from 2 to 4 years [39].

3.5 Matte Smelting of Galena

In the hearth process, lead is obtained by the roast-reduction [40] of sintered galena concentrate at 800–900 °C. Air blown into the furnace hearth through the tuyeres oxidizes lead sulfide, which interacts with the remaining lead sulfide according to the reaction:

$$2PbO\ (l) + PbS\ (l) = 3Pb\ (l) + SO_2(g) \tag{3.48}$$

Since the above reaction leading to the formation of metallic lead has resemblance with the production of blister copper in the Pierce–Smith side-blown converter due to the interaction of Cu_2O formed by blowing the white metal (Cu_2S), the hearth process is categorized under matte smelting. The free energy change for the reaction is negative at temperatures above 960 °C, smelting is carried out at relatively higher temperatures (1100–1200 °C) in excess of air to achieve a reasonable rate of production. The metallic lead trickles through the charge to the bottom. Major steps in extraction of lead from galena concentrate is shown in Appendix: A.2.

In a blast furnace, a charge consisting of sinter (obtained from the Dwight Lyod sintering machine), coke, limestone, and quartz is smelted in excess of air at 1100–1200 °C. The resulting products separate in four distinct layers depending on specific gravity. These are collected in the furnace hearth, from top to bottom, slag (sp. gr. 3.6), matte containing copper and other elements (sp. gr. 5), speiss ($FeAs_4$ + impurities: sp. gr. 6), and lead bullion (sp. gr. 11). The bullion is refined for removal of impurities and recovery of precious metals.

In recent years, the environment protection laws directing/regulating control of emission of lead fumes into the atmosphere have encouraged development of cleaner and less energy-intensive processes. In this regard the Ausmelt TSL process is quite attractive. Both high (50–75% Pb) and low (<50% Pb)-grade galena concentrate can be treated by the Ausmelt process [23] to produce lead bullion in two stages. Smelting of high-grade concentrate in the first stage produces lead bullion in contact with high lead molten slag containing 60–75% lead at relatively low temperature (950–1000 °C). The fume containing about 11% lead is recycled to the furnace. Sphalerite (ZnS), generally associated with galena (PbS), also gets oxidized to ZnO. The overall reactions are:

$$PbS\ (l) + O_2(g) = Pb\ (l) + SO_2(g) \tag{3.49}$$

$$2ZnS\ (s) + 3O_2(g) = 2ZnO\ (s) + 2SO_2(g) \tag{3.50}$$

The slag containing a significant amount of zinc and lead is subjected to further reduction and fuming in another furnace to transfer most of the lead to the bullion. This is the second stage of operation. When the lead level of the slag approaches 2–5%, reduction of ZnO starts producing zinc vapors. The slag from the second-stage operation is discarded because it contains much less lead and zinc.

Smelting of low-grade concentrate does not produce lead bullion directly due to the high-temperature requirements to obtain fluid slag and low activity of lead in the system. In this case, the first stage involves fuming of lead due to the high vapor pressure of PbS. A high sulfur potential in the bath is ensured by smelting the concentrate under reducing conditions. In the second step, the fumed lead obtained as a fine oxide/sulfate dust containing 70% lead is pelletized and reduced with coal lump to produce bullion and a discard slag. The metal is tapped continuously or periodically.

Ausmelt technology has gained prominence in the lead–zinc industries because of the development of efficient and flexible pyrometallurgical reactors with better environmental performance by reducing number of intermittent tapping by introducing a three-stage smelting/slag reduction/slag cleaning process. The process under operation at Hindustan Zinc Ltd. [41] produces 50,000 tons of lead bullion per year by smelting about 85,000 tons of lead concentrate containing 60% lead, 4% zinc, and 0.15% silver. A thorough knowledge of thermodynamics of the Pb–PbS–PbO system concerning reaction (3.49) and other competitive reactions (mentioned below) has been useful in the development of the three-stage process:

$$PbS\ (l) + O_2(g) \rightarrow [Pb]_{bullion}(l) + SO_2(g)$$

$$PbS\ (l) + 3/2\ O_2(g) \rightarrow (PbO)_{slag}(l) + SO_2(g) \qquad (3.49a)$$

$$PbS\ (l) \rightarrow PbS\ (g) \qquad (3.49b)$$

$$PbO\ (l) \rightarrow PbO\ (g) \qquad (3.49c)$$

$$[Pb]\ (l) \rightarrow Pb\ (g) \qquad (3.49d)$$

Thus, lead can enter in any phase (metal, slag, fume) but its concentration in slag and fume can be controlled by restricting the competitive reactions (3.49a–3.49d). Lead in the bullion is maximized by adjusting the operating variables. Low temperature reduces lead entry into the fume whereas high partial pressure of oxygen will reduce fume formation by lowering the activity of PbS, which has the highest vapor pressure. However, higher p_{O_2} will increase lead concentration in the slag by forming PbO according to the reaction (3.49a).

In the modified process [41], reduction of PbO, accumulated in the slag, is carried out by lead concentrate containing PbS instead of carbon by generating conditions for the reaction (3.48) to proceed in the forward direction. Production of lead by the old hearth process is based on this principle.

In the slag cleaning step, lead and zinc are removed from the slag. This is achieved by reducing PbO and ZnO by addition of coal. Temperature is increased to 1300 °C to promote fuming. Lead and zinc in the fume get re-oxidized in the post combustion zone. The modified process aims to discard slag with 0.5 and 3.0 wt% of lead and zinc, respectively.

The current problem of inadequate supply of Pb–Zn concentrate due to lower mining output and higher demand of lead and zinc metals has forced smelters to develop a suitable technology to recover lead and zinc from scrap, residues, and fumes along with the primary feed (the concentrate). In this context, the TSL Technology may be considered as an alternative because it can process a variety of feed materials at low capital and operating costs with minor environmental pollution. Furthermore, it can be easily incorporated into the existing flow sheet. After charging the solid feed from the furnace top, the partial pressure of oxygen can be precisely controlled in steps to generate conditions to separate valuables and nonvaluables into metal and slag, respectively, and volatiles into fumes The appropriate knowledge [42] of the slag phase equilibria, activities of Pb, Zn, PbO, and ZnO in the multicomponent slag system: SiO_2–CaO–FeO–PbO–ZnO and kinetics of transfer of species from the slag to the metal (and vice versa) has been useful in development of the process.

The technology invented by Queneau and Schuhmann [43, 44] and developed by Lurgi has gained prominence as the QSL process. It is a continuous and direct lead-making process that allows carrying out roast-reduction smelting of galena concentrate and carbon reduction of lead oxide slag [45] in a single reactor (a long horizontal-cylindrical converter). A bath of molten lead bullion is maintained below the slag layer contained in the converter. Green pellets prepared from a mixture of concentrate, recycle flue dust, and fluxes are charged into the oxidation zone of the reactor. There is rapid oxidation of sulfide in the molten bath at 950°–1000 °C by submerged injection of oxygen to form lead bullion, lead oxide slag, and sulfur dioxide gas. The lead oxide slag continuously flows through the reduction zone of the vessel and is reduced by submerged injection of pulverized coal. In this zone temperature gradually increases from 1000 °C to 1250 °C. The secondary bullion joins the primary bullion and is tapped continuously. The slag from the reduction zone passes through a settling zone and continuously tapped and water granulated.

3.6 Smelting of Pentlandite (Nickel–Copper–Iron Sulfide)

3.6.1 Matte Smelting of Pentlandite

The pentlandite [(NiFe)$_9$S$_8$], a nickel ore containing 1–3% Ni, is a mixture of sulfides of nickel and copper. Mineral processing produces a nickel sulfide concentrate containing sulfides of copper, iron, and cobalt, together with silver, platinum, and some arsenide and siliceous gangue. Both pyrometallurgical and hydrometallurgical methods have been adopted for production of nickel from sulfide/oxide ores. A flow sheet based on pyrometallurgical route practiced by the International Nickel Company of Canada for treatment of pentlandite is presented in Appendix: A.3a. Currently, about 50% of the world's sulfide concentrates are smelted by the Outotec flash smelting process [46] to produce nickel matte containing Ni_3S_2, Cu_2S, and FeS. Major amount of FeS is oxidized to FeO and SO_2, and FeO is slagged off with silica whereas SO_2 joins the flue gases. Basically, similar characteristics of sulfides of copper and nickel pose problems in their separation, which is achieved in several steps.

3.6.2 Converting of Nickel Matte

The normal extraction procedure of partial roasting, smelting, and converting as adopted in treatment of chalcopyrite renders pentlandite concentrate into a product which cannot be easily separated. Contrary to the production of blister copper by $Cu_2S–Cu_2O$ interaction $(Cu_2S + 2Cu_2O = 6Cu + SO_2)$ and lead bullion by PbS–PbO interaction $(PbS + 2PbO = 3Pb + SO_2)$, Ni_3S_2–NiO interaction does not produce nickel because the free energy change for the reaction $[Ni_3S_2$ (matte) $+ 4NiO$ (formed during partial blowing of the matte) $= 7Ni + 2SO_2]$ is positive $(\Delta G° = +21$ kJ at 1200 °C). Instead, a Ni_3S_2–Cu_2S matte and Cu–Ni alloy containing precious metals are produced by the converting operation. During the converting operation as per normal practice, heat is derived by the oxidation of FeS present in the matte to FeO, which is slagged off by the addition of quartz. The three products Ni_3S_2, Cu_2S, and Cu–Ni alloy are separated by a cumbersome procedure comprising of extremely slow controlled cooling (from 1100 °C to 400 °C in 3 days), grinding, magnetic separation to collect Cu–Ni alloy, followed by froth flotation to separate Ni_3S_2 and Cu_2S. For production of nickel, Ni_3S_2 may be cast into anodes and directly electrolyzed in a diaphragm cell separating anolyte and catholyte in order to prevent deposition of impurities, namely, Co and Fe, which are close to Ni in the electrochemical series. Precious metals join the anode slime along with sulfur. While following the pyrometallurgical route, Ni_3S_2 is roasted to NiO $(Ni_3S_2 + 7/2O_2 = 3NiO + 2SO_2)$, which is then reduced with carbon or hydrogen into crude nickel [NiO + C (or H_2) = Ni + CO (or H_2O)]. The impure nickel is refined by the carbonyl process (see Chapter 10).

Nickel matte containing Ni_3S_2, Cu_2S, and FeS cannot be directly blown into nickel in a Pierce–Smith side-blown converter. The conversion of molten copper matte into blister copper in the horizontal side-blown converters (discussed in Sect. 3.2.1) practiced throughout the world became a challenge to nickel metallurgists. Earlier, nonferrous metallurgists dreamed and tried to achieve the direct conversion of molten nickel sulfide matte into metallic nickel but all the efforts were a waste only. Hence, there was a strong opinion in the metallurgical world that blowing of nickel sulfide to nickel in a converter was not feasible [47, 48]. Nevertheless, the dream was kept alive by INCO's Research Department in Canada. INCO pursued this apparent desire intermittently, for a period of about 40 years. Based on some promising studies conducted in the early 1940s, it was finally realized that for successful operation the fundamental requirement was to use a turbulent bath as generated in a top-blown rotary converter. Application of this principle resulted in conversion of nickel sulfide to nickel containing less than 0.02% S. Tonnage tests were conducted during 1959–1960 in an experimental Kaldo converter using oxygen of 96% purity. The combined effect of turbulence and pure oxygen at high temperature of 1600 °C made the process autogenous even in the 3 ton unit employed. Thus, it has taken long time to understand that Ni_3S_2–NiO interaction leading to the formation of metallic nickel is endothermic and hence does not take place at the converting temperature of 1250 °C at which blister copper is produced in a Pierce–Smith converter by Cu_2S–CuO interaction. Development of Top-Blown Rotary Converter (TBRC) for direct conversion of nickel matte into nickel is the outcome of about 40 years of extensive research [49, 50] carried out at the Noranda Research Centre and the International Nickel Company in Canada.

3.6.2.1 Physicochemical Aspects of Direct Conversion of Molten Nickel Sulfide into Nickel

The essential difficulty encountered in converting nickel sulfide is the formation of solid nickel oxide. The appearance of massive quantities of this refractory material (melting point over 1900 °C) in a conventional side-blown converter brings the process to a halt. If the characteristic dry, NiO-rich slag is allowed to build up while surface blowing, oxygen efficiency is reduced, thereby causing the bath to cool and precipitate still more NiO until the blowing is stopped. Nickel oxide has rather limited thermodynamic stability at high temperature. At 1650 °C, approximately 3% CO in a CO–CO_2 gas mixture will theoretically reduce NiO to liquid nickel at $a_{NiO}=$ 0.6. Yet it is difficult to reduce this dry slag by a controlled atmosphere because NiO forms a number of high melting materials, for example, silicates, aluminates, spinels, and solid solutions like NiO–MgO with the refractory lining. These compounds and solid solutions increase the stability of NiO. It is important to note that the stable refractory oxides physically interfere with the reduction process.

Hence, successful conversion of nickel sulfide into nickel requires strict control of the amount of NiO formed. The mechanism of conversion can be discussed in terms of the following reactions:

$$\tfrac{1}{2}\,O_2(g) = [O] \tag{3.50}$$

$$Ni\,(l) + \tfrac{1}{2}\,O_2(g) = NiO\,(s) \tag{3.51a}$$

$$Ni\,(l) + [O] = NiO\,(s) \tag{3.51b}$$

$$[S] + O_2(g) = SO_2(g) \tag{3.52a}$$

$$[S] + 2[O] = SO_2(g) \tag{3.52b}$$

$$[S] + 2NiO\,(s) = 2\,Ni\,(l) + SO_2(g) \tag{3.52c}$$

$$Ni_3S_2(l) + 4NiO\,(s) = 7\,Ni\,(l) + 2SO_2(g) \tag{3.53}$$

For conversion of Ni_3S_2 to metallic nickel the following points must be considered:

1. The tendency to form NiO cannot be avoided because reaction (3.51a) is spontaneous at 1650 °C even when $a_{Ni}= 0.6$ and $p_{O_2}= 0.35$ mm Hg.
2. If reactions (3.52a) and (3.52b) are very much faster as compared to (3.51a) and (3.51b), they would effectively prevent the formation of NiO by preferentially consuming the available oxygen. If this is not the case, the only remaining hope

is that reaction (3.52c) would proceed at a rate comparable to the rate of formation of NiO and thus prevent undue accumulation of NiO.

3. The relative kinetics of these reactions have not been studied but it is clear that at some point in the process, reactions (3.52a, 3.52b, and 3.52c) would become diffusion limited with respect to sulfur while reactions (3.51a and 3.51b) could proceed without hindrance.

4. It is therefore inevitable that the system eventually becomes saturated with oxygen and continued addition of significant quantities of free oxygen results in the accumulation of prohibitive amounts of NiO.

5. When the oxygen containing blast is stopped, further desulfurization may be achieved only by reactions (3.52b) and (3.52c). A hot stream of neutral or even slightly reducing gas may be used to remove SO_2 from the system.

6. From a thermodynamic viewpoint reaction (3.52c) governs the final elimination of sulfur. For analysis, reactions (3.52c) and (3.53) should be considered simultaneously.

7. The degree to which reaction (3.52c) will desulfurize the melt depends on the temperature and p_{SO_2}.

8. Experience shows that desulfurization proceeds rapidly down to the neighborhood of 1% S and then slows down extensively.

9. Table 3.1 illustrates the rapid drop in p_{SO_2} at sulfur content below 1% and the beneficial effect of high temperature at low S content.

10. It is clear that final desulfurization at atmospheric pressure by reaction (3.52b) or (3.52c) is practicable only above 1600 °C.

Table 3.1 Calculated equilibrium values of p_{SO_2} over Ni–S–O melts saturated with NiO[a]

Approximate wt % S	Calculated p_{SO_2} (atm)		
	1500 °C	1600 °C	1700 °C
0.1	0.005	0.0 13	0.025
0.5	0.025	0.063	0.11
1	0.05	0.13	0.19
2	0.1	0.21	–

[a]Reproduced from P. E. Queneau et al. [50] with the permission of The Minerals, Metals & Materials Society

When converting has just begun and sulfur content of the bath is still high (20%) it is possible to operate at a temperature of about 1380 °C. The temperature must be increased as sulfur is eliminated so as to reach at least 1600 °C when sulfur has decreased to below about 5%; otherwise, NiO will accumulate. It is possible and advantageous to derive most of the required heat from combustion of sulfur. It is estimated that reaction (3.52a) releases about 1480 kcal kg^{-1} of sulfur oxidized at 1650 °C. If the necessary oxygen is supplied as air at 15.5 °C about 1930 kcal are required to heat it to 1650 °C resulting in a net deficit of 450 kcal kg^{-1} of sulfur oxidized. On the other hand, if 95% commercial O_2 is used, a net surplus of 1040 kcal kg^{-1} of sulfur is realized. Hence, the use of tonnage oxygen makes the process commercially attractive. If final desulfurization is to be affected by reaction

(3.52b) or (3.52c), the heat must be supplied by an external source, as reaction (3.52c) consumes about 2200 kcal kg^{-1} of sulfur eliminated. In addition to keeping the system sufficiently hot, it is important that the molten bath be well mixed.

The conversion of white metal to blister copper is much less critical in this respect because the miscibility gap in the Cu–Cu_2S system ensures that a phase of roughly constant sulfur content is presented (or exposed) to the blast throughout most of the process. As the copper-rich phase is produced, it sinks to the bottom of the converter and remains away from the blast, thus minimizing oxidation of the metal. Contrary to this, when nickel sulfide is converted to nickel, the bath becomes a single phase of decreasing sulfur content. If it is not thoroughly mixed, the region of the bath exposed directly to the blast may be temporarily depleted in sulfur, permitting reactions (3.51a) and (3.51b) to proceed preferentially. The precipitation of the finely dispersed NiO increases the viscosity of that part of the bath and further reduces the rate of mixing. The converter is rotated in order to improve the interaction between Ni_3S_2 and NiO (reaction 3.53) leading to the formation of nickel. Thus, the converter is very much similar to the Kaldo converter employed for steel production. In this way the Rotary Top-Blown Converter (TBRC) that made a place in the steel industry had a greater impact in nonferrous industry.

It is rather unfortunate that despite extensive researches conducted during 1940–1980 with enormous expenses on the development of TBRC at the Noranda Research Center the process failed to attain commercial status in other countries. Still, the conventional method of extraction of nickel from pentlandite incorporating partial roasting, smelting as discussed above and presented in Appendix: A.3a, is practiced on commercial scale.

Treatment of nickel sulfide concentrate by Ausmelt TSL process is carried out in two steps [23, 46]. The first step produces a low-grade matte and a discardable slag whereas the second step treats the matte to either a high-grade matte (>70% Cu/Ni) with low iron and high sulfur or a high-grade matte with low iron and sulfur.

3.7 Rate of Smelting and Converting Processes

Chaubal and Sohn [51] studied the effect of temperature on the overall kinetics of oxidation of chalcopyrite in the absence of heat and mass-transfer effects up to1150 K. According to them the pore-blocking model was applicable with an activation energy of 71 kJ mol^{-1} in the temperature range: 754 K–873 K and 215 kJ mol^{-1} below 754 K and the kinetics of oxidation in the initial stage was dominated by the rate of vaporization of sulfur. However, oxidation of the decomposition product above 873 K followed power-law kinetics. The rate of vaporization of sulfur adopted power-law kinetics with an activation energy of 208 kJ mol^{-1}. From their studies, Chaubal and Sohn [51] have concluded that the rate of oxidation of chalcopyrite follows first-order rate kinetics with respect to oxygen concentration and is inversely proportional to the square root of the particle size over the entire range of temperature studied.

From the preceding paragraphs in Sects. 3.2–3.6, it is clear that the major reactions in sulfide smelting and converting steps are the oxidation of sulfur and iron. As these process steps are carried out at sufficiently high temperatures, the chemical reactions are very fast, and hence, the overall rate of the process is invariably controlled by mass transfer of reactants between different phases. On the other hand, in the bath smelting processes designed for operation under gas–liquid mass transfer, the rate is controlled by the rate of supply of oxygen-containing gas which is dependent on factors, such as heat generation and removal, erosion of refractories, gas injection, and melt splashing. These considerations led to the development of larger, intense, and continuous pyrometallurgical processes involving continuous flow of materials contrary to the older batch processes working under near-equilibrium conditions.

3.8 Problems [52–54]

Problem 3.1
A copper ore having the following composition (wt%) on dry basis: Cu-8.8, S-36.6, SiO_2-19.0, Al_2O_3-5.6, Fe-29.7, and CaO-0.3, was roasted down to a sulfur content of 6.8%. Assuming that (i) 5% of the copper is converted to CuO and the remainder remains un-oxidized as Cu_2S, (ii) the iron oxidized forms only Fe_2O_3 and the rest remains as FeS, and (iii) the flue gases coming out of the roaster analyze: SO_2-2.61, SO_3-0.27, CO_2-0.12, O_2-17.20, and N_2-79.0% (by volume). Calculate for 1000 kg of raw ore:

(a) The weight and analysis of the roasted product.
(b) The volume of the gases.
(c) The volume of air supplied and the percent excess that of theoretically required.

Solution
(a) Let x be the weight of the roasted product.
   Hence, amount of S in the product $= 0.068\ x$ kg
   Amount of Cu that gets converted into CuO (1000 kg base) $= 0.05 \times 88 = 4.4$ kg
   $\therefore$ Amount of CuO $= \frac{80}{64} \times 4.4 = 5.5$ kg
   Hence the amount of Cu as $Cu_2S = 88 - 4.4 = 83.6$ kg
   $\therefore$ Amount of $Cu_2S = \frac{160}{128} \times 83.6 = 104.5$ kg
   The rest of S will report in FeS and will determine the weight of FeS.
   Hence, the S remaining in $Cu_2S = 104.5 - 83.6 = 20.9$ kg
   Since the roasted product contains $0.068\ x$ kg of S
   $\therefore$ Amount of S in FeS $= 0.068\ x - 20.9$
   $\therefore$ Amount of FeS $= \frac{88}{32} \times (0.068\ x - 20.9) = 0.187\ x - 57.5$
   Hence, amount of Fe $= \frac{56}{88} \times (0.187\ x - 57.5) = 0.119\ x - 36.6$
   $\therefore$ Fe as $Fe_2O_3 = 297 - (0.119\ x - 36.6) = 333.6 - 0.119\ x$
   $\therefore$ Amount of $Fe_2O_3 = \frac{160}{112} \times (333.6 - 0.119\ x) = 477 - 0.170\ x$

Since sum of all the compounds in the roasted ore $= x$

$\therefore x = 5.5 + 104.5 + (0.187\, x - 57.5) + (477 - 0.170\, x) + 190 + 56 + 3$

Hence, $x = 792$ kg

$\therefore$ Composition of the roasted product:

$CuO = 5.5$ kg $= 0.7\%$

$Cu_2S = 104{,}5$ kg $= 13.2\%$

$FeS = 0.187\, x - 57.5 = 90.5$ kg $= 11.4\%$

$Fe_2O_3 = 477 - 0.170\, x = 342.5$ kg $= 43.2\%$

$SiO_2 = 190$ kg $= 24.0\%$

$Al_2O_3 = 56.0$ kg $= 7.1\%$

$CaO = 3.0$ kg $= 0.4\%$.

(b) Since the amount of S entering the gases is known, the volume of gases may be
calculated from their S content. The flue gases contain 2.88% $(SO_2 + SO_3)$, each
of which contains 32 kg of S in 22.4 m^3 of the gas.

$\therefore$ S in each m^3 of gas $= \frac{32}{22.4} \times 0.288 = 0.0411$ kg

Hence, the total weight of S entering the gases $= 366 - (0.068 \times 792) = 312.2$ kg

$\therefore$ Cubic meters of dry gas $= \frac{312.2}{0.0411} = 7610$ Ans.

(c) Amount of nitrogen in gases $= 7610 \times 0.798 = 6070$ m^3

$\therefore$ Volume of air supplied $= \frac{6070}{0.79} = 7690$ m^3

The volume of excess air may be calculated from the free oxygen content of the
gases.

O_2 in gases $= 7610 \times 0.172 = 1309$ m^3

$\therefore$ Excess air supplied $= \frac{1309}{0.21} = 6233$ m^3

Air required $= 7690 - 6233 = 1457$ m^3

$\therefore$ Percent excess air supplied $= \frac{6233}{1457} = 427\%$ Ans.

Problem 3.2

In the first stage of blow 30 tons matte containing 42% Cu was charged into a
Pierce–Smith copper converter. The slag carrying 28% SiO_2, 63% FeO, and 4%
CuO was generated by adding an ore as the flux having 7% Cu, 16% Fe, 5% S, and
49% SiO_2. After pouring the first slag additional matte of the same weight as the FeS
which was oxidized from the first matte was charged. If the time of the blister–
forming stage is 2 h, calculate the following:

1. The total weight of the flux used and the total weight of the slag produced.
2. The weight of the blister copper formed, assuming it to be as pure Cu.
3. The cubic meters of blast used.
4. The total blowing time and the volume of the blast supplied per minute.

(atomic weights: Cu-64, O-16, Fe-56, S-32, Si-28)

Solution

Weight of Cu_2S in the first charge $= 30,000 \times 0.42 \times \left(\frac{160}{128}\right) = 15,750$ kg ($= 12,600$ kg Cu)

∴ Weight of FeS in the first charge $= 30,000 - 15,750 = 14,250$ kg ($=$ wt of the second matte).

Wieght of Cu_2S in the second matte charged $= 14,250 \times 0.42 \times \left(\frac{160}{128}\right) = 7480$ kg ($= 5985$ kg Cu)

∴ Weight of FeS in the second matte charged $= 14,250 - 7480 = 6770$ kg

Total weight of FeS $= 14,250 + 6770 = 21,020$ kg

∴ Weight of FeO $= 21,020 \times \left(\frac{72}{88}\right) = 17,190$ kg

Available SiO_2 in the flux: flux contains 7% Cu and 16% Fe.

Equivalent $Cu_2S = 0.07 \times \left(\frac{160}{128}\right) = 0.0875$, and FeO $= 16 \times \left(\frac{72}{56}\right) = 0.206$

Hence, SiO_2 for FeO $= 0.206 \times \left(\frac{28}{63}\right) = 0.092$

SiO_2 available $= 0.49 - 0.092 = 0.398$

SiO_2 required for FeO in the matte $= 17,190 \times \left(\frac{28}{63}\right) = 7640$ kg

Weight of the flux used $= \frac{7640}{0.398} = 19,200$ kg Ans (1)

Weight of the slag made$= \frac{19,200 \times 0.49}{0.28} = 33,600$ kg Ans. (1)

Total weight of copper in the matte $12,600 + 5985 = 18,585$ kg

Cu in the flux $= 19,200 \times 0.07 = 1340$ kg

∴Total Cu in the matte and flux $= 18,585 + 1340 = 19,925$ kg

CuO in the slag $= 33,600 \times 0.04 = 1344$ kg

∴ Cu in the slag $= 1344 \times \left(\frac{64}{80}\right) = 1075$ kg

Hence, weight of the blister produced $= 18,585 + 1340 - 1075$
$= 18,850$ kg Ans. (2)

Flux contains 5% S and 7% Cu.

S with Fe in the flux $= 0.05 - 0.07 \times \left(\frac{32}{160}\right) = 0.05 - 0.014 = 0.036$

Weight of FeS $= 0.036 \times \left(\frac{88}{32}\right) \times 19,200 = 1900$ kg

∴Total FeS oxidized $= 14,250 + 6770 + 1900 = 22,920$ kg

Formation of SO_2 gas: $2FeS + 3\,O_2 = 2FeO + 2SO_2$

and $Cu_2S + 3O_2 = 2CuO + 2SO_2$

Total O_2 required in slag-forming stage $= \frac{3 \times 22.4}{176} \times 22,920 + \frac{2 \times 22.4}{128} \times 1070$
$= 9125$ m^3

In the blister-forming stage: $Cu_2S + O_2 = 2Cu + SO_2$

O_2 required in blister-forming stage $= \frac{22.4}{128} \times 18,850 = 3300$ m^3

Grand total of O_2 required $= 9125 + 3300 = 12,425$ m^3

∴ Blast used $= \frac{12,425}{0.21} = 59,170$ m^3 Ans. (3)

Total time of blow $= \frac{12,425}{3300} \times 2 = 7.53$ h

∴ Volume of the blast supplied per minute $= \frac{59,170}{7.53 \times 60} = 131$ m^3 Ans. (4)

Problem 3.3

When air blow is stopped in the Pierce–Smith converter at 1200 °C blister copper contains 1 wt% oxygen. What is the residual sulfur in the blister if the partial pressure of SO_2 in the melt is one atmosphere? The free energy changes for relevant reactions are given below:

$\frac{1}{2}S_2(g) + O_2(g) = SO_2(g), \Delta G_1^o = -60,870$ cal.

$O_2(g) = 2[\%O]_{Cu}, \Delta G_2^o = -27,730$ cal.

$\frac{1}{2}S_2(g) = [\%S]_{Cu}, \Delta G_3^o = -19,720$ cal.

Solution

Reaction to be considered: $[\%S]_{Cu} + 2[\%O]_{Cu} = SO_2(g)$

$$\Delta G^o = \Delta G_1^o - \Delta G_2^o - \Delta G_3^o = -60,870 - (-27,730) - (-19,720)$$
$$= -13,420 \text{ cal}$$
$$= -RT\ln K$$

$\ln K = \frac{\Delta G^o}{RT} = \frac{13,420}{1.987 \times 1473} = 4.585$

$K = 98.02$

For the above equation, $K = \frac{P_{SO_2}}{[\%S].[\%O]^2} = \frac{1}{[\%S].[\%O]^2}$

Residual sulfur: $[\%S] = \frac{1}{K.[\%O]^2} = \frac{1}{98.02.[1]^2} = 0.0102\%$ Ans.

Problem 3.4

Toward the end of the second stage of converting copper matte at 1150 °C, both solid FeO and Cu_2O begin to float on the surface of blister copper. Calculate the lowest attainable level of iron (in wt%) in the molten copper at 1150 °C.

   Given:-

$Cu_2O(s) = 2 \text{ Cu (l)} + \frac{1}{2}O_2(g), \Delta G_1^o = 46,700 + 3.92 \, T. \log T - 34.1 \, T$ cal

$FeO (s) = Fe (\gamma) + \frac{1}{2}O_2(g), \Delta G_2^o = 63,310 - 15.62 \, T$ cal

$\log \gamma_{Fe(s)}^o \text{ (in Cu)} = \frac{4430}{T} - 1.41$

Solution

The reaction under question: $Cu_2O(s) + Fe(\gamma) = FeO(s) + 2 \text{ Cu (l)}$

$\Delta G^o = \Delta G_1^o - \Delta G_2^o = -16610 + 3.92 \, T. \log T - 18.48 \, T$ cal

   At 1150 °C (1423K), $\Delta G^o = -25,318$ cal, $K = 7744$ and $\gamma_{Fe}^o = 50.5$

$K = \frac{a_{FeO}.a_{Cu}^2}{a_{Cu_2O}.a_{Fe}} = 7744$

Since FeO and Cu_2O float as solids and in blister copper wt% Cu > 99.7% Cu, $a_{FeO} = a_{Cu_2O} = a_{Cu} = 1$

$$\therefore a_{Fe} = \frac{1}{K} = 0.000129$$

As $a_{Fe} = \gamma_{Fe}^{o}.x_{Fe}, x_{Fe} = \frac{0.000129}{50.5}$ (i.e., concentration of Fe in Cu is extremely low)

At infinitely dilute concentration of the solute Fe, wt% Fe is negligible as compared to the wt% of Cu ~100% and according to Eq. (1.78a) we can write:

$$wt\%Fe = \frac{x_{Fe}.100.M_{Fe}}{M_{Cu}}$$

$$\therefore wt\%Fe = \frac{x_{Fe} \times 100 \times 55.85}{65.54}$$

$$= \frac{0.000129 \times 100 \times 55.85}{50.5 \times 63.54} = 2.2 \times 10^{-4}$$

Thus the lowest attainable level of iron in the molten copper at 1150 °C is 2.2 10^{-4} % . Ans.

Problem 3.5

The Raoultian activity coefficient of PbO (γ_{PbO}^{o}) in FeO–SiO$_2$–PbO slag at infinite dilution is 0.10, whereas the Raoultian activity coefficient of FeO in a slag containing 55% FeO, 7% FeO$_{1.5}$, 37% SiO$_2$, and some other unspecified oxides (including PbO with average molecular weight of 70) is 0.76. This slag is in equilibrium with a matte containing 10% FeS, 89% Cu$_2$S, and 1.0% PbS at 1200 °C. Considering the matte to be an ideal molecular solution of the FeS–Cu$_2$S–PbS system, calculate the % Pb content of the slag from the following data:

Pb (l) + ½ O$_2$ (g) = PbO (l), ΔG^o = −46,930 + 20.20 T cal
Pb (l) + ½ S$_2$ (g) = PbS (l), ΔG^o = −37,580 + 19.12 T cal
Fe (γ) + ½ O$_2$ (g) = FeO (l), ΔG^o = −54,890 + 10.55 T cal
Fe (γ) + ½ S$_2$ (g) = FeS (l), ΔG^o = −24,250 + 5.50 T cal

At. wt Pb-207.21, Fe-55.85, Cu-63.54, S-32.06, Si-28.09.

Solution

Assuming that slag and matte do not mix, FeS of the matte interacts with PbO of the slag at the interface to convert PbO into PbS which moves into the matte. Likewise, when FeO of the slag interacts with PbS of the matte to convert PbS into PbO. This PbO will move into slag.

The slag composition is required in terms of mole fraction. Consider 100 g of slag.

$$x_{FeO} = \frac{\text{number of moles of FeO}}{\text{number of moles of (FeO} + \text{FeO}_{1.5} + \text{SiO}_2 + \text{MO)}}$$

$$= \frac{55/71.85}{\frac{55}{71.85} + \frac{7}{79.85} + \frac{37}{60.09} + \frac{1}{70}} = \frac{0.7655}{1.4832} = 0.5161$$

Therefore,

$$a_{FeO} = \gamma_{FeO}\, x_{FeO} = 0.76 \times 0.5161 = 0.3922 \tag{1}$$

(This activity is a function of slag composition only.) In order to calculate % Pb in the slag, a_{PbO} is expressed according to Henry's law in dilute solution: $a_{PbO} = k\,(\%\,Pb)_{slag}$

$$
\begin{aligned}
x_{PbO} &= \frac{\%PbO/223.21}{\dfrac{\%FeO}{71\,85} + \dfrac{\%FeO_{1.5}}{79.85} + \dfrac{\%SiO_2}{60.09} + \dfrac{\%MO}{70}} \\
&= \frac{\%PbO}{223.21 \times 1.4832}
\end{aligned}
\tag{2}
$$

On the other hand, we can express:

$$(\%PbO)_{slag} = (\%Pb)_{slag} \times \left(\frac{207.21 + 16}{207.21}\right)$$

Hence Eq. (2) can be modified as:

$$
\begin{aligned}
x_{PbO} &= \frac{(\%PbO)_{slag}}{223.21 \times 1.4832} = \frac{(\%Pb)slag}{223.21 \times 1.4832} \times \left(\frac{207.21 + 16}{207.21}\right) \\
&= \frac{(\%Pb)slag}{207.21 \times 1.4832}
\end{aligned}
$$

$$a_{PbO} = \gamma_{PbO}^{o}\cdot x_{PbO} = 0.10 \times \frac{(\%Pb)_{slag}}{207.21 \times 1.4832}$$

$$(\%Pb)_{slag} = 3073\, a_{PbO} \tag{3}$$

Similarly, we can calculate mole fractions of FeS and PbS in the matte. Consider 100 g of matte (contains 10% FeS, 89% Cu_2S, and 1% PbS).

Number of moles of FeS = 10/87.91 = 0.1137
Number of moles of Cu_2S = 89/159.14 = 0.5593
Number of moles of PbS = 1/239.27 = 0.0042

$$\therefore x_{FeS} = \frac{\text{number of moles of FeS}}{\text{number of moles of } (FeS + Cu_2S + PbS)} \tag{4}$$

$$= 0.1137/0.6777 = 0.1679 = a_{FeS}$$

$$\text{and } x_{PbS} = 0.0042/0.6772 = 0.0063 = a_{PbS} \tag{5}$$

(since matte forms an ideal solution, $a_i = x_i$)

The free energy change for the sulfide-oxide interaction at the interface:

PbS (l, in matte) + FeO (l, in slag) = FeS (l, in matte) + PbO (l, in slag)

$\Delta G^o = \Delta G^o_{FeS} + \Delta G^o_{PbO} - \left(\Delta G^o_{PbS} + \Delta G^o_{FeO}\right)$

$\qquad = [-24{,}250 + 5.50\ T + (-46{,}930 + 20.20\ T)] - [-37{,}580 + 19.12\ T +$
$\qquad\quad (-54{,}890 + 10.55\ T)]$

$\qquad = 21{,}290 - 3.97\ T$ cal

$\Delta G^o_{1473} = 21290 - 3.97 \times 1473 = 15{,}442$ cal

$\qquad = -RT\ln K$

$\ln K = -\dfrac{15{,}442}{1.987 \times 1473} = -5.276$

$K = 5.11 \times 10^{-3}$

and

$$K = \frac{a_{FeS} \cdot a_{PbO}}{a_{PbS} \cdot a_{FeO}} = 5.11 \times 10^{-3} \qquad (6)$$

Substituting the values of $a_{FeO} = 0.3922$ (1), $a_{FeS} = 0.1679$ (4) and $a_{PbS} = 0.0063$ (5) in Eq. (6)

$$\frac{0.1679 \times a_{PbO}}{0.0063 \times 0.3922} = 5.11 \times 10^{-3}$$

$$a_{PbO} = 7.52 \times 10^{-5} \qquad (7)$$

From Eqs. (7) and (3) we get

$(\%Pb)_{slag} = 3073 \times a_{PbO} = 3073 \times 7.52 \times 10^{-5} = 0.2312$ Ans.

References

1. Sohn, H. Y. (2014). Principles of copper production. In *Treatise on process metallurgy. Vol. 3, Industrial processes, Part A* (pp. 534–591). Oxford/Waltham: Elsevier. (Section 2.1.1).
2. Sohn, H. Y., Kang, S., & Chang, J. (2005). Sulfide smelting fundamentals, technologies and innovations. *Minerals and Metallurgical Processing, 22*, 65–76.
3. Shamsuddin, M., & Sohn, H. Y. (2019). Constitutive topics in physical chemistry of high-temperature nonferrous metallurgy – A Review: Part 1. Sulfide roasting and smelting. *JOM, 71* (9), 2353–2365.
4. Moore, J. J. (1990). *Chemical metallurgy* (2nd ed.). Oxford: Butterworth-Heinemann. (Chapter 7).
5. Gilchrist, J. D. (1980). *Extraction metallurgy* (2nd ed.). New York: Pergamon. (Chapter 12).
6. Biswas, A. K., & Davenport, G. W. (1976). *Extractive metallurgy of copper* (1st ed.). New York: Pergamon. (Chapter 4).
7. Compos, R., & Torres, I. (1993). Caletones Smelter: Two decades of technological improvements. In O. A. Landolt (Ed.), *Extractive metallurgy of copper, nickel and cobalt. Vol. II, Copper and nickel smelter operations* (pp. 1441–1460). Warrendale: TMS.
8. Sohn, H. Y., & Ramachandran, V. (1998). Advances in sulfide smelting-technology, R&D, and education. In J. A. Asteljoki & R. L. Stephens (Eds.), *Sulfide smelting '98, current and future practices* (pp. 3–37). Warrendale: TMS.

9. Newman, C. J., Probert, T. I., & Weddick, A. J. (1998). Kennecott Utah copper smelter modernization. In J. A. Asteljoki & R. L. Stephens (Eds.), *Sulfide smelting '98, current and future practices* (pp. 205–215). Warrendale: TMS.

10. Suzuki, Y., Suenaga, C., Ogasawara, M., & Yasuda, Y. (1998). Productivity increase in flash smelting furnace operation at Saganoseki Smelter & Refinery. In J. A. Asteljoki & R. L. Stephens (Eds.), *Sulfide smelting '98, current and future practices* (pp. 587–595). Warrendale: TMS.

11. Bureggemann, M., & Caba, E. (1998). Operation of the CONTOP process at the ASARCO El Paso Smelter. In J. A. Asteljoki & R. L. Stephens (Eds.), *Sulfide smelting '98, current and future practices* (pp. 159–166). Warrendale: TMS.

12. Mackey, P. J., & Tarasoff, P. (1983). New and emerging technologies in sulphide smelting. In H. Y. Sohn, D. B. George, & A. D. Zunkel (Eds.), *Advances in sulfide smelting* (pp. 399–426). Warrendale: TMS.

13. Alvarado, R., Achurra, G., & Mackay, R. (1998). Present and future situation of the Teniente process. In J. A. Asteljoki & R. L. Stephens (Eds.), *Sulfide smelting '98, current and future practices* (pp. 493–501). Warrendale: TMS.

14. Mounsey, E. N. (1995). Economic and technical evaluation for Ausmelt process systems for copper bearing materials. In W. J. Chen, C. Diaz, A. Luraschi, & P. J. Mackey (Eds.), *Proceedings of copper 95. Vol. IV, Pyrometallurgy of copper* (pp. 189–204). Montreal: CIM.

15. Marczeski, W. D., & Aldrich, T. L. (1986). *Retrofitting Hayden plant to flash smelting* (Paper No. A86-65). Warrendale: TMS.

16. Carr, H., Humphris, M. J., & Longo, A. (1997). The smelting of bulk Cu-Ni concentrates at the Inco Copper Cliff smelter. In C. Diaz, I. Holubec, & C. G. Tan (Eds.), *Proceedings of the Nickel-Cobalt 97 international symposium, Vol. III, Pyrometallurgical operations, environment, vessel integrity in high-intensity smelting and converting processes*, Metallurgical Society, CIM, Montreal, pp 5–16.

17. Kojo, I. V., Jokilaakso, A., & Hanniala, P. (2000). Flash smelting and converting furnaces: A 50 year retrospect. *JOM, 52*(2), 57–61.

18. Jones, D. M., Cardoza, R., & Baus, A. (1999). 1999 rebuild of the BHP San Manuel Outokumpu flash furnace. In D. B. George, W. J. Chen, P. J. Mackey, & A. J. Weddick (Eds.), *Copper 99-Cobre 99 proceedings of the fourth international conference. Vol. V, Smelting operations and advances* (pp. 319–334). Warrendale: TMS.

19. Davenport, W. G., King, M., Schlesinger, M., & Biswas, A. K. (2002). *Extractive metallurgy of copper* (4th ed.). Oxford: Elsevier Science Ltd. (Chapters 5–10).

20. Mackey, P. J., & Compos, R. (2001). Modern continuous smelting and converting by bath smelting technology. *Canadian Metallurgical Quarterly, 40*(3), 355–375.

21. Harris, C. (1999). Bath smelting in the Noranda process reactor and the El Teniente process converter compared. In D. B. George, W. J. Chen, P. J. Mackey, & A. J. Weddick (Eds.), *Copper 99-Cobre 99, Proceedings of the fourth international conference. Vol. V, Smelting operations and advances* (pp. 305–318). Warrendale: TMS.

22. Torres, W. E. (1998). Current Teniente converter practice at the SPL Ilo smelter. In J. A. Asteljoki & R. L. Stephens (Eds.), *Sulfide smelting '98, current and future practices* (pp. 147–157). Warrendale: TMS.

23. Mounsey, E. N., & Robilliard, K. R. (1994). Sulfide smelting using Ausmelt technology. *JOM, 46*(8), 58–60.

24. Floyd, J. M. (2005). Converting an idea into a worldwide business commercializing smelting technology. *Metallurgical and Materials Transactions B: Process Metallurgy and Materials Processing Science, 36B*, 557–575.

25. Shi, Y. F. (2006). Yunnan copper's Isasmelt-successful smelter modernization in China. In F. Kongoli & R. G. Reddy (Eds.), *International symposium on sulfide smelting. Vol. 8, Sohn international symposium* (pp. 151–162). Warrendale: TMS.

26. Peretti, E. A. (1948). An analysis of the converting of copper matte. *Discussion Faraday Society, 4*, 179–184.

27. Prevost, Y., Lepointe, R., Levac, C. A., & Beaudoin, D. (1999). First year of operation of the Noranda continuous converter. In D. B. George, W. J. Chen, P. J. Mackey, & A. J. Weddick (Eds.), *Copper 99-Cobre 99, Proceedings of the fourth international conference. Vol. V, Smelting operations and advances* (pp. 269–282). Warrendale: TMS.

28. Newman, C. J., Collins, D. N., & Wedick, A. J. (1999). Recent operation and environmental control in the Kennecott smelter. In D. B. George, W. J. Chen, P. J. Mackey, & A. J. Weddick (Eds.), *Copper 99-Cobre 99. Proceedings of the fourth international conference. Vol. V, Smelting operations and advances* (pp. 29–45). Warrendale: TMS.

29. Kojo, I. V., & Storch, H. (2006). Copper production with Outokumpu flash smelting an update. In F. Kongoli & R. G. Reddy (Eds.), *International symposium on sulfide smelting. Vol. 8, Sohn International Symposium* (pp. 225–238). Warrendale: TMS.

30. Goto, M., & Hayashi, M. (1998). *The Mitsubishi continuous process.* Tokyo: Mitsubishi Material Corporation. www-adm@mmc.co.jp

31. Goto, M., Oshima, I., & Hayashi, M. (1998). Control aspects in the Mitsubishi continuous process. *JOM, 50*(4), 60–65.

32. Worner, H. K. (1968). Continuous smelting and refining by the WORCRA process. In *Proceedign of the symposium on advances in extractive metallurgy,* April 1967, Institute of Mining and Metallurgy, pp. 245–257.

33. Schnalek, F., Holeczy, J., & Schmiedl, J. (1964). A new continuous converting of copper matte. *JOM, 16*(5), 416–419.

34. Mills, L. A., Hallett, G. D., & Newman, C. J. (1976). Design and operation of the Noranda process continuous smelter. In J. C. Yannpoulos & J. C. Agarwal (Eds.), *Extractive metallurgy of copper. Vol. I, Pyrometallurgy and electrolytic refining* (pp. 458–487). Warrendale: TMS.

35. Czernecki, J., Smieszek, Z., Miczkowski, Z., Dobrzanski, J., & Warrnuz, M. (1999). Copper metallurgy at the KGHM Polska Miedz S.A.- present state and perspectives. In D. B. George, W. J. Chen, P. J. Mackey, & A. J. Weddick (Eds.), *Copper 99-Cobre 99. Proceedings of the fourth international conference. Vol. V, Smelting operations and advances* (pp. 189–203). Warrendale: TMS.

36. Hunt, A. G., Day, S. K., Shaw, R. G., Montgomerie D., & West, R. C. (1999). Start up and operation of the #2 direct-to-copper flash furnace at Olympic Dam. In *Proceeding of the ninth international congress,* Australia, June 6–12, 1999.

37. Peuraniemi, E. J., & Lahtinen, M. (2006). Outokumpu blister flash smelting process. In F. Kongoli & R. G. Reddy (Eds.), *International symposium on sulfide smelting. Vol. 8, Sohn international symposium* (pp. 303–312). Warrendale: TMS.

38. Yazawa, A., Takeda, Y., & Waseda, Y. (1981). Thermodynamic properties and structure of ferrite slags and their process implications. *Canadian Metallurgical Quarterly, 20*(2), 129–134.

39. Taniguchi, T., Matsutani, T., & Sato, H. (2006). Technological innovations in the Mitsubishi process to achieve four years campaign. In F. Kongoli & R. G. Reddy (Eds.), *International symposium on sulfide smelting. Vol. 8, Sohn international symposium* (pp. 261–274). Warrendale: TMS.

40. Sevryukov, N., Kuzmin, B., & Chelishchev, Y. (1960). *General metallurgy* (B. Kuznetsov, Trans.). Moscow: Peace Publishers (Chapter 19 and 20).

41. McClelland, R., Hoang, J., Lightfoot, B., & Dhanavel, D. (2006). Commissioning of the Ausmelt lead smelter at Hindustan Zinc. In F. Kongoli & R. G. Reddy (Eds.), *International symposium on sulfide smelting. Vol. 8, Sohn international symposium* (pp. 163–171). Warrendale: TMS.

42. Hughes, S., Reuter, M. A., Baxter, R., & Kaye, A. (2008). *Ausmelt technology for lead and zinc processing.* In Lead and Zinc 2008, Johannesburg, The Southern African Institute of Mining and Metallurgy, pp 147–161.

43. Queneau, P. E., & Schuhmann, R., Jr. (1974). The QSL oxygen process. *JOM, 26*(8), 14–16.

44. Queneau, P. E. (1989). The QSL reactor for lead and its prospects for Ni, Cu and Fe. *JOM, 41* (12), 30–35.

45. Fischer, P., & Maczek, H. (1982). The present status of development of the QSL-lead process. *JOM, 34*(6), 60–64.
46. Makinen, T., & Taskinen, P. (2006). The state of the art in nickel smelting: Direct Outokumpu nickel technology. In F. Kongoli & R. G. Reddy (Eds.), *International symposium on sulfide smelting. Vol. 8, Sohn International Symposium* (pp. 313–325). Warrendale: TMS.
47. Newton, J. (1967). *Extractive metallurgy*. London: Wiley Eastern Pvt. Ltd.. (Chapter 7).
48. Boldt, J. R. & Queneau, P. E. (1967). *Winning of nickel* (Section C, pp. 191–386). London: Mathuen & Co. Ltd.
49. Saddington, R., Curlook, W., & Queneau, P. E. (1966). Tonnage oxygen for nickel and copper smelting at copper cliff. *JOM, 18*(4), 440–452.
50. Queneau, P. E., O'Neill, C. E., Illis, A., & Warner, J. S. (1969). Some novel aspects of the pyrometallurgy and vapometallurgy of nickel, Part I: Rotary top-blown converter. *JOM, 21*(7), 35–45.
51. Chaubal, P. C., & Sohn, H. Y. (1986). Intrinsic kinetics of the oxidation of chalcopyrite particles under isothermal and nonisothermal conditions. *Metallurgical Transactions B, 17*, 51–60.
52. Kubaschewski, O., & Alcock, C. B. (1979). *Metallurgical thermochemistry* (5th ed.). Oxford: Pergamon.
53. Curnutt, J. L., Prophet, H., McDonald, R. A., & Syverud, A. N. (1975). *JANAF thermochemical tables*. Midland: The Dow Chemical Company.
54. Barin, I., Knacke, O., & Kubaschewski, O. (1977). *Thermochemical Properties of Inorganic Substances, (supplement)*. New York: Springer.

Chapter 4
Metallurgical Slag

Along with valuable metals, pyrometallurgical methods of extraction also produce an appreciable proportion of a relatively unwanted material called slag. The slag comprising simple and/or complex compounds consists of solutions of oxides from gangue minerals, sulfides from the charge or fuel, and, in some cases, halides added as flux. Slag being immiscible and lighter than the metallic phase, it covers the metallic bath. The specific gravity of slag ranges between 3 and 4 as compared to 5.5 for sulfide matte and 7–8 for iron and steel. The slag cover protects the metal and matte from oxidation and prevents heat losses due to its poor thermal conductivity. In an electric furnace, slag may be employed as a heat resistor. It protects the melt from contamination from the furnace atmosphere and from the combustion products of the fuel. In pyrometallurgy, slags play a very important role in carrying out a number of physical and chemical functions.

In primary extraction, slag accepts gangue and unreduced oxides, whereas in refining, they act as a reservoir of chemical reactant(s) and an absorber of extracted impurities. In order to achieve these objectives, slag must possess certain optimum level of physical properties, such as low melting point, low viscosity, low surface tension, high diffusivity, and chemical properties, such as basicity, oxidation potential, and thermodynamic properties. The required properties of slag are controlled by the composition and structure. Hence, understanding of the physicochemical properties of slags has become extremely essential to meet a steadily increasing demand of better-quality steels in terms of lower impurity contents (ultralow sulfur, as low as 0.001 wt%S in line-pipe and hydrogen-induced corrosion-resistant steel), better internal quality (i.e., inclusion free), and mechanical properties in terms of strength, toughness, and workability under extreme forming conditions.

© The Minerals, Metals & Materials Society 2021
M. Shamsuddin, *Physical Chemistry of Metallurgical Processes, Second Edition*,
The Minerals, Metals & Materials Series, https://doi.org/10.1007/978-3-030-58069-8_4

4.1 Structure of Oxides

As most of the slag contains oxides as a major fraction, knowledge about the structure of pure oxides becomes essential in understanding the structure of slag. Majority of slags are formed by solution of mixed oxides and silicates, sometimes with aluminates, phosphates, and borates. Relative dimensions of cations and anions and the type of bonds between them are important factors in controlling the structure of pure oxides.

In solid oxides, metallic cations are surrounded by oxygen anions in a three-dimensional network. According to Pauling's [1] first law, in a close-packed structure, each cation should be surrounded by a maximum number of oxygen anions called the coordination number (CN). This number depends on the relative sizes and charges of cations and anions. The atomic radii of the elements in each column of the periodic table increase from top to bottom because the number of electron shells increases. But in a row, the atomic radii decrease from left to right since each electron of the given shell is attracted by a larger number of protons in the nucleus. Removal of electron(s) from atoms produces cations whereas addition produces anions. In the case of ions, the radius does not depend only on the radius of the corresponding atom but also on its charge. The radii of cations are much smaller than those of anions. Table 4.1 lists the radii of some common ions.

From Table 4.1 it is evident that on losing three electrons, the Fe^{3+} cation with more charge is smaller than the ion Fe^{2+} that is formed by losing two electrons. Thus, the R_c/R_a ratio as well as the CN control the structural arrangement of the ions. Hence, from simple geometrical considerations, the ionic structure of the oxide can be established. Table 4.2 illustrates the structural dependency on the CN and R_c/R_a ratio. Oxides with larger cations (from Ca^{2+} to Mg^{2+}) have an octahedral structure with a CN of 6, whereas oxides with smaller cations (Al^{3+}, Si^{4+}, and P^{5+}) have a

Table 4.1 Ionic radii[a]

Cations	Ca^{2+}	Mn^{2+}	Fe^{2+}	Fe^{3+}	Mg^{2+}	Al^{3+}	Si^{4+}	P^{5+}
R_c (Å)	0.93	0.80	0.75	0.60	0.65	0.50	0.41	0.35
Anions	O^{2-}	S^{2-}	F^-					
R_a (Å)	1.40	1.84	1.36					

[a]From Fundamentals of Metallurgical Processes by L. Coudurier et al. [2], © 1978, p 184, Pergamon Press Ltd. Reproduced with the permission of Pergamon Press Ltd

Table 4.2 CN, R_c/R_a ratio and structure of solid oxides[a]

Structure	CN	R_c/R_a	Examples
Cubic	8	1–0.732	–
Octohedral	6	0.732–0.414	CaO, MgO, MnO, FeO
Tetrahedral	4	0.414–0.225	SiO_2, P_2O_5, Al_2O_3
Triangular	3	0.225–0.155	–

[a]From Fundamentals of Metallurgical Processes by L. Coudurier et al. [2], © 1978, p 185, Pergamon Press Ltd. Reproduced with the permission of Pergamon Press Ltd

Fig. 4.1 Silica tetrahedron of four O^{2-} ions and one Si^{4+} ion (From Chemical Metallurgy by J. J. Moore [3], © 1990, p. 157, Butterworth Heinemann Ltd. Reproduced with the permission of Butterworth Heinemann Ltd.)

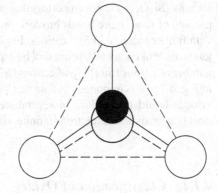

(a)   (b)

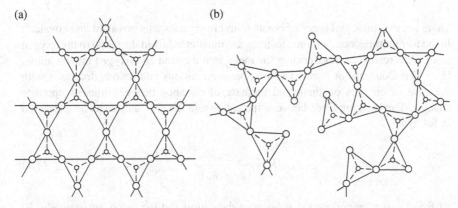

Fig. 4.2 Structure of silica. (**a**) solid and (**b**) liquid (From Fundamentals of Metallurgical Processes by L. Coudurier et al. [2], © 1978, p 186, Pergamon Press Ltd. Reproduced with the permission of Pergamon Press Ltd.)

tetrahedral structure. In the case of SiO_2, four O^{2-} ions provide the frame of the tetrahedron and the smaller Si^{4+} ion is situated within the frame (as shown in Fig. 4.1). Since the neighboring cations (Si^{4+}) are mutually repellent, according to Pauling's second law, the interval between two Si^{4+} ions should be maximum. Hence, the joining of two silica tetrahedra must occur at each vertex to produce a three-dimensional hexagonal network structure (Fig. 4.2a). In this arrangement of joining, each silicon atom has bonding with four atoms of oxygen and each oxygen atom has bonds with two atoms of silicon. In the solid state, this results in a structural formula of $(SiO_2)_n$ or simply SiO_2. The major covalent bond between silicon and oxygen is very strong, which makes silica a refractory material. Silica is a common constituent of most slags.

From the hexagonal network structure of silica, it is evident that two tetrahedra join at the vertex by one oxygen atom (one O^{2-} ion). During melting, the network of silica is destroyed and in the liquid phase the bonds between ions are broken by thermal agitation. With increasing temperature, bonds are broken one by one with gradual decrease in viscosity. Initially, with only a few bonds broken, large anions such as $(Si_4O_{13}^{10-})$ are formed together with a small amount of Si^{4+} cations. With passage of time, more bonds break to produce smaller anions, for example, $Si_3O_{10}^{8-}$ with further increase in Si^{4+} cations. Finally, at a very high temperature when all the joints are broken, all tetrahedra will be separated from each other to produce an equal number of anions (SiO_4^{4-}) and cations (Si^{4+}). The structure of molten silica is shown in Fig. 4.2b. It is important to note that the passage from completely solid silica to a viscous liquid takes place in several steps with no well-defined fusion point but merely a gradual change from infinite viscosity to measurable values.

4.1.1   Classification of Oxides

There are two principal types of bonds found in crystals: electrovalent and covalent. In the former type, one or more electrons are transferred from the metal to the oxygen atom. This results in transforming the metal into a cation and oxygen into an anion, O^{2-}. The Coulomb or attractive force between cations and anions decreases with decrease of charges on them and increase of distance between ions of opposite charges. The bonding force between the cation and the O^{2-} anion can be expressed as follows:

$$F = \frac{z^+ z^- e^2}{(R_c + R_a)^2} \tag{4.1}$$

where z^+ and z^- represent the valence of the cation and the anion, respectively, R_c and R_a are radii of cations and anions, and e is the electron charge.

The alkali metals with larger ionic radii, which belong to the first column of the periodic table, carry a smaller number of unit charge compared to the elements in the subsequent columns. Hence, the electrovalent bonding force according to Eq. 4.1 is relatively small. The bond strength further decreases because it is shared between a larger number of oxygen anions due to larger size of alkali cations and, hence, larger CN. The bonding force or the attractive force between cations and anions (which is proportional to the ratio: $z/(R_c + R_a)^2$) is listed in Table 4.3.

The central elements between alkali metals and halogens are of intermediate size. As they neither lose nor gain peripheral electrons easily as compared to elements located at the extremities, bonding with oxygen is mainly covalent. The bond strength between atoms forming a covalent molecule is very large, and high temperature is required to destroy such bonds. In case of smaller bond strength, covalent oxides such as CO_2 and SO_2 are gaseous. However, oxides exhibit varying proportions of both ionic and covalent bonding in slags. The ionic fraction of the bond shown in Table 4.3 decreases from sodium oxide to phosphorus oxide. Hence, each oxide has an ionic bond fraction and a covalent bond fraction. The ionic fraction happens to be a measure of the tendency to dissociate into simple ions in the liquid state. From the table, it is also evident that the fraction decreases approximately at

Table 4.3 Type of bonding and bond fraction between cations and O^{2-} [a]

Oxide	$\frac{z}{(R_c+R_a)^2}$	Ionic fraction of bond	Coordination number		Nature of the Oxide
			Solid	Liquid	
Na_2O	0.18	0.65	6	6–8	
BaO	0.27	0.65	8	8–12	
SrO	0.32	0.61	8		Network breakers
CaO	0.35	0.61	6		or
MnO	0.42	0.47	6	6–8	Basic oxides
FeO	0.44	0.38	6	6	
ZnO	0.44	0.44	6		
Mgo	0.48	0.54	6		
BeO	0.69	0.44	4		
....	....	....	....	....	....
Cr_2O_3	0.72	0.41	4		
Fe_2O_3	0.75	0.36	4		Amphoteric oxides
Al_2O_3	0.83	0.44	6	4–6	
....	....	....	....	....	....
TiO_2	0.93	0.41	4		Network formers
SiO_2	1.22	0.36	4	4	or
P_2O_5	1.66	0.28	4	4	Acid oxides

[a]From Fundamentals of Metallurgical Processes by L. Coudurier et al. [2], © 1978, p 187, Pergamon Press Ltd. Reproduced with the permission of Pergamon Press Ltd

the same rate as the attractive force between the ions increases. In the last group of oxides, namely, TiO_2, SiO_2, and P_2O_5, bonding is mainly covalent and the electrovalent proportion is strong due to small cations carrying higher charge with a CN of 4. On account of these factors, the bonding between cations and O^{2-} anions is strong. These simple ions combine to form complex anions such as SiO_4^{4-} and PO_4^{3-}, leading to the formation of a stable hexagonal network in slag systems. Hence, they are classified as "network formers" or "acid oxides." For example:

$$SiO_2 + 2O^{2-} = SiO_4^{4-} \tag{4.2}$$

$$P_2O_5 + 3O^{2-} = 2\left(PO_4^{3-}\right) \tag{4.3}$$

The first group of oxides in Table 4.3 with high ionic fraction form simple ions on heating beyond the melting point or when incorporated into a liquid silicate slag. For example:

$$CaO = Ca^{2+} + O^{2-} \tag{4.4}$$

$$Na_2O = 2Na^+ + O^{2-} \tag{4.5}$$

Each O^{2-} ion is capable of breaking one of the four bonds between two tetrahedra by putting itself at the joint as shown in Fig. 4.3. As they destroy the hexagonal network of silica by breaking the bond, they are called "network breakers" and also as "basic oxides."

Oxides of metals exhibiting variable valences, for example, Fe^{2+} and Fe^{3+}, show both basic as well as acidic characteristics. As the cation of the lower valence state happens to be larger, the metal–oxygen bonding in the corresponding oxide will be more electrovalent (ionic) and hence this oxide will be more basic than an oxide of higher valence. For example, out of the two oxides of iron, FeO is basic and Fe_2O_3 is acidic. Oxides such as Fe_2O_3, Cr_2O_3, and Al_2O_3 are known to be amphoteric [2] due to their dual characteristics because they behave like acids in basic slag and as bases in acid slag.

The scale of attraction between elements and oxygen given in Table 4.3 is a function of the size of ions and their charges. It should not be mistaken for affinity of the element for oxygen. It must be noted that affinity is a thermodynamic property of the molecule and not of the crystal.

4.2   Structure of Slag

It is well known that most of the slags are silicates. When a basic oxide is incorporated into the hexagonal network of silica, it forms two simple ions as presented according to Eqs. 4.4 and 4.5. Formation of O^{2-} ions breaks the three-dimensional Si-O network, which has been demonstrated in Fig. 4.3. The fraction of basic oxide, expressed as a O/Si ratio, plays an important role in destroying the number of Si-O joints. The effect of the addition of a basic oxide on the structure of silicates is evident from Table 4.4. Depending on the O/Si ratio of the silicates, a variety of structures are produced. For example, kaolinite, a hydrated silicate of aluminum, $Al_2 Si_2O_5(OH)_4$ or $Al(OH)_3.AlO(OH).2SiO_2$ with O/Si = 5/2, has a lamellar structure in which the elementary silica tetrahedra are united in a

Fig. 4.3 Schematic representation of breaking of a common oxygen bond in the silica tetrahedra in the presence of a basic oxide (From Chemical Metallurgy by J. J. Moore [3], © 1990, p 159, Butterworth Heinemann Ltd. Reproduced with the permission of Butterworth Heinemann Ltd.)

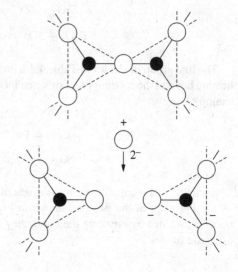

Table 4.4 Role of O/Si ratio in the structure of silicates[a]

O/Si	Formula	Structure
2/1	SiO_2	Silica tetrahedra form a perfect three dimensional hexagonal network
5/2	$MO.2SiO_2$	One vertex joint in each tetrahedron breaks to produce a two-dimensional lamellar structure.
3/1	$MO.SiO_2$	Two vertex joints in each tetrahedron break to produce a fibrous structure
7/2	$3MO.2SiO_2$	Three vertex joints in each tetrahedron break
4/1	$2MO.SiO_2$	All the four joints break

[a]From Fundamentals of Metallurgical Processes by L. Coudurier et al. [2], © 1978, p 189, Pergamon Press Ltd. Reproduced with the permission of Pergamon Press Ltd

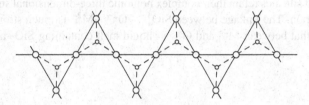

Fig. 4.4 Fibrous structure of a pyroxene (From Fundamentals of Metallurgical Processes by L. Coudurier et al. [2], © 1978, p 189, Pergamon Press Ltd. Reproduced with the permission of Pergamon Press Ltd.)

two-dimensional plane. The broken vertices of the tetrahedra combine with aluminum octahedra. On destruction of two vertices in each tetrahedron pyroxene, MgO. SiO_2 with O/Si = 3/1, exhibits a fibrous structure as shown in Fig. 4.4.

All the silica tetrahedra are separated from each other by Fe^{2+} ions on breakage of all the four vertices in solid fayalite, Fe_2SiO_4 (with O/Si = 4), which is an iron silicate. Unlike silica, silicates have a well-defined fusion point with moderate viscosity. With the weaker electrovalent bonds between simple cations and complex anions, SiO_4^{4-} break first. With increase of temperature, the solid silicate structure completely disappears. Structures of solid and molten fayalite are shown in Fig. 4.5a and b, respectively.

(a)   (b)

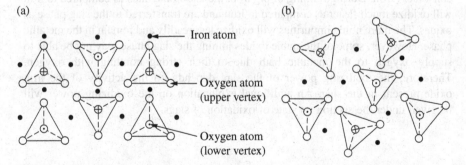

Iron atom

Oxygen atom (upper vertex)

Oxygen atom (lower vertex)

Fig. 4.5 Structure of fayalite: (**a**) solid (**b**) liquid. (From Fundamentals of Metallurgical Processes by L. Coudurier et al. [2], © 1978, p 190, Pergamon Press Ltd. Reproduced with the permission of Pergamon Press Ltd.)

While analyzing the structure of aluminate and phosphate slags, one should note that dissociation of Al_2O_3, P_2O_5, and B_2O_3 provides Al^{3+}, P^{5+} and B^{3+} and O^{2-} ions. These cations can substitute Si^{4+} in the silica tetrahedral network if the anion/cation ratio is adjusted to maintain electro-neutrality. Replacement of Si^{4+} with Al^{3+} reduces the electron field strength and the attraction between Al^{3+} and O^{2-}, whereas such a replacement with P^{5+} increases the attraction between P^{5+} and O^{2-} [3]. This leads to the decrease in the activity of basic oxides on addition of Al_2O_3, SiO_2, and P_2O_5.

Basic oxides show high mobility in liquid state due to reduction in the force of attraction compared to that of the solid state. These basic oxides are more ionic in the liquid state. In contrast, ionic characteristics of silicates do not increase in liquid state because liquid silicates retain their complex nonionic three-dimensional structure for a longer duration. The linkage between $SiO_4^{4-} - O^{2-} - M^{2+}$ is much stronger [3] as compared to that between M^{2+} and O^{2-} in liquid melt containing SiO_2 and a basic oxide, MO.

4.3 Properties of Slag

The knowledge of the following chemical and physical properties of slag is useful in optimizing conditions for recovery of the metal values and transfer of impurities from the metal to the slag phase during extraction and refining.

4.3.1 Oxidizing Power of Slag

Oxidizing power means the ability of the slag to take part in smooth transfer of oxygen from and to the metallic bath. The metallic bath in iron and steelmaking contains impurities that may be either more or less noble compared to the main constituent iron. During refining of pig iron, the less noble ones as compared to iron will oxidize much faster as compared to iron and are transferred to the slag phase as oxides. The more noble impurities will oxidize less readily and remain in the metallic phase. However, some of the stable oxides joining the slag phase may not be able to supply oxygen to the metallic bath due to their strong bonding with oxygen. Therefore, the oxidizing power of the slag depends on the activity of the iron oxide present in the slag. An additional explanation on the oxidizing power will be given under the section on state of oxidation of slags.

4.3.2 *Basicity of Slag*

The concept of basicity requires an understanding about the nature of the different oxides present in the slag. From the above discussion, it is clear that oxides are mainly divided into two groups: acids and bases. In aqueous solutions, base accepts one or more protons whereas acid provides proton(s). In slag systems, an acid oxide forms a complex anion by accepting one or more O^{2-} anions whereas a basic oxide generates O^{2-} anion(s):

$$\text{base} \rightleftarrows \text{acid} + O^{2-} \tag{4.6}$$

For example, SiO_2 (and P_2O_5, CO_2, SO_3 etc.) is an acid oxide because it accepts O^{2-} anions to form a complex silicate anion as per the reaction:

$$(SiO_2) + 2\left(O^{2-}\right) \rightleftarrows SiO_4^{4-} \tag{4.7}$$

On the other hand, basic oxides such as CaO, Na_2O, MnO, and so on generate O^{2-} anions:

$$(MnO) \rightleftarrows Mn^{2+} + O^{2-} \tag{4.8}$$

The amphoteric oxides such as Al_2O_3, Cr_2O_3, and Fe_2O_3 behave as acids in the presence of base(s) or bases in the presence of acid(s) according to the following chemical reaction (4.9) and (4.10), respectively:

$$(Al_2O_3) + \left(O^{2-}\right) \rightleftarrows 2\left(AlO_2^{-}\right) \tag{4.9}$$

$$(Al_2O_3) \rightleftarrows 2\left(Al^{3+}\right) + 3\left(O^{2-}\right) \tag{4.10}$$

A slag containing just sufficient O^{2-} anions to make each tetrahedron of the acid oxide independent of the rest is called a neutral slag. A slag containing 33.3 mol% SiO_2, corresponding to the composition $2CaO.SiO_2$ in the $CaO\text{-}SiO_2$ system is neutral. From Eqs. 4.7 and 4.4, it is clear that one molecule of SiO_2 is neutralized by two molecules of CaO by forming two cations of calcium (Ca^{2+}) and one complex silicate anion $\left(SiO_4^{4-}\right)$ which in turn forms one molecule of calcium orthosilicate ($2CaO.SiO_2$). Thus, depending on the percentage of SiO_2 being more than, equal to, or less than 33.3 mol%, the slag is acid, neutral, or basic. Similarly, three molecules of CaO will neutralize one molecule of P_2O_5 (and Al_2O_3) by forming calcium orthophosphate ($3CaO.P_2O_5$) and calcium orthoaluminate ($3CaO.Al_2O_3$), respectively.

Different methods have been suggested to express basicity index. Ionic theory [2, 3] expresses basicity as the excess of O^{2-} anions in 100 g of slag:

$$n_{O^{2-}} = n_{CaO} + n_{MgO} + n_{MnO} + \ldots - 2n_{SiO_2} - n_{Al_2O_3} \ldots - 3n_{P_2O_5} \qquad (4.11)$$

The degree of depolymerization of the silica network can be expressed as the ratio of the number of nonbridging oxygen atoms to the number of tetrahedrally coordinated atoms of silicon (i.e., NBO/T) [4, 5]. Thus, basicity may be considered as a slag structure index [4, 5] because it is dependent on the number of free oxygen anions ($n_{O^{2-}}$), which is related to the activity of free oxygen anions in the slag. As $a_{O^{2-}}$ cannot be measured experimentally and the ionic scale is not useful in industrial practice, basicity index has been defined in several different ways. The index is generally defined as the ratio of the total weight percent of the basic oxides to the total weight percent of the acid oxides. In a binary slag, for example, CaO–SiO_2, the basicity index (b) is given as:

$$b = wt\%CaO / wt\%SiO_2 \qquad (4.12)$$

Depending on the composition of the slag, the index may take into account the difference in the relative strength between bases. The basicity index [2] of a complex slag consisting of CaO, MgO, SiO_2, and P_2O_5, employed in dephosphorization of steel, is estimated according to the expression:

$$b = \frac{wt\%CaO + \frac{2}{3} wt\%MgO}{wt\%SiO_2 + wt\%P_2O_5} \qquad (4.13)$$

MgO is a weaker base as compared to CaO. Moore [3] has suggested the following expression for calculation of the index:

$$b = \frac{\text{no. of moles of basic oxides} - 3 \times \text{no. of moles of } (Al_2O_3 + P_2O_5)}{2 \times \text{no. of moles of } SiO_2}$$
$$(4.14)$$

The ratio, b, is so adjusted that it is less than, equal to, or greater than unity for acid, neutral, or basic slag, respectively.

For a complex slag comprising CaO, MgO, SiO_2, and P_2O_5, the basicity index may be further modified on the assumption that concentration of CaO and MgO are equivalent on molar basis. Converting this concentration in weight percent, CaO equivalent of MgO becomes $1.4 \times wt\% MgO = wt\% CaO$. On molar basis, $\frac{1}{2} P_2O_5$ is equivalent to SiO_2; hence, in terms of $wt\% SiO_2$, the equivalent of P_2O_5 would be $0.84 \times \%P_2O_5 = \%SiO_2$. Thus, basicity index [6] is given as:

$$b = \frac{wt\%CaO + 1.4 \, wt\%MgO}{wt\%SiO_2 + 0.84 \, wt\%P_2O_5} \qquad (4.15)$$

As composition of most of metallurgical slag is close to neutral, the amphoteric oxides are not included while calculating basicity.

4.3.3 Sulfide Capacity of Slag

Slag plays an important role in extraction and refining of metals, particularly in iron and steelmaking. From the viewpoint of sulfur removal from pig iron, the chemistry of sulfur in silicate slags becomes interesting. Sulfide is soluble in silicate melts but elemental sulfur does not dissolve to any appreciable extent. The chemical reaction leading to transfer of sulfur may be represented as:

$$\frac{1}{2}S_2(g) + (O^{2-}) = \frac{1}{2}O_2(g) + (S^{2-}) \tag{4.16}$$

$$K = \frac{(a_{S^{2-}})}{(a_{O^{2-}})}\left\{\frac{p_{O_2}}{p_{S_2}}\right\}^{\frac{1}{2}} = \frac{(x_{S^{2-}} \cdot \gamma_{S^{2-}})}{(x_{O^{2-}})}\left\{\frac{p_{O_2}}{p_{S_2}}\right\}^{\frac{1}{2}} \tag{4.17}$$

where p_{O_2} and p_{S_2} represent the partial pressure of oxygen and sulfur in the gaseous phase, respectively. From the above equation it may be concluded that for a fixed pressure of sulfur in the gas phase, the sulfide sulfur decreases as oxygen pressure increases. The sulfur affinity of a slag, presented as molar sulfide capacity is defined by the equation:

$$C_S' = (x_{S^{2-}})\left\{\frac{p_{O_2}}{p_{S_2}}\right\}^{\frac{1}{2}} = K.\frac{(x_{O^{2-}})}{(\gamma_{S^{2-}})} \tag{4.18}$$

Richardson and Fincham [7] were first to introduce the concept of sulfide capacity of a slag for assessing the extent of desulfurization according to the reaction (4.16). A more useful term weight percent sulfide capacity [6, 7] for a technologist is defined as:

$$C_S = (\%S)\left\{\frac{p_{O_2}}{p_{S_2}}\right\}^{\frac{1}{2}} \tag{4.19}$$

Thus, under similar conditions, a slag with a high C_S will definitely hold sulfur more strongly than the other with a low C_S and, hence, will prove to be a better desulfurizer in a metallurgical process. Richardson [8] has further pointed out that the sulfide capacity of a slag is proportional to the thermodynamic activity of the basic oxide in the slag. Considering the fact that the sulfide capacity of SiO_2, Al_2O_3, TiO_2, and CaF_2 is close to zero, Nilsson et al. [9] have defined a modified sulfide capacity function in the following manner:

$$C_S^M = \frac{C_S}{a_{M_xO}} = \frac{\text{wt\%S}}{a_{M_xO}} \left\{ \frac{p_{O_2}}{p_{S_2}} \right\}^{\frac{1}{2}} \tag{4.20}$$

The modified function, characteristic of each binary slag system, has been helpful in developing mathematical models [10] which can estimate the sulfide capacities in multicomponent systems. In view of the above-mentioned importance of the sulfide capacity data in steelmaking, Seetharaman and coworkers [11–17] have determined sulfide capacities of many binary, ternary, and multicomponent oxide melts. In general, sulfide capacity enhances with increase of temperature and basicity. The sulfide capacity of CaO–SiO_2–MgO–Al_2O_3–TiO_2 slag [14] decreases with the increase of TiO_2 content at higher temperatures, whereas the effect was insignificant at a lower temperature. MgO beyond 5% increases the sulfide capacity of this slag. MgO and MnO increase the sulfide capacities of CaO–Al_2O_3–SiO_2–MgO, CaO–Al_2O_3–SiO_2–MnO and CaO–Al_2O_3–SiO_2–MgO–MnO slags [15] in a low SiO_2 concentration range. Both basicity and temperature enhance the sulfide capacities of CaO–MgO–Al_2O_3–SiO_2–CrO_x slags [17] in the temperature range 1550–1625 °C. The sulfide capacity increases by the addition of CrO_x up to 5% but decreases on further increase of the CrO_x content.

4.3.4 Electrical and Thermal Conductivity

Molten silica is a poor electrical conductor. However, its conductivity increases to a great extent by addition of basic oxides, for example, CaO, FeO, or MnO as flux. This increase is due to the formation of ions. The conductivity values serve as a measure of degree of ionization of the slag. The electrical conductivity of slags depends on the number of ions present and the viscosity. Thus, conductivity will be greater in the liquid state and further increases with the temperature.

In general, thermal conductivity of slags is very low but heat losses are much higher due to convection. The knowledge about thermal conductivity is essential, not only in development of heat and mass-transfer models but also in optimization of the pyrometallurgical processes. The addition of basic oxides (network breakers) to silicate melts consisting of polymeric silicate anions, SiO_4^{4-} gradually causes disintegration of the silicate network. However, the addition of Al_2O_3 (an oxide of amphoteric characteristic) to a basic slag polymerizes the slag melt. Mostaghel et al. [18] have reported that the effective thermal conductivity of the industrial iron-based slag increases by the addition of Al_2O_3 up to 15%. This is attributed to the formation of strong Al–O covalent bonds. Kang and Morita [19] have carried out systematic thermal conductivity measurements of CaO–Al_2O_3–SiO_2 slags. Erickson and Seetharaman [20] have observed a significant increase in the effective thermal conductivity on addition of Al_2O_3 to a slag with a constant CaO/SiO_2 molar ratio and also on addition of SiO_2 to a slag with the CaO/Al_2O_3 molar ratio of 2.59, but there was very little effect of SiO_2 in a slag of 4.42 molar ratio.

4.3.5 Viscosity

Viscosity being one of the most important properties of slag melts does not only control the heat and mass transfer but also affects the kinetics of reactions at the slag/matte/metal interfaces [21, 22]. It depends on the composition of the polymeric complex ions present in the slag system. The viscosity, η, of a slag of the given composition decreases exponentially with increase of temperature according to the Arrhenius equation:

$$\eta = A \, \exp \, (E\eta/RT) \tag{4.21}$$

where A is a constant and $E\eta$ is the activation energy for viscous flow of the slag. However, decrease is small due to the large value of activation energy. $E\eta$ decreases rapidly on the addition of a basic oxide or halide as flux in molten silica. Basic oxides or halides with large ionic bond fraction are more effective in reducing viscosity than those with smaller bond fraction by breaking bonds between the silica tetrahedra. The first addition of flux to silica to the extent of 15 % has a far greater effect in decreasing the activation energy. This trend for the addition of CaO, Na$_2$O, and CaF$_2$ in silica is shown in Fig. 4.6. Na$_2$O and CaF$_2$ have a greater ionic bond fraction as compared to CaO.

Silica serves as a flux for basic slags in the same way as basic oxides do in acid slags. Figure 4.7 presents the effect of CaO/SiO$_2$ ratio and temperature on viscosities of acid and basic blast furnace slags of various compositions. From these figures it is noted that the lowest viscosity is found with a CaO/SiO$_2$ ratio of approximately 1.35, which is the eutectic composition at this temperature. It is also inferred that viscosity

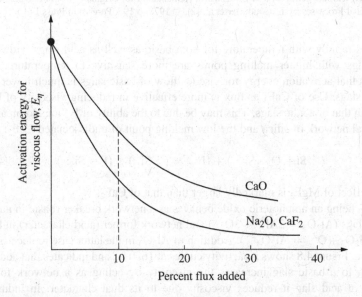

Fig. 4.6 Effect of the addition of a flux on the activation energy of a slag (From Fundamentals of Metallurgical Processes by L. Coudurier et al. [2], © 1978, p 191, Pergamon Press Ltd. Reproduced with the permission of Pergamon Press Ltd)

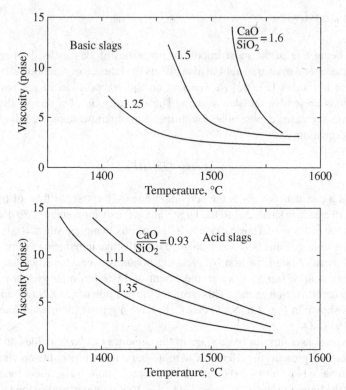

Fig. 4.7 Effect of CaO/SiO$_2$ ratio and temperature on the viscosity of acid and basic blast furnace slags (From Chemical Metallurgy by J. J. Moore [3], © 1990, p 171, Butterworth Heinemann Ltd. Reproduced with the permission of Butterworth Heinemann Ltd. Originally from Fundamentals of Metallurgical Processes by L. Coudurier et al. [2], © 1978, p 192, Pergamon Press Ltd.)

decreases rapidly with temperature for both basic as well as acid slags. However, basic slags with higher melting points are more sensitive to temperature. This indicates that activation energy for viscous flow of basic slags is much lower than for acid slags. Use of CaF$_2$ as flux is more effective in reducing viscosity of basic slags than that of acidic slags. This may be due to the ability of F$^-$ ions to break the hexagonal network of silica and the low melting point of undissociated CaF$_2$:

$$(\vdots Si - O - Si \vdots) + (F^-) = (\vdots Si^-) + (F - Si \vdots) \qquad (4.22)$$

The effect of MgF$_2$ is markedly larger than that of CaF$_2$.

Al$_2$O$_3$ being an amphoteric oxide behaves as a network breaker (basic in nature) in acid slags (Al$_2$O$_3$ = 2Al^{3+} + 3O^{2-}) and network former (acid character) in basic slags (Al$_2$O$_3$ + O^{2-} = Al$_2$O$_4^{2-}$). Addition of Al$_2$O$_3$ in the latter case replaces silica tetrahedra. Figure 4.8 shows isoactivity curves at 1600 °C and indicates that addition of Al$_2$O$_3$ to a basic slag increases the viscosity by acting as a network former whereas in acid slag it reduces viscosity due to its dual character. In industrial

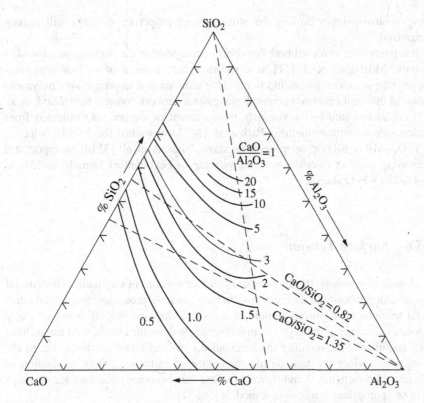

Fig. 4.8 Isoviscosity curves for CaO–SiO$_2$–Al$_2$O$_3$ slags at 1600 °C, viscosity in poise and composition in weight percent (From Fundamentals of Metallurgical Processes by L. Coudurier et al. [2], © 1978, p 194, Pergamon Press Ltd. Reproduced with the permission of Pergamon Press Ltd.)

practice if viscosity decreases slowly with rise in temperature, the slag is called a long slag (acid slags) whereas in the case of short slag (basic), viscosity decreases rapidly. According to Moore [3] a reduction in the viscosity of acid slags due to increase of temperature and addition of basic oxides is related to the electrostatic force of attraction between cation and anion complexes. A stronger force lowers the sensitivity of the slag to reduce viscosity by these changes. Surface properties of slags are affected by different additions. Viscosity is controlled by the movement of large anions. During steelmaking, highly viscous slag, for example, in the acid Bessemer process, does not take part in refining reactions; it only acts as a sink for oxide products. Such a dry slag can be easily separated from the steel melt. In all other steelmaking processes, slags with low viscosity are fluid and act as sink as well as take part in refining processes. Such thin or wet slags are helpful in improving the mass transport and, consequently, the rates of reaction.

Since the knowledge of viscosity is essential for modeling of high-temperature pyrometallurgical processes, extensive researches have been directed during the past few decades [23–32] to collect data on viscosities of slags of different compositions over a range of temperatures. It is surprising to note that despite exhaustive work, many controversies regarding the structure and properties of slags still remain unresolved.

It is interesting to record here the effect of composition on viscosity of a few slag systems. Mostaghel et al. [21] have reported that viscosity of an industrial zinc-copper slag increases by addition of 5–15% alumina due to progressive polymerization of the melt because alumina behaves as a network former. Kowalczyk et al. [33] conducted studies on viscosity measurement of copper slag obtained from chalcocite concentrate smelting. Park et al. [34] have studied the viscous behavior of FeO_x–Al_2O_3–SiO_2 copper smelting slags. Vidacak et al. [35] have reported a decreasing trend in viscosity with the increase of FeO content from 10 to 33% in Al_2O_3–CaO–FeO slags.

4.3.6 Surface Tension

In addition to viscosity, the knowledge of surface tensions of slag melts is also useful in optimizing mass-transfer conditions in bath smelting processes. Slag/metal interfacial tensions have played significant roles in enhancing the efficiency of many processes concerning metal production. Studies on dynamic interfacial tensions have been useful in understanding the mechanisms of slag–metal reactions due to the changes in surface tensions during the course of extraction. Thus, generation of accurate thermochemical and thermophysical data becomes necessary for development of appropriate extrapolation models [36, 37].

Surface properties of a slag and metal bath play an important role in iron and steelmaking. The high rates of reaction in basic oxygen converters is due to the physical conditions of the metal, slag, and gaseous phases in the converter. The theories regarding rapid reaction rates rely heavily on the formation of slag–metal emulsion and slag foams, leading to the creation of the large required reaction surface. Both these phenomena are governed by the surface properties [6] of the oxide and metal bath, which are controlled by composition and temperature. The most important feature of emulsion and foam is the considerable increase of the interfacial area [38] between the two phases leading to the high rate of reaction.

As surface tension is the work required to create a unit area of the new surface, the necessary energy for emulsifying a liquid or a gas in another liquid increases with increasing surface tension value. In a similar manner, energy is liberated when the interfacial area decreases. The decrease in the interfacial area is associated with a corresponding decrease in the total surface energy. The destruction of an emulsion is always a spontaneous process, called self-destruction. Hence, a low interfacial tension favors both the formation and retention of emulsion. On this basis, slag/metal and slag/gas systems are not suitable for emulsification because of the high equilibrium slag/metal interfacial tension [6], 800–1200 erg cm^{-2} as compared to about 40 erg cm^{-2} for a water/mineral oil interface. The value is also high for a slag/gas interface (400–600 erg cm^{-2}) as compared to 73 erg cm^{-2} for a water/gas interface. However, the slag/metal interfacial tension is considerably lowered to 1/100 of the equilibrium value due to mass transfer.

From the surface tension data for pure oxide [6], it is found that ionic octahedral oxides with high CN and NaCl structure, in liquid state, have very low viscosity and high surface tension; for example, for Fe_xO in equilibrium with solid iron at 1420 °C, $\eta = 0.3$ p and $\sigma = 585$ erg cm^{-2}; for Al_2O_3 at 2100 °C, $\eta = 0.5$ p and $\sigma = 690$ erg cm^{-2}; and for CaO at the melting point, $\eta < 0.5$ p and $\sigma > 585$ erg cm^{-2}. On the other hand, liquid oxides with low coordination of central atom with an oxygen atom (tetrahedral oxides–network formers) have high viscosity and low surface tension; for example, for SiO_2, $\eta = 1.5 \times 10^5$ p at 1942 °C and $\sigma = 307$ erg cm^{-2} at 1800 °C; for B_2O_3, $\eta = 300$ p at 1000 °C and $\sigma = 82$ erg cm^{-2} at 1000 °C. Addition of SiO_2 or P_2O_5 to a basic oxide lowers [3] the surface tension due to the absorption of a thin layer of anions, that is, SiO_4^{4-} and PO_4^{3-} on the surface. Vishkarev et al. [39] have reported lowering of surface tension of Fe_xO by excess oxygen. Thus, excess oxygen seems to be moderately surface-active in FeO.

4.3.7   Diffusivity

There is very limited information about diffusion coefficients in liquid oxides. The interdiffusion coefficient of oxygen with FeO has been established as 4×10^{-4} cm^{2}s^{-1} at 1550 °C. It decreases drastically with the increase of Fe^{3+} concentration due to the formation of $Fe^{3+} - O^{2-}$ complexes. Addition of network-forming covalent oxides, such as SiO_2, P_2O_5, and B_2O_3 in oxides such as FeO, NiO, CaO, and MnO reduces conductance, which is controlled by the movement of small cations. Diffusivity data are scarce as well as conflicting. Silicon diffuses very slowly because it is locked covalently in the large slow-moving anions [8]. Aluminum is the next slowest to diffuse as it is built into the anions. Calcium being the smallest ion is expected to diffuse most rapidly but oxygen diffuses much faster than calcium. Richardson [8] has reported that it is possible that oxygen moves faster by splitting away from and recombining with the slow moving silicate ions; however, one expects the larger oxygen ion to move slowly as compared to calcium. The self-diffusion coefficients of Si, Al, and Ca in $CaO-Al_2O_3-SiO_2$ melt are consistent [40].

4.4   Constitution of Metallurgical Slags

From the preceding paragraphs, we understand that slags are molten mixture of oxides and silicates, sometimes phosphates and borates. As the major constituents of iron blast furnace slag are SiO_2, CaO, and Al_2O_3, the slag system is represented by a ternary phase diagram of $SiO_2-CaO-Al_2O_3$. The melting points of SiO_2, CaO, and Al_2O_3 lie between 1700 and 2600 °C but are lowered by addition of other components. Melting points below 1400 °C are found in two regions (i) mixtures with 40–70% SiO_2 and 10–20% Al_2O_3 and (ii) mixtures with about equal quantities of lime and alumina and about 10% SiO_2. Most iron blast furnace slags fall within the first region, whereas the second region corresponds to the lime aluminate slags formed in the smelting of high alumina ores with lime. In addition to the three main constituents, SiO_2, CaO, and Al_2O_3, the iron blast furnace slags may contain variable amounts of magnesia, manganese oxide, and titanium oxides. The behavior of magnesia and manganese oxide is similar to that of lime but they have lesser affinity for SiO_2 and Al_2O_3.

There is no appreciable change in the melting point of a slag with high SiO_2 on replacement of some lime by equal amounts of MgO or MnO. However, the melting point decreases in case of slag with lower SiO_2 [41]. For example, the melting point of Ca_2SiO_4 is lowered by several hundred degrees by replacing about 10% CaO by the same amount of MgO. Based on this finding, dolomite is occasionally added to lower the melting point in blast furnaces to produce a slag with about 10% MgO. MnO also lowers the melting point in the orthosilicate range but it is less effective as compared to MgO. However, addition of TiO_2 in slags exhibits an entirely different behavior. With CaO, TiO_2 (melting point 1842 °C) forms a stable compound, $CaTiO_3$, which is partially immiscible with molten SiO_2. Addition of TiO_2 to $CaSiO_3$ slag lowers the melting point, extending a low-melting-slag area to a eutectic composition with 80% TiO_2 and 20% CaO melting at 1460 °C. In smelting of ilmenite, titania slags are generated by reduction of iron.

The major constituents of steelmaking and nonferrous slags are SiO_2, CaO, and FeO but some Fe_2O_3 is also present [2, 42]. In refining of steel and production of copper, lead, tin, and nickel by pyrometallurgical methods, conditions are much more oxidizing as compared to what prevails in the iron blast furnace. Hence, a part of iron after oxidation joins the slag phase. In addition, the slag may contain variable amounts of MgO, MnO, Al_2O_3, P_2O_5, ZnO, and so on, depending upon the metal production under question. Acid steelmaking slag often contains some MnO, whereas MgO and P_2O_5 are present in basic steelmaking slags. Copper and lead slags may contain some ZnO and tin slags contain PbO. All slags may contain some Al_2O_3. MnO, MgO, and ZnO, which behave like CaO and form silicates. P_2O_5 like SiO_2 is acidic in nature and forms stable compounds with lime and is less stable with FeO. In the ternary system CaO–FeO–P_2O_5, high stability of $Ca_3P_2O_8$ and $Ca_4P_2O_9$ gives rise to a range of liquid immiscibility between liquid calcium phosphate and liquid iron oxide.

The Kennecott-Outotech flash converter [43] and Mitshibishi continuous converter [44] for production of blister copper use silica-free slags based on CaO flux. The reasons for the use of calcium ferrite slag has been explained in Sects. 3.3.2 and 3.3.3. In the Mitsubishi process for direct conversion of concentrates into metallic copper, CaO-based CaO–Cu_2O–FeO–Fe_3O_4 slag [45] plays an important role in the converting operation because SiO_2-based slag cannot absorb magnetite and wustite formed under the high oxygen potential required to form metallic copper. The appropriate knowledge [46] of the slag-phase equilibria, activities of Pb, Zn, PbO, and ZnO in the multicomponent slag system, SiO_2–CaO–FeO–PbO–ZnO, and kinetics of transfer of species from the slag to the metal (and vice versa) have been useful in development of the Ausmelt process for the production of lead from low- as well as high-grade galena concentrates.

From a discussion point of view, the SiO_2–CaO–FeO ternary system seems interesting in steelmaking. Presently, major production of steel is based on basic processes, and hence, it will be more appropriate to discuss some aspects of physical chemistry of basic steelmaking slags in the following paragraphs.

4.4.1 State of Oxidation of Slag

The oxidizing or reducing power of a slag refers to its capacity to participate in the transfer of oxygen to and from the metallic bath. In iron and steelmaking, the metallic bath contains iron as the principal component and other elements that are more or less noble than iron. More noble elements will oxidize less readily than iron during the refining of pig iron and, consequently, will be found in the metallic phase. Less noble elements are easier to oxidize as compared to iron and will be found to a large extent in the slag as oxides. However, these oxides may be so stable that they are not able to supply oxygen to the metallic bath and, hence, the oxidizing power of the slag depends on the activity of the iron oxide present in the slag. The equilibrium between oxygen dissolved in the metal and iron oxide dissolved in the slag is explained as:

$$(FeO) \ (l) = [Fe] \ (l) + [O] \tag{4.23}$$

Where [ ] denotes the element dissolved in the metallic phase and ( ) represents the constituent/ion present in the slag.

$$K = \left[a_{Fe} \right] \frac{[a_O]}{(a_{FeO})} = \frac{[a_O]}{(a_{FeO})} \text{(in a low alloy steel melt } [a_{Fe}] = 1) \tag{4.24}$$

Thus, the activity of oxygen in the melt, $[a_O]$ is proportional to the activity of FeO in the slag (a_{FeO}). At 1600 °C liquid iron saturated with pure FeO dissolves 0.23% O, that is, when $a_{FeO} = 1$.

$$\text{Hence,} \ K = [a_O] = [f_O]wt\%O = [f_O][\%O] \tag{4.25}$$

$$(\text{assuming } [f_O] = 1), \ K = [a_O] = [\%O] = 0.23 \ (\text{maximum solubility}) \tag{4.26}$$

In the presence of slag-forming oxides, solubility of oxygen decreases and $[\%O]$ is accordingly modified as:

$$K = \frac{[a_O]}{(a_{FeO})} \tag{4.27}$$

$$[a_O] = K.a_{(FeO)} = 0.23.a_{FeO} \tag{4.28}$$

Thus, from Eqs. 4.26 and 4.28 we can write:

$$[\%O] = 0.23.a_{FeO} \tag{4.29}$$

where a_{FeO} is the activity of FeO in the coexisting slag relative to pure liquid FeO.

The oxygen content of steel estimated from slag composition, assuming slag–metal equilibrium is about 2–3 times [47] more than determined by vacuum fusion analysis of the metal sample. In all steelmaking processes, the oxygen content of the steel during refining or at tap is high when the total (FeO) content of the slag is high. This simply indicates the possibility of oxidation of Fe^{2+} into Fe^{3+}. If the slag is not in equilibrium with the metal, due consideration should be given to this aspect. For a given oxygen potential, the ratio Fe^{3+}/Fe^{2+} changes markedly with the composition of the slag. For example, at 1550 °C the ratio [47] in different melts is as follows:

1. in case of lime-saturated melts ≈ 0.33,
2. silica-saturated iron silicate melt ≈ 0.02 and
3. complex basic slag < 0.3.

However, in actual steelmaking practice, the ratio varies between 0.3 and 0.5. This high ratio indicates that the state of oxidation in steelmaking slags is much higher than that anticipated from slag–metal equilibrium. For the BOF and BOH practices [47], the oxygen content of the steel, estimated from the Fe_2O_3 content of the slag, is within 0.1–0.4% and for BOH roof lancing, it is 0.4–1.0%. This may be compared with the value of 0.23% in liquid iron saturated with liquid iron oxide. When the calculated value is greater than 0.23, the slag must be oxidizing to the metal.

On the basis of the evidence cited above, it is concluded that the oxygen activity of steelmaking slags is much greater [47] than that of the steel during refining and at the time of tapping. This noticeable difference in the state of oxidation of slag and metal is, of course, a desirable feature for rapid oxidation of impurities during steelmaking without over oxidation of the steel. In the bottom-blown Bessemer process, the FeO content of the slag is in the range of 6–12%, which is about half that for BOH and BOF processes. This difference is not surprising because in BOH and BOF furnaces, slag is exposed to a more severe oxidizing atmosphere. The difference in the state of oxidation of the slag in these processes has an important bearing on the dephosphorization of steel.

The activity data have been determined for a variety of slag compositions and have been usefully compiled in the form of isoactivity curves in the pseudoternary systems. The ternary phase diagram [48] $(CaO + MnO + MgO) - (SiO_2 + P_2O_5) - (FeO + Fe_2O_3)$ as shown in Fig. 4.9, can be used to know a_{FeO} in the normal range of steelmaking temperatures. For comparison, an alternative diagram [49] of $CaO–SiO_2–FeO$ is also given in Fig. 4.10. Both the figures indicate that for any given FeO content, its activity

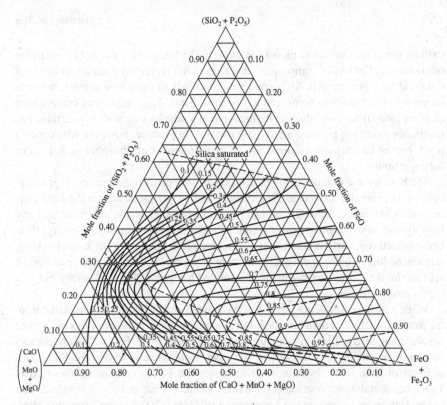

Fig. 4.9 Isoactivity lines of FeO in complex slag (CaO + MnO + MgO)—(SiO$_2$ + P$_2$O$_5$)—(FeO + Fe$_2$O$_3$) (Reproduced from E. T. Turkdogan [47] with the permission of Association for Iron & Steel Technology. Originally from E. T. Turkdogan and J. Pearson [48] with the permission of Journal of Iron & Steel Institute)

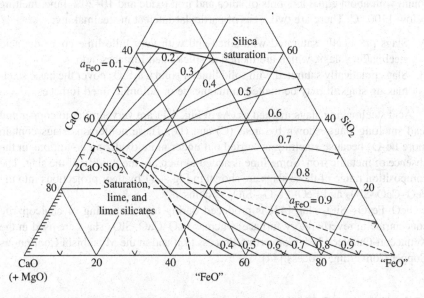

Fig. 4.10 Isoactivity lines of FeO in CaO–FeO–SiO$_2$ system at 1600 °C (Reproduced from J. F. Elliott [49] with the permission of The Minerals, Metals & Materials Society)

shall be maximum for the composition close to the line joining the FeO apex to the orthosilicate $2CaO.SiO_2$ composition due to the high preferential attraction between CaO and SiO_2. Increase in SiO_2 concentration tends to form iron silicate, whereas increase in CaO tends to form ferrite, Ca_2SiO_4. Thus, a_{FeO} reduces as composition changes away from the orthosilicate line on either side. A slag with $b = 2$, thus, has maximum oxidizing power for any FeO content. In practice, however, slags with a much higher basicity are required for effective removal of phosphorus and sulfur during refining.

The basicity and oxidizing power of a slag are independent properties. In practice, basicity is generally related to lime content and oxidizing power to the FeO concentration in the slag although the latter is also a base. Thus, the slag can be of any basicity and any oxidizing power provided requirements regarding viscosity, thermal conductivity, surface tension, and so on are properly met. In basic steelmaking, slag can be highly basic and highly oxidizing or highly basic and highly reducing. In acid steelmaking, the slag being silica saturated has practically no basicity but it is highly oxidizing.

When a ferrous slag (CaO–SiO_2–FeO) is melted in equilibrium with metallic iron, the iron is mostly in a divalent state, but some trivalent iron is also present. A pure wustite melt in equilibrium with iron contains about 10% Fe_2O_3. Fe_2O_3 content decreases on addition of SiO_2 and becomes insignificant beyond Fe_2SiO_4. But addition of lime to the wustite melt increases the Fe_2O_3 content to about 20%. Thus, silica stabilizes divalent iron oxide, whereas lime stabilizes trivalent iron oxide. Hence, at a steelmaking temperature of 1600 °C, the range of molten slags extends from pure wustite melt across the diagram [48] to compositions around $CaSiO_3$, and borders on the acid side by saturation with solid silica and on the basic side by saturation with solid lime and solid Ca_2SiO_4. The lowest melting points are found with about equal amounts of silica and iron oxide and 10–20% lime, melting below 1100 °C. There are two areas of particular interest in steelmaking:

A. Slags practically saturated with silica and with very little lime cover the acid steelmaking slags, which are melted in silica lined furnaces, and
B. Slags practically saturated with solid lime or solid Ca_2SiO_4 cover the basic steelmaking slags. It may be melted in magnesite or dolomite-lined furnaces.

Acid steelmaking slags marked as (A) in Fig. 4.11 are very similar to copper and lead smelting slags, shown by areas (C) and (D). These nonferrous slags contain more Fe_2O_3 because smelting is carried out under more oxidizing conditions in the absence of metallic iron. Some lime is added to increase the fluidity of the slag. The composition range of these slags are shown in Fig. 4.11 by their projections into the FeO–CaO–SiO_2 and FeO–Fe_2O_3–SiO_2 systems.

FeO–Fe_2O_3–SiO_2 slag [42] is produced during matte smelting of chalcopyrite concentrate in reverberatory furnace, whereas FeO–CaO–SiO_2 slags are used in the Kennecott-Outotec Flash Converting Process [43] and in the Mitsubishi Continuous Copper Converting Process [44].

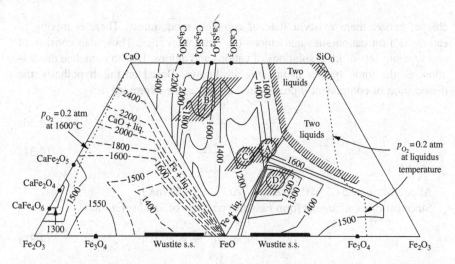

Fig. 4.11 Liquidus isotherms in the system: SiO_2–CaO–FeO–Fe_2O_3 (From Principles of Extractive Metallurgy by T. Rosenqvist [41], © 1974, p 337, McGraw-Hill Inc. Reproduced with the permission of McGraw-Hill Book Co.)

4.5 Slag Theories

Activities of various constituents present in the slag are required for characterization of the slag. As experimental measurement is difficult at high temperature due to the corrosive nature of slag systems, several theories have been developed for the estimation of activities and other thermodynamic properties of the slag constituents. All the theories are based on certain assumptions which limit their application in actual slags. The important theories may be classified into ionic and molecular theories. However, no theory is universally applicable in all types of slags. Hence, the activity of a component needs to be determined experimentally by a method based on the equilibrium between metal, slag, and atmosphere.

4.5.1 Ionic Theories

Herashymenko [50, 51], Temkin [52], Flood [53, 54], and Masson [55, 56] have made important contributions toward the development of ionic theories of slag. Only a brief mention of each will be given here; details can be obtained from references [2, 3, 50–56].

4.5.1.1 Temkin Theory

This theory is based on the assumptions that components of slag solutions are completely dissociated into ions. There is no interaction between ions of the same charge; hence, there exists a state of complete randomness. There is mixing of cations (i^+) on cation sites and anions (j^-) on anion sites. Thus, slag consists of two separate sets of ideal solutions of cations and anions. Due to complete dissociation, all the ionic species in the slag are known. Based on this hypothesis, the dissociation of component ij present in the slag can be expressed as:

$$ij = i^+ + j^- \tag{4.30}$$

$$K = \frac{a_{i^+} \cdot a_{j^-}}{a_{ij}} \tag{4.31}$$

At equilibrium, $\Delta G^o = 0$, $\therefore K = 1$ ($\Delta G^o = - RT\ln K$)

Since cations and anions form two different sets of ideal solutions:

$$a_{i^+} = x_{i^+} \text{ and } a_{j^-} = x_{j^-} \tag{4.32}$$

$$\therefore K = \frac{x_{i^+} \cdot x_{j^-}}{a_{ij}} = 1 \tag{4.33}$$

$$\text{or } a_{ij} = x_{i^+} \cdot x_{j^-} \tag{4.34}$$

It is assumed that the above relation holds good at various molar fractions of the component ij present in the slag solution. In general, Ca^{2+}, Mg^{2+}, Fe^{2+}, SiO_4^{4-}, PO_4^{3-}, $Al_2O_4^{2-}$, $Fe_2O_5^{4-}$, O^{2-}, and so on are present in slag solutions. The fraction of the O^{2-} anion and activity of the oxide in solution can be calculated by making use of Eqs. 4.11 and 4.34, respectively. Application of these equations will be illustrated in some problems.

4.5.1.2   Flood Theory

Flood, Forland, and Grjotheim [53, 54] have further extended Temkin's theory by considering equilibria between ions or compounds dissolved in the slag and elements dissolved in the metallic phase. It assumes ideal behavior for ions in slag and nonideal for elements in metal. For example, the sulfur–oxygen equilibrium between slag and metal may be represented in the following manner:

$$[S] + (O^{2-}) = (S^{2-}) + [O] \tag{4.35}$$

$$K = \frac{(a_{S^{2-}}) \cdot [a_O]}{[a_S] \cdot (a_{O^{2-}})} \tag{4.36}$$

As slag and metal phases exhibit ideal and nonideal behavior, respectively, K is expressed as:

$$K = \frac{(x_{S^{2-}}).[f_O.\%O]}{[f_S.\%S].(x_{O^{2-}})} = K'g(f) \tag{4.37}$$

where $K' = \frac{(x_{S^{2-}}).[\%O]}{(x_{O^{2-}}).[\%S]}$ (the apparent equilibrium constant)

$$\text{and } g(f) = \frac{[f_O]}{[f_S]} \tag{4.38}$$

when Henry's law is obeyed:

$$[f_O] = 1 = [f_S] \tag{4.39}$$

$$\text{hence } K = K' \tag{4.40}$$

Moore [3] has discussed the application of Flood's Theory in complex slags containing sulfides and oxides of Na and Ca and in the phorphorus–oxygen equilibrium between metal and slag in iron and steelmaking.

4.5.1.3 Masson theory

In order to calculate the activity of a basic oxide in a silicate slag, Masson [55, 56] has developed a theory to account for more complex anions compared to SiO_4^{4-}. Masson considers slags as complex solutions containing polymeric silicate anions. The character and quantity of basic oxide present in the slag control the degree of polymerization. Thus, in a highly basic slag, silica is present mainly as SiO_4^{4-} whereas less basic slag contains $SiO_4^{4-}, Si_2O_7^{6-}, Si_3O_{10}^{8-} \ldots\ldots \ldots\ldots Si_nO_{(3n+1)}^{2(n+1)-}$ anions in equilibrium with each other. The equilibria between these complex anions can be expressed as follows:

$$SiO_4^{4-} + SiO_4^{4-} = Si_2O_7^{6-} + O^{2-} \quad K_1 = \frac{x_{Si_2O_7^{6-}}.x_{O^{2-}}}{x_{SiO_4^{4-}}x_{SiO_4^{4-}}} \tag{4.41}$$

$$Si_2O_7^{6-} + SiO_4^{4-} = Si_3O_{10}^{8-} + O^{2-} \quad K_2 = \frac{x_{Si_3O_{10}^{8-}}.x_{O^{2-}}}{x_{Si_2O_7^{6-}}.x_{SiO_4^{4-}}} \tag{4.42}$$

$$Si_3O_{10}^{8-} + SiO_4^{4-} = Si_4O_{13}^{10-} + O^{2-} \quad K_3 = \frac{x_{Si_4O_{13}^{10-}}.x_{O^{2-}}}{x_{Si_3O_{10}^{8-}}.x_{SiO_4^{4-}}} \tag{4.43}$$

In the above equations K_1, K_2, K_3 are corresponding equilibrium constants. The mole fraction of each of the subsequently generated polymeric silicate anion by addition of a silica tetrahedron can be estimated as follows by assuming, $K_1 = K_2 = K_3 = K$. Hence,

$$x_{Si_2O_7^{6-}} = \frac{K \cdot x_{SiO_4^{4-}}}{x_{O^{2-}}} x_{SiO_4^{4-}} \tag{4.44}$$

$$x_{Si_3O_{10}^{8-}} = \frac{K \cdot x_{Si_2O_7^{6-}}}{x_{O^{2-}}} x_{SiO_4^{4-}} = K^2 \left[\frac{x_{SiO_4^{4-}}}{x_{O^{2-}}}\right]^2 \cdot x_{SiO_4^{4-}} \tag{4.45}$$

$$x_{Si_4O_{13}^{10-}} = \frac{K \cdot x_{Si_3O_{10}^{8-}}}{x_{O^{2-}}} x_{SiO_4^{4-}} = K^3 \left[\frac{x_{SiO_4^{4-}}}{x_{O^{2-}}}\right]^3 \cdot x_{SiO_4^{4-}} \tag{4.46}$$

where K is the composite equilibrium constant and, hence, the total silicate anions may be obtained as

$$\sum N_{\text{silicate anions}} = x_{SiO_4^{4-}} + x_{Si_2O_7^{6-}} + x_{Si_3O_{10}^{8-}} + \ldots\ldots\ldots x_{Si_nO_{(3n+1)}^{2(n+1)-}} \tag{4.47}$$

$$= x_{SiO_4^{4-}} \left[1 + \left[K\frac{x_{SiO_4^{4-}}}{x_{O^{2-}}}\right] + \left[K\frac{x_{SiO_4^{4-}}}{x_{O^{2-}}}\right]^2 + \left[K\frac{x_{SiO_4^{4-}}}{x_{O^{2-}}}\right]^3 + \ldots..\right] \tag{4.48}$$

$$= \frac{x_{SiO_4^{4-}}}{1 - K\left[\frac{x_{SiO_4^{4-}}}{x_{O^{2-}}}\right]} \tag{4.49}$$

(assuming $y = \left[K\frac{x_{SiO_4^{4-}}}{x_{O^{2-}}}\right]$ and summing the series: $1 + y + y^2 + y^3 + \ldots.. + y^n = \frac{1}{(1-y)}$ if $y < 1$)

According to Temkin theory:

$$\sum N_{\text{silicate anions}} = 1 - x_{O^{2-}} \tag{4.50}$$

(when there are only silicate and O^{2-} ions present)

From Eqs. 4.49 and 4.50 we get

$$(1 - x_{O^{2-}}) \left[1 - \frac{K x_{SiO_4^{4-}}}{x_{O^{2-}}}\right] = x_{SiO_4^{4-}}$$

$$\therefore x_{SiO_4^{4-}} = \left[\frac{1 - x_{O^{2-}}}{1 - K + K/x_{O^{2-}}}\right] \tag{4.51}$$

Form the chemical analysis of slag the mole fraction of SiO_2 can be expressed as:

$$x_{SiO_2} = \frac{n_{SiO_2}}{n_{total\ oxide}}$$

$$= \frac{n_{SiO_2}}{n_{SiO_2} + n_{MO(free)} + n_{MO(silicate)}} \qquad (4.52)$$

From the knowledge of the equivalent number of moles of SiO_2 required to produce each of the above complex silicate polymeric anions as well as the total number of oxides (SiO_2 + basic oxides) Eq. 4.52 may be expressed as:

$$x_{SiO_2} = \frac{x_{SiO_4^{4-}} + 2\, x_{Si_2O_7^{6-}} + 3\, x_{Si_3O_{10}^{8-}} + \cdots \cdots}{x_{O^{2-}} + 3\, x_{SiO_4^{4-}} + 5\, x_{Si_2O_7^{6-}} + 7\, x_{Si_3O_{10}^{8-}} + \cdots \cdots} \qquad (4.53)$$

By using appropriate chemical reactions such as (4.41), (4.42), (4.43) and assuming a composite equilibrium constant, K, instead of K_1, K_2, K_3 and so on, fractions of other silicate anions can be calculated. A general expression [3] may be written as:

$$x_{Si_nO_{(3n+1)}^{2(n+1)-}} = K \left[\frac{x_{SiO_4^{4-}}}{x_{O^{2-}}} \right]^{n-1} \cdot x_{SiO_4^{4-}} \qquad (4.54)$$

Combining the above Eqs. 4.41, 4.42, 4.43, 4.51 and 4.53 and making use of suitable expressions of the sum of infinite series $(1 + 2y + 3y^2 + \cdots \cdots = \frac{1}{(1-y)^2}$ and $3 + 5y + 7y^2 + 9y^3 + \cdots \cdots = \frac{3-y}{(1-y)^2}$, one can evaluate $x_{O^{2-}}$ by the expression relating x_{SiO_2} (by chemical analysis) with K and $x_{O^{2-}}$:

$$x_{SiO_2} = \frac{1}{3 - K + \frac{x_{O^{2-}}}{1 - x_{O^{2-}}} + \frac{K(K-1)}{\frac{x_{O^{2-}}}{1 - x_{O^{2-}}} + K}} \qquad (4.55)$$

If K is known, $x_{O^{2-}}$ can be calculated. Hence, activity of the basic oxide, MO can be calculated by the Temkin relationship: $(a_{MO} = x_{M^{2+}} \cdot x_{O^{2-}})$.

There is reasonable agreement between the activity values calculated for a number of basic oxides, CaO, FeO, MnO, and the experimentally determined values. However, the theory is not applicable to all slag systems due to oversimplified assumption of the composite constant ($K = K_1 = K_2 = K_3$ etc.). The equilibrium constant for more complex polymeric anions are definitely different from those for $Si_2O_7^{6-}$ and $Si_3O_{10}^{8-}$. Further, this theory considers only linear chains but some cyclic chains may exist with more complex polymeric anions. Similarly, to the Temkin and Flood theories, Masson's theory also assumes ideal behavior for cations and anions separately. The possibility of ion interaction cannot be ignored in actual solutions.

4.5.2 Molecular Theory

According to Schenck [57], the molecular theory assumes ideal behavior for all the molecules present in slag. Without forming ions, simple oxides such as CaO, MnO, FeO, Fe_2O_3, Al_2O_3, and SiO_2 either associate to form complex molecules such as Ca_2SiO_4, $CaAl_2O_4$, and $Ca_4P_2O_9$ or remain as free compounds. This makes it necessary to consider all possible molecules existing in the slag system. For example, in a ternary slag system of $CaO–MnO–SiO_2$, eight molecules MnO, CaO, $CaSiO_3$, Ca_2SiO_4, $MnSiO_3$, Mn_2SiO_4, $Ca_2Si_2O_6$, and $Ca_4Si_2O_8$ may exist. This requires a set of the following eight equations to calculate the mole fraction of each compound as under:

$$1.\ MnO + SiO_2 = MnSiO_3; K_1 = x_{MnSiO_3}/x_{MnO}.x_{SiO_2} \tag{4.56}$$

$$2.\ 2MnO + SiO_2 = Mn_2SiO_4;\ \ K_2 = x_{Mn_2SiO_4}/x_{MnO}^2.x_{SiO_2} \tag{4.57}$$

$$3.\ CaO + SiO_2 = CaSiO_3;\ \ K_3 = x_{CaSiO_3}/x_{CaO}.x_{SiO_2} \tag{4.58}$$

$$4.\ 2CaO + SiO_2 = Ca_2SiO_4; K_4 = x_{Ca_2SiO_4}/x_{CaO}^2.x_{SiO_2} \tag{4.59}$$

$$5.\ 2CaO + 2SiO_2 = Ca_2Si_2O_6; K_5 = x_{Ca_2Si_2O_6}/x_{CaO}^2.x_{SiO_2}^2 \tag{4.60}$$

$$6.\ 4CaO + 2SiO_2 = Ca_4Si_2O_8; K_6 = x_{Ca_4Si_2O_8}/x_{CaO}^4.x_{SiO_2}^2 \tag{4.61}$$

$$7.\ Ca_4Si_2O_8 = Ca_2Si_2O_6 + 2CaO; K_7 = x_{Ca_2Si_2O_6}.x_{CaO}^2/x_{Ca_4Si_2O_8} \tag{4.62}$$

$$8.\ Ca_4Si_2O_8 = 2(Ca_2SiO_4);\ \ K_8 = x_{Ca_2SiO_4}^2/x_{Ca_4Si_2O_8} \tag{4.63}$$

Assuming ideal behavior for all the compounds participating in the above equations the equilibrium constants have been expressed in terms of mole fractions. From the mass balance the number of moles of CaO, MnO, and SiO_2 can be expressed as:

$$n_{CaO} = n_{CaO(free)} + n_{CaSiO_3} + 2n_{Ca_2SiO_4} + 2n_{Ca_2Si_2O_6} + 4n_{Ca_4Si_2O_8} \tag{4.64}$$

$$n_{MnO} = n_{MnO(free)} + n_{MnSiO_3} + 2n_{Mn_2SiO_4} \tag{4.65}$$

$$x_{SiO_2} = n_{SiO_2(free)} + n_{MnSiO_3} + n_{Mn_2SiO_4} + n_{CaSiO_3} + n_{Ca_2SiO_4} + 2n_{Ca_2Si_2O_6}$$
$$+ 2n_{Ca_4Si_2O_8} \tag{4.66}$$

Although the molecular theory is applicable in certain slag systems, assumption of ideal solution behavior is not true in all the slags. Furthermore, the equilibrium constants relating the mole fractions of different molecules are not often known with a certain degree of accuracy. Winkler and Chipman [58] have applied this theory to calculate activities of CaO and FeO in a complex slag used in dephosphorization of steel.

4.6 Problems

Problem 4.1

For a slag containing (weight percent) CaO – 50%, MnO – 8%, FeO – 25%, and SiO_2 – 17% calculate (a) the excess base of slag assuming species $2MO.SiO_2$, (b) oxygen ion concentration of the slag, and (c) activity of FeO

Solution
Slag composition

Species	wt%	mol%
CaO	50	54.6
MnO	8	6.9
FeO	25	21.2
SiO_2	17	17.3

a. Excess base $= CaO + FeO + MnO - 2(SiO_2) = (54.6 + 6.9 + 21.2) - 2 \times 17.3 = 48.1$ mol

b. $n_{O^{2-}} = n_{CaO} + n_{FeO} + n_{MnO} - n_{SiO_2}$

$= 0.546 + 0.212 + 0.069 - 2 \times 0.173$

$= 0.481$

$x_{O^{2-}} = \dfrac{n_{O^{2-}}}{n_{O^{2-}} + n_{SiO_4^{4-}}} = \dfrac{0.481}{0.481 + 0.173} = 0.888$

c. $x_{Fe^{2+}} = \dfrac{0.212}{0.827} = 0.256$

$a_{FeO} = x_{Fe^{2+}} \cdot x_{O^{2-}} = 0.888 \times 0.256 = 0.227$ Ans.

Problem 4.2

Calculate the activity of MnO in a slag of the following composition (molar fraction):

CaO–0.34, MgO–0.11, FeO–0.31, MnO–0.14, SiO_2–0.02, Al_2O_3–0.04, and Fe_2O_3–0.04.

Solution
According to Temkin's theory activity of a basic oxide in a highly basic slag is given by the expression:

$$a_{MnO} = x_{Mn^{2+}} \cdot x_{O^{2-}}$$

$$x_{Mn^{2+}} = \frac{n_{MnO}}{n_{MnO} + n_{CaO} + n_{MgO} + n_{FeO}}$$

$$= \frac{0.14}{0.14 + 0.34 + 0.11 + 0.31} = \frac{0.14}{0.90} = 0.155$$

SiO_4^{4-}, $Al_2O_4^{2-}$ and $Fe_2O_5^{4-}$ anions are formed by the reactions (as the slag is highly basic amphoteric oxides Al_2O_3 and Fe_2O_3 act as acids, that is network former).

$$SiO_2 + 2O^{2-} = SiO_4^{4-}$$
$$Al_2O_3 + O^{2-} = Al_2O_4^{2-}$$
$$Fe_2O_3 + 2O^{2-} = Fe_2O_5^{4-}$$

$$n_{O^{2-}} = n_{MnO} + n_{CaO} + n_{MgO} + n_{FeO} - 2n_{SiO_2} - n_{Al_2O_3} - 2n_{Fe_2O_3}$$
$$= 0.14 + 0.34 + 0.11 + 0.31 - 2 \times 0.02 - 0.04 - 2 \times 0.04 = 0.74$$

$$x_{O^{2-}} = \frac{n_{O^{2-}}}{n_{O^{2-}} + n_{SiO_4^{4-}} + n_{Al_2O_4^{2-}} + n_{Fe_2O_5^{4-}}}$$
$$= \frac{0.74}{0.74 + 0.02 + 0.04 + 0.04} = \frac{0.74}{0.84} = 0.88$$
$$\therefore a_{MnO} = x_{Mn^{2+}} \cdot x_{O^{2-}} = 0.155 \times 0.88 = 0.137 \quad \text{Ans.}$$

Problem 4.3
Calculate the activity of MgO in a slag of the following mole fractions:
    CaO–0.44, MgO–0.10, FeO–0.12, Na$_2$O–0.08, Fe$_2$O$_3$–0.06, Al$_2$O$_3$–0.08, SiO$_2$–0.02, and P$_2$O$_5$–0.10.

Solution
As the slag is highly basic all the species are completely dissociated into ions. Acidic oxide such as SiO_2 and P_2O_5 as well as amphoteric oxides such as Fe_2O_3 and Al_2O_3 will form ions such as SiO_4^{4-}, PO_4^{3-}, $Fe_2O_5^{4-}$ and $Al_2O_4^{2-}$.

According to Temkin rule: $a_{MgO} = x_{Mg^{2+}} \cdot x_{O^{2-}}$

$$x_{Mg^{2+}} = \frac{n_{MgO}}{n_{MgO} + n_{CaO} + n_{FeO} + \frac{1}{2}n_{Na_2O}}$$
$$= \frac{0.10}{0.10 + 0.44 + 0.12 + 0.04} = \frac{1}{7}$$

SiO_4^{4-}, $Al_2O_4^{2-}$ and $Fe_2O_5^{4-}$ anions are formed by the reactions
$$SiO_2 + 2O^{2-} = SiO_4^{4-}$$
$$P_2O_5 + 3O^{2-} = 2PO_4^{3-}$$
$$Al_2O_3 + O^{2-} = Al_2O_4^{2-}$$
$$Fe_2O_3 + 2O^{2-} = Fe_2O_5^{4-}$$

$$n_{O^{2-}} = n_{CaO} + n_{MgO} + n_{FeO} + n_{Na_2O} - 2n_{SiO_2} - 2n_{Fe_2O_3} - n_{Al_2O_3} - 3n_{P_2O_5}$$
$$= 0.44 + 0.10 + 0.12 + 0.08 - 2 \times 0.02 - 2 \times 0.06 - 0.08 - 3 \times 0.10$$
$$= 0.74 - 0.54 = 0.20$$

$$x_{O^{2-}} = \frac{n_{O^{2-}}}{n_{O^{2-}} + n_{SiO_4^{4-}} + n_{Fe_2O_5^{4-}} + n_{Al_2O_4^{2-}} + n_{PO_4^{3-}}}$$

$$= \frac{0.20}{0.20 + 0.02 + 0.06 + 0.08 + 0.05} = \frac{20}{41}$$

$$\therefore a_{MgO} = x_{Mg^{2+}} \cdot x_{O^{2-}} = \frac{1}{7} \times \frac{20}{41} = \frac{20}{287} = 0.07 \text{Ans.}$$

Form the above problems, it is clear that Temkin's theory is applicable in highly basic slags when all the ions in the slag are known. Only in strongly basic slags are simple SiO_4^{4-} anions formed. In acid slags, there is a possibility of formation of complex anions, such as $Si_2O_7^{6-}$, $Si_3O_{10}^{8-}$, and so on.

Problem 4.4

A slag of the following fraction molar composition: MgO–0.25, FeO–0.25, CaO–0.10 and SiO_2–0.40 contains higher order polymeric silicate anions. Assuming the equilibrium constant for the formation of these silicate anions to be unity, calculate the mole fraction of oxygen and silicate ions in the slag.

Solution

$x_{SiO_2} = 0.4$ and $K = 1$; since higher silicate anions are present, according to Masson's theory (Eq. 4.55) we have:

$$x_{SiO_2} = \frac{1}{3 - K + \dfrac{x_{O^{2-}}}{1 - x_{O^{2-}}} + \dfrac{K(K-1)}{x_{O^{2-}}/\left(1 - x_{O^{2-}}\right) + K}}$$

$$0.4 = \frac{1}{3 - 1 + x_{O^{2-}}/\left(1 - x_{O^{2-}}\right)}$$

$$x_{O^{2-}} = 0.33$$

$$\therefore x_{silicate\ ions} = 1 - 0.33 = 0.67 \text{ Ans.}$$

Problem 4.5

If $K_{1-1} = K_{1-2} = \cdots \cdots = K_{1-n} = 0.163$ for a binary silicate melt containing MO and SiO_2, calculate, (a) the ionic fraction of O^{2-} in the melt at the maximum ionic fraction of $x_{SiO_4^{4-}}$ and (b) the mole fraction of SiO_2 in the melt under the condition mentioned in part (a).

Solution

a. To determine $x_{O^{2-}}$ when $x_{SiO_4^{4-}}$ has its maximum value, we may use Eq. 4.51, derived earlier:

$$x_{SiO_4^{4-}} = \frac{1 - x_{O^{2-}}}{1 + K\left(\frac{1}{x_{O^{2-}}} - 1\right)}$$

Differentiating $x_{SiO_4^{4-}}$ with respect to $x_{O^{2-}}$ and equating the result to zero to calculate $x_{O^{2-}}$ at the maximum of $x_{SiO_4^{4-}}$:

$$\frac{dx_{SiO_4}}{dx_{O^{2-}}} = \frac{-1\left[1 + K\left(\frac{1}{x_{O^{2-}}} - 1\right)\right] - (1 - x_{O^{2-}})\left[\frac{-K}{(x_{O^{2-}})^2}\right]}{\left[1 + K\left(\frac{1}{x_{O^{2-}}} - 1\right)\right]^2} = 0$$

$$\text{or } (K - 1)\, x_{O^{2-}}^2 - 2Kx_{O^{2-}} + K = 0$$

$$\therefore x_{O^{2-}} = \frac{2K \pm \sqrt{4K^2 - 4(K - 1)K}}{2(K - 1)} = \frac{K \pm \sqrt{K}}{K - 1}$$

Since $K = 0.163$ (given), $x_{O^{2-}} = 0.288$ and -0.677 (invalid)

$$\therefore x_{O^{2-}} = 0.288 \text{ when } x_{SiO_4^{4-}} \text{ is maximum Ans.}$$

b. At the maximum value of $x_{SiO_4^{4-}}$ the binary silicate has an orthosilicate composition of $2MO.SiO_2$, that is, $x_{SiO_2} = 0.333$. However, it can also be obtained from Eq. 4.55 by substituting $K = 0.163$ and $x_{O^{2-}} = 0.288$

$$x_{SiO_2} = \frac{1}{3 - K + \frac{x_{O^{2-}}}{1 - x_{O^{2-}}} + \frac{K(K-1)}{x_{O^{2-}} / (1 - x_{O^{2-}}) + K}}$$

$$x_{SiO_2} = 0.333 \text{ Ans.}$$

Problem 4.6

Derive a relationship between the activity of PbO and the mole fraction of PbO for $PbO-P_2O_5$ melts when $K_{1-1} = K_{1-2} = \ldots\ldots\ldots = K_{1-n} = K = 0$.

Solution

The relationship between a_{PbO} and x_{PbO} in $PbO-P_2O_5$ melts when $K_{1-1} = K_{1-2} = \ldots\ldots\ldots = K_{1-n} = K = 0$

According to Temkin $a_{PbO} = x_{Pb^{2+}} . x_{O^{2-}}$

Considering the reaction $2PO_4^{3-} = P_2O_7^{4-} + O^{2-}$

$$K = \frac{x_{P_2O_7^{4-}} . x_{O^{2-}}}{x_{PO_4^{3-}}^2} = 0$$

Thus, the melt consists of only cations: Pb^{2+} and anions: PO_4^{3-} and O^{2-}

$$\therefore x_{Pb^{2+}} = 1 \text{ and } x_{P_2O_7^{4-}} + x_{O^{2-}} = 1$$

$$a_{PbO} = x_{Pb^{2+}} . x_{O^{2-}} = x_{O^{2-}}$$

$$x_{O^{2-}} = \frac{n_{O^{2-}}}{n_{O^{2-}} + n_{PO_4^{3-}}}$$

From the ionic theory $n_{O^{2-}} = n_{PbO} - 3n_{P_2O_5}$ and $n_{PO_4^{3-}} = 2n_{P_2O_5}$

$$\therefore a_{PbO} = x_{O^{2-}} = \frac{n_{O^{2-}}}{n_{O^{2-}} + n_{PO_4^{3-}}}$$

$$= \frac{n_{PbO} - 3n_{P_2O_5}}{n_{PbO} - 3n_{P_2O_5} + 2n_{P_2O_5}}$$

$$= \frac{n_{PbO} - 3n_{P_2O_5}}{n_{PbO} - n_{P_2O_5}}$$

Dividing numerator and denominator by $n_{PbO} + n_{P_2O_5}$, we get

$$a_{PbO} = \frac{x_{PbO} - 3x_{P_2O_5}}{x_{PbO} - x_{P_2O_5}} \quad (x_{P_2O_5} = 1 - x_{PbO})$$

$$= \frac{x_{PbO} - 3(1 - x_{PbO})}{x_{PbO} - (1 - x_{PbO})} = \frac{4x_{PbO} - 3}{2x_{PbO} - 1} \quad \text{Ans.}$$

Problem 4.7
(a) Show that the activities of salts AY_2 and BX (where A and B are, respectively, divalent and univalent cations and X and Y are univalent anions) in a salt mixture: $AY_2 - BX$, are given as $a_{AY_2} = x_{A^{2+}} . x_{Y^-}^2$ and $a_{BX} = x_{B^+} . x_{X^-}$, where $x_{A^{2+}}, x_{Y^-}, x_{B^+}$ and x_{X^-} are ionic fractions.
(b) Construct an activity (a_{AY_2}, a_{BX} and a_{BY}) vs. mole fraction ($x_{AY_2} = 1.0$ to $x_{BX} = 1.0$) diagram for the Temkin ideal salt mixtures.

Solution
(a) According to Temkin's theory, fused salt mixtures of AY_2 and BX are solutions, which are completely dissociated into ions. There is no interaction between ions of the same charge and the state may be that of complete randomness. Hence, solution of salts may be assumed to be constituted of two ideal solutions; one of cations: A^{2+}, B^+ and another of anions: X^-, Y^-.

The heat of mixing of an ideal solution is zero, that is, $\Delta H^{M, \text{id}} = 0$ and in case of random mixing, entropy of mixing, $\Delta S^{M, \text{id}} = - R \sum x_i \ln x_i$.
Hence for the above system, we have:

$$\Delta S^{M,\text{id}} = -R \left(n_A \ln x_A + n_B \ln x_B + n_X \ln x_X + n_Y \ln x_Y \right) \tag{1}$$

$$\Delta G^{M,\text{id}} = \Delta H^{M,\text{id}} - T\Delta S^{M,\text{id}} = RT(n_A \ln x_A + n_B \ln x_B + n_X \ln x_X + n_Y \ln x_Y) \tag{2}$$

The mole fractions of anions and cations in the above system of salt mixture, AY_2 and BX, comprising two ideal solutions can be expressed as:

$$x_A = \frac{n_A}{n_A + n_B}, x_B = \frac{n_B}{n_A + n_B}, x_X = \frac{n_X}{n_X + n_Y}, x_Y = \frac{n_Y}{n_X + n_Y}$$

$$\therefore \Delta G^M = RT\left[n_A \ln\left(\frac{n_A}{n_A + n_B}\right) + n_B \ln\left(\frac{n_B}{n_A + n_B}\right) + n_X \left(\frac{n_X}{n_X + n_Y}\right) + n_Y \ln\left(\frac{n_Y}{n_X + n_Y}\right)\right]$$

$$\tag{3}$$

The partial molar free energy of mixing of the salt AY_2 can be expressed as:

$$\overline{G}^M_{AY_2} = \left(\frac{\partial \Delta G^M}{\partial n_{AY_2}}\right) = RT \ln a_{AY_2} \tag{4}$$

It may be noted that $dn_{AY_2} = \frac{1}{2}dn_Y = dn_A$

$$\therefore \left(\frac{\partial \Delta G^M}{\partial n_{AY_2}}\right)_{n_B,n_X} = \left(\frac{\partial \Delta G^M}{\partial n_A}\right)_{n_B,n_X,n_Y} + \left(\frac{\partial \Delta G^M}{\frac{1}{2}\partial n_Y}\right)_{n_B,n_X,n_A}$$

$$\text{or} \left(\frac{\partial \Delta G^M}{\partial n_{AY_2}}\right)_{n_B,n_X} = \left(\frac{\partial \Delta G^M}{\partial n_A}\right)_{n_B,n_X,n_Y} + 2\left(\frac{\partial \Delta G^M}{\partial n_Y}\right)_{n_B,n_X,n_A} \tag{5}$$

Differenting Eq. 3 with respect to n_A at constant n_B, n_X, n_Y

$$\left(\frac{\partial \Delta G^M}{\partial n_A}\right)_{n_B,n_X,n_Y} = RT\left[\ln\left(\frac{n_A}{n_A + n_B}\right) + n_A \frac{d}{dn_A}\left\{\ln\left(\frac{n_A}{n_A + n_B}\right)\right\} + n_B \frac{d}{dn_A}\left\{\ln\left(\frac{n_B}{n_A + n_B}\right)\right\}\right]$$

$$\tag{6}$$

On differentiation of appropriate terms, we get:

$$\frac{d}{dn_A}\left[\ln\left(\frac{n_A}{n_A + n_B}\right)\right] = \frac{n_B}{n_A}\frac{1}{(n_A + n_B)}$$

$$\frac{d}{dn_A}\left[\ln\left(\frac{n_B}{n_A + n_B}\right)\right] = -\frac{1}{(n_A + n_B)}$$

Substituting these values in Eq. 6 we get:

$$\left(\frac{\partial \Delta G^M}{\partial n_A}\right)_{n_B,n_X,n_Y} = RT\left[\ln\left(\frac{n_A}{n_A + n_B}\right) + n_A \cdot \frac{n_B}{n_A}\frac{1}{(n_A + n_B)} - n_B \cdot \frac{1}{(n_A + n_B)}\right]$$

$$= RT \ln\left(\frac{n_A}{n_A + n_B}\right) = RT \ln x_A \tag{7}$$

Similarly,

$$\left(\frac{\partial \Delta G^M}{\partial n_Y}\right)_{n_A, n_B, n_X} = RT\left[\ln\left(\frac{n_Y}{n_X + n_Y}\right) + n_Y \frac{d}{dn_Y}\left\{\ln\left(\frac{n_Y}{n_X + n_Y}\right)\right\} + n_X \frac{d}{dn_Y}\left\{\ln\left(\frac{n_X}{n_X + n_Y}\right)\right\}\right] \tag{8}$$

We have: $\frac{d}{dn_Y}\left[\ln\left(\frac{n_Y}{n_X + n_Y}\right)\right] = \frac{n_X}{n_Y}\frac{1}{(n_X + n_Y)}$

and $\frac{d}{dn_Y}\left[\ln\left(\frac{n_X}{n_X + n_Y}\right)\right] = -\frac{1}{(n_X + n_Y)}$

Substituting these values in Eq. 8 we get:

$$\left(\frac{\partial \Delta G^M}{\partial n_Y}\right)_{n_A, n_B, n_X} = RT\left[\ln\left(\frac{n_Y}{n_X + n_Y}\right) + n_Y \cdot \frac{n_X}{n_Y}\frac{1}{(n_X + n_Y)} - n_X \cdot \frac{1}{(n_X + n_Y)}\right]$$

$$= RT \ln\left(\frac{n_Y}{n_X + n_Y}\right) = RT \ln x_Y \tag{9}$$

From Eqs. 5, 7 and 9 we get:

$$\left(\frac{\partial \Delta G^M}{\partial n_{AY_2}}\right)_{n_B, n_X} = RT \ln x_A + 2RT \ln x_Y = RT \ln x_A x_Y^2 \tag{10}$$

From Eqs. 4 and 10 we can write: $a_{AY_2} = x_{A^{2+}} \cdot x_{Y^-}^2 = x_A \cdot x_Y^2$

Following the same procedure and noting that for the salt BX we can write:

$$\overline{G}_{BX}^M = \left(\frac{\partial \Delta G^M}{\partial n_{BX}}\right)_{n_A, n_Y} = RT \ln a_{BX} \tag{11}$$

and $dn_{BX} = dn_B = dn_X$

and $$\left(\frac{\partial \Delta G^M}{\partial n_{BX}}\right)_{n_A, n_Y} = \left(\frac{\partial \Delta G^M}{\partial n_B}\right)_{n_A, n_Y, n_X} + \left(\frac{\partial \Delta G^M}{\partial n_X}\right)_{n_A, n_B, n_Y} \tag{12}$$

Differentiating Eq. 3 with respect to n_B at constant n_A, n_X, n_Y.

$$\left(\frac{\partial \Delta G^M}{\partial n_B}\right)_{n_B, n_X, n_Y} = RT\left[n_A \frac{d}{dn_B}\left\{\ln\left(\frac{n_A}{n_A + n_B}\right)\right\} + \ln\left(\frac{n_B}{n_A + n_B}\right) + n_B \frac{d}{dn_B}\left\{\ln\left(\frac{n_B}{n_A + n_B}\right)\right\}\right] \tag{13}$$

On differentiation of appropriate terms, we get:

$$\frac{d}{dn_B}\left[\ln\left(\frac{n_A}{n_A + n_B}\right)\right] = \frac{n_B}{n_A}\frac{1}{(n_A + n_B)}$$

$$\frac{d}{dn_B}\left[\ln\left(\frac{n_B}{n_A + n_B}\right)\right] = -\frac{1}{(n_A + n_B)}$$

Subtituting these values in Eq. 13 we get:

$$\left(\frac{\partial \Delta G^M}{\partial n_B}\right)_{n_A, n_X, n_Y} = RT\left[\ln\left(\frac{n_B}{n_A + n_B}\right) + n_A \cdot \frac{n_B}{n_A} \cdot \frac{1}{(n_A + n_B)} - n_B \cdot \frac{1}{(n_A + n_B)}\right]$$

$$= RT \ln\left(\frac{n_B}{n_A + n_B}\right) = RT \ \ln x_B \tag{14}$$

Similarly,

$$\left(\frac{\partial \Delta G^M}{\partial n_X}\right)_{n_A, n_B, n_Y} = RT \ \ln x_X \tag{15}$$

From Eqs. 12, 14 and 15, we can write:

$$\left(\frac{\partial \Delta G^M}{\partial n_{BX}}\right)_{n_A, n_Y} = RT \ \ln x_B + RT \ \ln x_X = RT \ \ln x_B x_X \tag{16}$$

Comparing Eqs. 11 and 16 we can write:

$$a_{BX} = x_B \cdot x_X$$

(b) The formation of the salt BY in equilibrium with A^{2+}, B^+, X^- and Y^- ions present in the fused salt mixture may be considered because B^+ and Y^- are monovalent. Following the above procedure, the activity of BY can be obtained in terms of their ionic fractions.

$$\overline{G}^M_{BY} = \left(\frac{\partial \Delta G^M}{\partial n_{BY}}\right)_{n_A, n_Y} = RT \ln a_{BY} \tag{17}$$

$$\text{and} \left(\frac{\partial \Delta G^M}{\partial n_{BY}}\right)_{n_A, n_Y} = \left(\frac{\partial \Delta G^M}{\partial n_B}\right)_{n_A, n_Y, n_X} + \left(\frac{\partial \Delta G^M}{\partial n_Y}\right)_{n_A, n_Y, n_X}$$

$$= RT \ \ln x_B + RT \ \ln x_Y = RT \ \ln x_B x_Y$$

$$\therefore a_{BY} = x_B \cdot x_Y$$

Further, we want to calculate the activities of AY_2, BX and BY in the salt mixture. According to Temkin we can write:

$$a_{BX} = x_B \cdot x_X$$

$$= \left(\frac{n_B}{n_A + n_B}\right)\left(\frac{n_X}{n_X + n_Y}\right)$$

$$= \left(\frac{n_{BX}}{n_{AY_2} + n_{BX}}\right)\left(\frac{n_{BX}}{n_{BX} + 2n_{AY_2}}\right) = x_{BX}\left(\frac{n_{BX}}{n_{BX} + 2n_{AY_2}}\right)$$

Dividing numerator and denominator by $(n_{AY_2} + n_{BX})$ we get:

$$a_{BX} = x_{BX} \left[\frac{\frac{n_{BX}}{n_{AY_2}+n_{BX}}}{\frac{n_{BX}}{n_{AY_2}+n_{BX}} + 2\left(\frac{n_{AY_2}}{n_{AY_2}+n_{BX}}\right)} \right] = x_{BX} \left[\frac{x_{BX}}{x_{BX} + 2x_{AY_2}} \right] = \left(\frac{x_{BX}^2}{1 + x_{AY_2}} \right) \quad (18)$$

Making use of Eq. 18, we calculate and list the value of a_{BX} for different values of x_{AY_2}:

x_{AY_2}	0	0.05	0.1	0.2	0.3	0.4	0.5	0.6	0.7	0.8	0.9	1.0
a_{BX}	1	0.860	0.736	0.533	0.377	0.257	0.167	0.10	0.053	0.022	0.005	0

Similarly, $a_{BY} = x_B \cdot x_Y$

$$= \left(\frac{n_B}{n_A + n_B} \right) \left(\frac{n_Y}{n_X + n_Y} \right)$$

$$= \left(\frac{n_{BX}}{n_{AY_2} + n_{BX}} \right) \left(\frac{2n_{AY_2}}{n_{BX} + 2n_{AY_2}} \right) = x_{BX} \left(\frac{2n_{AY_2}}{n_{BX} + 2n_{AY_2}} \right)$$

Dividing numerator and denominator by $(n_{AY_2} + n_{BX})$ we get:

$$a_{BY} = x_{BX} \left[\frac{\frac{2n_{AY_2}}{n_{AY_2}+n_{BX}}}{\frac{n_{BX}}{n_{AY_2}+n_{BX}} + 2\left(\frac{n_{AY_2}}{n_{AY_2}+n_{BX}}\right)} \right] = x_{BX} \left[\frac{2x_{AY_2}}{x_{BX} + 2x_{AY_2}} \right]$$

$$= x_{BX} \left(\frac{2x_{AY_2}}{1 + x_{AY_2}} \right) \quad (19)$$

Making use of Eq. 19 we calculate and list the value of a_{BY} for different values of x_{AY_2}:

x_{AY_2}	0	0.1	0.2	0.3	0.4	0.5	0.6	0.7	0.8	0.9	1.0
a_{BY}	0	0.164	0.267	0.323	0.343	0.333	0.300	0.247	0.178	0.095	0

and $a_{AY_2} = x_A \cdot x_Y^2$

$$= \left(\frac{n_A}{n_A + n_B} \right) \left(\frac{n_Y}{n_X + n_Y} \right)^2$$

$$= \left(\frac{n_{AY_2}}{n_{AY_2} + n_{BX}} \right) \left(\frac{2n_{AY_2}}{n_{BX} + 2n_{AY_2}} \right)^2 = x_{AY_2} \left(\frac{2n_{AY_2}}{n_{BX} + 2n_{AY_2}} \right)^2$$

Dividing numerator and denominator by $(n_{AY_2} + n_{BX})$ we get:

$$a_{AY_2} = x_{AY_2} \left[\frac{\frac{2n_{AY_2}}{n_{AY_2}+n_{BX}}}{\frac{n_{BX}}{n_{AY_2}+n_{BX}} + 2\left(\frac{n_{AY_2}}{n_{AY_2}+n_{BX}}\right)} \right]^2 = x_{AY_2} \left[\frac{2x_{AY_2}}{x_{BX} + 2x_{AY_2}} \right]^2$$

$$= \left[\frac{4x_{AY_2}^3}{(1 + x_{AY_2})^2} \right] \tag{20}$$

Making use of Eq. 20 we calculate and list the value of a_{AY_2} for different values of x_{AY_2}:

x_{AY_2}	0	0.05	0.1	0.2	0.3	0.4	0.5	0.6	0.7	0.8	0.9	1.0
a_{AY_2}	1	0.0005	0.003	0.022	0.064	0.131	0.222	0.338	0.475	0.632	0.808	1.0

Variation of activity of AY_2, BX, and BY with composition is shown in Fig. 4.12.

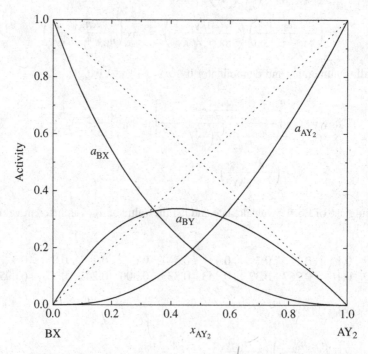

Fig. 4.12 Variation of activity of AY_2, BX and BY with composition

Problem 4.8

Calculate the activities of $CaCl_2$ and KBr in a salt mixture containing 40 mol% KBr.

Solution

Based on the above derivation, we can write expressions for the activity of $CaCl_2$, KBr, and KCl in the $CaCl_2$–KBr salt mixture as:

$$a_{CaCl_2} = \left[\frac{4x_{CaCL_2}^3}{(1 + x_{CaCl_2})^2}\right], \quad a_{KBr} = \left(\frac{x_{KBr}^2}{1 + x_{CaCl_2}}\right) \text{ and } a_{KCl} = x_{KBr}\left(\frac{2x_{CaCl_2}}{1 + x_{CaCl_2}}\right)$$

In the given problem $x_{KBr} = 0.4$ and $x_{CaCl_2} = 0.6$

$$a_{CaCl_2} = \left[\frac{4 \times (0.6)^3}{(1 + 0.6)^2}\right] = 0.3375$$

$$a_{KBr} = \left(\frac{(0.4)^2}{1 + 0.6}\right) = 0.10$$

$$a_{KCl} = 0.4\left(\frac{2 \times 0.6}{1 + 0.6}\right) = 0.30 \text{ Ans.}$$

Problem 4.9

Calculate the activity of NaCl, NaBr and KCl in a fused salt mixture: NaCl-KBr containing 40 mol% NaCl.

Solution

According to the Temkin rule the fused salt mixture contains Na^+, K^+, Cl^-, and Br^- ions:

$$a_{NaCl} = x_{Na^+}.x_{Cl^-}$$

$$= \left(\frac{n_{Na^+}}{n_{Na^+} + n_{K^+}}\right)\left(\frac{n_{Cl^-}}{n_{Cl^-} + n_{Br^-}}\right)$$

$$= \left(\frac{n_{NaCl}}{n_{NaCl} + n_{KBr}}\right)\left(\frac{n_{NaCl}}{n_{NaCl} + n_{KBr}}\right) = x_{NaCl}.x_{NaCl} = x_{NaCl}^2$$

Similarly, $a_{KBr} = x_{KBr}^2$

$$a_{NaBr} = x_{Na^+}.x_{Br^-}$$

$$= \left(\frac{n_{Na^+}}{n_{Na^+} + n_{K^+}}\right)\left(\frac{n_{Br^-}}{n_{Br^-} + n_{Cl^-}}\right)$$

$$= \left(\frac{n_{NaCl}}{n_{NaCl} + n_{KBr}}\right)\left(\frac{n_{KBr}}{n_{KBr} + n_{NaCl}}\right) = x_{NaCl}.x_{KBr}$$

Similarly, $a_{KCl} = x_{NaCl}. x_{KBr}$
In the given mixture: $a_{NaCl} = x_{NaCl}^2 = 0.4^2 = a_{KBr} = 0.16$
and $a_{NaBr} = x_{NaCl}. x_{KBr} = 0.4 \times 0.6 = a_{KCl} = 0.24$ Ans.

References

1. Pauling, L. (1960). *The nature of the chemical bonds*. Ithaca: Cornell University Press.
2. Coudurier, L., Hopkins, D. W., & Wilkomirski, I. (1978). *Fundamentals of metallurgical processes*. Oxford: Pergamon. (Chapter 6).
3. Moore, J. J. (1990). *Chemical metallurgy* (2nd ed.). Oxford: Butterworth Heinemann. (Chapter 5).
4. Mills, K. C. (1993). The influence of structure on the physicochemical properties of slags. *ISIJ International, 33*(1), 14–155.
5. Wang, L. J., Hayashi, M., Chou, K. C., & Seetharaman, S. (2012). An insight into slag structure from sulfide capacities. *Metallurgical and Materials Transactions B: Process Metallurgy and Materials Processing Science, 43B*, 1338–1343.
6. Geiger, G. H., Kozakevitch, P., Olette, M., & Riboud, P. V. (1975). Theory of BOF reaction rates. In R. D. Pehlke, W. F. Porter, P. F. Urban, & J. M. Gaines (Eds.), *BOF steelmaking* (Vol. 2, pp. 191–321). New York: Iron & Steel Society, AIME. (Chapter 5).
7. Richardson, F. D., & Fincham, C. J. B. (1954). Sulphur in silicate and aluminate slag. *Journal of Iron and Steel Institute, 178*, 4–14.
8. Richardson, F. D. (1974). *Physical chemistry of melts in metallurgy* (Vol. 1). London: Academic Press. (Chapter 2).
9. Nilsson, R., Seetharaman, S., & Jacob, K. T. (1994). A modified sulphide capacity function. *ISIJ International, 34*, 876–882.
10. Sichen, D., Nilsson, R., & Seetharaman, S. (1995). A mathematical model for estimation of sulphide capacities of multicomponent slags. *Steel Research, 11*, 458–462.
11. Nzotta, M. M., Sichen, D., & Seetharaman, S. (1998). Sulphide capacities in multi-component slag systems. *ISIJ International, 38*, 1170–1179.
12. Nzotta, M. M., Nilsson, R., Sichen, D., & Seetharaman, S. (1999). A study of the sulphide capacities of iron oxide containing slags. *Metallurgical and Materials Transactions B: Process Metallurgy and Materials Processing Science, 30B*, 909–920.
13. Nzotta, M. M., Andreasson, M., Jönsson, P., & Seetharaman, S. (2000). A study on the sulphide capacitites of steelmaking slags. *Scandinavian Journal of Metallurgy, 29*, 177–184.
14. Shankar, A., Görnerup, M., Lahiri, A. K., & Seetharaman, S. (2006). Sulphide capacity of high alumina blast furnace slags. *Metallurgical and Materials Transactions B: Process Metallurgy and Materials Processing Science, 37B*, 941–947.
15. Taniguchi, Y., Sano, N., & Seetharaman, S. (2009). Sulphide capacities of CaO-Al₂O₃-SiO₂-MgO-MnO slags in the temperature range 1673–1773 K. *ISIJ International, 49*(2), 156–163.
16. Taniguchi, Y., Wang, L. J., Sano, N., & Seetharaman, S. (2012). Sulphide capacities of CaO-Al₂O₃-SiO₂ slags in the temperature range 1673–1773K. *Metallurgical and Materials Transactions B: Process Metallurgy and Materials Processing Science, 43B*, 477–484.
17. Wang, L. J., Wang, Y. X., Chou, K. C., & Seetharaman, S. (2016). Sulfide capacities of CaO-MgO-Al₂O₃-SiO₂-CrOx slags. *Metallurgical and Materials Transactions B: Process Metallurgy and Materials Processing Science, 47B*, 2558–2563.
18. Mostaghel, S., Matsushita, T., Samuelsson, C., Björkman, B., & Seetharaman, S. (2012). Influence of alumina on physical properties of an industrial zinc–copper smelting slag: Part 2 – Apparent density, surface tension and effective thermal diffusivity. *Transactions. Institution of Mining and Metallurgy, 122*, 49–55. https://doi.org/10.1179/1743285512Y.0000000015.
19. Kang, Y., & Morita, K. (2006). Thermal conductivity of the CaO-Al₂O₃-SiO₂ systems. *ISIJ International, 46*(3), 420–426.
20. Eriksson, R., & Seetharaman, S. (2004). Thermal diffusivity measurements of some synthetic CaO-Al₂O₃-SiO₂ slags. *Metallurgical and Materials Transactions B: Process Metallurgy and Materials Processing Science 35B*, 461–469.
21. Mostaghel, S., Matsushita, T., Samuelsson, C., Björkman, B., & Seetharaman, S. (2012). Influence of alumina on physical properties of an industrial zinc–copper smelting slag: Part

1 – Viscosity. *Transactions. Institution of Mining and Metallurgy, 122*, 42–48. https://doi.org/ 10.1179/1743285512Y.0000000029.

22. Kucharski, M., Stubina, N. M., & Toguri, J. M. (1989). Viscosity measurement of molten Fe–O–SiO$_2$, Fe–O–CaO–SiO$_2$, and Fe–O–MgO–SiO$_2$ slags. *Canadian Metallurgical Quarterly, 28* (1), 7–11.

23. Seetharaman, S., Mukai, K., & Sichen, D. (2005). Viscosities of slags-an overview. *Steel Research International, 76*(4), 267–278.

24. Zhang, L., & Jahanshahi, S. (1998). Review and modelling of viscosity of silicate melts: Part I. Viscosity of binary and ternary silicates containing CaO, MgO, and MnO. *Metallurgical and Materials Transactions B: Process Metallurgy and Materials Processing Science, 29B*, 177–186.

25. Zhang, L., & Jahanshahi, S. (1998). Review and modeling of viscosity of silicate melts: Part II. Viscosity of melts containing iron oxide in the CaO–MgO–MnO–FeO–Fe$_2$O$_3$–SiO$_2$ system. *Metallurgical and Materials Transactions B: Process Metallurgy and Materials Processing Science, 29B*, 187–195.

26. Seetharaman, S., & Sichen, D. (1997). Viscosities of high temperature systems – A modelling approach. *ISIJ International, 37*(2), 109–118.

27. Shu, Q., & Zhang, J. (2006). A semi-empirical model for viscosity estimation of molten slags in CaO–FeO–MgO–MnO–SiO$_2$ systems. *ISIJ International, 46*(11), 1548–1553.

28. Nakamoto, M., Lee, J., & Tanaka, T. (2005). A model for estimation of viscosity of molten silicate slag. *ISIJ International, 45*(5), 651–656.

29. Kondratiev, A., Hayes, P. C., & Jak, E. (2006). Development of a quasi-chemical viscosity model for fully liquid slags in the Al$_2$O$_3$–CaO– 'FeO'–MgO–SiO$_2$ system. Part 1. Description of the model and its application to the MgO, MgO–SiO$_2$, Al$_2$O$_3$–MgO and CaO–MgO sub-systems. *ISIJ International, 46*(3), 359–367.

30. Kondratiev, A., Hayes, P. C., & Jak, E. (2006). Development of a quasi-chemical viscosity model for fully liquid slags in the Al$_2$O$_3$–CaO– 'FeO'–MgO–SiO$_2$ system. Part 2. A review of the experimental data and the model predictions for the Al$_2$O$_3$–CaO–MgO, CaO–MgO–SiO$_2$ and Al$_2$O$_3$–MgO–SiO$_2$ systems. *ISIJ International, 46*(3), 368–374.

31. Kondratiev, A., Hayes, P. C., & Jak, E. (2006). Development of a quasi-chemical viscosity model for fully liquid slags in the Al$_2$O$_3$–CaO– 'FeO'–MgO–SiO$_2$ system. Part 3. Summary of the model predictions for the Al$_2$O$_3$–CaO–MgO– SiO$_2$ system and its subsystems. *ISIJ International, 46*(3), 375–384.

32. Kondratiev, A., Hayes, P. C., & Jak, E. (2008). Development of a quasi-chemical viscosity model for fully liquid slags in the Al$_2$O$_3$–CaO– 'FeO'–MgO–SiO$_2$ system. The experimental data for the 'FeO'–MgO–SiO2, CaO–'FeO'–MgO– SiO$_2$ and Al$_2$O$_3$–CaO–'FeO'–MgO– SiO$_2$ systems at iron saturation. *ISIJ International, 48*(1), 7–16.

33. Kowalczyk, J., Mroz, W., Warczok, A., & Utigard, T. A. (1995). Viscosity of copper slags from chalcocite concentrate smelting. *Metallurgical and Materials Transactions B: Process Metallurgy and Materials Processing Science, 26B*, 1217–1223.

34. Park, H.-S., Park, S. S., & Sohn, I. (2011). The viscosity behavior of FeOt– Al$_2$O$_3$–SiO$_2$ copper smelting slags. *Metallurgical and Materials Transactions B: Process Metallurgy and Materials Processing Science, 42B*, 692–699.

35. Vidacak, B., Sichen, D., & Seetharaman, S. (2001). An experimental study of the viscosities of Al$_2$O$_3$–CaO-"FeO" slags. *Metallurgical and Materials Transactions B: Process Metallurgy and Materials Processing Science, 32B*, 679–684.

36. Jakobsson, A., Nasu, M., Mangwiru, J., Mills, K. C., & Seetharaman, S. (1998). Interfacial tension effects on slag/metal reactions. *Philosophical Transactions of the Royal Society of London Series A, 356*, 995–1001.

37. Seetharaman, S., Teng, L., Hayashi, M., & Wang, L. (2013). Understanding properties of slags. *ISIJ International, 53*(1), 1–8.

38. Shamsuddin, M., & Sohn, H. Y. (2020). Constitutive topics in physical chemistry of ironmaking and steelmaking: A review. *Trends in Physical Chemistry, 19*, 33–50.

39. Vishkarev, A. F., Dragomir, J., Din-Fen, U., & Yavoiski, V. J. (1965) Role of oxygen on surface tension of Fe$_x$O. In *Surface phenomena in melts, Symposium, Nalchik, Sept, 1964*, Published by Kabardino Balkarskoe Istatelstvo, Nalchik, pp. 327–332.

40. Towers, H., & Chipman, J. (1957). Diffusion of calcium and silicon in a lime-alumuna-silica slag. *Transactions of the Metallurgical Society, AIME, 209*, 769–773.

41. Rosenqvist, T. (1974). *Principles of extractive metallurgy*. New York: McGraw-Hill. (Chapter 11).

42. Sohn, H. Y. (2014). Principles of copper production. In *Treatise on process metallurgy. Vol 3, Industrial processes, Part A* (pp. 534–591). Oxford and Waltham: Elsevier. (Section 2.1.1).

43. Kojo, I. V., & Storch, H. (2006). Copper production with Outokumpu flash smelting an update. In F. Kongoli & R. G. Reddy (Eds.), *International Symposium on sulfide smelting. Vol. 8, Sohn international symposium* (pp. 225–238). Warrendale: TMS.

44. Goto, M., Oshima, I., & Hayashi, M. (1998). Control aspects in the Mitsubishi continuous process. *JOM, 50*(4), 60–65.

45. Yazawa, A., Takeda, Y., & Waseda, Y. (1981). Thermodynamic properties and structure of ferrite slags and their process implications. *Canadian Metallurgical Quarterly, 20*(2), 129–134.

46. Hughes, S., Reuter, M. A., Baxter, R., & Kaye, A. (2008). *Ausmelt technology for lead and zinc processing*. In Lead and Zinc 2008, Johannesburg, The Southern African Institute of Mining and Metallurgy, pp. 147–161.

47. Turkdogan, E. T. (1975). Physical chemistry of oxygen steelmaking thermchemistry and thermodynamics. In R. D. Pehlke, W. F. Porter, P. F. Urban, & J. M. Gaines (Eds.), *BOF steelmaking* (Vol. 2, pp. 1–190). New York: Iron & Steel Society, AIME. (Chapter 4).

48. Turkdogan, E. T., & Pearson, J. (1953). Activity of constituents of iron and steelmaking slags, Part 1, iron oxide. *Journal of Iron and Steel Institute, 173*, 217–223.

49. Elliott, J. F. (1955). Activities in the iron oxide-silica-lime system. *Transactions of the American Institute of Mining and Metallurgical Engineers, 203*, 485–488.

50. Herashymenko, P. (1938). Electrochemical theory of slag-metal equilibria. *Transactions of the Faraday Society, 34*(2), 1245–1257.

51. Herashymenko, P. & Speight, G. E. (1950). Ionic theory of slag-metal equilibria, Part I – Derivation of fundamental relationships. *Journal of Iron and Steel Institute, 166*, 169–183. Part II – Application to the basic open hearth process. *Journal of Iron and Steel Institute, 166*, 269–303.

52. Temkin, M. (1945). Mixtures of fused salts as ionic solutions. *Acta Physicochimica URSS, 20*, 411–420.

53. Flood, H., & Grjotheim, K. (1952). Thermodynamic calculation of slag equilibria. *Journal of Iron and Steel Institute, 171*, 64–70.

54. Flood, H., Forland, T., & Grjotheim, K. (1953). *Physical chemistry of melts* (pp. 46–59). London: Institute of Mining and Metallurgy.

55. Masson, C. R. (1965). An approach to the problem of ionic distribution in liquid silicates. *Proceedings. Royal Society of London, A287*, 201–221.

56. Whiteway, S. G., Smith, I. B., & Masson, C. R. (1970). Activities and ionic distribution in liquid silicates: application of polymer theory. *Canadian Journal of Chemistry, 48*, 1456–1465.

57. Schenck, H. (1945). *Physico-Chemistry of steelmaking* (p. 455). London: BISRA Translation.

58. Winkler, T. B., & Chipman, J. (1946). An equilibrium study of distribution of phosphorus between liquid iron and basic slags. *Transactions of the American Institute of Mining and Metallurgical Engineers, 167*, 111–133.

Chapter 5
Reduction of Oxides and Reduction Smelting

A number of metals are produced from oxide minerals. Iron, manganese, chromium, and tin are exclusively obtained from oxide ores. Rich ores, for example, hematite, pyrolusite, chromite, and so on can be directly reduced whereas lean ores are first concentrated by communication, classification, jigging, and tabling. Sulfide ores, such as galena and sphalerite concentrates, are first roasted to oxides for subsequent reduction to metals. Carbonate minerals, such as dolomite, are calcined prior to reduction into magnesium. However, oxides of noble metals like Ag_2O, PtO, and PdO are decomposed to Ag, Pt, and Pd at temperatures of 200 °C, 500 °C, and 900 °C, respectively.

All the oxides obtained either by concentration of oxide ores/minerals or by roasting of sulfide concentrates are reduced by means of a suitable reducing agent. Physical characteristics like melting and boiling points, vapor pressure, and so on of the metal and relative stability of its oxide as compared to the oxides of impurity metals and gangue mineral oxides are major factors in the selection of the reducing agents. In general, carbon, carbon monoxide, and hydrogen are the reducing agents [1] of commercial and industrial importance. In special cases, as in the production of very reactive metals such as uranium, thorium, beryllium, zirconium, titanium, and so on, metals having high affinity for oxygen, such as magnesium or calcium may be employed as reducing agents.

5.1 Reduction Methods

The method of reduction depends on the nature of the ore and the metal to be produced. Various techniques may be summarized as follows:

Blast Furnace Reduction The lumpy ore/concentrate requiring a high reduction potential is smelted in a blast furnace with coal/coke by blowing preheated air through tuyeres located at the base of the shaft or stack of the furnace. Combustion

M. Shamsuddin, *Physical Chemistry of Metallurgical Processes, Second Edition*,
The Minerals, Metals & Materials Series, https://doi.org/10.1007/978-3-030-58069-8_5

produces highly reducing CO gas. Thus coal/coke acts as an indirect reducing agent ($MO + CO = M + CO_2$). The direct reduction between coke and the metal oxide takes place in the lower part of the stack. The reducibility of metal oxides in blast furnace smelting is decided by CO/CO_2 ratio in the system. Coke not only provides the necessary heat required for reduction of oxides it also supports the entire burden of the stack of the blast furnace due to its high strength even at high temperature. The high melting oxides are smelted in blast furnaces of circular cross section fitted with tuyeres that are evenly spaced around the base of the stack. The charge comprising of ore lump, coke, and flux fed from the top of the furnace gets dried while descending from the stack and is subsequently reduced by the uprising reducing gas. A blast furnace of rectangular cross section with tuyeres located only along the two long sides is employed for production of lower boiling point metals like lead and zinc. This arrangement generates less extensive heat at the tuyere and decreases the metal loss through volatilization.

The most notable example is the production of pig iron by reduction of hematite with coke. Lead and zinc are also produced from the roasted calcine of galena and sphalerite by reduction in blast furnaces. Cuprite and calcined azurite and malachite can be reduced in rectangular blast furnaces, but generally, oxide ores of copper are treated by hydrometallurgical methods.

Electric Arc Furnace Smelting Ferroalloys are manufactured by smelting of ores in electric arc furnaces in the presence of carbon, for example, ferromanganese from pyrolusite (MnO_2), ferrosilicon from silica sand (SiO_2), and ferrochrome from chromite ($FeO.Cr_2O_3$).

Reverberatory Smelting Tin is produced by smelting cassiterite concentrate (SnO_2) in a reverberatory smelting furnace in the presence of carbon.

Reduction in Retort In Pidgeon process, calcined dolomite ($CaO.MgO$) is reduced with ferrosilicon at 1200 °C under a reduced pressure of 10^{-4} mm Hg and magnesium vapor is collected at the cooler end of the retort. In the old and obsolete retort process, after dead roasting of sphalerite, zinc oxide was reduced with carbon at 1100 °C.

Metallothermic Reduction Ferroniobium is manufactured by reduction of Nb-rich concentrate [mineral: niobite-tantalite $(FeMn)(NbTa)_5O_6$] with aluminum in a thermit process in which some preheat is necessary. Welding and repair of railway track by thermit process, in which high-grade hematite is reduced with aluminum, is well known. The Kroll [2] process is very popular for production of titanium sponge by reduction of vapors of $TiCl_4$ with liquid magnesium. Uranium, thorium, zirconium, and beryllium are obtained by reduction of their respective fluorides by magnesium and/or calcium.

Gaseous Reduction Hydrogen and water gas have been used successfully on commercial scale in the production of metals. The reduction of iron oxide with hydrogen [$FeO(s) + H_2(g) = FeO(s) + H_2O(g)$] becomes more favorable with increase of temperature. A convenient method is to use a mixture of hydrogen and carbon monoxide (H_2 + CO) known as "water gas" which may be produced by the reaction:

$$H_2O(\text{steam}) + C(s) = H_2(g) + CO(g) \tag{5.1}$$

In the Wiberg–Soderfors process [3], sponge iron is produced in a shaft kiln at 1100 °C by passing water gas (H_2 + CO) over iron ore or concentrate. As there is no liquid slag formation at this temperature, the impurities are not separated. The resultant solid product containing the gangue oxides named "sponge iron" is subsequently melted in an electric furnace for steelmaking.

In the carbonyl process, crude nickel is allowed to react with carbon monoxide gas at 100–200 atm pressure to form gaseous nickel carbonyl [$Ni(CO)_4$] that decomposes into pure nickel at atmospheric pressure. The reaction is represented as:

$$Ni\,(\text{crude}-s) + 4\,CO\,(g) \rightleftarrows Ni(CO)_4\,(g) \tag{5.2}$$

On the industrial scale, tungsten and molybdenum are produced by reduction of their oxides with hydrogen gas according to the reactions:

$$WO_3(s) + 3H_2(g) = W(s) + 3H_2O(g) \tag{5.3}$$
$$MoO_3(s) + 3H_2(g) = Mo(s) + 3H_2O(g) \tag{5.4}$$

Since there are equal number of gaseous molecules on both sides of the chemical equations, the rate of reduction is independent of the total pressure of the system. Although the standard free energy changes for these reactions are positive, the efficiency of reduction improves with increase of temperature as well as by maintaining large excess of hydrogen gas in the system.

Electrothermic Reduction   Stable oxides like Al_2O_3 and MgO can be reduced with carbon at temperatures above 2000 °C and 1840 °C, respectively. The smelting processes at such high temperatures for production of aluminum and magnesium would be uneconomical due to the high cost of fuel and the difficulty in procuring suitable refractory containers serving for number of heats. Further, the metal produced will be highly reactive and hence may pick up oxygen from refractories and from oxides of carbon. Aluminum will react with carbon to form carbide. This would result in a product mixture of Al–Al_2O_3–Al_4C_3. In order to overcome these difficulties, aluminum is extracted by electrolysis of Al_2O_3 dissolved in fused mixture of fluorides of NaF and AlF_3 and magnesium by electrolysis of $MgCl_2$ dissolved in fused chlorides, instead of the conventional carbon reduction methods.

5.2 Thermodynamics of Reduction of Oxides

Since the knowledge of thermodynamics and reaction kinetics is essential in the development of extraction processes for production of different metals, physico-chemical principles involved in selection of the technique and the reducing agent will be discussed in this section. The principles underlying metallothermic reduction will be explained with the aid of Ellingham diagram and the role of CO/CO_2 and H_2/H_2O ratio in gaseous reduction of oxides will be discussed separately.

5.2.1 Metallothermic Reduction

Metal, M can be obtained by reducing the metal oxide, MO_2 with a reducing agent R according to the reaction:

$$MO_2(s,l) + R(s,l) = M(s,l) + RO_2(s) \qquad (5.5)$$

For the above reaction to proceed in the forward direction, RO_2 must be more stable than MO_2, that is, the free energy of formation of RO_2 should be more negative as compared to that of MO_2. The change in free energy of the reaction (5.5) should be negative enough to get a large value of the equilibrium constant that is related to the free energy change, ΔG^o by the expression: $\Delta G^o = -RT \ln K$. A large equilibrium constant will generate a high proportion of M and RO_2 when the reaction reaches equilibrium. For example, the high negative value of ΔG^o for the reduction of Fe_2O_3 by aluminum [Fe_2O_3 (s) + Al (l) = 2 Fe (l) + Al_2O_3 (s)] at 1200 °C produces Fe and Al_2O_3. This forms the basis of the aluminothermic reduction of hematite ore, which has been used extensively for the repair of railway tracks. The reaction is initiated by igniting a magnesium ribbon embedded in the preheated charge of hematite and aluminum contained in a steel-capped vessel. Since the reaction is highly exothermic the heat generated is sufficient to produce molten iron. The technique popularly known as thermite or aluminothermic process has been extensively used to produce manganese, chromium, and ferroalloys such as ferroniobium and ferrovanadium.

In order to discuss the thermodynamics of reaction (5.5), the following three reactions may be considered:

$$MO_2(s,l) = M(s,l) + O_2(g) \qquad (5.6)$$

$$R(s,l) + O_2(g) = RO_2(s) \qquad (5.7)$$

$$M(s,l) + O_2(g) = MO_2(s) \qquad (5.8)$$

Making use of the first law of thermodynamics, ΔG^o for reaction (5.5) can be obtained by the sum of Eq. 5.6 and 5.7 or by the difference between ΔG^o for reactions (5.7) and (5.8).

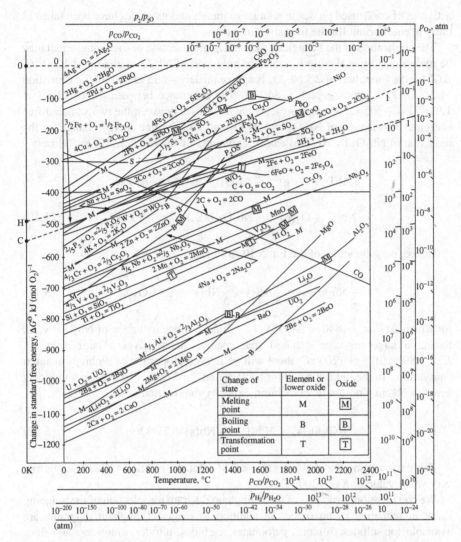

Fig. 5.1 Ellingham diagram for oxides. (From Chemical Metallurgy by J. J. Moore [5], © 1990, p 85, Butterworth Heinemann Ltd. Reproduced with the permission of Butterworth Heinemann Ltd)

The standard free energy changes in the formation of oxides (i.e., the relative stability of oxides or relative affinity of different metals for oxygen) may be represented graphically as a linear function of temperature. This idea was first conceived by Ellingham [4]. The ΔG^o vs T plots published by him in 1944 have been popularized as Ellingham diagrams (Fig. 5.1). In the Ellingham diagram, the standard free energy of formation of each oxide accounts for one mole of oxygen with the clear understanding that the distance between two curves at any temperature gives directly the free energy change accompanying the reduction of the oxide represented by the upper of the two curves by the metal whose oxide formation is represented by the lower curve at the chosen temperature. In these diagrams, activities of condensed phases in relation to metals and its oxides have been taken as unity at equilibrium for the formation of oxides.

The application of the Ellingham diagram in the selection of a reducing agent may be understood by analyzing Fig. 5.1 for reduction of Nb_2O_5 with Al. At 1200 °C, ΔG^o for the formation of 2/5 Nb_2O_5 is approximately -535 kJ and for the formation of $2/3Al_2O_3$ is approximately -840 kJ. The difference between the two lines at 1200 °C is -305 kJ, which represents the difference between the two standard free energies of formation of Nb_2O_5 and Al_2O_3 and also the free energy change for the reduction of Nb_2O_5 by Al. This can be explained by the reactions [6] as follows:

$$\frac{4}{3}Al(l) + O_2(g) = \frac{2}{3}Al_2O_3(s), \Delta G^o = -840 \text{ kJ} \tag{5.9}$$

$$\frac{4}{5}Nb(s) + O_2(g) = \frac{2}{5}Nb_2O_5 (s), \Delta G^o = -535 \text{ kJ} \tag{5.10}$$

Hence, on subtraction of the Eq. 5.10 from Eq. 5.9 we get:

$$\frac{2}{5}Nb_2O_5(s) + \frac{4}{3}Al(l) = \frac{4}{5}Nb(s) + \frac{2}{3}Al_2O_3(s) \tag{5.11}$$

for which $\Delta G^o = -840 - (-535) = -305$ kJ. Since reduction of Nb_2O_5 by Al, having a large negative standard free energy change gives a large value of $K = 1.89 \times 10^{13}$ at 1200 °C, there will be a large proportion of niobium metal at equilibrium. In the light of a similar background, lead cannot be used as a reducing agent to obtain niobium from niobium oxide by the reaction:

$$\frac{2}{5}Nb_2O_5(s) + 2Pb(l) = \frac{4}{5}Nb(s) + 2PbO(s) \tag{5.12}$$

The reason being the positive value of ΔG^o for the reaction (5.12) leading to a very small value of K at 1200 °C.

We can therefore make use of the Ellingham diagram for selection of the reducing agent in reduction of an oxide. In addition to oxides, Ellingham diagrams are available for sulfides, sulfates, carbonates, carbides, nitrides, chlorides, and fluorides. The shape and slope of the curves, which are of special significance, are governed by the entropy change accompanying each reaction. The slope of the ΔG^o vs T curves is the standard entropy change of the reaction, expressed as $\frac{\partial \Delta G^o}{\partial T} = -\Delta S^o$. ΔS^o and ΔH^o are temperature-dependent in the Gibbs–Helmholtz equation:

$$\Delta G^o = \Delta H^o - T\Delta S^o \tag{5.13}$$

From Fig. 5.1, it is evident that as long as the reactants and products remain in the same physical state, the variation of ΔS^o and ΔH^o with T is small and curves are virtually straight lines. At low temperature, a majority of reactions show this trend.

Most of the lines have roughly the same positive slope due to the increase in state of order of the system on the loss of 1 mol of oxygen gas. Hence, ΔS^o for most of reactions is approximately the same negative value. The slope of the line, ΔS^o will become more positive; that is, ΔS^o will be smaller at the melting and boiling points of the metal marked as M and B, respectively, as shown in Fig. 5.1. ΔS^o will be larger and hence the slope more negative when oxide changes phase. For example, while considering the line for oxidation of zinc:

$$2\,Zn(l) + O_2(g) = 2\,ZnO(s) \tag{5.14}$$

one finds a slight increase in the slope of the line at the melting point and a large increase at the boiling point of zinc. However, the two lines representing the formation of oxides of carbon demonstrate important exceptions to the general tendency of having positive slopes in Fig. 5.1.

$$C(s) + O_2(g) = CO_2(g); \Delta H^o_{298} = -397 \text{ kJ} \tag{5.15}$$

$$2C(s) + O_2(g) = 2CO(g); \Delta H^o_{298} = -222 \text{ kJ} \tag{5.16}$$

As one mole of gas appears on either side of Eq. 5.15, there is only a small change in entropy accompanying the reaction and hence ΔS^o for reaction (5.15) is $+0.8 \text{ J K}^{-1}$ and the line in Fig. 5.1 is almost horizontal. On the other hand, increase in the number of gas molecules accompanying reaction (5.16), ΔS^o is about $+170 \text{ J K}^{-1}$ and the corresponding line for the reaction has a pronounced negative slope $(-\Delta S^o)$. The significance difference in slope of the two lines representing the formation of CO_2 and CO is of great importance in metal extraction. The two lines intercepting at 720 °C indicate that CO is more stable as compared to CO_2 at above 720 °C. Thus, it is noted from Fig. 5.1 that CO becomes more and more stable as compared to metal oxides with increase of temperature, and therefore, carbon will reduce more and more metal oxides at higher temperature. The figure indicates that standard free change is negative for the reduction of stable oxides such as Al_2O_3, MgO, and TiO_2 with carbon above 2000 °C, 1840 °C, and 1730 °C, respectively. But smelting processes above such high temperatures would be highly expensive because of fuel requirements and the difficulty of procuring refractory containers that have to serve for a sufficiently longer time. Further, the metal produced would be highly reactive to pick up oxygen from oxide refractories and oxides of carbon. It may also react with carbon to form carbides like TiC and Al_4C_3. On account of these difficulties, aluminum is extracted by electrolysis of Al_2O_3 dissolved in fused cryolite, magnesium by electrolysis of $MgCl_2$ dissolved in fused chlorides, and titanium by reduction of $TiCl_4$ vapors by liquid magnesium, rather than by the conventional carbon reduction.

However, Ellingham diagram must be referred with due care and proper understanding because it has been constructed for reactions taking place under standard conditions, that is, unit activities of reactants or products. Further, it does not account for the kinetics of the reaction. The problem of deviation from unit activity is illustrated by the reduction of MgO with silicon (Pidgeon Process). According to Fig. 5.1, a positive free energy change ($\Delta G^o = +280$ kJ at 1200 °C) for the reaction:

$$2MgO(s) + Si(s) = 2\,Mg(g) + SiO_2(s) \qquad (5.17)$$

suggests that there is no chance of using silicon as a reducing agent to produce magnesium from magnesia. However, according to the van't Hoff isotherm, the actual free energy change accompanying reaction (5.17) is given as:

$$\Delta G = \Delta G^o + RT \ln K = \Delta G^o + RT \ \ln \left(\frac{p_{Mg}^2 . a_{SiO_2}}{a_{MgO}^2 . a_{Si}} \right) \qquad (5.18)$$

On lowering down p_{Mg} and a_{SiO_2} sufficiently, ΔG becomes negative, even if ΔG^o is positive. The Pidgeon process for the commercial production of magnesium from calcined dolomite (CaO.MgO) lowers p_{Mg} to about 10^{-4} atm. The activity of silica in the orthosilicate slag ($2CaO.SiO_2$) is automatically lowered to less than 0.001. The strong attraction of CaO for SiO_2 reduces the possibility of loss of MgO as magnesium silicate. In this process calcined dolomite (CaO.MgO) obtained by decomposition of dolomite, a mixed carbonate of magnesium and calcium ($CaCO_3.MgCO_3$):

$$CaCO_3.MgCO_3 \overset{800°C}{\rightarrow} CaO.MgO + 2CO_2 \qquad (5.19)$$

is reduced with silicon in a retort according to the reaction:

$$2(CaO.MgO)(s) + Si(s) = 2Mg(g) + 2CaO.SiO_2(l) \qquad (5.20)$$

Dolomite is cheaper as compared to magnesite ($MgCO_3$) and occurs more abundantly. In the Pidgeon process, ferrosilicon is used instead of pure silicon. Formation and dissolution of FeO in the ternary oxide melt: $CaO.FeO.SiO_2$ decreases the liquidus temperature of the slag. Magnesium vapors evolved are condensed in massive form without reoxidation. This may be compared with the formidable task of reduction of MgO with carbon at 1900 °C.

$$MgO(s) + C(s) \overset{1900°C}{\rightarrow} Mg(g) + CO(g) \qquad (5.21)$$

CO evolved along with magnesium vapor reoxidizes the magnesium on cooling. This requires shock cooling of magnesium vapors with cold hydrogen in order to prevent the formation of finely divided pyrophoric magnesium powder, which is difficult to handle.

From the above example, it is clear that when activities of reactants and products differ significantly from unity, the van't Hoff isotherm provides a better understanding of the thermodynamic feasibility of the reaction as compared to the standard free energy change and the Ellingham diagram. This has been demonstrated with the aid

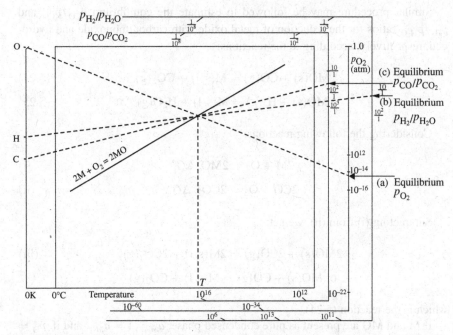

Fig. 5.2 Nomographic scales for the estimation of equilibrium p_{O_2}, equilibrium p_{H_2}/p_{H_2O}, and equilibrium p_{CO}/p_{CO_2} ratio for reactions: (a) $2\,M + O_2 = 2MO$, (b) $MO + H_2 = M + H_2O$, and (c) $MO + CO = M + CO_2$, respectively. (From Chemical Metallurgy by J. J. Moore [5], © 1990, p 40, Butterworth Heinemann Ltd. Reproduced with the permission of Butterworth Heinemann Ltd.)

of a problem (No. 5.10). The use of the diagram was further extended by Richardson and co-workers [7–9] who added nomographic scales shown in Fig. 5.2, which allows gas composition to be read off for different reactions. For example, consider oxidation of the metal, M into its oxide, MO:

$$2M(s,l) + O_2(g) = 2MO(s) \qquad (5.22)$$

If M and MO are pure (i.e., there is no mutual solubility of one into another, $a_M = 1 = a_{MO}$) the equilibrium partial pressure of oxygen, p_{O_2} can be calculated from the known value of ΔG^o:

$$\Delta G^o = -RT \ln\left(\frac{1}{p_{O_2}}\right) = 2.303\ RT\ \log\left(p_{O_2}\right) \qquad (5.23)$$

Log p_{O_2} (hence p_{O_2}) can be estimated by constructing a nomographic scale around ΔG^o vs T diagram. Since $\Delta G^o = 0$, when $p_{O_2} = 1$ atm (Eq. 5.23), the equilibrium p_{O_2} values radiate from the point 'O' on the ΔG^o axis because $\Delta G^o = 0$ at 0 K. In order to estimate the value of equilibrium p_{O_2} at a particular temperature, T marked on the ΔG^o vs T line corresponding to Eq. 5.22 is extended to intersect the nomograph (Fig. 5.2). Antilog of the point of intersection reading as $\log p_{O_2}$ will give p_{O_2}.

Similar procedure may be followed to estimate the equilibrium p_{CO}/p_{CO_2} and p_{H_2}/p_{H_2O} ratios for the reduction of metal oxide with carbon monoxide and hydrogen, respectively, according to the reactions:

$$MO(s) + CO(g) = M(s,l) + CO_2(g) \tag{5.24}$$

$$MO(s) + H_2(g) = M(s,l) + H_2O(g) \tag{5.25}$$

Considering the following reactions:

$$2M + O_2 = 2MO, \Delta G_i^o \tag{i}$$

$$2CO + O_2 = 2CO_2, \Delta G_{ii}^o \tag{ii}$$

Subtracting (i) from (ii), we get:

$$2MO(s) + 2CO(g) = 2M(s,l) + 2CO_2(g) \tag{iii}$$

$$\text{or } MO(s) + CO(g) = M(s,l) + CO_2(g)$$

which is the reaction (5.24).

If M and MO are present as pure condensed phase, $a_M = 1 = a_{MO}$, and if $p_{O_2} = 1$ atm, the equilibrium p_{CO}/p_{CO_2} ratio for reaction (iii) would be same as that for reaction (ii) because both the reactions are in equilibrium in the presence of O_2 at one atm. Thus for the reaction (iii):

$$\Delta G_{iii}^o = -RT \ln \left(\frac{p_{CO_2} \cdot a_M}{p_{CO} \cdot a_{MO}} \right)^2 = \Delta G_{ii}^o - \Delta G_i^o$$

$$= 4.606 \, RT \, \log \left(\frac{p_{CO}}{p_{CO_2}} \right) \tag{5.26}$$

If $p_{CO}/p_{CO_2} = 1, \Delta G_{ii}^o = 0$,   the   line   corresponding   to   the   reaction: $2CO + O_2 = 2CO_2$ can be extrapolated to the point "C" on the ΔG^o axis at 0 K. The equilibrium p_{CO}/p_{CO_2} ratio can be estimated for any reaction concerning reduction of pure condensed metal oxide with carbon monoxide by constructing another nomographic scale around ΔG^o vs T diagram. The equilibrium p_{CO}/p_{CO_2} ratio at the desired temperature, T marked on the ΔG^o vs T line corresponding to Eq. 5.24 is extended to intersect the nomograph (Fig. 5.2). Antilog of the point of intersection reading as $\log p_{CO}/p_{CO_2}$ will give p_{CO}/p_{CO_2}. By following a similar procedure third nomographic scale may be constructed to estimate the equilibrium p_{H_2}/p_{H_2O} ratio from the line radiating from the point "H" on the ΔG^o axis at 0 K.

5.2.2 Thermal Decomposition

The free energy of formation of oxides is invariably negative within the temperature range (0–2200 °C) considered in the Ellingham diagram (Fig. 5.1). Hence, decomposition of the oxide is not feasible. The oxide will spontaneously decompose at temperature above which the oxide formation line crosses the horizontal line, where $\Delta G^o = 0$. At the point of intersection, that is, where $\Delta G^o = 0$, there is equal possibility for the formation of oxide from the metal as well as for the decomposition of the oxide into metal. The lines corresponding to the following reactions:

$$4Ag(s) + O_2(g) = 2Ag_2O(s) \tag{5.27}$$

$$2Pt(s) + O_2(g) = 2PtO(s) \tag{5.28}$$

$$2Pd(s) + O_2(g) = 2PdO(s) \tag{5.29}$$

$$2Ni(s) + O_2(g) = 2NiO(s) \tag{5.30}$$

intersect $\Delta G^o = 0$ line at 200 °C, 500 °C, 900 °C, and 2400 °C, respectively. The reverse reaction, that is, decomposition is favored above each of these temperatures. The decomposition temperature is defined as the temperature at which the oxide formation line crosses the $\Delta G^o = 0$ line. This method of obtaining metals is called thermal decomposition.

Generally, oxides are reduced by means of a reducing agent, for example, C, CO, and H_2. These reducing agents are of great industrial importance and may be produced from raw materials like coal, oil, and natural gas. A large number of metals are produced from their oxides by reduction with C and/or CO, for example, iron, manganese, chromium, and tin exclusively from oxides; lead and zinc from oxides obtained by roasting of their sulfide minerals. The more noble metals like silver, mercury, and palladium are obtained by thermal decomposition of Ag_2O, HgO, and PdO, respectively. The more reactive metals such as uranium and thorium are produced by metallothermic reduction of their oxides. Metals like calcium and magnesium having higher affinity for oxygen are employed as reducing agents.

5.2.3 Reduction with Carbon Monoxide in the Presence and Absence of Carbon

The reduction reactions with carbon and carbon monoxide are controlled by the chemical kinetics and the related equilibria. For the reaction:

$$MO(s) + CO(g) = M(s, l, g) + CO_2(g), K = p_{CO_2}/p_{CO} \tag{5.24}$$

to be feasible, the standard free energy change should be negative. If M and MO are not mutually soluble, $a_M = 1 = a_{MO}$, the equilibrium constant, K is expressed in terms of the gas ratio: p_{CO_2}/p_{CO}. The corresponding gas ratio is shown in Fig. 5.3 for a number of metals. The gas ratio is a function of temperature only. From the figure, it is noted that the ratio varies from about 10^5 for the reduction of Cu_2O to Cu and Fe_2O_3 to Fe_3O_4 to 10^{-4} or less for the reduction of MnO and SiO_2. Even lower ratios are required for the reduction of Al_2O_3 and MgO.

The figure also shows the gas ratio, p_{CO_2}/p_{CO} for the Boudouard reaction:

$$C(s) + CO_2(g) = 2CO(g), K = p_{CO}^2/p_{CO_2} \qquad (5.31)$$

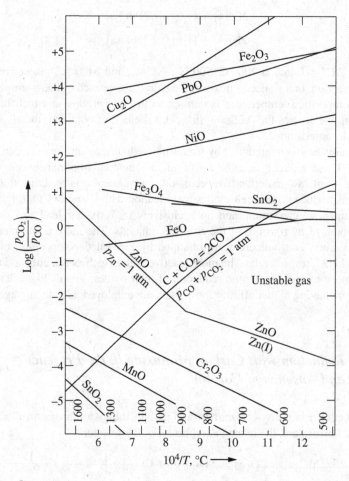

Fig. 5.3 $\log \frac{p_{CO_2}}{p_{CO}}$ vs. $\frac{1}{T}$ for the reduction of various oxides. (From Principles of Extractive Metallurgy by T. Rosenqvist [1], © 1974, p 266, McGraw-Hill Inc. Reproduced with the permission of McGraw-Hill Book Co.)

Here, the ratio, $\frac{p_{CO_2}}{p_{CO}} = \frac{p_{CO}}{K}$, is a function of p_{CO} and of the total pressure $(p_{CO_2} + p_{CO})$ due to the increase in the number of gaseous molecules in the reaction (5.31). The Boudouard reaction (5.31) may be disregarded if reduction is taking place in the absence of solid carbon. Under such a situation, reduction of the oxide will take place when the $\frac{p_{CO_2}}{p_{CO}}$ ratio [1] in the gaseous mixture is less than the equilibrium value for the concerned metal–metal oxide. Thus, a gas ratio between 10^5 and 10^2, that is, at very small concentration of CO in the gaseous mixture will reduce Cu_2O, PbO, and NiO. In practical sense it means that if reduction is initiated with pure CO, the entire amount of CO will be converted into CO_2. On the other hand, a gas almost free of CO_2 (i.e., $\frac{p_{CO_2}}{p_{CO}} = 10^{-5}$) is required for the reduction of MnO and SnO_2. Hence, the reaction will stop as soon as minute amounts of CO_2 are formed in the pure CO that is used initially. Thus, reduction of Cr_2O_3, MnO, and SiO_2 with CO is practically impossible.

In the presence of solid carbon in the gaseous mixture, the two reactions (5.24) and (5.31) take place simultaneously. At equilibrium, simultaneous reactions between MO, M, and C will take place at the temperature where the curves for the two reactions intersect. From Fig. 5.3, it is clear that at about 610 °C, SnO_2, Sn, and C may be in equilibrium with a gaseous mixture of $p_{CO_2} + p_{CO}$ at 1 atm. This means SnO_2 may be reduced by carbon at above 610 °C, when $p_{CO_2}/p_{CO} = 3$. Similarly, Fe_3O_4 may be reduced to FeO at above 650 °C and FeO to Fe above 700 °C. MnO and SiO_2 will be reduced with carbon at a total pressure of one atmosphere (i.e., $p_{CO_2} + p_{CO} = 1$ atm) in the temperature range of 1400–1600 °C. A gas ratio, $p_{CO_2}/p_{CO} = 10^{-4}$ is necessary for the reduction of MnO at 1500 °C. For reactions taking place above the equilibrium temperature, the gas mixture has a value intermediate between the values for reactions (5.24) and (5.31) and very close to the value for the reaction that has the highest reaction rate. Thus, it is wrong to say that the reduction with carbon is represented by reaction: MO + C = M + CO. This is applicable in case of reduction of the most stable oxides, which essentially requires a very low concentration of CO_2 in the gaseous mixture. From Fig. 5.3 it is evident that even at low temperatures carbon curve does not intersect the curves of NiO and other noble metal oxides. The enthalpy of reactions may be calculated from slope of the curves in Fig. 5.3 by van't Hoff equation. The reduction of less noble metal oxides with CO is endothermic while it is exothermic for the reduction of relatively noble metals. As the reaction between C and CO_2 (5.31) is highly endothermic, the reduction of all metal oxides with carbon is endothermic. The enthalpy of reaction increases with increasing stability of oxides. Thus, both high temperature and large amount of heat are required in reduction of stable oxides with carbon. In other words, there is extensive increase in fuel and energy requirements while reducing more stable oxides.

The reduction of ZnO presents an exceptional case due to the formation of liquid as well as gaseous zinc depending on temperatures. When liquid zinc is formed, the gas ratio is a function only of temperature. The equilibrium is disturbed on the formation of zinc vapor at a higher temperature according to the following reaction:

$$ZnO(s) + CO(g) = Zn(g) + CO_2(g) \tag{5.32}$$

The gas ratio for the above reaction is given by the expression:

$$\frac{p_{CO_2}}{p_{CO}} = \frac{K}{p_{Zn}} \tag{5.33}$$

In Fig. 5.3, the gas ratio p_{CO_2}/p_{CO} plotted for $p_{Zn}= 1$ atm intersects the curve for liquid zinc at its boiling point, 907 °C; that is, liquid and gaseous zinc are in equilibrium. The gas ratio will be displaced upward by one logarithmic unit for a partial pressure of 0.1 atm. At this pressure the two curves will intersect at about 740 °C, which is the dew point of zinc vapor at 0.1 atm. The thermodynamic condition for the reduction of ZnO is depicted in Fig. 5.3. Figure 5.4 gives the gas ratio, p_{CO_2}/p_{CO} at different partial pressures of zinc, $p_{Zn}= 0.1, 0.5, 1.0$, and 10 atm, together with the curve for reduction of ZnO to liquid zinc. The latter intersects the curves for different vapor pressures of zinc at temperatures where liquid zinc and zinc vapor are in equilibrium. The Boudouard reaction (5.31) is also shown in Fig. 5.4 with the gas ratio [1] for $p_{CO}= 0.1, 0.5, 1$, and 10 atm. Both the equilibria represented by Eqs. (5.32) and (5.31) must be satisfied for continuous reduction of ZnO by carbon. This means reduction will take place at the point of intersection of the two curves for reactions (5.31) and (5.32). Since the number of atoms of zinc and oxygen are equal in reduction of ZnO with carbon, partial pressures of Zn, CO, and

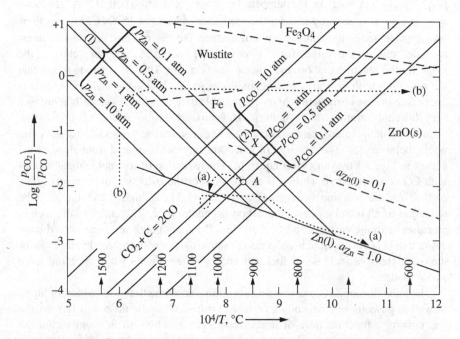

Fig. 5.4 Equilibrium gas ratios for reduction of ZnO(s) to Zn(l) at different activities of liquid zinc and partial pressures of zinc vapors and for the Boudouard reaction at different partial pressures of CO gas. (From Principles of Extractive Metallurgy by T. Rosenqvist [1], © 1974, p 307, McGraw-Hill Inc. Reproduced with the permission of McGraw-Hill Book Co.)

CO_2 are stoichiometrically related as: $p_{Zn} = p_{CO} + 2p_{CO_2}$. According to the Phase Rule, there is only one degree of freedom, either temperature or pressure and there are three components: Zn, O, and C with one restriction as $n_{Zn} = n_O$ and three phases: ZnO, C, and a gaseous phase. Assuming that p_{CO_2} is small compared to p_{CO}, we have $p_{Zn} \approx p_{CO} \approx \frac{1}{2}p_{tot}$. Figure 5.4 shows that for a total reaction pressure of 1 atm, curves for $p_{Zn} = 0.5 = p_{CO}$ intersect at 920 °C (point A). This is the lowest temperature at which continuous reduction of ZnO with solid carbon starts at 1 atm. At the point of intersection, A, the CO_2/CO ratio is about 1.2×10^{-2}, that is, p_{CO_2} is about 0.6×10^{-2} atm which is in agreement with the assumption of p_{CO_2} being low as compared to p_{CO}.

Reactions (5.31) and (5.32) are reversible; hence on cooling, the resultant gaseous mixture containing Zn, CO, and CO_2 may form ZnO. Since the reversion of reaction (5.32) is much faster compared to the reversion of 5.31, due precaution has to be taken to prevent the oxidation of zinc vapors during cooling by CO_2. During cooling the presence of even as low as 1% CO_2 in the gas mixture containing 50% each of Zn and CO will reoxidize nearly 2% of zinc. However, on lowering the CO_2 concentration to 0.1% in the presence of excess carbon at high temperatures, only 0.2% of zinc gets reoxidized. Even small amount of ZnO formed may adversely affect the process because it acts as nuclei for the condensation of zinc vapors. The zinc droplets covered with thin layer of ZnO (called blue powder) will prevent further coalescence. The amount of blue powder increases with increasing CO_2 content in the flue gas. Condensation of zinc will start when the saturation pressure of liquid zinc is equal to the partial pressure of zinc in the gas. The condensation occurs at a temperature of about 840 °C for a zinc pressure of 0.5 atm. To achieve 99% condensation at a total pressure of 1 atm at 600 °C, the saturation pressure of zinc should be about 0.01 atm. Due to supercooling in industrial processes, zinc condenser is operated at about 500 °C with metallic zinc seeding.

On further heating and subsequent cooling and condensation the gas ratio in the retort process follows the path indicated by the curve (a)–(a) in Fig. 5.4. However, the gas ratio in the Imperial Smelting blast furnace follows the path (b)–(b). At the point of intersection, X (intersection of the Boudouard line) reduction of ZnO (s) to a brass takes place with $a_{Zn\ (l)} = 0.1$ and $p_{CO} = 1$ atm at 890 °C. The highly endothermic simultaneous reduction [1] of zinc oxide with carbon according to the reaction

$$ZnO(s) + C(s) = Zn\ (g) + CO\ (g), \Delta H^o = 57 \text{ kcal}(238.5 \text{ kJ}) \tag{5.34}$$

presents a serious problem in providing the large amount of heat to the reaction mixture. Together with the added heat content of the reaction products at 1000 °C and heat of vaporization of zinc, a total heat of 90 kcal (376.6 kJ) per mole or about 1350 kcal (5648 kJ) per kilogram of zinc is required. In the retort process, the required heat generated in an outer combustion chamber is transferred by conduction through the retort wall into the charge. This arrangement not only puts a limit on the rate of production but also poses additional problems during cooling and condensation of the flue gases in the condenser by the release of large quantities of heat, approximately 400 kcal (1674 kJ) per kilogram of zinc. Removal of this heat in the condenser poses problems in the design of the condenser. The problem of condensing without oxidation of the zinc vapor was resolved by the Imperial Smelting Corporation by developing specially designed lead splash condensers and blast furnaces capable of smelting mixed ZnO–PbO sinters.

5.2.4 Reduction with Hydrogen

Hydrogen is expensive and hence used under special conditions. However, the main reason for its limited applications is the low exothermicity of reactions when oxides are reduced with hydrogen. Figure 5.5 shows the p_{H_2O}/p_{H_2} ratio for reduction of various metals with hydrogen. Figures 5.3 and 5.5 have close resemblance with respect to the shape and relative position of different metal–metal oxide reduction equilibria. While comparing the relative merits of C, CO, and H_2 reduction reactions,

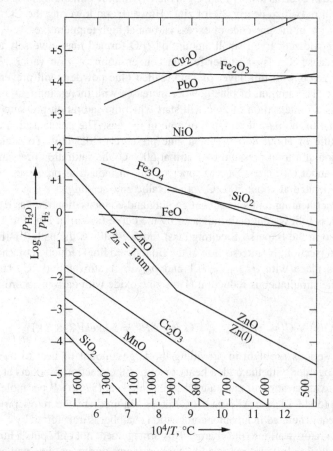

Fig. 5.5 $\log \frac{p_{H_2O}}{p_{H_2}}$ vs. $\frac{1}{T}$ for the reduction of various oxides. (From Principles of Extractive Metallurgy by T. Rosenqvist [1], © 1974, p 269, McGraw-Hill Inc. Reproduced with the permission of McGraw-Hill Book Co.)

exothermicity of the following two reactions should be considered in addition to reactions (5.15) and (5.16).

$$2CO(g) + O_2(g) = 2CO_2(g), \Delta H^o{}_{298} = -572 \text{ kJ} \tag{5.35}$$

$$2H_2(g) + O_2(g) = 2H_2O(g), \Delta H^o{}_{298} = -486 \text{ kJ} \tag{5.36}$$

The above heat changes indicate that the reduction of MO with H_2 will be less exothermic than the reduction by CO. In case a mixture of CO and H_2 is used as the reductant, the reactions will have to satisfy the water gas equilibrium according to the following reaction:

$$CO(g) + H_2O(g) = CO_2(g) + H_2(g) \tag{5.37}$$

Any shift in the reaction to the right at low temperatures means that CO is a better reducing agent than hydrogen at lower temperatures. Natural gas consisting mainly of methane can be used in reduction of oxides:

$$4MO(s) + CH_4(g) = 4M(s,l) + CO_2(g) + 2H_2O(g) \tag{5.38}$$

Although methane is a reducing agent of considerable significance, its decomposition to soot and hydrogen above 500 °C makes its direct use problematic. In fact, carbon and hydrogen generated due to the decomposition reduce the metal oxide in their own way.

The relative positions of the lines concerning reactions (5.15), (5.16), and (5.36) in the Ellingham diagram (Fig. 5.1) suggest that hydrogen is a better reducing agent than carbon and carbon monoxide, respectively, at temperatures below 650 °C and above 800 °C. However, for the reduction of iron oxide, hydrogen presents better reduction kinetics in the temperature range of 1150–1350 °C [10–13]. In certain cases, a mixture of both the gases is used.

5.3 Kinetics of Reduction of Oxides

The progress of gas–solid reactions like roasting and reduction of ores is complicated due to the fact that chemical reactions proceed along with temperature rise of the ore particles. A reaction cannot occur at all until the temperature is high enough at the reaction interface. Most of the processes are carried out under counter flow conditions in shaft furnaces or rotary kilns. The charge and the combustion/reduction gases move in opposite directions so that at any point the gas is only a little hotter than the surface of the charge. In a gas–solid reaction of the type MO(s) + A(g) = M (s) + AO(g), the reduction takes place in a number of steps:

1. Transfer of A, from the bulk gaseous phase to the outer surface of the solid.
2. Diffusion of A to the reaction interface.

3. Chemical reaction at the interface (adsorption of reaction species, chemical reaction and desorption of reaction product species).
4. Diffusion of AO away from the interface into the bulk of the gas phase.

The overall rate of reaction depends on whether the product M forms a porous layer and the relative kinetics of the chemical reaction at the MO/M interface and of the diffusion through the product layer. Szekely et al. [14] have discussed these cases in detail.

The rate of reaction can be expressed in terms of the fraction of solid reacted (f), which is defined, using a spherical solid without change in the overall size as an example, as

$$f = \frac{\omega_0 - \omega}{\omega_0 - \omega_f} = \frac{4/3\pi r_0^3 - 4/3\pi r^3}{4/3\pi r_0^3} = 1 - \frac{r^3}{r_0^3} \qquad (5.39)$$

where ω_o, ω_f, and ω represent, respectively, the original weight, final weight, and weight of the solid at any time t, and r_o and r, are, respectively, the initial radius and radius of the unreacted solid at any time t. If k and c, are, respectively, the rate constant based on the consumption rate of the gaseous reactant and the molar concentration of reactant gas; the conversion versus time for a system with a first-order reaction in which the diffusion through the product layer is much faster than the interfacial reaction is given by

$$1 - (1 - f)^{1/3} = \frac{bkc}{r_0 \rho} t \qquad (5.40)$$

where b is the number of moles of the solid reactant reacted by one mole of the gaseous reactant.

When the diffusion through the solid product layer controls the overall rate, the conversion versus time is described by [14].

$$\frac{6bD_e(c - c^{eq})}{\rho r_0^2} t = 1 + 2(1 - f) - 3(1 - f)^{2/3} \qquad (5.41)$$

where D_e stands for effective diffusivity through the product solid layer.

When the reactant solid contains porosity, the expression of the overall rate is more complex, and the reader is referred to a monograph on the subject by Szekely et al. [14].

The kinetics of the reaction is governed by the nature of the solid reaction product (porous or nonporous) formed on the reacting solid. In case of a porous film, there is no resistance to the reagents reaching the interface and the rate will not be affected by the reaction product. On the other hand, if the film is nonporous, the reagent has to diffuse through this protective film before it reaches the interface. The kinetics of reaction will markedly differ in this case. The rate of reaction gradually decreases as thickness of the product layer increases because diffusion path increases. The rate further decreases due to the decrease in the area of the unreacted core/reaction product interface.

Jander [15], Gnistling and Brounshtein [16], and Valensi [17, 18] have proposed models on the direct reduction of metal oxides with solid carbon. They have postulated that the gaseous intermediate reaction $[C(s) + CO_2(g) = 2\ CO(g)]$ was the rate-controlling step. As Jander's equation refers to the plane surface, it will be applicable for a sphere of very large radius as compared to the thickness of the product layer. Further, it is applicable only at early stages of the reaction when the volume of the reaction product is equal to the volume of the original material. The approximation made in Jander's equation does not work well beyond about 50% conversion because volume changes with the progress of the reaction. The progressive decrease in the area of the reaction interface in actual practice is taken into account in the models proposed by Gnistling and Brounshtein [16] and Valensi [17]. These are briefly summarized below.

5.3.1   Gnistling and Brounshtein Simplified Model

According to Fick's law, we can write:

$$J = -AD\frac{dc}{dr} = 4\pi r^2 D\frac{dc}{dr} \tag{5.42}$$

where J is the number of molecules of the reagent diffusing in time t through the product layer (Fig. 5.6). A and D represent the area of cross section of the reaction front and diffusivity through the product solid layer, respectively. Integration of Eq. 5.42 gives:

$$\int_{c_i}^{c} dc = -\frac{J}{4\pi D} \int_{r_1}^{r_0} \frac{dr}{r^2} \tag{5.43}$$

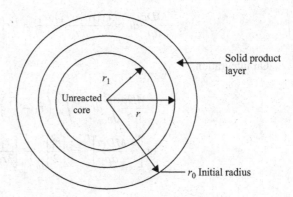

Fig. 5.6 Formation of a solid product (porous/nonporous) layer

$$\therefore c - c_i = -\frac{J}{4\pi D}\left(\frac{r_0 - r_1}{r_0 r_1}\right)$$ (5.44)

where c_i and r_1 are the concentration of the reagent at the interface and the radius of the unreacted core, respectively. For a diffusion-controlled process, $c_i = 0$,

$$\therefore J = -4\pi D\left(\frac{r_0 r_1}{r_0 - r_1}\right)c$$ (5.45)

We consider a simple case where c is constant

$$\text{Fraction reacted}\quad f = \frac{\frac{4}{3}\pi r_0^3 - \frac{4}{3}\pi r_1^3}{\frac{4}{3}\pi r_0^3} = 1 - \left(\frac{r_1}{r_0}\right)^3$$ (5.46)

$$\therefore r_1 = r_0(1 - f)^{1/3}$$ (5.47)

The number of moles of unreacted solid present at any time t is.
$N = \frac{4}{3}\pi r_1^3 \frac{\rho}{M}$ (M and ρ represent molecular weight and density of the reactant)
Since we can write: $\frac{dN}{dt} = \frac{dN}{dr_1} \cdot \frac{dr_1}{dt}$

$$\therefore \frac{4/3\pi\rho}{M}3r_1^2\frac{dr_1}{dt} = \frac{4\pi\rho r_1^2}{M}\frac{dr_1}{dt}$$ (5.48)

But the rate of change of N, $\left(\frac{dN}{dt}\right)$ is proportional to the flux of the material, J, diffusing through the spherical shell of thickness $(r_0 - r_1)$. Therefore, from Eqs. (5.45) and (5.48), we can write:

$$J = -4\pi D\left(\frac{r_0 r_1}{r_0 - r_1}\right)c = \alpha\, 4\pi\frac{\rho}{M}r_1^2\frac{dr_1}{dt}$$ (5.49)

where α is stoichiometry factor:

$$\therefore -\frac{MDc}{\alpha\rho}dt = \frac{r_1(r_0 - r_1)dr_1}{r_0} = \left(r_1 - \frac{r_1^2}{r_0}\right)dr_1$$ (5.50)

On integration,

$$-\frac{MDc}{\alpha\rho}\int_0^t dt = \int_{r_0}^{r_1}\left(r_1 - \frac{r_1^2}{r_0}\right)dr_1$$

$$\text{or}\left[-\frac{MDct}{\alpha\rho}\right]_0^t = \left[\frac{r_1^2}{2}\right]_{r_0}^{r_1} - \left[\frac{r_1^3}{3r_0}\right]_{r_0}^{r_1}$$ (5.51)

$$\text{or } -\frac{MDct}{\alpha\rho} = \frac{r_1^2}{2} - \frac{r_0^2}{2} - \frac{r_1^3}{3r_0} + \frac{r_0^3}{3r_0}$$

$$= \frac{1}{2}r_1^2 - \frac{1}{6}r_0^2 - \frac{1}{3}\frac{r_1^3}{r_0} \tag{5.52}$$

Substituting r_1 in terms of f we get:

$$-\frac{MDc}{\alpha\rho}t = \frac{1}{2}r_0^2(1-f)^{2/3} - \frac{1}{6}r_0^2 - \frac{1}{3}\frac{r_0^3}{r_0}(1-f) \tag{5.53}$$

$$= \frac{1}{2}r_0^2(1-f)^{2/3} - \frac{1}{6}r_0^2 - \frac{1}{3}r_0^2(1-f) \tag{5.54}$$

$$\text{or } -\frac{MDc}{\alpha\rho r_0^2}t = \frac{1}{2}(1-f)^{2/3} - \frac{1}{6} - \frac{1}{3}(1-f) \tag{5.55}$$

$$\text{or } \frac{2MDc}{\alpha\rho r_0^2}t = -(1-f)^{2/3} + \frac{1}{3} + \frac{2}{3}(1-f) \tag{5.56}$$

$$= 1 - \frac{2}{3}f - (1-f)^{2/3} \tag{5.57}$$

Thus, a plot of $\left[1 - \frac{2}{3}f - (1-f)^{2/3}\right]$ against t should be a straight line. This equation was found more appropriate as compared to Jander's equation but still it fails after about 90% reaction. This aspect has been considered in the model proposed by Valensi [17].

5.3.2   Valensi Model

Figure 5.7 shows the formation of a nonporous product layer. If the reagent is supplied continuously, c remains constant and then the rate of change of ω is proportional to J, the flux of reagent diffusing through the spherical shell of thickness $(r_2 - r_1)$.

According to Fick's law of diffusion $J = -AD\frac{dc}{dr} = -4\pi r^2 D\frac{dc}{dr}$.

If c_i is the concentration of the reagent at the interface, on integration:

$$\int_{c_i}^{c} dc = -\frac{J}{4\pi D}\int_{r_1}^{r_2}\frac{dr}{r^2} \tag{5.58}$$

$$c - c_i = -\frac{J}{4\pi D}\left(\frac{r_2 - r_1}{r_1 r_2}\right) \tag{5.59}$$

$$\text{or } J = -4\pi D\left(\frac{r_1 r_2}{r_2 - r_1}\right)(c - c_i) \tag{5.60}$$

Fig. 5.7 Formation of nonporous reaction products

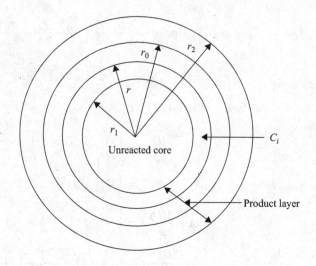

Unreacted core

C_i

Product layer

For a diffusion controlled process, $c_i = 0$

$$\therefore J = -4\pi D \left(\frac{r_1 r_2}{r_2 - r_1} \right) c \tag{5.61}$$

If ρ and M represent respectively, the density and molecular weight of the reactant (R), and the product (P) the weight of the unreacted core, ω can be expressed as: $\omega = \frac{4}{3}\pi r_1^3 \rho_R$ and thus, the number of molecules of the reactant is given as:

$$N = \frac{\omega}{M_R} = \frac{4}{3}\pi r_1^3 \frac{\rho_R}{M_R} \tag{5.62}$$

$$\frac{dN}{dt} = \frac{4\pi r_1^2}{M_R} \rho_R \frac{dr_1}{dt} \tag{5.63}$$

$$J = \alpha \frac{dN}{dt} \quad (\alpha \text{ is the stoichiometry factor}) \tag{5.64}$$

Combining Eqs. 5.61, 5.63, and 5.64, we can write:

$$-4\pi D \left(\frac{r_1 r_2}{r_2 - r_1} \right) c = \alpha \frac{4\pi r_1^2 \rho_R}{M_R} \frac{dr_1}{dt} \tag{5.65}$$

$$\therefore -\frac{M_R D c \, dt}{\alpha \rho_R} = \frac{(r_2 - r_1) r_1}{r_2} dr_1 = \left(r_1 - \frac{r_1^2}{r_2} \right) dr_1 \tag{5.66}$$

Since the number of molecules consumed = number of molecules formed ×
stoichiometry factor, we can write:

$$\frac{(\frac{4}{3}\pi r_0^3 - \frac{4}{3}\pi r_1^3)\rho_R}{M_R} = \frac{(\frac{4}{3}\pi r_2^3 - \frac{4}{3}\pi r_1^3)\rho_P}{M_P} . \alpha \tag{5.67}$$

$$\frac{1}{\alpha}\frac{M_P/\rho_P}{M_R/\rho_R} = \frac{r_2^3 - r_1^3}{r_0^3 - r_1^3} \tag{5.68}$$

If V_P and V_R represent volume of the product and the reactant, respectively, we
have: $\boxed{\frac{1}{\alpha}\frac{M_P/\rho_P}{M_R/\rho_R} = \frac{1}{\alpha}\frac{V_P}{V_R} = z}$

$$\therefore z = \frac{r_2^3 - r_1^3}{r_0^3 - r_1^3} \tag{5.69}$$

$$\text{or } r_2^3 - r_1^3 = z(r_o^3 - r_1^3) \tag{5.70}$$

$$\text{or } r_2^3 = z r_0^3 + r_1^3(1 - z) \tag{5.71}$$

$$\text{or } r_2 = \left[z r_0^3 + r_1^3(1 - z)\right]^{1/3} \tag{5.72}$$

Noting that if $z = 1$, $r_2 = r_o$, substituting the value of r_2 in Eq. 5.66 from 5.72
we get:

$$-\frac{M_R Dc}{\alpha \rho_R}dt = \left(r_1 - \frac{r_1^2}{\left[z r_0^3 + r_1^3(1 - z)\right]^{1/3}}\right)dr_1 \tag{5.73}$$

Integrating from the limits r_0 to r_1 for $t = 0$ and $t = t$

$$-\frac{M_R Dc}{\alpha \rho_R}t = \frac{1}{2}r_1^2 + \frac{z r_0^3 - \left[z r_0^3 + r_1^3(1 - z)\right]^{2/3}}{2(1 - z)} \tag{5.74}$$

$$\text{or } -2(1 - z)\frac{M_R Dc}{\alpha \rho_R}t = (1 - z)r_1^2 + z r_0^3 - \left[z r_0^3 + r_1^3(1 - z)\right]^{2/3}$$

$$z r_0^3 + 2(1 - z)\frac{M_R Dc}{\alpha \rho_R}t = \left[z r_0^3 + r_1^3(1 - z)\right]^{2/3} - (1 - z)r_1^2 \tag{5.75}$$

since fraction reacted, $f = \frac{\omega_0 - \omega}{\omega} = \frac{\frac{4}{3}\pi r_0^3 \rho_R - \frac{4}{3}\pi r_1^3 \rho_R}{\frac{4}{3}\pi r_0^3 \rho_R} = 1 - \left(\frac{r_1}{r_0}\right)^3 \therefore r_1 = r_0(1 - f)^{1/3}$.

Substituting the value of r_1 in terms of f in to Eq. 5.75, we get:

$$zr_0^2 + 2(1-z)\frac{M_R Dc}{\alpha\rho_R}t = \left[zr_0^3 + (1-z)r_0^3(1-f)\right]^{2/3} - (1-z)r_0^2(1-f)^{2/3}$$

(5.76)

$$= r_0^2[z + (1-z)(1-f)]^{2/3} - (1-z)r_0^2(1-f)^{2/3} \quad (5.77)$$

Dividing by r_0^2, we get:

$$z + 2(1-z)\frac{M_R Dc}{\alpha\rho_R r_0^2}t = [z + (1-z)(1-f)]^{2/3} - (1-z)(1-f)^{2/3} \quad (5.78)$$

$$= [z + 1 - z - f + zf]^{2/3} - (1-z)(1-f)^{2/3} \quad (5.79)$$

$$= [1 + (z-1)f]^{2/3} + (z-1)(1-f)^{2/3} \quad (5.80)$$

Thus, a plot $[1 + (z-1)f]^{2/3} + (z-1)(1-f)^{2/3}$ vs t should give a straight line. Equation 5.80 has been found to be valid till 100% of the reaction time.

5.3.3   Other Mathematical Models

In recent years, Sohn and coworkers [10–13] conducted detailed studies on the reduction kinetics of hematite particles (average particle size 21 μm) with hydrogen [10] and carbon monoxide [11] in the temperature range of 1150–1350 °C and 1200–1350°C, respectively under different partial pressures of hydrogen and carbon monoxide with the prime objective of developing a novel flash ironmaking process [13]. They have reported more than 90% reduction of hematite concentrate particles with hydrogen at 1300 °C in 3 s of residence time whereas the same order of reduction was obtained with CO at 1350 °C in 5 s. The rate of reduction with CO is slow but still fast enough for the flash ironmaking process. In both cases the reaction was found to be strongly temperature-dependent and followed the first-order rate kinetics with respect to the partial pressure of hydrogen and carbon monoxide with the activation energy of 214 and 231 kJ mol^{-1}, respectively.

The above studies will be useful in understanding the complex kinetics of the reduction of hematite with the gaseous mixture of H_2 and CO [generally, produced by reforming the natural gas: CH_4 (g) + H_2O (steam) = CO (g) + 3 H_2 (g)], not only in the flash ironmaking process but also in direct reduction of iron oxide without melting. In direct reduction, oxygen from the hematite concentrate is removed by H_2 and CO in steps by the following heterogeneous chemical reactions [19]:

$$3Fe_2O_3(s) + H_2(g) = 2Fe_3O_4(s) + H_2O(v) \quad (5.81)$$

$$Fe_3O_4(s) + H_2(g) = 3FeO(s) + H_2O(v) \qquad (5.82)$$

$$FeO(s) + H_2(g) = Fe(s) + H_2O(v) \qquad (5.83)$$

$$3Fe_2O_3(s) + CO(g) = 2Fe_3O_4(s) + CO_2(g) \qquad (5.84)$$

$$Fe_3O_4(s) + CO(g) = 3FeO(s) + CO_2(g) \qquad (5.85)$$

$$FeO(s) + CO(g) = Fe(s) + CO_2(g) \qquad (5.86)$$

However, the possibility of occurrence of the following important homogeneous reactions between the gaseous species present in the reducing gas mixture cannot be avoided:

$$2CO(g) \leftrightarrows C(s) + CO_2(g) \qquad (5.87)$$

$$H_2O(v) + CO(g) \leftrightarrows H_2(g) + CO_2(g) \qquad (5.88)$$

$$3H_2(g) + CO(g) \leftrightarrows CH_4(g) + H_2O(v) \qquad (5.89)$$

During the past three decades, considerable attention has been focused on the development of mathematical models to assess the rate controlling step in the reduction process. The available models for interpretation of kinetics of reduction of oxides may be categorized into three groups [19]: (i) one interface shrinking core model, (ii) three interfaces shrinking core model, and (iii) grain model.

On account of its simplicity, "one interface shrinking core model" has been employed in interpretation of reduction of metal oxides. For example, the reduction of wustite with H_2 and/or CO as depicted in Fig. 5.8 can be divided into four steps (as mentioned in Sect. 5.3). The slowest among all is considered to be the rate-controlling step. However, investigators have different views on the matter of the absolute control of the rate of reduction by a single step. For example, MacKewan [20] and Themelis and Gauvin [21] are of the opinion that the reduction of iron oxide is controlled to a major extent by chemical reaction whereas Bogdandy [22] considers it to be by the gaseous transport control, and Lu [23] has suggested the pore diffusion together with first-order chemical reaction as the rate-controlling step. On the other hand, Spitzer et al. [24, 25] have developed a model based on assumptions of isothermal and isobaric conditions to demonstrate diminishing possibility of the absolute control of the rate of reduction of the oxide pellet by a single step. Instead,

Fig. 5.8 Schematic diagram of one interface shrinking core model: reduction of dense hematite pellet

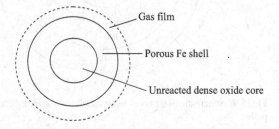

they suggested for simultaneous consideration of diffusion and chemical reaction in the rate-controlling step.

The "three interfaces shrinking core model" is presented in Fig. 5.9. The figure shows the reduction of hematite to iron with H_2/CO gas by the formation of three successive interfaces: hematite/magnetite, magnetite/wustite, and wustite/iron of varying thicknesses moving outward. The three successive interfaces make the situation complex due to the difference in flux of species entering and leaving the front and consumption of reactants. The gaseous product follows a similar sequence in the opposite direction. The reduction mechanism proposed by Edstrom [26] considers solid-state diffusion of ferrous ions through the wustite layer according to the reactions (5.83) and (5.86) which are summation of sub-reactions ($FeO = Fe^{2+} + O^{2-}$, $O^{2-} + CO = CO_2 + 2e$, and $Fe^{2+} + 2e = Fe$). Thus, removal of oxygen from the wustite surface increases the concentration of Fe^{2+}. A part of these ferrous ions together with electrons migrate to nucleation sites to deposit as iron whereas the remaining ferrous ions and electrons diffuse across the wustite and magnetite layers to react with magnetite and hematite to form wustite and magnetite, respectively, according to the reactions:

$$Fe^{2+} + 2e + Fe_3O_4(s) = 4FeO(s) \tag{5.90}$$

$$Fe^{2+} + 2e + 4Fe_2O_3(s) = 3Fe_3O_4(s) \tag{5.91}$$

Although the experimental data can be easily fitted in the above model having a number of parameters, the assumption of sharp and clear cut interfaces between different solid phases points out toward a fundamental mistake. There seems a rare possibility of having sharp interfaces due to the simultaneous chemical reactions and diffusion in formation of porous sub-oxides and metallic iron.

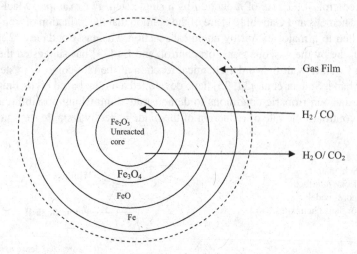

Fig. 5.9 Schematic diagram of three interfaces shrinking core model: reduction of dense hematite pellet

Fig. 5.10 Schematic representation of grain model for reaction of gases with porous oxide consisting of solid spheroidal grains of uniform radius

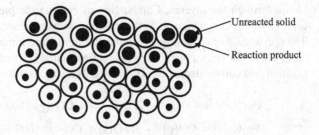

Unreacted solid

Reaction product

As simultaneous chemical reactions and diffusion during the reduction of an oxide pellet cannot be avoided, the three interfaces shrinking core model has very limited application in identifying the rate-controlling step. A number of investigators realized this difficulty in the reduction of porous and nonporous pellets. Tien and Turkdogan [27] developed a mathematical model by considering two separate zones in a pellet, one of the completely reduced layer of iron and the other layer of the partially reduced iron oxides. However, this concept happens to be a special case of the "grain model" (depicted in Fig. 5.10) of gas–solid reactions introduced by Szekely and coworkers [14, 28] which can be easily accommodated within the more advanced version of the mathematical model developed by Sohn [29–31]. Tien and Turkdogan [27] have considered grains of uniform size maintaining their initial physical structure throughout the reaction. Further, they have assumed that reaction of each grain proceeds like microscopic shrinking core without considering gaseous diffusion through the product layer of the grains.

5.4 Commercial Processes

In this section, a few commercial processes based on reduction of oxides and reduction smelting will be discussed. When primary metals are obtained in molten state during reduction of oxides with suitable reducing agents, the process is known as the reduction smelting. The term "smelting" stands for any metal production process that goes through a molten state.

5.4.1 Production of Iron

The production of iron presents the most important example of the reduction of oxide and reduction smelting. Iron is primarily extracted from hematite (Fe_2O_3) by reduction with coke in the blast furnace of circular cross section. This design generates more intensive heat. For the production of one ton of pig iron 1.6–2 ton of iron ore (50–60% Fe), 0.35–0.5 ton of coke and 0.4 ton of lime stone are charged in the blast furnace from top and about 1.4 ton of air (approximately 1000 Nm^3) is blown through the tuyeres. Combustion of coke in the preheated air at the tuyere level produces CO_2 which on further reaction with coke forms CO ($C + CO_2 = 2CO$). The uprising CO gas coming into contact with iron ore lumps reduces Fe_2O_3, Fe_3O_4, and FeO in the stack region (i.e., indirect reduction). Simultaneously, lime decomposition and carbon deposition (Boudouard) reactions also take place:

$$3Fe_2O_3(s) + CO(g) \rightarrow 2Fe_3O_4(s) + CO_2(g) \ (400 - 600\,^\circ C) \tag{5.92}$$

$$Fe_3O_4(s) + CO(g) \rightarrow 3FeO(s) + CO_2(g) \ (600 - 800\,^\circ C) \tag{5.93}$$

$$2CO(g) \rightarrow C(s) + CO_2(g) \ (\text{Boudouard reaction}: 540 - 650\,^\circ C) \tag{5.94}$$

$$FeO(s) + CO(g) \rightarrow Fe(s) + CO_2(g) \ (800 - 1100\,^\circ C) \tag{5.95}$$

$$CaCO_3(s) \rightarrow CaO(s) + CO_2(g) \ (800 - 900\,^\circ C) \tag{5.96}$$

The following slag forming reaction takes place at 1100 °C:

$$CaO(s) + SiO_2(s) = CaSiO_3(s) \tag{5.97}$$

The gangue oxides, such as P_2O_5, MnO, and SiO_2 present in the ore and FeO, are directly reduced by carbon at a high temperature (1200–1800 °C) existing in the hearth (direct reduction). The molten iron produced in the hearth in contact with coke at high temperature dissolves carbon and picks up phosphorus, manganese, and silicon.

$$FeO(s) + C(s) \rightarrow Fe(l) + CO(g) \tag{5.98}$$

$$C(\text{in coke}) \rightarrow [C] \tag{5.99}$$

$$P_2O_5(s) + 5C(s) \rightarrow 2[P] + 5CO(g) \tag{5.100}$$

$$MnO(s) + C(s) \rightarrow [Mn] + CO(g) \tag{5.101}$$

$$SiO_2(s) + 2C(s) \rightarrow [Si] + 2CO(g) \tag{5.102}$$

The Boudouard reaction leading to the carbon deposition [$2CO(g) \rightarrow C\ (s) + CO_2\ (g)$] at the iron–refractory interface takes place in the upper part of the stack where temperature varies from 550 to 650 °C. The finely deposited carbon on refractory wall takes part in direct reduction of iron oxide to a limited extent. In order to control the Mn, Si, P, and S contents of the hot metal, adequate slag basicity and hearth temperature need to be maintained. The effect of temperature can be analyzed by the following equations relating free energy and temperature [5, 32] for reactions (5.101) and (5.102) respectively:

$$\Delta G^o = 290300 - 173\ T \quad J\ mol^{-1} \tag{5.103}$$

$$\Delta G^o = 593570 - 396\ T \quad J\ mol^{-1} \tag{5.104}$$

Since ΔG^o decreases with increasing T (hearth temperature), the forward reactions leading to Mn and Si pick up in the metal is favored at a higher temperature. As the activity of acidic SiO_2 decreases in the presence of a basic oxide (CaO), the free energy change for reaction (5.102) under nonstandard conditions, according to the van't Hoff isotherm, can be expressed as:

$$\Delta G = \Delta G^o + RT \ln \left(\frac{[a_{Si}] \cdot p_{CO}^2}{(a_{SiO_2}) \cdot [a_C^2]} \right) \qquad (5.105)$$

And it is seen that ΔG becomes positive with a decrease of a_{SiO_2} in the basic slag CaO–SiO_2 to the extent that the following reverse reaction is favored:

$$[Si] + 2CO(g) \rightarrow (SiO_2) + 2[C] \qquad (5.106)$$

In a similar manner, the distribution of other elements may be discussed. But the situation is not straight forward in the case of manganese. Manganese pick up is influenced by temperature but follows a complex trend due to the basic nature of MnO in the presence of the acidic oxide SiO_2 as well as the basic oxide CaO in the slag. The distribution of sulfur in slag and metal depends on the formation of stable calcium sulfide according to the reaction [5]:

$$[FeS] + (CaO) \rightarrow (CaS) + [FeO] \qquad (5.107)$$

$$\Delta G^o = 72000 - 38 \, T \quad J \, mol^{-1} \qquad (5.108)$$

Equation 5.108 indicates that sulfur transfer from metal to slag is favored by increasing the hearth temperature. In addition to temperature, basic slag and reducing conditions favor desulfurization. Since transfer of sulfur from the metal to the slag requires reducing conditions, it is easier to control sulfur to the desired level in the blast furnace during ironmaking rather during steelmaking. Optimum conditions for desulfurization as derived by considering ionic reactions will be discussed in Chap. 7.

5.4.1.1 Recent Trends in Ironmaking

The production rate and furnace efficiency of the blast furnace have been improved by the use of preheated blast, oxygen enrichment in the blast, injection of hydrocarbon and steam into the blast, and high top pressure. These facilities have been adopted with the prime objective of reducing the coke rate (i.e., the amount of coke consumed for production of one ton of pig iron). Since reduction of iron ore in the blast furnace requires very high-grade coke, reduction in coke consumption will lower down the cost of production of iron. However, while implementing any advancement, pros and cons of indirect versus direct reduction reactions must be analyzed in light of differences in their thermo-chemical nature. For example, preheated air obtained by passing air through a heated refractory network in hot blast stoves needs less coke to generate the same amount of heat. In a similar manner preheated air enriched with oxygen also reduces the coke rate together with reduction/decrease in volume of the blast furnace gas ascending through the stack. The reduction in gas volume decreases the amount of heat transferred from the ascending gas to the charge. This results in lowering down the proportion of indirect reduction which has to be compensated by the direct reduction taking place in the hearth. In view of the nature of these two reactions, oxygen enrichment has to be maintained below 9% [5].

Any increase in pressure of the blast furnace gas at top of the stack will reduce the pressure difference between the gas at the stack and at the tuyere level. This reduction in pressure difference increases the residence time of the gas in the stack due to slower movement of the ascending gas. Thus, gas gets more contact time with the ore lump and hence reduces the coke rate and increases the rate of production of pig iron.

Combustion of the injected hydrocarbon into the stack of the blast furnace produces two reducing gases: carbon monoxide and hydrogen:

$$2CH_4 + O_2 = 2CO + 4H_2, \Delta H^o_{298} = -71.4 \text{ kJ} \tag{5.109}$$

CO and H_2 are better reducing agents for metal oxides, above and below 800 °C, respectively. Hence, both together improve the efficiency of the indirect reduction. Since combustion of coke is much more exothermic [5] than that of hydrocarbon ($2C + O_2 = 2CO$, $\Delta H^o_{298} = -222 \text{ kJ}$), a higher increase in blast temperature is required in case of hydrocarbon injection in order to maintain the hearth temperature. The commonly used hydrocarbons, for example, natural gas, atomized oil, and pulverized coal are injected just above the tuyeres. Steam injection through the tuyere also generates CO and H_2 but decreases the hearth temperature due to the highly endothermic [5] nature of the reaction (H_2O (steam) + C (s) = H_2 (g) + CO (g), $\Delta H^o_{298} = 132 \text{ kJ}$). This necessitates high blast preheat and/or oxygen enrichment to achieve the required hearth temperature.

Modern blast furnaces equipped with the above facilities produce over 10,000 tons of pig iron per day (hot metal) containing 3.5–4% C, 1–2% Si, 1% Mn, 0.2–2.5% P, and 0.04–0.2% S. Although currently the rate of production is very high, blast furnaces suffer from the disadvantages of poor control over composition (variation in composition of the pig iron from tap to tap) and of being uneconomical to developing countries with limited requirements as well as high capital and operating costs in addition to the requirement of costly high-grade coking coal as reductant. These limitations led to the development of two important alternatives to the blast furnace namely (i) direct reduction (DR) and (ii) direct smelting reduction (DSR) processes.

DR processes, based on the reduction of oxide with gases or carbon in solid state, produce solid iron in the form of sponge (popularly known as sponge iron and also referred as Direct Reduced Iron, DRI). It may be noted that the hot briquetted form of DRI is known as hot briquetted iron (HBI). The reactors employed for direct reduction processes are rotary kilns, low shaft furnace (similar to the bottom two third of the blast furnace), static bed reactors, and fluidized bed furnaces. The rotary kiln uses solid reducing agent, for example, coke breeze, coal dust, char, or anthracite, whereas other reactors use a mixture of $CO + H_2$ for reduction of iron ore pellets. The working temperatures of all these units (except the fluidized bed furnaces) range between 1000 and 1200 °C. The fluidized bed furnace is operated at a lower temperature (750 °C) due to the higher reducibility of hydrogen and efficient heat transfer in the fluidized bed. The lower temperature operation avoids fusion of particles. All these processes are almost DR processes with very little indirect reduction reaction. Even $CO + H_2$ act as a direct reducing agent without any intermediate reaction. Solid sponge iron produced by these units is of inferior purity as compared to the pig iron obtained from the blast furnace but has a better control over composition. Sponge iron is used as feed to electric furnaces for steelmaking.

In recent years, several direct smelting reduction processes have emerged as competitors to blast furnaces for production of iron and steel. Smelting reduction processes, like COREX and FINEX, incorporate both melting and reduction. The DSR processes can work with inferior quality raw materials compared to blast furnaces; for example, low-grade fine ores and iron-bearing plant wastes as the iron oxide feed and noncoking coal, carbon-bearing fines, and other relatively inexpensive materials containing carbon and hydrogen as reductants. These processes require lesser ore preparation (only by pelletizing and sintering) and materials handling, consume lower energy, provide better pollution control, and work at lower operating costs. In addition to these advantages, direct smelting processes can be installed at low capital cost and made continuous for iron and steel production because the molten iron produced is similar to the hot metal (pig iron) obtained from the blast furnace and, hence, can be fed directly to the steelmaking units. These advantages have led to the development of a large number of DSR processes [33] throughout the world. The Ausmelt [34] submerged lance technology discussed in Sect. 3.1.2.2 for sulfide smelting seems to have a promising future in direct smelting reduction of iron ores.

Both the gas-based DR and DSR processes use reformed gas. The natural gas can be reformed with steam or oxygen as per reactions:

$$CH_4 + H_2O = CO + 3H_2 \text{(endothermic)} \tag{5.110}$$

$$CH_4 + \frac{3}{2}O_2 = CO + 2H_2O \text{ (exothermic)} \tag{5.111}$$

The water gas reaction ($CO + H_2O = CO_2 + H_2$) is enhanced in the presence of suitable catalyst (nickel or platinum) at temperatures above 700 °C. The required amount of hydrogen is generated by reaction (5.110) at higher temperature and lower pressure according to Le Chatelier principle. The proportion of CO, H_2, CO_2, and H_2O can be adjusted by temperature control. Reformed gaseous mixture [33] generally contains about 85–90% CO + H_2 (with equal proportion of both) and about 12–15% CO_2 + H_2O (in the ratio of 4:1).

In general, SR processes utilize two reactor units, one for the pre-reduction of the oxide ore in the solid state and the other for the removal of the remaining oxygen by liquid phase reactions (smelting reduction vessel). Pre-reduction of hematite and magnetite to wustite is done by gases (CO and/or H_2) in the temperature range 850–1050 °C. This is followed by the reduction of molten FeO by CO or carbon in the smelting reduction step. The carbon that reacts with FeO in this stage may be solid carbon or carbon dissolved in molten iron. There is considerable increase in the rate of reduction due to enhancement in the contact between the liquid and other phases. In this way, the formation of liquid phase makes the SR process efficient.

Although in recent years, several direct smelting reduction processes have emerged as competitors to blast furnaces and direct reduction processes for production of iron, kinetics of the process is still not well understood. An attempt has been made here to outline the main features of reactions in smelting reduction. The pre-reduced ore produced in the pre-reduction unit according to the reactions (5.92), (5.93), and (5.95) is transferred to the smelting reduction vessel. In this vessel coal, lime stone and silica are charged and oxygen/air is blown. Coal is gasified and a slag (CaO–SiO_2–FeO) of variable FeO content is formed by the dissolution of FeO in SiO_2–CaO. The coal gasification and reduction of iron oxide present in the slag produce excessive volume of carbon monoxide and hydrogen. Gases bubbling through the slag give rise to the formation of foam. The smelting reduction of iron oxide rich slag in liquid state is either by CO or solid carbon or carbon dissolved in liquid iron bath in the temperature range of 1250–1650 °C, according to the following reactions:

$$(FeO) + \{CO\} = Fe(l) + \{CO_2\} \tag{5.112}$$

$$(FeO) + C\ (s) = Fe(l) + \{CO\} \tag{5.113}$$

$$(FeO) + [C] = Fe(l) + \{CO\} \tag{5.114}$$

In slag–metal reactions, brackets [], (), and {} are used for the constituents contained, respectively, in metal, slag, and the gaseous phases.

Chemical reactions in the part of smelting reduction involving the melt, solid, and gas are highly complicated and not well understood in terms of how much each reaction contributes toward the overall reduction [35]. The foam formation further complicates the mechanism of smelting reduction. As a consequence, the kinetics analysis of this process poses a daunting task, and presents a challenging but rich ground for further research for increased understanding and improvement of the process.

In Sect. 5.3, it has been mentioned that more than 90% reduction of hematite concentrate particles can be achieved with hydrogen at 1300 °C in 3 s and with CO at 1350 °C in 5 s [10–13]. These findings led to the development of a novel Flash Ironmaking Technology (IFT) at the University of Utah [13] to produce iron from magnetite concentrate, ranging from several microns to 100 microns without undergoing through pelletizing and sintering processes. IFT has a promising future due to low energy consumption and its effectiveness in reducing greenhouse gas emissions. Reduction can be carried out either by pure hydrogen gas or mixture of $H_2 + CO$, generated by partial oxidation of natural gas in the flash ironmaking reactor. The greenhouse gas emissions can be reduced up to 97% if pure hydrogen is used as the fuel and reductant whereas use of natural gas it is reduced up to 39–51% [13]. The industrial flash ironmaking reactor was designed after developing the rate equations for the reduction of magnetite concentrate fines with H_2 and CO gas mixture in the temperature range of 1200–1600 °C. In order to develop an industrial process, a laboratory-scale Utah Flash Reactor and a large-scale bench reactor were designed and installed for better understanding of the process. Based on these experimental results, the research group at the University of Utah aims commercialization of the process to achieve its great advantage of minimizing the energy requirements and CO_2 emissions.

5.4.2 Production of Zinc

Discussion in Sect. 5.2.3 clearly demonstrates that supply of larger amount of heat to the reaction mixture (ZnO + C) has been the major problem in production of zinc by the older retort processes. In order to supply the required quantity of heat, retorts have to be heated from outside by attaching external combustion chambers. This additional facility not only restricts the size of the retort (and thereby the rate of production) but also causes problems by releasing large amount of heat during cooling and condensation of the flue gases in the condenser. The horizontal as well as vertical retort processes work essentially on the same principle. These problems were sorted out by the Imperial Smelting Corporation by developing a blast furnace capable of smelting of mixed ZnO–PbO sinters. Thus, Imperial Smelting Process has improved the economics of zinc extraction with simultaneous production of lead.

Prior to the development of the Imperial Smelting Process, galena and sphalerite concentrates were smelted separately after dead roasting. The roasted lead oxide was smelted with coke and flux in a blast furnace to produce a crude lead bullion whereas zinc oxide was reduced with coke in horizontal or vertical retorts. In the blast furnace smelting of mixed ZnO–PbO sinters, obtained by blast roasting of galena and sphalerite concentrates, the following reduction reactions of PbO with CO and C need to be analyzed in addition to the reactions (5.31), (5.32), and (5.34).

$$PbO(s) + CO(g) = Pb(l) + CO_2(g), \quad \Delta G^{\circ}_{773} = -147 \text{ kJ} \tag{5.115}$$

$$PbO(s) + C(s) = Pb(l) + CO(g), \quad \Delta G^{\circ}_{773} = -76 \text{ kJ} \tag{5.116}$$

The above reactions together with values of free energy changes demonstrate that the reduction of PbO with CO at 500 °C not only provides a greater driving force than reduction with coke, but also indicate that the reaction (5.115), being a solid–gas reaction, will be kinetically faster than a solid-solid reaction (5.116). In addition, we should note that (i) the reaction (5.34: $ZnO + C = Zn + CO$) between ZnO and coke at 1000 °C produces zinc vapor with a free energy change of -29 kJ and (ii) reaction (5.32: $ZnO + CO = Zn + CO_2$) between ZnO and CO with a positive standard free energy change of $+76$ kJ at 1000 °C will proceed only if the partial pressure of CO is high. From Fig. 5.3, it is clear that ZnO can only be reduced at 1000 °C with a gas mixture having p_{CO_2}/p_{CO} ratio of less than 0.02. The gaseous zinc produced by reaction (5.32) will get re-oxidized during cooling when temperature goes below 950 °C because of higher stability of ZnO than CO_2 at 950 °C. Therefore, the zinc vapor must condense in a reducing atmosphere. For this reason, horizontal as well as vertical retorts are heated externally to prevent the oxidation of CO into CO_2 (at above 720 °C, CO is more stable as compared to CO_2). This reduces the thermal efficiency of the process and makes the horizontal retort a batch process. The zinc vapor is condensed and collected at a controlled distance from the reaction chamber. Vertical retorts are operated on a continuous basis with zinc oxide briquettes and coke being charged from the top and zinc vapor being removed rapidly and condensed in a liquid zinc bath at the top of the retort.

Simultaneous reduction of mixed ZnO–PbO sinter, obtained by roasting-cum-sintering (blast roasting) of galena and sphalerite concentrates, has become possible with the development of Imperial Smelting Furnace. The presence of FeO (gangue in the concentrate) in ZnO–PbO sinters further complicates the reduction reactions in the Imperial Smelting Process that produces molten lead bullion and gaseous zinc. The bullion is collected in the hearth of the blast furnace, and zinc vapor is rapidly shock cooled in a condenser on passing through the top of the furnace. Since zinc metal is produced as a vapor and lead as a liquid and unreduced FeO has to be transferred into the slag, the requirements of the Imperial blast furnace differ from those of lead and iron blast furnaces. On account of the simultaneous production of lead and zinc, the operational conditions in the Imperial furnace have some similarity to those of the lead and iron blast furnace practices. However, analogy with either is not close, some differences are fundamental. For example, in Fig. 5.3, we find that CO_2/CO ratio for reduction of PbO, ZnO, and FeO varies from one to another. As ZnO is more stable than PbO, the reduction of the former requires stronger reducing conditions than the latter. In the Imperial Smelting Process, gases generated in the furnace by the combustion of coke for reduction of ZnO are so reducing that lead oxide is reduced without any significant reduction of iron oxide. Therefore, one has to set conditions for the simultaneous reduction of PbO and ZnO without reducing FeO because the CO_2/CO ratio is of similar order in zinc as well as in iron blast furnaces. The reduction of iron oxides (in the burden and in the slag) has to be prevented because solid iron will badly affect the gas flow and affect the reduction equilibria.

A flow sheet showing various steps in extraction of zinc from sphalerite concentrate is presented in Appendix: A.4a. The mixed calcine, PbO–ZnO is charged with preheated coke and limestone flux at the top of the furnace and preheated air at 850–

950 °C is injected through the tuyeres located at the bottom of the stack. The reduction reactions occur in the stack of the furnace in three distinct zones:

1. Zinc oxide is reduced with coke according to the reaction (5.34) in the zinc reduction zone, located in the bottom third of the stack. This is the hottest part of the stack. A small amount of ZnO is also reduced by CO in this zone.
2. Lead oxide is reduced with CO according to the reaction (5.115) in the top third of the stack. Since the reactions (5.35) and (5.115) are highly exothermic there is no further coke requirement. CO is produced by the reduction of ZnO with coke and by combustion of coke in lower portion of the stack. This is a significant factor in the overall economics of the process.
3. The middle zone of the stack is known as the equilibrium zone, where ZnO is reduced with the uprising CO gas formed at the tuyeres level, that is, in the zinc reduction zone. CO is oxidized to CO_2.

The reoxidation of zinc vapors to ZnO [Zn (g) + CO_2 (g) = ZnO (s) + CO (g)] by CO_2 is prevented by maintaining a temperature higher than 1000 °C in the stack of the blast furnace by burning CO in preheated air, injected at a controlled rate at top of the stack. Since the combustion of CO to CO_2 is highly exothermic, a temperature above 1000 °C is easily attained in the stack. This ensures that the temperature is maintained above 1000 °C. The Imperial Smelting Process employs a specially designed lead splash condenser to prevent the reversion of zinc to ZnO by "shock cooling" of the zinc vapor present in the flue gas with a relatively high CO_2 concentration. The condenser is divided into four sections, each section containing two four-baffled rotors immersed in molten lead. Liquid lead is splashed as tiny droplets in the condenser by the rotating blades. This action compels the flue gas containing zinc vapor from the blast furnace to follow a zigzag path through the splashed droplets. In this way the gas is thoroughly scrubbed and shock cooled to less than 500 °C at the exit from a temperature of 1000 °C at entrance. This procedure causes condensation and dissolution of more than 95% of the zinc vapor in lead droplets. Lead containing 2.65% zinc pumped at 550 °C into a jacketed launder is cooled to 450 °C. As only 2.2% zinc is soluble in molten lead at 450 °C, the excess liquid zinc floating above the dense liquid lead is separated. The lead containing 2.2% zinc is re-circulated to the condenser for further operation. One of the most important features of the Imperial Process includes the high efficiency of the lead splash condenser despite high CO_2/CO ratio in the flue gas (14% CO_2 and 10–12% CO). Satisfactory zinc elimination and condensation have been achieved without any deposition of ZnO on the wall of the condenser. This permits a reasonably low fuel-to-zinc ratio (1.1 ton of zinc produced per ton of carbon consumed). The high efficiency of the Imperial lead splash condenser is related to the use of liquid lead for shock cooling of zinc vapors because lead can be more effectively splashed into tiny droplets due to its lower vapor pressure than zinc at the working temperature.

The gas leaving the blast furnace (or leaving the charge) contains 5.9% Zn, 11.3% CO_2, 18.3% CO, balance N_2. According to Morgan and Lumsden [36], about 40% ZnO is reduced in the shaft and remaining 60% in the tuyere zone. They have further reported that the gas leaving the tuyere has the composition 3.5% Zn, 10.7% CO_2, and 18.5% CO. This gives a CO_2/CO ratio of 0.57. A gas of this composition would not reduce iron oxide from the slag or present in the burden. At 1250 °C, the equilibrium ratio for the reduction of iron from wustite is 0.31.

5.4.3   Production of Tungsten and Molybdenum

Sheelite ($CaWO_4$) and Wolframite [$(Fe(Mn)WO_4$] are the two major sources of tungsten whereas molybdenum occurs as molybdenum sulfide, MoS_2 in the mineral molybdenite. Owing to their high melting points, major steps in flow sheets of tungsten and molybdenum extraction involve hydrometallurgical methods. However, molybdenite is roasted prior to leaching. Finally, both the metals are obtained by hydrogen reduction of their oxides. Thus, production of tungsten and molybdenum presents a combination of pyrometallurgical and hydrometallurgical techniques.

On the industrial scale [37, 38], tungsten and molybdenum are produced by the reduction of tungstic oxide (WO_3) and molybdic oxide (MoO_3) with hydrogen. In the extraction of tungsten from sheelite ($CaWO_4$) and wolframite [$(Fe(Mn)WO_4$] concentrates [37], first yellow tungstic oxide (WO_3) is obtained by the decomposition of the intermediate product ammonium paratungstate [5 $(NH_4)_2O.12WO_3.11H_2O$]. Its reduction takes place in the following stages:

$$2WO_3(s) + H_2(g) = W_2O_5(s) + H_2O(g) \tag{5.117}$$

$$W_2O_5(s) + H_2(g) = 2WO_2(s) + H_2O(g) \tag{5.118}$$

$$WO_2(s) + 2H_2(g) = W(s) + 2H_2O(g) \tag{5.119}$$

A higher temperature and a higher hydrogen concentration in the gas are required in every later stage of reduction. During operation, the nickel boats loaded with tungstic oxide move counter-currently to the flow of purified hydrogen gas in a steel tube of 50–100 mm diameter, kept in the electrically heated tube furnace. It is allowed to pass gradually through the zone of increasing temperature (maximum 850–860 °C) and to expose it gradually with drier hydrogen until its arrival at the water-cooled discharge end. The reduction is facilitated by the catalytic adsorption of dissociated hydrogen molecules on the surface of tungsten metal. However, the catalytic effect is extensively retarded by the presence of moisture. The maximum temperature of reduction is 850–860 °C. The output of the tube furnace depends on the capacity of the boats and their rate of movement through the tube, the higher the rate of movement the coarser the tungsten powder. An ordinary boat holds 50–180 g and travels at 50–180 cm h^{-1}. In general, reduction of tungstic oxide is carried out in two stages. The first stage produces the brown oxide (WO_2) at 720 °C. The brown oxide is mixed with a fresh charge of yellow oxide (WO_3) for the second stage reduction at 860 °C. A flow sheet showing various steps in extraction of tungsten from scheelite/wolframite concentrate is presented in Appendix: A.5.

The moisture content of the gas, which is affected by temperature, plays an important role in controlling the particle size of the reduced powder. Finer powder is obtained by maintaining a high rate of hydrogen flow, which is useful in removing water vapor generated during reduction. The depth of oxide in the boat affects the moisture content in the gas phase. Coarsening of the powder can be avoided by adjusting the temperature in different zones, separately so as to have gradual reduction from lower to higher temperature.

To produce anhydrous molybdic oxide (MoO_3), a rich molybdenite concentrate is first roasted at a temperature below 650 °C. In order to prevent melting and volatilization of MoO_3, temperature of roasting should not exceed 650 °C due to local overheating. The calcine is digested in an ammonia solution. Copper and iron present in the leach liquor as impurities are removed as sulfides in stages by treating the solution with ammonium sulfide. Molybdic acid (H_2MoO_4) is precipitated from the purified solution by hydrochloric acid treatment. Finally, the molybdic acid is calcined at 400–450 °C [38] in a muffle furnace to obtain anhydrous molybdic oxide (MoO_3).

The mechanism of reduction of molybdenum trioxide to molybdenum metal with hydrogen gas follows a path similar to the three-stage [37] reduction of tungstic oxide. The three oxides of molybdenum are reduced according to the following reactions:

$$2MoO_3(s) + H_2(g) = Mo_2O_5(s) + H_2O \ (g) \qquad (5.120)$$

$$Mo_2O_5(s) + H_2(g) = 2MoO_2(s) + H_2O \ (g) \qquad (5.121)$$

$$MoO_2(s) + 2H_2(g) = Mo(s) + 2H_2O \ (g) \qquad (5.122)$$

The first step is carried out at 450 °C to prevent the melting of the partially reduced MoO_3 at about 550 °C. The second step at 1100 °C and the last step require higher temperatures and lower water vapor contents in the gaseous mixture.

5.5 Problems [39–41]

Problem 5.1
Determine the temperatures at which nickel oxide can dissociate under (i) standard conditions (ii) a pressure of 10^{-4} mm Hg. Given that:

$$Ni(s) + ½ O_2(g) = NiO(s), \quad \Delta G^o = -62650 + 25.98T \ cal.$$

Solution
i. Under standard conditions, $p_{O_2} = 1$.

For the dissociation reaction:

$$2NiO(s) = 2Ni(s) + O_2(g), \Delta G^\circ = 125300 - 51.96T, \text{cal.}$$

Since Ni and NiO are pure solids, $a_{Ni} = 1 = a_{NiO}$, $K = p_{O_2} = 1$.

$$\Delta G^\circ = -RT \ln K = 0$$

$$\therefore 125300 - 51.96\, T = 0$$

$$T = \frac{125300}{51.96} = 2411\, K = 2138\,^\circ C \text{ Ans}$$

ii. $p_{O_2} = 10^{-4}$ mm Hg $= \frac{10^{-4}}{760}$ atm $= K$

$$\Delta G^\circ = -RT \ln K = -1.987 \times T \times \ln\left(\frac{10^{-4}}{760}\right) = 31.48T$$

$$\therefore 31.48T = 125300 - 51.96\, T$$

$$T = 1502K = 1229\,^\circ C \quad \text{Ans}$$

Problem 5.2
The standard free energy change for the reaction: $NiO(s) + CO(g) = Ni(s) + CO_2(g)$ at 1125 K is -5147 cal. Would an atmosphere of 15% CO_2, 5% CO, and 80% N_2 oxidize nickel at 1125 K?

Solution

$$NiO(s) + CO(g) = Ni(s) + CO_2(g), \text{ at } 1125K, \quad \Delta G^\circ = -5147 \text{ cal}$$

$$\Delta G^\circ = -RT \ln K = -5147 \text{ cal}$$

$$\ln K = \frac{5147}{1.987 \times 1125} = 2.303$$

$$K = 10$$

For the above reaction, $K = \frac{p_{CO_2} \cdot a_{Ni}}{p_{CO} \cdot a_{NiO}} = \frac{p_{CO_2}}{p_{CO}} = 10$ $(a_{Ni} = 1 = a_{NiO})$.

$$\therefore p_{CO} = 0.1 \cdot p_{CO_2}$$

In the given atmosphere, $p_{CO} = \frac{5}{15} \cdot p_{CO_2} = 0.33 \cdot p_{CO_2}$

This means the atmosphere under question is more reducing, and hence will not oxidize nickel at 1125 K.

Problem 5.3
What is the maximum partial pressure of moisture which can be tolerated in H_2–H_2O mixture at 1 atm total pressure without oxidation of nickel at 750 $^\circ C$? Given that:

$$Ni(s) + \tfrac{1}{2}O_2(g) = NiO(s), \Delta G_1^o = -58450 + 23.55T \text{ cal}$$

$$H_2(g) + \tfrac{1}{2}O_2(g) = H_2O(g), \Delta G_2^o = -58900 + 13.1T \text{ cal}$$

Solution

Reaction under question: $Ni(s) + H_2O(g) = NiO(s) + H_2(g)$.

$\Delta G_R^o = \Delta G_1^0 - \Delta G_2^o = 450 + 10.45T$ cal.

$T = 750 + 273 = 1023$ K

$$\therefore \Delta G_{1023}^0 = 450 + 10.45 \times 1023 = 11140$$
$$= -RT \ln K$$

$$\ln K = -\frac{11140}{1.987 \times 1023}$$
$$= -5.4804$$

$$K = 0.0041677 = \frac{p_{H_2}}{p_{H_2O}} (a_{Ni} = 1 = a_{NiO})$$

$$p_{H_2} = 0.0041677 \cdot p_{H_2O} \tag{1}$$

and

$$p_{H_2} + p_{H_2O} = 1 \tag{2}$$

from 1 and 2 we get:

$$p_{H_2O} = 0.99585 \text{ atm}$$

$$p_{H_2} = 0.00415 \text{ atm}$$

Hence, maximum tolerable partial pressure of moisture is 0.99585 atm i.e., 99.6%.

Problem 5.4

Calculate the maximum moisture content that can be tolerated in the gaseous mixture of H_2–H_2O for reduction of WO_3 by hydrogen gas in a system maintained at a total pressure of one atmosphere and 400, 700, and 1000 K. Assuming that WO_3 and W are present as pure solids, comment on the utilization efficiency of hydrogen.

a. Does the efficiency improve with increasing temperature?

b. What will be the effect of increasing the total pressure in the system on this efficiency?

Given that:

1. $WO_3(s) = W(s) + 3/2O_2(g), \Delta G_1^o + 201500 + 10.2T \log T - 91.7T$ cal

2. $H_2(g) + \tfrac{1}{2}O_2(g) = H_2O(g), \Delta G_2^o = -58900 + 13.1T$ cal

Solution

The free energy change, ΔG^o for the required reaction: $WO_3(s) + 3H_2(g) = W(s) + 3H_2O(g)$,

$$\Delta G^o = \Delta G_1^o + 3\,\Delta G_2^o = 24800 + 10.2\,T\,\log\,T - 52.4\,T \quad cal$$

i.

$$\Delta G^o \text{ at } T = 400\,K, \quad \Delta G_{400}^o = 14460 \text{ cal.}$$

ii.

$$\Delta G^o \text{ at } T = 700\,K, \quad \Delta G_{700}^o = 8440 \text{ cal.}$$

iii.

$$\Delta G^o \text{ at } T = 1000\,K, \quad \Delta G_{1000}^o = 2800 \text{ cal}$$
$$\Delta G^o = -RT\ln K$$

at $T = 400$ K, $K_{400} = 1.253 \times 10^{-8}$.
$T = 700$ K, $K_{700} = 2.317 \times 10^{-3}$
$T = 1000$ K, $K_{1000} = 0.244$.

$$K = \left\{\frac{p_{H_2O}}{p_{H_2}}\right\}^3 \text{ or } \left\{\frac{p_{H_2O}}{p_{H_2}}\right\} = (K)^{1/3}$$

$$\left\{\frac{p_{H_2O}}{p_{H_2}}\right\} = K_{400}^{1/3} = \left(1.253 \times 10^{-8}\right)^{1/3}$$

$$= 2.323 \times 10^{-3} \tag{1}$$

$$p_{H_2} + p_{H_2O} = 1 \tag{2}$$

From Eqs. 1 and 2 we get:

$$p_{H_2} = 0.9981 \text{ atm and } p_{H_2O} = 0.0019 \text{ atm}$$

$\therefore$ The maximum permissible moisture content in the gaseous mixture of H_2-H_2O at 400 K is 0.19%.

By similar calculations we find that the maximum permissible moisture content in the gaseous mixture at 700 K is 11.66% and at 1000 K is 38.5%.

a. Thus, we find that the maximum permissible moisture content in the gaseous mixture of H_2–H_2O in reduction of WO_3 increases with increase of the reduction temperature. This demonstrates that the efficiency of utilization of hydrogen increases with increase of temperature. Hence, reduction should be carried out at higher temperature.

b. Since the number of moles of gases, H_2 on the reactant side and H_2O vapors on the product side are equal, the efficiency of hydrogen utilization will remain unaffected with the increase of total pressure of the system.

Problem 5.5

Using the data given below, calculate the precise temperature at which wustite decomposes into metallic iron and magnetite: $4FeO(s) \rightarrow Fe(s) + Fe_3O_4(s)$, where "FeO" refers to wustite saturated with metallic iron and magnetite. Which of the two oxides will be formed first at 500 °C, if pure iron is oxidized by gradual increase of partial pressure of oxygen from $p_{O_2} = 0$ to higher?

$$Fe\,(s) + \frac{1}{2}O_2(g) = FeO(s), \qquad \Delta G_1^o = -62952 + 15.493\ cal$$

$$3FeO\,(s) + \frac{1}{2}O_2(g) = Fe_3O_4(s), \quad \Delta G_2^o = -74538 + 29.373\ cal$$

Solution

Wustite is a nonstoichiometric solid solution of metallic iron and magnetite (Fe_3O_4). The stoichiometric FeO does not exist in the phase diagram. Wustite can exist stably only above 560 °C, below which it decomposes to metallic iron and magnetite. At the decomposition temperature, three phases of metallic iron, wustite, and magnetite can coexist. This temperature is called the triple point. At the decomposition temperature (triple point):

$$\Delta G_1^o = \Delta G_2^o, \therefore T = \frac{11586}{13.88} = 834.7\ K = 561.7\,°C$$

Thus, wustite decomposes into metallic iron and magnetite at 561.7 °C. It means below this temperature wustite does not exist stably.

In order to know whether magnetite (Fe_3O_4) or wustite (FeO) is formed first at 500 °C (773 K) with gradual increase of p_{O_2}, we must have the values of standard free energy of formation of the two oxides with one mole of oxygen. These may be obtained from the given data:

$$\frac{3}{2}Fe(s) + O_2(g) = \frac{1}{2}Fe_3O_4(s)$$

$$\Delta G_3^o = \frac{1}{2}(3\,\Delta G_1^o + \Delta G_2^o) = -131697 + 37.926\ T \tag{3}$$

$$2Fe\,(s) + O_2(g) = 2FeO(s)$$

$$\Delta G_4^o = 2\,\Delta G_1^o = -125904 + 30.986\ T \tag{4}$$

For both the reactions (3 and 4) $K = \frac{1}{p_{O_2}}$, hence $\Delta G^o = -RTlnK = RTlnp_{O_2}$ ($T = 773$ K).

$\therefore p_{O_2} = 1.13 \times 10^{-29}$ atm for reaction (3): magnetite formation

and $p_{O_2} = 1.49 \times 10^{-29}$ atm for reaction (4): wustite formation

Thus, reaction (3) has a lower p_{O_2} than reaction (4). In other words, when we start the oxidation of pure iron by increasing gradually the partial pressure from $p_{O_2} = 0$ to higher, the first iron oxide to form at 500 °C would be magnetite, and not wustite.

Problem 5.6

Some investigators based on their microscopic studies reported that Cu_2O underwent thermal decomposition into CuO and Cu at 375 °C. Making use of the following thermodynamic data, show that the old microscopic observation was in error. In reality, Cu_2O does not decompose.

$$2Cu(s) + \frac{1}{2}O_2(g) = Cu_2O(s), \Delta G_1^o = -39855 + 17.041\, T\ cal$$

$$Cu_2O(s) + \frac{1}{2}O_2(g) = 2CuO(s), \quad \Delta G_2^o = -31347 + 22.643\, T\ cal$$

Solution

Consider the free energy change for the formation of CuO(s) from pure elements:

$$Cu(s) + \frac{1}{2}O_2(g) = CuO(s)$$

$$\Delta G_3^o = \frac{1}{2}\left(\Delta G_1^o + \Delta G_2^o\right) = -35601 + 19.842\, T\ cal. \tag{3}$$

Decomposition of Cu_2O into CuO and Cu can be represented by the reaction:

$$Cu_2O(s) = CuO(s) + Cu(s)$$

$$\Delta G_4^o = \Delta G_3^o - \Delta G_1^o = -RT\ lnK_4 \tag{4}$$

$$\text{and } K_4 = \frac{a_{CuO} \cdot a_{Cu}}{a_{Cu_2O}}$$

At the decomposition temperature, three solid phases (Cu, Cu_2O, and CuO each of at unit activity) coexist, hence.

$$K_4 = \frac{a_{CuO} \cdot a_{Cu}}{a_{Cu_2O}} = \frac{1 \times 1}{1} = 1$$

$$\therefore \Delta G_4^o = -RT\ ln\ K_4 = -RT \ln\left(1\right) = 0$$

$$\text{hence } \Delta G_3^o - \Delta G_1^o = 0 \text{ and } \therefore \Delta G_3^o = \Delta G_1^o$$

$$\therefore -35601 + 19.842\, T = -39855 + 17.041\, T$$

$$\text{or } T = -1519\, K.$$

As a temperature of -1519 K is not feasible, $Cu_2O(s)$ does not decompose into metallic Cu and CuO. Thus, the microscopic observation was erroneous.

Problem 5.7

In the basic open hearth process, the reaction of manganese in the bath with iron oxide (FeO) in the slag attains a condition very close to true equilibrium. The steel contains 0.065 at% Mn and the slag analyzes (by wt%) as FeO-76.94%, Fe_2O_3– 4.15%, MnO-13.86%, MgO-3.74%, SiO_2–1.06% and CaO-0.25%. Calculate the value of equilibrium constant and standard free energy change for the above reaction at 1655 °C, assuming that the slag and Fe–Mn system behave ideally at this temperature. Neglect the effect of other metalloids, present in the steel.

Given that: molecular weights of FeO: 71.85, Fe_2O_3:159.70, MnO: 70.94, MgO: 40.32, SiO_2: 60.09, and CaO: 56.08.

Solution

The reaction may be represented as:

$$(FeO)_{slag} + [Mn]_{metal} = (MnO)_{slag} + [Fe]_{metal}$$

$$K = \frac{(a_{MnO}) \cdot [a_{Fe}]}{(a_{FeO}) \cdot [a_{Mn}]}$$

Assuming ideal behavior in the slag and in Fe–Mn system at 1655 °C, we can write:

$$K = \frac{(x_{MnO}) \cdot [x_{Fe}]}{(x_{FeO}) \cdot [x_{Mn}]}$$

Since the other metalloids present are negligible and Mn content is small, mole fraction of Fe may be assumed to be one.

$$\therefore K = \frac{(x_{MnO})}{(x_{FeO}) \cdot [x_{Mn}]}$$

Analysis of the slag needs to be converted into mole fraction:

constituents	wt%	mol wt	g mol
FeO	76.94	71.85	1.0710
Fe_2O_3	4.15	159.70	0.0259
MnO	13.86	70.94	0.1954
MgO	3.74	40.32	0.0927
SiO_2	1.06	60.09	0.0176
CaO	0.25	56.08	0.0044

In slag, iron and oxygen are also present as Fe_2O_3. Hence, mole fraction of Fe_2O_3 should be converted in terms of FeO to obtain the total value of x_{FeO}. In order to produce FeO, the following reaction may be considered:

$$Fe_2O_3(s) \rightarrow 2FeO(s) + \frac{1}{2}O_2(g)$$

Thus, one mole of Fe_2O_3 gives 2 mol of FeO.

∴Total number of gmol of FeO = gmol of FeO + 2 × gmol of Fe_2O_3
$$= 1.0710 + 2 \times 0.0259 = 1.1228$$

Total gmol of constituents in the slag = 1.1228 + 0.1954 + 0.0927 + 0.0176 + 0.0044 = 1.4329.

$$\therefore Total(x_{FeO}) = \frac{1.1228}{1.4329} = 0.7835$$

$$(x_{MnO}) = \frac{0.1954}{1.4329} = 0.1363$$

$$[x_{Mn}] = \frac{0.065}{100} = 0.065 \times 10^{-2}$$

$$\therefore K = \frac{(x_{MnO})}{(x_{FeO}) \cdot [x_{Mn}]} = \frac{0.1363}{0.7835 \times (0.065 \times 10^{-2})} = 267.8 \text{ Ans.}$$

$$\Delta G^o = -RT \ln K = -1.987 \times 1928 \times \ln(267.8) = -21416 \text{ cal Ans.}$$

Problem 5.8

In the study of carbothermic reduction of Al_2O_3 to form Al (l), liquid aluminum is contained in a graphite crucible and equilibrated with CO gas at 2300 K. The total pressure over the system is 1 atm. (a) Calculate the partial pressures of Al (v), Al_2O (g), and CO (g) in the system for the metal being aluminum ($a_{Al} = 0.8$) plus carbon. Other vapor species may be ignored. (b) What would the partial pressures of Al (v), Al_2O (g), and CO (g) for the case when $a_{Al} = 0.1$? Given that:

$$Al(l) = Al(v), \Delta G_1^o = 72000 - 25.9\, T, \text{cal mol}^{-1} \tag{1}$$

$$2Al(l) + \frac{1}{2}O_2(g) = Al_2O(g), \Delta G_2^o = -44100 - 10.13\, T, \text{cal mol}^{-1} \tag{2}$$

$$C(s) + \frac{1}{2}O_2(g) = CO(g), \Delta G_3^o = -28500 - 19.91\, T, \text{cal mol}^{-1} \tag{3}$$

$$C(s) + O_2(g) = CO_2(g), \Delta G_4^o = -94800 - 0.02\ T, \text{cal mol}^{-1} \quad (4)$$

Solution

a. $Al(l) = Al(v), \Delta G_1^o = 72000 - 25.9\ T, \text{cal mol}^{-1}$.
$T = 2300$ K, $\Delta G_1^o = 12430 = -RT \ln K$
For Eq. 1.

$$K = \left[\frac{p_{Al(v)}}{a_{Al(l)}}\right] \quad (i)$$

$$\ln K = \frac{\Delta G^o}{-RT} = -\frac{12430}{1.987 \times 2300} = -2.7199$$

$$K = 0.06588 = 0.06588 = p_{Al}^o\ (a_{Al} = 1)$$

From Eq. (i), $p_{Al\ (v)} = K.a_{Al} = p_{Al}^o.a_{Al} = 0.06588 \times 0.8 = 0.05271$ atm Ans.

In order to calculate the partial pressures of Al_2O (g) and CO, consider the following reaction:

$$C(s) + Al_2O(g) = 2Al(v) + CO(g) \quad (5)$$

for reaction (5),

$$\Delta G_5^o = 2\Delta G_1^o + \Delta G_3^o - \Delta G_2^o$$

$$= 159600 - 61.58\ T\ \text{cal}$$

At 2300 K, $\Delta G_5^o = 17966$ cal $= -RT \ln K_5$.

$$\ln K_5 = -\frac{17966}{1.987 \times 2300} = -3.971$$

$$\therefore K_5 = 1.962 \times 10^{-2}$$

$$K_5 = \frac{p_{Al}^2 \cdot p_{CO}}{p_{Al_2O} \cdot a_C} \text{(graphite, } a_C = 1)$$

Hence, $p_{Al_2O} = \dfrac{p_{Al}^2 \cdot p_{CO}}{1.962 \times 10^{-2}} \quad (6)$

Ignoring other vapor species we can write as:

$$p_{Al} + p_{Al_2O} + p_{CO} = 1 \tag{7}$$

$$\therefore p_{Al} + \frac{p_{Al}^2 \cdot p_{CO}}{1.962 \times 10^{-2}} + p_{CO} = 1, \left(p_{Al\,(v)} = 0.05271\ \text{atm} \right)$$

Hence we get, $0.05271 + 0.1416.\ p_{CO} + p_{CO} = 1$.

$$\therefore p_{CO} = 0.837\ \text{atm Ans.}$$

and

$$p_{Al_2O} = 1 - p_{Al} - p_{CO} = 1 - 0.05271 - 0.837 = 0.11\ \text{atm}\quad \text{Ans.}$$

b. When $a_{Al} = 0.1$.

$$p_{Al} = p_{Al}^o \cdot a_{Al} = 0.06588 \times 0.1 = 6.588 \times 10^{-3}\text{atm}$$

Following a similar procedure and substituting $p_{Al} = 6.588 \times 10^{-3}\text{atm}$ in Eq. 7, we get:

$p_{CO} = 0.9913$ atm and $p_{Al_2O} = 2.14 \times 10^{-3}$ atm   Ans.

Problem 5.9

ZnO is reduced with carbon in a retort at 950 °C and condensed in a condenser at 450 °C. If the reaction product, zinc vapors enter the condenser at 950 °C and leave at 450 °C, assuming equilibrium conditions, calculate the efficiency of zinc recovery. Given that:

$$\log p(\text{mm Hg}) = -\frac{6620}{T} - 1.255\ \log\ T + 12.34$$

Solution

Since the reduction temperature is above the boiling temperature of zinc (907 °C) $p_{Zn} = 760$ mm Hg and at 450 °C (723 K), p_{Zn} can be calculated from the given expression:

$$\log p\ (\text{mm Hg}) = -\frac{6620}{723} - 1.255 \log\ (723) + 12.34 = -0.4045$$

$$p_{Zn} = 0.394\ \text{mm Hg}$$

The reduction reaction under question is: ZnO (s) + C(s) = Zn(g) + CO(g).

Gases leaving the condenser at 1 atm will have $\left[\frac{0.394}{760 - 0.394} \right] = 5.19 \times 10^{-4}$, mole of zinc per mole of CO.

$$\therefore \text{zinc loss} = 0.0519\%$$

Therefore, efficiency of zinc recovery = 99.95% Ans.

Problem 5.10

In the Pidgeon's process for the manufacture of magnesium, calcined dolomite is reduced with ferrosilicon according to the equation:

$$2MgO.CaO(s) + Si(s) = 2Mg(g) + 2CaO.SiO_2(s)$$

Given that:

$$2MgO(s) + Si(s) = 2Mg\,(g) + SiO_2(s)$$

$$\Delta G_1^o = 152600 + 11.37T \log\,T - 99.18T \text{ cal}$$

$$2CaO(s) + SiO_2(s) = 2\,CaO.SiO_2(s)$$

$$\Delta G_2^o = -30200 + 1.2T \text{ cal}$$

Evaluate the partial pressure of magnesium vapor at 1200 °C and discuss the feasibility of the process.

Solution

For the reduction reaction under question: $2MgO.\,CaO(s) + Si(s) = 2Mg(g) + 2CaO.$
$SiO_2(s)$ the standard free change is obtained as:

$$\Delta G^o = \Delta G_1^o + \Delta G_2^o = 122400 + 11.37\,T \log\,T - 100.38\,T \; (T = 1473\,K)$$
$$\Delta G_{1473}^o = 122400 + 11.37 \times 1473 \log\,1473 - 100.38 \times 1473$$
$$= 122400 + 53061 - 147860$$
$$= +27601 \text{ cal}$$

For the above reaction, $K = p_{Mg}^2$ (activities of solid components being unity).
$$\Delta G^o = -RT \ln K = -RT \ln p_{Mg}^2 = -4.575 \times T \times 2 \log p_{Mg}$$
$p_{Mg} = 0.00896$ atm
since $p_{Mg} < 1$ atm and $a_{SiO_2} = 0.001$ (in $2CaO.SiO_2$) slag, consider van't Hoff isotherm:

$$\Delta G = \Delta G^o + RT \ln\,K = 27601 + RT \ln \left[\frac{p_{Mg}^2 \cdot a_{SiO_2}}{a_{MgO} \cdot a_{Si}}\right]$$

In the Pidgeon process, reduction is carried out at 1200 °C under a vacuum of the order of 10^{-4} atm.

$$\therefore p_{Mg} = 10^{-4} \text{ atm.}$$

in $2MgO.CaO$, $a_{MgO} = 1$, and $a_{Si} = 1$

$$\therefore \Delta G = 27601 + 1.987 \times 2.303 \times 1473 \log \left[\frac{\left(10^{-4}\right)^2 \times 10^{-3}}{1 x 1} \right]$$

$$= 27600 - 74145 = -46545 \text{ cal}$$

Since ΔG is negative, reaction is feasible under the reduced pressure of 10^{-4} atm at 1200 °C in the presence of a basic oxide (CaO).

Problem 5.11

(Elliott, J. F. (1980–81) Personal communication, Massachusetts Institute of Technology, Cambridge, Mass., USA)

Zinc vapors can be produced from zinc oxide in the presence of carbon according to the following reactions:

$$ZnO(s) + CO(g) = Zn\ (v) + CO_2(g) \tag{i}$$

$$CO_2(g) + C(s) = 2CO(g) \tag{ii}$$

a. Plot the following lines on a graph of $\log p_{CO_2}/p_{CO}$ as the ordinate and $1/T$ as the abscissa:

$$Zn\ (l), a_{Zn} = 1; Zn\ (l), a_{Zn} = 0.1.$$

$$p_{Zn} = 1\ \text{atm}, p_{Zn} = 0.1\ \text{atm}$$

$$p_{CO} = 1\ \text{atm}, p_{CO} = 0.1\ \text{atm}$$

b. Consider a system which initially contains only ZnO and carbon. The total pressure is maintained at 1 atm. What is the lowest temperature at which reactions (i) and (ii) can be simultaneously in equilibrium? What are the related pressures of Zn (v), CO, and CO_2? Plot where these conditions are on your diagram (in the temperature range from 700 to 1500 °C).

Note: You are advised to take the necessary data from literature.

Solution

a. For the vaporization of zinc:

$$Zn\ (l) = Zn\ (v) \tag{1}$$

$$\log p\ (\text{mm Hg}) = -\frac{6620}{T} - 1.255 \log T + 12.34$$

$$\log p(\text{atm}) = \log p(\text{mm Hg}) - \log 760$$

$$\therefore \log p(\text{atm}) = -\frac{6620}{T} - 1.255 \log T + 9.459$$

In terms of free energy change for the reaction (1):

$$\Delta G_1^o = -RT \ln p_{Zn} = -4.575 \ T \ \log p_{Zn}$$

$$= -4.575 \ T \left\{ -\frac{6620}{T} - 1.255 \ \log \ T + 9.459 \right\}$$

$$= 30290 + 5.74 \ T \ \log T - 43.284 \ T \ \text{cal}$$

Decomposition reaction : $ZnO(s) = Zn(v) + \frac{1}{2}O_2(g)$ \hfill (2)

$$\Delta G_2^o = 115420 + 10.35 \ T \ \log T - 82.68 \ T \ \text{cal}$$

Other required reactions and their free energy values:

$$C(s) + O_2(g) = CO_2(g) \hfill (3)$$

$$\Delta G_3^o = -94200 - 0.2 \ T \ \text{cal}$$

$$C(s) + \frac{1}{2}O_2 = CO(g) \hfill (4)$$

$$\Delta G_4^o = -26700 - 20.95 \ T \ \text{cal}$$

$$CO(g) + \frac{1}{2}O_2 = CO_2(g) \hfill (5)$$

$$\Delta G_5^o = \Delta G_3^o - \Delta G_4^o = -67500 - 20.75 \ T \ \text{cal}$$

Reduction of ZnO into liquid and gaseous zinc:

$$ZnO(s) + CO(g) = Zn \ (l) + CO_2(g) \hfill (6)$$

$$\Delta G_6^o = 17630 + 4.61 \ T \ \log \ T - 18.35 \ T \ \text{cal}$$

$$ZnO(s) + CO(g) = Zn(v) + CO_2(g) \hfill (7)$$

$$\Delta G_7^o = \Delta G_2^o + \Delta G_5^o = 47920 + 10.35 \ T \ \log \ T - 61.63 \ T \ \text{cal}$$

From the above equations, we can get various p_{CO_2}/p_{CO} ratios for different lines:

A.   For Zn (l) lines, that is, Eq. 6, where $K_6 = \frac{p_{CO_2}.a_{Zn(l)}}{p_{CO}.a_{ZnO(s)}}$ and as $a_{ZnO(s)} = 1$.

$$\frac{K_6}{a_{Zn \ (l)}} = \frac{p_{CO_2}}{p_{CO}}$$

$$\Delta G_6^o = 17630 + 4.61\, T\, \log\, T - 18.35\, T$$

$$= -4.575\, T\, \log\, K_6 = -4.574\, T\, \log\left[\frac{p_{CO_2}.a_{Zn(l)}}{p_{CO}}\right]$$

$$\therefore \log\left[\frac{p_{CO_2}.a_{Zn(l)}}{p_{CO}}\right] = -\frac{3852.7}{T} - 1.007\, \log\, T + 4.01$$

$$\text{or } \log\left[\frac{p_{CO_2}}{p_{CO}}\right] = -\frac{3852.7}{T} - 1.007\, \log\, T + 4.01 - \log a_{Zn\,(l)} \tag{8}$$

p_{CO_2}/p_{CO} ratios calculated from Eq. 8 at different temperatures and activities of liquid zinc, $a_{Zn} = 1$ and $= 0.1$, are listed below:

T,K	$10^4/K$	K_6	$\dfrac{K_6}{a_{Zn(l)}}$	$\log\left[\dfrac{p_{CO_2}}{p_{CO}}\right]$	$\dfrac{K_6}{a_{Zn(l)}}$	$\log\left[\dfrac{p_{CO_2}}{p_{CO}}\right]$
			$a_{Zn(l)} = 1$		$a_{Zn(l)} = 0.1$	
973	10.28	1.09×10^{-3}	1.09×10^{-3}	-2.96	1.09×10^{-2}	-1.96
1073	9.32	2.32×10^{-3}	2.32×10^{-3}	-2.64	2.32×10^{-2}	-1.64
1173	8.53	4.28×10^{-3}	4.28×10^{-3}	-2.37	4.28×10^{-2}	-1.37
1273	7.86	7.14×10^{-3}	7.14×10^{-3}	-2.15	7.14×10^{-2}	-1.15
1373	7.28	1.10×10^{-2}	1.10×10^{-3}	-1.96	1.10×10^{-2}	-0.96
1473	6.79	1.58×10^{-2}	1.58×10^{-2}	-1.80	1.58×10^{-1}	-0.80
1573	6.36	2.18×10^{-2}	2.18×10^{-2}	-1.66	2.18×10^{-1}	-0.66
1673	5.98	2.87×10^{-2}	2.87×10^{-2}	-1.54	2.87×10^{-1}	-0.54
1773	5.64	3.65×10^{-2}	3.65×10^{-2}	-1.44	3.65×10^{-1}	-0.44

B. For Zn (v) lines, that is, Eq. 7, where $K_7 = \frac{p_{CO_2}.p_{Zn}}{p_{CO}.a_{ZnO(s)}}$ and as $a_{ZnO(s)} = 1$.

$$\frac{K_7}{p_{Zn}} = \frac{p_{CO_2}}{p_{CO}}$$

$$\Delta G_7^o = 47920 + 10.35\, T\, \log\, T - 61.63\, T \text{ cal.}$$

$$= -4.575\, T\, \log\, K_7 = -4.574\, T\, \log\left[\frac{p_{CO_2}.p_{Zn}}{p_{CO}}\right]$$

$$\therefore \log\left[\frac{p_{CO_2}.p_{Zn}}{p_{CO}}\right] = -\frac{10472}{T} - 2.262\, \log\, T + 13.47$$

$$\text{or } \log\left[\frac{p_{CO_2}}{p_{CO}}\right] = -\frac{10472}{T} - 2.262\, \log\, T + 13.47 - \log p_{Zn} \tag{9}$$

p_{CO_2}/p_{CO} ratios calculated from Eq. 9 at different temperatures and partial pressures of zinc, $p_{Zn} = 1$ atm and $= 0.1$ atm, are listed below:

T,K	10^4/K	K_7	$\dfrac{K_7}{a_{Zn(l)}}$	$\log\left[\dfrac{p_{CO_2}}{p_{CO}}\right]$	$\dfrac{K_7}{a_{Zn(l)}}$	$\log\left[\dfrac{p_{CO_2}}{p_{CO}}\right]$
			$p_{Zn} = 1$ atm		$p_{Zn} = 0.1$ atm	
973	10.28	8.87×10^{-5}	8.87×10^{-5}	-4.05	8.87×10^{-4}	-3.05
1073	9.32	7.14×10^{-4}	7.14×10^{-4}	-3.15	7.14×10^{-3}	-2.15
1173	8.53	3.96×10^{-3}	3.96×10^{-3}	-2.40	1.66×10^{-1}	-0.78
1273	7.86	1.66×10^{-2}	1.66×10^{-2}	-1.78	1.66×10^{-1}	-0.78
1373	7.28	5.55×10^{-2}	5.55×10^{-2}	-1.26	5.55×10^{-1}	-0.26
1473	6.79	1.56×10^{-1}	1.56×10^{-1}	-0.81	1.56	0.193
1573	6.36	0.38	0.38	-0.42	3.8	0.58
1673	5.98	0.83	0.83	-0.08	8.3	0.92
1773	5.64	1.64	1.64	0.21	16.4	1.215

C. For CO line consider reaction : $CO_2(g) + C(s) = 2CO(g)$ (10)

$$\Delta G_c^o = 2\Delta G_4^o - \Delta G_3^o = 40800 - 41.70\, T = -4.575\, RT \log K_c$$

$$\log K_c = -\frac{8916}{T} + 9.113$$

$$K_c = \frac{p_{CO}^2}{p_{CO_2}} \text{ or } \frac{p_{CO_2}}{p_{CO}} = \frac{p_{CO}}{K}$$

or $$\log\left[\frac{p_{CO_2}}{p_{CO}}\right] = -\log K_c + \log p_{CO} = \frac{8911}{T} - 9.113 + \log p_{CO}$$ (11)

p_{CO_2}/p_{CO} ratios calculated from Eq. 11 at different temperatures and partial pressures of CO, $p_{CO} = 1$ atm and $= 0.1$ atm, are listed below:

T,K	10^4/K	K_C	$\dfrac{p_{CO}}{K}$	$\log\left[\dfrac{p_{CO_2}}{p_{CO}}\right]$	$\dfrac{p_{CO}}{K_c}$	$\log\left[\dfrac{p_{CO_2}}{p_{CO}}\right]$
			$p_{CO} = 1$ atm		$p_{CO} = 0.1$ atm	
973	10.28	0.89	1.124	0.051	0.112	-0.95
1073	9.32	6.36	0.157	-0.804	0.016	-1.80
1173	8.53	32.50	0.031	-1.51	0.0031	-2.51
1273	7.86	128.56	7.78×10^{-3}	-2.11	7.78×10^{-4}	-3.11
1373	7.28	416.23	2.40×10^{-3}	-2.62	2.40×10^{-4}	-3.62
1473	6.79	1148.85	8.70×10^{-4}	-3.06	8.70×10^{-5}	-4.06
1573	6.36	2785.12	3.59×10^{-4}	-3.44	3.59×10^{-5}	-4.44
1673	5.98	6081.28	1.64×10^{-4}	-3.79	1.64×10^{-5}	-4.79
1773	5.64	12150	8.23×10^{-5}	-4.08	8.23×10^{-6}	-5.08

b. As stated in the problem, system contains only ZnO and C. O comes from ZnO (s) according to the reaction (7): $ZnO(s) + C(s) = Zn(v) + CO(g)$.

Thus, $n_{Zn} = n_O$.

Considering the formation of CO and CO_2 gases according to reaction (10), we can write:

$p_{Zn} = p_{CO} + 2p_{CO_2}$, since total pressure, $P_T = 1$ atm

$$p_{Zn} + p_{CO} + 2p_{CO_2} = 1$$

Assuming p_{CO_2} to be very small, we can write: $p_{Zn} = p_{CO} = 0.5$, and the lowest temperature at which the given chemical reactions (7) and (10) would be at equilibrium can be calculated by equating $\log\left[\frac{p_{CO_2}}{p_{CO}}\right]$ values derived in Eqs. 9 and 11, respectively, for reactions (7) and (10):

$$\log\left[\frac{p_{CO_2}}{p_{CO}}\right] = -\frac{10472}{T} - 2.262 \log T + 13.47 - \log p_{Zn}$$

$$= \frac{8911}{T} - 9.113 + \log p_{CO}$$

when $p_{Zn} = p_{CO} = 0.5$,

$$-\frac{10472}{T} - 2.262 \log T + 13.47 - \log(0.5) = \frac{8911}{T} - 9.113 + \log(0.5)$$

$$\text{or} -\frac{19383}{T} - 2.262 \log T = -22.583 + 2\log(0.5) = -23.186$$

$$\therefore T = 1193 \text{ K} = 920\,^{\circ}\text{C Ans.}$$

The relevant plots are shown in Fig. 5.11.

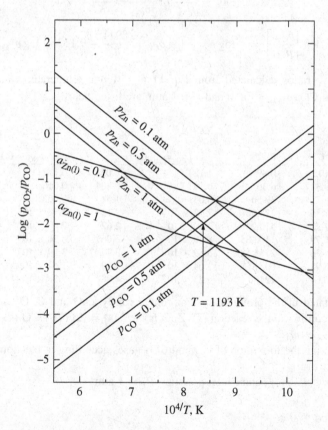

Fig. 5.11 $\log\left(\frac{p_{CO_2}}{p_{CO}}\right)$ vs. $\frac{1}{T}$ at different activities of zinc and partial pressures of zinc vapor and carbon monoxide

Problem 5.12

Show by suitable calculations why is it possible to operate the Zn–Pb blast furnace with such conditions that zinc oxide in burden is reduced at 1200 °C and a total pressure of 1 atm but iron oxide is not? Assume that the flue gas contains 6.5% zinc vapors (volume basis).

Given that: $Zn(g) + \frac{1}{2}O_2(g) = ZnO(s)$

$$\Delta G^o = -115420 - 10.35 \, T \, \log \, T + 82.38T \, cal$$

$$Fe(s) + \frac{1}{2}O_2(g) = FeO(l)$$

$$\Delta G^o = -62600 + 15.18 \, T \, cal$$

Solution

At $1473 \, K : \Delta G^o_f(ZnO) = -115420 - 48300 + 121345 = -42375 \, cal$
$\Delta G^o_f(FeO) = -62600 + 22360 = -40240 \, cal$

The reactions concerning the problem are:

$$ZnO(s) + CO(g) = Zn(g) + CO_2(g)$$
$$Fe(s) + CO_2(g) = FeO(l) + CO(g)$$

The overall reaction: $ZnO(s) + Fe(s) = FeO(l) + Zn(g)$, $K = p_{Zn}$

$$\Delta G^o_R = \Delta G^o_f(FeO) - \Delta G^o_f(ZnO)$$
$$= -40240 - (-42375) = +2135 \, cal$$
$$\Delta G = \Delta G^o_R + RT \, \ln p_{Zn}$$
$$= 2135 + 1.987 \times 1473 \times \ln 0.065$$
$$= 2135 - 8000 = -5865 \, cal$$

Since ΔG is negative, ZnO is reduced but not FeO.

References

1. Rosenqvist, T. (1974). *Principles of extractive metallurgy.* New York: McGraw-Hill. (Chapter 9 and 10).
2. Gilchrist, J. D. (1980). *Extraction metallurgy* (2nd ed.). New York: Pergamon. (Chapter 12).
3. Bogdandy, L. V., & Engell, H. J. (1971). *The reduction of iron ores.* Berlin: Springer Verlag. (Chapter 3).
4. Ellingham, H. J. T. (1944). Reducibility of oxides and sulphides. *Journal of the Society of Chemical Industry Transations, 63,* 125–133.

5. Moore, J. J. (1990). *Chemical metallurgy* (2nd ed.). Oxford: Butterworth-Heinemann. (Chapter 7).
6. Shamsuddin, M., & Sohn, H. Y. (2019). Constitutive topics in physical chemistry of high-temperature nonferrous metallurgy – A review: Part 2. Reduction and refining. *JOM, 71*(9), 3266–3276.
7. Richardson, F. D., & Jeffes, J. H. E. (1948). The thermodynamics of substances of interest in iron and steelmaking, part I-oxides. *Journal of the Iron and Steel Institute, 160*, 261–270.
8. Richardson, F. D., Jeffes, J. H. E., & Withers, G. (1950). The thermodynamics of substances of interest in iron and steelmaking, Part II-compounds between oxides. *Journal of the Iron and Steel Institute, 166*, 213–234.
9. Richardson, F. D., & Jeffes, J. H. E. (1952). The thermodynamics of substances of interest in iron and steelmaking, part III-sulphides. *Journal of the Iron and Steel Institute, 171*, 165–175.
10. Chen, F., Mohassab, Y., Jiang, T., & Sohn, H. Y. (2015). Hydrogen reduction kinetics of hematite concentrate particles relevant to a novel flash ironmaking process. *Metallurgical and Materials Transactions B: Process Metallurgy and Materials Processing Science, 46B*, 1133–1145.
11. Chen, F., Mohassab, Y., Zhang, S., & Sohn, H. Y. (2015). Kinetics of the reduction of hematite concentrate particles by carbon monoxide relevant to a novel flash ironmaking process. *Metallurgical and Materials Transactions B: Process Metallurgy and Materials Processing Science, 46B*, 1716–1172.
12. Fan, D.-Q., Sohn, H. Y., & Elzohiery, M. (2017). Analysis of the reduction rate of hematite concentrate particles by H_2 or CO in a drop-tube reactor through CFD modeling. *Metallurgical and Materials Transactions B: Process Metallurgy and Materials Processing Science, 48B*, 2677–2684. https://link.springer.com/content/pdf/10.1007%2Fs11663-017-1053-2.pdf.
13. Sohn, H. Y., Elzohiery, M. and Fan, D-Q. (2019) Development of flash ironmaking technology. In Petrova, V. M., (Ed.) *Advances in engineering research* (Vol. 26, pp. 23–106). New York (ISBN: 987-1-715-5 (eBook); ISSN: 2163-3932): Nova Science Publishers, Hauppauge, (Chapter 2).
14. Szekely, J., Evans, W., & Sohn, H. Y. (1976). *Gas-solid reactions.* New York: Academic Press. (Chapters 3, 4 and 5).
15. Jander, W. (1927). Reaktionen in fasten Zustand bei hohern Temperaturen. *Zeitschrift für Anorganische und Allgemeine Chemie, 163*, 1–30. 31–52.
16. Gnistling, A. M., & Brounshtein, B. I. (1950). The diffusion kinetics of reactions in spherical particles. *Journal of Applied Chemistry USSR, 23*, 1327–1338.
17. Valensi, G. (1935). Kinetics of oxidation of metallic spherules and powders. *Bulletin de la Société Chimique de France, 5*, 668–681.
18. Habashi, F. (1970). *Principles of extractive metallurgy, Vol. 1, General principles.* New York: Gordon & Breach (Chapter 7).
19. Ghadi, A. Z., Valipour, M. S., Vahedi, S. M., & Sohn, H. Y. (2020). A review on the gaseous reduction of iron oxide pellets. *Steel Research International, 91*(1), 1900270. https://doi.org/10.1002/srin.201900270.
20. McKewan, W. M. (1961). Reduction kinetics of magnetite in H_2-H_2O-N_2 mixtures. *Transactions of the Metallurgical Society, AIME, 221*, 140–145.
21. Themelis, W. H., & Gauvin, N. J. (1963). A generalized rate equation for reduction of iron oxides. *Transactions of the Metallurgical Society, AIME, 227*, 290–300.
22. von Bogdandy, L. & Engell, H. J. (1971) *The reduction of iron ores: Scientific basis and technology.* Berlin/Heidelberg: Springer. https://doi.org/10.1007/978-3-662-10400-2.
23. Lu, W. K. (1963) The general rate equation for gas-solid reactions in metallurgical processes. *Transactions of the Metallurgical Society, AIME 227*, 203–206.
24. Spitzer, R. H., Manning, F. S., & Philbrook, W. O. (1966). Mixed-control reaction kinetics in the gaseous reduction of hematite. *Transactions of the Metallurgical Society, AIME, 236*, 726–742.

25. Spitzer, R. H., Manning, F. S., & Philbrook, W. O. (1966). Generalized model for the gaseous, topochemical reduction of porous hematite spheres. *Transactions of the Metallurgical Society, AIME 236*, 1715–1724.

26. Edstrom, J., O. (1953) The mechanism of reduction of iron oxide. *Journal of the Iron and Steel Institute. 175*, 289–304.

27. Tien, R. H., & Turkdogan, E. T. (1972). Gaseous reduction of iron oxides: Part IV. Mathematical analysis of partial internal reduction-diffusion control. *Metallurgical Transactions, 3*, 2039–2048.

28. Sohn, H. Y., & Szekely, J. (1972). A structural model for gas-solid reactions with a moving boundary-III. A general dimensionless representation of the irreversible reaction between a porous solid and a reactant gas. *Chemical Engineering Science, 27*, 763–778. https://doi.org/10.1016/0009-2509(72)85011-5.

29. Sohn, H. Y. (1978). The law of additive reaction times in fluid-solid reaction. *Metallurgical Transactions B, 9*, 89–96. https://doi.org/10.1007/BF02822675.

30. Sohn, H. Y. (1991). The coming of age of process engineering in extractive metallurgy. *Metallurgical Transactions B, 22*, 737–754. https://doi.org/10.1007/BF02651151.

31. Sohn, H., Y. (2014). *Reaction engineering models*. Treatise Process Metall., Oxford, UK and Waltham, MA, USA: Elsevier, pp 758–810.

32. Bodsworth, C. (1963). *Physical chemistry of iron and steel manufacture*. Longmans Green & Co. Ltd., London (Chapter 12, 13, 14).

33. Ghosh, A., & Chatterjee, A. (2008). *Ironmaking and steelmaking*. Prentice Hall India Pvt. Ltd., New Delhi, (Chapter 13 and 14).

34. Floyd, J. M. (2005). Converting an idea into a worldwide business commercializing smelting technology. *Metallurgical and Materials Transactions B: Process Metallurgy and Materials Processing Science, 36B*, 557–575.

35. Shamsuddin, M. & Sohn, H. Y. (2019). Constitutive topics in physical chemistry of ironmaking and steelmaking: *A review. Trends in Physical Chemistry, 19*, 33–50.

36. Morgan, S. W. K., & Lumsden, J. (1959). Zinc blast furnace operation. *JOM, 11*(4), 270–275.

37. Shamsuddin, M. & Sohn, H. Y. (1981, February 22–24) *Extractive metallurgy of tungsten, Proceedings of the symposium on extractive metallurgy of refractive metals*, AIME, Chicago, pp. 205–230.

38. Sevryukov, N., Kuzmin, B. & Chelishchev, Y. (translated: Kuznetsov B.) (1960). *General metallurgy*. Moscow: Peace Publishers, (Chapter 19 and 20).

39. Kubaschewski, O., & Alcock, C. B. (1979). *Metallurgical thermochemistry* (5th ed.). Oxford: Pergamon.

40. Curnutt, J. L., Prophet, H., McDonald, R. A., & Syverud, A. N. (1975). *JANAF thermochemical tables*. Midland: The Dow Chemical Company.

41. Barin, I., Knacke, O., & Kubaschewski, O. (1977). *Thermochemical properties of inorganic substances, (supplement)*. New York: Springer.

Chapter 6
Interfacial Phenomena

A molecule in a bulk of material is pulled from all the sides whereas a molecule at the surface experiences no pull from the top and is pulled only from the sides and below. Bonding between molecules takes place with lowering of free energy. The bond energies between particles are negative because energy is evolved on bond formation. Therefore, atoms, molecules, or ions on the surface of a liquid or solid possess higher energies than those located in the bulk of the substance. The energy of the surface increases with the increase of the surface area. Hence, the surface free energy can be defined as the energy required to create 1 m^2 of new surface. This energy can be expressed in Joule per square meter (or ergs per square cm). As the molecules in solids are rigid, no such effect is realized. Surface free energy is a characteristic property of the liquid phase making a free surface. Since reduction in the surface area reduces the free energy, the surface of the liquid will tend to contract. Liquids tend to form spherical drops because spheres have the least surface area to volume ratio among all geometrical shapes. The tensional force causing the contraction of the free liquid surface is known as surface tension. The surface tension is defined as the force in newtons acting at right angles to a line 1 m long on the surface. The work done in producing 1 m^2 of new surface is equivalent to the work done in stretching the surface by 1 m along a 1 m line against the surface tension and this is the surface free energy, σ. Thus, the surface tension σ, newton per meter (N m^{-1}), is numerically equal to the surface free energy J m^{-2}. Surface tension is generally used with reference to liquid, whereas surface free energy and interfacial energy are used with reference to liquids and solids, respectively. Solids definitely have surface free energy, but due to their rigid surfaces, they are neither smooth nor free to allow the solid to fill the shape of a container, and therefore, the concept of surface tension is not applicable in solids. Table 6.1 lists the values of surface tension of a few liquids.

On the surface layer, atoms/molecules are bonded in two dimensions only. The bonds which are free often attract impure elements. The surface-active agents are preferentially absorbed on the surface and reduce the surface tension [1]. Thus, impurities in metals reduce the surface tension (Table 6.1 refers to the surface

M. Shamsuddin, *Physical Chemistry of Metallurgical Processes, Second Edition*, The Minerals, Metals & Materials Series, https://doi.org/10.1007/978-3-030-58069-8_6

Table 6.1 Surface tension of a few liquid metals

Metal	Temperature (°C)	$(N\,m^{-1})$ or $J\,m^{-2}$
Aluminum	660	0.914
Antimony	630.5	0.367
Beryllium	1283	1.390
Bismuth	671	0.378
Cadmium	321	0.570
Calcium	865	0.361
Chromium	1875	1.700
Copper	1083	1.303
Gold	1063	1.169
Iron	1550	1.788
Steel (0.4% C)	1600	1.560
Lead	327	0.444
Lead	1000	0.401
Mercury	20	0.480
Mercury	40	0.466
Silver	995	1.128
Silver	1120	0.923
Tin	232	0.531
Tin	1000	0.497
Zinc	650	0.750

tension of iron and steel). The surface tension of metals does not vary much with temperature.

Since most of the metals are extracted and refined in the liquid state and then solidified, surface tension plays a prominent role in nucleation of embryos during the course of solidification, which is probably the most important transformation in the field of metallurgy. As mixing in solid state takes place only by diffusion that is quite slow, most alloys are mixed in liquid state by convection and diffusion and then solidified to produce homogeneous starting materials. In Chap. 4, we noticed that slag with proper surface tension/interfacial energy and viscosity by control of temperature and composition plays an important role in the extraction and refining of metals. Many reactions of interest to metallurgists are hetrogeneous. For example, (i) reactions between two liquids, that is, transfer of species from liquid metal to slag (and vice versa) in steelmaking, (ii) reactions between a solid and a liquid, namely, leaching, precipitation of pure Al_2O_3 from sodium aluminate solution by seeding in the Bayer's process, precipitation of nickel and cobalt powders from ammoniacal solution by reduction with hydrogen gas at high pressure in the Sheritt Gordon process, removal of deoxidation products, and solid impurities from liquid melts, (iii) reactions between a gas and a solid, namely, reduction of oxides by $CO–CO_2$ gaseous mixture in blast furnaces, roasting of sulfides, and calcination of carbonates, and (iv) reactions between a gas and a liquid, such as nucleation of gas bubbles in liquid melts. Interfacial energy between respective interfaces plays an important role in all these reactions. In addition, we find a very effective and significant role of interfacial phenomena in the formation of foam and emulsion in basic oxygen steelmaking, Kennecott-Outotec Flash Converting and Mitsubishi Continuous Converting processes for copper production, fused salt electrolysis of halides, anode effect during electrolysis of Al_2O_3 dissolved in cryolite, froth flotation, comminution, penetration of metal in cavities of the mold walls, and so on. In the following sections, some applications of interfacial phenomena will be illustrated with suitable worked-out examples.

6.1 Precipitation

Precipitation of a solid from a liquid melt or aqueous solution and metallurgical transformation whether it is a change in state or a phase change within one of the three states are governed by processes of nucleation and growth. However, there is vast literature on kinetics of nucleation and growth during solidification of liquid metals, but very little is known about precipitation of nonmetallic phases (inclusions) in deoxidation of steel and precipitation of metal powders from aqueous solutions. There are several theories [2–4] on the kinetics of nucleation and growth. In the classical theory, the work done (ω) in forming a spherical nucleus of radius (r) is given by:

$$\omega = 4\pi r^2 \sigma + \frac{4}{3}\pi r^3 \left(\frac{\Delta G}{v}\right), \tag{6.1}$$

where σ is the interfacial tension between the matrix and the new phase, ΔG is the difference in free energy between the matrix and the new phase, and v is the molar volume of the new phase. There is a certain radius, r^c known as the critical nucleus radius, associated with the maximum free energy [2, 5] (Fig. 6.1). If $r < r^c$, the system lowers its free energy by dissolution of the solid. On the other hand, when $r > r^c$, the free energy of the system decreases as the solid grows. The unstable solid particles with $r < r^c$ are known as clusters or embryos while stable particles with $r > r^c$ are referred to as nuclei. When $r = r^c$, $d\omega/dr = 0$, the critical nucleus radius can be obtained by differentiating Eq. 6.1:

$$r^c = -\frac{2\sigma v}{\Delta G} \tag{6.2}$$

The corresponding work, ω^c (= ΔG^c the free energy change to form the nucleus of critical size) is obtained by substituting the value of r^c in Eq. 6.1.

$$\omega^c = \frac{16\pi\sigma^3 v^2}{3(\Delta G)^2} \tag{6.3}$$

Fig. 6.1 Free energy change for homogenous nucleation of a spherical nucleus of radius r

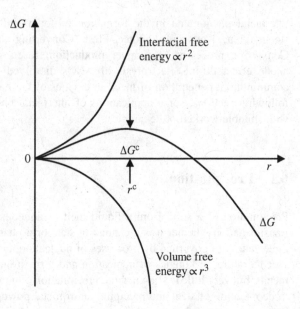

ΔG^c is temperature dependent and can be regarded as the necessary activation energy for the formation of the nucleus. The formation of the precipitate in this manner without artificial simulation [6] is known as homogeneous nucleation. Hence, ΔG^c_{hom} is the minimum work required to form the nucleus of critical size, r^c, homogeneously from the melt. The rate of formation of nuclei [7], I (per cm^3 of melt per second) homogeneously is expressed as:

$$I = A \, \exp\left(\frac{\Delta G^c_{hom}}{kT}\right) \tag{6.4}$$

where A is the frequency factor, k is the Boltzmann's constant and T is the temperature in degree absolute. Turpin and Elliott [7] have estimated the values of A as 10^{26} and 10^{30}, respectively, for Al_2O_3 and FeO in molten slags. The rate of formation of nuclei, I, varies enormously with supersaturation. The effect of ΔG^c_{hom} as given in Eq. 6.4 is so dominant that it overwhelms the frequency factor. This is visualized from the fact that $I = 1$ or 10,000 (per cm^3 of matrix per second) makes no significant difference on the rate of nucleation. The value of supersaturation necessary for homogeneous nucleation, at $I = 1$ is designated as ΔG^c_{hom}. If $I = 1$, Eq. 6.4 gives, $\Delta G^c_{hom} = -kT\ln A$. Hence, from Eqs. 6.3 and 6.4 we get:

$$\Delta G^c_{hom} = -2.7\nu\left(\frac{\sigma^3}{kT \log A}\right)^{\frac{1}{2}}. \tag{6.5}$$

Turpin and Elliott [7] have suggested a method to estimate oxygen concentration for supersaturation in Fe–O–Si (Al) solutions by using thermodynamic properties of

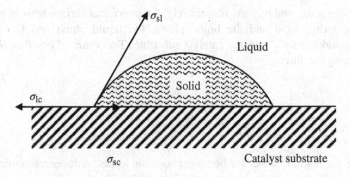

Fig. 6.2 Heterogenous nucleation on a planar substrate

the system and the interfacial tension between the melt and the oxide formed. Supersaturation might be achieved by addition of deoxidizers (Si or Al) or oxygen to the liquid melt. Alternatively, supersaturation may be achieved by freezing the melt because oxygen segregates in the residual liquid to a greater extent than do either Si or Al. It can also be attained by simultaneous additions of oxygen and the deoxidizers.

Supersaturation of the liquid melt can be obtained by first equilibrating it with pure oxide(s) at a high temperature, T_{eq}, and then rapidly cooling it to a lower temperature, T. This temperature difference ($T_{eq} - T$) is required to achieve super-saturation. This is known as the degree of undercooling in the case of solidification of pure metals and is expressed as ($T_f - T$) where T_f is the temperature of fusion. The driving force of solidification (activation barrier against nucleation) can be obtained directly from $\Delta G^o = \Delta H^o - T\Delta S^o$ as:

$$\Delta G^c_{hom} = \Delta H - T\Delta S = \Delta H_f - T.\frac{\Delta H_f}{T_f} = \frac{\Delta H_f}{T_f}[T_f - T] = \frac{\Delta H_f.\Delta T}{T_f}, \qquad (6.6)$$

where ΔH_f is the heat of fusion. Thus, the activation barrier is proportional to the degree of undercooling (i.e., $\Delta G^c_{hom} \propto \Delta T$). However, the high energy barrier required for homogeneous nucleation is drastically reduced due to the catalytic action of crevices and projections on the walls of the container in the presence of solid impurities in the melt. Hence, there is more likelihood of heterogeneous nucleation. The critical free energy barrier, ΔG^c_{het}, for heterogeneous nucleation in case of formation of a spherical cap on a planar substrate (catalyst) is modified as

$$\Delta G^c_{het} = \Delta G^c_{hom}.f(\theta), \qquad (6.7)$$

where $f(\theta)$, the shape factor, for the contact angle, θ made by the nucleus on the substrate (Fig. 6.2) is expressed as

$$f(\theta) = \frac{1}{4}\left(2 - 3\cos\theta + \cos^3\theta\right). \qquad (6.8)$$

In Fig. 6.2, σ_{sl}, σ_{lc}, and σ_{sc}, are, respectively, the interfacial surface tension between the precipitating solid and the liquid phase, the liquid phase and the catalyst substrate, and the solid and the catalyst substrate. The value of cos θ is obtained by balancing the forces:

$$\sigma_{lc} = \sigma_{sc} + \sigma_{sl} \cos \theta \qquad (6.9)$$

$$\therefore \cos \theta = \frac{(\sigma_{lc} - \sigma_{sc})}{\sigma_{sl}}. \qquad (6.10)$$

Thus, the nucleation barrier for heterogeneous nucleation reduces tremendously at lower values of θ, for example, $f(\theta) = 10^{-4}$ when $\theta = 10°$. Significant reductions are also possible at higher values of θ, for example, $f(\theta) = 0.02$ at $\theta = 30°$, and $f(\theta) = 0.5$ at $\theta = 90°$. Heterogeneous nucleation can also be achieved by seeding in the liquid solution. In the Bayer process, alumina from bauxite ore after crushing to about 100 mesh is dissolved in a hot solution of caustic soda under a pressure of 10–25 atm at 100–160 °C leaving aside the insoluble oxides in red mud. The sodium aluminate solution after separation from the mud is diluted and slowly cooled to precipitate aluminum as hydroxide. Precipitation is initiated by "seeding" the solution with freshly precipitated fine $Al(OH)_3$.

In order to obtain a fine-grained structure, a large number of nuclei have to be provided by addition of a grain refiner in the liquid metal. The grain refiner is either insoluble in the metal and forms a suitable nucleating surface or reacts with the metal to form the nucleating phase. Titanium carbide having a similar crystal structure, lattice parameter and bonding to that of aluminum is added to aluminum alloys to lower down the interfacial tension. It also lowers the degree of supercooling, ΔT, necessary to achieve a rapid rate of nucleation. The role of interfacial energy is very much evident in solidification of eutectic alloys and its modification. For example, the mode of growth is modified by addition of small quantities of sodium in an Al–Si eutectic alloy melt. Adsorption of sodium on the surface of nuclei lowers the interfacial energy. Another important example is that of modification of the grey cast iron by addition of magnesium or cerium to form a spherical graphite eutectic of austenite and graphite. In solid state transformations, heterogeneous nucleation may occur on nucleation sites provided by dislocations at grain boundaries, twin boundaries, slip planes and sub-grain boundaries, or by nonmetallic inclusions such as oxides or nitrides. The increased hardening effect in precipitation hardening of alloys is due to reduction in the interfacial energy between its parent phase and the nucleating phase [1].

In the Sheritt Gordon process [8], metallic nickel is precipitated from the leach liquor (obtained by leaching the partially roasted pentlandite concentrate with ammonia) by blowing hydrogen gas at 30 atm and at 175 °C. Metallic nickel powder is added as seeds to start the precipitation reaction. Addition of certain organic compounds affects the rate and mode of precipitation of the nucleus. Anthraquinone accelerates the rate of nickel precipitation by lowering the activation energy of the process, which is apparently due to the increase of the effective surface area of the solid particles. This happens to be one of the most important developments in extractive metallurgy during early 1950s [8, 9].

In the following sections, important applications of interfacial phenomena in nucleation of CO bubbles from liquid melt during steelmaking (in open hearth and electric furnace), role of interfaces in slag–metal reactions, high rate of refining in basic oxygen furnaces due to the formation of emulsion, and froth flotation of minerals will be discussed.

6.2   Nucleation of Gas Bubbles in a Liquid Metal

The pressure (P_b) on a spherical gas bubble of radius r located at a depth of h in a liquid metal of density ρ and surface tension σ is given by the expression:

$$P_b = P_a + \rho g h + \frac{2\sigma}{r} \tag{6.11}$$

where P_a is the atmospheric pressure over the surface of the liquid metal and g is the acceleration due to gravity. Metallostatic pressure over the bubble is given by $\rho g h$. $2\sigma/r$ is the term for the pressure required to maintain the metal/gas interface of the bubble against the surface tension. While considering the possibility of a gas bubble nucleating with a radius of 10^{-8} m at a depth of 1 m below the surface of molten steel of surface tension 1.5 J m^{-2} and density 7.1 g cm^{-3}, the required bubble pressure would be 2972 atm.

$$P_b = 1 + \frac{7100 \times 9.81 \times 1}{1.01 \times 10^5} + \frac{2 \times 1.5}{10^{-8} \times 1.01 \times 10^5}$$

$$= 1 + 0.69 + 2970.3 = 2972 \text{ atm}$$

The extremely high value of P_b rules out the possibility of homogenous nucleation of gas bubbles in liquid metal. Hence, for removal of a gaseous component, a gas/metal interface must be provided by (i) introduction of gas bubbles into the metal, (ii) vigorous stirring action, or (iii) gas trapped in crevices in the refractory wall of the furnace. For example, introduction of argon bubbles into liquid steel in vacuum degassing chamber (RH process – Chap. 8) increases the rate of removal of hydrogen by providing gas/metal interface in the bulk of the melt and stirring action.

In the refining of steel, oxidation of Mn, Si, and P takes place at the slag–metal interface. But due to the difficulty in nucleation of CO bubbles, oxidation of carbon practically does not take place at the slag–metal interface. Oxidation of carbon according to the reaction:

$$[C] + [O] = CO(g) \tag{6.12}$$

takes place at the gas–metal interface because it does not require nucleation of gas bubbles. The rate of diffusion of carbon and oxygen to the gas/metal interface is much higher in pneumatic processes compared to the open hearth furnace. Introduction of fresh air bubbles in bottom-blown Bessemer converter through the metal increases the total surface area of the gas/metal interface. Vigorous blowing also improves the transport of reactants and provides a stirring action. As a result, high decarburization rates up to 10% carbon per hour are possible. Similarly, the formation of slag–gas–metal emulsion in the basic oxygen steelmaking enhances decarburization rate as compared to that in the Bessemer process by providing fresh gas–metal interface. In the absence of gaseous medium, C–O reaction is very slow in the open hearth furnace, which can be explained by considering the physics of formation of the carbon monoxide bubble. The pressure inside the CO bubble of radius r in liquid metal of surface tension, σ given by Eq. 6.11 can be represented as:

$$p_{CO} = p_{atm} + \rho_{Fe}gh + \frac{2\sigma}{r} \tag{6.13}$$

In order to homogeneously form the CO bubbles of nearly 10^{-9} m radius in the melt, the pressure within the bubble should be of the order of 10^4–10^5 atm. This makes it clear that homogeneous nucleation would be unlikely because supersaturation of carbon and oxygen required to be in equilibrium with these pressures corresponding to nuclei of reasonable size are out of the real possibility. For the decarburization reaction (Eq. 6.12) the equilibrium partial pressure of CO at the gas/metal interface can be calculated by the expression:

$$p_{CO} = Kf_C [\%C] f_O[\%O] \tag{6.14}$$

But the bath can never be practically so supersaturated with carbon and oxygen that p_{CO} (equilibrium) is of the order of 10^4–10^5 atm at steelmaking temperatures. This requirement totally rules out the chances of homogenous nucleation of CO bubbles in the melt during refining. There seems a possibility of nucleation of CO bubbles at the slag–metal or refractory–metal interface. The possibility of nucleation at the former interface is almost negligible compared to that at the solid refractory–metal interface. Though molten metal always rests against the refractory walls having crevices, the high surface tension of molten steel prevents the crevices from wetting. The gas–metal interfaces provide the necessary interface for C–O reaction to occur and desorption of the gaseous product in the pore space.

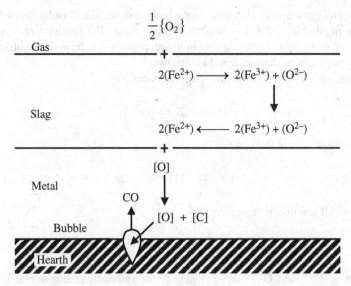

Fig. 6.3 Schematic representation of oxygen transfer from the furnace atmosphere through the slag and metal to a carbon monoxide bubble

The C–O reaction can continue at the pore space (crevice) until p_{CO} inside the bubble reaches the p_{CO} (equilibrium). As reaction progresses p_{CO} (bubble) increases and the bubble grows to attain hemispherical shape (bubble radius becomes equal to the crevice radius). Further, the increase in radius after attaining the hemispherical shape will decrease the pressure term, $\frac{2\sigma}{r}$ which means decrease in the internal pressure of the bubble (p_{CO}). This will favor the C–O reaction (6.12) in the forward direction. Any growth beyond the hemispherical shape makes the bubble more and more unstable. The bubble finally detaches from the refractory wall and moves upward through the melt. While ascending, the pressure term ($p_{atm} + \rho g h$) decreases and r increases. This will also push the C–O reaction to the right. The C–O reaction continues by repetition of the same process at the crevices if other requirements are fulfilled.

At crevices smaller than the critical pore size, p_{CO} (bubble) becomes equal to p_{CO} (eqm) before the bubble attains a hemispherical shape. Such crevices do not actively participate in the nucleation process because bubbles do not grow on such inactive sites. There is a minimum unwetted crevice size for nucleation of CO bubbles for a fixed value of $a_C \cdot a_O$ [i. e., wt % C. wt % O]. For the prevailing conditions in open hearth and electric arc furnaces, the unwetted crevices greater than 0.01 mm are the most effective nucleation sites for C–O reaction. Mechanism of oxygen transport from gas to slag has been demonstrated in Fig. 6.3.

As oxygen is easily available at the gas–metal interface in the pneumatic processes, the rate of reaction is affected by the diffusion of carbon to the interface. However, the vigorous stirring increases carbon transport to achieve very high rate of decarburization. Contrary to this, in open hearth and electric furnaces, oxygen has to diffuse from the furnace atmosphere through the slag and metal layers to reach the crevice–metal (gas–metal) interface. Since atomic/molecular dissolution of oxygen in slag is negligible, it has to dissolve in ionic form. The mechanism of oxygen transfer [10] from gas to metal via slag in hearth processes to form CO bubbles at the refractory wall can be demonstrated as follows:

Dissolution of oxygen at the gas–slag interface:

$$\frac{1}{2}\{O_2\} + 2e = (O^{2-}) \tag{6.15}$$

Oxidation of iron in the slag:

$$2(Fe^{2+}) = 2(Fe^{3+}) + 2e \tag{6.16}$$

The overall reaction is represented as:

$$2(Fe^{2+}) + \{\frac{1}{2} O_2\} = 2(Fe^{3+}) + (O^{2+}) \tag{6.17}$$

Migration of Fe^{3+} and O^{2-} from gas–slag phase interface to slag–metal interface completes the cycle:

$$2(Fe^{3+}) + O^{2+} = 2(Fe^{2+}) + [O] \tag{6.18}$$

The cycle is repeated and thus the dissolved oxygen in the metal diffuses to the crevice–metal (gas–metal) interface to combine with carbon. The rate of diffusion of oxygen can be increased by avoiding the diffusion path across the slag layer by addition of hematite (Fe_2O_3) heavier than the slag but lighter than the metal. While settling at the slag–metal interface, Fe_2O_3 supplies oxygen according to the following endothermic reaction; known as 'oreing' of the slag:

$$Fe_2O_3(s) \rightarrow 2(FeO) + [O] \tag{6.19}$$

Oreing increases the rate of decarburization to 0.6% carbon per hour from 0.12% to 0.18% carbon per hour under the normal condition prevailing in hearth furnaces, that is, oxygen coming from the furnace atmosphere. The marginal increase in decarburization rate by oreing is due to endothermic nature of the overall reaction:

$$2/3\ Fe_2O_3(s) + 2\ [C] = 4/3[Fe] + 2\{CO\}, \Delta H = 65\ kcal \tag{6.20}$$

The above endothermic reaction can be transformed into exothermic by flushing low-pressure oxygen over the slag surface to achieve dC/dt up to 3% carbon per hour.

$$[C] + [O] = \{CO\}. \quad \Delta H = -89\ kcal \tag{6.12a}$$

In order to further increase dC/dt in open hearth, oxygen lancing of the bath during refining period was practiced.

6.2.1 Role of Interfaces in Slag–Metal Reactions

The slag–metal and gas–metal interfaces play dominant roles in the transfer of solutes in the refining of molten pig iron. The partitioning [11] of slag (phase 1) and metallic bath (phase 2) is represented as:

Interface	Bulk phase No.1
	Stagnant boundary layer of phase 1
	Stagnant boundary layer of phase 2
	Bulk phase No.2

In general, three steps [11] may be involved in chemical reactions taking place in the above heterogeneous system: (i) diffusion of reactants across the boundary layer (s) to the interface, (ii) chemical reaction at the interface, and (iii) diffusion of products across the corresponding stagnant boundary layer(s) into the bulk phase. At steelmaking temperatures, the chemical reaction (step 2) is very fast, hence, to a major extent, the refining reactions are normally controlled by mass transport (either step 1 or 3). According to the Fick's law of diffusion, the rate of mass transport across the stagnant boundary layer is given by:

$$J = D_i \frac{(c_b - c_i)}{\delta} A \text{ g s}^{-1} \tag{6.21}$$

where D_i, c_b, c_i, δ, and A stand for diffusion coefficient, concentration in the bulk and at the interface, thickness of the stagnant boundary layer, and the interfacial area, respectively.

The thickness of the boundary layer depends on the slag composition and viscosity. Stirring decreases the thickness of the boundary layer and improves the mass transport. The refining reactions in various commercial processes are controlled by either the slag–metal interface or the gas–metal interface or by both types of interfaces. For example, refining takes place at slag–metal interfaces in open hearth and electric furnace without oxygen lancing and Kaldo converter, whereas in acid Bessemer process, it takes place at the gas–metal interfaces. In a number of furnaces and converters, such as open hearth and electric furnace with oxygen lancing, Thomas, rotor, LD, OLP, OBM, Pierce Smith side-blown converter for blister copper production, and top-blown rotary converter for nickel production, refining reactions are controlled by slag–metal as well as gas–metal interfaces (i.e., mixed control).

In steelmaking, oxidation of silicon, manganese, and phosphorus, which takes place at the slag–metal interface is controlled by diffusion across the stagnant boundary of the slag and is accelerated by vigorous stirring. In commercial practice, slag–metal system may be stirred vigorously by bottom blowing in Bessemer, OBM, and combined-blown processes, by carbon boil in open hearth and electric furnaces, by rotation of the vessel in Kaldo and Rotor processes, by supersonic oxygen jet in LD converter. Stirring not only reduces the stagnant boundary layer thickness but also increases the interfacial area. Further, viscosity is reduced by raising the temperature and by addition of suitable fluxes to improve the mass transport in the slag phase.

6.3   Emulsion and Foam

During the past seven decades, extensive research studies have been conducted on the structure, properties and ionic theories of slags at various pioneering institutions including the university of Oslo, Norway, Massachusetts Institute of Technology, U. S. Steel, USA, Imperial College of Science and Technology, UK, Max Plan Institute, Germany, Royal Institute of Technology, Sweden as well as in Japan and Canada. Despite substantial research investigations on slags, many of its important functions still remain mysterious, controversial and unresolved. This is particularly true in the case of the high rate of refining in BOF steelmaking processes. The reason for which may be accounted, as most of the investigations have been conducted under steady-state conditions, whereas a more complex behavior is found in actual dynamic operations [12]. However, a thorough knowledge of slag properties [13], such as basicity, surface tension, viscosity, and so on are essential for optimization of steelmaking reactions.

The high decarburization rate in basic oxygen converter comparable to that in the Bessemer process is obtained by the formation of slag–metal interfaces and emulsion. The slag–metal–gas emulsion helps in providing an ever fresh gas–metal interface, which accounts for the large required reaction surface necessary for enhancing the reaction rates. In these processes, the formation of slag foam poses one of the most difficult operating problems. Formation of both emulsion and foam is governed by the surface properties of slag and metal, which are finally controlled by composition and temperature. Before discussing the physical conditions and requirements for their formation, we must understand the difference between emulsion and foam [14]. Small droplets of one liquid or gas bubbles embedded in another liquid is known as emulsions if the neighboring droplets (liquid–liquid emulsion) or bubbles (gas–liquid emulsion) are separated by large distances to permit free movement of drops or bubbles. In the case of foams, the adjacent bubbles cannot move independently because the total volume of thin films separating the bubbles is larger as compared to the volume of the liquid medium in a gas/liquid emulsion. A tremendous increase in the interfacial area between the two phases happens to be the most significant contribution of any emulsion or foam. For example, the specific interfacial area of the emulsified liquid consisting of droplets of 0.4 mm radius [14] is about 10^5 cm^2 per cm^3.

As the interfacial energy (surface tension) is the energy required to create new unit surface area, the amount of energy needed for emulsifying a liquid or a gas in another liquid will increase with increasing surface tension. This means, the energy requirement for emulsification will decrease at lower interfacial tension, and hence, destruction of emulsion is going to be a spontaneous process and the decrease of interfacial areas will lead to liberation of energy. Thus, the formation as well as retention of an emulsion is favored by a low interfacial tension. It is important to note that from an energetic viewpoint, the slag–metal and gas–slag systems are not suitable for emulsification because of high surface tension [14] of the former lying between 800–1200 erg cm^{-2} (compared to 40 erg cm^{-2} for water/mineral oil interface) and that for the latter system lying between 400 and 600 erg cm^{-2} (compared to 72 erg cm^{-2} for water/gas and 20–60 erg cm^{-2} for organic liquid/gas system). However, in case of the slag/metal interface, it may be reduced to 1/100 of its equilibrium value by rapid mass transfer brought out by supersonic oxygen jet or by stirring with suitable means.

As regards to the destruction of emulsions, two factors have to be kept in mind: (i) the large difference in densities of metal and gas emulsions in slag and (ii) the coalescence of small droplets or bubble into larger units, which separate easily from the liquid medium. The destruction of foam may be hindered by any factor that may stop or slow down coalescence. High viscosity may stabilize both emulsions and foams. Viscosity of steelmaking slag controlled by composition, is generally low but increases due to precipitation of solid particles. These particles may stick to metal drops or gas bubbles and the process of coalescence may be slowed down. In general, the presence of surface-active agents such as CaS, Cr_2O_3, P_2O_5, V_2O_5, and so on make the foam more stable. The suspended second-phase particles such as Cr_2O_3 [15], CaO, MgO, and $2CaO.SiO_2$ [16] stabilize the foam and are more effective on foaming as compared to their effect in changing viscosity and surface tension. Carbonaceous material such as coke and charcoal particles play important roles in controlling slag foaming in bath smelting as well as in other steelmaking processes. Sulfur, which is a strong surface-active agent in liquid iron [17], increases the contact angle, that results in larger bubbles and less stable foam. According to Zhang and Fruehan [18], the slag foaming can be totally suppressed by covering only 15–20% of the slag surface with either of charcoal or coke particles. Based on their studies, they have concluded that the anti-foaming effect is primarily due to the nonwetting characteristic of the carbonaceous material with slag. Occasionally, emulsification of metal in slag may prove nuisance by the formation of abundant stable foams due to swelling and overrunning of the slag.

The formation and stability of gas emulsions depend on the structure and properties of slags. The octahedral oxides like FeO, Al_2O_3, CaO, and MgO exhibit very low viscosity and high surface tension. Contrary to this, tetrahedral oxides like SiO_2, P_2O_5, and B_2O_3 exhibit high viscosity and low surface tension (Table 6.2). The slag–metal interface directly concerned with the formation and destruction of metal emulsion happens to be the most important interface. It is also the most complex due to the absorption of ions SiO_4^{4-} and PO_4^{3-} coming from the slag phase and O^{2-}

Table 6.2 Viscosity and surface tension of selected oxides[a]

Oxides	Temperature (°C)	Viscosity (poise)	Surface tension (erg cm^{-2})
Fe$_x$O (in equilibrium with solid Fe)	1420	0.3	585
Al$_2$O$_3$	2100	0.5	690
CaO	m.p.	<0.5	>585
MgO	m.p.	<0.5	>585
SiO$_2$	1800	1.5×10^5	307
P$_2$O$_5$	100	–	60
B$_2$O$_3$	1000	300	82

[a]Reproduced from G. H. Geiger et al. [14] with the permission of Association for Iron & Steel Technology

and S^{2-} from the metal. Notable stability of foam in FeO–SiO$_2$ system begins with a melt containing about 40 wt% SiO$_2$, goes through a sharp maximum at approximately 60 wt% SiO$_2$ and falls to zero at higher silica contents. In all slag systems, stability is related to composition. Since most slags react with the metal bath, nonequilibrium conditions and mass transfer across the metal–slag interface are characteristic features of any steelmaking process, resulting in a drastic decrease in the interfacial tension. However, the decrease is beneficial for emulsification.

Following the benefits of interfacial phenomena, serious attention has been paid on the measurement of interfacial tension and correlation of the data with the overall kinetics of slag–metal reactions, even in the extraction of nonferrous metals [19, 20]. The emulsion of molten copper, matte, slag, and gas formed in the Mitsubishi Continuous Converter [21, 22] by blowing air along with CaCO$_3$ flux and O$_2$-enriched air (through different pipes) on the surface of the molten bath facilitates solid–liquid-gas reactions leading to the formation of copper and slag at much higher rate as compared to the rate of blister formation in the conventional Pierce Smith copper converter.

6.4 Froth Flotation

In Sects. 6.2 and 6.3, we have noticed that there is enormous increase in the surface area due to the introduction of gas bubbles in the Bessemer converter and formation of emulsions in the basic oxygen converter. In both the cases, we start with liquid metal–gas system but as refining progresses, we have two liquids (molten metal and molten slag) and gas (blown as air or oxygen and CO formed). We have discussed the role of interfacial tension in the nucleation of CO bubbles and formation of emulsions and foam. In froth flotation, we deal with systems comprising of a solid (mineral), a liquid (water), and a gas (air).

Equation 6.11 indicates that the pressure inside a bubble located at a certain depth depends on the interfacial tension between air and the liquid. Thus, the energy required to form an air bubble decreases with the decrease of surface tension. Addition of soap lowers surface tension by adsorbing at the air/water interface polar groups in water and nonpolar hydrocarbon groups on the air-side of the interface. In froth flotation, nonwetting mineral particles are attached to the surface of air bubbles rising through the water. In this way, they are separated from mineral particles which are wetted and consequently sink to the bottom. The froth formed by the bubbles runs off to the top of the flotation cell and breaks. The floated mineral particles can be recovered by filtration. Since the life of an air bubble in water is about 0.01 s [23], air bubbles burst and drop the mineral particles before reaching the surface of pure water. This makes it essential to stabilize the froths by introducing some surface-active organic chemicals known as frothers. The froths containing the mineral particles are skimmed off. Pine oil and higher alcohols are commonly used as frothers on the commercial scale. Small quantities of frother, pine oil (0.05 kg ton^{-1} of ore in a pulp) are added to produce stable froth on the surface of the pulp. Frothers are heteropolar substances like soaps, which lower the interfacial energy by adsorbing at the air/water interface. From the recovery viewpoint, frother should produce persistent froth with sufficient strength to carry mineral particles to the top. At the same time, it should break immediately on leaving the cell to avoid overflooding of the plant with froth in a very short time. In pine oil based on terpineol $C_{10}H_{17}OH$, the nonpolar hydrocarbon group is oriented on the air side of the interface. On the other hand, the polar hydroxyl group is attracted by hydrogen from the water molecules. In froth flotation, wettability of mineral particles plays an important role.

The interfacial tension between air and the mineral, σ_{AM} can be lowered by adding a substance which gets adsorbed at the mineral/air interface. Such a substance called "collector," when added, adsorbs on one mineral leaving no effect on others. For example, use of potassium ethyl xanthate as a collector in flotation of galena leaving silica is well known. Although water wets both silica and galena, addition of a small quantity of xanthate (25 g ton^{-1} of ore) raises the contact angle from 0° to 60° on the galena whereas the contact angle on the silica remains almost zero. The negatively charged sulfur atom of the polar group of the heteropolar xanthate anion, $-S-CS-O-C_2H_5$ is attracted to leave Pb^{2+} ions in the galena, PbS. The outward facing nonpolar ethyl group, C_2H_5, is weakly attracted to the water molecules compared to galena. These conditions raise the value of the interfacial tension between water and the mineral, σ_{WM}. Due to weaker water/xanthate attraction compared to the air/xanthate attraction, σ_{AM} is lowered and contact angle, θ, is raised. Silica remains wetted by water and sinks to join the tailings because of very poor adsorption on it. On the other hand, galena floating at the top is collected as a concentrate. Like xanthates used as collectors for sulfides, fatty acids are used for oxides, hydroxides and carbonates, and amines for silicates. An optimum value of σ_{WA} is important. Drastic lowering of σ_{WA} by excessive addition of frother will jeopardize the collector action of xanthate. Thus, there is an optimum frother addition to obtain the effective collector action. The mineral particle will not penetrate the air/water interface on excessive addition of collector and hence will not float.

Thus, the concentration of mineral by flotation is essentially based on the ability of solid surfaces to be wetted by water, which is controlled by the relative values of the interfacial energies: mineral–water, mineral–air and water–air. A larger value of the first compared to the other two prevents the mineral from wetting and the mineral is categorized as hydrophobic. On the other hand, the mineral is known to be hydrophilic if the mineral–air interfacial energy is the largest. Hence, when air is blown through the aqueous pulp of ground minerals, hydrophobic mineral particles will get attached to the air bubbles and rise and hydrophilic minerals will sink. In addition to frothers and collectors, some chemicals are used as modifiers. They modify the action of collectors by either increasing or decreasing the interfacial tension between the mineral particles and the air bubbles. Modifiers increasing or decreasing the interfacial energy are called depressors and activators, respectively, and are useful in developing selective froth flotation processes. In addition to these, chemical additives used to control pH of the solution are known as regulators [23] or conditioners [24]. Lime is commonly used to prevent precipitation of compounds drastically affecting the separation efficiency. These chemical reagents may be added to the pulp before or during the flotation process. For efficient flotation, most mineral particles are ground between 200 and 500 μm and diluted with water to pulp containing 25–45% solid.

6.5   Other Applications

In the following paragraphs, some *typical applications of interfacial phenomena* will be very briefly outlined.

As discussed above, low interfacial tension is also useful in crushing and grinding of minerals. The energy required to deform minerals and overcome friction increases tremendously due to an increase in surface area during comminution.

The shape of the boundary [1] is governed by the interfacial energy at the boundary between two solid phases in an alloy as well as between grains of the same metallic phase. The interfacial energy [25] also controls the shape of grains. The grain boundary energy reduces due to rearrangement of the metallic structure to produce grains of shape with a minimum ratio of grain boundary area to volume. A minor phase adopts a spherical shape in the grains of a structure consisting mainly of a major phase. A fairly long period of annealing of a steel structure consisting of a lamellar eutectoid of ferrite and cementite at about 650 °C results in spheroidization of cementite lamellae.

When metal and ceramic powders compacted at high pressure are sintered at moderate temperatures, the possibility of obtaining sound objects depends on the reduction of surface free energy [26] of the structure. This happens due to the diffusion of atoms or molecules at the point of contact between particles forming a junction. This enables the pores to become spheroidal and to contract by vacancy diffusion leading to the production of a sound object.

During electrolysis of Al_2O_3 dissolved in fused cryolite (Na_3AlF_6), as the Al_2O_3 content of the bath decreases, the electrolyte gradually loses its wetting ability due to increase in surface tension. This leads to accumulation of gas bubbles resulting in the increase of resistance to current at the interface between the anode and the electrolyte. Electric arc strikes wherever the gas film is temporarily broken. This is known as the anode effect [27]. The wetting ability of the electrolyte is restored by the addition of fresh alumina, which helps in attaining the proper interfacial tension of the bath. As a result, the gas film is removed from the anode and the voltage across the cell attains the normal value.

In electrolytic production of magnesium, surface tension and viscosity of highly viscous pure molten $MgCl_2$ is reduced to a desired value [28] by the addition of NaCl and KCl. A fused salt halide mixture containing 10% $MgCl_2$, 45% $CaCl_2$, 30% NaCl, and 15% KCl is used for a smooth running of the electrolytic cell. Excessive surface tension is undesirable because it retards the coalescence of the magnesium globules that collect at the surface of the chloride. Low surface tension facilitates the coalescence of metallic globules that collect at the surface of the cathode. Too low a surface tension is undesirable because magnesium that floats to the surface would break through the electrolyte film and is therefore exposed to atmospheric oxygen. The excessive decrease in the interfacial tension of the electrolyte due to the low $MgCl_2$ content is counteracted by $CaCl_2$, which like all alkaline chlorides, increases the surface tension. Hence, proper surface tension of the electrolytic cell has to be maintained by adjusting the bath composition.

In all the above cases, we have seen the beneficial effects of lowering of interfacial tension. In the following two examples, we find that a higher interfacial tension is required to achieve the goal:

In sand castings, objects with rough surfaces are produced when metal penetrates the cavities between the sand particles by wetting the sand. The extent of wettability of the mold material depends on the surface tension of the metal to be cast. For hemispherical cavities, the metallostatic pressure head that is required to force the metal in the cavity is directly proportional to the surface tension and inversely proportional to the radius. The relation will change according to the shape of the cavity. Definitely, a larger value of surface tension will prevent the mold cavities from wetting and hence reduces the penetration of the molten metal into the cavities.

It has been established that the liquid deoxidation product (FeO–MnO–SiO_2 slag) formed by deoxidation of liquid steel with ferromanganese and ferrosilicon floats out to the surface of molten steel within a ladle holding time of 20 min [10] at a velocity proportional to the square of their radii. According to Stoke's law, in the laminar movement of particles in a fluid depending on the densities of the fluid and solid, larger particles may rise to the surface or sink through a liquid metal faster than small particles. However, Rosegger [29] has reported that the rate of removal of the deoxidation product depends on its composition as well as on its size. Thus, relatively small alumina-rich particles formed by deoxidation with aluminum: 2 [Al] + 3[O] = (Al_2O_3) separates more rapidly than the larger silicate particles formed by deoxidation with silicon. Although alumina is denser than silica, the small difference in density does not account for this anomaly. According to Rosegger

[29], the rate of removal depends on the interfacial tension between the solid particles and the liquid steel. The surface tension of FeO increases with the presence of alumina whereas it decreases by addition of SiO_2 and MnO. A higher interfacial tension between liquid steel and solid Al_2O_3 particles provides a small attractive force between them. This means a small dragging force opposes the motion of alumina. Due to higher surface tension, liquid steel is not attracted to alumina. These facts are sufficient to explain the increased rate of removal of deoxidation product when steel is deoxidized with aluminum. On the other hand, a low interfacial tension due to SiO_2 and MnO (when steel is deoxidized with Si and Mn) causes a large dragging force and slows down the rate of removal because steel is attracted to these oxides. Thus, aluminum not only produces a stable oxide but also presents advantages over other deoxidizers in the removal of the deoxidation products.

6.6 Problems [30–32]

Problem 6.1
(Elliott, J. F., 190–1981, personal communication, Massachusetts Institute of Technology, Cambridge)

Assuming homogeneous nucleation of $Al_2O_3(s)$ from an Fe–Al–O melt at 1600 °C calculate:

(a) The spersaturation ratio of oxygen for the case where the interfacial tension for iron in contact with alumina is 2.0 J m^{-2}. The aluminum content of the melt is 0.03%.
(b) The critical size of nucleus that would form.
(c) Determine, if it would be possible also for this melt to nucleate FeO (l) for which the interfacial tension between the oxide and slag is 0.4 J m^{-2}.

Frequency factors of Al_2O_3 and FeO are, respectively, 10^{26} and 10^{30} and their molar volumes are 25 and 16 cm^3 mol^{-1}.

Solution
Homogeneous nucleation in Fe–Al–O melt at 1600 °C

$$\sigma_{Al_2O_3-Fe} = 2.0 \text{ J m}^{-2}, \%Al = 0.03$$

(a) Reaction under consideration:

$$2[Al] + 3[O] = Al_2O_3(s)$$

$\Delta G = \Delta G_{eq} - \Delta G_{ss}$ (eq and ss stand respectively, for equilibrium and supersaturation)

$$\Delta G = -RT \ln \left[\frac{[\%O]_{ss}^3 [\%Al]_{ss}^2}{[\%O]_{eq}^3 [\%Al]_{eq}^2} \right]$$

In both cases of supersaturation and equilibrium, neglecting interaction coefficients

$$f_O = 1 = f_{Al}.$$

ΔG_{hom}^c can be calculated from Eq. 6.5:

$$\Delta G_{hom}^c = -2.7\nu \left(\frac{\sigma^3}{kT \log A}\right)^{1/2}$$

k (Boltzmann constant) $= 1.38 \times 10^{-23}$ J K^{-1}

$\sigma = 2.0$ J m^{-2}, $T = 1873$ K, $A = 10^{26}$ for Al$_2$O$_3$, $\nu = 25$ cm^3 mol$^{-1} = 25 \times 10^{-6}$ m^3 mol^{-1}

$$\Delta G_{hom}^c = -2.7(25 \times 10^{-6} \text{ m}^3 \text{ mol}^{-1}) \left[\frac{(2.0 \text{ J m}^{-2})^3}{(1.38 \times 10^{-23} \text{ J K}^{-1})(1873) \log (10^{26})}\right]^{1/2}$$

$$= -2.33 \times 10^5 \text{ J mol}^{-1}$$

also,

$$\Delta G_{hom}^c = -RT \ln \left(\frac{[\%O]_{ss}^3 [\%Al]_{ss}^2}{[\%O]_{eq}^3 [\%Al]_{eq}^2}\right)$$

assuming $[\%Al]_{ss} = [\%Al]_{eq} = $ constant

$$\left(\frac{[\%O]_{ss}}{[\%O]_{eq}}\right)^3 = \exp\left(\frac{-\Delta G_{hom}^c}{RT}\right).$$

Hence, the supersaturation ratio:

$$\frac{[\%O]_{ss}}{[\%O]_{eq}} = \left[\exp\left(\frac{-\Delta G_{hom}^c}{RT}\right)\right]^{1/3}$$

$$= \left[\exp \frac{-(-2.33 \times 10^5 \text{ J mol}^{-1})}{(8.314 \text{ J mol}^{-1}\text{K}^{-1})(1873 \text{ K})}\right]^{1/3} = 146.5 \text{ Ans.}$$

(b) The critical size of nucleus is obtained from Eq. 6.2:

$$r^c = -\frac{2\sigma\nu}{\Delta G_{hom}^c}$$

$$= \frac{-2(2 \text{ Jm}^{-2})(25 \times 10^{-6} \text{ m}^3 \text{ mol}^{-1})}{-2.33 \times 10^5 \text{ J mol}^{-1}}$$

$$= 4.29 \times 10^{-10} \text{m} = 4.29°\text{Å Ans.}$$

(considering only the formation of Al$_2$O$_3$ not FeO. Al$_2$O$_3$)

(c) Whether it is possible to nucleate FeO in the slag.

$$[Fe] + [O] = [FeO](l)$$

$$\Delta G = \Delta G_{eq} - \Delta G_{ss}$$

$$\Delta G = -RT \ln \left[\frac{[\%O]_{ss}[\%Fe]_{ss}}{[\%O]_{eq}[\%Fe]_{eq}} \right].$$

Assuming $(\%Fe)_{ss} = (\%Fe)_{eq} = $ constant,

$$\Delta G = -RT \ln \left[\frac{[\%O]_{ss}}{[\%O]_{eq}} \right]$$

$\left[\frac{[\%O]_{ss}}{[\%O]_{eq}} \right]$ is the supersaturation ratio.

For calculating ΔG^c_{hom} for FeO (l)

$$\sigma_{FeO\text{-}Al} = 0.4 \text{ J m}^{-2}, \quad v_{FeO} = 16 \text{ cm}^3 \text{ mol}^{-1} = 16 \times 10^{-6} \text{ m}^3 \text{ mol}^{-1}, A_{FeO} = 10^{30},$$

$T = 1873$ K, considering Eq. 6.5:

$$\Delta G^c_{hom} = -2.7 \times \left(16 \times 10^{-6} \text{ m}^3 \text{ mol}^{-1} \right) \left[\frac{(0.4 \text{ J m}^{-2})^3}{(1.38 \times 10^{-23} \text{J K}^{-1}).1873.(\log 10^{30})} \right]^{1/2}$$

$$= -1.24 \times 10^4 \text{ J mol}^{-1}$$

Hence, the supersaturation ratio:

$$\left[\frac{[\%O]_{ss}}{[\%O]_{eq}} \right] = \exp \left(\frac{-\Delta G^c}{RT} \right)$$

$$= \exp \left(\frac{-1.24 \times 10^4 \text{ J mol}^{-1}}{8.314 \text{ J mol}^{-1}\text{K}^{-1} \times 1873 \text{ K}} \right) = 2.2.$$

As the supersaturation ratio of oxygen for FeO (l) nucleation is less than that of $Al_2O_3(s)$ nucleation, the melt should nucleate FeO (l). Ans.

Problem 6.2

What should be the interfacial tension between the molten metal–catalyst interface to reduce the activation barrier 100 times less as compared to that required for homogeneous nucleation in order to precipitate nuclei of the solid metal from the molten metal? The interfacial tension between the solid–liquid interface and the solid–catalyst interface are 0.28 and 0.13 J m^{-2}, respectively.

Solution

According to Fig. 6.2 and using the same symbols for solid, liquid, and catalyst, we have the following equation showing balancing of different forces:

$$\sigma_{lc} = \sigma_{sc} + \sigma_{sl} \cos \theta.$$

σ_{lc}, σ_{sc}, and σ_{sl} represent the interfacial tension between the liquid–catalyst, the solid–catalyst, and the solid–liquid interfaces, respectively.

$$\Delta G^c_{het} = \Delta G^c_{hom} f(\theta) = \Delta G^c_{hom} \cdot \frac{1}{100} \left[\text{given} : f(\theta) = \frac{1}{100} \right],$$

where $f(\theta)$, the shape factor, for the contact angle, θ made by the nucleus on the substrate (Fig. 6.2), is expressed as

$$f(\theta) = \frac{1}{4} \left(2 - 3 \cos \theta + \cos^3 \theta \right) = \frac{1}{100}$$

Solution of the above equation gives $\cos \theta = 0.8821$, $\theta = 28.10°$

$$\therefore \sigma_{lc} = \sigma_{sc} + \sigma_{sl} \cos \theta = 0.13 + 0.28 \cos \theta = 0.13 + 0.28 \times 0.8821$$
$$= 0.377 \textbf{ Ans.}$$

Problem 6.3

A spherical bubble of CO is within Fe–C–O melt at 1540 °C. The local pressure in the melt outside the bubble is 1.5 atm. The melt contains 0.1% C and the surface tension is 1.4 J m^{-2}.

(a) Calculate the pressure within the bubbles of 0.01 μm, 0.1 μm, 0.01 mm, 0.1 mm, 1 mm, and 1 cm radii.
(b) What is the equilibrium oxygen content of the melt for each of the bubble sizes in part (a)? given that $[\%C] + [\%O] = CO(g)$, $\log K = \frac{1168}{T} + 2.07$.
(c) Would it be necessary to make correction for CO_2 in the gas phase in your calculations in part (a) and (b)? Why? Explain.

Given that: $[\%C] + CO_2(g) = 2CO(g)$, $\Delta G° = 33280 - 30.38 \, T$ cal

Solution

(a) Spherical bubbles of CO in Fe–C–O melt at 1540 °C
Reaction: $[\%C] + [\%O] = CO(g)$
The total pressure inside the bubble of radius r is given by the expression:
$P_b = P_s + \frac{2\sigma}{r}$ (where P_s is the local pressure in the melt outside the bubble,
$P_s = P_{atm} + \rho_{Fe}gh = 1.5$ atm), $\sigma = 1.4$ J m^{-2}

$$\therefore P_b = 1.5 + \frac{2\sigma}{r}$$

$$(P_b - 1.5) = \frac{2\sigma}{r}$$

conversion factor

$1 \text{ atm} = 1.033 \times 10^4 \text{ kg m}^{-2}$

$1 \text{ J} = 0.102 \text{ kg m}$

$0.01 \ \mu m = 10^{-8} m, \ 0.1 \ \mu m = 10^{-7} m, \ 0.01 \ mm = 10^{-5} m.$

The total pressure $P_b \ (= p_{CO})$ can be calculated by the expression:

$(P_b - 1.5) = \frac{2\sigma}{r}$ for different radii, $(P_b - 1.5) \ 1.033 \times 10^4 = \frac{2 \times 1.4 \times 0.102}{r}.$

Radius	0.01 μm	0.1 0 μm	0.01 mm	0.1 mm	1 mm	1 cm
P_b (atm)	2766	278	4.26	1.78	1.53	1.503

(b) $[\%C] + [\%O] = CO(g) \quad \log K = \frac{1168}{T} + 2.07$

at 1540 °C (1813 K), $\log K = \frac{1168}{1813} + 2.07 = 2.714$

$K = 518$

$K = \frac{p_{CO}}{[a_C][a_O]} = \frac{p_{CO}}{[\ f_C.\%C][\ f_O.\%O]}.$

Assuming the validity of Henry's law: $f_C = 1 = f_O$

$K = \frac{p_{CO}}{[\%C][\%O]}.$

Assuming % C to be constant, $[\%C] = 0.1$

$[\%O] = \frac{p_{CO}}{518 \times 0.1} = (0.0193) \ p_{CO}.$

The equilibrium oxygen content of the melt calculated for different sizes of bubbles is listed below:

Radius	0.01 μm	0.1 0 μm	0.01 mm	0.1 mm	1 mm	1 cm
$p_{CO(atm)}$	2766	278	4.26	1.78	1.53	1.503
[%O]	53.4	5.37	0.082	0.034	0.0295	0.0290

(c) Whether it is necessary to correct for CO_2 (g) in bubbles

$$P_{bubble} = p_{CO} + p_{CO_2} = p_T$$

$[\%C] + CO_2(g) = 2CO(g), \ \Delta G^\circ = 33280 - 30.38 \ T$

at 1540 °C, $\Delta G^\circ = 33280 - 30.38 \times 1813 = -21800 \text{ cal} = -RT \ln K$

$K = 424.74$

$[\%C] = 0.1$

$K = 424.74 = \frac{p_{CO}^2}{[\%C] p_{CO_2}}$

$\therefore 424.74 = \frac{p_{CO}^2}{[0.1](p_T - p_{CO})}$

$42.47 \ (p_T - p_{CO}) = p_{CO}^2$

$$p_{CO}^2 + 42.47 \cdot p_{CO} - 42.47 \cdot p_T = 0$$

$$p_{CO} = \frac{-42.47 + \sqrt{(42.47)^2 - 4(1)(-42.47 p_T)}}{2}$$

p_{CO} and p_{CO_2} calculated at different bubble pressures (i.e., p_T from the above table) are listed below:

Radius	0.01 μm	0.1 0 μm	0.01 mm	0.1 mm	1 mm	1 cm
p_T(atm)	2766	278	4.26	1.78	1.53	1.503
p_{CO}(atm)	322	89.5	3.903	1.713	1.48	1.455
p_{CO_2}(atm)	2444	188.5	0.357	0.067	0.05	0.048

We do not have to correct for p_{CO_2} in part (a) as the total bubble pressure $(p_{CO} + p_{CO_2})$ was calculated. However, in part (b), it is necessary when bubbles are smaller as p_{CO_2} is quite high and thus it will affect the oxygen content of the melt.

Problem 6.4

A spherical bubble of CO is within the Fe–C–O melt at 1600 °C. The local pressure of the melt outside the bubble is 1.48 atm. The melt contains 0.2% C and 0.02% O and has a surface tension of 1.5 J m^{-2}. What is the critical pore size for nucleation of CO bubble?

Given that: [%C] + [%O] = CO(g) $\Delta G° = -4830 - 9.75\,T$ cal

Solution

At 1873 K, $\Delta G° = -4830 - 9.75 \times 1873 = -23091$ cal $= -RT \ln K$

$K = 495.2$

$p_{CO} = K$ [%O]. [%C] $= 495.2 \times 0.2 \times 0.02 = 1.98$ atm

$P_b = P_s + \frac{2\sigma}{r}$ $\qquad P_b = 1.98$ atm

$(P_b - P_s) = \frac{2\sigma}{r}$ $\qquad P_s = 1.48$ atm

$(1.98 - 1.48) \times 1.033 \times 10^4 = \frac{2 \times 1.5 \times 0.102}{r}$

$r = 59.2 \times 10^{-6}$ m $= 59$ μm $= 0.06$ mm Ans.

Problem 6.5

A spherical bubble of carbon monoxide is within the Fe-C-O melt at 1600 °C. The local pressure outside the bubble is 2 atm. The melt contains 0.2% C and 0.03% O and its surface tension is 1.5 J m^{-2}. From the following data

$$[\%C] + [\%O] = CO(g), \ \log K = \frac{1168}{T} + 2.07.$$

calculate the most probable site for nucleation among the bubble of 1 cm, 1 mm, 0.1 mm, 0.05 mm, 1 μm, and 0.1 μm radius.

Solution

Formation of spherical bubble of CO in Fe-C-O melt at 1873 K

According to the reaction: $[\%C] + [\%O] = CO(g)$
The pressure inside the bubble is given by

$P_b = P_s + \frac{2\sigma}{r}$ local pressure, $P_s = P_{atm} + \rho_{Fe}gh = 2$ atm
$\sigma = 1.5$ J m^{-2}, 1 J $= 0.102$ kg m, 1 atm $= 1.033 \times 10^4$ kg m^{-2}
$P_b - P_s = \frac{2\sigma}{r}$
$(P_b - 2) \times 1.033 \times 10^4 = \frac{2 \times 1.5 \times 0.102}{r} = \frac{0.306}{r}$

(i) $r = 0.1$ μm $= 10^{-7}$m

$(P_b - 2) \times 1.033 \times 10^4 = \frac{0.306}{10^{-7}}$
$P_b = 298$ atm

(ii) $r = 1$ μm $= 10^{-6}$ m
$(P_b - 2) \times 1.033 \times 10^4 = \frac{0.306}{10^{-6}}$
$P_b = 31.62$ atm

Similarly,

(iii) when $r = 0.05$ mm $= 5 \times 10^{-5}$ m
$P_b = 2.592$ atm
(iv) $r = 0.1$ mm $= 10^{-4}$ m
$P_b = 2.296$ atm
(v) $r = 1$ mm $= 10^{-3}$ m
$P_b = 2.0296$ atm
(vi) $r = 1$ cm $= 10^{-2}$ m
$P_b = 2.003$ atm

Given that $[C] = 0.2\%$ and $[O] = 0.03\%$
For the reaction $[C] + [O] = CO$ (g)
$\log K = \frac{1168}{1873} + 2.07 = 2.6936$
$\therefore K = 493.85$
By definition, $K = \frac{p_{CO}}{[\%C][\%O]}$,
that is, p_{CO} (equilibrium) $= K [\%C].[\%O] = 493.85 \times 0.2 \times 0.03 = 2.963$ atm.
Summary of the above calculations:

Bubble radius	0.1 μm	1 μm	0.05 mm			0.1 mm	1 mm	1 cm
P_b (atm)	298	31.62	2.592 close to the equilibrium $p_{CO} = $ 2.963 atm			2.296	2.030	2.003

Therefore, the probable site for the nucleation of CO bubble is the crevice of 0.05 mm radius. **Ans.**

Problem 6.6
A spherical bubble of CO is within Fe–C–O melt at 1577 °C. The total pressure outside the bubble is 2 atm. The melt contains 0.2% C and 0.03% O and its surface

tension is 1.5 J m^{-2}. What is the critical pore size for nucleation and growth of CO bubble?

Given that: [C] + [O] = CO (g), $\Delta G° = -4830 - 9.75\ T$ cal, $f_c = 1 = f_o$

Solution

At 1850 K, $\Delta G° = -4830 - 9.75 \times 1850 = -22867.5$ cal

$\Delta G° = -RT \ln k$

$K = 503.3$

p_{CO} (equilibrium) $= K f_C f_o$ [%O]. [%C] $= 503.3 \times 0.2 \times 0.03 = 3.02$ atm

for critical size, P_b (total) $= p_{CO}$ (equilibrium)

$P_b = P_s + \frac{2\sigma}{r}$

$(P_s = 2$ atm)

$\therefore (3.02 - 2) \times 1.033 \times 10^4 = \frac{2 \times 1.5 \times 0.102}{r} = \frac{0.306}{r}$

$r = 29 \times 10^{-6}$m

$= 29$ μm **Ans.**

OR

$P_b = P_s + \frac{2\sigma}{r}$ (1 atm $= 1.013 \times 10^5$ Pascal)

$P_b = p_{CO} = 3.02$ atm

$3.02 \times 1.013 \times 10^5 = 2 \times 1.013 \times 10^5 + \frac{2 \times 1.5}{r}$

or $(3.02 - 2)\ 1.013 \times 10^5 = \frac{3}{r}$

$r = \frac{3 \times 10^{-5}}{1.02 \times 1.013} = 2.9 \times 10^{-5} = 29 \times 10^{-6} = 29$ μm Ans.

Problem 6.7

(Elliott, J. F., 1980–1981, personal communication, Massachusetts Institute of Technology, Cambridge)

A bubble containing 0.05 m mol of argon forms a jet of the gas at the bottom of a ladle of steel. This bubble rises through the metal bath and finally bursts at the surface. The metal at 1600 °C contains 0.05%C, 5 ppm H, 0.004% N and 0.07% Si. The bath has been oxidized with silicon previously. Calculate the diameter and composition of the bubble at the bottom of the ladle and at the point of surface just before the bubble bursts. Assume that the gas in the bubble is at equilibrium with the melt. The depth of the metal is 2 m and the density of the metal is 7.1 g cm^{-3}. For calculation assume that the bubbles are hemispherical. Given that:

$\frac{1}{2} H_2(g) = [H]$ (ppm), $K_H = 24.3$

$\frac{1}{2} N_2(g) = [N]$ (ppm) $K_N = 0.045$

[%C] + [%O] = CO(g), $K_{CO} = 500$

$SiO_2(s) = [\%Si] + 2[\%O]$, $K = 2.82 \times 10^{-5}$

$e_C^C = 0.22$, $e_C^O = -0.097$, $e_C^N = 0.111$, $e_C^{Si} = 0.10$

$e_O^C = -0.13$, $e_O^O = -0.2$, $e_O^N = 0.057$, $e_O^{Si} = -0.14$

$e_N^C = 0.13$, $e_N^O = 0.05$, $e_N^N = 0$, $e_N^{Si} = 0.047$

$2 [H_{ppm}] + [\%O] = H_2O$ (g), $K = 0.006$

[%C] + 2[%O] = CO$_2$ (g), $K = 421$

Solution

Oxygen content of the deoxidized melt can be calculated from the following equations:

$SiO_2(s) = [\%Si] + [\%O]$, $a_{SiO_2} = 1$

$K = [\%Si] [\%O]^2 = 2.82 \times 10^{-5}$, $[\%Si] = 0.07$

$\therefore [\%O] = \left(\frac{2.82 \times 10^{-5}}{0.07}\right)^{1/2} = 0.02$

$[\%C] + [\%O] = CO$ (g), $K_{CO} = 500$, $\%C = 0.05$, $\%O = 0.02$

$\therefore p_{CO} = K_{CO}[\%C]. [\%O] = 500 \times 0.05 \times 0.02 = 0.5$ atm

In Fe–C–O–N–Si system:

$\log f_C = \%C. e_C^C + \%O.e_C^O + \%N.e_C^N + \%Si.e_C^{Si}$

$= 0.05 \times 0.22 + 0.02 \times (-0.097) + 0.004 \times 0.111 + 0.07 \times 0.10 = 0.0165$

$\therefore f_C = 1.04$

$\log f_O = \%O. e_O^O + \%C.e_O^C + \%N.e_O^N + \%Si.e_O^{Si}$

$= 0.02(-0.2) + 0.05 (-0.13) + 0.004 \times 0.057 + 0.07(-0.14) = -0.020072$

$\therefore f_O = 0.955$

After correction:

$p_{CO} = K_{CO}. f_C. [\%C]. f_O. [\%O] = 500 \times 1.04 \times 0.05 \times 0.955 \times 0.02 = 0.497$ atm

$\frac{1}{2} H_2(g) = [H]$ (ppm), $K_H = 24.3$

$K_H = \frac{H_{ppm}}{p_{H_2}^{1/2}} = \frac{5}{p_{H_2}^{1/2}} = 24.3$

$p_{H_2} = 0.042$ atm

$\frac{1}{2} N_2$ (g) $= [N]$ (ppm) $K_N = 0.045$

$K_N = \frac{\%N}{p_{N_2}^{1/2}} = \frac{0.004}{p_{N_2}^{1/2}} = 0.045$

$p_{N_2} = 0.008$ atm

$\log f_N = \%N. e_N^N + \%C.e_N^C + \%O.e_N^O + \%Si.e_N^{Si} + \%H.e_N^H$

$= 0 + 0.05 \times 0.13 + 0.02 \times 0.05 + 0.07 \times 0.047 + 0 = 0.01079$

$f_N = 1.025$

$K_N = \frac{f_N \%N}{p_{N_2}^{1/2}} = 0.045$

$p_{N_2} = \left(\frac{f_N.\%N}{0.045}\right)^2 = \left(\frac{1.025 \times 0.004}{0.045}\right)^2 = 0.0083$ atm

Water vapors: $2[H_{ppm}] + [\%O] = H_2O(g)$

$K = \frac{p_{H_2O}}{[H_{ppm}]^2 [f_O\%O]} = 0.006$

$p_{H_2O} = 0.006 \times [H_{ppm}]^2 [f_O\%O] = 0.006 \times 5^2 \times 0.955 \times 0.02 = 0.0029$ atm

$CO_2 : [\%C] + 2[\%O] = CO_2(g)$, $K = 421$

$K = \frac{p_{CO_2}}{[f_C.\%C].[f_O.\%O]^2} = 421$

$\therefore p_{CO_2} = 421.[f_C.\%C]. [f_O.\%O]^2 = 421 \times 1.04 \times 0.05 \times (0.955 \times 0.02)^2 = 0.008$ atm.

Thus, the equilibrium bubble holds the following gases at different partial pressures:

N_2-0.008, H_2-0.042, CO-0.497, CO_2-0.008, H_2O-0.003 atm ($P_{gases} = 0.558$ atm)

$$P_{Total} = p_{Ar} + 0.558$$

At the bottom of the ladle total pressure, $P_T = \rho gh + p_{atm}$

$\rho gh = 7.1(\text{gcm}^{-3}).\ 981(\text{cm s}^{-2}).\ 200\ (\text{cm})$
$\quad = 1393020\ (\text{dyne cm}^{-2})$
$\quad = \frac{1393020}{1.013 \times 10^6}$ atm $= 1.375$ atm
$\therefore P_T = 1.375 + 1 = 2.375$ atm
$\therefore p_{Ar} = 2.375 - 0.558 = 1.817$ atm

Volume of the hemispherical bubble $= \frac{2}{3}\pi r^3 = \frac{nRT}{P}$ ($n = 0.05$ m mol)

$v_b = \frac{0.05 \times 10^{-3} \times 0.082 \times 10^3 \times 1873}{1.817} = 4.23$ cc ($R = 0.082 \times 10^3$ cc atm deg mol^{-1})
$r = \sqrt[3]{\frac{4.23 \times 3}{2\pi}} = 1.26$ cm

$\therefore$Diameter of the bubble at the bottom of the ladle $= 2.52$ cm.
At the top of the ladle, $P_T = 1$ atm

$p_{Ar} = 1 - 0.558 = 0.442$ atm
$v_b = \frac{0.05 \times 10^{-3} \times 0.082 \times 10^3 \times 1873}{0.442} = 17.374$ cc
$r = \sqrt[3]{\frac{17.374 \times 3}{2\pi}} = 2.02$ cm
$\therefore$diameter of the bubble at top of the ladle $= 4.04$ cm.

At the top of the melt, $P = 1$ atm, $p_{Ar} = 0.442$ atm but the partial pressure of N_2, H_2, CO, CO_2 and H_2O are the same as at the bottom of the ladle.
Hence, the gas composition at different position would be as follows:

Species	Bottom		Top	
	p_i, atm	%	p_i, atm	%
Ar	1.817	76.505	0.442	44.2
N_2	0.008	0.337	0.008	0.8
H_2	0.042	1.768	0.042	4.2
CO	0.497	20.926	0.497	49.7
CO_2	0.008	0.337	0.008	0.8
H_2O	0.003	0.127	0.003	0.3
Total	2.375	100	1.00	100

Problem 6.8
Consider the effect of pressure in the above problem. How would answers be affected by having a pressure over the surface of the metal (a) 1 atm, (b) 0.001 atm, and (c) 10^{-6}atm. Make your comparison quantitatively.

Solution

(a) The conditions are the same as in the above problem, that is, the diameter of the bubble at top of the ladle = 4.04 cm and composition as shown in the table.

(b) Pressure at the surface (top) of the melt = 0.001 atm

P_T (at the bottom) = 1.375 + 0.001 = 1.376 atm

$P_T = p_{Ar} + P_{gases}$ (as per calculation in problem 6.7, $P_{gases} = 0.558$ atm)

∴ $p_{Ar} = 1.376 - 0.558 = 0.818$ atm

$$v_b = \frac{0.05 \times 10^{-3} \times 0.082 \times 10^3 \times 1873}{0.818} = 9.39 \text{ cc}$$

$$r = \sqrt[3]{\frac{9.39 \times 3}{2\pi}} = 1.65 \text{ cm}$$

Diameter of the bubble = 3.3 cm.

At the top of the ladle, since the total of the gases without argon ($P_{gases} = 0.558$ atm) exceeds the pressure at the surface of the melt, the bubble will be unstable and will burst after growing rapidly in size near the surface ($r \approx \infty$).

(c) Conditions are similar to that in part (b). The difference in pressure of 10^{-3} and 10^{-6} atm will have essentially no effect on behavior of the bubble.

Gas composition:

Bottom			Top	
Species	p_i, atm	%	p_i, atm	%
Ar	0.817	59.42	Equilibrium is not possible	
N_2	0.008	0.58		
H_2	0.042	3.05		
CO	0.497	36.15		
CO_2	0.008	0.58		
H_2O	0.003	0.22		
Total	1.375	100		

Based on the above calculations it may be concluded that the effectiveness of vacuum degassing is pronounced because of the large surface of the liquid exposed to the vacuum. In the previous problem, there will be turbulent boiling of the liquid but the vacuum degassing gives better kinetics of degassing.

Problem 6.9

The free energy change for the decarburization reaction: [%C] + [%O] = CO (g) is given as $\Delta G^\circ = -22400 - 39.6\ T$ J. If the steelmaking temperature is 1600 °C,

(a) What is the equilibrium concentration of oxygen in the melt when p_{CO} is 1 atm and the carbon content is 0.1%?

(b) What oxygen content in the melt having 0.1% C is required if it is necessary to form bubbles of 0.1 mm diameter to generate carbon boil? The ambient pressure at the location of the bubble is 1.5 atm. Surface tension of the melt is 1.2 J m^{-2}.

(c) The melt is in contact with pieces of slag in which a_{FeO} is 0.02 relative to pure liquid FeO. What is the equilibrium carbon content of the liquid iron if the CO pressure is (i) 1 atm and (ii) 0.01 atm.

Given that: FeO (l) = Fe(l) + [%O], $\Delta G^o = 120100 - 52.3\,T$ J

(d) What are the equilibrium oxygen concentrations in the melt for each of the conditions in part (c).

Assume that the activities of carbon and oxygen are equal to the weight percent concentration in the melt in all the cases.

Solution

$$[\%C] + [\%O] = CO(g), \quad \Delta G^o = -22400 - 39.6\,T \text{ J}$$

At 1873K, $\Delta G^o = -96570.8$ J $= -RT \ln K$

$$\ln K = \frac{96570.8}{8.314 \times 1873} = 6.2 \tag{1}$$

$$K = \frac{p_{CO}}{a_C.a_O} = \frac{p_{CO}}{[\%C].[\%O]} \tag{2}$$

(a) $p_{CO} = 1$ atm, %C $= 0.1$
From Eqs. 1 and 2 we get:

$$\ln \left[\frac{[\%C].[\%O]}{p_{CO}}\right] = -6.2$$

$$\therefore \left[\frac{[\%C].[\%O]}{p_{CO}}\right] = 0.002$$

$$\therefore [\%O] = \frac{0.002 \times 1}{0.1} = 0.02 \quad \text{Ans.}$$

(b) $p_{CO} = \frac{2\sigma}{r} + p_{surr}$ (1 atm. $= 1.013 \times 10^5$ N m^{-2})

$p_{surr} = 1.5$ atm, $\sigma = 1.2$ J m^{-2}, $r = \frac{0.01}{2}$ cm $= \frac{10^{-4}}{2}$ m

$$\therefore p_{CO} = \left\{\frac{2 \times 1.2 \times 2}{10^{-4}} \left(\frac{\text{N m}^{-1}}{\text{m}}\right) \cdot \frac{1}{1.013 \times 10^5} \left(\frac{\text{atm}}{\text{N m}^{-2}}\right)\right\} + 1.5 = 1.973 \text{ atm}$$

Since $\left[\frac{[\%C].[\%O]}{p_{CO}}\right] = 0.002$, $[\%C] = 0.1$

$$\therefore [\%O] = \frac{0.002 \times 1.973}{0.1} = 0.04 \text{ Ans.}$$

(c) FeO(l) = Fe(l) + [%O], $\Delta G^o = 120, 100 - 52.3\,T$, J

At 1873 K, $\Delta G^o = 22142.1$ J

$$= -8.314 \times 1873 \times \ln \left[\frac{a_{Fe}.[\%O]}{a_{FeO}}\right]$$

$$\ln \left[\frac{a_{Fe}.[\%O]}{a_{FeO}}\right] = -1.422$$

$$\therefore \left[\frac{a_{Fe}.[\%O]}{a_{FeO}}\right] = 0.24, \quad a_{FeO} = 0.02 \text{ (given) and assuming } a_{Fe} = 1$$

$$\therefore [\%O] = 0.24 \times 0.02 = 0.0048$$

(c.i) From part (a) we have

$$\left[\frac{[\%C].[\%O]}{p_{CO}}\right] = 0.002, \quad p_{CO} = 1 \text{ atm}, \quad [\%O] = 0.0048$$

$$\therefore [\%C] = \frac{0.002}{0.0048} = 0.415$$

(c.ii) $p_{CO} = 0.01$ atm, $[\%O] = 0.0048$

$$\therefore [\%C] = \frac{0.002 \times 0.01}{0.0048} = 0.00415 \quad \text{Ans.}$$

(d) The equilibrium oxygen concentration for both the conditions in part (c) are the same

That is, $[\%O] = 0.0048$ Ans.

References

1. Parker, R. H. (1978). *An introduction to chemical metallurgy* (2nd ed.). Oxford: Pergamon. (Chapter 6).
2. Cahn, J. W., & Hilliard, J. E. J. (1959). Free energy of a non-uniform system. III. Nucleation in a two-component incompressible fluid. *The Journal of Chemical Physics, 31*, 688–699.
3. Becker, R., & Doring, W. (1935). Kinetische Behandlung der Keimbildung in Ubersattingten Dampfen. *Annln der Physik, 24*, 719–731.
4. Vomer, M., & Weber, A. (1926). Keimbildung in Ubersattingten Gebilden. *Zeitschrift für Physikalische Chemie, 119*, 277–291.
5. Porter, D. A., & Easterling, K. E. (2004). *Phase transformations in metals and alloys*. Boca Raton: CRC, Taylor & Francis. (Chapter 2).
6. Bodsworth, C. (1994). *The extraction and refining of metals*. Boca Raton: CRC, Taylor & Francis. (Chapter 3).
7. Turpin, G. M., & Elliott, J. F. (1956). Nucleation of oxide inclusions in iron melts. *Journal of the Iron and Steel Institute, 24*, 217–224.
8. Evans, D. J. I. (1968). Production of metals by gaseous reduction from solutions, process and chemistry. In *Advances in extractive metallurgy, Proceedings of the symposium, Institution of Mining and Metallurgy*, London, April 17–20, 1967, (pp 831–907)
9. Meddings, B., & Mackiw, V. N. (1964). Gaseous reduction of metals from aqueous solutions. In M. E. Wadsworth, & F. T. Davis (Eds.), *Unit processes in hydrometallurgy*. Met. Soc. AIME, Dallas, February 25–28, 1963, Gordon & Breach, New York, Vol. 24, Group B, pp 345–384.
10. Ward, R. G. (1963). *An introduction to physical chemistry of iron and steelmaking*. London: Edward Arnold. (Chapter 13).
11. Tupkary, R. H., & Tupkary, V. R. (1998). *An introduction to modern steelmaking* (6th ed.). Delhi: Khanna Publishers. (Chapter 14).
12. Kapilashrami, A., Görnerup, M., Lahiri, A. K., & Seetharaman, S. (2006). Foaming of slags under dynamic conditions. *Metallurgical and Materials Transactions B: Process Metallurgy and Materials Processing Science, 37B*, 109–119.
13. Seetharaman, S., Teng, L., Hayashi, M., & Wang, L. (2013). Understanding the properties of slags. *ISIJ International, 53*, 1–8.
14. Geiger, G. H., Kozakevitch, P., Olette, M., & Riboud, P. V. (1975). Theory of BOF reaction rates. In R. D. Pehlke, W. F. Porter, P. F. Urban, & J. M. Gaines (Eds.), *BOF steelmaking* (Iron & steel technology) (Vol. 2, pp. 191–321). New York: AIME. (Chapter 5).
15. Swisher, J. H., & McCabe, C. L. (1964). Cr_2O_3 as a foaming agent in CaO-SiO_2 slags. *Transactions Metallurgical Society, AIME, 230*, 1669–1675.
16. Ito, K., & Fruehan, R. J. (1989). Study on the foaming of CaO- SiO_2-FeO slags. *Metallurgical Transactions B, 20B*, 509–514.

17. Zhang, Y., & Fruehan, R. J. (1995). Effect of the bubble size and chemical reactions on slag foaming. *Metallurgical and Materials Transactions B: Process Metallurgy and Materials Processing Science, 26B*, 803–812.

18. Zhang, Y., & Fruehan, R. J. (1995). Effect of carbonaceous particles on slag foaming. *Metallurgical and Materials Transactions B: Process Metallurgy and Materials Processing Science, 26B*, 813–819.

19. Utigard, T., & Toguri, J. M. (1985). Interfacial tension of aluminum in cryolite melts. *Metallurgical Transactions B, 16*, 333–338.

20. Jiang, R., & Fruehan, R. J. (1991). Slag foaming in bath smelting. *Metallurgical Transactions B, 22B*, 481–489.

21. Goto, M., Oshima, I., & Hayashi, M. (1998). Control aspects in the Mitsubishi continuous process. *JOM, 50*(4), 60–65.

22. Goto, M., & Hayashi, M. (1998). *The Mitsubishi continuous process*. Tokyo: Mitsubishi Material Corporation, www-adm@mme.co.jp

23. Moore, J. J. (1990). *Chemical metallurgy* (2nd ed.). Oxford: Butterworth-Heinemann. (Chapter 7).

24. Rosenqvist, T. (1974). *Principles of extractive metallurgy*. New York: McGraw-Hill. (Chapter 7).

25. Smith, C. S. (1964). Some elementary principle of polycrystalline microstructures. *Metallurgical Reviews, 9*, 1–48.

26. Kingery, W. D. (1960). *Introduction of ceramics*. New York: Wiley. (Chapter 5).

27. Thonstad, J., Fellner, P., Haarberg, G. M., Hives, J., Kvande, H., & Sterten, A. (2001). *Aluminium electrolysis* (3rd ed.). Aluminium-Verlag, Marketing & Kommunikation GmbH. (Chapter 7).

28. Sevryukov, N., Kuzmin, B., & Chelishchev, Y. (1960). *General metallurgy* (B. Kuznetsov, Trans). Moscow: Peace Publishers. (Chapter 17).

29. Rosegger, R. (1960, July 15). Use of aluminium in deoxidation. *Iron and Coal Trades Review*, pp. 131–140.

30. Kubaschewski, O., & Alcock, C. B. (1979). *Metallurgical thermochemistry* (5th ed.). Oxford: Pergamon.

31. Curnutt, J. L., Prophet, H., McDonald, R. A., & Syverud, A. N. (1975). *JANAF thermochemical tables*. Midland: The Dow Chemical Company.

32. Barin, I., Knacke, O., & Kubaschewski, O. (1977). *Themochemical properties of inorganic substances, (supplement)*. New York: Springer.

Chapter 7
Steelmaking

Steel is an alloy of iron and one or more element(s), namely carbon, nickel, chromium, manganese, vanadium, molybdenum, tungsten, and so on. Chemically, steels may be classified in two groups: plain carbon steels and alloy steels. The former comprises the alloys of iron and carbon, whereas the latter contains one or more elements in addition to carbon. The alloying elements improve the mechanical, magnetic and electrical properties, as well as the corrosion resistance of steels. Impurities like Si, Mn, S, P, Al, and O are invariably present in steels due to their association in pig iron obtained by reduction smelting of iron ore with coke and lime in the blast furnace. Essentially, steelmaking is the conversion of molten pig iron (hot metal) containing variable amounts of 4.0–4.5% carbon, 0.4–1.5% silicon, 0.15–1.5% manganese, 0.05–2.5% phosphorus (normally between 0.06 and 0.25%), and 0.15% sulfur (normally between 0.05 and 0.08%) to steel containing about 1% of controlled amount of impurities by preferential oxidation. Alternatively, steel can be produced from solid sponge iron obtained by solid-state reduction of iron ore in the shaft furnace or retort. Thus, basically two routes are adopted in the production of steels. The first one employs the basic oxygen furnace (BOF – LD/Q-BOP/Hybrid converters) for treatment of hot metal, and the second route uses the electric arc furnace (EAF) to treat steel scrap/sponge iron or direct reduced iron (DRI). Electric arc or induction furnaces are generally used in the production of alloy steels. Pig iron contains a total of about 10% of C, Si, Mn, P, S, and so on as impurities, whereas sponge iron contains gangue oxides of iron ore, such as Al_2O_3, SiO_2, CaO, and MgO. The amount and number of impurities depend on the quality of the iron ore, coke, and lime stone used in smelting. The molten pig iron is refined to molten steel under oxidizing conditions using iron ore and/or oxygen. On the other hand, scrap and sponge iron are melted in electric furnaces and refined for steel production.

Steelmaking is a process of selective oxidation of impurities, which is reverse of ironmaking (carried out under reducing atmosphere). In principle, it is similar to the fire refining of nonferrous metals (particularly blister copper and lead bullion), but the end product is an alloy, not a pure metal. However, the process of conversion of

© The Minerals, Metals & Materials Society 2021
M. Shamsuddin, *Physical Chemistry of Metallurgical Processes, Second Edition*,
The Minerals, Metals & Materials Series, https://doi.org/10.1007/978-3-030-58069-8_7

pig iron/sponge iron into steel is a complex process. The overall process includes a number of steps, namely, charge preparation, melting, refining, tapping, deoxidation, decarburization, alloying, teeming, casting, stripping, and so on. Except sulfur all other impurities are oxidized during steelmaking. These oxides are eliminated either as a gas (C $\rightarrow$ CO/CO$_2$) or as a liquid oxide product known as slag. The slag acts as an absorber of oxidized impurities by control of basicity, which is achieved by adjusting the chemical composition of the slag. As removal of sulfur requires reducing conditions, it is carried out in the blast furnace. In case the amount of sulfur, silicon, and phosphorus in the pig iron is beyond the normal level, these elements are removed outside the blast furnace (prior to charging in the steelmaking units) after tapping the hot metal in the ladle. Such a process of removal by treatment with different reagents is known as external desulfurization, desiliconization, and dephosphorization.

7.1 Steelmaking Processes

Prior to the development of the Bessemer process, steel was made on small scale by cementation and crucible processes. Bulk production of steel started soon after the Bessemer process was invented in 1856 in the United Kingdom. Since then a number of processes have been developed. It would be appropriate to give here only a very brief account on these processes. For details, readers are advised to go through references [1–4].

7.1.1 Bessemer Process

The process developed by Henry Bessemer to refine low-phosphorus pig iron by blowing air in acid-lined converter is known as the Bessemer process. In 1879, Sydney G. Thomas modified the Bessemer process for refining pig iron containing sulfur and high phosphorus by forming basic slag in a basic lined converter. This process is known as the basic Bessemer process and also as the Thomas process after the name of the inventor. Both acid and basic Bessemer processes employed converters of conico-cylindrical shape fitted with detachable bottom. The converter mounted on trunnions could be rotated through 360°. After bringing the converter in horizontal position, it was charged with blast furnace hot metal and air was blown through tuyeres located at the bottom. The converter was then slowly moved to a vertical position. The impurities (Si, Mn, and P) present in the hot metal are oxidized to form oxides, SiO$_2$, MnO, and P$_2$O$_5$, and slagged off. CO and CO$_2$ formed by oxidation of carbon join the flue gas and released into the atmosphere from the mouth of the converter. After refining (i.e., at the end of the blow), the converter is rotated to tap the metal and slag.

As the Bessemer process of steelmaking was extremely fast, taking only 20–25 min for a 20–25 ton vessel, it was adopted for mass production of steel on account of its speed, as well as lower cost. However, despite being fast and cheap, the process was completely abandoned after serving the steel industry for more than 100 years until a few decades after the Second World War due to (i) loss of heat generated through exothermic oxidation in heating the undesirable nitrogen from room temperature to 1600 °C, (ii) much higher residual nitrogen (~0.012%) causing stretcher strain in Bessemer steel compared to 50–60 ppm in steel produced by the open-hearth process, (iii) frequent removal of the bottom due to severe erosion of poor quality material available then for preparation of the detachable bottom, and (iv) incapability in refining hot metal containing high silicon and medium phosphorus in a single step. Hence, steel industries were attracted to the newly developed open-hearth process that could produce low nitrogen steel. Presently, Bessemer process is only of historical significance as it is credited with starting the mass production of steel.

7.1.2 Open-Hearth Process

In 1861, open-hearth furnace, based on heat-regenerating principles for achieving a temperature of 1600 °C, was developed by Siemens in Germany and Martin in France for production of steel using varying proportion of hot metal and steel scrap. The technique was known as Siemens-Martin or the Open-Hearth process due to the shallow hearth of the refining vessel. The capacity of open-hearth furnaces was generally very high: up to 500 tons per heat. It was provided with a number of doors in the front wall for charging steel scrap, hot metal, lime, and iron ore. During the course of heating, these doors were also used for inspection, sampling, and addition. Different types of refractories were used in various parts of the furnace. For example, hearth made of steel is covered with asbestos sheet having successive layers of porous fireclay bricks, magnesia, and finally working lining by ramming magnesia. The heat was generated by combustion of liquid and gaseous fuel with air through burners provided in the side walls. The air for combustion was preheated in regenerators to produce a flame temperature above 1600 °C. Silicon, carbon, manganese, and phosphorus are oxidized by oxygen supplied by iron ore and atmospheric oxygen. Before tapping, partial deoxidation was carried out in the furnace with limited alloy addition to reduce oxygen potential.

The open-hearth process introduced in 1865 contributed to the major production of steel throughout the world because it could produce various grades of quality steel with low residual nitrogen having good mechanical properties. The process was not only extremely slow taking 6–8 h per heat but was also associated with the requirement of external heat and problems of construction and maintenance of the roof. Attempts were made to increase productivity and reduce fuel consumption by injecting oxygen through consumable lances inserted at the slag–metal interface or through water-cooled lances inserted in the roof and positioned over the slag surface.

Despite extensive efforts, the open-hearth process could not compete with the rate of production achieved by modern basic oxygen converters which take only 30–40 min per heat for a 250-ton converter. As a result, it was replaced by oxygen steelmaking processes after serving the industry for about a century.

7.1.3 Electric Arc Furnace (EAF) Process

Electric arc and induction furnaces developed by Paul Heroult (1899) and Ferranti (1877) have emerged as the major alternative for the manufacture of steel from steel scrap. A number of mini steel plants producing less than 1 million of steel annually are essentially based on EAF. In most of the plants, sponge iron is charged along with steel scrap (which is in short supply). However, recent trend of integrating EAF with mini blast furnaces has prompted the use of varying proportions of hot metal, steel scrap, and sponge iron. Generally, basic electric steelmaking practice is adopted due to the presence of sulfur and phosphorus in the charge (hot metal/steel scrap/ sponge iron).

The furnace made of steel shell lined with refractories has three electrodes entering from the roof. The unit is provided with mechanisms to tilt the furnace and to move up and down the electrodes. The roof is lined with silica or high alumina bricks. The furnace shell is lined with fireclay bricks to provide thermal insulation. In the hearth, the fireclay bricks are backed by a few layers of fired magnesite bricks. Joint-free working hearth surface is prepared by ramming dolomite or magnesite. These three layers of refractories make the hearth strong enough to bear mechanical and thermal stresses and withstand the corrosive action of slag and metal at high temperatures. High temperature is generated by striking the electrodes against the charge. Operation of the basic EAF steelmaking consists of charging, melting, refining, and finishing. The charge consisting of sponge iron/steel scrap, lime/ limestone, iron ore, and coke is dropped from baskets after raising the electrodes and swinging the roof. Hot metal (if used) is poured through the door. Dephosphorization is carried out by making oxidizing slag. In case desulfurization is desired, a reducing slag is made after draining the oxidizing slag completely.

7.1.4 Top-Blown Basic Oxygen Converter Process

Availability of oxygen on tonnage scale at reasonable cost by the Linde-Frankle process opened new way for the development of faster steelmaking processes. The top-blown basic oxygen converter process, popularly known as the LD process, is the outcome of the development work carried out at the two towns Linz and Donawitz in Austria during late 1940s and early 1950s. This led to the beginning of modern steelmaking. In his patent, Henry Bessemer had mentioned about the benefit of blowing pure oxygen (in place of air), which was commercially not available in those days. In addition to the cost factor, it was not practically possible to blow pure oxygen through the bottom of the converter designed by Henry Bessemer due to severe corrosion problems in tuyeres. Although, at present, Bessemer processes (both acid as well as basic) have been completely abandoned Henry Bessemer will always be remembered for his invention and pioneering contributions to the steel industry, which laid the foundation for the subsequent development of oxygen converter processes.

A basic oxygen converter (capacity ranging from 100 to 400 tons) is a pear-shaped vessel made of steel plate lined with either pitch-bonded dolomite or impregnated fired magnesite. However, in recent years, the life of the refractory lining has been largely increased by the slag splashing technique. For splashing, a fraction of slag left in the converter after tapping is conditioned with dolomite. The mixture is then splashed in various parts of the converter by injecting nitrogen gas at different flow rates through the existing oxygen lance positioning it at different levels. Viscous slag thus attached to the working refractory lining reduces its consumption and increases life. Scrap, hot metal, flux, ferroalloys, and iron ore are charged from the top of the vessel. In order to produce a basic and reactive slag to achieve effective dephosphorization as well as desulfurization in an LD converter, reasonably good quality lime is charged. Lime with low silica is preferred because SiO_2 reduces the reactivity of CaO by forming $2CaO.SiO_2$ when the charge gets heated. About 3.5% MgO in lime seems beneficial, whereas 5% MgO slows down the formation of dicalcium silicate slag. For early formation of slag during the blow, a soft burnt lime of uniform size ranging from +8 to −40 mesh with low moisture content and low loss on ignition is required. High purity oxygen (99.5% pure) is blown at supersonic speed (flow rate 550–600 m^3 min^{-1} for about 15–20 min for a 160–180 ton converter) [1] through a water-cooled lance inserted through the mouth of the converter. The position of the lance fitted with copper nozzle is automatically adjusted in the bath according to the flow rate of oxygen. Initially, lances having single hole were used. Multihole lances were developed to blow large volume of oxygen in bigger vessels. Larger number of holes enhance the slag–metal reaction and increase productivity. In order to check the bath composition, a metal sample is taken after stopping the blow, raising the lance and tilting the vessel. Low-silicon hot metal is desired to reduce the volume of the slag produced by neutralization of silica (produced by oxidation of silicon) by lime. A silica-rich slag produced by blowing high-silicon hot metal is highly corrosive to the basic lining. Hence, high-silicon hot metal obtained from the blast furnace has to be first desiliconized in a ladle before charging into the converter. Lime is added before start of the blow, and fluorspar is fed to achieve the desired fluidity of the slag. Steel scrap, iron ore containing lesser gangue, and mill scale are charged to cool the overheated bath. In recent years, use of DRI in suitable sizes (3–15 mm) is steadily increasing. Steel is tapped into a ladle through the tap hole located in the nose of the converter by rotating it in the opposite side. The slag floating on the bath is prevented from entering into the ladle along with the metal by using a pneumatic slag stopper at the tap hole or a slag arresting device into the converter. After tapping steel, converter is rotated almost by 180° from upright position to allow the slag to flow into a slag ladle. As and when desired, some slag is retained in the converter for splashing. After deoxidation of the steel with ferromanganese, ferrosilicon, and aluminum, alloying elements are added in the ladle. The nitrogen content of the LD steel varies from 0.003 to 0.005% on account of blowing high-purity oxygen. About 125 m^3 of flue gas (generated per ton of liquid steel produced), loaded with dust (up to 20 kg ton^{-1} of liquid steel) containing appreciable amount of CO is subjected to cleaning before storage.

LD process was modified by CNRM in Belgium to refine high-phosphorus hot metal (as high as 2%, i.e., Thomas-grade pig iron) produced in Germany, France, Belgium, Luxenburg, and United Kingdom. The modified process, known as LDAC or OLP is very similar to LD in design and operation. The charge in LDAC contains less scrap as compared to that used in LD converter because excess lime has to be used for slagging large amount of phosphorus present in the hot metal. Hence, in order to accommodate larger amount of slag generated in this process, a larger vessel having 20% excess volume over LD converter is required. One third of the total lime required is charged in lumpy form before pouring the hot metal in the converter. The remaining two third is blown as lime powder along with oxygen gas. Slag formation is facilitated by the use of powdered lime. Foam formation which controls the overall process is affected by the lance height and oxygen/lime flow rate. In this process, dephosphorization ends well before decarburization due to simultaneous blowing of lime with oxygen. Putting the lance at the highest position oxygen blow is started. After passage of 5 min of oxygen blow, powdered lime (1–2 mm size) is fed through the lance at a predetermined rate. The lance is then lowered to control the foam. The blow is stopped after 15 min by which time carbon is brought to 1.5–1.7% and phosphorus to 0.2% and bath temperature reaches 1600–1690 °C. The slag rich in P$_2$O$_5$ is completely drained out. Blow is resumed again after fresh addition of coolant, flux, and so on and continued for another 5–8 min. In this way, phosphorus is reduced to 0.02%. For further reduction, three slag practice is adopted. The third slag is retained in the converter for the next heat.

A similar process in principle and practice named as oxygen lime process (OLP) was independently developed by IRSID in France. The process based on IRSID design is known as OLP, otherwise popularly known as the LDAC.

7.1.5 *Rotating Oxygen-Blown Converter Process*

The rotating oxygen-blown converter was developed concurrently and independently when LD process was developed in Austria. The refractory lining and oxygen lancing in this process are similar to those used in LD vessel, but the converter is rotated during refining. Since carbon monoxide evolved during refining is burnt in the converter, the rotation protects the lining from being overheated. Rotation increases the rate of refining by facilitating mixing of the slag, metal, and gas. Under this category, there are two processes, namely, Kaldo and Rotor.

The Kaldo process was developed to refine high phosphorus (~2%) hot metal under the leadership of Professor Bo-Kalling of Sweden. The first commercial plant installed at Domnarvet in 1954 started production in 1956. By 1967, there were 10 plants in Sweden, France, United Kingdom, United States, Japan, and so on. The Kaldo converter, originally developed for refining Thomas-grade pig iron, could successfully treat all qualities of hot metals. In size, shape, and refractory lining, the converter is similar to the LD vessel. The converter placed in a cradle and mounted on trunnions can be tilted for charging, tapping, slagging, and so on. During blowing, it is held at an angle of 16–20° to the horizontal and rotated at a maximum speed of 30 rpm. The accessories in Kaldo plant for charging scrap, flux and hot metal, lancing, tapping, slagging, and so on are very much similar to those in the LD converter. After lining, the effective volume of the Kaldo converter is reduced to about 0.5 m^3 ton^{-1} capacity as compared to 0.75 m^3 ton^{-1} of the LD vessel. The volume has to be increased for higher phosphorus content of hot metal to accommodate additional volume of slag that would be generated. Heat generated by combustion of CO inside the converter is transferred to the metal via the lining by radiation. Refractory consumption in Kaldo converter is higher than in the LD converter because rotation increases wear of the lining; for example, the lining life is 60–100 heats against 200–300 heats for similar material in LD vessel.

In 1952, at Oberhausen in Germany, a rotary furnace was employed (after replacing the burner with oxygen lance) for trials with the idea to partially refine hot metal for subsequent production of steel in an open-hearth furnace. But a successful trial established that the rotary furnace itself could be employed for the production of steel from charge containing scrap and hot metal. This method of production of steel in a rotary furnace was named as the Rotor Process. Several units of 100–120 ton capacity were set up in Germany, United Kingdom, and South Africa. In this process, a long cylindrical vessel (length four times the diameter) with openings at both ends is fitted with two lances. The primary lance for refining inserted in the slag–metal interface blows oxygen into the bath. The secondary lance is employed for supply of oxygen or mixture of oxygen and air above the surface of the bath to burn carbon monoxide (gas evolving due to C–O reaction) to carbon dioxide. Rotation of the vessel around its horizontal axis at a speed of 0.2–4.0 rpm protects the lining from overheating and also facilitates heat transfer by radiation and conduction. The refractory lining in the rotor is similar to that in Kaldo and LD converters, but life is shorter due to rotation. The converter can be lifted in vertical plane up to 90° and also can be turned for charging and tapping. The Rotor Process, however, did not achieve much success in the production of steel.

7.1.6 Bottom-Blown Oxygen Converter Process

Even after commercialization of the LD process, many European countries were refining high-phosphorus hot metal in the conventionally bottom-blown basic Bessemer converters (Thomas Process). Since air containing nearly 80% nitrogen was blown in these converters, the resulting steel contained as high as 0.015% nitrogen, which adversely affected the mechanical properties. Furthermore, the drastically reduced thermal balance of the Bessemer process in heating large volumes of nitrogen from room temperature to 1600 °C restricted the use of scrap to a certain maximum. Hence, attempts were made to reduce nitrogen content in steel by blowing oxygen-enriched air. Use of 40% oxygen in the blast-reduced nitrogen from 0.015 wt% to 0.001 wt% and permitted 20% scrap in the charge instead of 8–10%. The enriched oxygen blow drastically reduced life of the bottom refractories. The exothermic reaction between oxygen and molten iron not only increases temperature at the tuyere tip but also forms molten FeO that is highly corrosive and has a tendency to form low-melting compounds with oxide ingredients of the refractory lining. Thus, tuyeres have to be repaired/replaced frequently. Pure oxygen is blown from the top in LD converters and from sides in Kaldo and rotor processes. Among these, LD became popular and dominated the steel industry and was modified as LDAC/OLP to refine high-phosphorus pig iron produced in Western European countries. The modification offered some solution but industries were interested in developing a process wherein pure oxygen could be blown from the bottom of the Bessemer-type converter.

It was also realized that composition and temperature of the hot metal bath in the LD converter are not uniform in the presence of solid scrap in the charge although it is stirred by the top jet and evolution of carbon monoxide formed due to violent C–O reaction at 1600 °C. Constant and steady efforts to overcome the problems of slopping and concentration and temperature gradients in top-blown oxygen converters in refining high-phosphorus hot metal (1.5–2%) led to the development of bottom-blowing processes in 1960s. The process named OBM [oxygen-bolden blasen (bottom-blown) Maxhuette] was developed in Germany in 1967 and was commissioned for the first time at Maximillianshutte Iron and Steel Company. In 1969, the LWS process was developed in France by the joint efforts of three companies, Laire, Wendel, and Strunck. In the United States, the OBM process is known as the Q-BOP (quick or quiet-bath oxygen process). The OBM converter is basically a Bessemer-like vessel fitted with a special bottom, in which entire oxygen is introduced along with lime powder through the stainless steel tuyeres inserted in the magnesia-lined bottom. As blowing pure oxygen through the bottom generates intensive heat, tuyeres are protected by injecting methane or propane gas and fuel oil, in the OBM/Q-BOP and LWS processes, respectively, through the outer pipe surrounding the inner oxygen pipe. Endothermic decomposition of the gas/oil on entering into the melt derives heat from the steel bath and thus protects tuyeres from overheating. Hydrogen generated by dissociation gets dissolved in liquid steel. Hence, nitrogen is bubbled through the melt to remove hydrogen before tapping. The bottom-blown processes operate under conditions close to equilibrium. In the absence of top lancing system, these units require less tall buildings. This aspect encouraged steelmakers in the mid-1970s to install bottom-blown converters in the then existing open-hearth shops at lower investment cost as compared to the total investment on basic oxygen converters. It was also easier to switch over from Bessemer- to OBM-type converters without much expenditure.

In the OBM process, tuyeres are fixed in only one half of the converter bottom, which ensures upward flow of metal in this half and downward in the other half of the vessel without tuyeres. This arrangement provides sufficient turbulence to achieve adequate slag–metal contact. The design minimizes damage and repair of tuyeres because scrap can be charged in the part without tuyeres. It also increases the capacity of the vessel for the same inner volume (about 40% increase compared to the Bessemer converter) because hot metal can be filled almost half the bottom area in inclined position.

7.1.7 Hybrid/Bath-Agitated/Combined-Blown Process

Although supersonic oxygen jet impinges the molten bath from top in an LD converter, the kinetic energy is not very effectively converted to produce sufficient stirring in the bath. As a result, the temperature and concentration in the top-blown converters are not uniform. Re-blowing reduces productivity and poses problems in the manufacture of ultralow carbon steel. Bottom-blowing reduces temperature and concentration gradients and at one time it appeared that the OBM process would supersede LD steelmaking. Though bottom-blowing offers perfect stirring that is required for steelmaking and works under near equilibrium conditions, it suffers from high refractory consumption, tedious and cumbersome repairing and mainte-nance of the bottom, and high hydrogen and nitrogen pick up in steel. Based on extensive researches in the 1970s, it was established that good stirring and better mass transport could be obtained even by mild bottom-blowing. This led to the development of "bath-agitated process" (BAP). In this process, small amount of nitrogen or argon is introduced through symmetrically located bottom tuyeres (4–8 in numbers). Pure oxygen and powdered lime required for complete refining are blown from the top, as well as bottom. Addition of ore or scrap depends on the composition and temperature of the hot metal. Blowing is completed in three stages. When gaseous phase is introduced from both ends (top as well as bottom), the process is called the hybrid process. This is also known as combined-blown process. In recent years, these processes have gained prominence throughout the world.

As regards to the nomenclature of different processes, one should note that basic Bessemer, open-hearth, and EAF processes are known as the conventional pro-cesses, whereas oxygen-blown converter processes: top-blown (LD, LDAC, Kaldo, and rotor), bottom-blown through special tuyeres (Q-BOP/OBM and LWS), and the combined-blown (hybrid process) are classified as basic oxygen furnace (BOF) processes. Bessemer and BOF processes are also known as pneu-matic processes because air/oxygen is blown at high pressure. These processes are autogenous due to the highly exothermic nature of oxidation of impurities: carbon, silicon, manganese, and phosphorus, which generate sufficient heat to raise the temperature of the bath well above the melting point of steel. In EAF, the main source of heat is electric power, whereas in open-hearth furnace, liquid or gaseous fuel is used. At present, BOF and EAF together account for the major production of primary steel throughout the world. The open-hearth process, the major producer of steel a few decades ago, has presently disappeared from the scene.

Conversion of pig iron containing about 10 wt% of carbon, silicon, manganese, phosphorus, and so on to steel containing about 1 wt% of controlled amount of impurities by preferential oxidation as well as production of steel from steel scrap, sponge iron/DRI are referred as "primary steelmaking." Along with the principal gaseous oxidizing agent, oxygen, some iron ore is added in the converter. Oxidation of carbon forms gaseous products CO and CO_2, whereas oxidation of other impurities, as referred earlier, form SiO_2, MnO, P_2O_5, and so on, which are fluxed with lime and fluorspar to generate a basic slag (CaO as the major content). The role of slag in refining, as well as its structure, properties, and theories, have been discussed in Chap. 4.

In recent years, demand for high-quality steel with extremely low sulfur and phosphorus has enforced further *refining, desulfurization, degassing*, and so on along with *deoxidation* of the liquid steel tapped in the ladle. These operations are collectively known as 'secondary steelmaking'. Presently, such units have become an essential part of steel melting shops of any integrated steel plant. In order to produce a better quality steel at the primary level itself, pretreatment of the hot metal tapped from the blast furnace is carried out in the ladle with different reagents to reduce S, Si, and P before charging into the steelmaking units. Manufacture of stainless steel may be considered under secondary steelmaking because the EAF is used mainly as a melting unit where very little decarburization is carried out; for adequate decarburization, argon–oxygen decarburization vessel is employed.

In this chapter, the physical chemistry of sulfur, phosphorus, silicon, manganese, and carbon reactions have been discussed to derive optimum conditions for desulfurization, dephosphorization, and also for recovery of silicon and manganese. These physicochemical principles are applicable in general to all the steelmaking processes. A brief account on different processes with salient features of chemical reactions showing sequence of removal of impurities during refining has been included in this chapter. Deoxidation, degassing, and stainless steelmaking will be discussed with the aid of problems in Chap. 8.

7.2 Physicochemical Principles

As mentioned above, the process of steelmaking is based on the principle of fire refining. Elements having higher affinity for oxygen are preferentially oxidized and slagged off. Considering the fact that pig iron contains carbon, silicon, manganese, sulfur, and phosphorus as major impurities, it would be appropriate and interesting to discuss these reactions in order to set optimum conditions for removal/recovery of these elements, as the case may be.

7.2.1 Sulfur Reactions

Despite having a very low boiling point, an appreciable amount of sulfur remains in the hot metal due to its strong attraction with iron. Making use of interaction coefficients for the effect of various elements on the activity coefficient of sulfur in iron, activity of sulfur can be calculated by the expression:

$$\log f_S = 0.29 \times \%P + 0.11 \times \%C + 0.063 \times \%Si + 0.035 \times \%Al + 0.0097 \times \%W + 0.0027 \times \%Mo +$$
$$0.0026 \times \%Co - 0.0984 \times \%Cu - 0.011 \times \%Cr - 0.013 \times \%Nb - 0.016 \times \%V - 0.026 \times \%Ca -$$
$$0.027 \times \%O - 0.28 \times \%S - 0.952 \times \%Zr$$

$$(7.1)$$

Sulfur exists in slag as CaS to the extent of a few percent. Distribution of sulfur between slag and metal can be discussed according to the reaction:

$$[S] + (O^{2-}) = [O] + (S^{2-}) \qquad (7.2)$$

$$K = \frac{[a_O] \cdot (a_{S^{2-}})}{[a_S] \cdot (a_{O^{2-}})} \qquad (7.3)$$

Since the concentration of the impurity element in the melt is low, we can express the activity according to the Henry's law, $a_i = f_i$. wt % i, and applying Temkin theory of ionic slag melt, Eq. 7.3 may be expressed as:

$$K = \frac{[f_O . \%O] \cdot (x_{S^{2-}})}{[f_S . \%S] \cdot (x_{O^{2-}})} \qquad (7.4)$$

Desulfurizing index, D_S (also known as the partition coefficient), which is the ratio of the amount of sulfur in the slag to that in the metal, may be written as:

$$D_S = \frac{(x_{S^{2-}})}{[\%S]} = \frac{K \cdot (x_{O^{2-}}) \cdot [f_S]}{[f_O . \%O]} = K' \frac{(x_{O^{2-}})}{[\%O]} \qquad (7.5)$$

In Eq. 7.5, $(x_{O^{2-}})$ is related to the amount of CaO and other basic oxides that control the basicity (b) of the slag. [%O] in the hot metal depends on the FeO content of the slag, which controls the oxygen potential. The variation of the index with basicity and (%FeO) in the slag is shown in Fig. 7.1. The whole range [5] of FeO found in blast furnace, electric arc, and open-hearth furnaces have been indicated in the figure. The plot emphasizes the dominant role of the iron oxide content on desulfurization. Increasing basicity promotes desulfurization. The highly reducing slag typical of blast furnace has low FeO up to 0.1 mol%, whereas slag from an electric furnace contains approximately 1 mol% FeO. For a fixed basicity, the index is

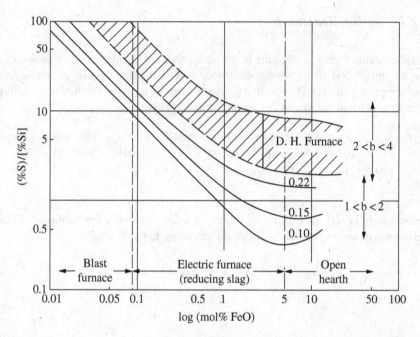

Fig. 7.1 Effect of basicity (b = wt% CaO/wt% SiO$_2$) and iron oxide content on desulfurization (Reproduced from R. N. Rocca et al. [5] with the permission of The Minerals, Metals & Materials Society)

inversely proportional to the (FeO) content in the slag or the oxygen dissolved in the metal [%O]. Assuming $x_{S^{2-}} = (\%S)$, from Eq. 7.5, we can write:

$$\frac{(\%S)}{[\%S]} \propto \frac{1}{[\%O]} \propto \frac{1}{(\%FeO)}$$

For a fixed FeO content of the slag: $\frac{(\%S)}{[\%S]} \propto (x_{O^{2-}}) \propto (\%CaO)$

Hence, for efficient removal of sulfur, high basicity and low oxygen potential are essential. Since K is favored at low temperature, thermodynamically lower temperature favors desulfurization but from kinetic viewpoint higher temperature is required. Equation 7.5 also concludes that higher activity coefficient of sulfur in the hot metal $[f_S]$ favors desulfurization. Hence, removal of sulfur from pig iron is relatively easier than from steel because $[f_S]$ increases in the presence of usual impurities in hot metal.

The relative desulfurization ability of basic cations can be visualized from Flood [6, 7] equation:

$$\log K' = -1.4 N'_{Ca^{2+}} - 1.9 N'_{Fe^{2+}} - 2.0 N'_{Mn^{2+}} - 3.5 N'_{Mg^{2+}} + 1.63 N'_{Na^+} + \dots\dots$$

$$(7.6)$$

where N' and K' stand for the electrically equivalent ionic fraction and equilibrium quotient, respectively. The relative contribution of different basic oxides to desulfurization can be estimated from the numerical coefficients of the Eq. 7.6. Based on such calculations, it has been estimated that the desulfurizing power of various cations relative to calcium (as unity) are in the ratio [6]:

$$Na(1070) : Ca(1.00) : Fe(0.325) : Mn(0.25) : Mg(0.0075)$$

The contribution of Na_2O has been included for comparison. But soda ash is not used because of its extremely corrosive action on the furnace lining. Hence, CaO makes the major contribution to desulfurization by its influence in producing high value of K. Although FeO may help in desulfurization as a basic oxide, the oxidizing potential of the slag is adversely affected. Negligible contribution of Mg^{2+} points out the difficulty in achieving adequate desulfurization in the blast furnace operating with high magnesia burden.

As steelmaking is carried out under oxidizing conditions (except electric arc furnaces—reducing slag period), the efficiency of desulfurization is very low. The value of the index ranges from 50 to 100 under blast furnace conditions and drops to 5–10 under basic steelmaking conditions. In slow processes like open hearth, 50% of the sulfur present in the charge may be eliminated. Thus, a maximum of 0.06–0.10% sulfur can be tolerated in the pig iron in such slow processes to produce finished steel with less than 0.005% sulfur. Due to insufficient time available for the transfer of sulfur from the metal to slag, there is no appropriate desulfurization in faster processes such as LD, OBM, and Hybrid. Hence, sulfur content of the molten pig iron must be below 0.05%. As sulfur content in steel is becoming more and more stringent, it should be less than 0.04%. Conditions in an electric furnace are favorable for desulfurization under reducing period. From the above discussion, it is evident that blast furnace has more favorable conditions for desulfurization. Hence, it is better to remove sulfur in the blast furnace. A pig iron containing high sulfur should be desulfurized outside the blast furnace before charging into the steelmaking furnace. This is known as external desulfurization of the hot metal.

In addition to the slag–metal equilibria considered above regarding sulfur distribution, there is a possibility of sulfur transfer from the furnace atmosphere to the slag–metal system according to the reaction [6]:

$$\frac{1}{2}\{S_2\} + (O^{2-}) = \frac{1}{2}\{O_2\} + (S^{2-}) \tag{7.7}$$

The equilibrium constant for the reaction (7.7) may be expressed as:

$$K = \frac{(a_{S^{2-}})}{(a_{O^{2-}})}\left\{\frac{p_{O_2}}{p_{S_2}}\right\}^{1/2} \tag{7.8}$$

The value of K cannot be estimated accurately because of the limitations of measuring the partial pressure of oxygen at the slag–metal interface in the working furnace. According to the law of mass action, sulfur transfer from the furnace atmosphere to the slag can be minimized by maintaining highly oxidizing atmosphere in the gas phase. This means the fuel must be burnt in excessive air or oxygen.

Making use of Flood's additive function, the modified equilibrium constant [7] can be expressed as

$$\log K' = 2.26 N'_{Ca^{2+}} + 2.76 N'_{Fe^{2+}} + 2.86 N'_{Mn^{2+}} + 4.36 N'_{Mg^{2+}} \qquad (7.9)$$

For lower sulfur in the slag phase, K' should be higher. CaO, FeO, and MnO seem to be equally effective in increasing the value of K', whereas MgO is more effective due to higher coefficient. Flushing of slag surface with oxygen will help transfer of sulfur from the slag to the furnace gas, but it is not practicable. Thus, removal of sulfur from metal to gas via slag is not of much significance; however, in open-hearth furnace, there is some desulfurization in this manner. Based on the above discussion, optimum conditions for maximum desulfurization can be summarized from the equation of Desulfurizing Index: $D_S = \frac{\left(x_{S^{2-}}\right)}{[\%S]} = \frac{K \cdot \left(x_{O^{2-}}\right) \cdot [f_S]}{[f_O \cdot \%O]}$

  (i) Basic slag gives high $(x_{O^{2-}})$, and hence favors desulfurization.
 (ii) High lime content—CaO makes the major contribution to desulfurization by its influence in producing high value of K.
(iii) Low iron oxide content—FeO below 2% as in the blast furnace or in the EAF under reducing conditions favors desulfurization. FeO has little effect in more oxidizing slag because increasing FeO increases both [wt% O] and K in an approximately compensating manner.
 (iv) Low temperature favors a high value of K.
  (v) High C, Si, and P in pig iron increase f_S and decrease f_O and for this reason blast furnace metal is more easily desulfurized than steel. It is therefore advantageous to remove sulfur in the blast furnace itself.
 (vi) High MgO in slag increases the value of K, and hence favors desulfurization.
(vii) High p_{O_2} in furnace atmosphere is beneficial for desulfurization.

In addition to equilibrium factors, the rate of approach to equilibrium must be reasonably rapid, which can be achieved by maneuvering the following:

  (i) Using a slag of appropriate fluidity and avoiding highly acid slag or operations below the liquidus temperature.
 (ii) Adding fluorspar (CaF_2) that not only increases fluidity but also increases the rate of desulfurization.
(iii) Stirring the bath by CO_2 evolution in open-hearth, mechanical stirring in rotor process or gas bubbling in all the pneumatic processes.
 (iv) Oxidizing atmosphere which is achieved by a high air/fuel ratio or by direct impingement of oxygen on the gas-slag interface.

Normally, it is believed that sulfur is transferred to the steelmaking slag in the form of sulfide, but recent investigations [8] have produced evidences for the presence of calcium sulfate in the slag during basic oxygen steelmaking. This may be due to the highly oxidizing conditions prevailing in basic oxygen converters during refining of the pig iron. The formation of CaS and $CaSO_4$ can be expressed according to the reactions: $[S] + (CaO) + [Fe] = (CaS) + (FeO)$ (under reducing conditions); $[S] + (CaO) + 3(FeO) = 3[Fe] + (CaSO_4)$ (under oxidizing conditions). In recent years, theoretical and industrial studies have been conducted on desulfurization using high- and low-basicity slags [9–11].

7.2.2 *Phosphorus Reactions*

Making use of the interaction coefficients for the effect of various elements on the activity coefficient of phosphorus in iron, the activity of phosphorus can be estimated by the expression:

$$\log f_P = 0.13 \times \%C + 0.13 \times \%O + 0.12 \times \%Si + 0.062 \times \%P + 0.024 \times \%Cu + 0.028 \times \%S$$
$$+0.006 \times \%Mn - 0.0002 \times \%Ni - 0.03 \times \%Cr$$

$$(7.10)$$

Practically, all the phosphorus present in the ore gets reduced along with iron in the blast furnace and joins the pig iron because the free energy of formation FeO, Fe_2O_3, Fe_3O_4, and P_2O_5 are of similar order. This is very much evident from the iron and phosphorus lines corresponding to the formation of oxides of iron and phosphorus in the Ellingham diagram, shown in Fig. 5.1. The two lines can be widely separated in the presence of strong and excess flux, lime [6]. ΔG^o vs T plots under standard and non-standard conditions are shown in Fig. 7.2. The activity of P_2O_5 in steelmaking slag of basicity 2.4 (containing even 25% P_2O_5) is reduced drastically to 10^{-15}–10^{-20}. Thus, for effective removal of phosphorus, basic steelmaking processes have to employ slag of high basicity. If basicity falls, phosphorus may revert to the metal phase. In acid steelmaking processes, it is not possible to remove phosphorus because the slag is almost saturated with silica.

Alternatively, one can visualize that for phosphorus to oxidize in preference to iron at 1600 °C, the free energy of formation of P_2O_5 should be lower than that of FeO, that is, $\Delta G_{P_2O_5} < \Delta G_{FeO}$. Hence, according to the van't Hoff isotherm under non-standard conditions for the reaction: $2[P] + 5[O] = (P_2O_5)$, we can write:

$$\Delta G_{P_2O_5} = \Delta G^o_{P_2O_5} + RT \ln \frac{(a_{P_2O_5})}{[a_P]^2 \cdot [a_O]^5} \qquad (7.11)$$

In order to satisfy the desired condition: $\Delta G_{P_2O_5} < \Delta G_{FeO}$, the variable term in Eq. 7.11 must be made negative by lowering down the activity of P_2O_5, which is possible by using a basic slag. Figure 7.2 indicates that $(a_{P_2O_5})$ is drastically lowered in the presence of lime by producing a basic slag according to the reaction:

$$4CaO + P_2O_5 \rightarrow Ca_3(PO_4)_2 + \text{extra CaO} \qquad (7.12)$$

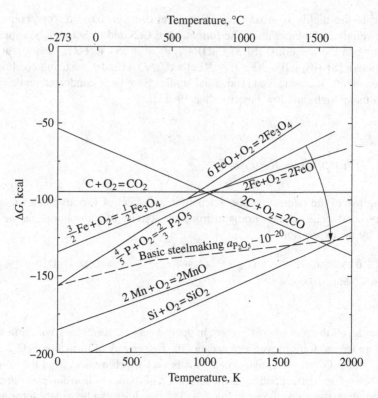

Fig. 7.2 ΔG^o vs T plots under standard and nonstandard conditions (From An Introduction to Physical Chemistry of Iron and Steelmaking by R. G. Ward [6], © 1962, p 124, Edward Arnold. Reproduced with the permission of Edward Arnold)

Activity of P_2O_5 can also be reduced by increasing the oxidation potential of the hot metal, which is achieved by adding iron ore or mill scale or by blowing oxygen. Effect of temperature [12] on ΔG^o for the reaction: $2[P] + 5[O] = (P_2O_5)$, expressed as $\Delta G^o = -683000 + 580\ T\ \text{J mol}^{-1}$, indicates that the negative value of ΔG^o increases with decrease of temperature. Hence, formation of P_2O_5 (i.e., dephosphorization reaction) is favored at lower temperature. But the melt temperature has to be maintained at the desired level to achieve the appropriate rate of refining at adequate fluidity. The distribution of phosphorus between slag and metal can be represented as:

$$2[P] + 5(FeO) + 3(CaO) = (3CaO.P_2O_5) + 5[Fe]$$
$$\text{or } 2[P] + 5[O] + 3(O^{2-}) = 2(PO_4{}^{3-}) \tag{7.13}$$

$$K = \frac{\left(a^2_{PO_4^{3-}}\right)}{[a_P]^2.[a_O]^5.(a_{O^{2-}})^3} \tag{7.14}$$

Assuming Henrian behavior in the melt and applying Temkin rule for ionic melts in the slag, we can write:

$$K = \frac{\left(x^2_{PO_4^{3-}}\right)}{[f_P.\%P]^2.[f_O.\%O]^5.(x_{O^{2-}})^3} \qquad (7.14a)$$

The dephosphorization index, D_P (i. e., phosphorus partition ratio), which is the ratio of the amount of phosphorus in the slag to that in the metal, is given by the expression:

$$D_P = \frac{\left(x_{PO_4^{3-}}\right)}{[\%P]} = K'^{1/2}[\%O]^{5/2}.(x_{O^{2-}})^{3/2} \qquad (7.15)$$

Healy [14] has expressed the effect of temperature on the phosphorus partition ratio correlating it with lime and iron contents of the slag by the following relation:

$$\log \frac{(\%P)}{[\%P]} = \frac{22350}{T} + 0.08 \ (\%CaO) + 2.5 \log \ (Fe_{total}) - 16 \pm 0.4 \qquad (7.16)$$

D_P is higher for higher basicity index and higher oxidizing power of the slag. Contribution of basic ions in D_P according to Flood and Grjotheim [6, 7] are as follows:

$$\log K' = 21 \ N'_{Ca}{}^{2+} + 18 \ N'_{Mg}{}^{2+} + 13 \ N'_{Mn}{}^{2+} + 12 \ N'_{Fe}{}^{2+} \qquad (7.17)$$

Based on Eq. 7.17, the dephosphorizing ability of various cations is in the ratio of:

$$Ca : \ Mg : \ Mn : \ Fe$$
$$10^{21} : 10^{18} : 10^{13} : 10^{12}$$

Thus, the equilibrium quotient for a lime slag will be 10^3 times larger than that for a slag in which magnesia is substituted for lime, and 10^8 and 10^9 times, larger than for MnO and FeO slag, respectively. The change in the dephosphorization index $\left[D_P = \frac{\left(x_{PO_4^{3-}}\right)}{[\%P]} \right]$ will not be very significant as this ratio appears squared in the definition of the equilibrium quotient and hence the dephosphorization ability of various cations will be approximately in the ratio of:

$$Ca \quad Mg \quad Mn \quad Fe$$
$$30000 : 1000 : 3 \quad : 1$$

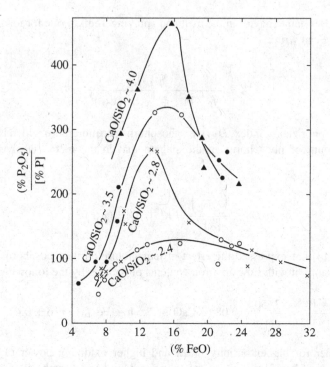

Fig. 7.3 Inter-relationship between dephosphorizing index, basicity, and FeO content (From An Introduction to Physical Chemistry of Iron and Steelmaking by R. G. Ward [6], © 1962, p 127, Edward Arnold. Reproduced with the permission of Edward Arnold. Originally from K. Balagiva et al. [13] with the permission of Journal of Iron & Steel Institute)

Thus, lime is 30 times more effective on molar basis and 20 times more effective on weight basis as compared to magnesia. Manganese and ferrous oxides make insignificant contributions. FeO is not only a base but is also effective as an oxidizing agent. Figure 7.3 shows the effect of basicity and FeO content of the slag on the dephosphorizing index, D_P. From the figure, it is clear that D_P increases with increase in the (FeO) content up to 15% due to the high oxidizing power. Beyond 15%, D_P decreases due to decrease in the lime proportion. In other words, the effect is neutral due to the loss of lime in the slag. Maximum dephosphorization is obtained when FeO content [6, 13] is around 15%. Dephosphorization is more effective at lower temperature because D_P increases with decrease of temperature. The soda ash is 100 times more effective compared to lime on molar basis, but it is avoided in practice due to its severe corrosive action on the furnace lining. The magnesia content of a basic steelmaking slag reaches an equilibrium with the lining and is hence not under control, and MnO depends on charge and is hence not much adjustable. The steel maker has the option of controlling lime, silica, and FeO. For charges containing high percentage of phosphorus, more than one slag is made to dephosphorize the metal bath to the desired level. In brief, high basicity, low temperature, and FeO content [13] around 15% favor dephosphorization of the hot metal by basic slags.

In the BOF or other oxygen top-blown practices, for example, LD, LDAC, Kaldo, and rotor, dephosphorization is realized in the early stages of refining when the carbon content of the bath is relatively high. This is due to the formation of a highly oxidizing basic slag in the early stages of oxygen blowing. However, because of the adverse effect of silica on $a_{P_2O_5}$ in the slag, the silicon in the charge should be maintained at less than 0.5% while blowing high-phosphorus pig iron. During refining of the high phosphorus charge, it is often necessary to employ a double-slag technique to ensure low residual phosphorus in the steel. With the low P charge, for example, 0.1% or less, single slag practice is adequate to produce steel with 0.01% P or less.

The optimum conditions for dephosphorization can be derived from the equation defining the index:

$$D_P = \frac{\left(x_{PO_4^{3-}}\right)}{[\%P]} = K'^{\frac{1}{2}}[\%O]^{\frac{5}{2}} \cdot (x_{O^{2-}})^{\frac{3}{2}}$$

(i) Basic slag gives a high value of $(x_{O^{2-}})$

(ii) High lime content—lime makes largest contribution to K'.

(iii) Soda ash and other alkali oxides are even stronger dephosphorizers but extremely corrosive to the refractory lining, hence not used.

(iv) Ferrous oxide close to 15%, although FeO increases [%O], it also decreases K' and thus optimum is about 15%.

(v) Low temperature gives a high value of K'.

From elementary mass transport considerations, dephosphorization will be accelerated by the use of fluid stag and by turbulence in the slag–metal system as in the case of desulfurization. From the above discussion, it is clear that basicity and oxygen potential are the two important factors in desulfurization, as well as in dephosphorization.

7.2.3 Silicon Reactions

One of the most important parameters in determining the quality of pig iron obtained from the blast furnace is the silicon content, which is closely watched before tapping. High-silicon pig iron is required in the acid steelmaking processes to make relatively acid slag to ensure longer life of the refractory lining. Oxidation of silicon also generates sufficient heat. However, basic steelmaking processes need low silicon iron because the entire amount of acid silica generated due to the oxidation of silicon has to be neutralized by lime to produce slag with basicity (CaO/SiO_2 ratio) between 2 and 4, which is needed for effective desulfurization, as well as dephosphorization. Pig irons containing high silicon together with high phosphorus and high sulfur are not suitable for acid, as well as basic steelmaking processes. Such pig irons are first subjected to external desiliconization to reduce the silicon content to a level suitable for basic steelmaking, which can remove phosphorus. In addition to its contributions in selection of acid or basic steelmaking process, silicon plays an important role in deoxidation of steel and also as an alloying element in acid-resistant and transformer steels. Depending on the steelmaking practice and the type of hot metal used, the silicon content of the charge varies from 0.5 to 1.5%.

Due to the strong attraction between iron and silicon, the Fe–Si system exhibits large negative deviation from the Raoult's law. The activity coefficient of silicon in iron in the presence of various elements can be calculated by the following expression incorporating interaction coefficients:

$$\log f_{Si} = 0.18 \times \%C + 0.11 \times \%Si + 0.058 \times \%Al - 0.058 \times \%S + 0.025 \times \%V + 0.014 \times$$
$$\%Cu + 0.005 \times \%Ni + 0.002 \times \%Mn - 0.0023 \times \%Co - 0.23 \times \%O \qquad (7.18)$$

Oxidation of silicon is an exothermic reaction and provides some of the heat necessary for increasing the temperature of the bath during blowing.

$$[Si] + 2[O] = SiO_2(s) \qquad (7.19)$$

$$K = \frac{a_{SiO_2}}{[a_{Si}][a_O]^2} = \frac{a_{SiO_2}}{[f_{Si}.\%Si][f_O.\%O]^2}$$

$$\therefore [\%Si][\%O]^2 = \frac{a_{SiO_2}}{f_{Si}.f_O^2.K} \qquad (7.20)$$

With increasing silicon content in iron the activity coefficient of oxygen decreases and that of silicon increases. Since silica is a very stable oxide once silicon is

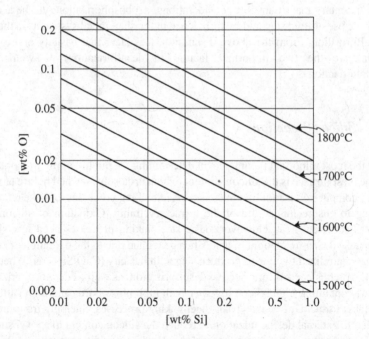

Fig. 7.4 Linear relationship between [%Si] and [% O] (Reproduced from J. Chipman [15] with the permission of Association for Iron & Steel Technology)

oxidized to SiO_2, the danger of its reversion does not arise. K has been found to be constant because of the compensating nature of a_O and a_{Si}. This constancy gives rise to a linear variation [15] of [%Si] with [%O] (Fig. 7.4). However, one should also consider the effect of [%Mn] on the variation of [%Si] with [%O] because hot metal contains Mn and the slag contains MnO. This will be discussed in the next section.

The extremely low activity of silica in basic steelmaking slag poses no danger of preferential reduction of silica like that of phosphorus removal. In basic steelmaking process, the silicon content of pig iron should be kept as low as possible to decrease the lime consumption with the prime objective of controlling the required basicity for phosphorus removal at a minimum slag volume. In case of high silicon entering the basic steelmaking furnace, either double slag practice or slag flushing has to be adopted. Alternatively, external desiliconization of the hot metal has to be done outside the blast furnace before charging it in a basic steelmaking furnace.

7.2.4 Manganese Reactions

Manganese is the second important element in determining the quality of pig iron. About 50–75% of the manganese in the burden gets reduced and joins the hot metal. Hence, the resulting pig iron contains 0.5–2.5% manganese. During steelmaking, a major amount of manganese is lost into the slag and very little is used to meet the specifications. Some heat is generated during steelmaking by the oxidation of manganese. Some manganese is required to control the deleterious effects of sulfur and oxygen and also for the improvement of mechanical properties of the steel.

The inter-relationship between Si, Mn, and O contents of iron in equilibrium with silica-saturated SiO_2–MnO–FeO slag, shown in Fig. 7.5, is of direct interest in steelmaking. From the figure, it is evident that (i) for a slag containing about 20% MnO at equilibrium, a maximum of 0.1% Mn is found in the metal at the steelmaking temperature and (ii) with increasing [%Mn] content of the metal, the (%MnO) content of the slag (containing 50% SiO_2 and the rest being FeO and MnO) increases, whereas the oxygen content of the metal decreases and silicon content increases. The activity coefficient of MnO is much higher in basic steelmaking slag compared to that of acid steelmaking. Hence, conditions for maximum recovery of manganese can be derived by considering the following equilibria:

$$(FeO) + [Mn] = (MnO) + [Fe]$$
$$\text{or } (Fe^{2+}) + [Mn] = (Mn^{2+}) + [Fe] \tag{7.21}$$

$$K = \frac{(a_{Mn^{2+}})[a_{Fe}]}{(a_{Fe^{2+}})[a_{Mn}]} = \frac{(x_{Mn^{2+}}) \cdot [f_{Fe} \cdot \%Fe]}{(x_{Fe^{2+}}) \cdot [f_{Mn} \cdot \%Mn]} \tag{7.22}$$

or

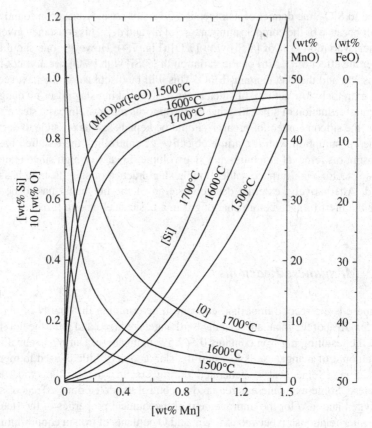

Fig. 7.5 Interrelationship of Si, Mn, and O contents of iron in equilibrium with silica-saturated slag: SiO_2–MnO–FeO (From An Introduction to Physical Chemistry of Iron and Steelmaking by R. G. Ward [6], © 1962, p 155, Edward Arnold. Reproduced with the permission of Edward Arnold)

$$K' = \frac{(x_{Mn^{2+}}) \cdot [\%_{Fe}]}{(x_{Fe^{2+}}) \cdot [\%_{Mn}]} \tag{7.23}$$

At equilibrium, the distribution of Mn in the slag and metal is given by the relation:

$$\frac{(x_{Mn^{2+}})}{[\%_{Mn}]} = K' \frac{(x_{Fe^{2+}})}{[\%_{Fe}]} \tag{7.24}$$

From the equation, it is apparent that the conditions for the highest possible recovery of Mn, that is, minimum slag–metal distribution ratio, are as follows:

(i) Minimum $(x_{Fe^{2+}})$, that is, a low FeO content in the slag.

(ii) Minimum K' requires a low SiO_2 content as evident from the following relation [6] showing the effect of various anions in the slag:

$$\log K' = 3.1 x_{SiO_4^{4-}} + 2.5 x_{PO_4^{3-}} + 2.4 x_{O^{2-}} + 1.5 x_{F^-} \qquad (7.25)$$

(iii) Low slag volume and a high temperature

(iv) High manganese in the charge

 (v) Partial or complete flushing of the slag

7.2.5   Carbon Reactions

During refining of steel, oxidation of silicon, manganese, and phosphorus takes place at the slag–metal interface. The oxidation of carbon practically does not take place at the slag–metal interface because of the difficulty of nucleation of CO bubbles there. C–O reaction takes place at the gas–metal interface since it eliminates the necessity of nucleating gas bubbles. During refining of steel, oxygen has to dissolve first in the bath before it reacts with the dissolved impurities. In the absence of other slag forming constituents, the maximum solubility of oxygen in liquid iron at 1600 °C is 0.23 wt%. Beyond this composition, liquid FeO is formed.

The oxidizing power of the slag is proportional to the FeO content of the slag, which in turn depends on the oxygen content of the metal. During steelmaking, the reaction between carbon and dissolved oxygen is of utmost importance. Generally, pig iron contains about 4 wt% carbon. The solubility of carbon in steel is affected by the presence of impurities and alloying elements. The solubility of carbon in iron decreases in the presence of V, Cr, Mn, and W, whereas it increases due to Co, Ni, Sn, and Cu. The combined effect of alloying elements on the activity coefficient of carbon can be estimated by the following equation incorporating the interaction coefficients:

$$\log f_C = e_C^C.\%C + e_C^V.\%V + e_C^{Cr}.\%Cr + e_C^{Mn}.\%Mn +$$
$$e_C^W.\%W + e_C^{Co}.\%Co + e_C^{Ni}.\%Ni + \ldots \qquad (7.26)$$

Oxidation of carbon may be considered by the reaction:

$$[C] + [O] = CO \qquad (7.27)$$

$$K = \frac{p_{CO}}{[a_C].[a_O]} = \frac{p_{CO}}{[f_C.\%C].[f_O.\%O]} \qquad (7.28)$$

$$\therefore [\%C].[\%O] = \frac{p_{CO}}{K.f_C.f_O} = \frac{p_{CO}}{K'} \qquad (7.29)$$

At any chosen pressure of CO, [%C] vs [%O] plot (Fig. 7.6) presents an inverse hyperbolic [6, 16] relationship. During oxidation, oxygen is continuously transferred from the slag to the bath, where it continuously reacts with carbon to form CO. The main resistance to the oxygen flow is the slag–metal and the metal-gas interfaces, whereas inside the steel bath the transfer of dissolved oxygen is very fast. The temperature dependence of the equilibrium constant [17], K has been reported as:

$$\log K = \frac{1056}{T} + 2.131 \tag{7.30}$$

and

$$K = \frac{p_{CO}}{[a_C] \cdot [a_O]} = \frac{p_{CO}}{[\%C] \cdot [\%O]} \tag{7.31}$$

In Eq. 7.31, activities of carbon and oxygen have been taken as being equal to [%C] and [%O] considering the solutions at infinite dilution.

The activity coefficient of carbon in iron increases with increasing carbon content and that of oxygen decreases with increasing carbon content. The net result is that the product [%C]·[%O] for a given p_{CO} decreases slightly [18] with increasing carbon content (Fig. 7.7). For [C] below 0.5% and at steelmaking temperature, the product [%C]·[%O] is 0.002 at $p_{CO} = 1$ atm. Since steelmaking is a dynamic process, the concentration of carbon and oxygen in the bulk metal

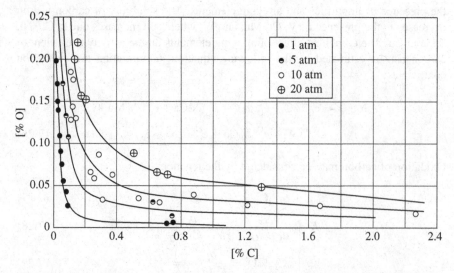

Fig. 7.6 Carbon and oxygen contents of molten iron in equilibrium with carbon monoxide at different pressures (From An Introduction to Physical Chemistry of Iron and Steelmaking by R. G. Ward [6], © 1962, p 88, Edward Arnold. Reproduced with the permission of Edward Arnold. Originally from S. Marshall and J. Chipman [16] with the permission of American Society for Metals)

Fig. 7.7 Variation of the product [%C]·[%O] with carbon content in iron at different temperatures at 1 atm pressure of carbon monoxide (Reproduced from E. T. Turkdogan [18] with the permission of Association for Iron & Steel Technology)

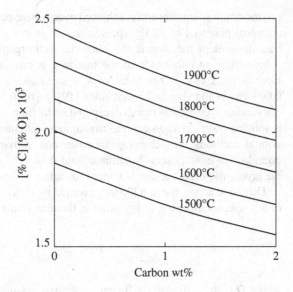

phase is not in equilibrium with the CO pressure prevailing in the bubbles. At the gas bubble–metal interface the reaction is close to equilibrium. Thus, it can be observed that the excess oxygen and carbon in the bulk metal phase is helpful in the transfer of the reactants by diffusion to the gas–metal interface in the violently stirred metal bath. From Sect. 4.4.1, it is clear that the oxygen content of the hot metal, [%O] increases with increase of (a_{FeO}) in the slag and decreases with increase of carbon content [%C] in the bath (Fig. 7.6). Thus, while considering the relations in Figs. 7.6 and 7.7 and the effect of (a_{FeO}) on [%O], it follows that in BOH and BOF practices the iron oxide contents of the slag increases with decreasing carbon in steel during refining and at tap. Hence, there is a general trend in the variation of slag composition with the carbon content of the metal.

At low [%C] content, the gas evolution from the melt that results from decarburization contains some CO_2 with CO as the major species. For reactions $\{CO\} + [O] = \{CO_2\}$ and $[C] + [O] = \{CO\}$ making use of K values, the gas composition can be computed. In practice, however, the gas composition in the converter remains essentially unchanged (about 10% CO_2 and 90% CO) during blowing. This gas composition is close to that for liquid iron (with 0.01% C at 1 atm total pressure and 1600 °C) in equilibrium with pure liquid iron oxide (i.e., $a_{FeO} = 1$).

7.2.6 Kinetics of Slag–Metal Reactions

Slag–metal reactions in refining of steel for removal of silicon, manganese, and phosphorus and in ironmaking for removal of sulfur belong to the category of heterogeneous liquid–liquid interfacial reactions. As temperatures are high enough in both ironmaking and steelmaking, reactions are generally very fast. Hence, a thermodynamic equilibrium is achieved everywhere particularly at the interface. The chemical potential of all the species present in the top level of the metal and the bottom level of the slag is the same. In well-organized convective transfer, the composition on either side of the interface is constant in the stagnant boundary layer of thickness, δ. The value of δ in steelmaking is supposed to be 0.015 and 0.003 cm in the slag and metal sides [19], respectively. In fluids, transport is by convection over long distances except when the fluid is very viscous where diffusion is much slower. Mechanical, electromagnetic, pneumatic, or hydraulic stirring are normal methods for achieving adequate convective transfer and to reduce the boundary layer thickness. In laminar fluid flow (i.e., parallel to the surface within the layer), the mass transfer is normal to surface by diffusion.

Diffusion across the thin film is governed by Fick's law. The flux, J $(g\ cm^{-2}\ s^{-1})$ of any species diffusing at any point in the direction x is given by:

$$J = -D \left(\frac{dc}{dx}\right) \tag{7.32}$$

where D is the diffusion coefficient of the species in the phase through which it is diffusing, and $\frac{dc}{dx}$ is the concentration gradient in the direction of x. The value of the diffusion coefficient depends on the nature of the diffusing species, as well as the medium, concentration level, and the temperature. However, variation with concentration is complicated, and the effect of temperature on D is expressed as:

$$D = D_o\ e^{-E/RT} \tag{7.33}$$

where D_o and E stand, respectively, for a constant and the activation energy for diffusion. E is the energy required to bring the diffusing atom or molecule into the intermediate position between the two lattice sites from which it is likely to fall into an empty site adjacent to that being vacated as it is likely to return to its original position. The activation energy of diffusion in metals is low (generally up to about 20 kcal mol^{-1}) and seems to depend mainly on the nature of the solvent. This value is lower than the activation energy for chemical reactions. The activation energy of diffusion in slag is considerably larger ($\sim$100 kcal mol^{-1}) due to the higher viscosity of the slag.

The rate of reaction at the interface of two liquid phases, for example, metal and slag, is affected by several factors. During smelting of iron ore in a blast furnace, the transfer of sulfur from metal to slag involves following three steps:

(i) Transport of reactants to the slag/metal interface, for example, in the transfer of sulfur from pig iron to the slag, oxygen ions in the slag (O^{2-}) and sulfur dissolved in the metal [S] are the two reactants.

(ii) Chemical reaction at the interface: $(O^{2-}) + [S] = (S^{2-}) + [O]$

(iii) Transport of the products such as sulfide ions in the slag (S^{2-}) and oxygen dissolved in the metal [O] away from the slag/metal interface.

The second step involving the chemical reaction that obeys the laws of chemical kinetics is temperature-dependent. According to the Arrhenius equation, the rate constant, k, is expressed as:

$$k = A \, e^{-E/RT} \qquad (7.34)$$

where A is the frequency factor. As smelting temperature is high, the rate of the chemical reaction will be high and hence the reaction at the interface will not be rate-controlling. On the other hand, if E, the activation energy for the reaction is large, the rate of the chemical reaction (step ii) will be as important as the mass transport in steps (i) and (iii) in controlling the overall rate of the heterogeneous reaction. In the example cited above the activation energy of 100 kcal mol^{-1} for the chemical reaction (step ii) involving the transfer of sulfur from metal to slag in ironmaking is high. Hence step (ii) seems to be rate-controlling.

The mass transfer occurring near the boundary of a fluid phase depends on the nature of the phase on the other side of the boundary. The transport of the species on both sides of the stagnant boundary layer becomes important. According to the two-film theory, a static boundary layer exists at both sides of the interface. Figure 7.8 shows the transfer of species from liquid metal to liquid slag. For the transport from the bulk of the metal phase to the metal interface, the amount of substance passing per unit time per unit area, that is, flux:

$$J = \frac{dm}{dt} \Big/ A \qquad (7.35)$$

Fig. 7.8 Transport in liquid slag and metal phases across the slag/metal interface, δ_m and δ_s are the effective boundary layer thickness in the metal and slag phases, respectively (From An Introduction to Chemical Metallurgy by R. H. Parker [19], © 1978, p 255, Pergamon Press Ltd. Reproduced with the permission of Pergamon Press Ltd)

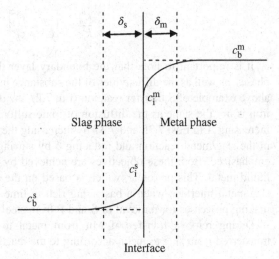

where, $\frac{dm}{dt}$ is rate of mass transfer and A is the area of the interface. If δ is the thickness of the stagnant boundary layer, the concentration gradient across the boundary is given by:

$$\frac{dc}{dx} = \frac{dc}{\delta} \tag{7.36}$$

From Eqs. 7.32, 7.35, and 7.36, we can write:

$$\frac{dm}{dt} = JA = -\left(\frac{DA}{\delta}\right)\left(c_b^m - c_i^m\right) = k_m^m\left(c_i^m - c_b^m\right) \tag{7.37}$$

where $k_m^m = \left(\frac{DA}{\delta}\right)$ is the mass transfer coefficient, c_i^m and c_b^m are the concentrations at the interface and in the bulk of the metal phase, respectively. In case there is no accumulation of the substance at the interface, the rate of mass transfer away from the slag interface may be expressed as:

$$\frac{dm}{dt} = k_m^s\left(c_i^s - c_b^s\right) \tag{7.38}$$

where c_i^s and c_b^s are the concentrations at the interface and in the bulk of the slag phase, respectively. At steady state, the two rates are equal.

$$\therefore k_m^m\left(c_i^m - c_b^m\right) = -k_m^s\left(c_i^s - c_b^s\right) \tag{7.39}$$

The change in sign accounts for the mass transfer in opposite directions relative to the interface.

$$\therefore \frac{k_m^m}{k_m^s} = \left[\frac{c_b^s - c_i^s}{c_i^m - c_b^m}\right] \tag{7.40}$$

It is important to note that the boundary layer thickness in the metal and slag phases, as well as the diffusivities of the substance in these phases, will differ. In the above example of S transfer reaction (Eq. 7.2), the diffusion of atomic sulfur in pig iron is not the same as the diffusion of ionic sulfur in the slag. The importance of increasing A in Eqs. 7.35 and 7.37 by increasing the area of the slag/metal interface in the slag/metal reaction and reducing δ by stirring in Eq. 7.37 need not be over emphasized. Both these objectives are achieved by introducing gas bubbles in the liquid metal. This principle, which is based on the reactions that take place at the slag/metal interface when a basic slag rich in lime is used, is applied in all steel-making processes for removal of S and P from steel.

During removal/transfer of Mn from metal to slag, oxygen has to be first transferred from slag to metal according to the reaction:

$$(O^{2-}) = [O] + 2e \tag{7.41}$$

The available oxygen in basic slag exists as ions (O^{2-}) because the basic slag consists of oxides completely dissociated into ions. The two electrons released in Eq. 7.41 are balanced as per the reaction:

$$(Fe^{2+}) + 2e = [Fe] \tag{7.42}$$

Thus, the overall reaction is

$$(FeO) = (Fe^{2+}) + (O^{2-}) = [Fe] + [O] \tag{7.43}$$

If manganese has to be transferred from the metal (pig iron) to the oxidizing slag (i.e., containing high FeO), the reaction would be:

$$[O] + [Mn] = (MnO) \tag{7.44}$$

This reaction may involve two stages: (i) the dissociation of FeO and (ii) the nucleation of MnO. But the theory of slag structure discussed in Chap. 4 confirms that ionization of FeO in steelmaking slag is a fast process. Being highly soluble in slag, nucleation of MnO may not occur at the slag/metal interface. Therefore, both these are not rate-controlling. The rate may be controlled either by the diffusion of oxygen to the interface, or Fe^{2+} through the slag, or Mn through the metal. Diffusion through slag is probably slower during most of the process and is hence rate-controlling. On the other hand, if MnO is to be reduced from the slag to the metal for recovery, the reaction sequence would be as follows:

$$(MnO) = (Mn^{2+}) + (O^{2-}) \tag{7.45}$$

$$(O^{2-}) = [O] + 2e \tag{7.46}$$

and

$$(Mn^{2-}) + 2e = [Mn] \tag{7.47}$$

This reaction stops if oxygen gets accumulated in the metal phase. For the above reactions to proceed, oxygen has to be removed by decarburization reaction: $[C] + [O] = CO$ (g). Since nucleation of CO involves a high activation energy, it is slow enough to control the rate of the overall reaction:

$$(MnO) + [C] = [Mn] + \{CO\} \tag{7.48}$$

This reaction is controlled by oxygen diffusion to sites where the evolution of CO proceeds easily. Reduction of SiO_2 in the hearth of the blast furnace is much slow and takes place in several steps:

(i) Dissociation of silica into ions: $2SiO_2 = SiO_4^{4-} + Si^{4+}$.
(ii) Spontaneous disintegration of silicate anions at the metal surface:

$$SiO_4^{4-} = [Si] + 4[O] + 4e$$

(iii) Dissolution of [Si] and [O] in the metal and neutralization of positive ions by
      the electrons released: $Si^{4+} + 4e = Si$.

For the above reactions to proceed smoothly, it is essential to avoid accumulation of
[O]. Hence, the reaction $[C] + [O] = CO$ (g) must occur.

The overall reaction being first order is likely to be controlled by diffusion of
oxygen to sites where CO is evolved or nucleated. But very high activation energy
for the C–O reaction (of the order of 250 kJ mol^{-1}), suggests that reaction is not
diffusion controlled. Instead, reaction is controlled by the rupture [20] of the high
energy Si-O bond in the second step of the chain of reactions.

7.3   Pre-treatment of Hot Metal

In recent years, the demand for high-quality steel with specific properties such as
improved strength, ductility, and toughness under severe forming conditions has
forced steelmakers to look at options to drastically reduce the impurities in steel, in
some steels even to the level of a few parts per million. Out of the main impurities
carbon, silicon, sulfur, and phosphorus present in the hot metal, carbon is reduced
from 4.0–4.5% to 0.03–0.04% at the end of oxygen blow in the converter. It can be
further reduced by vacuum degassing. Silicon gets eliminated almost completely
along with carbon oxidation. But varying amounts of sulfur and phosphorus remain
in liquid steel even at the end of the blow. As it is not always possible to reduce
sulfur and phosphorus simultaneously, it has become almost essential to charge hot
metal containing 0.005–0.010% each of sulfur and phosphorus to produce steel
products that have a good finished surface free from internal cracks. Thus, stringent
requirement of high-quality steel and factors arising due to the availability of poor
quality raw materials such as coke with high sulfur and iron ore with high phosphorus have forced the addition of one more step, that is, pretreatment between the blast
furnace and the steelmaking unit.

Pretreatment of hot metal reduces flux consumption, slag volume, iron loss, and
slopping and hence increases productivity because ideally speaking, steelmakers
have now to worry only about decarburization. Silicon, sulfur, and phosphorus can
be reduced by addition of appropriate reagents. The reagents are injected into the hot
metal as fine powder to increase the surface area in order to enhance the rate of
reaction. In this operation, hot metal is subjected to external treatments in ladles
(before charging the hot metal in steelmaking units) for removal of silicon, sulfur,

and phosphorus. Such treatments are known as external desiliconization, external desulfurization, and external dephosphorization, respectively.

7.3.1 External Desiliconization

Silicon is removed by injecting a mixture of mill scale, lime, fluorspar (CaF_2), sinter fines, iron ore/manganese ore fines along with oxygen gas to produce a neutral slag. In order to maintain the desired basicity of the slag during steelmaking the amount of flux proportionally increases with increasing silicon content in the hot metal. The resulting increase in slag volume requires extra heat for melting and fluxing during slag formation. This favors transfer of more FeO into the slag phase in order to maintain the required oxidizing power. External desiliconization lowers down flux consumption, as well as FeO loss. But it is not practiced under normal conditions in order to avoid problems in handling highly siliceous slag that causes severe attack on basic refractories. Hence, it is advisable to take necessary steps to produce low silicon hot metal in the blast furnace itself. For this purpose, the blast furnace has to be operated under acid burdening condition so that the basicity is maintained close to unity. Such a hot metal has less than 0.5% silicon and a slightly higher sulfur content that can be removed by external desulfurization.

7.3.2 External Desulfurization

From Sect. 7.2.1, it is evident that the optimum conditions necessary for desulfurization requires highly basic slag, reducing atmosphere, and high temperature, which prevail in blast furnaces. Even under these conditions if the resulting hot metal contains higher sulfur, it is treated in ladles for additional removal of sulfur before charging into steelmaking units. Desulfurization increases the productivity of the blast furnace and reduces slag volume, coke and flux consumption, and alkali build up on the furnace wall. It also facilitates production of low silicon pig iron. In order to produce steel for (i) efficient continuous casting unit, (ii) special steel plates, and (iii) manufacture of pipe lines free from hydrogen-induced cracking, the sulfur content must be brought down to 0.02%, 0.01%, and 0.001% (10 ppm), respectively. In order to achieve these specifications in steels, the initial sulfur in the hot metal should be lowered down to a minimum of 0.02–0.03%. This can be achieved by injecting a powder mixture [1] of calcium carbide, limestone, lime, and carbon in proportion (by weight), varying from 48 to 68% CaC_2, 31 to 9% $CaCO_3$, 17 to 22% CaO, and 4 to 1% C. Each constituent in the mixture contributes in its own way. On injection into the hot metal calcium carbide gets decomposed into calcium vapor and a graphite layer. The vapor instantaneously reacts with sulfur present in the hot metal and forms calcium sulfide. Lime increases the basicity of the slag formed, whereas dissociation of limestone provides CO_2 for agitation and helps in generating reducing atmosphere at the reaction sites (on reaction with carbon). Injection of a mixture of carbide and magnesium granules brings down the sulfur level to 0.01%. Soda ash is very effective due to its high sulfide capacity, but its use is restricted owing to its hazardous effects.

7.3.3 External Dephosphorization

Since almost the entire phosphorus present in the blast furnace burden joins the hot metal produced by reduction smelting, phosphorus has to be removed externally like silicon and sulfur when very low phosphorus steel needs to be produced. The optimum conditions for phosphorus removal as discussed in Sect. 7.2.2 have to be fulfilled for external dephosphorization in the ladle. However, it is not practiced because the demand for very low phosphorus steel is limited. Occasional requirements are managed by multiple slag practice at the primary steelmaking stage itself without bothering about minor loss of productivity.

7.3.4 Simultaneous Removal of Sulfur and Phosphorus

Although dephosphorization reduces the cost of hot metal by increased recycling of the slag to blast furnace, permits tapping of steel for continuous casting at higher temperature, and lowers down the final sulfur and phosphorus in high alloy steels, it is not practiced under normal conditions in bulk steel production. The reason being that dephosphorization is favored at lower temperature under oxidizing conditions in the presence of highly basic slag that is produced only after complete removal of silicon. Hence, an alternative method like blowing of lime powder along with oxygen (as practiced in the LDAC process) is adopted to refine high-phosphorus hot metal (1–2% P). Since desulfurization requires highly basic slag, reducing atmosphere, and high temperature, the simultaneous removal of sulfur and phosphorus is not feasible to the desired extent during steelmaking. However, based on recent researches, Japanese steelmakers [1] have suggested that simultaneous removal of sulfur and phosphorus can be achieved at high basicity (CaO/SiO_2) of 2 at temperature less than 1400 °C for a hot metal containing less than 0.25% silicon. Desiliconization in the first step and simultaneous removal of sulfur and phosphorus in the second step is carried out. It is important to note that simultaneous removal of silicon, sulfur, and phosphorus is not advisable due to high reagent consumption together with sudden temperature drop, problem in handling large volume of slag, and increased production cost.

Pig iron containing high percentage of carbon can be treated with sodium carbonate for excessive removal of silicon, sulfur, and phosphorus. Such a pretreatment will produce hot metal requiring decarburization by oxygen blowing during the steelmaking step by generating a small quantity of slag. The following reactions take place in soda ash treatment [4]:

$$Na_2CO_3 + [Si] = (Na_2O) + (SiO_2) + [C] \qquad (7.49)$$

$$Na_2CO_3 + [S] + 2[C] = (Na_2S) + 3CO \qquad (7.50)$$

$$5Na_2CO_3 + 4[P] = 5(Na_2O) + 2(P_2O_5) + 5[C] \qquad (7.51)$$

However, use of soda ash has been restricted in recent years because addition of Na_2CO_3 to hot metal generates dense fumes which pollute environment and slag disposal contaminates the ground water.

Pretreatment of hot metal will be more widely adapted in future not only due to increasing demand of lower level of sulfur and phosphorus in steel but also due to the deteriorating quality of raw materials, such as high phosphorus iron ore and high sulfur coke.

7.4   Chemistry of Refining

The physicochemical principles involved in deriving optimum conditions for desulfarization, dephosphorization, and recovery of silicon and manganese, discussed in Sect. 7.2, are invariably applicable in all the steelmaking processes. The chemistry of conversion of hot metal into steel is essentially the same for all the steelmaking processes, mentioned in Sect. 7.1. However, in this section, more emphasis has been given to basic oxygen and electric arc furnaces, which are currently employed for major production of steels. Basic oxygen furnace processes include top-blown oxygen, bottom-blown oxygen, combined-blown (hybrid/bath agitated), and rotating oxygen-blown converters. A brief discussion on Bessemer and open-hearth processes has also been included for comparison.

7.4.1   Important Chemical Reactions in Steelmaking

It is important to note that for effective refining of steel, the activities of impurities and oxygen should be high and those of product oxides should be low. The silicon and manganese contents of the hot metal can be adjusted in such a way that FeO and MnO formed during refining dissolves SiO_2 to produce a low melting ternary slag: FeO–MnO–SiO_2. But this slag does not dissolve P_2O_5 because FeO–MnO is a weak base. The addition of CaO and/or MgO forms a basic slag (CaO–FeO–MnO) of high basicity to absorb P_2O_5 without any chance of reversion of phosphorus. Sulfur removal to some extent is also possible by such a slag. Based on the above discussion, two important conclusions may be drawn for selecting a steelmaking process: (i) sulfur and phosphorus cannot be removed along with silicon and (ii) the removal of silicon should not be attempted simultaneously with phosphorus and to some extent sulfur.

The heat generated by the highly exothermic oxidation of carbon, silicon, manganese, sulfur, and phosphorus present in the hot metal during blowing in the BOF converters is sufficient to raise the temperature of the melt from the initial 1250–1350 °C (melting temperature of pig iron containing about 10% of impurities) to 1600–1650 °C (temperature required to produce molten steel containing about 1% impurities).

7.4.1.1 Top-Blown Basic Oxygen Converter Process

In LD converter, oxygen gas is blown at a pressure of 8–10 atmosphere with supersonic speed (1.5–2.2 times the speed of sound) through a convergent-divergent nozzle. The jet is retarded during its travel in the converter atmosphere and becomes subsonic at some distance from the nozzle. The length of the supersonic region (in which velocity of jet is more than the velocity of sound) is affected by the blowing speed and the ratio of the densities of the jet gas and the medium. Thus, some ambient medium is carried away along with the oxygen jet. Depending upon rate of blowing and lance height, jet may contain 60% O_2 and balance CO and CO_2 at the point where it strikes the bath. The amount of (CO + CO_2) increases from supersonic core to the subsonic region of the jet. Jet strike causes stirring in the bath though its kinetic energy is not sufficient to bring out proper mixing of the slag and metal. This results in a temperature and a concentration gradient, which are both responsible for ejection as well as slopping. As soon as oxygen is blown into the hot metal, iron is oxidized first, due to its abundance, although the free energy of formation of FeO is higher than those of SiO_2, MnO, P_2O_5, and CO. Some Fe_2O_3 is also formed:

$$2[Fe] + \{O_2\} = 2(FeO) \tag{7.52}$$

$$4(FeO) + \{O_2\} = 2(Fe_2O_3) \tag{7.53}$$

Thus, FeO formed together with iron ore/mill scale present in the charge makes iron rich slag during the early stage of the blow. [Si] and [Mn] present in the metal react with the slag constituents to form silicate slag according to the reactions:

$$2(FeO) + [Si] = (SiO_2) + [Fe] \tag{7.54}$$

$$(FeO) + [Mn] = (MnO) + [Fe] \tag{7.55}$$

$$(SiO_2) + 2(FeO) = (2\,FeO \cdot SiO_2) \tag{7.56}$$

$$(SiO_2) + 2(MnO) = (2\,MnO \cdot SiO_2) \tag{7.57}$$

(FeO) and (MnO) are released upon the dissolution of lime in FeO rich slag, according to the reactions:

$$(2 \, FeO \cdot SiO_2) + (2 \, CaO) = (2CaO \cdot SiO_2) + 2(FeO) \qquad (7.58)$$

$$(2 \, MnO \cdot SiO_2) + (2 \, CaO) = (2 \, CaO \cdot SiO_2) + 2(MnO) \qquad (7.59)$$

During blowing, the elimination of silicon and manganese start simultaneously, although the former gets eliminated within 3–5 min of blow, whereas manganese follows silicon only up to a certain point. Thereafter, it may remain the same for a while or revert to a limited extent toward the end of the blow. The carbon removal starts only after a few minutes of blowing. Dephosphorization goes concurrently along with decarburization but is completed well before carbon is completely removed. As the silicon content decreases the rate of decarburization increases. The rate increases from an initial low value to attain a peak in the middle of the blow and decreases thereafter in the third and last stage of the blow when carbon level lowers to 0.2% or less and emulsion collapses. During the initial period of slow decarburization, the FeO content of the slag increases up to 14–16%, whereas it decreases to 7–9% in the middle stage of the peak period due to the higher consumption of oxygen than supply. During the initial period of the blow, the [Mn] content in the bath decreases [reactions (7.55) and (7.57)] due to the formation of a manganese silicate (2 $MnO \cdot SiO_2$) slag. However, lime dissolution [reaction (7.59)] releases MnO, which gets reduced by carbon:

$$(MnO) + [C] = [Mn] + \{CO\} \qquad (7.60)$$

The above reaction during the extensive decarburization period increases the [Mn] content of the bath, which decreases again toward the end of the blow due to reoxidation when the decarburization rate becomes low. Figure 7.9, while showing the sequence of impurity removal during blowing in LD converters, clearly displays Mn hump [21] related to the change in oxidation/reduction pattern of the manganese reaction.

The reason for dephosphorization and decarburization proceeding simultaneously during the initial period of the blow is due to the formation of basic and oxidizing slag containing about 15% FeO. The efficiency of dephosphorization increases with an increase in slag basicity (optimum 3.5–3.6).

$$2 \, [P] + 5 \, (FeO) = (P_2O_5) + 5 \, [Fe] \qquad (7.61)$$

$$(P_2O_5) + 4(CaO) \rightarrow Ca_3(PO_4)_2 + excess \, CaO \qquad (7.62)$$

Despite the requirement of reducing condition, 6–8% S is removed in the LD converter in the presence of the oxidizing slag due to its highly basic nature and the high temperature of the bath (1650–1700 °C).

$$(FeS) + (CaO) = (CaS) + (FeO) \qquad (7.63)$$

$$(FeS) + [Mn] = (MnS) + [Fe] \qquad (7.64)$$

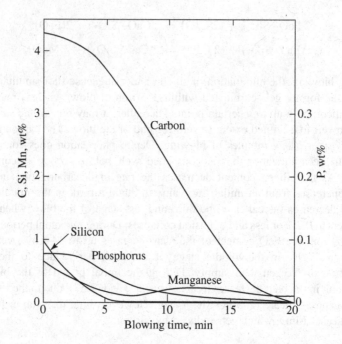

Fig. 7.9 Sequence of removal of impurities in LD converter (Reproduced from E. T. Turkdogan [21] 1974, Trans. Inst. Min. Metall., 83, C67, © Institute of Materials, Minerals and Mining, reproduced with the permission of Taylor & Francis Ltd.)

$$(FeS) + (MgO) = (MgS) + (FeO) \tag{7.65}$$

Mechanism of Refining Although the bath in the LD converter is stirred by the oxygen gas blown at a pressure of 8–10 atmosphere at supersonic speed (1.5–2.2 times the speed of sound) through a convergent-divergent nozzle, the slag and the metal are not properly mixed due to the lack of sufficient kinetic energy generated by the striking jet. The insufficient mixing gives rise to temperature and concentration gradients that cause ejection as slopping. The oxygen molecules striking the surface of the hot metal may lead to (i) the formation of CO by C–O interaction, (ii) formation of FeO, and (iii) [O] pick-up by the hot metal. The high activation energy involved in the nucleation of CO prevents the first reaction, whereas the third reaction is unlikely to take place due to the limited solubility of oxygen in iron (~ 0.23%). The total requirement of oxygen is about 3 wt% of the mass of the metallic bath for the oxidation of impurities, which cannot be achieved in 15 min of blowing. This suggests that oxygen is transferred to the metallic bath by the highly exothermic oxidation of iron to FeO (i.e., process ii) and its subsequent dissolution in the bath. The heat generated during process (ii) is adequate to increase temperature of the impact zone (known as the hot spot) to 2200–2500 °C. The hot spot was earlier considered to be the cause of the high rate of refining in the LD converter but now it has been confirmed that the high rate of reaction is in fact due to the formation of slag–metal–gas emulsion. As the jet velocity increases, the peripheral liquid is splashed in the converter atmosphere as droplets, which burn in oxygen to produce CO:

$$2[C] + \{O_2\} = 2\{CO\} \tag{7.66}$$

Since the droplets not only contain FeO and/or slag inclusions but also carbon and oxygen, FeO–C reaction leading to the formation of CO gas is favored. The highly viscous slag formed at the beginning of the blow (before complete dissolution of lime) helps in the formation of foam/emulsion. The formation of slag–metal–gas emulsion takes place from the two immiscible liquids (slag and metal), which cover the gas bubbles. Continuous generation of CO increases the volume of emulsion, which can persist until the FeO–C reaction continues in the metal droplets. Viscosity and surface tension play an important role in stabilizing the foam and emulsion. The emulsion increases the interfacial area of contact between the two phases to a great extent. It has been estimated that an apparent area of about 12–14 m^2 of a 100-ton LD converter increases to as high as 7500 m^2 [22] due to the formation of slag–metal–gas emulsion. The tremendous interfacial area increases the rate of decarburization up to 10–12 wt% C h^{-1}, comparable with the rate of refining achieved in Bessemer converters. Thus, there are two separate zones of refining in the LD vessel, namely, reaction in the bulk metallic phase and reaction in the emulsion. Major refining takes place after emulsion is formed. About 65–70% of the total carbon is removed through emulsion reaction. The slower bulk phase refining before the formation of emulsion becomes major toward the end after the collapse of the emulsion. As concentration of carbon in droplets held in emulsion approaches the metallic bath composition, the overall decarburization rate decreases because the emulsion cannot persist any longer due to the insufficient available volume of CO within the emulsion. From this stage onward, the rate of decarburization is not controlled by the supply of oxygen, but by the carbon content.

From Sect. 7.2.2, it is clear that efficient dephosphorization requires highly oxidizing and fluid slag, which is formed after 4–6 min of the blow. As dephosphorization proceeds concurrently with decarburization, efficient elimination of phosphorus and carbon demands early formation of such a slag. Since dephosphorization is extremely fast due to the large interfacial area and efficient mass transport in emulsion, phosphorus should be completely removed before the collapse of emulsion, failing which it has to be performed in the bulk phase. Considering the fact that dephosphorization in the bulk phase is extremely slow, it needs to be completed by the time carbon is reduced to 0.7–1.0%, that is, well before the collapse of emulsion, which starts at about 0.3% C.

The rate of refining is affected by the rate of oxygen flow, number of holes in the lance tip, lance height (distance between the lance tip and the bath surface), characteristics of the oxygen jet impinging the bath surface (pressure, velocity, etc.), bath temperature, slag characteristics, and so on. For example, increase in lance height decreases the rate of decarburization but increases the dephosphorization rate. Generally, lance height of 1.5–2.0 m is used at the start of the blow to favor early slag formation and of about 0.8–1.2 m for decarburization. The composition and temperature of the bath at the end of refining are greatly influenced by the basic design of the LD vessel, degree of combustion of waste gases, blowing time, and the volume/ basicity/viscosity of the slag. However, prior planning has to be made for effective dephosphorization well before decarburization. Some operators vary the oxygen pressure keeping the lance height constant during the entire period of blow, whereas others prefer altering the lance height at constant pressure of oxygen gas.

LDAC process follows an almost similar blowing procedure to that practiced in the LD converter. In this process also dephosphorization and decarburization are favored by increasing and decreasing the lance height, respectively. Charge consisting of scrap, iron ore, bauxite, and lumpy lime is fed into the hot converter containing some slag from the previous heat. Hot metal is then poured. In case the silicon content of the hot metal is low, some silica may be added for early formation of foamy slag. Keeping the lance at the highest position oxygen blow is started. After 5 min of blow, powdered lime (1–2 mm size) is fed through the lance at a predetermined rate. The lance is then lowered to control the foam. The blowing is stopped after 15 min by which time the bath temperature reaches 1650–1690 °C. The lime required for dephosphorization is charged in two or three steps with intermittent slagging. After the formation of the initial slag with lumpy lime, powdered lime is blown with oxygen through the lance in the hot spot (impact zone) to facilitate foam formation, which increases the rate of dephosphorization reaction (7.61/7.62). Foaming is controlled by maneuvering the lance height, flow rates of oxygen, and lime powder.

At the end of the first stage of blow, carbon and phosphorus are, respectively, lowered down to 1–1.5 wt% from the initial 3.8–4.0 wt% and to 0.2–0.4 wt% from 0.8–1 wt%. The first stage slag containing 60% CaO, 20–25% P_2O_5, 3–5% Fe (as oxide), and few percent of MnO and MgO is drained off and can be sold as fertilizer. After removal of the first slag, blow (oxygen and lime powder) is resumed again with fresh addition of coolant (iron ore, scrap) depending on the bath temperature. The second stage of blowing brings down carbon to 0.5 wt% and phosphorus to 0.1 wt% and bath temperature attains 1600 °C. The second slag contains 50–52% CaO, 10–15% P_2O_5, 10–15% Fe (as oxide), and a few per cent of MnO and MgO. After draining off part of this slag also, third-stage blowing is carried out to achieve the desired carbon level in steel. The duration of blow in the second and third stages is decided by the carbon level attained at the end of the first stage. The final slag (after the third stage), which is low in quantity containing about 20% Fe, may be retained in the converter for the next operation. LDAC process removes a major fraction of phosphorus present in the pig iron by draining two intermediate slags and thereby renders very effective desulfurization. Finally, it produces steel containing 0.02% P, low S, and 0.001–0.002% N with 99.5% pure oxygen blowing. However, the yield is slightly less and the duration of blow is longer (about 50–60 min) as compared to the LD process.

7.4.1.2   Bottom-Blown Oxygen Converter Process

The concept of the OBM converter was derived from the fact that blowing oxygen from the bottom would definitely improve the reaction kinetics compared with the air blown Bessemer converter. Immediately after charging scrap and hot metal, lime is blown with oxygen gas. The blow is divided into three parts. The first blow that lasts for 16–17 min produces high P_2O_5 slag (18–22% P_2O_5 and 12–15% FeO) and reduces carbon to 0.3 wt% and phosphorus to 0.08%. The second stage blow that lasts for about a minute (after charging fresh lime) reduces carbon to 0.1 wt% and phosphorus to 0.025 wt%. Finally, nitrogen is blown to reduce hydrogen dissolved. The resulting steel contains 0.004 wt% H and 0.005 wt% N. The process can produce different varieties of plain carbon steels from varying quality of hot metals.

It is well known that the rate of decarburization in the conventional air-blown Bessemer converter is extremely fast, taking only 20–25 min for a 20–25 ton vessel. The rate of decarburization (hence the overall refining) in the OBM process may be even faster compared to Bessemer because reaction kinetics is definitely affected by the presence of the large percentage of nitrogen in air blown. The slag–metal and gas–metal interfaces in the OBM converter attain near equilibrium condition due to stirring caused by bottom blowing. In the LD converter, oxygen is supplied to the metal through the slag layer. As the top blowing supersonic oxygen jet does not provide sufficient stirring, the slag is over-oxidized beyond equilibrium state. Inadequate stirring causes temperature and concentration gradients, which give rise to slopping and ejections. Bottom-blowing prevents building up of any such gradients and reduces iron loss in slag by 5 wt% as compared to top-blown converter processes. As a result, an OBM converter can easily produce very low carbon steels, which can be obtained in an LD process only at the cost of extra loss of iron in slag. Due to intensive stirring, OBM provides better partitioning of phosphorus and sulfur and offers high manganese recovery. Figure 7.10 shows the sequence of removal of impurities in an OBM converter. As blowing pure oxygen through the bottom generates intense heat, these injectors known as the Savard-Lee injectors [23] are protected by flowing methane or propane gas/fuel oil through the outer pipe surrounding the inner oxygen pipe. Endothermic decomposition of the gas/oil on entering the melt absorbs heat from steel bath near the injector tip and thus protects the injectors and furnace wall from overheating. Hydrogen generated by dissociation is dissolved in liquid steel. Hence, nitrogen is bubbled through the melt to remove hydrogen before tapping.

The OBM process is ideally suited for production of very low carbon steel (0.01–0.02 wt%). However, it faces problems like difficulty in feeding lime from the bottom, careful adjustment of flow rates of oxygen, hydrocarbon, and lime to achieve proper thermal balance at the tuyere junction, maintenance and repair of tuyeres, and so on. Despite these limitations the overall performance of the OBM process may be categorized as an efficient one. The advantages derived from bottom-blowing are listed below:

(i) Lower temperature and concentration gradient

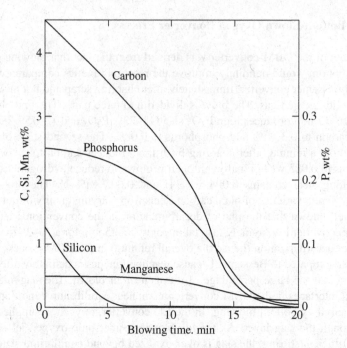

Fig. 7.10 Sequence of removal of impurities in OBM converter (Reproduced from E. T. Turkdogan [21] 1974, Trans. Inst. Min. Metall., 83, C67, © Institute of Materials, Minerals and Mining, reproduced with the permission of Taylor & Francis Ltd.)

  (ii) Reduction in FeO loss in slag results in better metallic yield

 (iii) Lesser slopping and dust formation

 (iv) Higher recovery of manganese and aluminum due to lower residual oxygen in the melt

  (v) Lower residual sulfur and phosphorus in steel on account of blowing powdered lime and improved agitation

 (vi) Production of ultra-low carbon steel without over-oxidation of the melt and slag

However, the following difficulties are encountered in bottom-blowing:

  (i) Highly specialized and skilled workers are needed to manufacture and maintain the bottom of the converter for injection of gas, as well as solid (oxygen and lime together).

 (ii) Relatively higher nitrogen content of the finished steel makes it unsuitable for deep drawing.

(iii) Each tuyere that pierces through the bottom functions as the heart of the converter and needs special attention in terms of heat balance around it during the blowing period. Depending upon the amount of oxygen blown, the coolant must be supplied in appropriate proportion so as to enable tuyeres to serve for longer duration.

7.4.1.3   Hybrid/Bath-Agitated/Combined-Blown Process

Better mixing of slag and metal in the hybrid process at all stages of blowing prevents slopping and ejections and eliminates temperature and concentration gradients in the bath. After charging scrap and ore and pouring hot metal, two-third of the lime powder and oxygen are blown from the top and bottom in the first stage of blowing. The first stage of blow reduces carbon to 1.5 wt% from the initial 3.5–4.0 wt% and phosphorus to 0.2–0.4 wt% from 1.8 wt%. The bath temperature gradually rises to 1600–1650 °C at the end of the blow due to exothermic oxidation of silicon, carbon, and iron. Oxidation of silicon starts first because SiO_2 is much more stable compared to the other oxides. In order to protect the basic lining of the vessel, lime is blown right from the beginning. After elimination of silicon, oxidation of carbon to CO starts and continues throughout the blow. Major phosphorus removal takes place at the end of the blow. The resulting slag contains about 60% CaO, 20–25% P_2O_5, and 3–5% Fe (as oxide). The non-foamy slag rich in P_2O_5 is partly drained off. The slag containing 60% CaO produced in the first stage of blowing favors dephosphorization whereas low FeO content enhances desulfurization.

After draining off part of the first slag some scrap and iron ore are charged depending on the temperature attained during the first stage of blowing. This is followed by the second stage blowing of lime powder with oxygen. The second stage blow produces metal containing 0.5% C and less than 0.1% P and a slag containing about 50–52% CaO, 15–20% P_2O_5, and 7–10% Fe (as oxide). A part of this slag is also drained off to remove additional phosphorus and sulfur. Blowing of oxygen and lime powder in the third and last stage is carried out to achieve the final desired level of carbon in the steel. The carbon level attained after the first stage blow decides the total duration of blow in the second and third stages. If low carbon steel has to be made, the small quantity of the final slag, which contains about 20% Fe (as oxide) is left in the converter for continuous operation.

The hybrid process makes manufacture of very low phosphorus and sulfur steel easier by draining off two intermediate slags. In this process, dephosphorization takes place even in the presence of carbon and a low phosphorus steel is produced by stopping refining at the desired carbon content. The basic intermediate slags with high partition coefficient (15–20 for the first slag and 10 for the second) favor desulfurization. Use of 99.5% pure oxygen finally produces steel with 0.001–0.002 wt% nitrogen. The sequence of removal of impurities in the hybrid converter is shown in Fig. 7.11. The change in slag composition during the course of blowing for the same converter is depicted in Fig. 7.12. During the initial stages of blow, slag temperature is higher (by about 100 °C) than that of the metal. But as blow progresses, this temperature difference decreases to about 30–50 °C, which is further brought down by blowing argon gas from the bottom tuyeres for about 2 min at the end of oxygen blow.

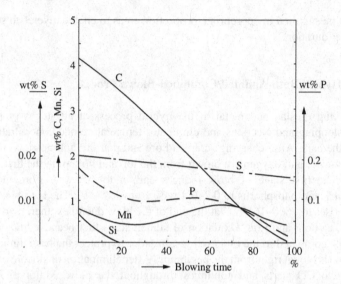

Fig. 7.11 Sequence of removal of impurities in a combined-blown converter (From Fundamentals of Steelmaking Metallurgy by B. Deo and R. Boom [24], © 1993, p 162, Prentice Hall International. Reproduced with the permission of the author, B. Deo)

Fig. 7.12 Change in slag composition in a combined-blown converter (From Fundamentals of Steelmaking Metallurgy by B. Deo and R. Boom [24], © 1993, p 162, Prentice Hall International. Reproduced with the permission of the author, B. Deo)

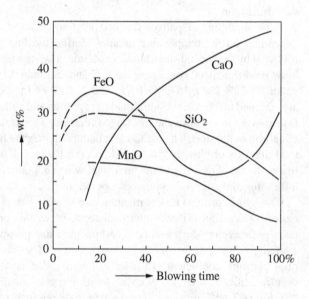

The post-blow inert gas stirring also helps in composition homogenization and dephosphorization because slag is highly basic and oxidizing due to high CaO and FeO contents. From Fig. 7.12, it is clear that CaO content of the slag increases continuously during the blow due to successive dissolution of lime. In the limited slag volume, initially the FeO concentration is high but it decreases as slag volume increases with the progress of blow. As the carbon content in the bath decreases, C–O reaction diminishes toward the end of the blow, the iron in the bath (rather than carbon) starts reacting with oxygen, thereby increasing the FeO content of the slag. Thus, the quantity of slag goes on increasing as more and more lime gets dissolved with the progress of the blow. In order to insure adequate refining and removal of impurities, CaO/SiO_2 ratio of at least 3.0 has to be maintained in the final slag. Hence, the amount of lime required depends mainly on the silicon content of the hot metal; for example, lime requirement increases from 60 kg ton^{-1} of liquid steel to 120 kg ton^{-1} as silicon in bath increases from 0.6 to 1.2%. The bath-agitated process offers benefits like:

(i) Reduction in slopping and FeO content in the slag as compared to the LD process
(ii) Better possibility of treating higher silicon hot metal
(iii) Better control of bath composition
(iv) Better recovery of manganese
(v) Production of ultra-low carbon steels without over-oxidation of the metal and slag
(vi) Better gas–slag–metal interactions as compared to the interactions occurring in LD converters

The combined top and bottom blowing in the hybrid process provides better mixing of slag and metal and prevents slopping and ejections. This combination also eliminates temperature and concentration gradients in the bath and generates conditions closer to equilibrium for slag–metal and gas–metal reactions in the converter. In this process, adequate mixing is achieved by bottom blowing (OBM process) whereas decarburization and dephosphorization are controlled by raising and lowering of the lance at a constant pressure of oxygen (LD process). In this way, the hybrid process derives benefits from both processes and thus has emerged as the improved version of LD and OBM. It is steadily gaining prominence on account of the above listed advantages. One can ascertain that no more new LD and OBM vessels will be installed. There is every likelihood that they will be replaced with hybrid converters with slight modifications depending upon available facilities.

7.4.1.4 Rotating Oxygen-Blown Converter Process

In the Kaldo process for converting operation lime, ore and fluxing material are charged in the hot vessel (in vertical position) containing some slag left over from the previous heat. After rotating the vessel in the horizontal position, first the scrap is charged and then the hot metal is poured. The converter is tilted to the blowing angle of 16–20° to the horizontal and lance is inserted through the hood at a proper angle. The speed of rotation of the vessel and the rate of flow of oxygen are increased slowly to their maximum value in about 5–7 min. Conditions are generated to oxidize the entire silicon, most of the manganese present in the charge and some of the iron. A basic, oxidizing and fluid slag is generated when lime gets dissolved in

SiO_2, MnO, and FeO formed by oxidation. After 18–20 min of blow when the entire solid charge becomes molten, the speed of rotation of the vessel and oxygen flow rate are reduced to about half of the starting value. Maximum dephosphorization takes place in this period and the blow is stopped after about 25 min, and P_2O_5 rich slag is drained off completely. The blow is resumed again at 6–8 rpm and low oxygen flow rate after addition of fresh lime and coolant (ore: depending on the bath temperature). The second blow is continued for a predetermined time in the case of two slag practice. When a very low phosphorous content is desired the second intermediate slag is discarded and blowing is continued again. This slag (generated after the third stage of blowing) is retained in the converter for the next heat. In the Kaldo process, refining is controlled by adjusting the operating variables, namely, speed of rotation of the converter, rate of flow of oxygen, pressure of oxygen, and lance angle. Slow speed of rotation helps in cooling the refractory lining heated by combustion of CO inside the vessel. At high speed (30–40 rpm), the metal drawn up along the wall falls as a shower of molten droplets through the gaseous atmosphere and is oxidized directly. It also generates an enormous slag–metal interface suitable for oxidation of silicon, manganese, and phosphorus as well as subsequent transfer of SiO_2, MnO, and P_2O_5 to the slag phase.

Since oxygen is supplied at an angle to the bath surface at a lower jet velocity (as compared to that in an LD vessel), formation of emulsion does not take place to any significant extent as it happens in an LD converter. The speed of rotation also subsides emulsion formation. Silicon and manganese are oxidized during the early part of the blow. Formation of basic, oxidizing, and fluid slag within a short time of 10–12 min due to the presence of slag from the previous heat and slow speed of rotation at the beginning favors high rate of dephosphorization. The initial slow rate of decarburization increases rapidly when metal droplets start falling in the converter atmosphere at high speed of rotation. Since decarburization and dephosphorization take place independently in the Kaldo converter, there is no binding as in the LD process that dephosphorization has to be completed well before collapse of the emulsion. In this process, a major amount of carbon is eliminated well before phosphorus removal. The rate of dephosphorization increases with (i) decrease of speed of rotation of the converter (ii) decrease of oxygen flow rate (iii) increase of lance height, and (iv) increase in lance angle. The decarburization rate increases with (i) increase in speed of rotation of the vessel (ii) increase in oxygen flow rate (iii) decrease in lance height, and (iv) decrease in lance angle.

The relative rates of decarburization and dephosphorization can be achieved by an appropriate combination of the above listed factors. Thus, the Kaldo process is more versatile compared to the LD process and can produce a large range of products using hot metal of varying composition. As the process is controlled by several operating variables, it can produce steels with low P, S, C, and N for extra deep-drawing applications. In addition to the production of carbon and alloy steels like in the LD process, it also produces fertilizer-grade slag as a by-product. Despite being a versatile process, it has not been commercially exploited due to high refractory consumption. The life of refractory lining in a Kaldo converter is nearly 60–100 heats as against 200–300 heats for similar materials in an LD vessel. Productivity decreases considerably because of limited availability of the converter on account of repair of the lining.

In the rotor process, a part or entire amount of the final slag from the previous heat is retained. The heat is started by charging some lime that is spread all over the lining by rotating the vessel. This is followed by charging the balanced lime, scrap, ore, and hot metal. After bringing the vessel to the blowing position, oxygen is blown through the primary lance at a constant flow rate of $70–85$ m^3 min^{-1}, while the vessel is under rotation. The blow of enriched air or pure oxygen, along with lime powder if required, through the secondary lance starts a little later. The rotor is blown for half the time from each end to maintain an even condition of the lining throughout the entire length. The first stage of blow takes nearly $20–24$ min ($10–12$ min from each end) and produces high P_2O_5 slag (fertilizer grade). Blowing is stopped and the vessel is rotated to separate slag and metal. The slag is drained off and blowing resumes after charging fresh lime and little sand to facilitate slag formation. In the rotor process, decarburization is favored by increasing the depth of the primary lance in the bath, whereas dephosphorization is enhanced by reducing the depth. The FeO content of the slag increases, that is, the slag becomes oxidizing when primary lance is withdrawn from the bath. The low speed employed in this process does not affect refining because rate of refining is controlled by varying oxygen pressures in both the lances.

The process, however, did not achieve much success as it could not compete with the LD process in terms of rate of production, capital, and operational cost. Further, fairly high consumption of refractory in the rotor process due to rotation and intensive heat generated on account of combustion of CO proved to be the main bottleneck in its adoption.

7.4.2 Electric Arc Furnace (EAF) Process

The charge consisting of sponge iron, steel scrap, lime/limestone, iron ore, and coke is dropped from baskets by raising the electrodes and swinging the roof. Hot metal (if used) is poured through the door. As soon as the arc is struck, melting begins and a pool of molten metal is formed. Oxidation of silicon, manganese, and phosphorus (into their respective oxides, SiO_2, MnO, and P_2O_5) in liquid medium leads to slag formation. The basicity of the slag increases with dissolution of lime in the slag. CO generated due to the oxidation of carbon stirs the bath and helps in flushing out the dissolved gases and floating out inclusions. Basic oxidizing slag favors dephosphorization according to reactions (7.61 and 7.62). The slag is drained off through the door after tilting the furnace to remove a major amount of the oxidized phosphorus. In order to maintain the basicity of the slag, fresh lime and iron ore are charged for further refining. Removal of phosphorus enhances the C–O reaction, which in turn produces frothing in the slag due to intensive carbon boil. Rise in the slag level due to frothing causes an overflow through the door sill. Discard of (P_2O_5) with the overflowing slag helps further dephosphorization and brings down the phosphorus level to $0.015–0.02\%$. After draining out the slag, the bath is first deoxidized in the furnace itself with ferrosilicon and ferromanganese. Finally, deoxidation is carried out with aluminum after tapping the steel melt in a ladle.

In order to meet the demand of restrictive sulfur content, a second reducing slag is made after draining the oxidizing slag completely. The reducing slag is formed by charging burnt lime, sand, and fluorspar. Addition of pulverized coke forms calcium carbide under the arc according to the reaction:

$$(CaO) + 3C = (CaC_2) + \{CO\} \tag{7.67}$$

(CaC_2) reacts with FeO in the slag:

$$(CaC_2) + 3(FeO) = (CaO) + 3Fe + 2\{CO\} \tag{7.68}$$

This favors diffusion of [FeO] from the metal to the slag for further reduction of (FeO) in the slag. Thus, carbide automatically reduces FeO content of the slag, which amounts to decrease in oxidation potential of the bath. The EAF slag containing 60–65% CaO, 18–22% SiO_2, 7–9% MgO, 1–2% CaC_2 and small amounts of FeO, CaS, MnO, Al_2O_3, and so on is reducing in nature, hence favors desulfurization.

7.4.3 Open-Hearth Process

Open-hearth furnace can treat varying proportions of steel scrap and hot metal to produce steel. In basic open-hearth process, iron ore, limestone or lime are added to generate basic oxidizing slag in order to remove phosphorus present in the hot metal. It is important to note that the silicon content of the hot metal, containing 0.22–0.30% phosphorus, should not exceed 1% so as to achieve the desired basicity of the slag. In open-hearth furnace, slag performs dual functions: (i) receives impurities and (ii) refines the medium. As per normal practice in steel plants, scrap, limestone, and iron ore were charged first. After heating the charge to a state of incipient fusion, hot metal was poured. Addition of limestone was advantageous because its decomposition gives rise to evolution of CO_2, which causes agitation in the bath. For melting, the charge burners were turned on in full. In the melt, silicon and manganese were oxidized as per reactions (7.54) and (7.55) which were facilitated by reaction (7.52). As free energy of formation of SiO_2 is much less compared to that of CO, oxidation of carbon normally does not occur in the melting down period. During melting, lime reacts with SiO_2 to form slag (reactions 7.58 and 7.59). In order to achieve the desired carbon content in steel, extra carbon in the form of petroleum/ anthracite coke or graphite block had to be added below the level of steel scrap, particularly when the charge consisted of a larger proportion of scrap.

As steelmaking is an oxidizing process, a steady and continuous supply of oxygen has to be maintained by dissociation of iron ore. Being an endothermic reaction dissociation of Fe_2O_3 at the slag–metal interface is slow due to limited supply of heat at the interface. Hence, supply of oxygen is increased by oxygen lancing into the bath (hot metal) beneath the slag level. The bath is agitated vigorously by evolution of CO due to C–O reaction at the steelmaking temperature. The CO evolution is known as carbon boil, which improves heat transfer, slag–metal interaction, and flotation of inclusions. Increased solubility of lime at higher temperature increases basicity of the slag. Viscosity is controlled by the addition of fluorspar (CaF_2). The two main slag–metal reactions, dephosphorization and desulfurization, are favored by highly basic slag. But the former requires oxidizing slag and low temperature, whereas the latter needs reducing slag and high temperature. As the basic open-hearth slag that contains about 15% FeO is highly oxidizing, it is suitable for dephosphorization; desulfurization takes place only to a limited extent. In order to avoid phosphorus reversion, excess oxygen is partially removed by addition of ferrosilicon and ferromanganese. Finally, deoxidation is carried out by aluminum after tapping the steel in a ladle.

7.4.4 Bessemer Process

As oxidation of carbon, silicon, manganese, sulfur, and phosphorus are highly exothermic reactions, the heat generated during blowing makes the Bessemer process self-sustaining and no extra fuel is required. The hot metal containing about 3.5–4.0% C, 0.04–0.05% S, 0.03–0.04% P, 2.0–2.5% Si, and 0.75–1.0% Mn was refined in acid Bessemer converter, whereas basic Bessemer was employed for the hot metal analyzing 3.0–3.5% C, 0.08–0.10% S, 1.8–2.5% P, 0.6–1.0% Si, and 1.0–2.5% Mn. Oxidation of silicon and phosphorus is the principal source of heat in acid as well as in basic Bessemer processes, respectively. Oxidation of carbon and manganese also provides some heat to make up part of the requirement. Oxidation of all these impurities raises the temperature of the hot metal from the initial 1250–1350 °C to 1600–1650 °C, suitable for refining molten steel. Oxidation of carbon produces long flames at the mouth of the converter. The C–O reaction controls the flame length, which diminishes when reaction ceases. However, oxidation of phosphorus continues for another 3–4 min. Due to larger mass, initially iron gets oxidized as soon as air is blown into the hot metal. In both acid and basic Bessemer processes, the FeO formed first reacts with silicon. This is followed by the oxidation of manganese present in the hot metal. Excess (FeO) and (MnO) react with (SiO_2) to form silicate slag (reactions: 7.56 and 7.57). Oxidation of carbon begins only after silicon and manganese present in the hot metal are completely oxidized: ([C] + (FeO) = [Fe] + {CO}).

In the Bessemer converter, the heterogeneous C–O reaction takes place at sites of nitrogen bubbles formed in the metallic bath, in which air is blown. CO that escapes through the mouth of the converter burns and produces flame. The flame is longer when C–O reaction is fast and gradually diminishes as reaction ceases. The converter is tilted for addition of deoxidizers (ferrosilicon and ferromanganese) and carbon in appropriate form for adjustment of the composition.

Operation of the basic Bessemer differs from that of the acid Bessemer in charging limestone before starting the blow. The sequence of removal of silicon, manganese, and carbon is similar to that followed in the acid Bessemer process; that is, oxidation of carbon starts after silicon and manganese have been completely removed. In basic process, oxidation of phosphorus starts when the flame due to burning of CO has almost diminished. This necessitates blowing to be continued for an additional 3–4 min. This duration is known as the *"after-blow period."* During this period, oxidation of phosphorus generates sufficient heat for lime dissolution in the slag. Phosphorus oxidized to P_2O_5 in the after-blow period reacts with (CaO) to form calcium phosphate (reactions: 7.61, and 7.62). The basic slag is helpful in desulfurization to some extent (reactions: 7.63, 7.64, and 7.65).

There is large variation in the slag produced by acid and basic processes. For example, acid Bessemer slag contained 57–67% SiO_2, 12.0–18.0% FeO, 12.0–18.0% MnO, 2.0–4.0% Al_2O_3, less than 1.0% FeO, CaO, and MgO in traces, whereas basic Bessemer slag analyzed 7.0–10% SiO_2, 12.0–17.0% FeO, 25.0–30.0% CaO, 18.0–22.0% P_2O_5, 10.0–15.0% MgO, 10.0% Al_2O_3, 5.0–6.0% Fe_2O_3, and MnO in traces.

It is interesting to note that the first revolution in the steel industry was observed in the 1850s by invention of the Bessemer pneumatic process. The second revolution took place in the 1950s by the availability of tonnage oxygen. The third revolution in steelmaking technology came through the hybrid process in the late twentieth century. The second revolution is important from the view point of thermodynamics of refining, whereas the third one representing near equilibrium conditions is based on the thorough understanding of thermodynamics and kinetics of refining reactions. The close equilibrium is attained without loss of production and quality of steel even by charging scrap and hot metal in certain proportions.

7.5 Problems [25–27]

Problem 7.1
A pig iron containing 0.1% S, 0.3% Si, 3.4% C, 0.8% Mn, and 1.8% P is desulfurized at 1300 °C according to the following two reactions:

1. $[S]_{1wt \%}$ + (CaO) (s) + [C] = (CaS) (s) + CO (g), ΔG_1^o = 23,090–25.14 T cal
2. $[S]_{1wt \%}$ + 2 (CaO) (s) + ½ [Si] = (CaS) (s) + ½ (2CaO.SiO$_2$) (s),
   ΔG_2^o = − 53,945 + 17.85 T cal

Assuming that iron is saturated with carbon and all solid compounds are pure and p_{CO} = 1 atm, show that desulfurization is favored in the presence of silicon, given that, f_S = 3.6 and f_{Si} = 4.5.

Solution
T = 1573 K,

$\Delta G_1^o = 23090 - 25.14\ T$ cal $= -RT\ln K_1, \Delta G_1^o = -16455.7, K_1 = 193.4$

$\Delta G_2^o = -53945 + 17.85\ T$ cal $= -RT\ln K_2, \Delta G_2^o = -25867, K_2 = 3930$

K_1 and K_2 for reactions (1) and (2), respectively, can be expressed as:

$$K_1 = 193.4 = \frac{a_{CaS} \cdot p_{CO}}{[a_S](a_{CaO})[a_C]} = \frac{1}{[\,f_S.\%S]}$$

(since CaS, CaO and $2CaO.SiO_2$ are pure solids, $a_{CaS} = 1 = a_{CaO} = a_{2CaO.SiO_2}$ and pig iron is saturated with C, $a_C = 1$ and $[a_S] = [f_S. \% S]$)

$$\therefore [\%S]_1 = \frac{1}{f_S.K_1} = \frac{1}{3.6 \times 193.4} = 0.00144$$

$$K_2 = \frac{a_{CaS} \cdot a_{2CaO.SiO_2}^{\frac{1}{2}}}{[a_S] \cdot (a_{CaO}^2) \cdot [a_{Si}]^{\frac{1}{2}}} = \frac{1}{[\,f_S.\%S] \cdot [\,f_{Si}.\%Si]^{\frac{1}{2}}}$$

$$\therefore [\%S]_2 = \frac{1}{K_2 f_S \cdot [\,f_{Si} .\%Si]^{\frac{1}{2}}}$$

$$= \frac{1}{3930 \times 3.6\,[4.5 \times 0.3]^{\frac{1}{2}}} = 0.00006$$

Hence, the presence of silicon in reaction (2) favors desulfurization.

Problem 7.2
(Elliott, J. F., 1980–1981, personal communication, Massachusetts Institute of Technology, Cambridge)

Using the interaction coefficients and thermodynamic data, compare the relative case with which sulfur can be removed from the following two compositions of liquid iron at 1600 °C.

Liquid iron	%C	%Si	%Mn	%S	%P
A	0.05	0.002	0.05	0.025	0.01
B	2.00	0.80	0.90	0.025	0.05

Given that:

$$[\%C] + [\%O] = CO(g), \Delta G^o = -22400 - 39.6\,T, J$$
$$e_S^C = 0.11, e_S^{Si} = 0.063, e_S^{Mn} = -0.026, e_S^S = -0.028, e_S^P = 0.29, \text{and } e_S^O = -0.27$$
$$e_O^C = -0.34, e_O^{Si} = -0.23, e_O^{Mn} = -0.083, e_O^S = -0.27, e_O^P = 0.13, \text{and } e_O^O = -0.20$$

Solution
The desulfurization reaction may be considered as follows:

$$[S] + (O^{2-}) = (S^{2-}) + [O] \tag{1}$$

$$K = \frac{[a_O] \cdot (a_{S^{2-}})}{(a_{O^{2-}}) \cdot [a_S]} = \frac{[\,f_O\%O] \cdot (x_{S^{2-}})}{(x_{O^{2-}}) \cdot [\,f_S\%S]} \tag{2}$$

$$\log K = \log \left(\frac{x_{S^{2-}}}{x_{O^{2-}}}\right) + \log [\%O] + \log f_O - \log [\%S] - \log f_S \qquad (3)$$

$\log \left(\frac{x_{S^{2-}}}{x_{O^{2-}}}\right)$ refers to the property of slag only, and further the above Eq. (3) can be written separately for two iron A and B, marked correspondingly and on equating the $\log K$, we get:

$$\log \left(\frac{x_{S^{2-}}}{x_{O^{2-}}}\right) + \log [\%O]^A + \log f_O^A - \log [\%S]^A - \log f_S^A$$

$$= \log \left(\frac{x_{S^{2-}}}{x_{O^{2-}}}\right) + \log [\%O]^B + \log f_O^B - \log [\%S]^B - \log f_S^B$$

$$\log \left[\frac{[\%S]^A}{[\%S]^B}\right] = \log [\%O]^A + \log f_O^A - \log f_S^A - \log [\%O]^B - \log f_O^B$$

$$+ \log f_S^B \qquad (4)$$

Assuming $p_{CO} = 1$ atm for the CO bubbles coming out of the melt.

$[\% C] + [\% O] = CO$ (g), $\Delta G^\circ = -22400 - 39.6\, T$, J

at $T = 1873$ K,

$\Delta G^\circ = -96570.8$ J $= -RT\ln K$

Assuming the validity of Henry's law: $a_i = $ wt $\% i$

$$K = \frac{p_{CO}}{a_C . a_O} = \frac{p_{CO}}{[\%C].[\%O]}$$

$$\therefore \ln K = \ln \left(\frac{p_{CO}}{[\%C].[\%O]}\right) = \frac{96570.8}{8.314 \times 1873} = 6.2015$$

$$\therefore \frac{p_{CO}}{[\%C].[\%O]} = 493.5$$

In iron A, $[\%C] = 0.05$

$$[\%O]^A = \frac{1}{493.5 \times [\%C]} = \frac{1}{493.5 \times 0.05} = 0.04053$$

In iron B, $[\%C] = 2.00$, $[\%O]^B = 0.00101$
From the value of interaction coefficients:

$$\log f_S^A = \%C.e_S^C + \%Si.e_S^{Si} + \%Mn.e_S^{Mn} + \%S.e_S^S + \%P.e_S^P + \%O.e_S^O$$
$$= 0.05 \times (0.11) + 0.002 \times (0.063) + 0.05 \times (-0.026) + 0.025 \times$$
$$(-0.028) + 0.01 \times (0.29) + 0.041 \times (-0.27) = -0.004544$$

$$\log f_S^B = 2.0 \times (0.11) + 0.80 \times (0.063) + 0.90 \times (-0.026) + 0.025 \times$$
$$(-0.028) + 0.05 \times (0.29) + 0.00101 \times (-0.27) = 0.2614$$

$$\log f_O^A = \%C.e_O^C + \%Si.e_O^{Si} + \%Mn.e_O^{Mn} + \%S.e_O^S + \%P.e_O^P + \%O.e_O^O$$
$$= 0.05 \times (-0.34) + 0.002 \times (-0.23) + 0.05 \times (-0.083) + 0.025 \times$$
$$(-0.27) + +0.01 \times (0.13) + 0.041 \times (-0.20) = -0.03526$$

Similarly, $\log f_O^B = -0.93897$

Substituting the values in Eq. (4)

$$\log \left[\frac{[\%S]^A}{[\%S]^B} \right] = \log [\%O]^A + \log f_O^A - \log f_S^A - \log [\%O]^B - \log f_O^B + \log f_S^B$$

$$= \log (0.04053) - 0.03526 - (-0.0045) - \log (0.00101) - (-0.93897) + 0.2614$$
$$= 2.7731$$

$$\therefore \left[\frac{[\%S]^A}{[\%S]^B} \right] = 593$$

Hence, after desulfurization [%S] in iron B = $\frac{1}{593}$ [%S] in iron A. Thus, better desulfurization (i.e., lower residual sulfur in the hot metal) can be achieved in the presence of larger amount of impurities (i.e., total of C, Si, Mn and P).

Problem 7.3

Calculate the pressure of CO and CO_2 in the gas bubbles formed from the reaction between carbon and oxygen in liquid iron at 1600 °C for melts containing (a) 1% C, (b) 0.1% C, and (c) 0.03% C. Assume: $a_i = [\%i]$ and $p_{CO} + p_{CO_2} = 1$ atm

Given that:

$$CO_2 (g) = [\%C] + 2 [\%O], \Delta G_1^o = 184,100 - 47.9 \, T \, J$$
$$[\%C] + [\%O] = CO (g), \Delta G_2^o = -22,400 - 39.6 \, T \, J$$

Solution

For the reaction under question:

$$CO_2(g) + [\%C] = 2CO(g), \Delta G^o = \Delta G_1^o + 2.\Delta G_2^o = 139300 - 127.1 T = -RT\ln K$$

and $K = \dfrac{p_{CO}^2}{p_{CO_2}.[\%C]}$, ΔG^o at $1873K = -98758.3$ J

Thus, we can write:

$$\Delta G^o = -98758.3 = -8.314 \times 1873 \times \ln\left[\frac{p_{CO}^2}{p_{CO_2}[\%C]}\right]$$

$$\therefore \ln\left[\frac{p_{CO}^2}{p_{CO_2}[\%C]}\right] = 6.342$$

$$\text{and}\left[\frac{p_{CO}^2}{p_{CO_2}[\%C]}\right] = 567.93$$

given $p_{CO} + p_{CO_2} = 1$ atm

$$\therefore \left[\frac{p_{CO}^2}{(1-p_{CO})\,[\%C]}\right] = 567.93$$

(a) $[\%C] = 1$ wt%, on substitution we get the following equation:

$$p_{CO}^2 + 567.93\,p_{CO} - 567.93 = 0$$

$$\therefore p_{CO} = 0.998 \text{ atm}$$

and $p_{CO_2} = 0.002$ atm

(b) similarly when $[\%C] = 0.1\%$, $p_{CO} = 0.983$ atm and $p_{CO_2} = 0.017$ atm
(c) when $[\%C] = 0.03\%$, $p_{CO} = 0.947$ atm and $p_{CO_2} = 0.053$ atm Ans.

Problem 7.4

Carbon in the liquid iron can reduce silica in the refractory by the reaction:
$2[\%C] + SiO_2(l) = [\%Si] + 2CO(g)$. The standard free energy of formation of silica is given as
$[\%Si] + 2[\%O] = SiO_2$ (l), $\Delta G^o = -594100 + 230\,T$ J

(a) What are the equilibrium concentrations of carbon and oxygen in the metal at 1600 °C and $p_{CO} = 1$ atm when $\%Si = 0.16$?
(b) What will happen to the composition of the metal in part (a) if the temperature of the system is raised to 1800 °C?

Assume that activities of species in the metal are equal to their compositions in wt%.

Solution

$$[\%Si] + 2[\%O] = SiO_2(l), \Delta G_1^o = -594100 + 230T \text{ J}$$

From the previous question we know that $[\%C] + [\%O] = CO$ (g),

$$\Delta G_2^o = -22400 - 39.6T \text{ J}$$

Therefore, ΔG^o for the reaction under question: $2[\%C] + SiO_2$ (l) $= [\%Si] + 2CO$ (g) is obtained as $\Delta G^o = 2\Delta G_2^o - \Delta G_1^o = 549300 - 309.2\,T$ J at 1873 K, $\Delta G^o = -29831.6$ J $= -RT\ln K$

$$\text{and } K = \frac{[\%\text{Si}].p_{CO}^2}{[\%\text{C}]^2.a_{SiO_2}}$$

$$\therefore \Delta G^o = -29831.6 \text{ J} = -8.314 \times 1873 \times \ln\left[\frac{[\%\text{Si}].p_{CO}^2}{[\%\text{C}]^2.a_{SiO_2}}\right]$$

$$\ln\left[\frac{[\%\text{Si}].p_{CO}^2}{[\%\text{C}]^2.a_{SiO_2}}\right] = 1.9157$$

$$\left[\frac{[\%\text{Si}].p_{CO}^2}{[\%\text{C}]^2.a_{SiO_2}}\right] = 6.792$$

$$\text{if } a_{SiO_2} = 1, \frac{[\%\text{Si}].p_{CO}^2}{[\%\text{C}]^2.} = 6.792$$

(a) $p_{CO} = 1$ atm, [% Si] = 0.16

$$[\%\text{C}]^2 = \frac{0.16 \times 1}{6.791}$$

$$[\%\text{C}] = 0.1535 \text{ wt\% Ans}$$

Consider equation: [C] + [O] = CO (g), $\Delta G_2^o = -22,400 - 39.6\,T$
at 1873 K, $\Delta G_2^o = -96570.8$ J $= -RT \ln K$, $K = 493.5 = \frac{p_{CO}}{[\%\text{C}].[\%\text{O}]}$

$$\therefore [\%\text{O}] = \frac{p_{CO}}{[\%\text{C}].493.5} = \frac{1}{0.1535 \times 493.5} = 0.013\% \text{ Ans.}$$

(b) when $T = 1800\,°C = 2073$ K,

$$\Delta G^o = 549300 - 309.2\,T = 549300 - 309.2 \times 2073 = -91671.6 \text{ J}$$

$$\therefore \ln\left[\frac{[\%\text{Si}].p_{CO}^2}{[\%\text{C}]^2.}\right] = \frac{91671.6}{8.314 \times 2073} = 5.319$$

$$\text{and } \left[\frac{[\%\text{Si}].p_{CO}^2}{[\%\text{C}]^2.}\right] = 204.168$$

when $p_{CO} = 1$ atm and [% Si] = 0.16
    [%C] = 0.028%
for the reaction: [% C] + [% O] = CO (g), at 2073 K, $\Delta G_2^o = -104490.8$ J

$$\therefore \ln\left[\frac{p_{CO}}{[\%C].[\%O]}\right] = \frac{104490.8}{8.314 \times 2073} = 6.063$$

$$\text{or } \frac{p_{CO}}{[\%C].[\%O]} = 429.6$$

when $p_{CO} = 1$ atm and $[\%C] = 0.028\%$

$$[\%O] = \frac{p_{CO}}{[\%C].429.6} = \frac{1}{0.028 \times 429.6} = 0.083$$

Hence, the composition of the metal at 2073 K would be as follows:

$$[\%Si] = 0.16 \text{ wt}\%, [\%C] = 0.028 \text{ wt}\%, \text{ and } [\%O] = 0.083 \text{ wt}\%. \text{ Ans}$$

Discussion With the increase of temperature from 1600 °C to 1800 °C, the free energy for the decarburization reaction decreases (for example, at 1600 °C, $\Delta G^0 = -96570.8$ J and at 1800 °C, $\Delta G^0 = -104490.8$ J). This means decarburization reaction becomes more feasible with increase of temperature; hence, lower carbon is attained in the melt.

Problem 7.5
In the basic open-hearth process, the reaction of manganese in the bath with iron oxide (FeO) in the slag attains a condition very close to true equilibrium. The steel contains 0.065 at % Mn and the slag analyzes (by wt%) as FeO-76.94%, Fe_2O_3-4.15%, MnO-13.86%, MgO-3.74%, SiO_2-1.06%, and CaO-0.25%. Calculate the value of equilibrium constant and standard free energy change for the above reaction at 1655 °C, assuming that the slag and Fe–Mn system behave ideally at this temperature. Neglect the effect of other metalloids present in the steel.

Given that: molecular weights of FeO: 71.85, Fe_2O_3:159.70, MnO: 70.94, MgO: 40.32, SiO_2: 60.09, and CaO: 56.08.

Solution
The reaction may be represented as:

$$(FeO)_{slag} + [Mn]_{metal} = (MnO)_{slag} + [Fe]_{metal}$$

$$K = \frac{(a_{MnO}).[a_{Fe}]}{(a_{FeO}).[a_{Mn}]}$$

Assuming ideal behavior in the slag and in Fe–Mn system at 1655 °C, we can write:

$$K = \frac{(x_{MnO}).[x_{Fe}]}{(x_{FeO}).[x_{Mn}]}$$

Since the other metalloids present are negligible and Mn content is small, mole fraction of Fe may be assumed to be 1.

$$\therefore K = \frac{(x_{MnO})}{(x_{FeO}) \cdot [x_{Mn}]}$$

Analysis of the slag needs to be converted into mole fraction:

Constituents	wt%	mol wt	g mol
FeO	76.94	71.85	1.0710
Fe$_2$O$_3$	4.15	159.70	0.0260
MnO	13.86	70.94	0.1954
MgO	3.74	40.32	0.0928
SiO$_2$	1.06	60.09	0.0176
CaO	0.25	56.08	0.0045

In slag, iron and oxygen are also present as Fe_2O_3. Hence, mole fraction of Fe_2O_3 should be converted in terms of FeO in order to obtain total value of x_{FeO}. Since the conversion takes place according to the reaction: $Fe_2O_3(s) \rightarrow 2FeO(s) + \frac{1}{2}O_2(g)$, 1 mol of Fe_2O_3 will give 2 mol of FeO.

$\therefore$Total number of g mol of FeO = g mole of FeO + 2 $\times$ g mol of Fe_2O_3

$$= 1.0710 + 2 \times 0.0260 = 1.1230$$

Total g mol of constituents in the
slag $= 1.1230 + 0.1954 + 0.0928 + 0.0176 + 0.0045 = 1.4333$

$$\therefore Total(x_{FeO}) = \frac{1.1230}{1.4333} = 0.7835$$

$$(x_{MnO}) = \frac{0.1954}{1.4333} = 0.1363$$

$$[x_{Mn}] = \frac{0.065}{100} = 0.065 \times 10^{-2}$$

$$\therefore K = \frac{(x_{MnO})}{(x_{FeO}) \cdot [x_{Mn}]} = \frac{0.1363}{0.7835 \times (0.065 \times 10^{-2})} = 267.7$$

$$\Delta G^o = -1.987 \times 1928 \times \ln(267.7) = -21414 \text{ cal} \quad \text{Ans}$$

References

1. Ghosh, A., & Chatterjee, A. (2008). *Ironmaking and steelmaking*. New Delhi: Prentice Hall India Pvt. Ltd. (Chapters 16 and 17).
2. Tupkary, R. H., & Tupkary, V. R. (2008). *An introduction to modern steel making* (4th ed.). New Delhi: Khanna Publishers.
3. Trubin, K. G., & Oiks, K. N. (1974). *Steelmaking open hearth and combined process*. Moscow: MIR Publishers.
4. Chakrabarti, A. K. (2007). *Steel making*. New Delhi: Prentice-Hall of India Pvt. Ltd. (Chapters 3,4 and 5).

5. Rocca, R. N., Grant, J., & Chipman, J. (1951). Distribution of sulfur between liquid iron and slags of low iron concentrations. *Transactions of the American Institute of Mining and Metallurgical Engineers, 191*, 319–326.
6. Ward, R. G. (1962). *An introduction to physical chemistry of iron and steel making*. London: Edward Arnold. (Chapter 11).
7. Flood, H., & Grjotheim, K. (1952). Thermodynamic calculation of slag equilibria. *Journal of Iron and Steel Institute, 171*, 64–70.
8. Basu, S., Seetharaman, S., & Lahiri, A. K. (2010). Thermodynamics of phosphorus and sulfur removal during basic oxygen steelmaking. *Steel Research International, 81*(11), 932–939.
9. Bora, D., Masanori, S., & Toshihiro, T. (2010). Sulphide capacity prediction of molten slag by using a neutral network approach. *ISIJ International, 50*(8), 1059–1063.
10. Toshihiro, T., Yumi, O., & Mitsuru, U. (2010). Trial on the application of capillary phenomenon of solid CaO to desulfurization of liquid Fe. *ISIJ International, 50*(8), 1071–1079.
11. Chen, S., Wang, X., He, X., Wang, W., & Jiang, M. (2013). Industrial application of desulfurization using low basicity refining slag in tire cord steel. *Journal of Iron and Steel Research International, 20*(1), 26–33.
12. Moore, J. J. (1990). *Chemical metallurgy* (2nd ed.). Oxford: Butterworth-Heinemann. (Chapter 7).
13. Balagiva, K., Quarrel, A. G. and Vajragupta, P. (1946) A laboratory investigation of the phosphorus reaction in the basic steelmaking process, *Journal of Iron and Steel Institute, 153*, 115–145.
14. Healy, G. W. (1970). Distribution of phosphorus in BOF process. *Journal of Iron and Steel Institute, 208*, 666–669.
15. Chipman, J. (1964). Physical chemistry of liquid steel. In G. Derge (Ed.), *Basic open hearth steel making* (pp. 640–714). New York: AIME. (Chapter 16).
16. Marshall, S., & Chipman, J. (1942). The carbon-oxygen equilibrium in liquid iron. *Transactions of the American Society for Metals, 30*, 695–741.
17. Turkdogan, E. T., Davis, L. S., Leake, L. E., & Gokcen, N. A. (1955). The reaction of carbon and oxygen in molten iron. *Journal of Iron and Steel Institute, 181*, 123–128.
18. Turkdogan, E. T. (1975). Physical chemistry of oxygen steelmaking thermchemistry and thermodynamics. In R. D. Pehlke, W. F. Porter, P. F. Urban, & J. M. Gaines (Eds.), *BOF steelmaking* (Vol. 2, pp. 1–190). New York: Iron & Steel Society, AIME. (Chapter 4).
19. Parker, R. H. (1978). *An introduction to chemical metallurgy* (2nd ed.). Oxford: Pergamon. (Chapter 7).
20. Gilchrist, J. D. (1980). *Extraction metallurgy* (2nd ed.). New York: Pergamon. (Chapter 12).
21. Turkdogan, E. T. (1974). Reflections on research in pyrometallurgy and metallurgical chemical engineering. *Transactions of the Institution of Mining and Metallurgy, Section C, 83*, C67–C82.
22. Geiger, G. H., Kozakevitch, P., Olette, M., & Riboud, P. V. (1975). Theory of BOF reaction rates. In R. D. Pehlke, W. F. Porter, P. F. Urban, & J. M. Gaines (Eds.), *BOF steelmaking* (Vol. 2, pp. 191–321). New York: Iron & Steel Society, AIME. (Chapter 5)
23. Shamsuddin, M., & Sohn, H. Y. (2020). Constitutive topics in physical chemistry of ironmaking and steelmaking: A review. *Trends in Physical Chemistry, 19*, 33–50.
24. Deo, B., & Boom, R. (1993). *Fundamentals of steelmaking metallurgy*. London: Prentice Hall International. (Chapter 5).
25. Kubaschewski, O., & Alcock, C. B. (1979). *Metallurgical thermochemistry* (5th ed.). Oxford: Pergamon.
26. Curnutt, J. L., Prophet, H., McDonald, R. A., & Syverud, A. N. (1975). *JANAF thermochemical tables, 1975 supplement*. Midland: The Dow Chemical Company.
27. Barin, I., Knacke, O., & Kubaschewski, O. (1977). *Themochemical properties of inorganic substances, (supplement)*. New York: Springer-Verlag.

Chapter 8
Secondary Steelmaking

Until the recent past, the mechanical properties and surface quality of steel produced by conventional processes, for example, open hearth, LD, OBM, and so on were up to the satisfaction of consumers in meeting their general requirements. However, in recent years, there has been a steadily increasing demand for better quality steels in terms of lower impurity contents, better surface finish, better internal quality (i.e., inclusion-free), and specific grain size as well as in their mechanical properties in terms of strength, toughness, and workability under extreme forming conditions. In order to fulfill the stringent demand of consumers it has become essential on the part of steel producers to drastically lower down the impurity level in steel, in some cases to a few parts per million. For example, alloy steel forgings, line-pipe steel, and HIC-resistant steel need ultralow sulfur, as low as 0.001% S (10 ppm). It is important to note that secondary treatment reduces the concentration of sulfur, oxygen, nitrogen, hydrogen, and nonmetallic inclusions. Thus, modern steelmaking is classified into two categories: "primary and secondary." The first category includes the major bulk of steelmaking processes that carry out total refining and melting (if required). In fast (primary steelmaking) processes where operations are completed within 60 min, the resulting steel may not always meet the desired specifications. Hence, the primary steelmaking processes are nowadays restricted to bulk steel production of ordinary quality steels for construction purposes. In order to adhere to the specific composition, molten steel from these units is processed in ladles. This ladle treatment is known as "secondary steelmaking," which in fact is the second stage of refining that is carried out for the final refining and finishing. By controlling the composition in this way, the deleterious effect of impurities on the mechanical properties is avoided and steel with better internal quality and ultralow carbon/sulfur/phosphorus can be produced. Nowadays, secondary steelmaking units have become an essential part of the integrated steel plants to supply sophisticated grade of steel for continuous casting. In recent years, a large number of research investigations [1–6] have been conducted on secondary refining to produce inclusion-free steels containing ultralow carbon, silicon, and phosphorus.

© The Minerals, Metals & Materials Society 2021   293
M. Shamsuddin, *Physical Chemistry of Metallurgical Processes, Second Edition*,
The Minerals, Metals & Materials Series, https://doi.org/10.1007/978-3-030-58069-8_8

Secondary steel refining homogenizes the temperature and composition of the steel bath; facilitates decarburization, desulfurization, dephosphorization, deoxidation, nitrogen and hydrogen removal, and attainment of the required teeming temperature; controls the shape of the inclusions; and improves the cleanliness of the steel. All these objectives cannot be achieved in one single unit. In order to fulfill these requirements, one or more units like inert gas purging (IGP), ladle furnace (LF), injection metallurgy (IM), vacuum degassing, and so on may be employed. In this chapter, a brief account on IGP, LF, and IM will be given and the physicochemical aspects of refining using synthetic slag, deoxidation, and degassing will be discussed. These units are employed according to the requirement of the desired quality of steels. The sequence of use of units differs from plant to plant. Stainless steelmaking, which involves melting in electric arc furnaces (EAF) and decarburization and alloying in ladles either under argon or vacuum, will also be discussed in this chapter.

8.1   Inert Gas Purging (IGP)

In 1933, the Perrin process was developed as a novel method for thorough mixing of slag and metal with the prime objective of effective removal of nonmetallic inclusions from steel before casting. In this process, liquid steel is allowed to fall in a short time as a fine stream over a synthetic nonoxidizing molten slag pool. The heavy turbulence due to enforcement, which causes the vigorous mixing of the slag and metal results in effective and efficient desulfurization as well as in deoxidation and scavenging of nonmetallic inclusions to produce finally cleaner low-sulfur steel. Better results were obtained by pouring the synthetic slag and the liquid refined steel simultaneously in a transfer ladle. In recent years, efficient mixing is brought out by purging molten steel with argon from the ladle bottom either through porous bricks or through a top lance immersed into the melt in an open ladle. The stirring caused by gas bubbles, not only helps in the homogenization of temperature and composition (better mixing of alloying elements) but also facilitates faster deoxidation and flotation of inclusions. IGP is used very effectively for efficient mixing of slag and metal in synthetic slag refining as well as in IM for desulfurization and dephosphorization. A high degree of deoxidation and desulfurization can be achieved by maintaining proper slag composition and by thorough stirring of the slag and metal by vigorous gas blowing. Nonmetallic inclusions are carried away by the rising argon bubbles to join the slag phase. Dissolved hydrogen in the metal diffuses into the bubbles which act as sites for C–O reaction. CO gas formed also diffuses into argon bubbles.

In recent years, the introduction of Sealed Argon Bubbling (SAB) and Capped Argon Bubbling (CAB) by Japanese has further improved the IGP technique. In SAB, an immersion box is dipped below the synthetic slag layer kept on the surface of the bath in such a way that it is located above the porous plug fitted at the bottom of the ladle. This arrangement protects the steel from atmospheric exposure by rising bubbles. In CAB, a cover is placed over the ladle. The Nippon Steel Corporation has introduced a new process known as CAS-OB (standing for Composition Adjustment by Sealed Argon Bubbling—Oxygen Bubbling) to minimize the loss of deoxidizers and alloy additions. The combination of high turbulence and stirring makes dissolution faster as well as raises the bath temperature due to exothermic oxidation of aluminum by oxygen blown in the melt.

8.2 Ladle Furnace (LF)

In recent years, ladle furnaces are used extensively to carry out most of the secondary refining economically. A simple ladle fitted with a bottom plug for purging/introducing argon and a lid with piercing electrodes works like an arc furnace for heating the steel bath. The lid (top cover) protects the bath from atmospheric oxidation. For evacuation, another lid with openings for additions and injection is attached to the ladle. Thus, in one unit that is equipped with heating facility, stirring, vacuum treatment, deoxidation, desulfurization, composition adjustment, synthetic slag refining, injection, and so on can be carried out with proper control of temperature. Slag composition can also be adjusted in the ladle furnace according to the requirement by adding CaO, CaF_2, and so on.

In the ladle furnace, steel can be refined to the desired adequate level by a synthetic slag. A basic slag consisting of CaO (50–55%), MgO (9–10%), SiO_2 (5–12%), Al_2O_3 (20–25%), FeO + MnO (1–2%), TiO_2 (0.4%), and some CaF_2 is generally used. The heating and stirring capabilities make the ladle versatile for (i) deoxidation either in vacuum or by addition of Mn, Si, Al, and so on (ii) faster melting and homogenization after alloy additions, (iii) effective desulfurization by forming suitable slag, (iv) elimination of nonmetallic inclusions, and (v) better alloy recovery.

8.3 Deoxidation

Iron ore is smelted under highly reducing conditions to produce hot metal (molten pig iron) that is refined under oxidizing conditions to make steel. In all steelmaking processes, either air or oxygen is blown or surplus air/oxygen is provided to facilitate quick oxidation of impurities. Under these conditions, oxygen easily gets dissolved in the steel melt. At 1600 °C, the maximum solubility of oxygen in pure iron is 0.23 wt%, which increases to 0.48% [7] at 1800 °C. Solubility is affected by impurities like C, Si, Mn, P, and so on and alloying elements, such as Ni, Co, Cr, V, and so on. The effect of alloying elements on the activity coefficient of oxygen

Fig. 8.1 Effect of alloying
elements on the activity
coefficient of oxygen in
liquid iron at 1600 °C
(Reproduced from J. F.
Elliott [8] and E. T.
Turkdogan [9] with the
permission of Association
for Iron & Steel
Technology)

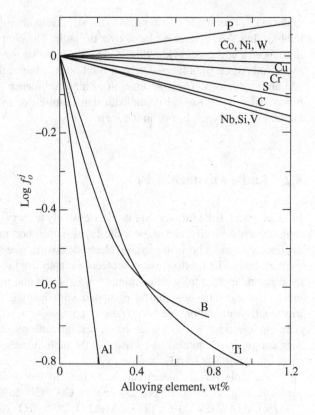

in liquid iron at 1600 °C is shown in Fig. 8.1. During steelmaking, as refining
progresses, the concentrations of impurities decrease, and there is a continuous
increase of dissolved oxygen in the steel melt. Thus, finally as high as 0.05–0.1%
[O] is dissolved in the melt. The excess oxygen is expelled during the solidification
of steel due to the low solid solubility of about only 0.003% (i.e., 30 ppm). Hence,
the production of sound castings warrants the removal of residual oxygen from the
refined steel. The process of removal of oxygen from the steel melt is known as
"deoxidation." Oxygen from the steel melt can be removed by the following two
methods:

1. Precipitation deoxidation, where the residual oxygen is allowed to react with
   elements having higher affinity for oxygen (compared to what iron has for
   oxygen) to form oxide products. These oxides are relatively more stable as
   compared to FeO and have very limited solubility in liquid steel. Being lighter
   than steel the oxide products rise to the top surface and can be easily removed.
   Precipitation deoxidation is practiced extensively because it is very effective in
   decreasing oxygen content of the steel.
2. Alternatively, the oxygen content of the refined steel can be reduced by vacuum
   treatment of the bath or by bubbling inert gas through the melt. Application of

vacuum as well as bubbling of inert gas lowers down the partial pressure of CO in the system and favors oxygen removal from the melt due to the formation of CO gas according to the reaction:

$$[C] + [O] = CO(g) \tag{8.1}$$

8.3.1 Choice of Deoxidizers

On addition of the appropriate amount of the deoxidizer in the steel bath, the reaction affecting the removal of some dissolved oxygen and the formation of the oxide of the most reactive metal present in the deoxidizer takes place. The reaction continues until equilibrium is attained between the oxygen in the bath and the deoxidizing element and the oxide formed. A generalized form of chemical equilibrium that deals with the deoxidation product in contact with the steel melt may be represented as:

$$x[M] + y[O] = M_xO_y(s, l) \tag{8.2}$$

for which the equilibrium constant is given as:

$$K = \frac{a_{M_xO_y}}{[a_M]^x[a_O]^y} \tag{8.3}$$

If the deoxidation product is pure solid, $a_{M_xO_y} = 1$ and since we are dealing with infinitely dilute solutions of deoxidizers and oxygen in the melt, in the simplest case, according to Henry's law, we can write $a_M = [\%M]$ and $a_O = [\%O]$. Then the solubility product, $K' = [\%M]^x[\%O]^y$ is constant at a given temperature. K' is known as the deoxidation constant of the element, M. However, in most cases, the deoxidation problem is not so simple. In order to account for any deviation from Henry's law ($a_i = f_i \cdot [\%i]$), Eq. 8.3 is modified as:

$$K = \frac{1}{f_M^x[\%M]^x f_O^y[\%O]^y} \tag{8.4}$$

Thus, the equilibrium constant, K and the deoxidation constant, K' are related as:

$$f_M^x \cdot f_O^y \cdot K' = \frac{1}{K} \tag{8.5}$$

When Henry's law is obeyed in infinitely dilute solution, f_M and $f_O \rightarrow 1$ as [% M] and [%O] $\rightarrow 0$, the two constants K and K' are simply the inverse of each other (i.e., $K' = \frac{1}{K}$).

The choice of deoxidizers depends on a number of factors based on thermodynamics and the physics of formation and removal of the oxide products. Thermodynamically, the best deoxidizing element (deoxidizer) should have the least amount of dissolved oxygen [O] left in equilibrium with its own lowest concentration in the steel melt. Aluminum, silicon, and manganese are used as common deoxidizers because they are reasonably cheap. Sometimes zirconium, titanium, and calcium are used in deoxidation of steel but they are costlier than common deoxidizers. Although rare earths, beryllium, and thorium can serve as deoxidizers because their oxides are more stable, the high cost factor encourages their application only in special cases. They are normally added in the steel bath after adequate deoxidation. Chromium, vanadium, and boron form relatively stable oxides resulting in the reduction in oxygen content of the bath but there is some loss of the added elements. Nickel, copper, cobalt, tin, lead, silver, niobium (columbium), and tantalum will not deoxidize the steel bath because their oxides are less stable as compared to FeO. In addition to the stability of the oxides, the activity of the elements and oxygen in the bath as well as the composition, purity, and nature (acid or basic) of the oxide formed must be given due consideration in understanding the behavior of the element toward oxygen. The residual content of the deoxidizer in steel after deoxidation should not adversely affect the mechanical properties of steel. Thus, it is essential that the residual deoxidizing element left must be controlled within the prescribed limit of the chemical specifications. The rate of deoxidation, that is, the formation of oxide products must be fast. Since kinetic data on deoxidation are very limited, thermodynamic consideration plays a major role in the selection of deoxidizers as well as in the estimation of residual content of the deoxidizers in steel at the end of deoxidation.

Zirconium being the most powerful deoxidizer is used in alloy steelmaking. The extent of deoxidation achieved by 8 wt% Si can be easily obtained by 0.7 wt% B or 0.1% Ti or 0.002% Al or 0.00003% Zr [7]. The deoxidizing power of a number of elements has been shown in Fig. 8.2 in terms of activities of oxygen and the deoxidizers (added elements). Activity instead of composition is preferred because the oxygen content in alloy steels will be strongly affected by the presence of alloying elements in the steel melt. Hence, the knowledge of the interaction coefficients [10] of all the elements present in the melt with oxygen and the deoxidizing element is essential for a precise estimation of the oxygen content of the bath on addition of the deoxidizer. Under normal conditions, when the steel melt is deoxidized by a strong deoxidizer like Al the concentration of oxygen is calculated by the activity of Al only after its equilibrium is reached. However, the case is not so simple when more than one deoxidizer is used.

8.3.2 Complex Deoxidizers

Deoxidation carried out by only one element is called "Simple Deoxidation" whereas addition of more than one element in the steel bath for this purpose is

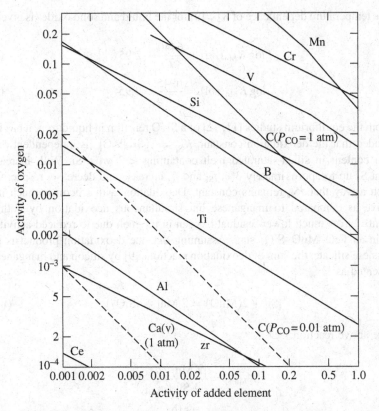

Fig. 8.2 Deoxidizing power of different elements at 1600 °C (Reproduced from J. F. Elliott [8] with the permission of Association for Iron & Steel Technology)

known as "Complex Deoxidation," Si + Mn, Si + Mn + Al, Ca + Si, and Ca + Si + Al are important complex deoxidizers. The use of complex deoxidizers may result in a different reaction product from that obtained when only one element is used. The resulting oxide products may form a solution in which the activities of each of these oxides may be reduced extensively. Lowering of activities increases the deoxidizing power of the element to a degree much above what would be achieved by using one element. Silicon and manganese present the most widely used combination of complex deoxidizer. These are added simultaneously in the form of ferroalloys in the steel bath held in a furnace or ladle. Deoxidation with manganese gives rise to the formation of liquid or solid solution of FeO and MnO, which may be represented as:

$$[\text{Mn}] + (\text{FeO})(\text{s}, \text{l}) = [\text{Fe}](\text{l}) + (\text{MnO})(\text{s}, \text{l}) \tag{8.6}$$

Since FeO and MnO form ideal solutions ($a_i = x_i$) and at infinite dilution $a_{\text{Mn}} = [\%\,\text{Mn}]$ and $a_{\text{Fe}} = 1$.

$$K_{\text{Mn}} = \frac{(x_{\text{MnO}})}{(x_{\text{FeO}})[\%\text{Mn}]} \tag{8.7}$$

The temperature dependence of K_{Mn} [9] for the liquid and solid oxides is given as:

$$\log K_{Mn}(\text{liq}) = \frac{6440}{T} - 2.95 \tag{8.8}$$

$$\log K_{Mn}(\text{sol}) = \frac{6945}{T} - 2.95 \tag{8.9}$$

From the equilibrium studies [11, 12] of a Si–O reaction in liquid iron, it has been concluded that the deoxidation constant, $K'_{Si} = [\%Si].[\%O]^2$ is independent of the silicon content in silica-saturated melt containing < 3 wt% Si. With increasing percent Si up to approximately 3%, f_{Si} and f_O increase and decrease, respectively, in such a way that K_{Si} remains constant. Deoxidation with silicon is much more effective as compared to manganese but simultaneous deoxidation by both the elements leaves much lower residual oxygen in the melt due to reduced activity of SiO_2 in an FeO–MnO–SiO_2 slag. Assuming that the deoxidation product is pure manganese silicate, the sum of deoxidation reactions [9] by silicon and manganese is represented as:

$$[Si] + 2(MnO) = 2[Mn] + (SiO_2) \tag{8.10}$$

For the above reaction:

$$K_{Mn,Si} = \frac{[\%Mn]^2 (a_{SiO_2})}{[\%Si] (a_{MnO}^2)} \tag{8.11}$$

$$\text{and} \quad \log K_{Mn.Si} = \frac{1510}{T} + 1.27 \text{ (Ref.9)} \tag{8.12}$$

The results based on such calculations are shown in Fig. 8.3. The figure highlights the role of manganese in enhancing the deoxidizing power of silicon although the effect decreases with increasing silicon content. From the figure, it is clear that the residual oxygen in the melt containing 0.05% Si decreases from 0.022 to 0.015 wt% when the manganese content is increased from zero to 0.8%. However, when the silicon content increases to 0.15% a similar increase in manganese in the melt reduces the residual oxygen from 0.013 to 0.009 wt%.

From the above equilibrium data, the amount of Si/Mn to be added in a steel melt to get the desired level of deoxidation can be calculated. The deoxidation product, solid silica, is formed above a critical ratio [9] of [%Si]/[%Mn] at a particular temperature. The critical Si and Mn contents of steel in equilibrium with silica-saturated $MnSiO_3$ slag after deoxidation at different temperatures have been presented in Fig. 8.4. From the figure, it is evident that at all temperatures for the metal compositions lying above the curve, manganese does not take part in the deoxidation reaction and solid silica is formed. On the other hand, for the metal composition lying below the curve, the deoxidation product is liquid manganese silicate whose composition is controlled by the ratio [%Si]/[%Mn] in the metal.

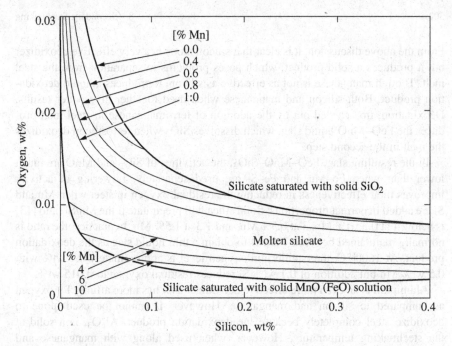

Fig. 8.3 Simultaneous deoxidation of steel by silicon and manganese at 1600 °C (Reproduced from J. F. Elliott [8] and E..T. Turkdogan [9] with the permission of Association for Iron & Steel Technology)

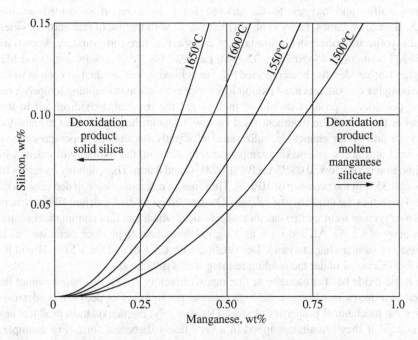

Fig. 8.4 Critical silicon and manganese contents of steel in equilibrium with manganese silicate slag after deoxidation at different temperatures (Reproduced from E. T. Turkdogan [9] with the permission of Association for Iron & Steel Technology)

From the above discussion, it is clear that silicon alone is a very effective deoxidizer but it produces a solid product, which poses problems in separation from the steel melt. Though manganese is not as effective as silicon, it produces a liquid deoxidation product. Both silicon and manganese when used together give better results. Deoxidation, first carried out by the addition of ferromanganese in steel melt produces the FeO–MnO liquid slag, which dissolves SiO_2 when ferrosilicon deoxidizes the melt in the second step.

In the resulting slag, FeO–MnO–SiO_2, the activities of SiO_2 and MnO are much lower than when Fe–Mn and Fe–Si are used separately. Lowering of activity improves their effectiveness in reducing the residual oxygen in steel when Mn and Si are added in correct proportions. Solid silica will precipitate if the Mn/Si ratio [13] is below 4 at 0.3 wt% Mn, 3 at 0.5% Mn, and 1.3 at 1.5% Mn. In practice, the ratio is normally maintained between 7 and 4 to obtain a thin liquid slag as the deoxidation product. At 1600 °C, the equilibrium oxygen level is approximately 0.1 wt% with 0.5 wt% Mn but addition of 0.1 wt% Si reduces residual oxygen to 0.015 wt%.

Aluminum is an even more effective deoxidizer as it has more affinity for oxygen as compared to silicon and manganese. However, it cannot be used alone to deoxidize steel completely because the deoxidation product, Al_2O_3, is a solid at the steelmaking temperature. However, when used along with manganese and silicon, alumina will dissolve in the liquid slag product of deoxidation. In order to deoxidize steel to a very low level of residual oxygen complex, deoxidizers containing alkaline earth [14] elements are used. The rare earth elements or alloys based on them are employed in conjunction with common deoxidizers to lower down sulfur and oxygen to the desired level. In a report submitted by the U.S. Bureau of Mines, Leary et al. [14] have dealt with the use of rare earth silicides and cryolite in molten steel. Similarly, a commercial rare earth mixture, known as "REM" containing 48–50% Ce, 32–34% La, 13–14% Nd, 4–5% Ps, and 0.6–1.6% higher lanthanides, has been reported [9]. For achieving low residual oxygen in steel, the complex deoxidizers must exhibit low vapor pressure at the refining temperature. The deoxidation product should be molten and the residual deoxidizer left in the steel at the end of the treatment must be low. Calcium having stronger affinity for oxygen and sulfur cannot deoxidize steel efficiently because its vapor pressure is above 1 atm at the steelmaking temperature. In addition, the solubility of calcium in liquid iron is very low: 0.032% Ca [9] at 1600 °C and 4 atm. The solubility increases to about 0.35% in the presence of 10% Si. Thus, use of calcium silicide in deoxidation of steel improves the efficiency of calcium. On dissolution in steel, calcium silicide reacts with oxygen to form molten calcium silicate slag, which can flux alumina inclusions. An alloy of Ca, Si, Al, and Ba is also a good deoxidizer to produce clean steel as it possesses similar characteristics. Deoxidation with Ca + Si and Ca + Si + Al will be further discussed under the heading refining with synthetic slag.

Boric oxide has the capacity to flux many refractory oxides but boron cannot be used as a deoxidizer because the high residual boron left after deoxidation deteriorates the mechanical properties of steel. Occasionally, the deoxidation products are beneficial if they remain entrapped in a very finely dispersed form. For example, very fine dispersion of Al_2O_3 particles without coagulation provides the possible nucleation sites during solidification of steel resulting in a very fine grain structure. Addition of zirconium prevents segregation of sulfides in free cutting steels. In order to avoid intergranular fracture, the dissolved nitrogen in steel is fixed up as harmless nitrides by the use of zirconium and titanium.

8.3.3 Vacuum Deoxidation

Carbon can serve as a very strong deoxidizer if the partial pressure of CO in the melting unit is reduced by applying vacuum or blowing inert gas. It is important to note that CO line in Fig. 8.2 goes below the line for aluminum if p_{CO} is reduced from 1 atm to 0.001 atm. Deoxidation of a steel bath in this manner lowers the oxygen concentration in the steel to such an extent that the bath reacts with the deoxidation products, slag, and refractories, and hence the effectiveness of vacuum deoxidation is limited. Advantages of lowering p_{CO} by either argon bubbling or vacuum treatment are made use of in decarburization of stainless steel in argon oxygen decarburization (AOD) and vacuum oxygen decarburization (VOD) processes. This will be discussed in the next section.

8.3.4 Deoxidation Practice

On the industrial scale, there are three methods of deoxidation. After refining, molten steel can be deoxidized either inside the furnace, called furnace deoxidation or during tapping in a ladle, called ladle deoxidation. For production of fine-grained steel or when inadequate deoxidation is required, a small part of total deoxidation is carried out in ingot molds, known as ingot deoxidation.

As deoxidation lowers the oxidizing potential of the bath there is a fair chance of reversion of the refining reactions if oxidized refining slag is present in contact with the metal. Stable oxides like SiO_2 and MnO are not prone to reversion in acid steelmaking processes. However, P_2O_5 in basic steelmaking is very easily reduced from the slag to the metal phase on drop of the oxygen potential. Hence, deoxidation can be completed inside the acid furnace in the presence of the refining slag in contact with the metal bath. However, in basic processes, the refining slag containing P_2O_5 has to be removed prior to deoxidation of the bath inside the furnace. Alternatively, deoxidation can be carried out in the ladle. In general, the refining slag is flushed off in basic processes and deoxidation may be carried out partly in the furnace and for a major part in the ladle. As products of deoxidation in the furnace get more time to reach the surface of the bath, furnace deoxidation is useful in the production of clean steel. In recent years, ladle furnaces equipped with heating and purging facilities are employed for effective deoxidation and efficient removal of deoxidation products.

Since the amount of oxygen dissolved in steel is related to the carbon content, three different types of ingots are teemed. Liquid steel containing less than 0.15% carbon contains enough dissolved oxygen. This steel is not deoxidized at all because the dissolved oxygen ensures rimming action during solidification. Since the dissolved oxygen reduces with increase of carbon content, steel containing 0.15–0.30% carbon requires partial deoxidation to produce semi-killed or balanced types of ingots. The high carbon steel containing more than 0.3% carbon is fully deoxidized or killed. In this context it is important to note that alloy steels are fully killed to get high recovery of alloying elements. From the recovery and loss point of view, alloying elements may be classified into four groups. Recovery of elements like Zn, Cd, and Pb is very poor due to their high vapor pressure at the steelmaking temperature and hence cannot be easily added into the steel melt. Elements like Cu, Ni, Co, Mo, and W can be recovered up to almost 100% because they do not get oxidized, and hence can be added at any stage of steelmaking. A good recovery of P, Mn, and Cr is possible only with proper care since they are partially oxidized during steelmaking, and hence are added in a partially deoxidized bath. Highly oxidizable elements like Al, Si, Ti, and Zr are added only after adequate deoxidation of the bath. As some alloying elements are lost during deoxidation, the bath recovery of such elements is poor.

8.3.5   Removal of Deoxidation Products

There are two different types of problems in deoxidation of steels. The first one is related to dissolution of deoxidizer(s) into molten bath, chemical reaction between the deoxidizing element(s) and the dissolved oxygen, and the nucleation and growth of the deoxidation product(s). The second one is concerned with the elimination of the deoxidation product(s) from the bath. The process with respect to the first aspect is fast. The latter, which requires further growth of the product(s) by coalescence and removal by flotation, is slow, and hence, rate limiting. Thus, the kinetics of removal of the deoxidation product(s) from the bath becomes more important from the viewpoint of production of cleaner steel as compared to the kinetics of deoxidation reactions. The mechanism of precipitation deoxidation needs to be understood in steps like (i) faster dissolution and homogenization of deoxidizer(s) in the bath in order to facilitate quicker oxide formation, (ii) formation of critical nuclei of the deoxidation product, (iii) growth of the deoxidation product, and (iv) removal of the product by flotation to improve cleanliness.

The mechanically entrapped oxide products in steel are called nonmetallic inclusions, which deteriorate the mechanical properties. The size, shape, distribution, and chemical composition of inclusions contribute effectively in controlling the properties of steel. This makes it essential to remove the deoxidation products from the steel melt to get clean steel. Thus, from a cleanliness point of view, a gaseous product of deoxidation would be most appropriate. Only carbon produces a gaseous deoxidation product under reduced pressure according to the reaction (8.1). As this reaction is not very favorable under standard conditions, and economics do not permit vacuum treatment of steel on a large scale, carbon cannot be used as a deoxidizer for the production of clean steel. Further, nucleation and growth of CO bubble have to be considered under vacuum in the light of interfacial phenomena discussed in Chap. 6. Deoxidizers other than carbon form liquid or solid products.

Formation of a solid deoxidation product will give rise to a new phase that will grow during the course of deoxidation and has to rise to the surface of the melt for elimination. Otherwise it will disperse in the melt and on solidification may be entrapped in steel as nonmetallic inclusions. For nucleation and growth of deoxidation products, the required interface may be provided by inhomogenities, for example, the formation of Al_2O_3/steel interface while deoxidizing steel with aluminum at the beginning. Final deoxidation may be carried by other deoxidizers. The rate of rise of the deoxidation product (v) in a quiet bath may be estimated from Stoke's law [15]:

$$v = \frac{2}{9} g r^2 \left(\frac{\rho_{liq} - \rho_{dp}}{\eta} \right) \tag{8.13}$$

where g, r, ρ_{liq}, ρ_{dp}, and η stand respectively, for acceleration due to gravity, radius of the deoxidation product, densities of the liquid metal and the deoxidation product, and the viscosity of the liquid metal. Bodsworth and Bell [16] have shown that r^2 factor plays an important role in controlling the time required for the particles to rise to the surface of the metallic bath. On the basis of Stoke's law, it can be demonstrated that particles of deoxidation product less than 0.001 cm radius will not move to the surface of the metallic bath in a usual ladle within the normal holding time of 20 min [13], whereas larger particles (radius > 0.01 cm) are completely eliminated. These figures emphasize the significance of coalescence of deoxidation products in the formation of particles of larger radii to facilitate rapid rise to the surface of the steel melt. Since coalescence of the deoxidation product is more likely in liquid state, deoxidation is often carried out to obtain liquid products. The liquid melt is never in still condition. Due to the presence of convection currents of appreciable strength, deoxidation products (particles) follow a zigzag path instead of moving vertically upward according to Stoke's law. During the process of zigzag movement, the particles grow in size due to coalescence and are finally eliminated. The rate of removal is also affected by the interfacial energy between the liquid metal and the deoxidation product. Wetting of the product by the melt in case of low interfacial energy will lower the rate of rise of the particle (hence removal) due to the dragging effect of liquid metal on the particles. Contrary to this, high interfacial energy will enhance the rate of removal of the product by lowering the dragging effect.

Turbulence caused by blowing inert gas increases the rate of flotation of solid particles (nonmetallic inclusions). Thus, IGP from the bottom of the ladle is beneficial in producing cleaner steel. The extent of deoxidation increases with increase of the gas flow rate. However, there is an optimum flow rate beyond which the deoxidation effect decreases. This may be due to the fact that excessive stirring brings back the floated oxide inclusions into the melt and also causes oxidation by air. Hence, the net deoxidation obtained is a balance between the rate of removal of oxide inclusions and the rate of oxidation.

8.4   Stainless Steelmaking

Ultralow carbon steels containing 12–30% chromium are classified as stainless steels. Depending upon the need, other alloying elements like nickel, copper, molybdenum, titanium, nitrogen, and so on may be added to develop/improve the desired mechanical properties. By and large, stainless steels contain 17–18 wt% chromium. Nickel increases the corrosion resistance and stabilizes the austenitic structure. Most widely used austenitic stainless steel contains 18% chromium, 8% nickel, and about 0.03% carbon. The next in use is the ferritic variety. Thus, it appears simple and straightforward to manufacture stainless steel by either adding chromium while decarburizing plain carbon steel bath or by melting stainless steel scrap and adjusting the composition. However, simultaneous control of chromium and carbon at the desired level encounters certain difficulties. Firstly, the main source of chromium, ferrochrome (50–70% Cr) containing varying amounts of carbon is available in three grades: low carbon (0.1% C), medium carbon (2% C), and high carbon (7% C) ferrochrome. Although high carbon grade is the cheapest, its use increases the carbon content of stainless steel. Use of the costly low carbon ferrochrome increases the cost of production. On the other hand, use of stainless steel scrap in the manufacture of stainless steel was discouraged earlier because it was not possible to melt the scrap in an EAF without carbon pick-up due to carbon electrodes. Hence, there was a huge stock of stainless steel lying as wasteful material. The economics of recovery of valuable alloying elements (Cr, Ni, etc.) present in stainless steel encouraged the steel industries to develop new processes of stainless steelmaking. This has been possible by understanding the physical chemistry of simultaneous oxidation of carbon and chromium at different temperatures.

8.4.1   Physicochemical Principles

The problem of increasing the chromium content of steel along with decarburization of the bath to the desired specifications was analyzed by comparing the relative stabilities of Cr_2O_3 and CO in the Ellingham diagram. Figure 5.1 shows that the lines corresponding to the formation of Cr_2O_3 and CO according to the reactions:

$$\frac{4}{3}Cr\,(s) + O_2(g) = \frac{2}{3}Cr_2O_3(s) \tag{8.14}$$

$$\text{and} \quad 2C(s) + O_2(g) = 2CO(g) \tag{8.15}$$

intersect at 1220 °C. Hence, the two reactions are in equilibrium at this temperature and the free energy change for the reaction:

$$2\,Cr(s) + 3CO(g) = Cr_2O_3(s) + 3C(s) \tag{8.16}$$

is zero at 1220 °C. Above this temperature, CO is more stable as compared to Cr_2O_3. Thus, carbon can be oxidized in preference to chromium at above 1220 °C. However, in practice, preferential oxidation of carbon in the Fe–Cr–Ni–C melt containing 10% Cr and 0.05% C occurs at above 1800 °C when $p_{CO} = 1$ atm. Even higher temperature is required for preferential oxidation of carbon with increasing percentage of chromium in the melt. This restricts the initial chromium and carbon in the bath while melting stainless steel scrap. Hence, the required chromium has to be made up by charging costly low carbon ferrochrome in the finishing stage. Since high temperature melting causes severe corrosion of the refractory lining, attempts were made to develop processes that can be carried out at lower temperatures with higher amount of stainless steel scrap and cheaper high carbon ferrochrome in the charge.

In the light of the above discussion, it may be concluded that the entire operation of stainless steelmaking is centered on the Fe–Cr–C system dealing with the prime objective to achieve the desired concentration of Cr and C in the bath by injecting oxygen. As chromium exhibits two valences, there is a lot of controversy over the forms of oxides present in the slag. Although divalent and trivalent ions (Cr^{2+}, Cr^{3+}) in oxidized states suggest the formation of CrO and Cr_2O_3, there is also evidence for the existence of Cr_3O_4 in oxide melts. Thus, in the Fe–Cr–C–O system, liquid metal coexists with liquid or solid oxide phases and a gas phase (mainly CO in a gaseous mixture). The chemical equilibria concerning the preferential oxidation of carbon from Fe–Cr alloys can be discussed in terms of reduction of Cr_3O_4, $FeCr_2O_4$ (chromite ore fed in the charge) and distorted spinnel ($Fe_{0.67}Cr_{2.33}O_4$). While discussing reduction of these oxides with carbon, Hilty and Kaveney [17] have reported a simplified equation relating [% Cr] and [% C] in the melt as a function of temperature at 1 atm pressure:

$$\log \frac{[\%Cr]}{[\%C]} = -\frac{13800}{T} + 8.76 \tag{8.17}$$

Considering the effect of p_{CO} he has further modified the above equation [17] as:

$$\log \frac{[\%Cr]}{[\%C]} = -\frac{13800}{T} + 8.76 - 0.925 \log p_{CO} \tag{8.18}$$

Equation 8.17 and the experimental data presented in Fig. 8.5 suggest that at any [% Cr] in the melt, temperature has to be increased to lower down carbon. Equation

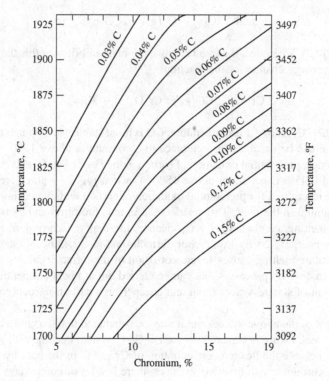

Fig. 8.5 Effect of temperature on chromium and carbon contents of oxygen saturated steel melt (Reproduced from D. C. Hilty and T. F. Kaveney [17] with the permission of Association for Iron & Steel Technology)

8.18 and Fig. 8.6 demonstrate that the same Cr and C level in the melt can be achieved at much lower temperature by lowering the partial pressure of carbon monoxide (p_{CO}) in the converter by blowing a mixture of oxygen and argon. From Fig. 8.5, it is evident that very high temperature is required for decarburization below 0.04% C in the melt containing 15% Cr when $p_{CO} = 1$ atm but the same can be achieved at much lower temperature when p_{CO} is reduced (Fig. 8.6). Toward the end of the operation, ferrosilicon (Fe–Si) is added to reduce chromium oxide held in the slag during the oxidizing period. Chromium transfer from the slag to the metallic bath improves the economics of the process, thermodynamic explanation for which is given below:

$$(Cr_2O_3) + \frac{3}{2}[Si] = 2[Cr] + \frac{3}{2}(SiO_2) \tag{8.19}$$

$$K = \frac{[a_{Cr}^2] \cdot \left(a_{SiO_2}^{3/2}\right)}{(a_{Cr_2O_3}) \cdot \left[a_{Si}^{3/2}\right]} = \frac{f_{Cr}^2 \cdot [\%Cr]^2 \cdot \left(a_{SiO_2}^{3/2}\right)}{(a_{Cr_2O_3}) \cdot f_{Si}^{3/2} \cdot [\%Si]^{3/2}} \tag{8.20}$$

Fig. 8.6 Effect of temperature and pressure on chromium and carbon contents of oxygen saturated steel melt (Reproduced from D. C. Hilty and T. F. Kaveney [17] with the permission of Association for Iron & Steel Technology)

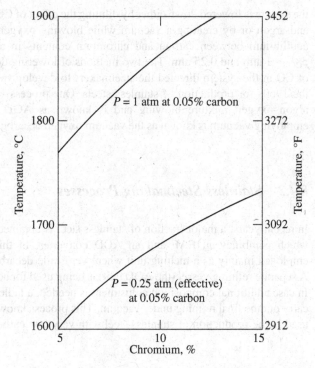

On rearranging the terms, we get:

$$\frac{(a_{Cr_2O_3})}{[\%Cr]^2} = \frac{f_{Cr}^2}{f_{Si}^{3/2}} \cdot \frac{1}{K} \cdot \frac{(a_{SiO_2}^{3/2})}{[\%Si]^{3/2}} \tag{8.21}$$

By approximation, the above equation may be simplified as:

$$\frac{(\%Cr)}{[\%Cr]} = K' \cdot \frac{(a_{SiO_2})}{[\%Si]} \tag{8.22}$$

As the activity of silica is extremely low in basic slags the above equation suggests that Cr can be easily transferred from the slag to the metallic bath by reducing Cr_2O_3 with ferrosilicon or aluminum.

According to the Le Chatelier principle, reaction (8.1) is more favorable in the forward direction with decrease of pressure. Hence, more carbon will get oxidized (i.e., decarburization or reduction in carbon content in steel) if the partial pressure of the system is lowered down either by diluting the effect of CO in a mixture of oxygen and argon or by creating a vacuum while blowing oxygen. Figure 8.6 shows the equilibrium between carbon and chromium contents in a stainless steel melt at $p_{CO} = 1$ atm and 0.25 atm. The two methods of lowering down the partial pressure of CO in the system directed the steelmakers to develop two different processes in the 1960s for production of stainless steels. One process was based on the use of argon–oxygen mixture blowing and is known as AOD and the other process employing vacuum is known as the vacuum oxygen decarburization (VOD) process.

8.4.2   Stainless Steelmaking Processes

In recent years, a major fraction of stainless steel is obtained by the duplex process, which combines an EAF and an AOD converter. In this process, the EAF is employed mainly as a melting unit where very little decarburization is carried out. A separate refining vessel (the AOD unit) is being used for adequate decarburization. In case additional composition adjustment is needed, a ladle furnace is employed to carry out the final refining under vacuum. This process known as Triplex Refining is used in the production of stainless steel with very low carbon and nitrogen.

8.4.2.1   AOD Process

The AOD converters are lined with basic magnesite bricks. The molten charge from the EAF is transferred into the converter and blown with oxygen or mixture of argon and oxygen from the top through a supersonic lance similar to the one used in LD converters. Blowing starts with an oxygen/argon mixture in the ratio of 3:1 when initial carbon content of the bath is high. As decarburization progresses, carbon decreases and the proportion of oxygen is gradually reduced to maintain an oxygen/argon ratio of 1:3 in the final part of the blow. CO evolved is burnt by the supersonic top lancing as in the case of the BOF process. This minimizes the proportion of CO in the flue gas and thus helps in raising the bath temperature. By the end of the first stage blow, the bath temperature rises to about 1700 °C, which is brought down to the required level by addition of appropriate amount of coolants, such as nickel, stainless steel scrap, and high carbon ferrochrome. At this stage, ferrosilicon and aluminum are added to reduce the chromium oxide held in the slag and thus chromium is transferred into the metallic phase. In case ultralow sulfur stainless steel is required, the first oxidizing slag is drained off and a fresh reducing slag is made under argon stirring. The entire operation takes about 2 h per heat and life of the refractory lining is about 80 heats.

8.4.2.2 VOD Process

The VOD unit consists of a vacuum tank, a ladle furnace with argon stirring, and a cover with lance. The charge is first melted in an EAF and transferred to the VOD vessel when carbon is reduced to 0.7–0.8%. After achieving the appropriate order of vacuum, oxygen blow from the top and argon bubbling from the bottom of the ladle are started. Argon bubbling helps in starting early decarburization. It is necessary as otherwise decarburization reaction is delayed due to lack of turbulence. Carbon is lowered to 0.02% at 15–18% chromium concentration by this technique at a temperature of 1600 °C. Toward the end, vacuum is broken and the bath is deoxidized with ferrosilicon and aluminum. This process can lower down carbon to about 0.02% with 15–18% chromium in the steel melt, and is hence very useful in the production of ultra-low carbon steel as well.

8.4.2.3 Direct Stainless Steelmaking Processes

When stainless steel is produced by reduction smelting of chromite ore with hot metal and ferroalloys, the technique is categorized as direct stainless steelmaking process. The technology is particularly useful in countries where stainless steel scrap and ferrochrome are either expensive or not available. In this context, the process developed by Kawasaki Steel Corporation, Japan, is worth mentioning. The hot metal is first dephosphorized in a ladle and then charged into a combined-blown converter. This is followed by charging of chromite ore, ferroalloy, and coke. The charge mixture is reduction smelted. The process is popularly known as SR–KCB (Smelting Reduction–Kawasaki Combined Blowing). For adjustment of carbon composition and alloy additions, the VOD vessel is finally made use of. In recent years, the following processes [18] have been developed for stainless steelmaking:

1. Metal Refining Process (MRP) based on OBM practice working under near equilibrium conditions was developed by Mannesmann Demag, Germany.
2. Cruesot-Loire-Uddeholm (CLU) Process jointly developed by Uddeholm, Belzium, and Cruesot-Loire, France, use a mixture of oxygen, steam, argon, and nitrogen for bottom blow and a mixture of oxygen, argon, and nitrogen for top blow.
3. K–BOP process developed by Kawasaki Steel Corporation, Japan, has been modified by Voest Alpine Industrie-Anlageneinbau, Austria and renamed as K-OBM-S.

8.5   Injection Metallurgy (IM)

Injection technique has been extremely useful in the production of ultralow sulfur steel. In this technique, the powder of a strong desulfurizing reagent along with a carrier inert gas is injected in the refined molten steel kept in a transfer ladle. The large interfacial area of contact of the fine powder brings out very efficient desulfurization in a short time of 8–10 min. Continuous injection may be carried out either by injecting reagent powder into molten steel along with argon flow through a lance immersed from top or by feeding reagent powder encased in a steel tube immersed in molten steel. Extensive desulfurization as well as deoxidation along with modification of inclusions are possible by addition of Ca–Si alloy into the steel melt to obtain the product with the desirable properties. IM has been successful in giving the required calcium treatment to line-pipe steel employed for transporting natural gas over long distances in arctic regions where pipes have to withstand high pressure, H_2S corrosion, and sub-zero temperature. Calcium boils at 1484 °C and its vapor pressure at the steelmaking temperature of 1600 °C is approximately 1.92 atm. It is alloyed with silicon to lower its vapor pressure so that it can be used as an effective desulfurizer (in Ca–Si alloy, if $a_{Ca} = 0.15$, $p_{Ca} = p_{Ca}^o \cdot a_{Ca} = 1.92 \times 0.15 = 0.288$ atm). Hence, injection of Ca–Si alloy eliminates the possibility of vapor formation. The efficiency of desulfurization can be further improved by injecting Ca–Si alloy powder at the deepest possible level in the melt. In this way, the problem of vaporization from the bath surface is avoided. Alternatively, Ca–Si powder encased in a steel tube is fed continuously into the melt by a machine. Increased surface area of the injected powder gives better desulfurization whereas feeding through the tube modifies inclusions.

By and large, under the current practice, the desulfurizing reagent is injected along with a strong deoxidizer to effectively carry out both desulfurization and deoxidation. It is important to note that steel melt must be deoxidized completely for effective desulfurization. In general, desulfurizing agents are also very effective deoxidizers, for example, Ca, Mg, and so on. The presence of a low or high amount of silicon and manganese in the bath does not effectively deoxidize the melts. Hence, the bath has to be deoxidized by at least aluminum (if not by any other strong deoxidizer). Aluminum in the form of a big block, or cube, or wire is placed deep in the bath in such a way that it travels in solid state well below the bath surface before melting and reacts with the dissolved oxygen. Though aluminum is costly, it is more effective when introduced in wire form.

When steel is deoxidized with aluminum, the deoxidation product, Al_2O_3, gets dispersed in the melt as nonmetallic inclusions and poses problems in continuous casting by forming alumina streaks in the product, which deteriorate the mechanical properties. In a similar manner the desulfurization product, CaS, may also remain in suspension thereby making nonclean/dirty steel. The problem caused by the dispersion/suspension of Al_2O_3 and CaS along with other inclusions may be reduced by introducing calcium or a Ca–Al wire in the melt at the end of desulfurization. This treatment changes the morphology of alumina–sulfide inclusions from an acicular shape to a globular shape, which does not cause any problem in rolling.

As nonmetallic inclusions (mostly oxides and some sulfides) in steels deteriorate the mechanical properties and corrosion resistance, considerable efforts have been made in recent years [1, 3, 5] to produce clean (inclusion-free) steels. In order to control/check inclusions generating from (i) precipitation deoxidation due to reaction between the deoxidizer and the dissolved oxygen, (ii) erosion of refractory lining, and (iii) slag particle entrapment, the following measures should be taken:

(a) Transfer of slag containing high FeO and MnO from the furnace to the ladle must be minimized.
(b) Liquid deoxidation products coalesce faster and hence removal becomes easier by rapid flotation of bigger particles. Use of complex deoxidizers is beneficial in this regard.
(c) Zigzag movement created by gas stirring is helpful in coalescence and flotation of the deoxidation products. However, very high gas flow is not advisable from a cleanliness viewpoint.

Since it is not possible to produce completely or nearly inclusion-free steel every attempt is made to modify the nucleation of inclusions to make them harmless by changing their shape and morphology. This is achieved by injecting calcium in molten steel. As the presence of dissolved gases in steel causes the formation of subsurface pinholes in continuous casting, the oxygen content in the melt is minimized by killing (deoxidizing) steel with aluminum. The oxidation product, Al_2O_3, causes nozzle clogging in continuous casting due to sticking of Al_2O_3 inclusion to the inner wall of the casting nozzle. This problem can be avoided by calcium treatment at the final stage in a ladle or tundish because it gives rise to the formation of a low-melting deoxidation product consisting of CaO and Al_2O_3, may be some SiO_2 also. As a powerful desulfurizer, calcium forms CaS that generates a liquid CaS–MnS solution with MnS. However, formation of low-melting compounds, $12CaO \cdot 7Al_2O_3$ and $3CaO \cdot Al_2O_3$ (melting points: 1455 and 1535 °C) are preferred.

A group of 14 elements with identical chemical properties, popularly known as rare earths (RE) are strong desulfurizers as well as deoxidizers. They are available as REM [9] containing about 50% Ce. REM can modify inclusions and also refine grains. However, they are not widely used like calcium due to the problem of process control. Addition of selenium and tellurium improves the machinability of steels containing sulfur by modifying inclusions in globular shape, which can be easily deformed during hot working.

8.6 Refining with Synthetic Slag

Although sulfur present in coke joins the hot metal, it is effectively removed during reduction smelting of the iron ore in the blast furnace by creating conditions (like basic and reducing slag containing high CaO and <1% FeO) favorable to desulfurization. The resulting hot metal still containing more than 0.05% sulfur is subjected to pretreatment in a ladle before charging into steelmaking units. The treatment known as external desulfurization has already been discussed in Sect. 7.3. Similarly, about 85–90% phosphorus present in iron ore joins the hot metal after smelting. If phosphorus content of the hot metal is more than the normal value, it may be removed by external dephosphorization. However, for the production of ultralow sulfur steel containing nearly 0.001% S, hot metal has to be further desulfurized by secondary refining using synthetic slag. In Sect. 7.2.1, for an ionic reaction (7.2): [S] + (O^{2-}) = [O] + (S^{2-}), the desulfurizing index (partition or distribution ratio), D_S, which is the ratio of amount of sulfur in slag to that in the metal has been expressed as:

$$D_S = \frac{(x_{S^{2-}})}{[\%S]} = \frac{K \cdot (x_{O^{2-}}) \cdot [f_S]}{[f_O \cdot \%O]} = K' \frac{(x_{O^{2-}})}{[\%O]}$$

As $(x_{O^{2-}}) \propto (\%CaO)$ assuming $(x_{S^{2-}}) = (\%S)$ at a fixed $(\%CaO)$ we can write:

$$D_S = \frac{(\%S)}{[\%S]} \propto \frac{1}{[\%O]} \propto \frac{1}{(\%FeO)}$$

For a fixed FeO content of the slag: $\frac{(\%S)}{[\%S]} \propto (x_{O^{2-}}) \propto (\%CaO)$

It has already been discussed in Sect. 7.2 that CaO is the most powerful desulfurizer among all the basic oxides present in iron and steelmaking slags. For better and efficient desulfurization, D_S should have a higher value that can be obtained by higher CaO in the slag and lower oxygen in the metal, that is, [% O]. It is important to note that $(x_{O^{2-}})$ increases with increase of $(\%CaO)$ in the slag, and $(\%FeO)$ in slag decreases with decrease of [%O] in the metal bath.

In the production of high-quality fine-grained steels containing less than 20 ppm [19] of oxygen and sulfur, calcium aluminate slags play specific roles in ladle deoxidation and desulfurization. The isotropic mechanical properties of steel are ensured by the formation of evenly dispersed fine globules of sulfide inclusions during solidification. These objectives can be achieved by a fine dispersion of tiny particles of liquid calcium aluminate in steel during the ladle treatment. However, the presence of alumina inclusions corresponding to the concentration of oxygen in the range of 15–25 ppm is objectionable in many applications. For example, tool life is shortened by the presence of abrasive alumina in the steel. For longer life and better machinability, the oxide inclusions should be dispersed in the form of finely divided particles of calcium aluminate, silicate, or aluminosilicate. For production of high-quality clean steel (steel with minimum nonmetallic inclusions), refining in ladles can be controlled in a better manner if the exact equilibrium state of reactions between liquid steel and calcium aluminate based slags is known.

Ozturk and Turkdogan [19] have discussed the theoretical and practical aspects of ladle oxidation and desulfurization of steel in the production of clean steel. They have reported that the desulfurization index (distribution or partition ratio), D_S, for the lime-saturated calcium aluminate slag is about 100 times [20] larger than that for the aluminate-saturated slag. It is important to note that the aluminate slag analyzing $50CaO–50Al_2O_3$, generally employed in ladle desulfurization has only average desulfurization capacity. Since alumina is generated when steel is deoxidized with aluminum, a lime-saturated slag having $(\%CaO)/(\%Al_2O_3)$ a ratio of 1.5 can be obtained by the addition of an appropriate amount of dry burnt lime in the commercially available pre-fused calcium aluminate. For such a lime-saturated slag in equilibrium with the deoxidized steel containing approximately 0.04% Al (dissolved) at approximately 1600 °C, the desulfurization index is approximately 1500. At the end of the ladle refining, if the concentration of sulfur in the lime-saturated slag reaches to 1% (about half of the S saturation level), the equilibrium residual sulfur content in the steel would be about 7 ppm. However, in order to achieve such a low residual sulfur, the plant trials demonstrated that the aluminate slag and the deoxidized steel must be thoroughly mixed by stirring the melt by argon injection, preferably by employing a porous plug at the bottom of the ladle.

There is a vital difference between pretreatment and secondary refining. The former is carried out in a ladle without any provision of bottom stirring whereas in the latter case the ladle is fitted with a bottom porous plug for argon purging. Further, in secondary refining, a lance is inserted in the melt from the top for gas stirring and injection of desulfurizing agent. Thus, in the latter case, the ladle is very similar to that used in IM, which is employed in secondary steelmaking for extensive desulfurization to produce ultralow sulfur steel. Alternatively, for general usage, treatment with synthetic slag on top of the molten steel in the ladle furnace brings out satisfactory desulfurization. In synthetic slag refining, some CaF_2, SiO_2, and Al_2O_3 are also added along with the major additives, CaO and Al. The efficiency of desulfurization can be improved by a slag consisting of CaO, Al_2O_3, and SiO_2 as the major components. Aluminum not only deoxidizes the melt but also favors desulfurization by forming a synthetic slag ($CaO–Al_2O_3–SiO_2$). Aluminum being a powerful deoxidizer helps to achieve low oxygen level in steel. The overall reaction can be summarized as:

$$(CaO) + [S] + \frac{2}{3}[Al] = (CaS) + \frac{1}{3}(Al_2O_3) \tag{8.23}$$

$$K = \frac{(a_{CaS}).(a_{Al_2O_3})^{1/3}}{(a_{CaO}).[a_S].[a_{Al}]^{2/3}} \tag{8.24}$$

Assuming Henrian behavior in the metallic bath and approximating $(a_{CaS}) = (\%S)$, we can write:

$$K = \frac{(\%S).(a_{Al_2O_3})^{1/3}}{(a_{CaO}).[f_S.\%S].[f_{Al}.\%Al]^{2/3}} \tag{8.25}$$

The desulfurization index (partition or distribution ratio) can be expressed as:

$$D_S = \frac{(\%S)}{[\%S]} = \frac{K(a_{CaO}) \cdot f_S \cdot f_{Al}^{2/3} \cdot [\%Al]^{2/3}}{(a_{Al_2O_3})^{1/3}} \qquad (8.26)$$

Hence, $D_S \propto \frac{(a_{CaO}) \cdot [\%Al]^{2/3}}{(a_{Al_2O_3})^{1/3}}$

Since (a_{CaS}) increases and $(a_{Al_2O_3})$ decreases with increase of wt% CaO in the slag, D_S increases with increase of wt% CaO in the slag and wt% Al in molten steel. The effect of temperature on K for the reaction (8.23) has been expressed by the following relation [19]:

$$\log K = \frac{16680}{T} - 4.965 \qquad (8.27)$$

From the expression, it is evident that desulfurization is favored at lower temperature because K (directly related to the value of desulfurization index, D_S) increases with decrease of temperature.

At the given temperature and concentration of aluminum dissolved in iron, D_S increases with increase of CaO in the slag. The D_S for the lime saturated slag was found to be two orders [19] of magnitude larger than that for the aluminate saturated melt. At a given temperature and CaO/Al$_2$O$_3$ ratio in the aluminate slag, the D_S increases with the two-third power of the concentration of aluminum in steel. However, it increases by a factor of approximately 4 on decrease of temperature from 1650 to 1500 °C.

8.7 Vacuum Degassing

During steelmaking, impurities like carbon, silicon, manganese, and phosphorus are eliminated but oxygen, nitrogen, and hydrogen dissolve in the steel melt. Due to atomic dissolution, the solubility of these diatomic gases in iron/steel is directly proportional to the square root of the partial pressure of the gas in equilibrium with the melt. It also depends on the composition of steel and temperature. The solubility of nitrogen and hydrogen in liquid iron is much higher than in solid iron. This large difference in solubility is of great significance in the production of sound ingots and castings. As solubility decreases with decrease of temperature, excess gases dissolved in steel are liberated during solidification. As a result, the concentration of the gas in the liquid near the solid/liquid interface increases and rapid build-up of such concentrations may form gas bubbles that give rise to the formation of skin or pinholes, blowholes, pipes, and so on in castings. The unsoundness caused by these cavities affects the mechanical properties of steel except when the gas evolution is controlled to produce ingots of semi-killed or rimming steels. Since solubility of nitrogen as well as hydrogen is more in gamma iron [21] as compared to that in delta

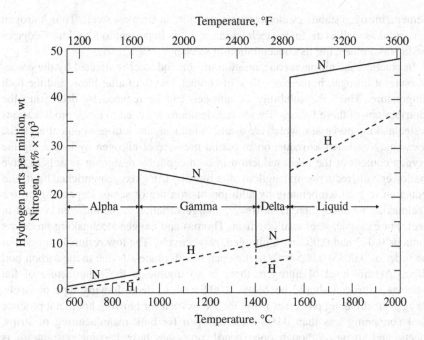

Fig. 8.7 Effect of temperature on the solubility of nitrogen and hydrogen in pure iron at 1 atm pressure of each gas (Reproduced from J. Chipman [21] with the permission of Association for Iron & Steel Technology)

iron (Fig. 8.7), steels forming the gamma phase [8] during solidification will have lesser tendency to form bubbles of hydrogen, or possibly nitrogen, than steels that solidify to form the delta phase. Steels containing more than 0.5% C or sufficient Mn/Ni belong to the former category. Vacuum degassing methods are employed for removal of dissolved oxygen, nitrogen, and hydrogen as well as for the production of clean steel. Removal of oxygen has been discussed in Sect. 8.3 under the heading deoxidation. In this section, emphasis is given to the problems concerning the solubility of nitrogen and hydrogen in steel and their removal.

8.7.1 Nitrogen in Iron and Steel

As steelmaking is carried out in open atmosphere, nitrogen gets a chance to enter in liquid steel. It may also be picked up from the raw material charged in the furnace and from the nitrogen coming in contact with the bath during melting and/or refining. In refined steel, the nitrogen content depends on the steelmaking process employed. Nitrogen is well known to cause yield-point phenomena and is hence responsible for stretcher-strain formation in deep-drawing and pressing operations. In cast steels, the presence of nitrogen as AlN causes intergranular fracture. On the other hand, AlN precipitation in wrought iron is helpful in restricting the grain growth. As an alloying element, nitrogen stabilizes the austenitic structure in stainless steels. Thus, nitrogen is harmful as well as useful in steelmaking and it is important to have knowledge of the factors influencing its concentration in steels.

In practice, the nitrogen concentration in iron and steel is affected by the partial pressure of nitrogen in the blast, time of contact, length of after blow, and the bath temperature. Thus, the solubility of nitrogen can be reduced by decreasing the contributions of these factors. These considerations have led to many modifications in steelmaking processes, which are useful in reducing the nitrogen content in steels. These may be listed as reduction in partial pressure of nitrogen by increasing the oxygen content of the blast, reduction in bath depth by designing vessels to have shallower bath, reduction in length of after blow by using oxygen-enriched blast, and reduction in bath temperature by addition of iron ore or scrap. Depending on the steelmaking practice, there is, however, a large variation in the nitrogen content in steel. For example, steel resulting from Thomas and oxygen steelmaking processes contains 0.022 and 0.002 wt% nitrogen, respectively. The low residual nitrogen of the order of 0.0035–0.005 wt% in the open hearth process is due to the carbon boil effect. At this level of nitrogen, there is no trouble in the manufacture of flat products. Nitrogen should be below 0.002 wt% in deep drawing type of steels. Oxygen steelmaking processes using 99.5% oxygen for blowing hot metal produce steel containing less than 0.002 wt% nitrogen for bulk manufacturing of strips, sheets, and so on. Although open hearth processes have become obsolete, it is interesting to note that low nitrogen steels can be produced by effective carbon boil. This phenomenon is also useful in reducing nitrogen in EAF steelmaking. CO bubbles provide flushing action and help in nitrogen removal during the boil.

If the nitrogen content of bubbles is assumed to reach equilibrium with the nitrogen in the steel melt, the nitrogen removed during any period of carbon elimination can be calculated. The equilibrium partial pressure of nitrogen in the bubbles is related to the solubility of nitrogen in steel, which can be explained by considering its atomic dissolution in the following manner:

$$\frac{1}{2}N_2 = [N] \tag{8.28}$$

$$K = \frac{[a_N]}{\sqrt{p_{N_2}}} \tag{8.29}$$

Since solubility of nitrogen is small, according to Henry's law, activity of the solute is proportional to concentration, that is, $[a_i] \propto [\%i]$ or $[a_i] = f_i [\%i]$. At infinitely dilute concentration, $f_i = 1$, hence $[a_i] = [\%i]$.

$$\therefore K = \frac{f_N \cdot [\%N]}{\sqrt{p_{N_2}}} \tag{8.30}$$

At 1600 °C and 1 atm pressure of nitrogen gas, 0.045% nitrogen dissolves in iron melt. At this low concentration, $f_N = 1$.

Thus, $K = 0.045$ (from Eq. 8.30)

$$\therefore p_{N_2} = \left[\frac{\%N}{0.045}\right]^2 \tag{8.31}$$

If the total pressure in the bubble is 1 atm, the volume fraction of N_2 (v_{N_2}) and CO (v_{CO}) in the bubble is in the volume ratio of

$$\left[\frac{\%N}{0.045}\right]^2 : 1 - \left[\frac{\%N}{0.045}\right]^2$$

$$\text{since} \left[\frac{\%N}{0.045}\right]^2 \ll 1$$

$$\left[\frac{\%N}{0.045}\right]^2 : 1$$

As 1 g mol of all gases occupy the same volume, N_2 and CO are in the weight ratio

$$\left[\frac{\%N}{0.045}\right]^2 \text{ mol wt } N_2 : \text{mol wt CO}$$

and the weight percent of nitrogen and carbon removed is in the ratio $\left[\frac{\%N}{0.045}\right]^2 28 : 12$

If the rate of C drop is $d[\%C]/dt$, the rate of nitrogen drop [13] will be

$$\frac{d[\%N]}{dt} = \frac{28}{12}\left[\frac{\%N}{0.045}\right]^2 \frac{d[\%C]}{dt} \tag{8.32}$$

But the rate of nitrogen removal estimated from the observed rate of carbon drop based on Eq. 8.32 gives rate 5–10 times higher than the observed values. This suggests that nitrogen within the bubbles may not be in equilibrium with the molten steel as assumed in the above derivation.

The solubility of nitrogen at 1 atm pressure of nitrogen in equilibrium with iron is shown in Fig. 8.7. There is slow rise in the solubility of nitrogen in solid iron with increasing temperature but at the melting point it increases very rapidly. It also rises in liquid steel but at a slow rate. Extensive studies have demonstrated that the solubility of nitrogen in iron is significantly influenced by alloying elements. The solubility [8, 22] increases in the presence of vanadium, niobium (columbium), tantalum, chromium, and manganese whereas it decreases due to carbon, silicon, nickel, cobalt, copper, and tungsten (Fig. 8.8). The effect of alloying elements (j) on the activity coefficient of nitrogen [8, 22] in liquid iron at 1 atm pressure and 1600 °C is presented in Fig. 8.9. Making use of the data in the figure, the interaction coefficient of nitrogen for the effect of the alloying elements (j) can be estimated.

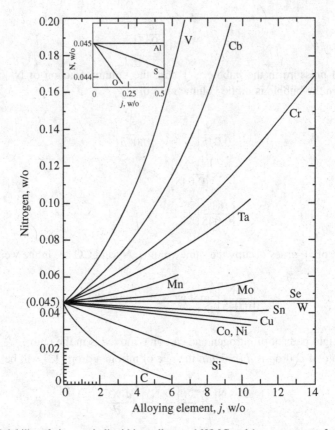

Fig. 8.8 Solubility of nitrogen in liquid iron alloys at 1600 °C and 1 atm pressure of nitrogen gas (Reproduced from J. F. Elliott [8] with the permission of Association for Iron & Steel Technology. Originally from R. D. Pehlke and J. F. Elliott [22] with the permission of The Minerals, Metals & Materials Society)

Thus, the solubility of nitrogen in alloy steels can be calculated by combining the binary data in the following equation:

$$\log f_N^j = e_N^N \cdot \%N + e_N^V \cdot \%V + e_N^{Cr} \cdot \%Cr + e_N^{Mn} \cdot \%Mn + \\ e_N^W \cdot \%W + e_N^{Ni} \cdot \%Ni + e_N^{Co} \cdot \%Co + e_N^{Si} \cdot \%Si + e_N^C \cdot \%C \tag{8.33}$$

8.7.2   Hydrogen in Iron and Steel

Though completely unwanted, hydrogen somehow enters into all steels. Without any increase in strength it decreases ductility of steels at all concentrations. On increasing the concentrations beyond 1.5 ml/100 g of steel, the appearance of hairline cracks seriously affects the mechanical properties of fully killed high-alloy steels. However, there is no such problem in rimming or semi-killed steels because the scavenging

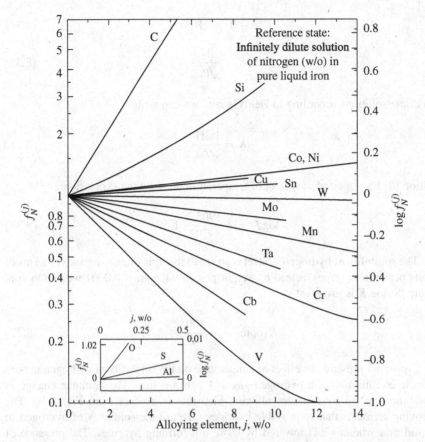

Fig. 8.9 Effect of alloying elements on the activity coefficient of nitrogen in liquid iron at 1600 °C (Reproduced from J. F. Elliott [8] with the permission of Association for Iron & Steel Technology. Originally from R. D. Pehlke and J. F. Elliott [22] with the permission of The Minerals, Metals & Materials Society)

action of gases reduces the hydrogen content below the safe level. At a concentration exceeding the solid solubility limit, hydrogen is rejected on solidification, which frequently results in ingot porosity and unsoundness, that is, the formation of blowholes and pinholes. In steelmaking, the primary source of hydrogen is moisture, which may come from slaked lime, moist air, wet scrap, wet or partially dried refractories (channels and runners), wet molds, and rusty charge. Hence, the best way to overcome problems with hydrogen would be to avoid charging wet materials in the furnace or prevent such materials coming into contact with liquid steel. A vigorous and prolonged boil can eliminate hydrogen to a major extent but vacuum treatment is the most effective method of reducing hydrogen from steel.

In iron, hydrogen dissolves atomically according to the reaction:

$$\frac{1}{2}H_2 = [H] \tag{8.34}$$

$$K = \frac{[H]}{\sqrt{p_{H_2}}} \tag{8.35}$$

In dilute solutions according to Henry's law, we can write:

$$K = \frac{[\%H]}{\sqrt{p_{H_2}}} \tag{8.36}$$

Elliott [8] has reported the effect of temperature on K by the expression:

$$\log K = -\frac{1905}{T} - 1.59 \tag{8.37}$$

The solubility of hydrogen in steel is so small that it is often expressed in terms of parts per million (ppm) instead of weight percent (1 ppm = 0.0001 wt%). On ppm scale [8] log K is given as:

$$\log K(\text{ppm}) = -\frac{1905}{T} + 2.41 \tag{8.38}$$

Figure 8.7 presents the effect of temperature on the solubility of hydrogen in pure iron in equilibrium with hydrogen gas at 1 atm pressure. The dramatic change in solubility caused by different alloying elements is evident from Fig. 8.10. The alloying elements that form stable hydrides increase the solubility of hydrogen in liquid iron whereas it is lowered by those not forming hydrides. The presence of niobium (columbium), tantalum, vanadium, titanium, manganese, and chromium increases the solubility whereas the presence of carbon, boron, silicon, aluminum, germanium, tin, copper, and cobalt decreases the solubility [8, 23].

The variation of the activity coefficient of hydrogen [8, 23] in iron with the alloying elements (j) is presented in Fig. 8.11, and the data can be made use of in estimating the interaction coefficients for the effect of alloying additions on the activity coefficient of hydrogen in steel. Thus, the solubility of hydrogen in alloy steels can be calculated from the interaction coefficients by the following equation:

$$\log f_H^j = e_H^H \cdot \%H + e_H^{Ta} \cdot \%Ta + e_H^{Ti} \cdot \%Ti + e_H^{Ni} \cdot \%Ni +$$
$$e_H^C \cdot \%C + e_H^{Si} \cdot \%Si + e_H^{Cr} \cdot \%Cr + e_H^{Co} \cdot \%Co \tag{8.39}$$

Water vapor coming in contact with steel or slag leads to the formation of hydrogen which gets dissolved in the melt as per reaction:

$$H_2O(g) = 2[H] + [O] \tag{8.40}$$

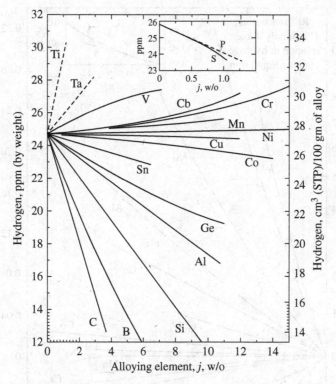

Fig. 8.10 Solubility of hydrogen in liquid iron alloys at 1592 °C and 1 atm pressure of hydrogen gas (Reproduced from J. F. Elliott [8] with the permission of Association for Iron & Steel Technology. Originally from M. Weinstein and J. F. Elliott [23] with the permission of The Minerals, Metals & Materials Society)

Since solubilities of hydrogen and oxygen in steel are very low, we can write:

$$K = \frac{[a_H]^2 \cdot [a_O]}{p_{H_2O}} = \frac{[\%H]^2 [\%O]}{p_{H_2O}} \tag{8.41}$$

On wt% scale, log K has been expressed [13] as:

$$\log K = -\frac{10390}{T} + 7.81 \tag{8.42}$$

Hence, at 1600 °C, $[\%H] = \dfrac{1.35 \times 10^{-3} \cdot p_{H_2O}^{1/2}}{[\%O]^{1/2}} \tag{8.43}$

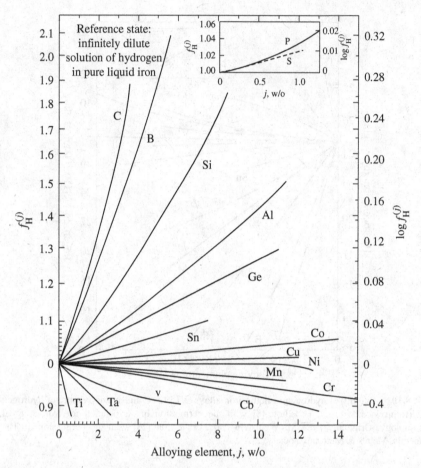

Fig. 8.11 Effect of alloying elements on the activity coefficient of hydrogen in liquid iron at 1592 °C (Reproduced from J. F. Elliott [8] with the permission of Association for Iron & Steel Technology. Originally from M. Weinstein and J. F. Elliott [23] with the permission of The Minerals, Metals & Materials Society)

8.7.3 Vacuum Treatment of Steel

The objectives of vacuum degassing include removal of hydrogen from steel to avoid long annealing treatment, removal of oxygen as carbon monoxide, and production of steels with very low carbon content (< 0.03 wt%). The principle is based on the usefulness of the Sievert's law relationship expressed by Eqs. 8.28 and 8.34. These equations suggest that according to the Le Chatelier principle, reactions can be directed in the reverse direction by subjecting the molten steel to vacuum. Hence, hydrogen, nitrogen, as well as the oxygen contents of steel can be lowered by vacuum treatment according to the reactions:

$$2[H] = H_2(g) \tag{8.44}$$

$$2[N] = N_2(g) \tag{8.45}$$

$$[C] + [O] = CO(g) \tag{8.46}$$

The effectiveness of vacuum treatment increases with increase of the surface area of liquid steel exposed to vacuum. For this purpose, metal is allowed to flow in the form of a thin stream or even fall as droplets to accelerate the degassing process. In general, the vacuum treatment is employed to reduce the hydrogen and oxygen contents of steel. Simultaneously, there is some reduction in nitrogen as well. However, this treatment is not aimed for nitrogen removal alone due to slow kinetics. The extent of removal of nitrogen under vacuum is limited to about 30%. On account of its technical importance and complexity of problems, the thermodynamics of solubility of nitrogen in steel and the kinetics of adsorption and desorption of nitrogen were studied in detail by Pehlke and Elliott [22, 24]. The rate of dissolution of nitrogen in steel is not much affected by alloying elements, such as Al, Nb (Cb), Cr, Ni, Si, and W. However, the surface active solutes, O and S, decrease the rate drastically by blocking the surface sites.

When argon is purged into molten steel, the gas bubbles start rising in the melt and gradually leave. During this movement, dissolved nitrogen and/or hydrogen diffuse(s) into the argon bubbles from the molten steel around. Depending on the rate of diffusion bubbles pick up dissolved gas(es) according to the laws of thermodynamic equilibria before leaving the melt. However, the rate of degassing will basically depend on the rate of purging of the inert gas. The possible steps [24] in transport of nitrogen from the bulk of molten iron/steel to the bulk of the gaseous phase (Fig. 8.12) may be listed as:

(i) Transport of atomic or dissolved nitrogen [N] from the bulk metal to the metal boundary layer (MBL).
(ii) Transport of the dissolved nitrogen [N] through the metal boundary layer (MBL) to the gas boundary layer (GBL).
(iii) Chemical reaction (formation of molecular nitrogen) at the MBL/GBL interface.
(iv) Transport (diffusion) of N_2 molecules in GBL.
(v) Transport of N_2 molecules from GBL to the bulk gas.

It is important to note that in a well-stirred melt, for example, in blowing inert gas in vacuum degassing process, or carbon boil in open hearth, or vigorous stirring in induction melting, step (i) is not likely to be rate controlling. Similarly, steps (iv) and (v) are unlikely to be rate controlling because they are affected by the flow rate of the carrier gas, which is generally very high. Thus, under these conditions either step (ii) or (iii) may be rate controlling.

The rate controlling step (iii), that is, the chemical reaction on the interface leading to the formation of molecular nitrogen is expressed as:

Fig. 8.12 Kinetics of transfer of dissolved nitrogen

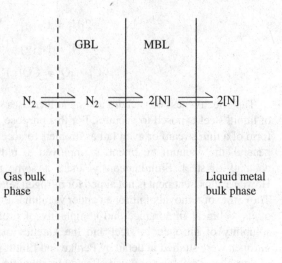

The rate of nitrogen removal is given as:

$$2N \; \overset{\overrightarrow{k}}{\underset{\overleftarrow{k}}{\rightleftharpoons}} \; N_2 \,(g) \tag{8.47}$$

The rate of nitrogen removal is given as:

$$-\frac{d[\%N]}{dt} = \frac{A}{v}\left(k[\%N]^2 - kp_{N_2}\right) \tag{8.48}$$

where A, v, and k are the interfacial area, volume of the melt, and rate constant, respectively. $p_{N_2} = 0$, when carrier gas used is pure argon.

$$\therefore -\frac{d[\%N]}{dt} = \frac{A}{v}\, k \,[\%N]^2 \tag{8.49}$$

$$\text{or} -\frac{d[\%N]}{[\%N]^2} = \frac{A}{v} k \, dt \tag{8.50}$$

On integration from 0 to t for the corresponding nitrogen content: $[\%N]_i$ to $[\%N]$ we get:

$$-\int_{[\%N]_i}^{[\%N]} \frac{d[\%N]}{[\%N]^2} = \frac{A}{v}\, k \int_0^t dt \tag{8.51}$$

$$\text{or} \quad \frac{1}{[\%N]} - \frac{1}{[\%N]_i} = \frac{A}{v}kt \quad (8.52)$$

In Fe–O–N system plot of $\left\{ \frac{1}{[\%N]} - \frac{1}{[\%N]_i} \right\}$ vs. $\frac{A}{v}$ t varies linearly although the slope varies with the concentration of the dissolved oxygen [O]. If Ar–N$_2$ mixture is used as a carrier gas, the partial pressure of nitrogen must be considered. Following important conclusions have been drawn from the above discussions:

1. Rate of desorption is proportional to the square of dissolved nitrogen concentration in the melt.
2. Desorption of nitrogen from liquid iron is a second order reaction.
3. Chemical reaction at the gas–metal interface controls the rate.
4. The decrease in rate of reaction is proportional to the partial pressure of nitrogen above the liquid melt.

The vacuum degassing methods for treatment of steel may be classified into three groups [25]:

8.7.3.1 Ladle Degassing

The teeming ladle filled with steel to one-fourth of its height is placed inside a vacuum chamber that is connected to a pumping system. The upper and lower parts of the ladle are usually lined with high alumina and fire bricks, respectively. In a similar way, the vacuum chamber cover and bottom are also lined to protect them from direct radiation and accidental failures. In order to make degassing effective, the steel melt is stirred either by bubbling argon (Sect. 8.1) or by electromagnetic induction. Gas purging follows just after evacuation begins. Introduction of gas for stirring provides an interface which facilitates degassing and also homogenizes the bath thoroughly. The fall of pressure causes vigorous reaction in the ladle and therefore the bath looks as if it is boiling. In some cases, steel is heated electrically to compensate for heat loss during degassing. A rough estimate is first made to see how much time is required to achieve the level of degassing at a particular vacuum. In general, pumping is carried out to attain the ultimate vacuum of 1–10 mm Hg, which is supposed to be adequate for degassing. There are number of methods under this category with varying combinations of stirring and heating. Ladle degassing is most widely used because it is the simplest of all the degassing methods. Along with degassing other metallurgical operations like deoxidation, decarburization, desulfurization, alloying, and melt homogenization can be carried out under vacuum.

8.7.3.2 Stream Degassing

In this case, molten steel is allowed to flow down under vacuum as a stream from the furnace to ladle to another ladle or a mold. A very high rate of degassing is achieved due to the large increase in surface area of molten steel in the form of falling droplets. In practice, various degassing methods, for example, ladle to mold, ladle to ladle, and so on are used to accomplish degassing by pouring steel from one vessel to another. In every method, degassing takes place when liquid steel droplets are exposed to vacuum. Thus, choice of a proper vacuum pump and vacuum chamber is important to achieve the adequate level of degassing. The chamber must be big enough to house the ladle or the mold(s). Currently, steam degassing is not much in use.

8.7.3.3 Circulation Degassing

Molten steel is circulated in the form of a stream exposed to vacuum. Alternatively, a small portion (10–15%) may be lifted and subjected to vacuum exposure. The former method is known as RH process after Ruhrstahl Heraus and the latter as DH process after Dortmund-Horder and is also known as a lifter degassing process. A brief description of both the methods is given below:

RH Degassing Process The process originally developed by Rheinstahl Heinrich Shutte at Hattingen, Germany, was modified by the Horohito Works of the Fuji Iron and Steel Co., Japan. Figure 8.13 shows the schematic arrangement of the RH degassing unit. A long cylindrical steel vessel lined with alumina is attached with two long legs, called snorkels. There are arrangements at the top for exhaust, alloy addition, and so on. Snorkels lined with high alumina dip into the molten steel contained in a ladle.

The chamber is preheated to 900–1000 °C prior to charging molten steel. The vessel is lifted and lowered to an appropriate level to move the snorkels in the ladle containing the steel. The chamber is evacuated and liquid steel rises in the chamber. The lifter gas is then introduced in the inlet snorkel at the point where the alumina tubular tip is attached to it. The melt is forced to rise in the vacuum chamber (the cylindrical steel vessel) through one snorkel due to reduced pressure and by injecting inert gas. In the chamber, liquid steel gets disintegrated into small droplets and degassed and flows down through another snorkel under gravity. In this way the melt is circulated. The degassed steel is cooler and heavier than that in the ladle and hence it forces the light undegassed steel upward thereby ensuring adequate mixing and homogeneity of the bath. The rate of degassing is controlled by adjusting the vacuum and the rate of flow of the lifter gas. The average rate of circulation is 12 ton min^{-1}. Twenty minutes are required to treat 100 tons of steel to bring a 90% reduction of hydrogen content. It is an efficient degassing process and used extensively.

If w ton of steel melt contained in a vessel is circulated to the vacuum chamber at the rate of r ton min^{-1}, the final (H_f) and the initial (H_{in}) hydrogen contents in the melt, after t min of circulation, are related according to the expression:

$$\ln \left[\frac{H_{in} - H_{eq}}{H_f - H_{eq}} \right] = \frac{r}{w}.t \qquad (8.53)$$

Fig. 8.13 RH degassing unit

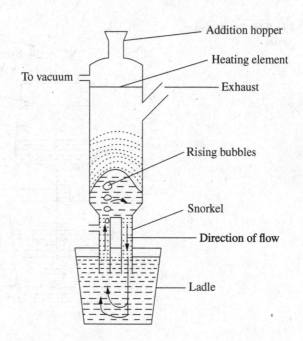

where H_{eq} is the equilibrium hydrogen content in the steel melt under the reduced pressure of p_{H_2} (i.e., the order of vacuum).

In recent years, the RH process has been modified by providing additional facility of oxygen blowing. The process known as RH-OB is useful for the production of ultralow carbon steel. The modern degassing processes are employed for deoxidation, desulfurization, alloying, melt homogenization, and faster removal of inclusions in addition to degassing and decarburization.

DH Degassing Process The process also known as the Lifter Degassing process was developed in Germany. In this process, about 10–15% of the total steel contained in a ladle is lifted and treated at a time under vacuum. The process is repeated a number of times until the required level of degassing is achieved. Figure 8.14 shows the arrangement of the DH degassing unit.

The vacuum chamber has a long leg (snorkel lined from both sides), which dips in the steel pool contained in a ladle. The chamber is lined with fire clay and alumina refractory bricks in the upper and lower parts, respectively. The vessel is provided with an exhaust, a hopper for alloy addition, and a heating device. As the atmospheric pressure would force the molten steel to rise to a height of 1.4 m in the chamber under vacuum, the length of the snorkel must be more than 1.4 m. Before starting the process, the chamber is preheated. The snorkel tip is allowed to dip into

Fig. 8.14 DH degassing
vessel

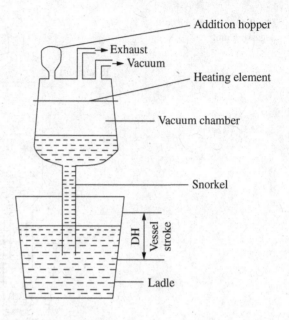

the liquid steel by lowering the preheated chamber in a tapping ladle. By moving the
evacuated chamber up and down, the molten steel is forced to enter the preheated
chamber so as to be exposed to the vacuum. The pattern of mixing is similar to that
of the RH process. The stroke of movement of the DH vessel is about 50–60 cm with
a cycle of 20 s. As only 10–15% of the molten steel is exposed to vacuum at a time,
7–10 cycles are needed to expose the total quantity of steel once. Thus, 20–30 cycles
are required to achieve the appropriate level of degassing in 15–20 min.

If $1/m$ part of the total quantity of steel taken in the DH degassing vessel is forced
to enter in the preheated vacuum chamber, the final hydrogen content H_f and the
initial hydrogen content H_i are related to number of times (n) the vessel is raised and
lowered by the expression [26]:

$$H_f = H_{eq} + \left(\frac{m-1}{m}\right)^n H_i \tag{8.54}$$

where H_{eq} stands for the equilibrium hydrogen content in the steel melt under the
reduced pressure p_{H_2} atm (i.e., the order of vacuum).

8.8   Problems [27–29]

Problem 8.1
Steel is desulfurized with a synthetic slag, calcium aluminate $(CaO–Al_2O_3)$, by adding aluminum. If sulfur in steel is reduced from 0.015% to 0.0025% by equilibrating 100 ton of steel with 1250 kg of slag, calculate the desulfurizing index/partition ratio.

Solution
Amount of sulfur transferred from the steel melt to the slag

$$= 100 \times 1000 \times (0.0150 - 0.0025) \times 10^{-2} = 12.5 \text{ kg}$$

$$\therefore \text{wt\% S in slag } (\%S) = \frac{12.5 \times 100}{1250} = 1.0\%$$

$$\therefore D_S = \frac{(\%S)}{[\%S]} = \frac{1}{0.0025} = 400 \qquad \text{Ans.}$$

Problem 8.2
Calculate the theoretical amount of pure aluminum required to deoxidize 100 ton of steel at 1600 °C, if oxygen has to be reduced from the initial value of 0.03 to 0.001 wt%. What is the deoxidation constant?

Given that: $2[\%Al] + 3[\%O] = Al_2O_3(s)$, $\Delta G^o = -294800 + 94.75 \ T$ cal

Solution
For the reaction: $2[\%Al] + 3[\%O] = Al_2O_3(s)$,   at 1873 K,

$$\Delta G^o = -294800 + 94.75 \times 1873 = -117333 \text{ cal} = -RT \ln K$$

$$K = 4.92 \times 10^{13}$$

$$\text{The equilibrium constant, } K = \frac{(a_{Al_2O_3})}{[a_{Al}^2] \cdot [a_O^3]}$$

As concentration of Al and O in the steel melt is very low we can assume that Henry's law $(a_i = \text{wt\% } i)$ is obeyed by both the solutes, $a_{Al_2O_3} = 1$

$$\therefore K = \frac{1}{[\%Al]^2 \cdot [\%O]^3} = 4.92 \times 10^{13}$$

$$\text{Deoxidation constant, } K' = \frac{1}{K} = 2.03 \times 10^{-14} \qquad \text{Ans.}$$

$$= [\%Al]^2 \cdot [\%O]^3$$

$$\therefore [\%Al]^2 = \frac{2.03 \times 10^{-14}}{[\%O]^3} = \frac{2.03 \times 10^{-14}}{[0.001]^3} = 2.03 \times 10^{-5}$$

$$[\%Al] = 4.50 \times 10^{-3}$$

Residual Al in the melt $= 100 \times 1000 \times \dfrac{4.50 \times 10^{-3}}{100} = 4.50 \;\; \text{kg}$

Amount of Al consumed (reacted) in removal of oxygen

$$= \frac{(0.03 - 0.001)}{100} \times 100 \times 1000 \times \frac{54}{48} = 32.625 \;\; \text{kg}$$

Total amount of Al required $= 4.50 + 32.625 = 37.125 \text{ kg}$ Ans.

Problem 8.3 (Elliott, J. F., 1980–1981, Personal Communication, Massachusetts Institute of Technology, Cambridge)

The scrap charge to a steelmaking system usually contains elements like aluminum, copper, nickel, and manganese. Some of these metals will remain principally in the bath at the end of the refining operation, some will be distributed between the slag and the bath and some will go completely into the slag.

(i) Set a criterion by which you may be able to determine into which of these categories a metal may fall. Assume that the bath of steel at the end of the heat and before deoxidation contains 0.1% C at 1600 °C. Make sure that your criterion accounts in some simple way for the oxidizing strength of the slag.
(ii) Rank the four metals mentioned above according to the three categories. Explain the reasoning behind your ranking.

Given that at 1600 °C

$$[\%C] + [\%O] = CO(g), \Delta G^o = -23100 \text{ cal}$$

$$Fe(l) + [\%O] = FeO(l), \Delta G^o = -5500 \text{ cal}$$

$$2[\%Al] + 3[\%O] = Al_2O_3(s), \Delta G^o = -101400 \text{ cal}$$

$$2[\%Cu] + [\%O] = Cu_2O(l), \Delta G^o = 40600 \text{ cal}$$

$$[\%Ni] + [\%O] = NiO(s), \Delta G^o = 30700 \text{ cal}$$

$$[\%Mn] + [\%O] = MnO(s), \Delta G^o = -12400 \text{ cal}$$

Solution

Considering the formation of oxides per atom of oxygen, the free energy of formation is listed below:

Oxide	ΔG^o (cal)
CO	−23,100
FeO	−5500
$Al_{2/3}O$	−33,800
Cu_2O	+40,600
NiO	+30,700
MnO	−12,400

These oxides are reduced by the dissolved carbon, [C] to form CO gas. In general, the reduction of oxides may be expressed as follows:

$$M_xO + [C] = xM + CO(g) \tag{1}$$

$$\log K = \log \left[\frac{a_M^x \cdot p_{CO}}{a_{M_xO} \cdot a_C} \right]$$

$$\text{or } \log a_{M_xO} = -\log K + \log p_{CO} + x \log a_M - \log a_C \tag{2}$$

Considering, 0.05% M in the metallic bath, $a_M = [\%M] = 0.05$, $\log(0.05) = -1.3$, $p_{CO} = 1$ atm, $a_C = [\%C] = 0.1$, Eq. 2 may be expressed as:

$$\log a_{M_xO} = -\log K + \log(1) + x \log(0.05) - \log(0.1)$$
$$= -\log K - 1.3x + 1.0 \tag{3}$$

$$Al_{2/3}O + [C] = 2/3\,Al(l) + CO(g) \tag{4}$$

$$\Delta G_r^o = \Delta G_{CO}^o - \Delta G_{Al_{2/3}O}^o = -23100 - (-33800) = 10700 \text{ cal}$$
$$= -RT\ln K = -4.575\,T\log K$$

$\log K = -\frac{10700}{4.575 \times 1873} = -1.25$, since for the reaction (4), we have $p_{CO} = 1$ atm, $a_{Al} = [\%Al] = 0.05$, $a_C = 0.1$, $x = 2/3$, from Eq. 3, we can write:

$$\therefore \log a_{Al_{2/3}O} = -\log K + 2/3 \log a_{Al} + 1.0 = -\log K + 2/3 \log(0.05) + 1.0$$
$$= -(-1.25) + 2/3(-1.3) + 1.0 = 1.3833$$

$$\therefore \log a_{Al_2O_3} = 3 \times 1.3833 = 4.15$$

Thus, $a_{Al_2O_3} = 14125$

$$Cu_2O + [C] = 2Cu + CO(g) \tag{5}$$

$$\Delta G_r^o = \Delta G_{CO}^o - \Delta G_{Cu_2O}^o = -23100 - 40600 = -63700 \text{ cal}$$
$$= -4.575 \times 1873 \times \log K$$
$$\therefore \log K = 7.434$$

For Eq. 5, we have: $p_{CO} = 1$ atm, $a_{Cu} = [\%Cu] = 0.05$, $a_C = 0.1$, $x = 2$. Hence, from Eq. 3, we can write:

$$\log a_{Cu_2O} = - \log K + 2(-1.3) + 1.0$$
$$= - 7.434 - 2.6 + 1.0 = -9.03$$
$$a_{Cu_2O} = 9.25 \times 10^{-10}$$

Similarly, for the reaction: $NiO + [O] = Ni + CO(g)$, $a_{NiO} = 2.64 \times 10^{-7}$ and for the reaction: $MnO + [O] = Mn + CO(g)$, $a_{MnO} = 0.028$.

Thus, the result, based on the criterion of activity of oxides in the slag, may be summarized as follows:

M_xO_y	$\log K$	$\log a_{M_xO_y}$	$a_{M_xO_y}$	M distributed in
Al_2O_3	−1.25	4.15	14,125	Slag
Cu_2O	7.43	−9.03	9.25×10^{-10}	Metal
NiO	6.28	−6.58	2.64×10^{-7}	Metal
MnO	1.25	−1.55	0.028	Slag/metal

From the values of activity of different oxides, it may be concluded that the metal will join the slag or the metallic bath. For example, aluminum present in the charge will get oxidized immediately and the oxide product, Al_2O_3, which attains a very high activity will enter in the slag phase. Extremely low values of activities of Cu_2O and NiO suggest that there is hardly any chance of oxidation of copper and nickel. Hence, both Cu and Ni will remain in the metallic phase, that is, pig iron. $a_{MnO} = 0.028$, suggests that Mn will get distributed partially in the metallic bath and partially in the slag.

Problem 8.4
180 tons of molten steel containing 0.2 wt% oxygen is deoxidized in an open hearth furnace at 1627 °C by adding 900 kg of ferrosilicon (analyzing 50 wt% Si and balance Fe). Assuming that Henry's law is obeyed for the solutes, calculate the equilibrium constant and the silicon and oxygen contents of the melt after deoxidation.

Given that:

$$[Si](in Fe) + 2[O](in Fe) = SiO_2(s), \Delta G^o = -142000 + 54.97T \text{ cal}$$

Solution

Since Henry's Law is obeyed for both the solutes (i.e., for Si and O)

$a_{Si} = [\%Si]$, $a_O = [\%O]$, SiO_2 being solid, $a_{SiO_2} = 1$

$$K = \frac{a_{SiO_2}}{[\%Si] \times [\%O]^2} = \frac{1}{[\%Si] \times [\%O]^2}$$

$$\Delta G^o = -RT \ln K = -142000 + 54.97\, T = -142000 + 54.97 \times 1900$$
$$= -37557$$
$$K = 20923$$

$\because$ 100 kg steel contains 0.2 kg oxygen

$\therefore$ Total oxygen in the steel melt before deoxidation $= \dfrac{0,2 \times 180 \times 1000}{100} = 360$ kg.

Amount of Si required $= \dfrac{28}{32} \times 360 = 315$ kg, for the reaction:

$$[Si] + 2[O] = (SiO_2)$$

but Si added $= \dfrac{900}{2} = 450$ kg (since ferrosilicon contains 50% Si)

$\therefore$ Excess Si (added) going into the melt as Si $= 450 - 315 = 135$ kg.

Total iron $= 180 \times 1000$ kg $+ 450$ kg (from ferrosilicon) $= 180450$ kg.

$$[\%Si] = \frac{135 \times 100}{180450 + 135} = 0.075 = a_{Si}(\text{in Fe}-\text{Si}-\text{O after deoxidation})$$

$$K = 20923 = \frac{1}{[\%Si][\%O]^2} = \frac{1}{0.075[\%O]^2}$$

$$[\%O]^2 = \frac{1}{0.075 \times 20923} = 0.0006373$$

$$\therefore [\%O] = 0.025244$$

and $[\%Si] = 0.075$

and equilibrium constant $= 20923$ \quad Ans.

Problem 8.5

Calculate the amount of ferrosilicon per ton of steel containing 0.3% C for deoxidation at 1600 °C to obtain a final oxygen content of 0.006%. Ferrosilicon contains 80% Si.

Given that:

$$CO(g) = [C] + [O], \quad K = 2.4 \times 10^{-3} \text{ and}$$

$$SiO_2(s) = [Si] + 2[O], \quad K = 2.82 \times 10^{-5}$$

$$e_O^C = -0.13, e_O^O = -0.2, e_{Si}^{Si} = 0.029, e_{Si}^C = 0.24$$

How much ferrosilicon will be required in deoxidation of 200 tons of steel?

Solution

The initial oxygen content in steel can be obtained by considering oxygen–carbon equilibrium. Since concentration of oxygen and carbon in steel is very low (i.e., solution is very dilute), we can assume the validity of Henry's law for dissolution of oxygen as well as carbon.

$$\text{Hence,} \quad a_O = [\%O] \quad \text{and} \quad a_C = [\%C] = 0.3$$

$$\text{for the reaction } CO(g) = [C] + [O]$$

$$K = \frac{a_C.a_O}{p_{CO}} = \frac{0.3 \times [\%O]}{1} = 0.0024$$

$$\therefore [\%O] = \frac{0.0024}{0.3} = 0.008\%$$

Consider Fe–O–C system

$$\log f_O = e_O^O.\%O + e_O^C.\%C$$

$$= (-0.2) \times 0.008 + (-0.13) \times 0.3 = -0.0406$$

$$\therefore f_O = 0.91$$

Considering deviation from Henrian behavior:

$$a_O = f_O.[\%O]$$

$$\therefore [\%O] = \frac{a_O}{f_O} = \frac{0.008}{0.91} = 0.009 \text{ (initial oxygen in steel before deoxidation)}$$

Thus, oxygen removed by deoxidation $= 0.009 - 0.006 = 0.003$

$$SiO_2(s) = [Si] + 2[O]$$

$$K = \frac{a_{Si}.a_O^2}{a_{SiO_2}} = 2.82 \times 10^{-5} (a_{SiO_2} = 1, \ SiO_2 \text{ being a solid})$$

$$\therefore a_{Si}.a_O^2 = 2.82 \times 10^{-5}$$

$$(a_O = f_O[\%O] = 0.91 \times 0.006 = 0.0055)$$

$$a_{Si} = \frac{2.82 \times 10^{-5}}{a_O^2} = \frac{2.82 \times 10^{-5}}{(0.0055)^2} = 0.932 = [\%Si]$$

Now consider Fe–C–Si system

$$\log f_{Si} = e_{Si}^{Si}.\%Si + e_{Si}^{C}.\%C = 0.029 \times 0.932 + 0.24 \times 0.3 = 0.099$$
$$\therefore f_{Si} = 1.256$$
$$[\%Si] = \frac{a_{Si}}{f_{Si}} = \frac{0.932}{1.256} = 0.7442 (\text{after deoxidation})$$

Change in [O] (i.e. [%O] removed) $= 0.009 - 0.006 = 0.003$

Corresponding amount of Si, removed as $SiO_2 = 0.003 \times \dfrac{28}{32} = 0.00263$

Amount of ferrosilicon added per ton of steel

$$= \left[\frac{0.00263}{100} \times \frac{1000}{0.8}\right] \text{ for } SiO_2 + \left[\frac{0.7442}{100} \times \frac{1000}{0.8}\right] \text{Si residual}$$
$$= 0.033 + 9.303 = 9.306 \text{ kg per ton of steel Ans.}$$

Amount of ferrosilicon required for 200 tons of steel

$$= 9.306 \times 200 = 1861.2 \text{ kg Ans.}$$

Problem 8.6

Calculate the residual oxygen content of liquid iron containing 0.10% Si in equilibrium with solid silica at 1600 °C.

Given that: $e_{Si}^{Si} = 0.32, e_{O}^{O} = -0.20, e_{Si}^{O} = -0.24, e_{O}^{Si} = -0.14$

$$Si(l) + O_2(g) = SiO_2(s), \quad \Delta G_1^o = -226500 + 47.50\,T \text{ cal}$$
$$O_2(g) = 2[O](\% \text{ in Fe}), \quad \Delta G_2^o = -55800 - 1.46\,T \text{ cal}$$
$$Si(l) = [Si](\% \text{ in Fe}), \quad \Delta G_3^o = -28500 - 6.1\,T \text{ cal}$$

Solution

In order to get the residual oxygen content of the liquid iron, one has to consider the formation of pure silica from the dissolved [Si] and [O] in liquid iron.

$$[Si] + 2[O] = SiO_2(s)$$
$$\Delta G_r^o = \Delta G_1^o - \Delta G_2^o - \Delta G_3^o = -142200 + 55.06\,T \text{ cal} \tag{1}$$
$$\Delta G_{1873}^o = -39073 \text{ cal} = -RT \ln K$$

$$K = \frac{(a_{SiO_2})}{[a_{Si}] \cdot [a_O^2]} \quad \text{In dilute solutions :} \quad a_i = f_i.\%i$$

$$= \frac{1}{[f_{Si}.\%Si] \cdot [f_O.\%O]^2} \quad \text{since } SiO_2 \text{ is a solid, } a_{SiO_2} = 1 \quad (2)$$

$$\ln K = -\frac{\Delta G_r^o}{RT} = \frac{39073}{1.987 \times 1873} = 10.4987$$

$$K = 36270 \tag{3}$$

From Eqs. 2 and 3, we get

$$\frac{1}{K} = [f_{Si}.\%Si] \cdot [f_O.\%O]^2 = \frac{1}{36270} = 2.76 \times 10^{-5} \tag{4}$$

$$\log f_{Si} = [\%Si].e_{Si}^{Si} + [\%O].e_{Si}^{O} = 0.1 \times 0.32 + [\%O] \times (-0.24)$$
$$= 0.032 - 0.24.[\%O] \tag{5}$$

$$\log f_O = [\%O].e_O^O + [\%Si].e_O^{Si} = [\%O] \times (-0.20) + 0.1 \times (-0.14)$$
$$= -0.20.[\%O] - 0.014 \tag{6}$$

Taking log of Eq. 4, we get:

$$\log f_{Si} + 2\log f_O + \log [\%Si] + 2\log [\%O] = \log (2.76 \times 10^{-5})$$

Substituting the values of $\log f_{Si}$ and $\log f_O$ from Eqs. 5 and 6, respectively, we get:

$$-0.032 - 0.24.[\%O] - 0.40.[\%O] - 0.028 + \log (0.10) + 2\log [\%O] = -4.5595$$
$$\text{or } 2\log [\%O] - 0.64.[\%O] = -4.5595 + 1.0 + 0.06 = -3.4995$$

Solving for [%O], we get [%O] = 0.018 Ans.

Problem 8.7 (Elliott, J. F., 1980–1981, Personal Communication, Massachusetts Institute of Technology, Cambridge)

An Fe–O–C melt contains 0.05 wt% C and 0.04 wt% O at 1600 °C. One kg of silicon metal is added to one ton of the melt. (i) How much SiO_2 is formed (% by weight of metal) as a result of this addition, and what is the final composition of the melt? The temperature is still 1600 °C and wt% C remains unchanged. (ii) The deoxidized melt is then cooled to 1536 °C. How much SiO_2 (wt%) is formed as a result of the change in temperature, and what is the final composition of the melt? The density of liquid iron is 7.2 gm cm^{-3}.

Given that: $SiO_2(s) = [\%Si] + 2[\%O]$, $\log K = -\frac{30400}{T} + 11.58$.

Solution

(i) For the given Fe–O–C melt at 1600 °C, C = 0.05 wt% and O = 0.04 wt%

There is no change in composition on addition of Si at 1600 °C.

$$SiO_2(s) = [\%Si] + 2[\%O], \qquad \log K = -\frac{30400}{1873} + 11.58 = -4.6506 \qquad (1)$$

$$x \qquad x \qquad 2x$$

That is, $28x$ [Si] reacts with $32x$ [O]

Initial wt% Si in the melt = $1/1000 \times 100 = 0.1$

$$\text{wt\% Si in the melt at 't'}_{eq} = 0.1 - 28x \qquad (2)$$

$$\text{wt\% O in the melt at 't'}_{eq} = 0.04 - 32x \qquad (3)$$

For the above reaction, $K = \frac{[\%Si].[\%O]^2}{a_{SiO_2}}$, neglecting the effect of interaction coefficients since [%O] as well as [%C] are very low in concentration.

$a_{SiO_2} = 1$ because SiO_2 is a solid, hence we can write:

$$\log K = -4.6506 = \log\left\{[\%Si].[\%O]^2\right\}$$

$\therefore [\%O]^2 = \frac{2.2354 \times 10^{-5}}{[\%Si]}$, substituting the values of [%O] and [%Si] from Eqs. 2 and 3, we get $[0.04 - 32x]^2 = \frac{2.2354 \times 10^{-5}}{[0.1 - 28x]}$

On simplification we get:

$$-28672x^3 + 174.08x^2 - 0.301x = -1.377 \times 10^{-4}$$

Solving by trial and error, $x = 8.23 \times 10^{-4}$

Hence, the final composition of the melt would be:

$$[\%C] = 5 \times 10^{-2}$$

$$[\%O] = 0.04 - \left(32 \times 8.23 \times 10^{-4}\right) = 1.372 \times 10^{-2}$$

$$[\%Si] = 0.10 - \left(28 \times 8.23 \times 10^{-4}\right) = 7.7 \times 10^{-2}$$

Hence, the final weight of Si in 1000 kg melt = $7.7 \times 10^{-2} \times 10 = 0.77$ kg

Therefore, weight of Si consumed to form $SiO_2 = 1.0 - 0.77 = 0.23$ kg

Amount of SiO_2 formed by 0.23 kg of Si = $\frac{60}{28} \times 0.23 = 0.493$ kg Ans.

(ii) In order to get the amount of SiO_2 formed and to calculate the final composition of the melt cooled to 1536 °C after deoxidation at 1600 °C, consider the equilibrium at 1809 K.

$$\log K = -\frac{30400}{1809} + 11.58 = -5.2249$$
$$\therefore K = 5.96 \times 10^{-6}$$
$$\therefore [\%O]^2 = \frac{5.96 \times 10^{-6}}{[\%Si]} \qquad (4)$$

whereas from part (i): $[\%O] = 0.01372 - 32x$ and

$$[\%Si] = 0.077 - 28x$$

Substituting the values of $[\%O]$ and $[\%Si]$ in Eq. 4, we get:

$$(0.01372 - 32x)^2(0.077 - 28x) = 5.96 \times 10^{-6}$$
$$\text{or} - 28672x^3 + 103.432x^2 - 7.286 \times 10^{-2}x = -8.52 \times 10^{-6}$$

Solving by trial and error, $x = 1.46 \times 10^{-4}$, hence, the final composition of the melt at 1536 °C would be

$$[\%O] = 0.01372 - 32 \times 1.46 \times 10^{-4} = 9.048 \times 10^{-3}$$
$$[\%Si] = 0.077 - 28 \times 1.46 \times 10^{-4} = 7.29 \times 10^{-2}$$
$$[\%C] = 5 \times 10^{-2}$$

$\therefore$ The final weight of Si in 1000 kg melt at 1536 °C $= \frac{7.29 \times 10^{-2}}{100} \times 1000 = 0.729$ kg
Weight of Si in the melt at 1600 °C $= 0.77$ kg
Hence, weight of Si consumed in forming $SiO_2 = 0.77 - 0.729 = 0.041$ kg
$\therefore$ Weight of SiO_2 formed as a result of the change of temperature from 1600 to 1536 °C $= 0.041 \times \frac{60}{28} = 0.088$ kg Ans

Problem 8.8 (Nagamori, M., 1979–1980, Personal Communication, University of Utah, Salt Lake City)

What is the effect of temperature on decarburization limit in the production of stainless steel? Assuming that the slag is saturated with chromite ($FeCr_2O_4$), demonstrate this aspect by calculating the minimum carbon that can be achieved at 1600 °C and 1700 °C from the following data:

$$Fe(s) + 2Cr(s) + 2O_2(g) = FeCr_2O_4(s) \quad \Delta G_1^o = -333550 + 74.2\,T \text{ cal}$$

$$C(s) + 0.5O_2(g) = CO(g) \qquad\qquad \Delta G_2^o = -26700 - 20.95\,T \text{ cal}$$

$$C(s) = [\%C](\text{in Fe}) \qquad\qquad\qquad \Delta G_3^o = 5400 - 10.1\,T \text{ cal}$$

$$Cr(s) = [\%Cr](\text{in Fe}) \qquad\qquad \Delta G_4^o = 5000 - 11.31\,T \text{ cal}$$

$$Fe(s) = Fe(l) \qquad\qquad\qquad\qquad \Delta G_5^o = 3587 - 1.995\,T \text{ cal}$$

How would your answer get modified by accounting the given interaction coefficients?

$$e_C^{Cr} = -0.024 \text{ and } e_C^C = 0.22$$

Solution
The extent of decarburization increases with increase of temperature.

As chromite is in equilibrium with carbon in steel, the following reaction may be considered for stainless steel production:

$$FeCr_2O_4(s) + 4[\%C] = Fe(l) + 2[\%Cr] + 4CO(g)$$

The standard free energy change (ΔG^o) for the above reaction can be obtained by making use of the above five equations:

$$\Delta G^o = \Delta G_5^o + 2\Delta G_4^o + 4\Delta G_2^o - \Delta G_1^o - 4\Delta G_3^o$$
$$= 218737 - 142.2\,T$$

At 1600 °C, $\Delta G^o = 218737 - 142.2 \times 1873 = -47604$ cal

$$K = 3.59 \times 10^5$$

The equilibrium constant, K for the reaction under question is given as:

$$K = \frac{a_{Fe} \cdot [\%Cr]^2 \cdot p_{CO}^4}{a_{FeCr_2O_4} \cdot [\%C]^4} = \frac{0.95 \times 5^2 \times 1}{1 \times [\%C]^4} = 3.59 \times 10^5$$

(assuming $a_{Fe} = 0.95$, $[\%Cr] = 5$, $p_{CO} = 1$ and $a_{FeCr_2O_4} = 1$)

$$[\%C]^4 = \frac{0.95 \times 25}{3.59 \times 10^5} = 6.62 \times 10^{-5}$$

$$\therefore [\%C] = 9.02 \times 10^{-2} = 0.09\% \ \left(\text{Henrian activity}, a_C^H\right)$$

Similarly, at 1973 K, the equilibrium constant, $K = 7.07 \times 10^6$

$$\text{and } [\%C]^4 = \frac{0.95 \times 25}{7.07 \times 10^6} = 3.36 \times 10^{-6}$$

$$\therefore [\%C] = 0.043\% \text{ (Henrian activity)}$$

Thus, we see that lower carbon can be achieved by increasing the temperature from 1600 to 1700 °C. Ans.

At 1600° C the activity coefficient of steel containing 0.09% C and 5% Cr will be

$$\log f_C = e_C^C.[\%C] + e_C^{Cr}.[\%Cr] = 0.22 \times 0.09 + (-0.024) \times 5 = -0.1002$$

$$\therefore f_C = 0.794$$

$$\text{As } a_C^H = f_C.[\%C]_{actual}$$

$$\therefore 0.09 = 0.794 \times [\%C]_{actual}$$

$$\text{or } [\%C]_{actual} = \frac{0.09}{0.794} = 0.1134\%C$$

$$\log f_C = e_C^C.[\%C] + e_C^{Cr}.[\%Cr] = 0.22 \times 0.1134 + (-0.024) \times 5 = 0.0951$$

$$\therefore f_C = 0.8034, \text{consequently,}$$

$$[\%C] = \frac{0.09}{0.8034} = 0.112\% \text{ C } (\approx \text{ the actual carbon, } 0.1134\% \text{ C})$$

Similarly, at 1700 °C, $[\%C]_{actual} = \frac{0.043}{0.794} = 0.054\%$ C

Thus, answer is modified by considering interaction coefficients in the steel melt but higher degree of decarburization is achieved at higher temperature, 1700 °C.

Problem 8.9

Calculate the rate of removal of nitrogen at 1600 °C from a steel melt containing 0.005 wt% N with a carbon boil which removes carbon at the rate of 0.015% per minute.

Solution

We can make use of Eq. 8.32 to calculate the rate of removal of nitrogen from the steel melt.

$$\frac{d[\%N]}{dt} = \frac{28}{12}\left[\frac{\%N}{0.045}\right]^2 \frac{d[\%C]}{dt}$$

$$= \frac{28}{12}\left[\frac{0.005}{0.045}\right]^2 \times 0.015$$

$$= \frac{28}{12} \times 0.012 \times 0.015$$

$$= 4.2 \times 10^{-4} \text{ wt\% per minute Ans.}$$

Problem 8.10

A large amount of liquid steel is processed at 1500 °C under reduced pressure of 130 Pa. Calculate the volume of gas liberated under these conditions when (i) the melt reacts with 14 g mol of water according to the reaction: H_2O (g) = H_2 (g) + [% O], (ii) the melt reacts with 20 kg of slag containing 20% FeO according to the reaction: (FeO) + [%C] = Fe(l) + CO(g).

Solution

(i)
$$P = 130\,Pa, T = 1773\ K$$

$$H_2O\ (g) = H_2\ (g) + [O](g) \tag{1}$$

Since pressure on top of the melt is very low, the reaction (1) will take place in the forward direction until the entire volume of water (14 g mol) is driven out.

14 g mol H_2O will produce 14 g mol of H_2 gas.

$$n = 14\ g\ mol, P = 130\ Nm^{-2}, R = 8.314\ JK^{-1}\ mol^{-1}$$

$$\therefore v = \frac{nRT}{P} = \frac{14 \times 8.314 \times 1773}{130} = 1587.5\ m^3$$

Thus, the volume of hydrogen gas liberated at 1500 °C and 130 Pa = 1587.5 m^3 Ans.

(ii)
$$(FeO) + [\%C] = Fe(l) + CO\ (g) \tag{2}$$

The above reaction will proceed unless entire amount of FeO present in the slag is consumed in the formation of CO gas.

$$\text{Amount of FeO in 20 kg of slag} = 20 \times \frac{20}{100} = 4\ kg$$

From reaction (2), it is clear that (55.85 + 16) g of FeO produces (12 + 16) g of CO gas.

That is, 1 g mol of FeO produces 1 g mol of CO

$$\left(\frac{4 \times 1000}{71.85}\right) \text{g mol of FeO produces} \left(\frac{4 \times 1000}{71.85}\right) \text{g mol of CO gas}$$

$$n_{CO} = \left(\frac{4\,000}{71.85}\right), T = 1773\ K, P = 130\ N\ m^{-2}$$

$$\therefore v = \frac{nRT}{P} = \frac{4000 \times 8.314 \times 1773}{71.85 \times 130} = 6312.6\ m^3$$

Thus, volume of CO gas liberated at 1500 °C and 130 Pa is 6312.6 m^3 Ans.

Problem 8.11

Why circulation method of degassing is preferred over ladle degassing? Justify your answer in the light of the following data:

(i) $\frac{1}{2}H_2(g) = [H](ppm)$, $\Delta G^o = 31973 - 44.36T$ J
(ii) Surface tension of the melt at 1600 °C = 1.6 J m^{-2}
(iii) Surrounding pressure in the melt at the location of the bubble of 10^{-8} m radius is 1.5 atm.

Solution

$$\Delta G^o = 31973 - 44.36 \times 1873 = -51113.3 \; J = -RT\ln K$$

$$\ln K = 3.2824, K = 26.64$$

$$\frac{1}{2}H_2(g) = [H](ppm)$$

$$K = \frac{[H](ppm)}{\sqrt{p_{H_2}}} = \frac{26.64}{\sqrt{1}} = 26.64 \; (p_{H_2} = 1 \text{ atm})$$

$$\therefore [H] \; (ppm) = K = \sqrt{p_{H_2}} = 26.64$$

$$r = 10^{-8} \text{ m, surrounding pressure} = 1.5 \text{ atm}$$

$$p_{H_2} = p_{surr} + \frac{2\sigma}{r} = 1.5 \times 10^5 \text{ Pa} + \frac{2 \times 1.6}{10^{-8}} J \, m^{-2}$$

$$= 1.5 \times 10^5 Pa + 3.2 \times 10^3 \times 10^5 Pa = 10^5(1.5 + 3200)P = 3201.5 \text{ atm}$$

$$\therefore [H](ppm) = 26.64\sqrt{3201.5} = 1507.3 \text{ ppm. Ans.}$$

Thus, we see that solubility of hydrogen in the melt decreases from bottom to the top of the ladle on account of the corresponding decrease of pressure. At depth where surrounding pressure on the bubble is 1.5 atm, solubility is 1507.3 ppm and at the top it is 26.64 ppm (i.e., at one atm pressure). Hence, circulation degassing will be faster and more effective compared to ladle degassing because under circulation the entire steel melt is gradually exposed to vacuum in the form of tiny droplets.

Problem 8.12

The surface tension of liquid iron at 1600 °C is 1.8 J m^{-2}

(i) What is the concentration of hydrogen in the melt when it is in equilibrium with hydrogen gas at 1 atm pressure?
(ii) What would be the concentration of hydrogen in the melt, if instead, it were in equilibrium with bubbles of pure hydrogen of diameter 0.1 μm? The total pressure in the melt at the location of the bubbles is 1 atm. Given that,

$$\frac{1}{2}H_2(g) = [H](ppm), \Delta G^o = 8720 - 11.02 \, T \text{ cal}$$

Solution

$T = 1873\,K, \Delta G^o = 8720 - 11.02 \times 1873 = -11920\,cal = -4.575\,T \log K$

$\log K = 1.3911$

$K = 24.6 = \dfrac{[H](ppm)}{\sqrt{p_{H_2}}}$

(i) $p_{H_2} = 1\,atm$

$\therefore [H](ppm) = K \cdot \sqrt{p_{H_2}} = 24.6$ Ans.

(ii) bubble diameter = 0.1 μm, $r = 0.05 \times 10^{-6}\,m$, 1 atm = 10^5 Pa

$$P_b = P_{surr} + \frac{2\sigma}{r} = 10^5\,Pa + \frac{2 \times 1.8 \ J\,m^{-2}}{0.05 \times 10^{-6}m} = 721 \times 10^5\,Pa = 721\,atm$$

$\therefore [H](ppm) = K \cdot \sqrt{p_{H_2}} = 24.6\sqrt{721} = 661\,ppm$ Ans.

Problem 8.13

With the aid of the following data, justify why circulation degassing is preferred over ladle degassing.

 (i) Surface tension of liquid steel at 1600 °C is 1.6 J m^{-2}
 (ii) $\frac{1}{2}H_2(g) = [H](ppm)$, $\Delta G^o = 31973 - 44.36T$ J
 (iii) Surrounding pressure in the melt at the location of the bubble of 10^{-9} m, 10^{-6} m and 0.05×10^{-6} radii is 2 atm.

Solution

$T = 1873\,K, \Delta G^o = 31973 - 44.36 \times 1873 = -51113.28 \ J = -RT\ln K$

$\ln K = 3.2824, K = 26.64$

For the reaction: $\frac{1}{2}H_2(g) = [H](ppm)$, $K = \frac{[H](ppm)}{\sqrt{p_{H_2}}}$

Since $p_{H_2} = 1\,atm$, $[H](ppm) = K \cdot \sqrt{p_{H_2}} = 26.64$

When bubble radius is 10^{-9} m

$$P_b = P_{surr} + \frac{2\sigma}{r} = 2 \times 10^5\,Pa + \frac{2 \times 1.6 \ J\,m^{-2}}{10^{-9}m} = 32002 \times 10^5 Pa = 32002\,atm$$

$\therefore [H](ppm) = K \cdot \sqrt{p_{H_2}} = 26.64\sqrt{32002} = 4766\,ppm$

Similarly, when $r = 10^{-6}$ m, [H] (ppm) = 155.3 ppm.
and when $r = 0.05 \times 10^{-6}$ m, [H] (ppm) = 675 ppm.

The solubility of hydrogen varies according to the radius of the bubble and the radius depends on the depth of its location. Hence, circulation degassing will be more effective for removal of hydrogen because in this process the entire steel melt is exposed to vacuum in the form of tiny droplets.

Problem 8.14

(i) 200 ton of steel containing 20 ppm of hydrogen is degassed at 1600 °C in RH vessel under vacuum of 100 μm by purging argon gas through tuyeres. If the rate of circulation of molten steel to the vacuum chamber is 16 tons per minute, calculate the final hydrogen content of the steel after 20 min of degassing operation.

Given that, $\frac{1}{2}H_2(g) = [H](ppm)$, $K_H = 24.6$ at 1600 °C

(ii) What would be the rate of circulation to achieve the final hydrogen content of 5 ppm in 15 min?

Derive the necessary expression for the above calculations.

Solution
In order to solve this problem, one should first derive the expression relating the rate of circulation of steel, time of degassing, initial and final hydrogen content in molten steel with the total quantity of steel taken in a ladle for degassing by RH process under the given order of vacuum in the reaction vessel. When molten steel is exposed to vacuum, molecular hydrogen comes out of steel according to the reaction: $[H] = \frac{1}{2}H_2(g)$. The equilibrium constant (K_H) for this reaction can be made use of to calculate the equilibrium hydrogen, $[H]_{eq}$ in steel under the reduced pressure (p_{H_2}), that is, the order of vacuum.

$$K_H = \frac{p_{H_2}^{1/2}}{[H]_{eq}}$$

$$[H]_{eq} = \frac{p_{H_2}^{1/2}}{K_H}$$

Now, consider the hydrogen balance using the following symbols:

Total amount of steel in the ladle $= w$ ton.
Rate of removal of hydrogen from steel $= w_1$ (g min^{-1})
Rate at which hydrogen is transported to the vacuum chamber $= w_2$ (g min^{-1})
Rate of circulation of steel in the vacuum chamber $= r$ (ton min^{-1})

$$\therefore w_1 = w \times 10^6 \times \left[\frac{dH}{dt} \times 10^{-6}\right] = w.\frac{dH}{dt} \tag{1}$$

$$w_2 = r \times 10^6 . \left[H - H_{eq} \right] \times 10^{-6} = r . \left[H - H_{eq} \right] \qquad (2)$$

Equating Eqs. 1 and 2 we get:

$$w . \frac{dH}{dt} = r . \left[H - H_{eq} \right]$$

$$\text{or} \quad \frac{dH}{H - H_{eq}} = \frac{r}{w} . dt \qquad (3)$$

If H_{in} and H_f are the initial and final hydrogen contents of steel, respectively, integration of Eq. 3 between the limits $t = 0$, $H = H_{in}$ and $t = t$, $H = H_f$ gives:

$$\ln \left[\frac{H_{in} - H_{eq}}{H_f - H_{eq}} \right] = \frac{r}{w} . t$$

(i) From the given equilibrium constant, we can calculate the equilibrium hydrogen in steel at $p_{H_2} = 100 \ \mu m = \frac{100 \times 10^{-3}}{760}$ atm. $= \frac{0.1}{760}$ atm

For the reaction: $\frac{1}{2} H_2(g) = [H](ppm)$, $K_H = \frac{H(ppm)_{eq}}{\sqrt{p_{H_2}}}$

$$H(ppm)_{eq} = K_H . \sqrt{p_{H_2}} = 24.6 \sqrt{\frac{0.1}{760}} = 0.2822 \ ppm$$

Since we know that $H_{in} = 20$ ppm, $H_{eq} = 0.2822$ ppm, and $t = 20$ min, we can calculate, H_f by substituting these values in equation: $\ln \left[\frac{H_{in} - H_{eq}}{H_f - H_{eq}} \right] = \frac{r}{w} . t$, derived above.

$$\ln \left[\frac{20 - 0.2822}{H_f - 0.2822} \right] = \frac{16}{200} \times 20 = 1.6$$

$$\therefore \left[\frac{20 - 0,2822}{H_f - 0.2822} \right] = 4.953$$

Hence, $H_f = 4.26$ ppm Ans.

(ii) Similarly, the rate of circulation (r) can be calculated to achieve $H_f = 5$ ppm

$$\ln \left[\frac{20 - 0.2822}{H_f - 0.2822} \right] = \ln \left[\frac{20 - 0.2822}{5 - 0.2822} \right] = 1.43 = \frac{r}{200} \times 15$$

$$r = 19.1 \ ton \ min^{-1} Ans.$$

Problem 8.15

Under similar conditions (as stated in Problem 8.14), repeat the above calculations of part (i) and (ii) for 200 ton of steel containing 0.7% C, 1.5% Mn and 1% Si at 1600 °C incorporating the given values of interaction coefficients.

$$e_H^H = 0, e_H^C = 0.06, e_H^{Mn} = -0.002, e_H^{Si} = 0.027.$$

Solution

$$\log f_H = \%H.e_H^H + \%C.e_H^C + \%Mn.e_H^{Mn} + \%Si.e_H^{Si}$$
$$= 0 + 0.7 \times 0.06 + 1.5 \times (-0.002) + 1 \times 0.027 = 0.066$$
$$\therefore f_H = 1.164$$

For the reaction: $\frac{1}{2}H_2(g) = [H](ppm)$, $K_H = 24.6$ at 1600 °C

$$K_H = \frac{[H](ppm)_{eq}}{\sqrt{p_{H_2}}}$$

$$[H](ppm)_{eq} = K_H.\sqrt{p_{H_2}} = 24.6.\sqrt{\frac{0.1}{760}} = 0.2822 \text{ ppm} = 0.2822 \times 10^{-4}\% = a_H^{Heq}$$

In extremely dilute solutions: $f_H = \frac{a_H^{Heq}}{[\%H]}$

$$[\%H](\text{in alloy}) = \frac{a_H^{Heq}}{f_H} = \frac{0.2822 \times 10^{-4}\%}{1.164} = 0.2424 \times 10^{-4} = 0.2424 \text{ ppm}$$

$$\ln\left[\frac{H_{in} - H_{eq}}{H_f - H_{eq}}\right] = \frac{r}{w}.t = \frac{16}{200} \times 20 = 1.6$$

(i) where $H_{in} = 20$ ppm and $H_{eq} = 0.2424$ ppm

$$\therefore \ln\left[\frac{H_{in} - H_{eq}}{H_f - H_{eq}}\right] = 1.6$$

$$\left[\frac{H_{in} - H_{eq}}{H_f - H_{eq}}\right] = 4.953 = \frac{20 - 0.2424}{H_f - 0.2424}$$

$$H_f = \frac{19.7576 + 4.953 \times 0.2424}{4.953} = \frac{20.9583}{4.953} = 4.23 \text{ ppm Ans.}$$

(ii) Value of r, when $t = 15$ min and $H_f = 5$ ppm

$$\therefore \ln \left[\frac{H_{in} - H_{eq}}{H_f - H_{eq}}\right] = \frac{r}{200} \cdot 15$$

$$\ln \left[\frac{20 - 0.2424}{5 - 0.2424}\right] = \frac{3r}{40}$$

$$\therefore r = \ln \left[\frac{19.7576}{4.7576}\right] \times \frac{40}{3} = 1.424 \times \frac{40}{3} = 19 \text{ ton min}^{-1} \quad \text{Ans}$$

Problem 8.16

The Dortmund-Horder degassing vessel holds 1/10 of total quantity of steel at 1600 °C.

(i) Develop an equation relating the lowering of hydrogen content to number of times the vessel is raised up and lowered down.

(ii) Estimate the number of times the vessel must be raised to bring down the hydrogen content to 1.5 ppm from an initial level of 6.5 ppm under a vacuum of 100 μm. At atmospheric pressure of hydrogen, steel dissolves 26 ppm of hydrogen at 1600 °C.

Solution

(i) Let H_i and H_{eq} be the initial and equilibrium (under vacuum) hydrogen content of steel, respectively, and H_L^1, H_L^2, H_L^3 ....be %H in steel after raising and lowering the vessel once, twice, thrice, and so on.

$$H_L^1 = \frac{1}{10} H_{eq} + \frac{9}{10} H_i$$

$$H_L^2 = \frac{1}{10} H_{eq} + \frac{9}{10} H_L^1 = \frac{1}{10} H_{eq} + \frac{9}{10} \left(\frac{1}{10} H_{eq} + \frac{9}{10} H_i\right)$$

$$= \frac{1}{10} H_{eq} + \frac{9}{10} \cdot \frac{1}{10} H_{eq} + \left(\frac{9}{10}\right)^2 H_i$$

and

$$H_L^3 = \frac{1}{10} H_{eq} + \frac{9}{10} H_L^2$$

$$= \frac{1}{10} H_{eq} + \frac{9}{10} \left[\frac{1}{10} H_{eq} + \frac{9}{10} \cdot \frac{1}{10} H_{eq} + \left(\frac{9}{10}\right)^2 H_i\right]$$

$$= \frac{1}{10} H_{eq} + \frac{9}{10} \cdot \frac{1}{10} H_{eq} + \left(\frac{9}{10}\right)^2 \cdot \frac{1}{10} H_{eq} + \left(\frac{9}{10}\right)^3 H_i$$

If $H_L^n (= H_f)$ is the hydrogen content of steel after raising up and lowering down the vessel n times,

$$H_L^n = \frac{1}{10}H_{eq} + \frac{1}{10} \cdot \frac{9}{10}H_{eq} + \frac{1}{10} \cdot \left(\frac{9}{10}\right)^2 H_{eq} + \ldots + \frac{1}{10} \cdot \left(\frac{9}{10}\right)^{n-1} H_{eq} + \left(\frac{9}{10}\right)^n H_i$$

$$= \frac{1}{10}H_{eq}\left[1 + \frac{9}{10} + \left(\frac{9}{10}\right)^2 + \ldots + \left(\frac{9}{10}\right)^{n-1}\right] + \left(\frac{9}{10}\right)^n H_i$$

$$= \frac{1}{10}H_{eq}\left[\frac{1}{1 - \frac{9}{10}}\right] + \left(\frac{9}{10}\right)^n H_i = H_{eq} + \left(\frac{9}{10}\right)^n H_i = H_f$$

(ii) Given that, $H_i = 6.5$ ppm and $H_f = 1.5$ ppm $= H_L^n$

$\frac{1}{2}H_2 = [H]$ (at one atmospheric pressure steel dissolves 26 ppm of hydrogen, that is, $[H] = 26$ ppm)

$$\therefore K = \frac{[H]}{\sqrt{p_{H_2}}} = \frac{26}{\sqrt{1}} = 26$$

$$p_{H_2} = 100 \text{ micron} = 100 \times 10^{-6}\text{m Hg} = 100 \times 10^{-3}\text{mm Hg} = \frac{0.1}{760} \text{ atm}$$

$$\therefore H_{eq} = K\sqrt{p_{H_2}} = 26 \times \sqrt{\frac{0.1}{760}} = 0.2982 \text{ ppm}$$

$$H_L^n = H_{eq} + \left(\frac{9}{10}\right)^n H_i = H_f$$

$$1.5 = 0.2982 + \left(\frac{9}{10}\right)^n .6.5$$

$$n = 17 \text{ times Ans.}$$

References

1. Yang, W., Wang, X., Zhang, L., Shan, Q., & Liu, X. (2013). Cleanliness of low carbon aluminium-killed steels during secondary refining processes. *Steel Research International, 84* (5), 473–489.
2. Bora, D., Masanori, S., & Toshihiro, T. (2010). Sulphide capacity prediction of molten slag by using a neutral network approach. *ISIJ International, 50*(8), 1059–1063.
3. Toshihiro, T., Yumi, O., & Mitsuru, U. (2010). Trial on the application of capillary phenomenon of solid CaO to desulfurization of liquid Fe. *ISIJ International, 50*(8), 1071–1079.
4. Gilbert, S., Monos, J. J., & Turkdogan, E. T. (1988) Ladle refining of steel using an exothermic synthetic slag. *Proceedings of the steelmaking conference, Toronto,* pp. 291–301.
5. Wang, L., Zhuo, X., & Zhano, J. (2003). Controlling inclusion composition in steelmaking process for tyre cord steel. *Journal of University of Science and Technology Beijing, 25*(4), 308–315.

6. Chen, S., Wang, X., He, X., Wang, W., & Jiang, M. (2013). Industrial application of desulfurization using low basicity refining slag in tire cord steel. *Journal of Iron and Steel Research International, 20*(1), 26–33.

7. Tupkary, R. H., & Tupkary, V. R. (2008). *An introduction to modern steelmaking* (7th ed.). Delhi: Khanna Publishers.

8. Elliott, J. F. (1985). Physical chemistry of liquid steel. In C. R. Taylor (Ed.), *Electric furnace steelmaking* (pp. 291–319). New York: Iron & Steel Society, AIME (Chapter 21).

9. Turkdogan, E. T. (1975). Physical chemistry of oxygen steelmaking thermochemistry and thermodynamics. In R. D. Pehlke, W. F. Porter, P. F. Urban, & J. M. Gaines (Eds.), *BOF steelmaking* (Vol. 2, pp. 1–190). New York: Iron & Steel Society, AIME (Chapter 4).

10. Lupis, C. H. P., & Elliott, J. F. (1966). Generalized interaction coefficients. Part I: Definitions. *Acta Metallurgica, 14*, 529–538.

11. Gokcen, N. A., & Chipman, J. (1952). Silicon-oxygen equilibrium in liquid iron. *Transactions of the Metallurgical Society, AIME, 194*, 171–182.

12. Turkdogan, E. T. (1965). Causes and effects of deoxidation occurring during cooling and solidification. *Transactions of the Metallurgical Society, AIME, 233*, 2100–2112.

13. Ward, R. G. (1962). *An introduction to the physical chemistry of iron and steelmaking.* London: Edward Arnold Ltd (Chapters 14, 15 and 16).

14. Leary, R. J., Coulehan, R. T., Tucker, H. A., & Wilson, W. W. (1968). *Rare earth silicides and cryolite in molten steel.* USBM, RI 7091.

15. Moore, J. J. (1990). *Chemical metallurgy* (2nd ed.). Oxford: Butterworth/Heinemann (Chapter 7).

16. Bodsworth, C., & Bell, H. B. (1972). *Physical chemistry of iron and steel manufacture* (2nd ed.). London: Longman (Chapters 4, 5 and 6).

17. Hilty, D. C., & Kaveney, T. F. (1985). Stainless steel making. In C. R. Taylor (Ed.), *Electric furnace steelmaking* (pp. 143–160). New York: Iron & Steel Society, AIME (Chapter 13).

18. Ghosh, A., & Chatterjee, A. (2008). *Ironmaking and steelmaking.* New Delhi: Prentice-Hall India (Chapter 21).

19. Ozturk, B., & Turkdogan, E. T. (1984). Equilibrium S distribution between molten calcium aluminate and steel part 1 CaS-CaO-Al$_2$O$_3$ melts equilibrated with liquid Fe containing Al and S. *Metal Science, 18*, 299–305.

20. Ozturk, B., & Turkdogan, E. T. (1984). Equilibrium S distribution between molten calcium aluminate and steel part 2 reaction of Si and Al in steel with calcium aluminate. *Metal Science, 18*, 306–309.

21. Chipman, J. (1964). Physical chemistry of liquid steel. In G. Derge (Ed.), *Basic open hearth steel making* (pp. 640–724). New York: AIME. (Chapter 16).

22. Pehlke, R. D., & Elliott, J. F. (1960). Solubility of nitrogen in liquid iron alloys, I thermodynamics. *Transactions of the Metallurgical Society, AIME, 218*, 1088–1101.

23. Weinstein, M., & Elliott, J. F. (1963). Solubility of hydrogen at one atmosphere in binary iron alloys. *Transactions of the Metallurgical Society, AIME, 227*, 382–391.

24. Pehlke, R. D., & Elliott, J. F. (1963). Solubility of nitrogen in liquid iron alloys, II kinetics. *Transactions of the Metallurgical Society, AIME, 227*, 844–855.

25. Bodsworth, C. (1990). *The extraction and refining of metals.* Boca Raton: CRC Press. (Chapter 5).

26. Shamsuddin, M., & Sohn, H. Y. (2020). Constitutive topics in physical chemistry of ironmaking and steelmaking: A review. *Trends in Physical Chemistry, 19*, 33–50.

27. Kubaschewski, O., & Alcock, C. B. (1979). *Metallurgical thermochemistry* (5th ed.). Oxford: Pergamon.

28. Curnutt, J. L., Prophet, H., McDonald, R. A., & Syverud, A. N. (1975). *JANAF thermochemical tables.* (1975). Midland: The Dow Chemical Company.

29. Barin, I., Knacke, O., & Kubaschewski, O. (1977). *Thermochemical properties of inorganic substances (supplement).* New York: Springer.

Chapter 9
Role of Halides in Extraction of Metals

The halide route in metal extraction has been found to be economically suitable for the large-scale production of many reactive and rare metals, for example, beryllium, calcium, magnesium, niobium, sodium, tantalum, thorium, titanium, uranium, vanadium, and zirconium. The refractory oxides of some of these metals can be reduced with carbon only at high temperatures. In most cases carbides are formed, decomposition of which is difficult. During reduction many of these metals dissolve large amounts of oxygen (even in solid state), which cannot be easily removed. For effective removal of oxygen, these oxides have to be reduced with calcium. However, whether reduction reaction is carried out between liquid calcium and solid/ liquid oxide, separation of the solid CaO (deoxidation product) is often troublesome. Reduction of fluoride/chloride by calcium or magnesium is a much easier operation where slag/metal separation is facilitated by the formation of either molten metal or molten slag. For example, reduction of uranium tetra fluoride by calcium produces liquid uranium whereas molten magnesium chloride slag is obtained when gaseous titanium tetra chloride or zirconium tetra chloride is reduced with magnesium. Generally, minerals do not occur as halides (fluoride, chloride, bromide, or iodide); however, conversion of stable oxides of rare/reactive metals (viz. Be, Ti, Zr, U, and Th) into halides offers an alternative route for metal extraction. As halides of reactive metals are less stable than their oxides, metals can be easily obtained by metallothermic reduction. Carbon may be used as a reducing agent to remove oxygen during chlorination of oxides of reactive metals like titanium, zirconium, and magnesium whereas UO_2 and ThO_2 are converted to fluorides by hydrofluorination. Beryl ($3BeO \cdot Al_2O_3 \cdot 6SiO_2$) is subjected to complex fluoride processing to convert beryllium into water-soluble sodium beryllium fluoride. Subsequently, the leach liquor is processed to obtain BeF_2. These chlorides and fluorides are reduced with magnesium or calcium for metal production.

On the possibilities of reduction of stable oxides of titanium, magnesium, zirconium, and beryllium with carbon, it is important to note that, under standard conditions, the free energy change for reduction of TiO_2, MgO, ZrO_2, and BeO with carbon becomes negative [1] (Fig. 5.1) at 1730, 1840, 2130, and 2320 °C,

© The Minerals, Metals & Materials Society 2021
M. Shamsuddin, *Physical Chemistry of Metallurgical Processes, Second Edition*,
The Minerals, Metals & Materials Series, https://doi.org/10.1007/978-3-030-58069-8_9

respectively. The carbothermic smelting processes at such high temperatures would be uneconomical due to high fuel consumption as well as the limited life of the refractory containers. The highly reactive nature of the metal produced, may pick up oxygen from oxide refractories as well as CO_2. In order to overcome these problems, titanium, zirconium, and beryllium are produced by reduction of $TiCl_4$ and $ZrCl_4$ vapors and glassy solid BeF_2 with liquid magnesium at appropriate temperatures, in the range 800–1200 °C. Magnesium is extracted by electrolysis of $MgCl_2$ dissolved in a fused $CaCl_2$–$NaCl$–KCl salt mixture. Beryllium, thorium, and zirconium can also be obtained by fused salt electrolysis of $BeCl_2$, $ThCl_4$, and K_2ZrF_6 dissolved in the mixture of inert salts of alkali or alkaline earth metals. The success of electrolysis depends on the fact that halides generally form more conducting melts with lower melting points. These metals cannot be electrolytically deposited from aqueous electrolytes due to hydrogen evolution at the cathode instead of metal deposition.

The stability of halides decreases in the following order: fluoride > chloride > bromide > iodide. Among all the halogens (fluorine, chlorine, bromine, and iodine) chlorine has been used extensively in many extraction processes. Fluoride and chloride routes have been employed in opening of refractory ores and chloride and iodide in refining of reactive metals. The popularly known reducing agent carbon cannot be used for reduction of chlorides (or other halides) since CCl_4 is less stable than many metal chlorides. Figure 9.1 shows the relative stability [1] in terms of free energy of formation of various halides and oxides. From the figure, it is evident that hydrogen can be used as a reducing agent for reduction of many metal chlorides because stability of HCl increases with increase of temperature. However, economics do not permit hydrogen reduction. Formation of iodides of many metals (e.g., Ti, Zr, Hf, and Th) at lower temperatures and their decomposition at higher temperatures into pure metal and iodine, provide a very effective method for preparation of ultra-pure metals. In brief, the success of a halide route in the extraction of reactive metals depends [2, 3] on purification by distillation due to the high vapor pressure of halides, low solid solubility of halides in metals, purification by crystallization due to difference in aqueous solubility, and electrolytic reduction of halides due to low melting point and high electrical conductivity.

Several metals form more than one chloride. The higher chlorides are generally more volatile and can be more easily reduced to lower chlorides by hydrogen. Various disproportionate reactions between chlorides of different valences (of the same metal) may be used in purification [3] and reduction of chlorides. Although halogen-based processes are feasible, halogens are costly reagents and they are often not generated cheaply [4] in the process. It has been used in the production of reactive and rare metals extensively but not for common metals. Chlorides are relatively hygroscopic, and hence, more costly fluorides are occasionally preferred. Fluorides are more stable and less volatile as compared to chlorides whereas iodides are less stable and more volatile than chlorides. Although bromine has properties common to those of chlorine and iodine, it is most expensive, and, hence, finds limited applications. Generally, chlorides are used.

The extent of use of halides in flow sheet varies from metal to metal. For example, extraction of beryllium, titanium, and zirconium is entirely based on halides from the

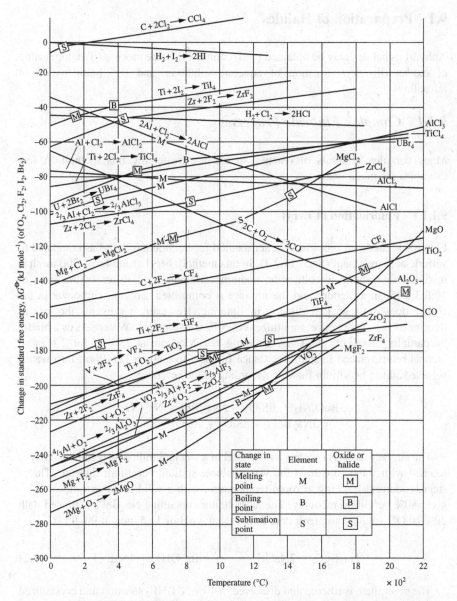

Fig. 9.1 $\Delta G°$ vs T diagram for the formation of some metal halides and oxides (From Chemical Metallurgy by J. J. Moore [1], © 1990, p 263, Butterworth Heinemann Ltd. Reproduced with the permission of Butterworth Heinemann Ltd)

opening of the ore/mineral to the production of metals. On the other hand, suitable intermediates like UF_4 and ThF_4 are produced by hydrofluorination of UO_2 and ThO_2, respectively, prior to a metallothermic reduction. In the following section, the preparation of halides and their reduction will be discussed briefly.

9.1 Preparation of Halides

Anhydrous halides may be obtained by (i) complex fluoride process (ii) halogenation of oxides (iii) crystallization of aqueous solutions, and (iv) halogenation of ferroalloys.

9.1.1 Complex Fluoride Processes

Many complex minerals of reactive metals are treated by this method. A few examples are given below:

9.1.1.1 Fluorination of Beryl

Fluoride breakdown provides a better method for treatment of beryl as compared to sulfuric acid treatment (Sect. 11.2.1). In this method, beryl ground to -200 mesh is mixed with sodium silicofluoride, sodium carbonate, and water in the ratio of 10:8:3:1 [5] in a blender and the mixture is compacted. Sodium carbonate in the charge does not take part directly in sintering. However, it prevents the loss of fluorine as SiF_4 and, hence, minimizes Na_2SiF_6 consumption. Water acts as a binder by partially dissolving Na_2CO_3. Sintering is carried out at 750 °C for 2 h in an inconel retort, placed inside a resistance furnace, to convert beryllium into water-soluble sodium beryllium fluoride according to the reaction:

$$3BeO.Al_2O_3.6SiO_2 + 2Na_2SiF_6 + Na_2CO_3$$
$$= 3Na_2BeF_4 + 8SiO_2 + Al_2O_3 + CO_2 \qquad (9.1)$$

The sintered mass is crushed and ground in a wet ball mill. The resultant slurry is leached with water to extract the water-soluble sodium beryllium fluoride. Three-stage, cocurrent leaching at room temperature, with a solid-to-water ratio of 1:3.5 gives 80% beryllium recovery. The leach liquor containing Na_2BeF_4 is treated with 20% NaOH solution to precipitate $Be(OH)_2$ at a pH of 11.5, and at 90–95 °C.

$$Na_2BeF_4 + 2NaOH \xrightarrow{90-95\ °C} Be(OH)_2 + 4NaF \qquad (9.2)$$

The precipitate is filtered and dissolved in 40% NH_4HF_2 solution and crystallized as $(NH_4)_2BeF_4$ by evaporation:

$$Be(OH)_2 + 2NH_4HF_2 \xrightarrow{RT} (NH_4)_2BeF_4 + 2H_2O \qquad (9.3)$$

The double fluoride is decomposed at 950 °C to get anhydrous BeF_2:

$$(NH_4)_2BeF_4 \xrightarrow{900\ °C} 2NH_4F(g) + BeF_2 \tag{9.4}$$

The filtrate containing fairly dilute sodium fluoride solution can be completely recovered at a pH of 4.5 as iron cryolite (Na_3FeF_6) [5] by the addition of $FeCl_3$ ($6NaF + FeCl_3 = Na_3FeF_6 + 3NaCl$) to make the process self-sustaining (see Appendix A.6). The iron cryolite can be used directly to replace sodium silicofluoride in the main fluorination reaction:

$$3BeO.Al_2O_3.6SiO_2 + 2Na_3FeF_6 \rightarrow 3Na_2BeF_4 + 6SiO_2 + Al_2O_3 + Fe_2O_3 \tag{9.5}$$

9.1.1.2 Fluorination of Zircon

Zircon ($ZrSiO_4$) is broken by sintering with potassium silicofluoride in a rotary furnace at 700 °C for about 4 h in the presence of potassium chloride to convert zirconium into water-soluble double fluoride according to the reaction:

$$ZrSiO_4 + K_2SiF_6 \xrightarrow{KCl} K_2ZrF_6 + 2SiO_2 \tag{9.6}$$

The sintered mass is crushed and ground to 100 mesh powder and leached with water containing 1% HCl for 2 h with stirring at 85 °C. The leach liquor is subjected to 16–18 fractional crystallization stages in a stainless steel vessel between temperatures of 100 and 19 °C for zirconium/hafnium separation to produce nuclear-grade zirconium. Alternatively, zircon is chlorinated to $ZrCl_4$ and subjected to hydrolysis to obtain $ZrOCl_2$, which is dissolved in ammonium thiocyanate and purified by solvent extraction to prepare a hafnium-free solution of zirconium. The purified solution is treated with potassium dihydrogen fluoride to precipitate K_2ZrF_6, which is dissolved in hot water and recrystallized and dried to yield pure K_2ZrF_6 crystals. K_2ZrF_6 crystals are dissolved in fused salt mixture of NaCl and KCl [6] for electrolytic extraction of zirconium.

9.1.2 Halogenation of Oxides

This is the most common method for the preparation of anhydrous halides. Some examples are given below:

9.1.2.1 Hydrofluorination

Uranium dioxide obtained after a series of extraction steps that include sulfuric acid digestion, ion-exchange concentration, solvent extraction purification, ammonium diuranate precipitation, and hydrogen reduction of UO_3, is converted to uranium tetrafluoride before metallothermic reduction (see Appendix A.7). Fluoride is preferred over oxide for metal production because the former can be easily reduced without oxygen pick up. UF_4 is produced through hydrofluorination of UO_2 powder at 450 °C in a screw-type stirred bed reactor [7], made of inconel, according to the reaction:

$$UO_2(s) + 4HF(g) = UF_4(s) + 2H_2O(g) \qquad (9.7)$$

For this purpose, UO_2 powder is fed from the top into an inclined tubular reactor while HF gas flows up from the bottom. The coaxial screw-stirring of the reactor tube helps to churn the down-flowing oxide powder repeatedly, thereby exposing fresh oxide surface for HF attack. The rpm of the screw and the oxide feeding rate are controlled in such a way that the solid received at the bottom of the reactor is completely UF_4.

A similar technique is employed for conversion of thorium dioxide [8] into thorium tetrafluoride (Appendix A.8). In this conversion process, a group of four horizontal stirred beds, one above the other, is used. The top one is for calcination in air to remove traces of nitric acid, water, and carbonaceous matter. The middle two are used for actual hydrofluorination. The lower bed is mainly for partial sintering and densification of the product in hydrogen fluoride. Each bed is 6.5 m long with a diameter of 40 cm. The lower ones are made of inconel and the upper one of stainless steel. The powder flows from a hopper at the top via a seal hopper between the calcination and hydrofluorination sections, to another hopper and feeding section at the bottom.

Fluidized bed reactors are also employed in hydrofluorination of UO_2 and ThO_2 powders.

9.1.2.2   Chlorination

As magnesium, titanium, and zirconium have high affinity for oxygen and their oxides are very stable; MgO, TiO_2, and ZrO_2 are chlorinated in the presence of carbon in a fluidized bed reactor or a shaft furnace at temperatures between 700 and 1000 °C according to the reactions:

$$2MgO(s) + C(s) + 2Cl_2(g) = 2MgCl_2(l) + CO_2(g) \qquad (9.8)$$

$$TiO_2(s) + C(s) + 2Cl_2(g) = TiCl_4(g) + CO_2(g) \qquad (9.9)$$

$$ZrO_2(s) + C(s) + 2Cl_2(g) = ZrCl_4(g) + CO_2(g) \qquad (9.10)$$

The starting materials are calcined magnesite (MgO), natural or synthetic rutile (TiO_2), or ZrO_2 containing less than 200 ppm hafnium obtained by processing zircon sand ($ZrSiO_4$) in several steps including alkali digestion with caustic soda at 500–600 °C and purification of the leach liquor by solvent extraction.

Chlorination of MgO Chlorination is done in an electric shaft furnace. The rate and completeness of chlorination, to a great extent, depends on the quality of magnesium oxide and reducing material used. Both MgO and the reducing agent must be pure, finely divided, thoroughly mixed, and briquetted. The shaft furnace consists of a steel cylindrical shell lined with refractory bricks. Two rows of carbon electrodes are inserted at the bottom of the furnace with three electrodes per row that are equally spaced along the circle. The electrodes of a row are shifted through 60° [9] in plan with respect to those of the other row. The space between the rows of electrodes is filled with carbon resistor blocks that are capable of producing a temperature up to 1000 °C. Chlorine gas is introduced into the furnace through three inlets located between the two rows of electrodes. The materials are charged into the furnace at the top through a double bell and hopper-feed chamber. The molten $MgCl_2$ is run off through a tap hole located at the bottom of the furnace.

The charge consists of briquettes produced by extrusion of proper mixture of MgO and coke and a suitable binder. The burden in the furnace during chlorination may be divided into three zones. In the top zone, which is 2 or 3 m deep, the charge is heated by flue gases to drive off moisture. At the bottom of this zone, chlorine that is not used in the main reaction zone is absorbed. The chlorination together with the fixation of oxygen evolved takes place in the middle or reaction zone. In this zone a temperature above the melting point of $MgCl_2$ (718 °C) has to be maintained to produce molten $MgCl_2$, which can easily trickle down so as to expose fresh MgO for chlorination. In order to avoid choking of the chlorinator at the bottom of the reaction zone, the charge must remain in lump form. The carbon resistor blocks filling the bottom zone do not only generate heat but also serve as a filter for molten $MgCl_2$. The MgO particles entrapped in $MgCl_2$ get chlorinated in this zone by the rising current of chlorine. For free flowing of molten $MgCl_2$ through the blocks, a temperature of 150–200 °C above the melting point of $MgCl_2$ is desired. The magnesium chloride is tapped every 3–4 h into ladles with tightly fitting lids, which are transported to electrolytic cells for metal production.

Chlorination of Natural or Synthetic Rutile The great difficulties encountered in the complete removal of oxygen from oxygen-bearing titanium ores have led to the development of extraction processes that start with oxygen-free material. Hence, manufacture of titanium tetrachloride of adequate purity happens to be the first critical step in the production of titanium metal. Presently, and, by far, a large amount of titanium metal is obtained by the reduction of titanium tetrachloride. $TiCl_4$ is prepared by the combined reduction and chlorination of rutile (natural or synthetic) according to reaction (9.9). More abundant ilmenite ($FeO \cdot TiO_2$) is not used due to high chlorine consumption and formation of large amounts of $FeCl_2$ and $FeCl_3$. The following three processes [10] have been developed for production of $TiCl_4$:

1. Chlorination of rutile (natural or synthetic) in a fluidized bed reactor
2. Chlorination of pelletized charge of rutile and coke in a shaft furnace

3. Chlorination by bubbling chlorine gas into a molten salt bath fed with a mixture of titania and coke powder

In the United States, $TiCl_4$ is commercially manufactured in fluidized bed chlorinators, operated at approximately 950 °C. In this process, petroleum coke containing greater than 98% carbon and rutile in the ratio of 1:1, are fed from the top and chlorine gas is inserted from the bottom of the chlorinator. In order to initiate the reaction at the beginning, oxygen/air is mixed with chlorine. Once chlorination is initiated, the rate of reaction (9.9) is controlled by adjusting the rate of supply of chlorine gas. All appropriate steps are taken to prevent carrying over of unreacted rutile and coke by the $TiCl_4$ gas formed.

In Russia and Japan, briquettes of coke and titania slag, obtained by electric furnace smelting of ilmenite ore, are chlorinated in shaft furnaces. For effective chlorination, titania is mixed and ground with coke in the presence of coal tar as a binder. The mixture is briquetted and the briquettes are heated in a kiln at 800 °C in the absence of air to remove volatile material. As the resulting product is porous and occupies a large volume, it is crushed and briquetted again and then fed into another chlorinator, operated at 900 °C by electric heating. Highly exothermic reaction (9.9) provides part of the heat necessary for chlorination.

The impurities present in the rutile are likewise chlorinated (Appendix A.9) resulting in impure $TiCl_4$ (97–99%) containing 1.5–2.5% $SiCl_4$, 0.02–0.002% $FeCl_3$, 0.1–0.3% Cl_2, and up to 0.15% VCl_5. $TiCl_4$ boils at 136.4 °C and evolves with CO, CO_2, and excess Cl_2. The resultant chlorides and oxychlorides may be divided into volatile compounds: $FeCl_3$ (boiling point: 319 °C), VCl_2 (164 °C), $VOCl_3$ (127 °C), $SiCl_4$ (57 °C), $COCl_2$, SCl_2, and HCl and nonvolatile compounds: $FeCl_2$, $MgCl_2$, $CaCl_2$, NaCl, and KCl. The nonvolatiles may form a coating around TiO_2 particles and hinder the process of chlorination. The volatile chlorides are carried away into the condenser as fine dusts. They distil over and condense with $TiCl_4$ and, hence, must be removed before $TiCl_4$ is reduced to metal. Chlorinator is a specially designed furnace to withstand highly corrosive conditions in a chlorine atmosphere in the temperature range of 700–1000 °C. The furnace is lined with graphite or silica. The product of chlorination is taken into a nickel-lined condenser.

In Russia, the use of fused salt bath of chlorides of alkali and alkaline earth elements has been reported for chlorination of rutile/titania slag containing appreciable amounts of CaO and MgO.

Chlorination of ZrO_2 As the reaction (9.10) leading to chlorination of ZrO_2 in the presence of carbon is exothermic, the equipment must ensure quick dissipation of heat to achieve a high rate of chlorination. The highly corrosive nature of chlorine and chloride at high temperatures (700–900 °C) restricts the choice of the material of construction and the type of equipment for large-scale production. Nickel and inconel can withstand chlorine up to 600 °C but in the production of $ZnCl_4$ involving temperatures of 700–900 °C, quartz or silica has to be used in the reaction zone. The chlorination (Appendix A.10) is facilitated if ZrO_2 and high-purity carbon are thoroughly mixed in the presence of a suitable binder (sugar solution) to obtain a thick paste in a kneading machine. The green compacts of cylindrical shape obtained by extrusion of the paste are dried in an oven at 100 °C and then coked at 800 °C in an inert atmosphere prior to chlorination. The coking operation decomposes the sugar to yield a hard and porous mass. The chlorinator consists of a silica tube clamped and held vertically between nickel flanges with a chlorine inlet chamber at the bottom and a feed chamber at the top. The tube is heated by an electric resistance furnace. A nickel condenser is used for collecting the vapors of zirconium tetrachloride. After charging the coked briquettes, the chlorinator is heated to the operating temperature of 850 °C under the flow of argon. Argon is replaced with chlorine as soon as the operating temperature is attained. Chloride is collected from the bottom by opening the condenser in polythene bags and stored in air-tight containers prior to its use in the reduction step.

In a similar manner, pure ThO_2 obtained by H_2SO_4 digestion of monazite, solvent extraction purification, and thermal denitration is chlorinated at 900 °C in the presence of coke [5].

9.1.3 Halogenation of Ferro-Alloys

The direct halogenation of niobium/tantalum and vanadium ores is not practiced on a commercial scale because ferro-niobium and ferro-vanadium are available on a large scale from the steel industry. Hence, for the production of relatively expensive niobium, tantalum, and vanadium, ferro-alloys can be used as raw materials in place of ore concentrates.

9.1.3.1 Fluorination of Ferro-Niobium

Ferro-niobium [5] is dissolved in 50% KOH solution at 100 °C. The filtrate containing potassium niobate and tantalate is treated with NaCl solution to precipitate sodium niobate and tantalate, which are converted to niobium and tantalum pentoxides by allowing it to react with ammonia at a pH of 5–6. The pentoxides are dissolved in hydrofluoric acid and subjected to solvent extraction purification to produce a tantalum-free solution of niobium fluoride that is treated with potassium fluoride solution to precipitate pure K_2NbF_7 at 0 °C for subsequent extraction of niobium metal.

9.1.3.2 Chlorination of Ferro-Niobium and Ferro-Vanadium

In practice, ferro-alloys with 70–80% V or Nb are subjected to chlorination. Vanadium is mainly converted to tetrachloride at 1000 °C whereas niobium forms niobium pentachloride at 550 °C:

$$V(s) + 2Cl_2(g) = VCl_4(g) \tag{9.11}$$

$$2Nb(s) + 5Cl_2(g) = 2NbCl_5(g) \tag{9.12}$$

In both the cases, along with VCl_4 and $NbCl_5$ gases, a large amount of $FeCl_3$ is generated. Small quantities of $AlCl_3$, $TiCl_4$, and so on are also formed. Since all the sources of niobium are associated with small quantities of tantalum and both niobium and tantalum and their compounds are chemically similar in nature, niobium and tantalum pentachlorides are formed in chlorination of ferro-niobium along with impurities of ferric chloride, tungsten penta, and hexachlorides, aluminum trichloride and titanium tetrachloride.

9.1.4   Crystallization from Aqueous Solution

Although magnesite ($MgCO_3$) and dolomite ($CaCO_3 \cdot MgCO_3$) are rich in magnesium content; the major production of magnesium comes from sea water containing about 0.13% magnesium in the form of chlorides such as carnallite ($MgCl_2 \cdot KCl \cdot 6H_2O$) and bischofite ($MgCl_2 \cdot 12H_2O$). In addition to the water of crystallization both carnallite and bischofite are associated with some more water in attached form. The electrolytic process for the production of magnesium requires the removal of the entire water prior to the electrolysis. Both carnallite and bischofite decompose into MgO and HCl on heating even at low temperatures. The residual MgO has to be minimized because it reacts with the graphite anodes. Accumulation of MgO in the electrolytic cell will reduce the overall energy efficiency of the process. Hence, preparation of anhydrous $MgCl_2$ with low residual MgO happens to be the most important step in the production of magnesium by the electrolytic method.

9.1.4.1   Dehydration of Carnallite

On rapid heating, carnallite melts in its water of crystallization [9] at 1 atm pressure and temperatures above 110–120 °C. On slow heating, however, all the molecules of water of crystallization may be driven off without melting. The first stage of dehydration is carried out in a rotary kiln at temperature not exceeding 600 °C, for 2 h to yield carnallite with 3–4% water and 1–2% MgO. The second stage of dehydration at 750–800 °C involves melting of the carnallite in a resistance furnace with steel electrodes through the carnallite charge. The molten carnallite is driven off into two mixers that are maintained at 780–850 °C by electric resistance heating. The molten and dehydrated carnallite containing about 50% $MgCl_2$ and 0.5–0.9% MgO is tapped and sent for electrolysis.

The method of dehydration has gone through several modifications in the recent past. For example, in the first step, the Dead Sea Magnesium Co. charges carnallite into a fluidized bed dryer, maintained at 130 °C, to remove the water of crystallization. Temperature of the dryer is slowly raised to 200 °C. By gradual heating, about 95% water is removed from the carnallite and some HCl is emitted. The resulting product containing 3–6% water and 1–2% MgO from the first stage of drying is fed to a chlorinator consisting of three chambers (maintained at 700–750 °C) for successive melting, chlorination, and settling. The carnallite melted in the first chamber is chlorinated with recycled chlorine produced in the electrolytic cell in the second chamber according to the reaction $MgO + C + Cl_2 \rightarrow MgCl_2 + CO/CO_2$. The chlorinated product is held for sufficient time in the third chamber to allow settling of the remaining MgO and other insoluble materials at the bottom and the clean molten $MgCl_2$ containing 0.2–0.6% MgO is transferred to the electrolytic plant for electrolysis.

9.1.4.2 Dehydration of Bischofite

Magnesium chloride forms hydrous crystals with 1, 2, 4, 6, 8, and 12 molecules of water of crystallization [9]. Each one is stable at a definite temperature and a change-over from one hydrate to another takes place in the following sequence:

$$MgCl_2.12H_2O \rightleftharpoons MgCl_2.8H_2O \text{ at } -19.4\,^{\circ}C \tag{9.13}$$

$$MgCl_2.8H_2O \rightleftharpoons MgCl_2.6H_2O \text{ at } -3.4\,^{\circ}C \tag{9.14}$$

$$MgCl_2.6H_2O \rightleftharpoons MgCl_2.4H_2O \text{ at } 17\,^{\circ}C \tag{9.15}$$

$$MgCl_2.4H_2O \rightleftharpoons MgCl_2.2H_2O \text{ at } 182\,^{\circ}C \tag{9.16}$$

$$MgCl_2.2H_2O \rightleftharpoons MgCl_2.H_2O \text{ at } 240\,^{\circ}C \tag{9.17}$$

On heating, crystals of hydrous magnesium chloride decompose at relatively low temperatures, and, hence, rapid heating may result in melting and foaming. Additionally, it may be noted that simple heating will not remove all the water of crystallization from hydrous magnesium chloride because at above 200 °C it hydrolyzes to form chloride. In the temperature range of 300–555 °C the following reaction takes place:

$$MgCl_2 \text{ (l)} + H_2O \text{ (g)} \rightleftharpoons Mg(OH)Cl(l) + HCl(g) \tag{9.18}$$

For which the equilibrium constant is given as:

$$K_1 = \frac{p_{HCl}}{p_{H_2O}} \tag{9.19}$$

Above 555 °C, MgO is formed according to the reaction:

$$MgCl_2(l) + H_2O\ (g) \rightleftharpoons MgO(s) + 2HCl\ (g) \qquad (9.20)$$

Accordingly, the equilibrium constant is modified as:

$$K_2 = \frac{p^2_{HCl}}{p_{H_2O}} \qquad (9.21)$$

From the above reactions it is clear that heating in air will not remove all the water of crystallization without decomposing the $MgCl_2$. In order to avoid decomposition of $MgCl_2$, the ratio of partial pressure of the HCl to the water vapors in the gaseous phase must be greater than the equilibrium constants, K_1 and K_2 of reactions (9.18) and (9.20) respectively, at a given temperature, that is,

$$\frac{p_{HCl}}{p_{H_2O}} > K_1, \text{ and } \frac{p^2_{HCl}}{p_{H_2O}} > K_2$$

The above facts suggest that dehydration of bischofite takes place in two stages. In the first stage the bischofite is slowly heated in a current of hot gases in a rotary kiln to obtain $MgCl_2$ with 1.5 or 2 molecules of water of crystallization. This operation must be very carefully carried out because bischofite melts at 106 °C. Melting of $MgCl_2$ and deposition of the product on the wall of the kiln retards the operation. As melting point of the charge increases with decrease of water content, the temperature has to be increased by heating with a counter current flow of flue gases.

The partial pressure of HCl in the gaseous phase has to be increased during the second stage of heating either by introducing HCl gas into the system or by generating it in the kiln from the anode gases of the electrolytic cell and carbon according to the reaction:

$$Cl_2(g) + H_2O(g) + C(s) = 2HCl(g) + CO(g) \qquad (9.22)$$

The second stage may be carried out either in a rotary kiln or in an arc furnace at a temperature above the melting point of anhydrous $MgCl_2$ to yield molten $MgCl_2$ containing not more than 1.0% MgO and 0.5% H_2O.

In recent years, although rotary kilns have been replaced with evaporators and fluid bed dryers, the basic principles of dehydration, as discussed above, remain the same. In the first stage of the Hydro Magnesium Process, magnesium chloride solution is heated in an evaporator to produce pure bischofite ($MgCl_2 \cdot 6H_2O$) containing 45–50% water. In the second stage of drying, which is carried out in the fluidized bed dryer, bischofite is treated with hot air to obtain $MgCl_2 \cdot 2H_2O$. The final drying is performed at 330 °C in the presence of HCl gas to produce anhydrous $MgCl_2$ containing about 0.1% MgO. Use of HCl prevents hydrolysis and reduces the residual MgO.

9.2 Purification of Chlorides

As chlorine is a very reactive reagent, chlorination of minerals, concentrates, and ferroalloys give rise to a number of chlorides. Hence, the resultant product has to be purified prior to metal production. Water-soluble chlorides may be purified by recrystallization or reprecipitation [3]. Ion exchange or solvent extraction method may also be employed. Nonvolatile chlorides may be collected as a separate fraction and volatile chlorides are separated by fractional distillation making use of the difference in boiling points.

Zirconium tetrachloride, the chlorination product of ZrO_2 obtained by calcination of zirconium hydroxide after solvent-extraction purification of alkali-digested liquor of zircon does not need any purification. $ZrCl_4$ can be directly reduced with magnesium. Chlorination products of natural rutile (96% TiO_2) and synthetic rutile (>90% TiO_2) containing SiO_2, Fe_2O_3, Al_2O_3, and so on have to be purified to remove $FeCl_3$, $SiCl_4$, $AlCl_3$, and so on. Since chlorine, as well as other halogens, are very reactive, gangue oxides present in the ores/minerals also are chlorinated along with oxides of the desired metals. This is the reason why pitchblende (oxide mineral of uranium), monazite (complex phosphate of thorium and rare earth), zircon sand ($ZrSiO_4$), niobite-tantalite (mixed ore of Nb and Ta), and so on are not chlorinated.

9.2.1 Purification of Titanium Tetrachloride

The gaseous product after chlorination of rutile is first subjected to dedusting treatment. $FeCl_3$ (boiling point: 319 °C) is allowed to deposit in a condenser maintained at 200 °C and then filtered. The principal impurities left at this stage are $AlCl_3$, $FeCl_3$, and $VOCl_3$. The first two are removed as sludge by scrubbing $TiCl_4$ with cold, recycled $TiCl_4$ followed by decantation and filtration. The volatile $VOCl_3$ is reduced to less volatile VCl_3 by H_2S. $TiCl_4$ thus purified is then subjected to fractional distillation to separate out most of the volatile chlorides and oxychlorides based on the differences in their boiling points. The $TiCl_4$ is obtained as a middle fraction, (boiling point 136.4 °C). The lower boiling material containing HCl, $COCl_2$, PH_3, CCl_4, and $SiCl_4$ is first removed. The heavy fraction is discarded as a slurry. In the second stage, $TiCl_4$ is driven off into a condenser to condense as a transparent colorless liquid. The purity of the final $TiCl_4$ is about 99.98% and it must be stored in mild steel containers under completely anhydrous condition in the presence of argon to avoid hydrolysis or the dissolution of atmospheric oxygen and nitrogen. A flow sheet showing the various steps in purification of $TiCl_4$ is presented in Fig. 9.2.

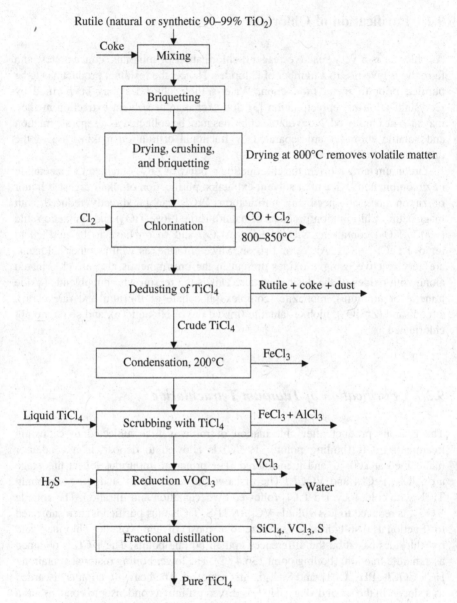

Fig. 9.2   Purification of TiCl$_4$ vapors

9.2.2   Purification of Niobium Pentachloride

Chlorides of niobium and tantalum with similar chemical characteristics may be separated by fractional distillation. The chloride mixture obtained by chlorination of ferro-niobium at 550 °C, containing 67.2% NbCl$_5$, 6.0% TaCl$_5$, 26.4% FeCl$_3$, and 0.4% WOCl$_4$ with respective boiling points of 254, 239, 315, and 228 °C is separated by fractional distillation into a niobium-rich fraction containing 99.98% $NbCl_5$, and 0.01% $TaCl_5$; a tantalum-rich fraction containing 93.2% $TaCl_5$, 0.6% $NbCl_5$, and 6.2% $WOCl_4$; and an iron-rich fraction with 98.7% $FeCl_3$, and 0.06% $TaCl_5$. The gaseous chlorination product is stored in a condenser maintained at 120 °C to condense $NbCl_5$, $TaCl_5$, and $FeCl_3$ and to distil vapors of $AlCl_3$ and $TiCl_4$. The resulting mixture of $FeCl_3$, $NbCl_5$, and $TaCl_5$ is purified by fractional reduction [4, 5] with hydrogen at 350 °C. $FeCl_3$ is reduced to $FeCl_2$ and removed as a solid. At 500 °C, $NbCl_5$ is reduced with hydrogen to solid $NbCl_3$ whereas the unreduced $TaCl_5$ is carried away with the gas stream.

9.2.3 Purification of Vanadium Tetrachloride

During chlorination of ferro-vanadium, VCl_3 and VCl_2 may be formed in addition to VCl_4 due to poor availability of chlorine. However, the temperature is controlled below 1000 °C by limiting the chlorine flow rate to prolong the furnace life. A similar procedure based on fractional distillation, which makes use of the difference in boiling points of various chlorides is adopted to obtain pure VCl_3 gas.

9.3 Metal Production

Reactive metals in pure state can be obtained by metallothermic reduction of halides using a suitable reducing agent or by electrolysis of fused salts containing mixture of fluorides or chlorides. Reduction of $TiCl_4$, $ZrCl_4$, VCl_3, BeF_2, UF_4, ThF_4, K_2TaF_7, and so on may be carried out by metals like magnesium, calcium, and sodium. However, magnesium, beryllium, thorium, and zirconium can be obtained, respectively, by fused salt electrolysis of $MgCl_2$ dissolved in $NaCl–KCl–CaCl_2$ melt, $BeCl_2$ dissolved in molten $NaCl$, $ThCl_4$ dissolved in $NaCl–KCl$ eutectic melt, and K_2ZrF_6 dissolved in $NaCl–KCl$ melt.

9.3.1 Metallothermic Reduction

Thermodynamic considerations help in the selection of the reducing agent. The free energy changes involved in the formation of halides and of oxides of some reactive metals with temperature are shown in Fig. 9.1. The data in the figure are useful for the selection of reducing agents on the basis of relative values of free energy of formation. In the extraction of reactive metals, reduction of halides is preferred over that of oxides with the prime objective of producing oxygen-free metals. Air has to be excluded from the reduction system because oxygen and nitrogen form solid solution of oxides and nitrides in the metals, making them hard and brittle [5]. To achieve this purpose an evacuated system or an inert atmosphere of argon or helium is employed. Of the halides, chlorides and fluorides are usually preferred over iodides or bromides on economic grounds, and between chlorides and fluorides, the latter are preferred because of the hygroscopic nature of the former.

The state of metal and slag after reduction depends entirely on the amount of heat generated from the reduction reaction. Sufficient availability of heat at the reduction temperature will melt both the metal and slag leading to clear separation of one from the other. The metal generally being heavier trickles through the slag and gets collected at the bottom of the vessel and can be removed as a solid ingot on cooling. In case insufficient heat is generated during reduction of oxides with calcium or magnesium, the metal is obtained as a powder in a matrix of the slag. The metal and slag are sintered together to some extent, particularly when the reactants are pressed together before reduction. The exact form of the reduced metal particles depends on whether the metal is molten and sufficient time is available for coalescence. If a flux like calcium chloride is used for calcium oxide slag and the product is held for a reasonable period of time after reduction, the metal may be recovered in the form of nearly spherical beads of appreciable size. In the absence of a flux, large agglomerates are obtained. On the other hand, when the temperature of reduction is neither sufficient to melt the metal nor sufficient for sintering of metal particles, it results in the formation of finely divided pyrophoric metal powder. When the heat of the reaction is sufficient to melt the slag but not the metal the resulting metal is obtained in the form of a sponge. For example, reduction of $TiCl_4$ and $ZrCl_4$ vapors by liquid magnesium produces solid titanium and zirconium sponge and liquid $MgCl_2$ slag. The low boiling points of rare/reactive metal chlorides help in sintering of the metal into sponge.

In order to achieve the highest possible temperature, normally adiabatic reaction conditions are preferred. Rapid reaction and thermal insulation of the system help in attaining adiabatic conditions. The vessel together with reactants is soaked at a temperature just below the required reduction temperature. Additional heat may be introduced into the system by using a booster in the charge. The common booster, iodine or sulfur may combine with excess of the reducing agent during the course of reduction. Too high a reduction temperature may cause rapid vaporization of either reactant or product leading to explosion. For example, sodium boiling at 800 °C may pose this problem. Excessive reaction between the container and the products or reactants must be minimized to avoid undue corrosion of materials of construction. In the following paragraphs, reduction of $TiCl_4$, UF_6, BeF_2, and so on with calcium, magnesium, and sodium will be discussed briefly.

9.3.1.1 Reduction of Uranium Tetrafluoride

Both magnesium and calcium can reduce UF_4 to yield liquid uranium in massive ingot form. However, calcium is preferred over magnesium due to the high heat of reduction, which is used to promote efficient slag-metal separation. Further, calcium also simplifies the operation as no pressure vessel is required, whereas owing to the high vapor pressure of magnesium at the reduction temperature, a pressure vessel is necessary. Magnesium is easily available in abundance at a cheaper rate but reduction reaction is less exothermic. Thus, external heat is required in this case for good slag/metal separation. Before reduction, the high-purity UF_4 powder is thoroughly mixed with calcium granules and the mixture is charged into a leak-tight tapered stainless bomb reactor, lined with dry CaF_2. The CaF_2 lining serves as a heat insulator as well as for containment of molten uranium. After charging, the bomb is tightly capped and lowered into a gas-heated furnace (already preheated to 1000 °C) to heat up to 700 °C within 4 h. The calciothermic reduction is triggered in an argon atmosphere by electrically igniting a magnesium ribbon embedded at the top of the charge. The reaction is instantaneous and spreads rapidly through the mass. The heat generated is sufficient to produce liquid uranium (melting point, 1135 °C) according to the reaction:

$$UF_4 \text{ (s)} + 2Ca \text{ (s)} = U(l) + 2\, CaF_2(s) \tag{9.23}$$

The liquid uranium coalesces and collects at the bottom of the reactor to form a solid ingot on cooling. The reactor vessel is lifted out of the furnace while still hot, and then cooled in air. Uranium ingot is removed by chiseling the slag (Appendix A.7). The ingot is washed, pickled with 40% nitric acid, rewashed, and subjected to infrared drying. Finally, it is melted in an induction furnace (3000 cps) at 200 °C above the melting point of uranium to about 1350 °C.

A similar procedure is followed in calciothermic reduction of thorium tetrafluoride (Appendix A.8).

9.3.1.2 Reduction of Titanium Tetrachloride

Both sodium and magnesium can be used to reduce $TiCl_4$ vapors (boiling point 136.4 °C) according to the reactions:

$$TiCl_4(g) + 4Na \text{ (l)} = Ti \text{ (s)} + 4NaCl \text{ (l)} + 130 \text{ kcal} \tag{9.24}$$

$$TiCl_4(g) + 2Mg \text{ (l)} = Ti \text{ (s)} + MgCl_2(l) + 225 \text{ kcal} \tag{9.25}$$

Thus, there are two routes for the reduction of titanium tetrachloride. The sodium-based process, known as the Hunter [11, 12] process, was first developed in 1910. This process was superseded by the Kroll process [13, 14], which uses liquid magnesium as the reductant. Although both the reactions (9.24 and 9.25) are exothermic, magnesium is preferred over sodium because the latter provides a very narrow range of working temperature (about 75 °C) due to the low boiling point of sodium (877 °C) and high melting point of NaCl (801 °C). Moreover, the slag, NaCl cannot be removed by vacuum distillation and leaching leads to the oxidation of the titanium sponge. The commercial use of calcium as a reducing agent has been discouraged on account of higher cost, scarcity, narrow working temperature range, and difficulty in $CaCl_2$ removal.

The use of magnesium as a reducing agent was considered on account of its relatively low melting point of 711 °C and relatively high boiling point of 1120 °C. This provides a working temperature range of 711–1120 °C. Additional advantages include easy availability on a large scale at a cheaper rate and convenient removal of Mg and $MgCl_2$ entrapped in titanium sponge by vacuum distillation. As reaction (9.25) is relatively more exothermic, $TiCl_4$ vapors can be easily reduced at 800–900 °C. Since magnesium reacts readily with the atmospheric oxygen to form MgO, it is given a preliminary treatment with dilute hydrochloric acid (pickling). The cleaned ingots are rinsed with cold water and dried.

The Kroll reactor made of stainless steel is charged with magnesium ingots in a proportion that is 10% above the stoichiometric requirement and sealed. The reactor is evacuated and filled with argon. After the chamber has been completely purged and filled with argon, magnesium is melted and held at 850 °C. This is followed by admission of $TiCl_4$ into the reactor at a rate sufficient to attain and maintain temperature of 900–1000 °C. $TiCl_4$ vaporizes and reacts with the molten magnesium (Appendix A.9). The reaction is allowed to proceed for 3 days [15]. In order to prevent contamination of the titanium that is produced, the product must be cooled under a protective atmosphere of an inert gas. The solid particles of reduced titanium sinter into a sponge-like mass, which is removed by drilling or chipping out by hand in large blocks. The removal must be done in dried atmosphere because titanium absorbs moisture. The liquid $MgCl_2$ slag is run off through a tap hole twice during the reduction and again at the end. The reactor is removed from the furnace, cooled to room temperature, and stored for 24 h under a small positive pressure of the inert gas, before opening in a dry room. The mixture of titanium sponge, residual $MgCl_2$, and excess magnesium is drilled out on a lathe in the form of 6–12 mm turnings. A jet of helium is kept ready to prevent the finely divided metal catching fire. The trapped Mg and $MgCl_2$ (as well as small amounts of $TiCl_3$ and $TiCl_2$) from the sponge are driven off by vacuum fractional distillation (0.01–0.1 mm Hg) at a higher temperature of 1000 °C for days to ensure removal of impurities from the titanium metal. It is a better method as there is no risk of hydrogen contamination as observed in leaching with dilute HCl, but distillation process is energy intensive with poor thermal efficiency. The titanium sponge analyzes as Mg: 0.5, Cl: 0.15, Fe: 0.1, H: 0.05 – 0.10, N up to 0.05, C: 0.05 (all in weight percent) and balance as Ti. After removal of $MgCl_2$, the sponge may be ground and pressed into ingots or melted in a special electric arc furnace and cast into ingots.

The powder may be pressed into steel molds at high pressure into bars and sintered in vacuum of the order of 10^{-4} mm Hg at 1000 – 1100 °C for 15 – 16 h. The powder metallurgy technique, however, suffers from low production rate and the difficulty of obtaining large sintered bars. The titanium produced analyzes as Ti: 99.8%, C < 0.05, H: 0.01, N $\leq$ 0.05, O < 0.2, Si: 0.07 (all in weight percent), and Mg in traces.

From the preceding paragraphs, it is clear that the Kroll process, incorporating a number of steps, namely, carbochlorination of TiO_2, purification $TiCl_4$ by vacuum fractional distillation, magnesiothermic reduction, removal of impurities (Mg and $MgCl_2$) trapped in the sponge by vacuum distillation, and electrolysis of $MgCl_2$ to get reductant Mg and Cl_2 for subsequent usage, is a batch process. The number of extracting steps involved in the production of titanium metal make the process very energy intensive, time consuming, and hence a very costly technique. Since the largest Kroll reactors in the world can produce about 10–15 tons of the sponge in a batch, a large number of reactors are required to install a titanium plant of capacity 10,000 tons per year [15]. The entire operation from chlorination to the shipment of the titanium sponge takes about 17 days [16]. Thus titanium in general, is twenty times more expensive than carbon steel and 4–5 times costlier than stainless steel [17]. In view of the high cost of production titanium has limited applications despite desirable properties like light weight, high strength, corrosion resistance, and bio-compatibility (most preferred for biomedical implants). In order to fulfil the industrial demand of titanium at lower price, in recent years, extensive researches have been directed to develop low-cost methods, which can replace the Kroll process. During the past several decades, a number of processes based on thermochemical and electrochemical principles have been proposed. Almost all the processes are at the laboratory or pilot plant stage with the prime objective of reducing the cost of production compared to that of the Kroll process together with the alternative of converting the batch process (Kroll) into a continuous one. A brief account of a few thermochemical processes will be given here. The electrolytic processes will be discussed in Chap. 12 on Electrometallurgy.

Proposed Thermochemical Processes A number of processes based on two approaches (i) reduction of $TiCl_4$ and (ii) reduction of TiO_2 have been proposed.

Under the **first category**, TiRO process developed at CSIRO, Australia produces commercial-purity titanium powder in two major steps. In the first step, $TiCl_4$ is allowed to react with magnesium powder in a fluidized bed reactor. The greatly increased rate of reaction during gas–solid fluidization reduces both the operating and capital costs [18]. The resultant titanium metal particles of 1.5 μm [19] uniformly get dispersed inside the larger spheroidal $MgCl_2$ particles (350 μm). In the second step, $MgCl_2$ is removed by vacuum distillation to obtain sintered porous spheroidal particles of titanium. In the continuous vapor phase reduction process [20] developed by the US Department of Energy, Albany Research Center, liquid $TiCl_4$ with argon gas (as a carrier) and magnesium wire are fed into a shaft reactor at 1000 °C to produce Ti, $MgCl_2$, and Mg powders, which are removed from the argon stream by an electronic precipitator. Vuuren et al. [21] have developed CSIR-Ti process to obtain titanium powder directly by metallothermic reduction of $TiCl_4$ in a molten salt.

In the past two decades, the Armstrong process [22–24] based on the principle of the reduction of $TiCl_4$ with sodium (Hunter process) has received considerable attention. In this process, $TiCl_4$ vapor is directly fed into the molten sodium stream whereas in the Hunter process, the vapor simply makes contact on the surface of the molten sodium. The difference in the method of operation makes the Armstrong process continuous. The products titanium and sodium chloride and the residual sodium are separated by filtration, distillation, and washing. Vapor phase Hunter process [16] has also been reported wherein sodium vapor is reacted with gaseous titanium tetrachloride [$TiCl_4(g) + 4Na(g) = Ti(s) + 4NaCl(s)$]. Nanoparticles of titanium encapsulated in NaCl are obtained. After removal of NaCl, the resulting titanium particles contain low oxygen.

In order to reduce the cost of production of titanium, under the **second category**, direct reduction of TiO_2 has been considered by a number of investigators because the cost of producing titanium from titanium tetrachloride is slightly more than half of the total cost of titanium sponge [16]. TiO_2 can be reduced with CaH_2 [25, 26] at 1100 °C according to the reaction:

$$TiO_2(s) + 2CaH_2(s) \rightarrow Ti\ (s) + 2CaO\ (s) + 2H_2(g) \qquad (9.26)$$

A commercial plant based on this principle is under operation in Russia but details are not available. In the preform reduction process (PRP) developed by Okabe et al. [27], the sintered (at 800 °C) preform made out of mixture of TiO_2 powder and a flux (CaO or $CaCl_2$) was allowed to react with calcium vapor in a stainless steel container at temperatures varying from 800 to 1000 °C for 6 h. Fine titanium powder of 99% purity was recovered by leaching the resultant mass with acid.

As it is not possible to lower down the oxygen content of titanium below 1% by magnesium reduction of TiO_2 at temperatures around 700 °C, two-step thermo-chemical reduction methods have been reported in literature to produce low-oxygen titanium powder. In the first step, TiO_2 is reduced to form Ti-O solid solution (TiO_2 + Mg $\rightarrow$ Ti$-$O + MgO). The remaining oxygen removal (deoxygenation) from the solid solution is carried out by calcium [28] (Ti$-$O + Ca $\rightarrow$ Ti + CaO). Use of solid/liquid/vapor calcium as well as Ca-saturated salt have been reported for deoxygenation of titanium and its alloys [15]. Okabe et al. [29] have used chemically active calcium dissolved in molten $CaCl_2$ at 1000–1200 °C for removal of oxygen from TiO_2 to obtain extra low-oxygen titanium. $CaCl_2$, acting as a flux, facilitates the reaction by decreasing the activity of CaO, the by-product. This method can deoxidize titanium to 20–60 ppm oxygen by the use of Ca-saturated $CaCl_2$ at 1000 °C. Xia et al. [30] have recently developed a new method, called "hydrogen-assisted magnesiothermic reduction" (HAMR) for the removal of oxygen from commercial titanium dioxide to produce metallic titanium powder in two steps. In the first step > 95% oxygen is removed by Mg reduction in a hydrogen atmosphere. This is followed by deoxygenation in the second step for additional reduction in oxygen with the prime objective of obtaining Ti powder that meets the standard specifications. According to the investigators [30], the emerging HAMR process has the potential to compete directly with the Kroll process. Based on thorough compositional analysis, they have further claimed that the product meets the ASTM B 299 standards, suitable to industries.

In the preceding paragraphs, a number of possibilities to produce low-cost titanium have been listed. However, it is difficult to ascertain how the proposed processes are going to handle the technical challenges of reducing the oxygen, chlorine, and nitrogen to extremely low level on tonnage scale. When it is difficult to lower down the oxygen content in the titanium sponge (Kroll process) below 0.2%, a serious attention has to be paid about the removal of these impurities to the required level from fine powders.

9.3.1.3 Reduction of Zirconium Tetrachoride

Kroll process is adapted in a similar manner for the production of zirconium sponge from $ZrCl_4$ vapors by reduction with liquid magnesium (Appendix A.10). However, there are differences in reduction of BeF_2, $NbCl_5/TaCl_5$, VCl_3, and K_2NbF_7/K_2TaF_7. Salient features are mentioned below. Details can be obtained from References [5, 10].

9.3.1.4 Reduction of Beryllium Fluoride

Glassy solid beryllium fluoride obtained by complex fluoride process (discussed in Sect. 9.1.1.1, Appendix A.6) is crushed to 3 mm size and reduced with magnesium according to the reaction:

$$BeFe_2(s) + Mg\ (l) = Be\ (s) + MgF_2(s) \qquad (9.27)$$

The stoichiometric quantities of mixed reactants are charged into a monolythic crucible containing fused $CaCl_2$ at 1000 – 1100 °C. The reaction takes place rapidly because it initiates at about 650 °C. The reaction products are heated to above the melting point of beryllium (1283 °C) and held for 2 h. Beryllium being lighter floats to the surface aggregating into small beads of about 12 mm size. Calcium chloride acting as a flux lowers the melting point of magnesium fluoride (1263 °C). The crucible containing the reaction products is cooled to 1050 – 1100 °C. The beads are quickly removed and placed in an argon-filled vessel that is then cooled to room temperature.

9.3.1.5 Reduction of Chlorides of Niobium/Tantalum and Vanadium

For production of niobium and tantalum, vapors of $NbCl_5$ and $TaCl_5$ are brought into contact with liquid magnesium contained in a crucible. $NbCl_5$ is distilled from a glass vessel into the reactor at a rate sufficient to maintain the crucible temperature at 800 °C without external heat to react with magnesium in excess of about 10%. The reactor is then cooled and dismantled. The resultant product containing Nb, Mg, and $MgCl_2$ is leached with water, dilute acid, and water again to recover niobium. A similar procedure is adopted in reduction of $TaCl_5$ vapors with liquid magnesium.

Vanadium trichloride is reduced with magnesium in a mild steel reactor heated by means of a gas-fired furnace. Magnesium after pickling, washing, and drying is melted in a crucible and then heated to 750–800 °C. VCl_3 is then allowed to react with the molten magnesium at a controlled rate so as to maintain the temperature between 750 and 850 °C without any additional heat from outside. After reduction the crucible is cooled and transported quickly for distillation at 920–950 °C for 8 h under continuous evacuation to remove Mg and $MgCl_2$.

9.3.1.6   Reduction of Potassium Heptafluoroniobate/Tantalate

Dry potassium heptafluoroniobate (K_2NbF_7) is reduced with sodium in either a closed bomb or an open reactor.

$$K_2NbF_7(s) + 5Na(l) = Nb(s) + 5\ NaF(s) + 2KF(s) \qquad (9.28)$$

Alternate layers of double fluoride and freshly cut cubes of sodium packed in the open reactor are covered with sodium chloride. NaCl acts as a trap for condensation of sodium vapors. The reaction is initiated within a few minutes from the top by a simple ring burner fitted around the reactor. The burner is then gradually lowered to allow the reaction to penetrate to the bottom of the reactor within 20 min. The initiation temperature is normally about 420 °C. After cooling for several hours, NaCl is recovered for reuse. The excess metallic sodium and potassium is destroyed by adding methylated spirit into the reactor. A similar procedure is adopted in reduction of potassium heptafluorotantalate (K_2TaF_7).

9.3.2   Fused Salt Electrolytic Process

The use of electric power for the production of rare metals may help in reducing the cost compared to the use of reducing agents like calcium, magnesium, and sodium. Further, the product is not contaminated with the reducing agents and any of its associated impurities. The commercial electrolytic processes for rare metal production are based on fused salt bath. Chlorides and fluorides of the reactive/rare metals dissolved in the mixture of inert salts of alkali and/or alkaline earth metals, are invariably electrolyzed. Production of rare and reactive metals by fused salt electrolytic process will be discussed in Chap. 12 on Electrometallurgy.

9.4   Purification

The following two processes for purification of reactive metals are based on halide routes:

9.4.1   Disproportionate Process

Reaction involving different valences is called "disproportionate reaction." This can be very effectively used in the preparation of ultrapure metals. Reactive metals like aluminum, beryllium, titanium, zirconium, and niobium can be purified by disproportionate reaction processes [31]. To some extent, the difference in valences, is taken advantage of in the formation and decomposition of halides of the same metal at different temperatures. For example, impure liquid aluminum can be vaporized as aluminum monochloride by passing aluminum trichloride gas into the molten bath at 1000 °C according to the reaction:

$$AlCl_3(g) + 2Al \ (l, impure) \xrightarrow{1000\ °C} 3AlCl \ (g) \qquad (9.29)$$

Then, solid or liquid aluminum can be recovered at lower temperature by reversing the reaction:

$$3AlCl \ (g) \xrightarrow[600-700\ °C]{cooling} 2Al \ (s/l, pure) + AlCl_3 \ (g) \qquad (9.30)$$

$AlCl_3$ gas regenerated is recycled for further reaction with impure aluminum. In order to produce aluminum metal of 99.999% purity, the disproportionate reaction should be carried out with crude aluminum having lower concentration of Ca, Ti, Mn, Mg, Zn, and Cu because these impurities cannot be effectively removed from commercial aluminum.

Preparation of ultrapure beryllium by this process involves the formation of beryllium chloride gas and sodium vapors at about 1200 °C in the first step from impure beryllium and sodium chloride vapors. The resultant product on cooling to about 900 °C, produces pure beryllium according to the reactions:

$$Be \ (s, impure) + 2 \ NaCl \ (g) \xrightarrow{1200\ °C} BeCl_2(g) + 2 \ Na \ (g) \qquad (9.31)$$

$$BeCl_2(g) + 2 \ Na \ (g) \xrightarrow{900\ °C} Be(s, pure) + 2 \ NaCl(l) \qquad (9.32)$$

The other examples are:

$$5NbCl_3(s) \rightarrow 2Nb \ (s, pure) + 3NbCl_5(g) \qquad (9.33)$$

$$4TiI_3(g) \rightarrow Ti \ (s, pure) + 3TiI_4(g) \qquad (9.34)$$

$$4ZrI_3(g) \rightarrow Zr \ (s, pure) + 3ZrI_4(g) \qquad (9.35)$$

9.4.2 Iodide Process

Out of all the halides, iodides are the least stable, and, hence, can be easily decomposed. The iodides of reactive metals such as titanium, zirconium, hafnium, and thorium, formed at lower temperatures can be decomposed at higher temperatures to prepare highly pure metals and the liberated iodine can be recycled. van Arkel process [5, 32, 33] based on this principle will be discussed in Chap. 10 on Refining under the heading ultrapurification.

9.5 Problems [34–36]

Problem 9.1
(Nagamori, M., 1979–1980, personal communication, University of Utah, Salt Lake City)
Develop suitable equations to express efficiency of chlorination of some oxides.

Solution
Let us consider chlorination of rutile and magnesia.

The chlorination of rutile:

$$TiO_2(s) + 2Cl_2(g) = TiCl_4(g) + O_2(g)$$

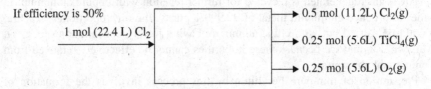

If efficiency is 50%   → 0.5 mol (11.2L) $Cl_2(g)$
1 mol (22.4 L) Cl_2 → 0.25 mol (5.6L) $TiCl_4(g)$
                         → 0.25 mol (5.6L) $O_2(g)$

According to Dalton's law, partial pressure of any gas (p_i) is given as $p_i = x \cdot P$, where x_i and P respectively, represent mole fraction and the total pressure. If v represents volume of the gas, from stoichiometric viewpoint of the reaction, we can write:

$$v_{TiCl_4} = v_{O_2} \tag{1}$$

$$\text{Mole faction of chlorine gas,} \, x_{Cl_2} = \frac{v_{Cl_2}^{out}}{v_{Cl_2}^{out} + v_{TiCl_4}^{out} + v_{O_2}^{out}} = \frac{v_{Cl_2}^{out}}{v_{Cl_2}^{out} + 2\,v_{O_2}^{out}} \tag{2}$$

$$\text{Stoichiometrically,} \, v_{O_2}^{out} = \frac{1}{2} \cdot v_{Cl_2}^{used} \tag{3}$$

$$\text{and} \, v_{Cl_2}^{out} = v_{Cl_2}^{in} - v_{Cl_2}^{used}$$

$$\text{or} \, v_{Cl_2}^{used} = v_{Cl_2}^{in} - v_{Cl_2}^{out} \tag{4}$$

Substituting Eq. 4 in Eq. 3, we get

$$v_{O_2}^{out} = \frac{1}{2}\left(v_{Cl_2}^{in} - v_{Cl_2}^{out}\right) \tag{5}$$

By definition efficiency of utilization can be expressed as:

$$E = \frac{v^{in}_{Cl_2} - v^{out}_{Cl_2}}{v^{in}_{Cl_2}}$$

$$\text{or } v^{out}_{Cl_2} = v^{in}_{Cl_2} - E.v^{in}_{Cl_2} \tag{6}$$

Substituting Eqs. 5 and 6 in Eq. 2, we get

$$x_{Cl_2} = \frac{v^{in}_{Cl_2} - E.v^{in}_{Cl_2}}{\left(v^{in}_{Cl_2} - E.v^{in}_{Cl_2}\right) + 2\left\{\frac{1}{2}\left[v^{in}_{Cl_2} - \left(v^{in}_{Cl_2} - E.v^{in}_{Cl_2}\right)\right]\right\}}$$

$$= \frac{1 - E}{1 - E + [1 - 1 + E]} = 1 - E \tag{7}$$

$$\text{Therefore}, E = 1 - x_{Cl_2} = 1 - \frac{p_{Cl_2}}{P} \tag{8}$$

Thus, the utilization efficiency of chlorine may be estimated if p_{Cl_2} and P are known. The value of p_{Cl_2} can be calculated from the equilibrium constant, whereas P may be chosen arbitrarily depending on the system as specified.

The chlorination of magnesia:

$$MgO\ (s) + Cl_2(g) = MgCl_2\ (l) + \frac{1}{2}\ O_2(g)$$

The chlorine balance may be represented as:

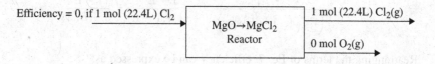

According to the Dalton's law of partial pressure: $p_i = x_i. P$ (where P is the total pressure and p_i and x_i respectively, stand for partial pressure and mole fraction of the component i.

$$\text{From stoichiometry}, v^{out}_{O_2} = \frac{1}{2} v^{used}_{Cl_2} \tag{1}$$

$$x_{Cl_2} = \frac{v_{Cl_2}^{out}}{v_{Cl_2}^{out} + v_{O_2}^{out}} \tag{2}$$

$$v_{Cl_2}^{used} = v_{Cl_2}^{in} - v_{Cl_2}^{out} \tag{3}$$

Substituting Eq. 3 in Eq. 1, we get:

$$v_{O_2}^{out} = \frac{1}{2}\left(v_{Cl_2}^{in} - v_{Cl_2}^{out}\right) \tag{4}$$

Combining Eqs. 4 and 2, we get:

$$x_{Cl_2} = \frac{v_{Cl_2}^{out}}{v_{Cl_2}^{out} + \frac{1}{2}\left(v_{Cl_2}^{in} - v_{Cl_2}^{out}\right)} = \frac{v_{Cl_2}^{out}}{\frac{1}{2}\left(v_{Cl_2}^{in} + v_{Cl_2}^{out}\right)} \tag{5}$$

On the other hand, utilization efficiency of $Cl_2(g)$ is given by the expression:

$$E = \frac{v_{Cl_2}^{in} - v_{Cl_2}^{out}}{v_{Cl_2}^{in}}$$

$$v_{Cl_2}^{out} = (1 - E)\, v_{Cl_2}^{in} \tag{6}$$

Substituting Eq. 6 in Eq. 5, we get

$$x_{Cl_2} = \frac{(1 - E)v_{Cl_2}^{in}}{\frac{1}{2}\left(v_{Cl_2}^{in} + (1 - E)v_{Cl_2}^{in}\right)} = \frac{1 - E}{\frac{1}{2}(1 + 1 - E)} = \frac{1 - E}{1 - E/2} \tag{7}$$

Rearranging the terms of Eq. 7, efficiency can be expressed as:

$$E = \frac{1 - x_{Cl_2}}{1 - \frac{1}{2}x_{Cl_2}} = \frac{1 - \frac{P_{Cl_2}}{P}}{1 - \frac{1}{2}\frac{P_{Cl_2}}{P}}$$

Problem 9.2

(a) Assuming aluminum-vaporization and aluminum-deposition take place in a reactor, derive general relationships to correlate the ingoing number of moles of $AlCl_3(g)$ and $AlCl(g)$ to the outgoing number of moles of $AlCl_3(g)$ and $AlCl$ (g) in a disproportionation reaction: $3AlCl(g) = 2Al(s) + AlCl_3(g)$.

(b) If the values of the equilibrium constants for the reaction are 1.293 and 6.337×10^8 at 1200 and 650 °C, respectively, and pure $AlCl_3(g)$ is injected under 1 atm pressure into a bath of molten aluminum at 1200 °C, calculate the chemical composition of the off-gas composed of $AlCl_3$ and $AlCl$. How many moles of aluminum are vaporized by the gas from the bath?

(c) When a gas mixture consisting of 0.62 mol $AlCl_3$ and 1.14 mol $AlCl$ is passed and equilibrated in a reactor maintained at 650 °C and 1 atm, calculate the chemical composition of the off-gas leaving the reactor. How many moles of aluminum is recovered in the reaction?

Solution

(a) Vaporization and deposition of aluminum are represented in the diagram below:

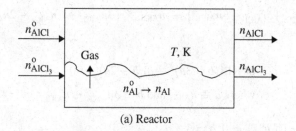

(a) Reactor

Let (i) n^o_{AlCl} and $n^o_{AlCl_3}$ moles of AlCl(g) and AlCl$_3$(g) enter in the reactor maintained at T K, and 1 atm pressure. (ii) n_{Al} number of moles of Al is purified from n^o_{Al} moles and (iii) n_{AlCl} and n_{AlCl_3} moles of gases come out of the reactor.

That is, we have a situation: $n^o_{Al}, n^o_{AlCl_3}, n^o_{AlCl} \rightarrow n_{AlCl}, n_{AlCl_3}, n_{Al}$

$$\text{Al balance}: n^o_{Al} + n^o_{AlCl_3} + n^o_{AlCl} = n_{AlCl_3} + n_{AlCl} + n_{Al} \tag{1}$$

$$\text{Cl balance}: 3n^o_{AlCl_3} + n^o_{AlCl} = n_{AlCl} + 3n_{AlCl_3} \tag{2}$$

$$n_T = n_{AlCl} + n_{AlCl_3} \tag{3}$$

There are three species and two mass balances; hence, one independent equation must be sought. According to the Dalton's law of partial pressure,

$$p_i = \frac{n_i}{n_T}.P \tag{4}$$

For the reaction: $3AlCl(g) = 2Al(l, s) + AlCl_3(g)$

$$K = \frac{p_{AlCl_3}.a^2_{Al}}{p^3_{AlCl}} \text{ (as Al is pure, } a_{Al} = 1)$$

$$= \frac{\left(n_{AlCl_3}/n_T\right).P.1^2}{\left(n_{AlCl}/n_T\right)^3 P^3} = \frac{n_{AlCl_3}.\, n^2_T}{n^3_{AlCl}.P^2}$$

$$\therefore K.n^3_{AlCl}.P^2 = n_{AlCl_3}.\, n^2_T \tag{5}$$

$$\text{From Eq. 1}\quad n_{Al} = n^o_{AlCl_3} + n^o_{AlCl} - n_{AlCl_3} - n_{AlCl} \tag{6}$$

$$\text{From Eq. 2}\quad n_{AlCl} = 3n^o_{AlCl_3} + n^o_{AlCl} - 3n_{AlCl_3} \tag{7}$$

Substituting Eq. 7 in Eq. 3 we get:

$$n_T = n_{AlCl_3} + \left(3n^o_{AlCl_3} + n^o_{AlCl} - 3n_{AlCl_3}\right) = 3n^o_{AlCl_3} + n^o_{AlCl} - 2n_{AlCl_3} \qquad (8)$$

Subtituting Eqs. 7 and 8 in Eq. 5, we get the required relationship:

$$KP^2\left(3n^o_{AlCl_3} + n^o_{AlCl} - 3n_{AlCl_3}\right)^3 - n_{AlCl_3} \times \left(3n^o_{AlCl_3} + n^o_{AlCl} - 2n_{AlCl_3}\right)^2$$
$$= 0 \qquad (9)$$

$$\text{or } KP^2(b - 3a)^3 - a\,(b - 2a)^2 = 0 \qquad (10)$$

$$\text{where } a \equiv n_{AlCl_3} \text{ and } b \equiv 3n^o_{AlCl_3} + n^o_{AlCl} \qquad (11)$$

when $n^o_{AlCl_3} = 1,\ n^o_{AlCl} = 0, b = 3$

(b) $T = 1200\ °C$, Substituting: $P = 1$ atm, $K = 1.293$, $n^o_{AlCl_3} = 1, n^o_{AlCl} = 0$ in
    Eq. 10, we get : $f(a) = 1.293\,(3 - 3a)^3 - a\,(3 - 2a)^2 = 0$

$a = 0.5, f\,(0.5) = 4.364 - 2.000 = 2.364$
$a = 0.7, f\,(0.7) = 0.943 - 1.792 = -0.849$
$a = 0.63, f\,(0.63) = 1.768 - 1.907 = -0.139$
$a = 0.61, f\,(0.61) = 2.071 - 1.933 = 0.138$

Thus $f\,(a) = 0$ when $a = 0.62 = n_{AlCl_3}$
Then from Eq. 7, $n_{AlCl} = 3n^o_{AlCl_3} + n^o_{AlCl} - 3n_{AlCl_3}$
$$= 3 \times 1 + 0 - 3 \times 0.62 = 1.14$$
From Eq. 6, the Al picked up by the gas is obtained as

$$n_{Al} = 3 - 0.62 - 1.14 = 1.24 \text{ mol}$$

The off-gas contains 1.14 mol of AlCl and 0.62 mol of AlCl$_3$ gases on vaporization of 1.24 mol of aluminum at 1200 °C. This is presented in reactor (b).

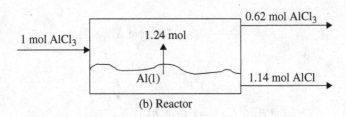

(b) Reactor

(c) $T = 650\ °C, P = 1$ atm, $K = 6.337 \times 10^8$, $n^o_{AlCl_3} = 0.62, n^o_{AlCl} = 1.14$

From Eq. 11, $b \equiv 3n^o_{AlCl_3} + n^o_{AlCl} = 3 \times 0.62 + 1.14 = 3$
To obtain a, substitute these values in Eq. 10

$$f(a) = 6.337 \times 10^8 (3 - 3a)^3 - a(3 - 2a)^2$$
$$= 1.711 \times 10^{10} (1 - a)^3 - a(3 - 2a)^2 = 0$$

if $a = 0.9996$, $f(0.9996) = 1.09504 - 1.0012 = +0.0938$
if $a = 0.9997$, $f(0.9997) = 0.0462 - 1.0009 = -0.539$

$$f(a) = 0 \text{ when } a = 0.99961 = n_{AlCl_3}$$

Then from Eq. 7, $n_{AlCl} = 3n^o_{AlCl_3} + n^o_{AlCl} - 3n_{AlCl_3}$

$$= 3 \times 0.62 + 1.14 - 3 \times 0.99961 = 0.0012$$

and from Eq. 6, $n_{Al} = n^o_{AlCl_3} + n^o_{AlCl} - n_{AlCl_3} - n_{AlCl}$

$$= 0.62 + 1.14 - 0.99961 - 0.0012 = 0.7592$$

The off-gas leaving the reactor contains 0.0012 mol of AlCl and 0.99961 mol of AlCl$_3$ gases and 0.7592 mol of pure aluminum is recovered at 650 °C. This is presented in reactor(c) shown below:

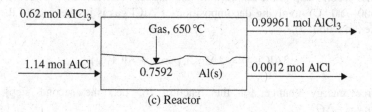

(c) Reactor

Problem 9.3
Making use of the following data, develop ΔG^o vs T equations for the two steps in purification of aluminum by disproportionate process:

$$2Al\,(s) + Cl_2(g) = 2\,AlCl\,(g), \Delta G^o_1 = -21360 + 11.75\,T\log T - 76.5\,T \text{ cal} \quad (1)$$

$$2Al\,(l) + Cl_2(g) = 2\,AlCl\,(g), \Delta G^o_2 = -25860 + 11.80\,T\log T - 71.9\,T \text{ cal} \quad (2)$$

$$Al\,(s) + 1.5\,Cl_2(g) = AlCl_3(g), \Delta G^o_3 = -138160 + 2.48\,T\log T + 4.75\,T \text{ cal} \quad (3)$$

$$Al\,(l) + 1.5\,Cl_2(g) = AlCl_3(g), \Delta G^o_4 = -140400 + 2.50\,T\log T + 7.05\,T \text{ cal} \quad (4)$$

Estimate the composition of the equilibrium gaseous phase at 1200 °C and 700 °C and comment on the feasibility of the two steps of the disproportionate process. Aluminum melts at 660 °C.

Solution
Since both the working temperatures (1200 and 700 °C) are above the melting point of aluminum (660 °C), we consider only reactions (2) and (4). In the first step of the disproportionate process, impure liquid aluminum is vaporized as aluminum monochloride by passing aluminum trichloride gas into the molten bath at 1200 °C according to the reaction:

$$AlCl_3(g) + 2Al(l, impure) = 3AlCl(g) \qquad (5)$$

The free energy change for the reaction (5), can be estimated from free energy values of reactions (2) and (4).

$$\begin{aligned} \Delta G_5^o &= 1.5 \, \Delta G_2^o - \Delta G_4^o \\ &= 1.5 \, (-25860 + 11.80 \, T \log T - 71.9 \, T) - (140400 + 2.50 \, T \log T + 7.05 \, T) \\ &= 101610 + 15.2 \, T \log T - 114.90 \, T \text{ cal} \end{aligned}$$

Since the second step in the disproportionate process carried out at lower temperature (600–700 °C) involving decomposition of AlCl gas is just reverse of the first step (i.e., reaction: 5):

$$3AlCl(g) = 2Al(s, l, pure) + AlCl_3(g) \qquad (6)$$

The free energy change for the reaction (6, i.e., the second step) will be: $\Delta G_6^o = -\Delta G_5^o$

$$\therefore \Delta G_6^o = -101610 - 15.2 \, T \log T + 114.90 \, T \text{ cal}$$

At 1200 °C, that is, 1473 K,

$$\begin{aligned} \Delta G_5^o &= 101610 + 15.2 \, T \log T - 114.90 \, T \\ &= 3297 \text{ cal} = -4.575T. \log K \\ K &= 0.3241 \end{aligned}$$

For Eq. 5, $K = \dfrac{p_{AlCl}^3}{p_{AlCl_3}} = 0.3241$

$$\therefore p_{AlCl_3} = 3.085.p_{AlCl}^3 \qquad (7)$$

And from the given conditions we have,

$$p_{AlCl_3} + p_{AlCl} = 1 \qquad (8)$$

From Eqs. 7 and 8, we have: $3.085.p_{AlCl}^3 + p_{AlCl} = 1$

On solving by trial and error we get: $p_{AlCl} = 0.533$ atm

$$p_{AlCl_3} = 1 - 0.533 = 0.467 \text{ atm}$$

Assuming ideal behavior, on volume basis the gaseous phase at 1200 °C contains 53.3% AlCl and 46.7% AlCl$_3$.

At 700 °C, that is, 973 K,

$$\Delta G_5^o = 101610 + 15.2 \times 973 \times \log(973) - 114.9 \times 973$$
$$= 34005 \text{ cal}$$
$$= -4.575\, T \log K$$
$$K = 23 \times 10^{-9}$$
$$K = \frac{p_{AlCl}^3}{p_{AlCl_3}} = 23 \times 10^{-9}$$

$$\therefore p_{AlCl_3} = \frac{1}{23 \times 10^{-9}} p_{AlCl}^3 = \frac{10^9}{23} p_{AlCl}^3 \qquad (9)$$

$$\text{and } p_{AlCl_3} + p_{AlCl} = 1 \qquad (8)$$

From Eqs. 9 and 8, we have: $\frac{10^9}{23}.p_{AlCl}^3 + p_{AlCl} = 1$

On solving by trial and error we get: $p_{AlCl} = 0.002842$ atm

$$\text{and } p_{AlCl_3} = 0.997158 \text{ atm}$$

Assuming ideal behavior, on volume basis the gaseous phase at 700 °C contains 0.28% AlCl and 99.72% AlCl$_3$.

From the above calculations it is clear that at 1200 °C, the proportion of AlCl gas is much higher than it is at 700 °C. This is consistent with the first step involved in refining of aluminum by the disproportionate process when AlCl gas is formed by passing AlCl$_3$ gas into impure molten aluminum at 1200 °C. On lowering the temperature to 700 °C, the second step involving decomposition of AlCl gas into pure Al and AlCl$_3$ gas takes place. Hence, proportion of AlCl in the gaseous mixture in the second step drastically decreases.

Problem 9.4

From the following data comment on the feasibility of production of titanium (i) by carbothermic reduction of TiO$_2$ at 1200 °C and (ii) via chlorination of TiO$_2$ at 1200 °C under reducing conditions and subsequent reduction of TiCl$_4$ with magnesium at 750 °C. Given that:

$$\text{Ti (s)} + O_2(g) = TiO_2(s), \Delta G_1^o = -223500 + 41.55\, T, \text{cal} \tag{1}$$

$$2\,C\,(s) + O_2(g) = 2CO\,(g), \Delta G_2^o = -53400 - 41.90\, T, \text{cal} \tag{2}$$

$$\text{Ti (s)} + 2Cl_2(g) = TiCl_4(g),$$
$$\Delta G_3^o = -180700 - 1.8\, T \log T + 34.65\, T, \text{cal} \tag{3}$$

$$\text{Mg (l)} + Cl_2(g) = MgCl_2(l),$$
$$\Delta G_4^o = -147850 - 13.58\, T \log T + 72.77\, T, \text{cal} \tag{4}$$

Solution

Route I: Carbothermic reduction of TiO_2 at 1200 °C:

$$TiO_2(s) + 2C\,(s) = Ti\,(s) + 2CO\,(g) \tag{5}$$

$$\Delta G_5^o = \Delta G_2^o - \Delta G_1^o = -53400 - 41.90\, T - (-223500 + 41.55\, T)$$
$$= 170100 - 83.45\, T$$

At 1473 K, $\Delta G_5^o = 170100 - 83.45\, T = 170100 - 83.45 \times 1473 = +47178$ cal

Since the free energy change for the carbothermic reduction of TiO_2 at 1200 °C is positive reaction will not be feasible.

Route II: Chlorination of TiO_2 at 1200 °C in the presence of carbon:

$$TiO_2(s) + 2C\,(s) + 2Cl_2(g) = TiCl_4(g) + 2CO\,(g) \tag{6}$$
$$\Delta G_6^o = \Delta G_2^o + \Delta G_3^o - \Delta G_1^o = -10600 - 1.8\, T \log T - 48.80\, T$$

At 1473 K, $\Delta G_6^o = -90883$ cal

Since the free energy change for chlorination of TiO_2 at 1200 °C under reducing conditions is negative, reaction is feasible.

Reduction of $TiCl_4$ (g) with Mg (l) at 750 °C:

$$TiCl_4(g) + 2Mg\,(l) = Ti\,(s) + MgCl_2(l) \tag{7}$$

$$\Delta G_7^o = 2\Delta G_4^o - \Delta G_3^o = -115000 - 25.36\, T \log T + 110.89\, T$$

At 1023 K, $\Delta G_7^o = -79646$ cal

Since the free energy change for reduction of $TiCl_4$ (g) with Mg (l) at 750 °C is negative, reaction is feasible.

Problem 9.5

What is the permissible limit of HCl gas in the gaseous mixture: H_2–HCl for reduction of chromium chloride by hydrogen at 1100 K? Given that:

$$H_2(g) + Cl_2(g) = 2HCl\,(g),\, \Delta G_1^o = -43540 + 1.98\,T \log T - 10.44\,T\,\text{cal} \quad (1)$$

$$Cr\,(s) + Cl_2(g) = CrCl_2\,(s),\, \Delta G_2^o = -93900 - 8.7\,T \log T + 55.0\,T\,\text{cal} \quad (2)$$

Solution

The reaction under consideration:

$$CrCl_2\,(s) + H_2(g) = Cr\,(s) + 2HCl\,(g), \quad (3)$$

$$\Delta G_3^o = \Delta G_1^o - \Delta G_2^o$$

$$= 50360 + 10.68\,T \log T - 65.44\,T\,\text{cal}$$

At 1100 K, $\Delta G_3^o = 14106$ cal

$$\therefore -4.575\,T \log K = 14106,\quad K = 0.001574$$

for reaction (3), $K = \frac{p_{HCl}^2}{p_{H_2}}$

$$\therefore \frac{p_{HCl}^2}{p_{H_2}} = 0.001574 \ \text{ or } p_{H_2} = \frac{1}{0.001574} p_{HCl}^2 = 635.32 \cdot p_{HCl}^2 \quad (4)$$

$$\text{and } p_{H_2} + p_{HCl} = 1 \quad (5)$$

From (4) and (5) we get: $635.32 \cdot p_{HCl}^2 + p_{HCl} - 1 = 0$

$$p_{HCl} = 0.039 \text{ atm}$$

$$p_{H_2} = 0.961 \text{ atm}$$

In the light of the above calculations it is evident that reduction of $CrCl_2$ by hydrogen will proceed in the forward direction only if the concentration of HCl gas in the gaseous mixture: H_2–HCl does not exceed beyond 3.9% by volume. Hence, HCl gas has to be removed by inserting fresh hydrogen gas into the reduction chamber.

References

1. Moore, J. J. (1990). *Chemical metallurgy* (2nd ed.). Oxford: Butterworth-Heinemann. (Chapter 7).
2. Coudurier, L., Hopkins, D. W., & Wilkomirsky, I. (1978). *Fundamentals of metallurgical processes* (1st ed.). Oxford: Pergamon Press Ltd. (Chapter 5).
3. Rosenqvist, T. (1974). *Principles of extractive metallurgy*. New York: McGraw-Hill. (Chapter 14).
4. Gilchrist, J. D. (1980). *Extraction metallurgy* (2nd ed.). Oxford: Pergamon. (Chapter 12).

5. Jamrack, W. D. (1963). *Rare metal extraction by chemical engineering techniques*. New York: Macmillan. (Chapter 2 and 8).
6. Mantell, C. L. (1960). *Electrochemical engineering* (4th ed.). New York: McGraw-Hill. (Chapter 5).
7. Harrington, C. D., & Ruehle, A. E. (1959). *Uranium production technology*. New York: D. van Nostrando. Inc. (Chapters 5–8).
8. Cuthbert, F. L. (1958). *Thorium production technology*. Reading: Addison-Wesley. (Chapter 5 and 6).
9. Sevryukov, N., Kuzmin, B., & Chelishchev, Y., (translated: Kuznetsov, B.) (1960). *General metallurgy*. Moscow: Peace Publishers. (Chapter 17, 18 and 21).
10. Minkler, W. W., & Baroch, E. F. (1981). The production of titanium, zirconium and hafnium. In J. K. Tien & J. F. Elliott (Eds.), *Metallurgical treatises* (pp. 171–189). Warrendale: Metallurgical Society, AIME.
11. Hunter, M. A. (1910). Metallic titanium. *Journal of the American Chemical Society, 32*, 330–336.
12. Forbath, T. P. (1958). Sodium reduction route yields titanium. *Chemical Engineering, 65*, 124–127.
13. Kroll, W. J. (1940). The production of ductile titanium. *Transactions of the Electrochemical Society, 78*, 35–47.
14. Kroll, W. J. (1955). How commercial Ti and Zr were born. *Journal of the Franklin Institute, 260*, 169–192. (Science Direct, Elsevier).
15. Zhang, Y., Fang, Z. Z., Sun, P., Zheng, S., Xia, Y., & Free, M. (2017). A perspective on thermochemical and electrochemical processes for titanium metal production. *JOM, 69*, 1861–1867.
16. Fray, D. (2008). Novel methods for the production of titanium. *International Materials Review, 53*, 317–325.
17. Fang, Z. Z., Paramore, J. D., Sun, P., Ravichandran, K. S., Zheng, S., Xia, Y., Cao, F., Koopman, M., & Free, M. (2018). Powder metallurgy of titanium- past, present and future. *International Materials Review, 63*, 407–459.
18. Brooks, G., Cooksey, M., Wellwood, G., & Goodes, C. (2007). Challenges in light metal production. *Transactions of the Institution of Mining and Metallurgy, Section C: Mineral Processing and Extractive Metallurgy, 116*, C25–C33.
19. Doblin, C., Chryss, A., & Monch, A. (2012). Titanium powder from the TiRO process. *Key Engineering Materials, 520*, 95–100.
20. Hansen, D. A., & Gerdemann, S. J. (1998). Producing titanium powder by continuous vapor phase reduction. *JOM, 50*, 56–58.
21. van Vuuren, D. S., Oosthuizen, S. J., & Heydenrych, M. D. (2011). Titanium production via metallothermic reduction of $TiCl_4$ in molten salt. *Journal of the Southern African Institute of Mining and Metallurgy, 111*, 141–147.
22. Wei, C., Yamamoto, Y., & Peter, W. H. (2010). Investigations of pressing and sintering process of CP-Ti powder made by Armstrong process. *Key Engineering Materials, 436*, 123–129.
23. Crowley, G. (2003). How to extract low cost titanium. *Advanced Materials and Processes, 161*, 25–27.
24. Gerdemann, S. J., Oden, L. L., & White, J. C. (1997). In B. Mishra & G. J. Kipouros (Eds.), *Proceedings of the 1997 materials week 'titanium extraction and processing'* (pp. 49–54). Indianapolis: TMS.
25. Borok, B. A. (1965) Lowering the cost of titanium. *Transaction of the Central Research Institute for Ferrous Metallurgy, 43*, 69–80.
26. Froes, F. H. (1998). The production of low-cost titanium powders. *JOM, 50*(9), 41–43.
27. Okabe, T. H., Oda, T., & Mitsuda, Y. (2004). Titanium powder production by preform reduction process. *Journal of Alloys and Compounds, 364*, 156–161.

28. Zhang, Y., Fang, Z. Z., Xia, Y., Huang, Z., Lefler, H., Zhang, T. Y., Sun, P., Free, M. L., & Gua, J. (2016). A novel chemical pathway for energy efficient production of titanium metal from upgraded titanium slag. *Chemical Engineering Journal, 286*, 517–527.

29. Okabe, T., Oishi, T., & Ono, K. (1992). Preparation and characterization of extra-low oxygen titanium. *Journal of Alloys and Compounds, 184*, 43–56.

30. Xia, Y., Fang, Z. Z., Zhang, Y., Lefler, H., Zhang, T. Y., Sun, P., & Huang, Z. (2017). Hydrogen- assisted magnesiothermic reduction (HAMR) of commercial TiO_2 to produce titanium powder with controlled morphology and particle size. *Materials Transactions, 56*(7), 355–360.

31. Shamsuddin, M., & Sohn, H. Y. (2019). Constitutive topics in physical chemistry of high-temperature nonferrous metallurgy – A review: Part 2. Reduction and refining. *JOM, 71*(9), 2366–2376.

32. Shelton, R. A. J. (1968). Thermodynamic analysis of the van Arkel iodide process. *Transactions of the Institution of Mining and Metallurgy, Section C, 77*, C32–C35.

33. Shelton, R. A. J. (1968). Rate of deposition of zirconium in the van Arkel iodide process. *Transactions of the Institution of Mining and Metallurgy, Section C, 77*, C113–C119.

34. Kubaschewski, O., & Alcock, C. B. (1979). *Metallurgical thermochemistry* (5th ed.). Oxford: Pergamon.

35. Curnutt, J. L., Prophet, H., McDonald, R. A., & Syverud, A. N. (1975). *JANAF thermochemical tables*. Midland: The Dow Chemical Company.

36. Barin, I., Knacke, O., & Kubaschewski, O. (1977). *Themochemical properties of inorganic substances, (supplement)*. New York: Springer-Verlag.

Chapter 10
Refining

In general, primary metals obtained through various processing routes are associated with some impurities. Ores, flux, and fuel are the main sources of impurities. As many physical, chemical, and mechanical properties of metals are influenced by the presence of impurities, it becomes essential to refine them for effective usage. A metal may be subjected to one or more refining methods depending on the chemical characteristics of the metal as well as those of the impurities. Broadly speaking, the objective of refining is to produce the metal as pure as possible and to recover precious metals such as gold and silver as by-products. On the other hand, steel-making is an example of refining to produce steel with controlled amount of impurities. Therefore, purity is a relative term, which is referred to be based on the usage of the metal. One property may not be affected by the presence of a particular impurity but another property may be significantly influenced by the same. For example, density of copper does not change with traces of oxygen, whereas dissolved oxygen reduces the electrical conductivity of copper to a large extent. Similarly, the presence of traces of hafnium in zirconium poses problems in the application of the latter when used as fuel canning material in nuclear reactors. Low-neutron absorption cross-section makes zirconium (0.15 barn per atom) a valuable material for fuel cans. As the neutron absorption cross-section of the co-occurring sister element, hafnium, is very high (115 barn per atom) it must be restricted below 200 ppm in zirconium for nuclear applications.

The common metals, namely, copper, lead, zinc, nickel, tin, aluminum, and so on produced in bulk are refined by different methods based on the difference in physical and chemical characteristics of the metal and impurities. During refining of high reactive metals such as uranium, thorium, titanium, and zirconium, it is beneficial to eliminate those impurities associated with the ore right in the beginning. In processing reactive metals, the compound obtained from the ore breakdown step [1] is purified by chemical separation techniques like selective distillation (titanium), chemical precipitation (uranium), ion exchange (uranium, zirconium), and solvent extraction (uranium, thorium, and zirconium, etc.). These separation techniques are discussed in the next chapter on hydrometallurgy. Thus, there are two basic aspects

M. Shamsuddin, *Physical Chemistry of Metallurgical Processes, Second Edition*,
The Minerals, Metals & Materials Series, https://doi.org/10.1007/978-3-030-58069-8_10

in the refining of reactive metals, namely, purification of compounds and refining of the primary metals.

10.1 Principle

Thermodynamically, the pure phase is unstable. It has an inherent tendency to pick up impurities from the surrounding environment because free energy of the pure phase is lowered by absorption of impurities. Hence, purification becomes more and more difficult as we approach the pure state because the purer phase is more prone to picking impurities. Knowledge of thermodynamics and kinetics are of utmost importance for generating favorable conditions in all refining techniques. It is more difficult to remove impurities held at lower chemical potential. During refining, the impure metal is brought into contact with slag or vacuum, which can pick impurities at lower chemical potential. Hence, knowledge about activities of the metal and its impurities would be helpful in formulating thermodynamic conditions for refining by a particular technique. Since chemical potential decreases with decrease in the concentration of the impurity, it becomes more and more difficult to achieve 100% pure metal. Thermodynamically, a metal with a number of impurities needs to be treated as a multicomponent solution, which requires adequate knowledge of interaction coefficients [2].

Difference in physical and chemical characteristics of the metal and the impurities are made use of in various refining processes. For example, a large difference in the extent of affinities of different metals for oxygen is employed in refining steel, copper, and lead bullion. Secondly, distribution of various elements in different phases [3] differs widely and these phases can be easily separated.

10.2 Methods of Refining

All the refining methods are classified into four main groups:

1. *Metal–slag processes*: In these processes, the less noble metal is oxidized and slagged off. This is commonly known as fire refining. Removal of carbon, silicon, manganese, and phosphorus from pig iron during steelmaking falls under this category. Copper from lead, bismuth, and tin can be removed by treating the crude with sulfur. Similarly, chlorination is useful in eliminating zinc from lead; zinc, copper, and lead from bismuth; lead from tin; and silver, copper, and zinc from gold.
2. *Metal–metal processes*: Liquation and zone refining are the important examples in this group. Zone refining is used for ultrapurification of some metals.
3. *Metal-gas processes*: Under this category, distillation, vacuum treatment, carbonyl decomposition, disproportionate reaction, and iodide decomposition form

important examples. Disproportionate reaction and iodide decomposition techniques are employed for ultra-purification of reactive metals.

4. *Miscellaneous group*: This group includes electrorefining of common metals like copper, lead, tin, and so on using aqueous electrolytes and of reactive metals like aluminum, magnesium, zirconium, and so on using fused salt electrolytes, and electrotransport method for ultra-purification.

It is important to note that no single process is adequate for refining all the metals or for removing all types of impurities. The choice of the refining technique, therefore, depends upon the characteristics of the metal, the nature of impurities, and often also upon the route employed for extraction. In addition to the above-mentioned methods, remelting techniques, namely, vacuum arc melting, electron-beam melting, and electroslag refining are employed for refining and consolidation of reactive metals.

10.2.1   Fire Refining

The degree of affinity of different metals for oxygen forms the basis of fire refining. In addition, it is useful to have knowledge of the activities of elements dissolved in the metallic phase and of oxides dissolved in the slag. During fire refining, the metal is oxidized by blowing air or oxygen or by addition of some oxide of the metal to be refined. For example, pig iron is oxidized with air or oxygen or iron ore and impure lead with air or lead oxide. The pig iron obtained from the blast furnace is saturated with carbon, and in addition, contains silicon, manganese, phosphorus, and a small quantity of sulfur. With the exception of sulfur, other impurity elements are removed by oxidation and subsequent dissolution of oxides in the slag during steelmaking. Slag plays an important role in refining of steel. Some sulfur is eliminated in steelmaking by gaseous phase reactions but major removal takes place in the blast furnace because desulfurization requires reducing atmosphere. Thus, steelmaking is primarily a process of converting pig iron containing approximately 10 wt% of carbon, manganese, silicon, and phosphorus into steel with about 1 wt% of some impurity elements.

The equipment employed in fire refining depends both on the metal in question and on the method of oxidation. Oxidation of impurities with air or oxygen in general and the above impurities in pig iron in particular is exothermic. In some cases, refining may be carried out without additional fuel. For example, in the Bessemer converter, where air is blown through molten pig iron, the heat evolved is sufficient to make up the heat losses and raise the temperature of the hot metal from 1300 °C to about 1600 °C, which is needed to melt steel. When oxygen is blown in an LD converter on top of the molten iron, a fair amount of cold scrap or some iron ore is charged to maintain the required steelmaking temperature of 1600 °C. Otherwise the bath temperature will rise to 1700 °C because the heat loss through the escaping gases in the absence of nitrogen is much less. Silicon, manganese, and phosphorus get oxidized and slagged. In addition to converters, hearth electric arc furnaces as well as ladles may be employed for refining. In the following sections, fire refining of pig iron, blister copper, and lead bullion are discussed.

10.2.1.1 Refining of Pig Iron

It has already been stated that steelmaking is the conversion of pig iron containing about 10 wt% of carbon, silicon, manganese, phosphorus, sulfur, and so on to steel with a controlled amount of impurities, that is, to the extent of about 1 wt%. With the exception of sulfur, removal of all other impurities is favored under oxidizing conditions. Oxidation of carbon generates only gaseous products whereas all other impurities generate solid products. In all the fire refining processes such as steelmaking or refining of blister copper and lead bullion, oxygen has to first dissolve in the metallic bath to react with the dissolved impurities. Liquid iron can dissolve 0.23 wt% oxygen in equilibrium with pure liquid FeO at 1600 °C in the absence of slag forming oxides, whereas in the presence of slag forming oxides the solubility of oxygen in steel decreases. At 1600 °C, solubility is expressed as $[\%O] = 0.23.a_{FeO}$, for the equilibrium reaction: FeO (l) = Fe (l) + [O]; where a_{FeO} is the activity of FeO in the slag. From Fig. 4.9, it is evident that for any given FeO content, its activity shall be maximum for the composition close to the line joining FeO apex to the orthosilicate: $2CaO.SiO_2$ composition [4] due to the high preferential attraction between CaO and SiO_2. SiO_2, MnO, and P_2O_5 formed by the oxidation of Si, Mn, and P, respectively, join the slag phase. The formation of these oxides can be facilitated by decreasing their activities, which is possible by providing oxides of opposite chemical character to serve as flux. For smooth separation of the oxide products, both the metal and the slag must be fluid enough to separate into two layers. The slag being lighter and immiscible, floats on top of the metal and, therefore, can be easily separated. As SiO_2 and P_2O_5 are acid oxides, a basic flux is required for the formation and easy removal of the slag. A higher proportion of basic flux will ensure prevention of the reverse reaction, particularly in the case of phosphorus removal. Silica being a very stable oxide, once it is formed, there is no chance of it reverting to silicon in the metallic bath. For effective refining, the activity of reactants (i.e., impurities) and oxygen should be high and that of product oxides should be low. However, it is possible to adjust silicon and manganese in the hot metal in such a way that during refining, FeO and MnO formed would be able to dissolve SiO_2 to produce a low-melting ternary slag, $FeO–MnO–SiO_2$. In this slag, P_2O_5 will not be fixed up because FeO–MnO is not a strong base. A strong basic slag is formed by the addition of CaO and/or MgO to absorb P_2O_5. In the presence of the highly basic slag, $CaO–FeO–P_2O_5$, there is no danger of phosphorus reversion. Sulfur removal to some extent is also possible by such a slag. Based on the above discussion, two important conclusions may be drawn with regard to selecting the steelmaking process. Firstly, sulfur, and phosphorus cannot be removed along with silicon as the impurity. Secondly, removal of silicon should not be tried along with phosphorus and to some extent sulfur. Removal of sulfur, carbon, silicon, manganese, and phosphorus has been discussed in detail in Chap. 7 on steelmaking.

10.2.1.2 Refining of Copper

The blister copper obtained from the Peirce-Smith converter contains large amount of impurities such as Fe: 0.01 – 0.04, Ni: 0.005–0.6, S: 0.01–0.05, O: 0.2–0.5, Zn: 0.01–0.05, As: 0.01–0.1, Sb: 0.01 – 0.1 (all in weight per cent) and also very small amounts of Pb, Se, Te, Au, and Ag. Copper produced by single-step smelting and continuous converting [5] contains 0.2–0.4% O and up to 1% S. Sulfur and oxygen dissolved in copper form SO_2 bubbles during solidification. It is estimated that 0.01 wt% S and 0.01 wt% O dissolved in copper would give rise to the formation of 2 cm^3 of SO_2 at 1083 °C per cm^3 of solidified copper. Hence, it has to be refined for commercial usage. There are two important methods of refining blister copper: fire refining and electrolytic refining. Fire refining does not remove/recover bismuth and precious metals. On the other hand, the electrolytic process not only produces high-purity copper but also improves economy of the process by successfully recovering silver and gold as by-products. Thus, the high capital investment on the electrolytic plant is fully compensated by the precious metal that is recovered and the quality of the product. In order to get a high-quality product, a major proportion of blister copper is first fire refined customarily prior to electrolytic refining with the prime objective of reducing the refining cost.

Fire refining of blister copper is carried out at 1150–1200 °C either in a rotary furnace (3–5 m diameter and 9–14 m long) or in a reverberatory furnace (12 – 15 m long, 5 m wide and 1 m deep) with a normal capacity of about 300–400 tons. The furnace is lined with magnesite or silica bricks and heated by burning oil, natural gas, or pulverized coal. Molten copper is charged into the furnace and then air is blown through pipes inserted into the molten copper. Owing to the larger mass, copper first gets oxidized to form cuprous oxide, which is distributed all over in the melt.

$$4Cu + O_2 \rightarrow 2Cu_2O. \tag{10.1}$$

The cuprous oxide formed spreads instantaneously throughout the molten copper and oxidizes, the impurities (M) having higher affinity for oxygen as compared to copper

$$Cu_2O + M \rightarrow 2Cu + MO. \tag{10.2}$$

At equilibrium, the dissociation pressure of the oxides of impurities and that of cuprous oxide are equal:

$$p_{O_2[Cu_2O]} = p_{O_2(MO)} \tag{10.3}$$

In addition to Cu_2O, oxide(s) of impurity metal(s) may take part in the oxidation of some impurities depending upon their affinity for oxygen. For example, M_1, having more affinity for oxygen than M_2, gets oxidized and removed as M_1O:

$$M_1 + M_2O \rightarrow M_1O + M_2. \tag{10.4}$$

The removal of the slag floating on the bath, formed collectively by oxides of impurity metals, facilitates further refining because of the decreased chemical potential of impurity oxides in it. Depending on the extent of affinity for oxygen in decreasing order, the sequence of removal from molten copper is as follows: Al, Si, Mn, Zn, Sn, Fe, Ni, As, Sb, Pb, and Bi.

Nickel, having lower affinity for oxygen than copper, can be removed from the blister at 1250 °C by generating conditions to get a negative free energy change for the reaction:

$$Cu_2O + Ni \rightarrow NiO + 2Cu \tag{10.5}$$

Since the standard free energy of the formation of Cu_2O is less than that of NiO, copper will get preferentially oxidized and lost in the slag. According to the van't Hoff isotherm, the free energy change under nonstandard conditions is given by

$$\Delta G = \Delta G^o + RT \ln \left(\frac{a_{NiO}}{a_{Cu_2O}} \right) \tag{10.6}$$

Hence, ΔG can be made negative by lowering the value of a_{NiO} by adjusting the slag composition. Silver and gold cannot be eliminated by fire refining because of their very weak affinity for oxygen.

The fire refining by air blowing reduces sulfur to approximately 0.002% at the expense of dissolving an appreciable quantity of oxygen in copper. The essential reaction in removal of S by air can be represented as

$$[S] + O_2(g) = SO_2(g) \tag{10.7}$$

And the corresponding dissolution of oxygen in copper takes place according to the reaction:

$$O_2(g) = 2[O] \tag{10.8}$$

The equilibrium relationship between [S] and [O] dissolved in copper is represented as:

$$[S] + 2[O] = SO_2(g) \tag{10.9}$$

$$K = \frac{p_{SO_2}}{[\%S].[\%O]^2} = 10^6 \text{ at } 1200\,^\circ\text{C(Reference 5)} \tag{10.10}$$

From the high value of the equilibrium constant, it is evident that sulfur elimination continues along with oxygen dissolution in copper even when [S] is reduced to 0.002% at $p_{O_2} \sim 0.2$ atm and $p_{SO_2} > 1$ atm. Finally, at the end of refining, the residual sulfur in the copper melt reduces to about 0.001 wt%, but [O] increases to about 0.6 wt%. During solidification, the decreasing solubility of oxygen in solid copper causes precipitation of the entire oxygen as solid Cu_2O to generate about 6 wt% of oxide inclusion. Since the electrical and mechanical properties of copper get significantly deteriorated in the presence of oxygen, molten copper is deoxidized by introducing a reducing gas or hydrocarbon after skimming off the slag. Most of the oxygen from the molten copper is removed by hydrocarbon according to the reactions:

$$C(s) + [O] = CO(g) \tag{10.11}$$

$$CO(g) + [O] = CO_2(g) \tag{10.12}$$

$$H_2(g) + [O] = H_2O(g) \tag{10.13}$$

The hydrocarbon blow is stopped when the oxygen content of the bath attains a value of 0.15%. In this way, copper oxide precipitation is minimized but hydrogen generated from the hydrocarbon dissolves in copper to a limited extent. Finally, the fire-refined copper containing approximately 0.002 wt% S, 0.15 wt% O, and 2×10^{-5} wt% H is cast into anodes to be refined to 99.99% purity by electrolytic refining, during which silver and gold are recovered.

10.2.1.3   Refining of Lead Bullion

Lead bullion containing approximately [6] 1.5 Cu, 1.2 Zn, 0.75 As, 0.5 Sb, 0.3 Sn, 0.4 Bi, 0.1 S, 0.2 Fe, 0.45 Ag, and 0.1 Au as impurities (all in weight per cent) is subjected to a series of operations to refine lead and to recover precious metals and other elements. For copper removal, the lead bullion is stirred with sulfur in a kettle at 330–350 °C. The resulting Cu_2S, being lighter and insoluble in lead rises to the surface as a dross consisting of Cu_2S, PbS, and mechanically entrapped molten lead. In order to minimize the loss of lead, the temperature is raised to 370 °C before skimming. Since the concentration of lead is much larger, PbS is first formed and gets dissolved in the molten lead because of high solubility of PbS in lead at the working temperature. Copper having higher affinity for sulfur as compared to lead forms Cu_2S according to the reaction:

$$[PbS] + 2[Cu] \rightarrow Pb\ (l) + Cu_2S\ (s) \qquad (10.14)$$

$$\text{At equilibrium, } K = \frac{1}{[a_{PbS}] \cdot [a_{Cu}^2]} \quad [a_{Cu_2S} = 1 = a_{Pb}] \qquad (10.15)$$

$$\text{or } K = \frac{1}{[\%S] \cdot [\%Cu]^2} \ \left(\text{assuming validity of Henry's law}\right) \qquad (10.16)$$

Hence, the residual copper in the bullion can be expressed as:

$$[\%Cu] = \frac{1}{\sqrt{K[\%S]}} \qquad (10.17)$$

Thus, higher the amount of sulfur dissolved in the molten lead (in the form of PbS) the lower the residual copper in lead after drossing. The sulfide carrying 95% Pb and 3% Cu is recycled.

The next step of fire refining, based on oxidation with air and litharge (PbO), is carried out in a reverberatory furnace at 350 °C. Impurities like zinc, arsenic, antimony, and tin, which have stronger affinity for oxygen as compared to lead form oxides that are insoluble in lead. Being amphoteric, these oxides are acidic to PbO and form zincates, arsenates, antimonates, and stannate, respectively, according to the reactions [7]:

$$Zn + 2PbO = PbO.ZnO + Pb \qquad (10.18)$$

$$2As + 6PbO = 3PbO.As_2O_3 + 3Pb \qquad (10.19)$$

$$2Sb + 4PbO = PbO.Sb_2O_3 + 3Pb \qquad (10.20)$$

$$Sn + 3PbO = PbO.SnO_2 + 2Pb \qquad (10.21)$$

The mixture of these compounds being lighter, floats on the surface of molten lead and is skimmed off [7, 8]. Oxidation can be speeded up by blowing air through the molten bullion. Removal of arsenic, antimony, and tin, results in softer or more ductile lead and hence the process is called *softening*. Silver and gold are not oxidized but they join the skim to which some molten lead droplets containing dissolved Ag and Au are entrained. Thus, loss of precious metals is the most serious drawback of fire refining of lead.

Alternatively, lead bullion is poured through a molten oxidizing mixture of soda salt containing $NaNO_3$, NaOH, and NaCl. This is known as the *Harris process*. In this treatment, lead is first oxidized and forms sodium plumbite with the alkaline melt at 700 °C:

$$5Pb + 2NaNO_3 + 8NaOH \rightarrow 5Na_2PbO_2 + N_2 + 4H_2O \qquad (10.22)$$

As tin, arsenic, and antimony have higher affinity for oxygen, they get oxidized by sodium plumbite and finally form oxysalts: Na_2SnO_3, Na_3AsO_4, and Na_3SbO_4. The overall reactions can be represented by the following chemical equations:

$$5Sn + 4NaNO_3 + 6NaOH \rightarrow 5Na_2SnO_3 + 2\,N_2 + 3H_2O \tag{10.23}$$

$$2As + 2NaNO_3 + 4NaOH \rightarrow 2Na_3AsO_4 + N_2 + 2H_2O \tag{10.24}$$

$$2Sb + 2NaNO_3 + 4NaOH \rightarrow 2Na_3SbO_4 + N_2 + 2H_2O \tag{10.25}$$

In this process, PbO acts as an oxygen carrier from the primary oxidant to the impurities. After all the impurities are oxidized, PbO starts accumulating in the melt as an orange–yellow product. The working temperature of the alkaline melt, $NaNO_3$–$NaOH$, is lowered by addition of NaCl to 420–450 °C. The alkaline melt becomes thicker as it picks up more impurities. On accumulating [7] about 18 wt% arsenic or 20 wt% tin or 30 wt% antimony, it is drained off from the chemical treatment tank to recover impurity elements and chemicals for reuse.

For removal of precious metals (Ag and Au), partially purified lead is treated with zinc at 450 °C. The excess zinc dissolved is removed by chlorination at 390 °C. Bismuth is removed as solid intermetallics by the addition of Ca–Mg alloy. Desilverizing, degolding, and debismuthizing are discussed under the heading of liquation.

Iron from crude tin is removed by blowing air or steam and skimming the dross floating on the surface of the tin melt. Lead can be recovered by injecting chlorine gas or adding stannous chloride in the tin melt:

$$Pb\,(l) + SnCl_2(l) \rightarrow Sn\,(l) + PbCl_2(l) \tag{10.26}$$

10.2.2 Metal–Metal Refining

Out of the two main processes under this category, liquation will be discussed in this section. Zone refining will be covered under the heading of ultra-purification.

10.2.2.1 Liquation

Removal of an impurity metal from the crude by selective melting is called liquation. This refining technique is based on three factors: liquid immiscibility, difference in melting points, and densities of the constituents in the alloy system. Metal impurity systems exhibiting considerable liquid immiscibility between solvent and solute elements with larger difference in melting points and densities may be refined by the liquation technique. The efficiency and degree of separation depend on the constitutional phase diagram of the metal-impurity system. In some systems, solubility of the impurity (solute) decreases drastically with decrease of temperature toward the melting point of the metal (solvent). A few selected systems [9] with these characteristic features have been presented in Fig. 10.1. The variation in the solubility of impurities in the crude melt can be used in refining/separation. For example, zinc and lead are partially miscible in liquid state. When crude zinc is kept in a furnace at a temperature just above the melting point of zinc (419.6 °C), two liquid layers, one rich in zinc and the other rich in lead are formed. The latter being heavier settles to the bottom of the furnace, and both these liquids can be tapped separately. Theoretical limit to this separation is 0.85 wt% Pb in zinc and the lead contains 1.3 wt% Zn. The complete immiscibility would be ideal for this type of refining. Iron and copper can be removed from lead by liquation because both have limited solubility in molten lead. Similarly, liquation can be employed to recover low-melting tin coated on high-melting iron by heating the tin-plated steel sheet to above the melting point of tin (about 400 °C) to drain out tin. In systems like Sn–As and Sn–Fe, while cooling, the excess solute separates out as intermetallic compounds of the two elements on exceeding the solubility limit.

Another aspect of liquation comprises removal of the impurity element by forming a high-melting compound with the metal, which is added to the molten crude. The compound is separated in solid form.

During smelting of the galena as well as mixed sphalerite–galena concentrates, the entire silver and gold are transferred in the lead bullion. The softened lead is stirred with 1–2 wt% zinc at 450 °C. Zinc forms a number of intermetallic compounds with Ag and Au, such as Ag_2Zn_3, Ag_2Zn_5, $AuZn$, Au_2Zn_5, and $AuZn_3$ with respective melting points of 665, 636, 725, 664, and 475 °C (higher than the melting point of lead). Being lighter than lead these compounds concentrate over the surface of the melt from where they are collected as silver/gold–zinc crust. This method of recovery of precious metals known as the *Parkes process,* is based on the principle that silver and gold dissolve more readily in zinc than in lead and mutual solubility of lead and zinc is low.

The lighter intermetallics floating on the surface of molten lead are skimmed off and treated separately for the recovery of precious metals. Since the solubility of zinc in lead is limited, the excess zinc also floats on the surface as a saturated solution of lead in zinc. Hence, a complex product consisting of silver, gold, lead, and zinc gets collected over the bath surface. The impurities like arsenic, antimony, and tin not only increase zinc consumption but also slow down the flotation of the crust. Thus, a clear separation from the underlying lead is suppressed. This highlights the importance of removal of As, Sb, and Sn prior to desilverizing. The Ag–Zn crust is treated for the recovery of silver, gold, zinc, and lead. After desilverizing, the lead containing 0.6–0.8 wt% Zn (the excess zinc dissolved in lead, added over the stoichiometric requirements) is treated with chlorine at 350–390 °C to remove zinc (*Betterton process*).

Bismuth present in lead can be very effectively removed by electrorefining. However, its concentration can be lowered down by the *Kroll–Betterton process.* In this process, dezinced lead is treated with Ca–Mg alloy at 350 °C to remove bismuth as intermetallics: Bi_3Ca (melting point 507 °C), Bi_2Ca_3 (928 °C), and

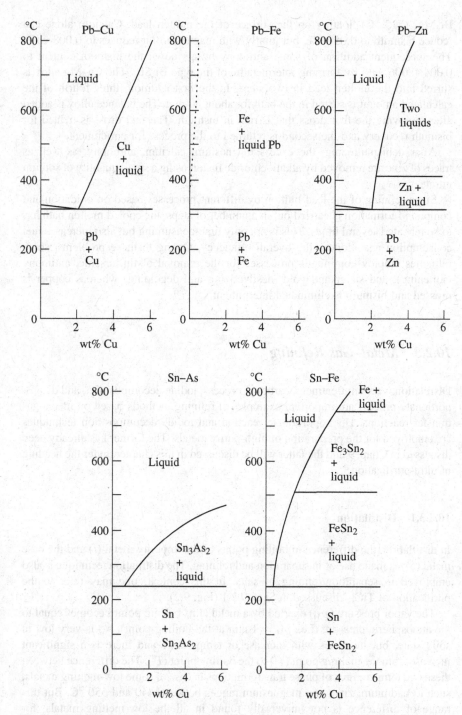

Fig. 10.1 Phase diagrams of some selected immiscible systems (From The Extraction and Refining of Metals by C. Bodsworth [9], © 1990, p 219, Taylor and Francis. Reproduced with the permission of Taylor and Francis Group, LLC)

Bi_2Mg_3 (715 °C) floating on the surface of the molten lead. Calcium alone can reduce bismuth to 0.05 wt%, but jointly with magnesium, it reduces to 0.008 wt%. The subsequent addition of some antimony brings down the bismuth content to 0.004–0.006 wt% by forming intermetallic of the type Bi_xSb_y. The Ca–Mg alloy is stirred into the molten lead in two steps. In the first addition, three-fourth of the calculated amount is stirred in the bath for about 30 min. The balance alloy is added after removing the first dross that is rich in bismuth. The first dross is treated for bismuth recovery and the second is returned to the process for enrichment.

After debismuthizing, the excess magnesium, calcium, antimony, as well as traces of zinc are removed by alkali–chloride fusion using a small quantity of sodium nitrate.

Since refining of the lead bullion by different processes based on oxidation and compound formation is carried out in a number of steps, the cooled molten bath has to be reheated several times. This is not only time-consuming but also increases fuel consumption and reduces the overall efficiency. Owing to these problems, some refineries employ continuous processes for the removal of tin, arsenic, antimony (softening), and silver and gold (desilverizing and degolding) whereas copper is drossed and bismuth is eliminated intermittently.

10.2.3   Metal–Gas Refining

Distillation, vacuum treatment, carbonyl process, iodide decomposition, and disproportionate process are important examples of refining methods based on metal–gas transfer reactions. Disproportionate reaction and iodide decomposition techniques are employed for the preparation of high-purity metals. The former has already been discussed in Chap. 9 and the latter will be discussed in this chapter under the heading of ultra-purification.

10.2.3.1   Distillation

In distillation, the difference in boiling points of the impurity metal (i) and the base metal (M) is made use of in separation and refining. The distillation technique is also employed in separation/refining of salts. In this context, one may refer to the purification of $TiCl_4$ discussed in Sect. 9.2.1 (Fig. 9.2).

The vapor pressure (p_M) exerted by a metal at its boiling point becomes equal to the atmospheric pressure (i.e., $p_M = 1$ atm at the boiling point). p_M is very low in solid state, but increases with increase of temperature and there is a significant increase from the melting point (T^f) to the boiling point (T^e). The difference between these two temperatures of phase transformations in case of some low-melting metals, such as cadmium, zinc, and magnesium ranges between 440 and 550 °C. But this range of difference is not universally found in all the low-melting metals, for example, in case of even very low-melting gallium (melting point: 29 °C), indium

(157 °C), and tin (232 °C), the difference is of the order of about 2000–2500 °C and for bismuth (melting point: 271 °C), lead (327 °C), and aluminum (660 °C), it is, respectively, 1356, 1410, and 1790 °C. The reason for this variation may be understood by analyzing the values of the latent heat of transformation of metals for changes from liquid to gas and solid to liquid. The temperature of evaporation of the metal species i, is related to the latent heat of evaporation according to the Clausius–Clapeyron equation:

$$\ln p_i = \frac{L_i^e}{RT} + C \tag{10.27}$$

where p_i is the vapor pressure of species i, at temperature T K, L_i^e the latent heat of evaporation, and C is a constant. It has to be noted that L_i^e increases with the group number of the periodic table; for example, the latent heat of evaporation of transition metals (chromium, molybdenum, and tungsten) is much higher than those of alkali (lithium, sodium, and potassium) and alkaline earth metals (barium, strontium, and calcium). However, the variation in the melting point with the latent heat of fusion is not as systematic as observed in the case of liquid–gas transformation. This is also evident from the constancy of the entropy of evaporation $\left(\Delta S^e = \frac{L^e}{T^e} = 22.4 \text{ cal K}^{-1} \text{ mol}^{-1}\right)$ according to Trouton's rule. However, the entropy of fusion $\left(\Delta S^f = \frac{L^f}{T^f} = 2.2 \text{ cal K}^{-1} \text{ mol}^{-1}\right)$ according to Crompton and Richards rule is not so strictly constant.

The temperature difference between melting and boiling points plays an important role on the pressure exerted by the metal at temperature 100 °C above the melting point. This is directly related to the rate of evaporation of the metal from the melt. Hence, at the selected temperature, element(s) exerting relatively higher vapor pressure can be removed completely by distillation. However, this cannot be taken for granted universally because the rate of evaporation may be enhanced or retarded by the nature of interaction between the impurity elements (i) and the base metal (M) in the melt. The partial pressure of a species, i, p_i in an ideal solution is proportional to its mole fraction x_i, (i. e., $p_i \propto x_i$, according to the Raoult's law). Hence, p_i can be expressed as:

$$p_i = p_i^o . x_i, \tag{10.28}$$

where p_i^o is the partial pressure of pure i at the temperature under consideration. Thus, in a solution of M–i containing 1 at % i, $x_i = 0.01$, the vapor pressure of the impurity above the melt would be 100 times less than p_i^o. In case of deviation, Eq. 10.28 is modified as:

$$p_i = p_i^o . a_i \tag{10.29}$$

where a_i is the activity of the component, i in the solution. This means in ideal solution, $a_i = x_i$, and to account for deviation from ideality, a term activity coefficient, γ_i is introduced as: $a_i = \gamma_i \cdot x_i$. γ_i may be >1 or <1, in case of positive or negative deviation, respectively. Hence, Eq. 10.29 is modified as:

$$p_i = p_i^o \cdot \gamma_i \cdot x_i \tag{10.30}$$

From this equation, it is evident that the vapor pressure of component, i is more or less than that expected according to the Raoult' s law in systems exhibiting respectively, positive ($\gamma_i > 1$) or negative ($\gamma_i < 1$) deviation from ideality. It must be noted that the forces of attraction between dissimilar atoms (M–i) is weaker and stronger than the forces of attraction between similar atoms (M–M or i–i), respectively, in case of positive and negative deviations. For example, in a Pb–Zn system showing positive deviation from ideality the vapor pressure of both the constituents, Pb and Zn is higher than that expected according to the Raoult's law. Hence, zinc boiling at lower temperature (boiling point 907 °C) can be easily removed from the lead. Contrary to this, removal of As and Sb from Fe melt is difficult because Fe–As and Fe–Sb systems exhibiting negative deviation due to stronger attraction between dissimilar Fe and As, and Fe and Sb atoms have a tendency to form intermetallic compounds that lower the vapor pressure of As and Sb over the melt.

However, refining by distillation is restricted to metals having boiling point below 1000 °C due to operating problems. Under normal atmospheric pressure, mercury (boiling point 357 °C), cadmium (765 °C), sodium (892 °C), and zinc (907 °C) may be refined by the distillation technique. A few more metals that have vapor pressure of the order of 10^{-3} atm at about 1000 °C may be refined by vacuum distillation. For example, metals like tellurium (boiling point 990 °C), magnesium (1105 °C), calcium (1487 °C), antimony (1635 °C), and lead (1740 °C) exerting vapor pressure of 10^{-3} atm at 509, 608, 803, 877, and 953 °C, respectively, can be distilled at 1000 °C under a vacuum of the order of 10^{-3} atm. However, high degree of separation is not possible by distillation due to the presence of lower mole fractions of a number of impurities in the melt together with varying nature of simultaneous interactions between impurity metals and the base metal and impurities. As a result, two or more metals having appreciable vapor pressure at the operating temperature may evaporate. Under such circumstances, separation is facilitated by selective condensation. The less volatile element condenses in the hottest part of the condenser whereas the more volatile element condenses in the cooler part.

The rate of evaporation of a species, i from the surface of the molten metal under equilibrium conditions, as derived by Langmuir, according to the kinetic theory of gases is expressed as:

$$\text{Rate of evaporation,} \quad r_i = p_i \left(\frac{M_i}{2\pi RT} \right)^{1/2} \tag{10.31}$$

where p_i is the vapor pressure of the species, i in the melt, R the gas constant and M_i the molecular weight of i. At constant temperature, the rate of evaporation is maximum when equilibrium is attained between the number of molecules leaving the melt surface and those condensing on the surface. Equation 10.31 assumes that all the molecules striking the surface, condense as well. Generally, in the majority of evaporation processes, transfer of vapors away from the liquid metal surface happens to be the rate-controlling step, not the rate of evaporation. On substituting the value of p_i from Eq. 10.29 in Eq. 10.31, we get:

$$r_i = p_i^o . a_i \left(\frac{M_i}{2\pi RT}\right)^{1/2}$$

(10.32)

Since the rates of evaporation and condensation are equal at equilibrium, Eq. 10.32 also denotes the rate of transfer of atoms/molecules from the melt to the gaseous phase. If the species i, as solute and the base metal M as solvent vaporize at a particular temperature, the relative rates of evaporation can be expressed as:

$$\frac{r_i}{r_M} = \frac{p_i^o a_i}{p_M^o a_M} \left(\frac{M_i}{M_M}\right)^{\frac{1}{2}}$$

(10.33)

where r_M, p_M^o, and a_M represent the rate of evaporation, vapor pressure under standard conditions, and activity of the metal M, respectively, at the chosen temperature. From Eq. 10.33 it is evident that the ratio r_i/r_M increases as the extent of vaporization (i.e., removal) of the solute, i increases without much loss of the solvent metal, M. This forms the basis of refining by distillation. Thus, selection of adequate distillation temperature is of utmost importance in order to remove most of the relatively high volatile impurities while retaining most of the solvent metal. This is called fractional distillation. Alternatively, the entire melt can be evaporated and the components separated by fractional condensation from the gaseous phase at appropriate temperatures. This procedure has been most successfully employed in the separation of $TiCl_4$ and $ZrCl_4$ from the resultant mass obtained by chlorination of rutile and zircon, respectively.

Based on the above discussion, it is concluded that the metal impurity systems with a larger difference in vapor pressures can be separated by single-stage distillation whereas a special technique known as rectification is needed for systems with small differences in vapor pressures. For example, the crude zinc (boiling point 907 °C) obtained from the blast furnace containing about 2–3 wt% Pb (boiling point 1740 °C) and 0.3–0.5 wt% Cd (boiling point 765 °C) is refined in a multistage rectification column consisting of 30–40 trays, operated in the temperature range of 800–1200 °C. Since, at the working temperature, vapor pressures of cadmium and lead are, respectively, much higher and much lower than that of Zn, both zinc and cadmium are distilled away, leaving lead in the molten state, which can be drained off. The zinc vapor is condensed while cadmium is retained in the gaseous state. In order to obtain a higher purity product, distillation has to be repeated several times, but at the cost of yield.

The rate of evaporation can be enhanced by reducing the gaseous phase above the melt or by providing a cooler surface for condensation of volatile species. The rate of removal is retarded by the absorption of surface-active solutes on the surface of the melt. The rate of evaporation may reduce to zero if the surface is completely covered by a slag layer. A reasonable rate of evaporation can be achieved by providing a large surface area per unit volume of the melt together with appropriate stirring. As melt experiences different pressures due to the variation in the hydrostatic pressure at different depths, it would be beneficial to maintain a thin layer of the metal to get a larger surface area.

10.2.3.2 Pyrovacuum Treatment

Metals obtained via the pyrometallurgical route that involves smelting, based on mainly slag–metal reactions in molten state, pick up oxygen, nitrogen, hydrogen, and carbon from the atmosphere, ore, flux, and fuel in addition to other metallic impurities. The problems caused by dissolved oxygen, nitrogen, and hydrogen and their removal from steel have been discussed in detail in Chap. 8 on "Secondary Steelmaking" under the headings: deoxidation and vacuum degassing. In vacuum degassing, the removal of carbon as carbon monoxide by the reaction between dissolved carbon and dissolved oxygen in steel has also been discussed. Though dissolved sulfur from the blister copper is removed by blowing air, conditions may be created to eliminate sulfur according to the reaction: $[S] + 2[O] \rightarrow SO_2(g)$ under vacuum. This will avoid the problems of removal of larger amount of oxygen, dissolved during air blowing, by hydrocarbon. Similarly, carbon picked up by reactive and rare metals like titanium, vanadium, tantalum, and so on can be removed by pyrovacuum treatment.

Impurities in reactive metals can be classified into two groups. The first group belongs to metallic impurities derived from co-occurring sister elements in the natural minerals, for example, Hf with Zr, Nb(Cb) with Ta, V with Ti. The second group consists of metallic and interstitial impurities picked up in the crude during extraction processes. Purification with respect to the first kind of impurities is carried out at an early stage in the extraction flow sheet, by techniques such as selective chemical reaction (i.e., carrier salt precipitation), fractional distillation/crystallization, ion exchange, and solvent extraction. These steps should in fact be termed as separation techniques rather than refining. These will be discussed in the chapter on Hydrometallurgy. The refining processes are concerned mainly with the treatment of the crude metals to the stage of final purity.

Reactive metals produced by metallothermic reduction of their oxides and halides are associated with residual excess reductants, by-products, and interstitial impurities picked up during reduction. High-vacuum and high-temperature techniques are most effective in refining these impure metals. Highly volatile impurities can be removed by simple evaporation. However, the removal of dissolved carbon and oxygen is achieved by special mechanisms such as carbon deoxidation and sacrificial deoxidation. Removal of volatile impurities is affected by selective evaporation of impurities. The extent of purification depends on the distribution coefficient, (k), which is the ratio of concentration of impurity in the metal to the concentration in the vapor phase. For effective purification, k should be much smaller than unity. If k is just smaller than unity, purification occurs with a considerable loss of the base metal. Purification by this mode is dependent on the vapor pressure of the impurity, which in turn depends upon temperature, its concentration in the melt and its rate of diffusion from the bulk to the surface and also on the vapor pressure of the base metal. For effective purification, the vapor pressure of the impurity should be more than 10^{-3} torr and should exceed the vapor pressure of the base metal by a factor of 10 or more. Impurities like Al, Mg, Ca, Fe, Si, and Mn are effectively removed from crude refractory metals because of their higher vapor pressure. Vanadium, niobium, tantalum, and molybdenum obtained by aluminothermic reduction [10] are purified by pyrovacuum technique to remove Al, Fe, Mn, O, and so on.

In sacrificial mode of refining, purification takes place via evaporation of suboxides of the metal to be refined. Sub-oxides are formed as a result of the reaction of oxygen in the metal with metal itself even at very low oxygen content. Carbon has been used as a deoxidizer for the removal of last traces of oxygen from refractory metals by formation of CO according to the reaction:

$$[C] + [O] = CO \, (g) \tag{10.34}$$

$$K = \frac{p_{CO}}{[a_C] \cdot [a_O]} = \frac{p_{CO}}{[f_C \cdot \%C] \cdot [f_O \cdot \%O]} \tag{10.35}$$

$$\therefore [\%C] \cdot [\%O] = \frac{p_{CO}}{K \cdot f_C \cdot f_O} = \frac{p_{CO}}{K'} \tag{10.36}$$

The product $[\%C] \cdot [\%O]$, known as the deoxidation constant represents the deoxidation tendency by carbon for a given metal. A low value of $[\%C] \cdot [\%O]$ signifies higher deoxidation tendency and is calculated by assuming the concentration of $[\%O]$ and $[\%C]$ equivalent to the concentration of oxide and carbide existing in equilibrium with the pure metal.

10.2.3.3   Carbonyl Process

By and large, removal of impurities by evaporation from solid crude metal is not effective due to very slow diffusion in solid state as compared to the rate of diffusion in liquid state. However, this limitation of slow rate of diffusion is eliminated when the entire mass of the crude metal is volatilized. Refining of crude nickel by the Mond carbonyl process is based on this principle.

During extraction of nickel, a sulfide ore [pentlandite $(NiFe)_9S_8$] containing about 1–3% Ni and some Co, Cu, and Fe is processed to obtain Ni_3S_2, which is dead roasted to nickel oxide. The oxide is reduced with carbon or hydrogen at 400 °C to impure nickel, which may contain 15–20% iron together with some cobalt and copper. After leaving the reducing chamber the crude nickel is brought into contact with CO gas in another reactor to form $Ni(CO)_4$ vapor at 50 °C and 1 atm pressure or liquid carbonyl at 50 °C and 20 atm. In this process some $Fe(CO)_5$ is also formed.

$$Ni(s) + 4\,CO\;(g) \rightleftarrows Ni(CO)_4(g) \qquad\qquad (10.37)$$

$$Fe(s) + 5\,CO(g) \rightleftarrows Fe(CO)_5(g) \qquad\qquad (10.38)$$

At 1 atm pressure [11] $Ni(CO)_4$ boils at 43 °C whereas $Fe(CO)_5$ boils at relatively higher temperature of 105 °C. Cobalt carbonyl, $Co_2(CO)_8$ has very low vapor pressure and copper and precious metals present in the crude do not form any carbonyl under these operating conditions.

Thus, if dry CO is passed over the crude nickel placed in a chamber at 80 °C, most of the nickel and only a small fraction of iron will volatilize as $Ni(CO)_4$ and $Fe(CO)_5$ and impurities will be left as solid residue. As 4 mol of CO are consumed during the formation of 1 mol of $Ni(CO_4)_4$, the free energy change of the reaction (10.37) is dictated by the large entropy contribution. Since reaction (10.37) is only slightly exothermic, it will proceed in the reverse direction at higher temperature (>100 °C) due to large entropy contribution. Hence, nickel carbonyl vapors directed in another chamber (decomposer) maintained at 160–180 °C decompose to produce pure nickel. On the other hand, $Fe(CO)_5$, which requires still higher temperature for decomposition, is retained in the gaseous phase. Decomposition of $Ni(CO)_4$ is facilitated by finely divided nickel powder kept in the decomposer, which acts as seed. With subsequent deposition, the particle size increases and seed grows with time to 99.99% pure Ni shots of about 6 mm diameter. The gaseous phase, mainly CO, contaminated with iron carbonyl from the decomposer is cooled to liquefy Fe $(CO)_5$ prior to recirculation over the crude nickel in the volatilizer.

The rate of reaction between solid nickel/iron and CO gas is slow at temperature near 50 °C but increases with increase of temperature. Thus, product carbonyl has greater chance of contamination at higher temperatures. According to the Le Chatelier's principle, increase of pressure will favor reactions (10.37) and (10.38) in the forward direction but the decomposition temperature of both the carbonyls will also increase. As 4 and 5 mol are consumed during the formation of $Ni(CO)_4$ and $Fe(CO)_5$, respectively, the increase of pressure is more effective in raising the decomposition temperature of the latter. In actual practice, on an industrial scale the crude nickel is allowed to react with CO gas at 50 °C and 20 atm to produce liquid Ni $(CO)_4$. The nickel carbonyl contaminated with iron carbonyl is subjected to thermal decomposition at temperatures between 205 °C and 250 °C to dissociate $Ni(CO)_4$ into pure nickel while retaining almost all the $Fe(CO)_5$ in the gaseous phase.

10.2.4   Miscellaneous Group

Electrorefining of common metals as well as reactive metals using aqueous and fused salt electrolytes, respectively, are discussed along with electrowinning in Chap. 12 on Electrometallurgy, and electrotransport method for the preparation of ultrapure metals will be covered in the next section on ultra-purification.

10.3 Ultra-Purification

High-purity metals are required in the field of electronics and fundamental research. The important methods, such as zone refining, electrotransport, and iodide decomposition will be discussed in this section. Disproportionate process has already been discussed in Sect. 9.4.1.

10.3.1 Zone Refining

This technique is based on the principle of fractional or selective crystallization of solid solutions and suitable intermediate phases. As it is used for ultra-purification, the impurity level in the metal to be refined must be low. During operation, a relatively thin molten zone is moved at rates of up to several mm per hour along the length of a solid bar. Impurities lowering the melting point of the basis metal are accumulated into this zone and removed to the end of the bar. On the other hand, impurities raising the melting point of the metal are concentrated into the solid and move toward the other end, in a direction opposite to that of the moving zone. A number of such passes are required during the operation for any appreciable removal of impurities. In order to get maximum segregation, the zone width and speed of travel are adjusted. The extent of refining depends on the relative solubility of the impurities in the liquid and solid phases at a particular temperature. The possibility of refining is better in case of a larger difference in solubility (i.e., farther apart the solidus and liquidus lines). The difference expressed as the partition, segregation, or distribution coefficient is the ratio of the solid to the liquid solubility. The distribution coefficient, k, may be estimated from equilibrium phase diagrams [12] (Fig. 10.2). Considering the solidus and liquidus as straight lines, k may be assumed to be constant against temperature changes. Thus, k approaches unity when liquidus and solidus are close. In such cases, there is little chance of refining. The chances of refining increase when k, is either <1 or >1.

The measure of purification is given by equilibrium distribution coefficient, k_o, defined as the ratio of the concentration of impurity in the solid c_s to that in the liquid c_l. According to Pfann [13], the distribution coefficient and concentration are related as:

$$c_s/c_o = 1 - (1 - k)e^{-kx/l} \tag{10.39}$$

where c_o is the initial concentration of solute in the solid, c_s is the solute concentration at a distance x, l is the length of the molten zone, and k the effective distribution coefficient is related to the equilibrium value [11], k_o as:

$$k = k_o/\left[k_o + (1 - k_o)e^{-fd/D}\right] \tag{10.40}$$

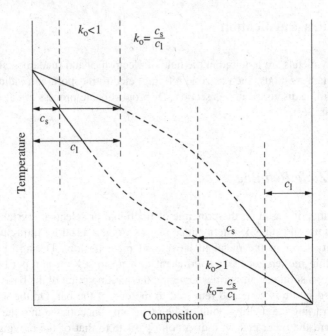

Fig. 10.2 Derivation of segregation coefficient

where, f, d and D are the freezing velocity, the thickness of the enriched or depleted solute layer at the solid/liquid interface, and the solute diffusion coefficient, respectively. For effective purification, k_o should be either far greater than unity or far less depending on the nature of the impurities. Impurities having $k_o < 1$ will be concentrated in the zone last to solidify and those with $k_o > 1$ will concentrate at the starting end. With the choice of suitable segregation coefficients along with nonequilibrium directional solidification, one can achieve very effective purification (Fig. 10.3), but it may turn up as a wasteful operation due to rejection of the impure portion after each melting and solidification.

Thus, maximum segregation can be achieved with both directional solidification and zone refining by generating conditions to prevent any significant diffusion in the solidifying metal.

Uniformity of the liquid depends on diffusion and to some extent on convection in the absence of stirring. Vertical arrangement of the bar provides better chance for convection in zone refining. Conventional zone refining can be carried out in a system holding the bar horizontally or vertically and moving either the container or the source of heat. Different methods of heating, for example, gas flames or induction can be used. In order to prevent the metal from oxidation, refining is carried out under vacuum or in an inert gas atmosphere. The first important application of zone refining was for the purification of germanium for use in transistors.

Fig. 10.3 Distribution of
solute impurity for various
values of segregation
coefficients

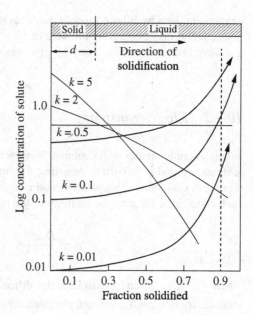

Initially it was developed for ultra-purification of low-melting metals such as tin,
zinc, cadmium, and so on. With advancements in the technique, it has been extended
to the purification of rare earth and reactive metals [14–18]. Each impurity may be
reduced to 1 part or less per million.

In order to prevent the metal from contamination by coming into contact with the
crucible, the floating zone technique was developed. This technique has facilitated
refining of refractory metals. In this technique, the metal bar is held vertically in a
container under controlled atmosphere without any side support. Surface tension
plays an important role in preventing the molten zone between the two solid portions
from collapsing. The presence of surface films on the metal and the application of an
electromagnetic field are also helpful in protecting the molten zone. A small portion
of the ingot supported at its ends is melted by induction or electron-beam bombardment. In the latter case, electrons are emitted from a heated tungsten loop. The
accelerated electrons by an applied potential are focused on to a localized region of
the specimen. Under these conditions, the efficiency of refining improves due to
various secondary effects. Removal of inclusions and self-slagging impurities of
different densities from the main metal takes place due to settling or flotation.
Controlled atmosphere (vacuum or a circulating gas) is helpful in the removal of
dissolved gases and volatile impurities.

The limitations of induction heating in generation of high temperatures in a
narrow zone required for refining refractory metals is eliminated by electron-beam
bombardment. Titanium, zirconium, niobium (columbium), and tantalum have been
purified by electron-beam floating zone refining. In zone-refined zirconium [10],

oxygen and iron have been brought down to 36 and 20 ppm from 140 to 150 ppm, respectively, under operating pressure of 10^{-8} torr. In titanium, oxygen, and nitrogen were brought down to 3 and 3.5 ppm, respectively, from 10 to 4.5 ppm.

10.3.2 Electrotransport

When a crude metal to be refined is subjected to a DC potential at elevated temperature and high current densities, the impurities of the alloy get transported relative to each other. This phenomenon is known as "Electrotransport" [19]. Under such conditions, the electron mobility of the interstitial solute can be expressed as:

$$\mu_i = \frac{D_i.e}{kT}\left(z_i - \frac{\partial e_i}{e}\right) \tag{10.41}$$

where D_i, k, z_i, e_i, and e stand for the diffusion coefficient, Boltzmann constant, valence, friction coefficient, and electronic charge, respectively. The term $\left(z_i - \frac{\partial e_i}{e}\right)$ is known as effective valence, z^* whose sign determines the direction of electrotransport of the constituents under the influence of an electric field. A negative sign indicates the predominance of electron friction force and suggests that electrotransport would be toward the anode. This phenomenon has been observed in case of group IV metals. Similarly, a positive sign of z^* indicates electrotransport toward the cathode, as observed in case of group V metals. In dilute solution, the flux of the solute, J_s (mol cm^{-2} s^{-1}) due to electrotransport is related to the concentration, c_i (mol cm^{-3}), the electron mobility, μ_i (cm^2 s^{-1} V^{-1}), and the electron field mobility, E (V cm^{-1}) by the expression:

$$J_s = c_i.\mu_i.E \tag{10.42}$$

In addition to electrotransport, there would be concentration diffusion $(-D_i dc_i/dx)$ in the opposite direction:

Thus the net flux, $\quad J = -D_i\frac{dc_i}{dx} - c_i.\mu_i.E \tag{10.43}$

Under the steady state, the net flux would be zero

$$\therefore D_i\frac{dc_i}{dx} = -c_i.\mu_i.E \tag{10.44}$$

The apparatus used for electrotransport refining consists of a vacuum chamber, a specimen holding device, and the arrangements for the DC supply. The specimen in the form of a wire or small diameter rod is refined under the vacuum of the order of 10^{-9} torr.

The main problems encountered are contamination from the surroundings and transport of impurities either from the electrodes or from the cooler ends of the specimen. These problems have been solved by maintaining an adequate leak-tight system and using high-purity tantalum adapters for minimizing the transport of impurities. The adapter at the anode end is usually made thicker, which remains considerably cooler and helps to prevent the flow of impurities from the anode to the specimen. The adapter at the cathode end is made thinner and kept at the same temperature as that of the specimen. Electrotransport refining [15, 20] of a number of rare earth, refractory and reactive metals, for example, gadolinium, zirconium, hafnium, vanadium, niobium (columbium), tantalum, and thorium have been successfully carried out in helium atmosphere. By this method carbon, nitrogen, and oxygen [10] in thorium have been reduced to less than 2, 0.5, and 1 ppm, respectively, from the starting level of 50, 35, and 80 ppm. However, electrotransport is an inefficient purification technique from the point of view of energy consumption. Ultimate purification to very low levels of interstitial impurities can be achieved if the initial purification is done either by iodide refining or electron-beam melting.

10.3.3 Iodide Decomposition

The iodide decomposition technique developed by van Arkel and de Boer [21] is a typical example of purification by chemical vapor deposition. Currently, it is extensively used for the production of high-purity rare and refractory metals [17, 22] metals. The process is based upon the reversibility of the reaction between a metal and iodine:

$$M\,(\text{crude}) + 2I_2(g) = MI_4(g) \qquad (10.45)$$

$$MI_4(g) = M\,(\text{pure}) + 2I_2(g) \qquad (10.46)$$

The formation of metal iodide takes place at a lower temperature, which decomposes at much higher temperature releasing iodine for reuse. The choice of iodine is based upon the fact that iodides are generally less stable as compared to fluorides, chlorides, and bromides. According to the Le Chatelier's principle, iodide formation (reaction 10.45) is favored at higher pressure, whereas the decomposition (reaction 10.46) is favored at a reduced pressure. From reactions (10.45) and (10.46), it is evident that there would be some increase in the volume on production of the pure metal because both the reactant iodine as well as the generated iodine are vapors. Although the metal production by decomposition of the iodide is favored at a reduced pressure, the pressure cannot be lowered to a very low level. Highly reduced pressure will not only restrict the throughput of the material for a particular size of the reactor but will also badly affect the formation of iodide from impure metal and

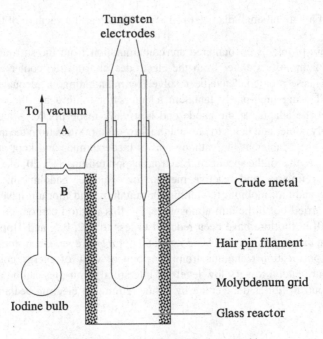

Fig. 10.4 Glass reactor for iodide decomposition

iodine, carried out in the same reactor. In practice, pressure of the order of 10^{-3} to 10^{-4} atm has been found to be adequate for the formation as well as decomposition of iodides at the recommended filament temperatures. The thermodynamic and kinetic aspects of the van Arkel iodide process have been discussed in detail by Shelton [23, 24]. He has concluded that the rate of deposition of zirconium is controlled by vapor transport process (i.e., diffusion of iodine from the filament) at ambient and filament temperatures. The rate is affected by the variation in the chemical reactivity of the feed material and the geometry of the reactor.

The process is carried out in a batch-type reactor made of glass or silica, as shown in Fig. 10.4. It is fitted with an incandescent filament with a provision for electric heating. A reactor made of inconel [1] has also been used for large-scale production of high-purity titanium and zirconium. Two ends of a long filament pass through suitable seals. A side arm holds iodine in a bulb with the provision for evacuation. The crude metal is packed in the annular space between the reactor wall and the perforated molybdenum screen. After evacuation, the system is sealed at point "A." Iodine is introduced in the reactor by heating the bulb and then the reactor is sealed at point "B." Iodine vapors react with the impure metal to form metal iodide that gets decomposed after reaching the hot filament. The liberated iodine reacts with the crude metal again to re-form the iodide vapor. This process is repeated until the entire operation is carried out. In this process, a continuous transportation of the metal takes place from low temperature to high temperature under a steady state of diffusion.

The filament of diameter as low as 2.5 μ made of tungsten or molybdenum is used for the deposition of the metal after decomposition of the metal iodide. It may be thicker if the metal being processed is employed as a filament. The filament is heated

Table 10.1 Bulb and filament temperature of iodide
refining of reactive metals

Metals	Bulb temperature, °C	Filament temperature, °C
Ti	200–525	1100–1400
Zr	250–400	1300
Hf	300–400	1600
Th	450	1300–1400

by electric current. Metal deposition in the form of adherent crystals increases the thickness of the filament. Hence, it is necessary to increase the current to maintain the required temperature for decomposition of the iodide vapor. This requirement of increase of current imposes a severe limitation on the maximum diameter of the final rod that can be obtained by this technique. A largest production unit can produce only rods of 25–50 mm diameter. Thus, for a higher rate of production, the length of the filament has to be increased rather than the diameter. The temperatures of the bulb and filament employed in the case of titanium, zirconium, hafnium, and thorium are listed in Table 10.1.

Conditions such as the formation of volatile iodide at low temperature and the subsequent easy decomposition at elevated temperatures, higher melting point of the metal compared to the dissociation temperature of the iodide, and very low vapor pressure of the metal at the decomposition temperature of the iodide facilitate refining by this method. If the melting point of the metal is lower than the decomposition temperature of the iodide (i.e., filament temperature) adherent crystals on the filament will not build up. There is every likelihood that the filament may be attacked by the molten metal. Thermodynamically, decomposition of the iodide (i.e., deposition of the pure metal) is favored at a reduced pressure, but there is a limitation on the reduction of pressure due to operational difficulties. This limit is set by the pressure at which the rate of evaporation of the metal (being purified and deposited on the filament) due to its own vapor pressure at the filament temperature becomes equal to the rate of deposition after decomposition. The rate increases with increase of the filament temperature up to a certain value but it decreases at very high temperature due to the loss of metal by evaporation. Hence, the melting point of the metal should be much higher than the filament temperature. The rate of deposition and the quality of the refined metal are affected by the nature of the metal, impurities in the starting metal, temperature of the crude metal, and the amount of iodine and crude metal charged in a vessel of certain geometry. The product obtained is in the form of large crystals [1] adherent to the filament in a packed form in a fairly regular fashion. There is a report of producing 6 nine grade Ti [25] by a new iodide process.

Generally, the yield per cycle is about 20% of the crude loaded in the reactor. In one cycle, the impurities can be reduced from 5000 to 1000 ppm and by multiple treatments to below 200 ppm. The major problem is due to the contamination of the metal from the residual gases in the vessel and due to the transportation of some metallic impurities from the vessel to the filament by iodine [26]. Effectiveness of the purification by this method is shown in Table 10.2.

Table 10.2 Effective purification by iodide process[a]

Metals	Type of feed material	Hot wire decomposition temperature, °C	Temperature of feed, °C	Analysis of metals (ppm)								
				–	Al	C	Fe	Hf	N	O	Si	Zr
Ti	Kroll sponge	1300	150	Before	<50	613	220	–	127	4800	665	–
				After	<50	105	75	–	40	150	195	–
Zr	Kroll sponge	1300	250	Before	50	132	95	170	41	1360	120	–
				After	10	115	<15	85	29	230	50	–
Hf	Ca-reduced vacuum sintered metal	1500–1800	280–300	Before	–	–	580	–	250	6900	–	13
				After	–	–	110	–	22	172	–	19

[a]Reproduced from C. V. Sundaram et al. [10] with the permission of The Indian Institute of Metals

The van Arkel process was employed for the first time to produce ultrapure titanium from titanium sponge obtained by the Kroll reduction process. During the refining of titanium, the crude is packed between the molybdenum grid and the reactor maintained at 200 °C to form TiI_4 at the maximum rate. At higher wall temperature, the rate of deposition on the filament decreases due to the formation of titanium di-iodide ($TiI_4 + I_2 = 2TiI_2$). At 400 °C, the rate of deposition approaches zero because the system contains only TiI_2 instead of TiI_4. At higher wall (bulb) temperature of 550 °C, the rate of decomposition of TiI_2, and thereby the deposition on the filament is faster than that achieved with TiI_4. Temperature of this order is not recommended in a glass reactor. However, higher bulb temperature prevents the deposited titanium from being contaminated with iron and silicon because iodides of these metals are unstable at higher temperature and hence do not enter the vapor phase. Relatively thicker titanium filament is used instead of very thin tungsten. The reactor is operated at a reduced pressure of few millimeters of mercury by maintaining the filament in the temperature range, 1100–1400 °C depending upon the requirement of the deposit. Larger crystals are deposited at 1300–1400 °C and smaller at only 1100 °C.

Zirconium sponge is also purified by following a similar procedure as adapted in the purification of titanium. In this case, the reactor wall is maintained at 200–400 °C to form ZrI_4 vapors that is made to decompose at the zirconium filament heated to 1300 °C. The reactor is operated under the vacuum of the order of 10^{-3} mm Hg. The element hafnium, which is chemically similar, is purified by forming HfI_4 vapors at 300–400 °C and decomposing the iodide at 1600 °C.

For the purification of thorium, the reactor bulb fitted with tungsten or thorium filament is first degassed at 500 °C under a reduced pressure of 10^{-4} mm Hg. Thorium iodide (ThI_4) is then allowed to form at 250–270 °C and then volatilized at 450 °C for decomposition into pure thorium and iodine at the filament heated to 1300–1400 °C. As thorium iodide is more stable compared to the iodides of titanium and zirconium, a higher temperature is required for its decomposition. During prolonged heating at 500–600 °C, formation of thorium tri-iodide (ThI_3) retards the rate of deposition of thorium on the filament because ThI_3 decomposes slowly due to its slower rate of vaporization than that of the ThI_4.

It is important to note that no refining technique is capable of removing all the impurities. Depending on the nature of the impurities and the metal (to be purified) different techniques have to be employed for the removal of different impurities. For example, a crude base metal, M contains impurities i, j, k, l, which are, respectively, more prone to forming stable oxide, relatively more volatile, nobler than the base metal, M, and more soluble in the selected electrolyte. In addition, i, j, and l are baser to M. Under such a condition, the following sequence of removal by different methods would be more justified:

(a) First of all, i should be removed as a dross by subjecting the crude metal, M to preferential oxidation.
(b) In the next step, the more volatile impurity, j can be removed by distillation at appropriate temperature.

(c) The resulting metal, M, after removal of i and j may be electrorefined. Since k is
    nobler than the parent metal M, it will collect at the bottom of the electrolytic
    tank. By virtue of its character, the impurity, l, will get dissolved in the
    electrolyte without being co-deposited on the cathode along with M.

10.4   Refining Along with Melting and Consolidation

Refractory and reactive metals like tantalum, niobium (columbium), tungsten,
molybdenum, titanium, and zirconium initially produced in the form of granules,
powder, or sponge are usually refined and consolidated by vacuum arc remelting or
electron-beam melting. The metal to be refined is first cast or forged in the form of an
electrode. In the case of metal powder, appropriate size of electrodes are made by die
pressing and welding.

After achieving, a vacuum of the order of 10^{-3} mm Hg in the *vacuum arc
remelting* furnace an electric arc is struck between the negative metal electrode
and a metal bath contained in the water-cooled copper crucible that serves as the
positive electrode. The metal electrode is melted and slowly lowered to maintain a
constant distance between the electrode tip and surface of the molten pool in the
crucible. Under vacuum melting, refining is assisted by several metallurgical operations, such as (i) removal of volatile impurities from the electrode and the bath
depending on temperature and pressure, (ii) removal of carbon and dissolved oxygen
by formation of gaseous carbon monoxide, (iii) deoxidation of the metal by vaporization of its oxide, (iv) removal of dissolved hydrogen in atomic form by formation
of molecular hydrogen gas according to the reaction: $[H] + [H] \rightarrow H_2$ (g), which is
favored under vacuum, and (v) uniform distribution of small inclusions in the melt.

In vacuum arc remelting, impurity pick up from the container is avoided by
holding the molten pool in a water-cooled copper mold.

The processing of refractive and reactive metals by *vacuum arc remelting* has led
to the development of *electron-beam melting*. In this technique, the kinetic energy of
the highly accelerated beam of electrons impinging on the metal is converted into
thermal energy. The resulting accumulated energy is large enough to melt the
material in due course. The operation is carried out under reduced pressure of less
than 10^{-3} mm Hg. Under this order of vacuum, volatile impurities can be selectively
removed and ingots of better purity along with other advantages of vacuum arc
remelting can be obtained. Electron-beam melting has been useful in improving the
purity of molybdenum, tungsten, and titanium in addition to its application in the
processing of tantalum, niobium, zirconium, and hafnium.

Plasma torch [27] has been developed to produce solar grade silicon. The
concentration of boron has been reduced from 5 ppm to <2 ppm by a technique
based on the combination of inductive plasma torch and electromagnetic
stirring [28].

In recent years, the *electroslag refining* technique, which was initially developed
for refining steel, has found application in purification and consolidation of a few

nonferrous metals. The metal to be refined is first cast or forged to make an electrode, one end of which is held in a refining molten slag. The slag is kept molten by passing heavy current. The heat generated is sufficient to raise the temperature of the electrode that dips in the slag to a value above its melting point. As a result, the electrode tip melts and the molten droplets fall through the reactive slag and get refined and solidified in the water-cooled mold. The process is continued by lowering the electrode. Refining takes place at the electrode tip–slag, droplet–slag, and metal pool–slag interfaces. In electroslag refining of steel and nickel alloys, calcium fluoride slags are generally used. CaO and Al_2O_3 have also been employed. Use of basic slag facilitates sulfur removal from the electrode metal. Large size nonmetallic inclusions are eliminated and finer inclusions are uniformly distributed in the refined ingots.

10.5 Problems [29–31]

Problem 10.1

Copper from the impure lead can be removed by adding PbS to the bath at 400 °C according to the reaction: PbS(s) + 2Cu(l) = Cu_2S(s) + Pb(l). If the standard free energy of the reaction at 400 °C is −19,500 cal, calculate the residual copper in the refined lead, assuming that (i) the sulfides are immiscible at low concentrations and (ii) Cu in low concentration forms a regular solution in lead at low concentration:

Given that, $\overline{H}_{Cu} = 1200$ cal mol^{-1}

Solution

The reaction under question: PbS (s) + 2Cu (l) = Cu_2S (s) + Pb (l),

$$\Delta G^o = -19,500 \text{ cal} = -RT \ln K, K = \frac{a_{Pb}}{a_{Cu}^2}$$

$$\log K = \frac{-19,500}{-4.575 \times 673} = 6.3333 = \log a_{Pb} - 2 \log a_{Cu} \qquad (1)$$

As the concentration of Cu is extremely low in the Pb-Cu system, $a_{Pb} = 1$,

$$\log a_{Cu} = -3.6667$$

For regular Pb-Cu alloys, $\overline{H}_{Cu} = RT \ln \gamma_{Cu} = 1200$ cal mol^{-1}

$$\therefore \log \gamma_{Cu} = \frac{1200}{4.575 \times 673} = 0.3897$$

$$a_{Cu} = \gamma_{Cu} \cdot x_{Cu}$$

$$\therefore \log x_{Cu} = \log a_{Cu} - \log \gamma_{Cu}$$
$$= -3.6667 - 0.3897 = -4.0564$$

$$x_{Cu} = 0.0000878$$

According to Eq. 1.78b,

$$wt\% \ Cu = \frac{x_{Cu} \times 100 \times 63.54}{207.21} = \frac{0.0000878 \times 100 \times 63.54}{207,21} = 0.0027\% \quad Ans.$$

Problem 10.2

(Nagamori, M., 1979–1980, personal communication, University of Utah, Salt Lake City)

Crude lead bullion containing Sn, Bi, and Cu is melted at 800 °C and oxidized by adding litharge slag. From the given data show that only tin can be removed successfully by drossing. Assuming that the reaction products are pure solid oxides of Pb, Sn, Bi, and Cu, calculate the minimum Sn content that can be achieved by this softening process. How are Bi and Cu eliminated from the crude lead? Given that:

$$Pb(1) + \frac{1}{2} O_2(g) = PbO(s), \Delta G_1^o = -53,300 + 25.70 \ T \ cal$$
$$Sn(1) + O_2(g) = SnO_2(s), \Delta G_2^o = -140,180 + 51.52 \ T \ cal$$
$$2Bi(1) + 3/2 \ O_2(g) = Bi_2O_3(s), \Delta G_3^o = -141,050 + 69.94 \ T \ cal$$
$$2Cu(s) + \frac{1}{2} O_2(g) = Cu_2O(s), \Delta G_4^o = -40,500 - 3.92 \ T \ logT + 29.5 \ T \ cal$$

atomic weights: Pb- 207.2, Sn- 118.7,

and at 800 K: $\gamma_{Sn}^o(\text{in Pb}) = 6.8$, $\gamma_{Cu}^o(\text{in Pb}) = 10$ and $\gamma_{Bi}^o(\text{in Pb}) = 0.39$

Solution

Assuming that the crude lead is covered with both PbO and MO (M = Sn, Bi, Cu). Interactions between M in the crude lead and the litharge (PbO) slag results in the following reactions:

$$PbO(s) + M \ (\text{in Pb}) = MO(s) + Pb(l)$$
At 1073 K, ΔG_1^o (PbO) = −25724 cal
ΔG_2^o (SnO$_2$) = −84899 cal
ΔG_3^o (Bi$_2$O$_3$) = −66004 cal
ΔG_4^o (Cu$_2$O) = −21594 cal

For the softening reactions:

$$2PbO + Sn = SnO_2 + 2Pb,$$
$$\Delta G_5^o = -84899 - 2 (-25724) = -33451,$$
$$K_5 = 6.52 \times 10^6$$

$$3PbO + 2Bi = Bi_2O_3 + 3Pb,$$
$$\Delta G_6^o = -66004 - 3(-25724) = +11168,$$
$$K_6 = 5.31 \times 10^{-3}$$

$$PbO + 2Cu = Cu_2O + Pb,$$
$$\Delta G_7^0 = -21594 - (-25724) = +4130,$$
$$K_7 = 1.44 \times 10^{-1}$$

Tin:

($a_{SnO_2} = 1 = a_{PbO}$; both SnO_2 and PbO are solids,
$a_{Pb} \approx 1$, crude lead contains more than 99% Pb)

$$K_5 = \frac{a_{SnO_2}.a_{Pb}^2}{a_{PbO}^2.a_{Sn}} = \frac{1}{a_{Sn}} = 6.52 \times 10^6$$

$$a_{Sn} = \frac{1}{6.52 \times 10^6} = 1.5 \times 10^{-7} = \gamma_{Sn}^0.x_{Sn} = 6.8.x_{Sn}$$

$x_{Sn} = 2.3 \times 10^{-8}$ (i.e., concentration of Sn in Pb is extremely low)

At infinitely dilute concentration of the solute Sn, wt% Sn is negligible as compared to the wt% of Pb ~100% and according to Eq. 1.78b,

$$\text{wt\% Sn} = \frac{x_{Sn}.100.M_{Sn}}{M_{Pb}}$$

$$\therefore \text{wt\% Sn} = \frac{x_{Sn} \times 100 \times 118.7}{207.2}$$

$$= \frac{2.3 \times 10^{-8} \times 100 \times 118.7}{207.2} = 1.3 \times 10^{-6}$$

Thus, the attainable lowest limit of tin in lead is 1×10^{-6}% Ans.

Bismuth:

$$K_6 = \frac{a_{Bi_2O_3}.a_{Pb}^3}{a_{PbO}^3.a_{Bi}^2} = \frac{1}{a_{Bi}^2} = 5.31 \times 10^{-3}$$
$$a_{Bi} = 13.7 \gg 1$$

This means that Bi content of the crude lead tries to become as high as 13.7 (a_{Bi}). But the crude lead cannot take up this much of bismuth. In other words, the elimination of Bi from the crude lead by oxidation is not possible. Conversely, Pb will get oxidized preferentially prior to the oxidation of bismuth.

Copper:

$$K_7 = \frac{a_{Cu_2O}.a_{Pb}}{a_{PbO}.a_{Cu}^2} = \frac{1}{a_{Cu}^2} = 0.144$$
$$a_{Cu} = 2.63 = \gamma_{Cu}^0.x_{Cu} = 10.x_{Cu}$$
$$x_{Cu} = 0.263$$

This means Cu content in Pb tries to become as high as 0.263 ($=x_{Cu}$). Since the Cu content in the initial crude lead is less than this, Pb will get oxidized preferentially prior to the oxidation of Cu from the Pb–Cu liquid.

In view of the above problems, calcium is added to eliminate Bi from Pb by forming Bi_2Ca_3 (Kroll-Betterton process). Sulfur is added to eliminate Cu by forming Cu_2S which is skimmed off from the surface of sulfur-treated Pb alloys.

Problem 10.3

One ton of zinc-free lead bullion containing 8 kg silver is treated with zinc at 800 K for silver recovery as pure solid Ag_2Zn_3. Calculate the amount of zinc to remove 97% silver from this bullion. The standard free energy change for the reaction: $2Ag$ (l) $+ 3Zn(l) = Ag_2Zn_3(s)$ is -30 kcal gatom^{-1}.

Given that: $(\gamma^o_{Zn})_{Pb} = 15$ and $(\gamma^o_{Ag})_{Pb} = 2$.

Atomic weights: Pb-207.21, Zn-65.37 and Ag-107.91

Solution

One ton of lead bullion contains 8 kg Ag and 992 kg Pb

Number of moles of Ag $= \frac{8}{107.91} = 0.07414$ kg mol/ton bullion.

Number of moles of Pb $= \frac{992}{207.21} = 4.78741$ kg mol/ton bullion.

The final Ag in the bullion $= (1 - 0.97) \times 0.07414 = 0.0022242$ kg mol/ton refined bullion

$\therefore x_{Ag}$ in the refined bullion $= \frac{0.0022242}{4.86155} = 4.575 \times 10^{-4}$

For the reaction: $2Ag(l) + 3Zn(l) = Ag_2Zn_3(s)$, $\Delta G^o = -30000$ cal gatom^{-1}

$\Delta G^o = -30000 = -RT\ln K$

$K = 1.57 \times 10^8$

$$= \frac{a_{Ag_2Zn_3}}{a^2_{Ag}.a^3_{Zn}} = \frac{1}{a^2_{Ag}.a^3_{Zn}}$$

$\therefore a^2_{Ag}.a^3_{Zn} = \frac{1}{1.57 \times 10^8} = x^2_{Ag}.\left(\gamma^o_{Ag}\right)^2.x^3_{Zn}.\left(\gamma^o_{Zn}\right)^3 = 6.37 \times 10^{-9}$

$x^3_{Zn} = \frac{6.37 \times 10^{-9}}{x^2_{Ag}.\left(\gamma^o_{Ag}\right)^2.\left(\gamma^o_{Zn}\right)^3} = \frac{6.37 \times 10^{-9}}{\left(4.575 \times 10^{-4}\right)^2.(2)^2.(15)^3} = 0.000002186$

$x_{Zn} = 0.012978$

$\therefore$ resdual zinc content in the bullion $= 0.012978 \times 4.78741 \times 65.37 = 4.059$ kg,

and zinc required for reaction with silver $= \frac{0.97 \times 8 \times 65.37}{2 \times 107.91} = 2.35$ kg. (assuming that Zn concentration is small and neglecting the residual silver in the bullion).

$\therefore$ Total zinc required for removal of 97% silver from the bullion

$= 4.06 + 2.35 = 6.41$ kg. Ans.

Problem 10.4

During the refining of crude lead, where the lead bullion is treated with zinc for the removal of silver and gold in zinc crust (Parkes process), the bullion contains an appreciable amount of zinc. Subsequently, chlorine gas is bubbled through the lead bullion at 400 °C to remove zinc (Batterton Process).

(a) Calculate the limiting zinc concentration in the chlorine-purified lead, assuming that molten chlorides containing 98 mol% $ZnCl_2$ and 2 mol% $PbCl_2$ above the Pb–Zn alloy form an ideal solution. Given that: $(\gamma^o_{Zn})_{Pb} = 15$

$Pb(l) + Cl_2(g) = PbCl_2(l), \Delta G^o_1 = -83600 - 14.12\ T\ \log\ T + 73.3\ T$ cal

$Zn(l) + Cl_2(g) = ZnCl_2(l), \Delta G^o_2 = -93950 + 27.35\ T$ cal

(b) What is the effect of temperature on refining with chlorine? Discuss with the aid of suitable calculations.

Solution

The refining reaction in the Batterton Process:

$$Zn(l) + PbCl_2(l) = ZnCl_2(l) + Pb(l)$$

$$\Delta G^\circ = \Delta G_2^\circ - \Delta G_1^\circ = -93950 + 27.35\,T$$

$$-(-83600 - 14.12\,T\,\log T + 73.3\,T\,)$$

$$= -10350 + 14.12\,T\,\log T - 45.95\,T\ \text{cal}$$

$$\Delta G^\circ = -14400\ \text{cal}$$

$$= -RT\ln K$$

at $T = 673\,K$, $K = 47494$

For the above reaction $K = \dfrac{a_{Pb} \cdot a_{ZnCl_2}}{a_{Zn} \cdot a_{PbCl_2}} = 47494$

Since chlorides form an ideal solution:

$$x_{ZnCl_2} = 0.98 = a_{ZnCl_2} \text{ and } x_{PbCl_2} = 0.02 = a_{PbCl_2}$$

As the first approximation for the lead bullion we may set $a_{Pb} = 1$, then a_{Zn} can be calculated in the following manner:

$$a_{Zn} = \frac{1 \times 0.98}{0.02 \times 47494} = 1.03 \times 10^{-3}$$

$$x_{Zn} = \frac{a_{Zn}}{\gamma_{Zn}^0} = \frac{1.03 \times 10^{-3}}{15} = 6.88 \times 10^{-5}$$

From Eq. 1.78b, we can calculate the residual zinc in the purified lead:

$$\%Zn = \frac{x_{Zn} \times 100 \times 65.37}{207.21} = \frac{6.88 \times 10^{-5} \times 100 \times 65.37}{207.21} = 0.00217 \text{ Ans.}$$

(b) In order to demonstrate the effect of temperature on the refining process, calculate K, a_{Zn}, x_{Zn} and $\%Zn$ at three different temperatures, 350 °C (623 K), 400 °C (673 K) and 500 °C (773 K). Following the above procedure carried out at 673 K, the results are listed below:

Temperature, K	K	a_{Zn}	x_{Zn}	wt % Zn
623	112,712	0.000435	0.000029	0.00091
673	47,494	0.00103	0.000069	0.00217
773	11,793	0.00415	0.000277	0.00874

From the above table it is evident that the value of K decreases with increase of temperature. This means the chlorination reaction: $Zn(l) + PbCl_2(l) = ZnCl_2(l) + Pb(l)$, is more favorable with decrease of temperature, that is refining is more effective at lower temperature. Table also shows that the residual zinc content decreases with decrease of temperature. In order to achieve the reasonable rate of refining, the optimum temperature of chlorination is about 390–400 °C (663–673 K).

Problem 10.5

Liquid copper containing silicon as an impurity is allowed to come into equilibrium with oxygen at a partial pressure of 10^{-14} atm. Calculate the minimum silicon content of the melt at 1200 °C assuming that copper will not oxidize under these conditions.

Given that: melting point, Cu: 1083 °C, Si: 1410 °C

Atomic weight, Cu: 63.54, Si: 28.09

$\gamma_{Si}^{\circ} = 0.006$ (standard state: pure liquid Si at 1200 °C).

$Si(l) + O_2(g) = SiO_2(s)$, $\Delta G^{\circ} = -897050 + 168.7\ T$ J mol^{-1} (25 – 1410 °C).

Solution

ΔG° at 1473 K $= -897050 + 168.7 \times 1473 = -648555$ J mol^{-1}

At equilibrium, $\Delta G^{\circ} = -RT\ln K = -648555$

$K = 9.988 \times 10^{22}$

and $K = \dfrac{a_{SiO_2}}{a_{Si(l)} \cdot p_{O_2}}$ since SiO_2 is a solid, $a_{SiO_2} = 1$

$$\therefore a_{Si(l)} = \frac{1}{K \cdot p_{O_2}} = \frac{1}{9.988 \times 10^{22} \times 10^{-14}}$$
$$= 1.001 \times 10^{-9}$$

As x_{Si} is very small, Henrian behavior can be assumed, hence we can write: $a_{Si} = \gamma_{Si}^{\circ} \cdot x_{Si}$

$$\therefore x_{Si} = \frac{a_{Si}}{\gamma_{Si}^{\circ}} = \frac{1.001 \times 10^{-9}}{0.006} = 1.669 \times 10^{-7}$$

wt% Si $= \dfrac{x_{Si} \times 100 \times M_{Si}}{M_{Cu}} = \dfrac{1.669 \times 10^{-7} \times 100 \times 28.09}{63.54} = 7.377 \times 10^{-6}$ Ans.

Problem 10.6

For the removal of zinc, a liquid Cd–Zn alloy is treated with excess solid CdO at 427 °C according to the reaction: $Zn(l) + CdO(s) = Cd(l) + ZnO(s)$. If solid ZnO and solid CdO exist as separate phases without any appreciable mutual solubility, calculate the equilibrium zinc concentration of the alloy coexisting with these oxides. The free energy of formation of solid ZnO and solid CdO at 427 °C, are −278,236 and − 184,514 J mol^{-1}, respectively. The activity coefficients of zinc and cadmium for the pure metals as reference states are given as:

$$\log \gamma_{Zn} = 0.87\,(1 - x_{Zn})^2 - 0.30\,(1 - x_{Zn})^3$$

$$\log \gamma_{Cd} = 0.42 \, (1 - x_{Cd})^2 + 0.30 \, (1 - x_{Cd})^3$$

Atomic weight: Zn-65.37 and Cd-112.41

Solution

The free energy change for the reaction: $Zn(l) + CdO(s) = Cd(l) + ZnO(s)$ is obtained as:

$$\Delta G_r^\circ = \Delta G_{ZnO}^\circ - \Delta G_{CdO}^\circ$$
$$= -278236 - (-184514) = -93722 \text{ J} = -RT\ln K$$
$$K = \frac{a_{ZnO} \cdot a_{Cd}}{a_{CdO} \cdot a_{Zn}} = \frac{a_{Cd}}{a_{Zn}} \quad \text{(as ZnO and CdO are solids, } a_{ZnO} = 1 = a_{CdO})$$
$$\ln K = \frac{\Delta G_r^\circ}{-RT} = \frac{-93722}{-8.314 \times 700} = 16.104$$
$$K = 9.86 \times 10^6$$
$$\therefore \frac{a_{Cd}}{a_{Zn}} = 9.86 \times 10^6 = \frac{\gamma_{Cd} \cdot x_{Cd}}{\gamma_{Zn} \cdot x_{Zn}}$$

$$\log \gamma_{Cd} - \log \gamma_{Zn} = 6.994 - \log \left(\frac{x_{Cd}}{x_{Zn}}\right) \tag{1}$$

Substituting the given values of activity coefficients of Cd and Zn in Eq. 1, we get:

$$0.42x_{Zn}^2 + 0.30x_{Zn}^3 - 0.87 \, (1 - x_{Zn})^2 + 0.30 \, (1 - x_{Zn})^3 = 6.994 - \log \left(\frac{1 - x_{Zn}}{x_{Zn}}\right)$$

On simplification we get:

$$0.45x_{Zn}^2 + 0.84x_{Zn} - 7.564 + \log \left(\frac{1 - x_{Zn}}{x_{Zn}}\right) = 0$$

Since x_{Zn} is very small x_{Zn}^2 and $0.84x_{Zn}$ can be neglected

$$\therefore \log \left(\frac{1 - x_{Zn}}{x_{Zn}}\right) = 7.564$$

and hence $\left(\frac{1 - x_{Zn}}{x_{Zn}}\right) = 3.66 \times 10^7$

or $x_{Zn} = 2.37 \times 10^{-8}$

$\text{wt\% Zn} = \frac{x_{Zn} \times 100 \times M_{Zn}}{M_{Cd}} = \frac{2.37 \times 10^{-8} \times 100 \times 65.37}{112.41} = 1.39 \times 10^{-6}$ Ans.

Problem 10.7

In the Mond carbonyl process, crude nickel is purified by volatilization and subsequent decomposition of $Ni(CO)_4$ according to the reactions: $Ni(s) + 4CO(g) \rightleftharpoons Ni(CO)_4(g)$. What are the effects of temperature and pressure on this reaction? Discuss these aspects by calculating percent volumes of CO and $Ni(CO)_4$ in CO–$Ni(CO)_4$ gaseous mixture at 100, 150, and 200 °C, at total pressures of 20, 50, and 100 atm. At what temperature can equal volume fractions of both the gases be obtained? The free energy change for the reaction is given as: $\Delta G^\circ = -33,500 + 95 \, T$ cal

Solution

For the reaction: $Ni(s) + 4 \, CO \, (g) \rightleftharpoons Ni(CO)_4(g)$, $\Delta G^\circ = -33500 + 95 \, T$ cal

At 100 °C (373 K), $\Delta G_{373}^\circ = -33500 + 95 \times 373 = +1935$ cal

$$K_{373} = 0.0735$$

$$K = \frac{p_{(NiCO)_4}}{p_{CO}^4} = 0.0735$$

$$p_{(NiCO)_4} = p_{CO}^4 \cdot 0.0735 \tag{1}$$

$$p_{(NiCO)_4} + p_{CO} = 20 \tag{2}$$

From (1) and (2) we get:

$$p_{CO}^4 \cdot 0.0735 + p_{CO} - 20 = 0 \tag{3}$$

On solving by trial and error, we get: $p_{CO} = 3.85$ atm and $p_{(NiCO)_4} = 16.15$ atm.

Assuming ideal behavior of gases: %CO $= 19.25$ and %Ni(CO)$_4$ $= 80.75$ (by volume).

Following the above procedure at different temperatures and pressures, values of K, partial pressures, and % volume composition of CO and Ni(CO)$_4$ gases in the CO–Ni(CO)$_4$ gaseous mixture, calculated, are listed in table below:

| Temperature (K) | K | Pressure (atm) | | | | | |
| | | 20 | | 50 | | 100 | |
		%CO	%Ni(CO)$_4$	%CO	%Ni(CO)$_4$	%CO	%Ni(CO)$_4$
100	0.0735	19.25	80.75	9.95	90.05	5.98	94.02
150	0.00035	61.0	39.0	34.92	65.08	21.74	78.26
200	0.0000052	96.4	3.6	77.06	22.94	54.39	45.41

Since the carbonyl reaction is only slightly exothermic, it proceeds in reverse direction even at 100 °C due to large contribution of entropy which makes ΔG°_{373} positive (= + 1935 cal). From the above table, it is quite evident that the equilibrium constant for the reaction: Ni (s) + 4CO(g) = Ni(CO)$_4$ decreases drastically from 0.0735 to 0.0000052 with the corresponding increase of temperature from 100° to 200 °C. Hence, the formation of Ni(CO)$_4$ decreases with increase of temperature.

This table clearly shows that Ni(CO)$_4$ content in the gaseous mixture decreases with increase of temperature at all pressures. According to the Le Chatelier's principle, formation of Ni(CO)$_4$ will be favored with increase of pressure because there is severe reduction in volume on the formation of 1 mol of Ni(CO)$_4$ by consuming 4 mol of CO gas. Table clearly demonstrates that the carbonyl content in the gaseous mixture increases with increase of pressure at all temperatures.

Assuming ideal behavior, gaseous mixture having equal volume fractions of CO and Ni(CO)$_4$ at a total pressure of 1 atm, the partial pressure of both the gases will be equal, hence $p_{CO} = p_{(NiCO)_4} = 0.5$ atm.

$$Ni(s) + 4CO(g) = Ni(CO)_4, \Delta G^{\circ} = -33500 + 95\,T \text{ cal} \tag{1}$$

$$K = \frac{p_{(NiCO)_4}}{p_{CO}^4} = \frac{0.5}{(0.5)^4} = 8$$

$$\Delta G^\circ = -RT \ln K = -4.575\, T \log (8) = -4.132\, T \tag{2}$$

From Eqs. 1 and 2:

$$\Delta G^\circ = -33500 + 95\, T = -4.132\, T$$

$$\text{or } 95\, T + 4.132\, T = 99.132\, T = 33500$$

$$T = 338\, K = 65\,^\circ C \quad \text{Ans.}$$

Problem 10.8

For the determination of heat for mixing of Cd–Zn alloys by liquid metal solution calorimetry, the alloy and the corresponding stoichiometric mechanical mixture were dissolved in high-purity tin contained in the calorimeter at 270 °C, maintained under a vacuum of 10^{-2} mm Hg. After 10 additions, the resulting alloy solvent analyzed 10 at %Cd, 10 at %Zn, and 80 at %Sn. In order to reuse the tin solvent in the calorimetric measurements what is the order of vacuum required to distil Cd and Zn completely (theoretically) from the ternary alloy melt at 600 °C?

Given that: $\gamma_{Sn} = 1.1$, $\gamma_{Cd} = 1.2$ and $\gamma_{Zn} = 1.5$.

Vapor pressures (in mm Hg) of tin, cadmium, and zinc are given as function of temperature by the following expressions:

$$\log (p_{Sn}) = -\frac{15500}{T} + 8.23$$

$$\log (p_{Cd}) = -\frac{5819}{T} - 1.257 \log T + 12.287$$

$$\log (p_{Zn}) = -\frac{6620}{T} - 1.255 \log T + 12.34$$

Solution

Vapor pressures of pure Sn, Cd, and Zn can be obtained at 600 °C by substituting $T = 873$ K in the given expressions:

$$\log (p_{Sn}^\circ) = -\frac{15500}{873} + 8.23 = -9.52, \quad p_{Sn}^\circ = 3 \times 10^{-10} \text{ mm Hg}$$

$$\log \left(p_{Cd}^{o}\right) = -\frac{5819}{873} - 1.257 \times \log (873) + 12.287 = 1.92,$$

$$p_{Cd}^{o} = 83.20 \text{ mm Hg}$$

$$\log \left(p_{Zn}^{o}\right) = -\frac{6620}{873} - 1.255 \times \log (873) + 12.34 = 1.07,$$

$$p_{Zn}^{o} = 11.75 \text{ mm Hg}$$

Hence, the partial pressures of Sn, Cd and Zn over the ternary alloy (Sn–Cd–Zn):

$$p_{Sn} = x_{Sn} \cdot \gamma_{Sn} \cdot p_{Sn}^{o} = 0.80 \times 1.1 \times 3 \times 10^{-10} = 2.64 \times 10^{-10} \text{ mm Hg}.$$

$$p_{Cd} = x_{Cd} \cdot \gamma_{Cd} \cdot p_{Cd}^{o} = 0.1 \times 1.2 \times 83.2 = 9.98 \text{ mm Hg}.$$

$$p_{Zn} = x_{Zn} \cdot \gamma_{Zn} \cdot p_{Zn}^{o} = 0.1 \times 1.5 \times 11.75 = 1.76 \text{ mm Hg}.$$

Thus, a vacuum of the order of 10^{-3} mm Hg would be sufficient to distil Cd and Zn completely from the ternary alloy melt at 600 °C. Ans.

References

1. Jamrack, W. D. (1963). *Rare metal extraction by chemical engineering techniques*. New York: Macmillan. (Chapters 2 and 8).
2. Lupis, C. H. P., & Elliott, J. F. (1966). Generalized interaction coefficients. Part I: Definitions. *Acta Metallurgica, 14*, 529–538.
3. Rosenqvist, T. (1974). *Principles of extractive metallurgy*. New York: McGraw-Hill. (Chapter 13).
4. Turkdogan, E. T., & Pearson, J. (1953). Activity of constituents of iron and steelmaking slags, Part I, Iron oxide. *Journal of the Iron and Steel Institute, 173*, 217–223.
5. Davenport, D. W., King, M., Schlesinger, M., & Biswas, A. K. (2002). *Extractive metallurgy of copper* (4th ed.). Oxford: Pergamon. (Chapters 15 and 16).
6. Moore, J. J. (1990). *Chemical metallurgy* (2nd ed.). Oxford: Butterworth Heinemann. (Chapter 7).
7. Sevryukov, N., Kuzmin, B., & Chelishchev, Y. (1960). *General metallurgy* (B. Kuznetsov, Trans.), Moscow: Peace Publishers,
8. Pehlke, R. D. (1973). *Unit processes in extractive metallurgy*. New York: American Elsevier Pub. Co
9. Bodsworth, C. (1990). *The extraction and refining of metals*. Boca Raton: CRC Press. (Chapter 5).
10. Sundaram, C. V., Garg, S. P., & Sehra, J. C. (1979) Refining of reactive metals, *Proceedings of international conference on metal sciences – Emerging frontiers*, Varanasi, India, November, 23–26, 1977, I.I.M. Calcutta, pp. 351–367.
11. Coudurier, L., Hopkins, D. W., & Wilkomirski, I. (1978). *Fundamentals of metallurgical processes*. Oxford: Pergamon. (Chapter 5).
12. Bailey, A. R. (1964). *A text book of metallurgy*. New York: Macmillan. (Chapter 8).
13. Pfann, W. G. (1959). *Zone melting*. New York: Wiley. (Chapters 2 and 3).
14. Fort, D., Jones, D. W., Beaudry, B. J., & Gschneidner, K. A., Jr. (1981). Zone refining of less common metals. *Journal of the Less Common Metals, 81*, 273–292.

15. Fort, D. (2001). Purification of the rare earth metals. In Y. Waseda & M. Isshiki (Eds.), *Purification process and characterization of ultra high purity metals* (pp. 155–177). Berlin: Springer. (Chapter 2).

16. Yuge, N., Hanazawa, K., & Kato, Y. (2004). Removal of metal impurities in molten silicon by directional solidification with electron beam heating. *Materials Transactions, 45*(3), 850–857.

17. Takeda, O., Uda, T., & Okabe, T. H. (2014). Rare earth, titanium group metals, and reactive metals production. In S. Seetharaman (Ed.), *Treatise on process metallurgy. Vol. 3, Industrial process, Part A* (pp. 995–1069). Oxford: Elsevier. (Chapter 2.9).

18. Maurits, J. E. A. (2014). Silicon production. In S. Seetharaman (Ed.), *Treatise on process metallurgy. Vol. 3, Industrial process, Part A* (pp. 919–948). Oxford: Elsevier. (Chapter 2.6).

19. Verhoeven, J. D. (1966). Electrotransport as a means of purifying metals. *JOM, 18*, 26–31.

20. Peterson, D. T., & Schmidt, F. A. (1972). Electrotransport of C, N and O in gadolinium. *Journal of the Less Common Metals, 29*, 321–327.

21. van Arkel, A. E., & de Boer, J. H. (1925). Darstellung von reinsen Titanium, Zirkonium, Hafnium und Thoriummetall. *Zeitschrift für anorganische und allgemeine Chemie, 148*, 345–353.

22. Yoshimura, Y., Oonishi, T., & Kuramoto, M. (1994). Purification of titanium by iodide process. *Materia Japan, 33*, 48–50.

23. Shelton, R. A. J. (1968). Thermodynamic analysis of the van Arkel iodide process. *Transactions Institution of Mining and Metallurgy, Section C, 77*, C32–C35.

24. Shelton, R. A. J. (1968). Rate of deposition of zirconium in the van Arkel iodide process. *Transactions Institution of Mining and Metallurgy, Section C, 77*, C113–C119.

25. Yoshimura, Y. (1996). Production of 6 nine grade Ti by a new iodide process. *Titanium Japan, (Japan Titanium Society), 44*, 71–73.

26. Shamsuddin, M., & Sohn, H. Y. (2019). Constitutive topics in physical chemistry of high-temperature nonferrous metallurgy – A review: Part 2. Reduction and refining. *JOM, 71*(9), 2366–2376.

27. Delannoy, Y., Alemany, C., Li, K.-I., Proulx, P., & Trassy, C. (2002). Plasma refining process to provide solar-grade silicon. *Journal of Solar Energy Materials and Solar Cells, 72*, 69–75.

28. Alemany, C., Trassy, C., Pateyom, B., Li, K.-I., & Delannoy, Y. (2002). Refining of metallurgical-grade silicon by inductive plasma. *Journal Solar Energy Materials and Solar Cells, 72*, 41–48.

29. Kubaschewski, O., & Alcock, C. B. (1979). *Metallurgical thermochemistry* (5th ed.). Oxford: Pergamon.

30. Curnutt, J. L., Prophet, H., McDonald, R. A., & Syverud, A. N. (1975). *JANAF thermochemical tables*. Midland: The Dow Chemical Company.

31. Barin, I., Knacke, O., & Kubaschewski, O. (1977). *Themochemical properties of inorganic substances, (supplement)*. New York: Springer.

Chapter 11
Hydrometallurgy

Hydrometallurgy refers to the extraction of metals and production of inorganic substances from minerals through aqueous solutions. Although the entire amount of iron, lead, tin, and antimony and major production of copper and nickel come from the pyrometallurgical route, hydrometallurgy plays an important role in the extraction of reactive and rare metals and some common metals (e.g., aluminum, uranium, thorium, zinc, copper, nickel, tungsten, and molybdenum). There has been significant increase in hydrometallurgical processing of copper ores/minerals at the mine site [1] during the past few decades. In some cases, both methods are involved. For example, sphalerite and molybdenite concentrates are roasted to oxides (ZnO and MoO_3) and dissolved in sulfuric acid and ammonia, respectively, and the resultant liquor after purification is electrolyzed and subjected to precipitation/ reduction for production of zinc and molybdenum. On the other hand, gold from its ore is dissolved in alkali metal cyanide in the presence of oxygen and is cemented with zinc and finally treated at high temperature. In rare metal extraction, a combination of hydrometallurgy and pyrometallurgy is often adopted for the production of a suitable intermediate as well as the metal. In hydrometallurgy, there are two major steps for obtaining the metal value. The first step, known as leaching or lixiviation, is carried out with the objective of bringing the metal into aqueous solution. In the second step, the metal is recovered from the leach liquor by cementation or precipitation by controlling the operating conditions/variables. In addition, it is necessary to purify the leach liquor from the insoluble residues and suspended solid particles by adding steps like solid–liquid separation and purification before the metal recovery. Hydrometallurgy is comparatively a newer technique compared to pyrometallurgy which is being practiced from ancient times. The former offers the following advantages over the latter:

1. Extraction of metals by the pyrometallurgical route requires high-grade ores or concentrates to economize the thermal energy in heating the gangue and the necessary flux for slagging. Thus, with deteriorating ore quality, higher grinding and beneficiation costs will be incurred in future on the treatment and handling

© The Minerals, Metals & Materials Society 2021
M. Shamsuddin, *Physical Chemistry of Metallurgical Processes, Second Edition*,
The Minerals, Metals & Materials Series, https://doi.org/10.1007/978-3-030-58069-8_11

of a larger proportion of the gangue for liberation of the mineral value. As the grade of ore is not so crucial in the hydrometallurgical method of extraction, it is going to be still profitable on the ore bodies, which are uneconomical to be treated by pyrometallurgy.

2. Contrary to pyrometallurgy, very little or no fuel is required in hydrometallurgical methods of extraction.

3. Hydrometallurgical methods can treat a variety of feeds like low-grade and complex ores and concentrates, speiss, tailings, residues, matte, and metal scrap by slight adjustment of process variables.

4. The equipment needed is relatively simple and inexpensive, and hence, hydrometallurgical plants can be erected at low capital cost.

5. As the capital cost involved in pyrometallurgical route is very high, sufficiently large ore reserves that can last for longer time have to be assured from an economic viewpoint. Contrary to this, hydrometallurgical plants can be installed for smaller ore bodies with low investment.

6. The reagent used during leaching or digestion can be easily regenerated. This adds to the economy of the process.

7. Precious metals like gold, silver, and so on can be easily recovered as a byproduct. Elemental sulfur can be recovered from the treatment of sulfide ores.

8. Many a time, metals obtained do not need refining. For example, zinc deposited after electrolysis of the purified leach liquor can be directly used for alloy making.

9. A large amount of gangue present in the ore remains unaffected during leaching, whereas it has to be slagged off in pyrometallurgical smelting.

10. There are mild corrosion problems in hydrometallurgy compared to those of expensive refractory lining in furnaces forcing periodic shutdown during replacement.

However, despite the flexibility of treating a wide variety of feed and producing the metal in powder form directly by chemical and electrolytic reduction of the purified leach liquor together with relatively easier control of instrumentation, hydrometallurgical processes suffer from two major drawbacks. At room or moderate temperatures, the rate of production is much slower compared to the pyrometallurgical methods of extraction. The use of lean and complex ores generates large volumes of dilute solutions that require large space for handling. Disposal of such effluents pollutes ground and water, if not the atmosphere.

From literature, it is evident that presently a large number of hydrometallurgical processes developed on the laboratory scale are available. These cannot be adopted in preference to the well-established pyrometallurgical methods without testing their economic feasibility on a larger scale. Despite the above-mentioned limitations, hydrometallurgy presents a better future in treatment of lean and complex ores of nonferrous metals and production of metals from scrap, slag, tailings, residues, etc. Since many materials would not respond to leaching with conventional lixiviants, new solvents and new techniques have to be developed. Unless a major breakthrough is achieved in these developments, pyrometallurgy will continue to be used in future years.

Generally, the following steps are incorporated in hydrometallurgical treatment of ores:

1. *Mineral beneficiation*: The ore has to be grounded to a very fine size for effective leaching. In case of treatment of sulfides and carbonates, roasting and calcinations, respectively, are necessary prior to leaching.
2. *Leaching*: Metal values are dissolved in suitable solvents by leaching the ore/concentrate.
3. *Solid–liquid separation*: Undissolved residues and suspended particles have to be removed by filtration.
4. *Concentration and purification*: Very dilute leach liquor has to be concentrated to a desired level depending on the requirements during recovery. Special techniques such as ion exchange and solvent extraction are employed in the extraction of rare metals.
5. *Recovery of metal value*: The metal may be precipitated from the leach liquor by control of pH and p_{O_2}, cementation, or blowing hydrogen gas at high pressure.

In the case of extraction of reactive metals like beryllium, zirconium, titanium, uranium, and thorium, a step called production of suitable intermediate is added. For example, in uranium extraction, magnesium or ammonium diuranate is precipitated by adding magnesia or ammonium hydroxide to the uranyl sulfate solution obtained after sulfuric acid digestion. In the following sections, methods of leaching, purification, and metal recovery with suitable examples will be discussed with major emphasis on the physical chemistry of the techniques.

11.1 Leaching

Leaching means dissolution of the metal value into a suitable reagent that may also dissolve many other metals present in the ore. Prior to leaching, the ore may be subjected to preliminary treatments like crushing, grinding, and concentration by mineral beneficiation methods, which will not be dealt here. Readers are advised to refer to books and journals available on this subject. Crushing and grinding of the ore to a particular size help to liberate the mineral particles that can be easily digested by the solvent. Grinding is generally not necessary for porous ores. For leaching purposes, products are divided into two groups, namely, slimes and sands. Slimes are finely ground products that tend to pack in a vat or tank and thus prevent the free circulation of liquid through the interstices of the ore bed. Sands are coarse products to permit circulation of solvents through the void space between the ore particles. Hence, slimes and sands are treated by agitation and percolation leaching methods, respectively. The leaching of low-grade ore is done on a very large scale. The great bulk of the treated ore requires a large amount of solvent. In such cases, for a profitable process, the solvent must be cheap and should be regenerative, if necessary. However, the cost of reagents for leaching the mineral containing large amount of valuable metal concentrated in a small bulk is not an important factor. Due to the small amount of metal in the low-grade ores, the leaching cost must be reduced by using cheap reagents in dilute solutions. Although most leaching processes can be accelerated by heating the solvent, heating is seldom practiced in large-scale leaching.

Some materials like oxidized copper and uranium ores and certain gold ores can be directly leached. Occasionally, ores are concentrated prior to leaching to reduce the bulk so as to economize the process by using less amount of solution. For example, gold ore is concentrated by flotation. The solvent to be used in leaching must be selective to the metal and not to the gangue, cheap and readily available in large quantities, regenerative, and quick in carrying out dissolution for commercial production. Acids, alkalis, and salts have been very effectively used in leaching, for example, sulfuric acid in leaching of uranium and oxidized copper ores and roasted zinc concentrate, sodium hydroxide in dissolution of bauxite, sodium carbonate solution in leaching of scheelite and oxidized uranium ores and in the presence of oxygen, sodium cyanide and potassium cyanide solution in leaching of gold and silver ores are being used on a commercial scale to bring metal values in the leach liquor according to the reactions (only a few selected ones):

$$2U_3O_8 + O_2 + 6H_2SO_4(aq) \rightarrow 6UO_2(SO_4)(aq) + 6H_2O \qquad (11.1)$$

$$ZnO + H_2SO_4(aq) \rightarrow ZnSO_4(aq) + H_2O \qquad (11.2)$$

$$Al_2O_3(bauxite) + 2NaOH\ (aq) \rightarrow 2NaAlO_2(aq) + H_2O \qquad (11.3)$$

$$CaWO_4(scheelite) + Na_2CO_3(aq) \rightarrow Na_2WO_4(aq) + CaCO_3 \qquad (11.4)$$

$$4Au + 8NaCN(aq) + O_2 + 2H_2O \rightarrow 4\ NaAu(CN)_2(aq) + 4NaOH \qquad (11.5)$$

The use of ferric salts [2] in dissolution of CuS and Cu$_2$S depends on the fact that they are reduced to ferrous salts, as evident by the following reaction:

$$Cu_2S + 2FeCl_3(aq) \rightarrow 2CuCl\ (aq) + 2FeCl_2(aq) + S \qquad (11.6)$$

Ferric sulfate also functions in a similar way. Oxide and carbonate minerals of copper can be easily dissolved in dilute sulfuric acid, but leaching of sulfides requires an oxidant in addition to the acid. The rate of leaching is enhanced tremendously in the presence of certain bacteria.

11.1.1   Leaching Methods

The choice of a particular technique depends on factors such as type of ore deposits, desired leaching rates, composition of the ore, nature of the gangue associated with the ore, and the subsequent separation and precipitation or extraction technique to be adopted. A brief outline of each method is given below.

11.1.1.1 In Situ Leaching

This method of leaching also known as solution mining is used when very low-grade ore is left out in the worked-out mines and also for recovery of metals from low-grade deep-seated ore deposits. In this method, ore bodies are fractured at the surface for penetration of the lixiviant inside the mine and the resultant leach liquor is pumped out to recover the metal value. In situ leaching has been successfully practiced for recovery of copper and uranium in the Western United States.

11.1.1.2 Heap Leaching

It is practiced by spraying a solvent over the ore lumps of less than 200 mm diameter, stacked in open atmosphere with the facility for drainage for collection of the leach liquor. The process is slow with low recovery.

11.1.1.3 Percolation Leaching

It is adopted for crushed ore of 6 to 10 mm size placed in large tanks by percolating a number of solutions in increasing concentration. For effective leaching, the ore should be course enough so that the leaching solution can move freely through the voids. The method is also known as sand leaching because of the use of coarse particles. The tanks made of wood and concrete and lined with lead or asphalt are used. To facilitate addition and withdrawal of leach solution and wash water, tanks are fitted with filter at the bottom.

11.1.1.4 Agitation Leaching

It is employed for ore fines ground to less than 0.4 mm diameter. This is also known as slime leaching. The slime and the leach solution are agitated in one or more agitators until the ore minerals have dissolved. Some agitators have mechanically driven paddles or elevators inside the agitation tank that facilitates continuous circulation of the pulp to achieve complete dissolution. For slime leaching, another type of tank known as pachuca in the form of a cylindrical vessel with a conical bottom fitted with a coaxial pipe, with both ends open, for introducing compressed air for agitation is more popular. In both types of tanks, the particles remain suspended in leach solution and are stirred mechanically or with jets of the compressed air. Though expensive, this is a faster and more efficient method as compared to the percolation leaching.

11.1.1.5 Pressure Leaching

Leaching is carried out in autoclaves at high pressure and relatively at higher temperature than possible in open leaching tanks. High-pressure leaching is advantageous when gaseous reagents, for example, oxygen and ammonia, are involved.

The amount of gas held in the solution increases with pressure. Equations 11.1 and 11.5 demonstrate that oxygen is necessary for dissolution of uranium oxide in dilute sulfuric acid and of gold and silver in cyanide solution. The rate of dissolution of gold in cyanide solution increases with increase of oxygen pressure, which optimizes at a certain value. It is possible to dissolve sulfides directly in acids or ammoniacal solution in the presence of oxygen at higher temperature and pressure. Thus, pressure leaching is advantageous because it permits a much higher concentration of gaseous reagents and higher operating temperature, which hastens the dissolution. In some cases, both these factors are important, or at least one is significant. It also prevents dissolution of gangue minerals. Regarding concentration of the gas, it is interesting to record some facts about the solubility of substances in superheated water. At the critical point of water [3] (i.e., at 374 °C and 218 atm) liquid water and steam become a single phase. Solubility of gases decreases with increase of temperature between the freezing point of water and the normal boiling point. Dissolved air and gas come out of the solution as water warms up and practically all gases are expelled at the boiling point of water. On the other hand, many salts show increased solubility in water with rise of temperature.

At higher temperature and pressure maintained in the autoclave, these effects are reversed. As water approaches the critical point, the solubility of salts and gases approaches zero and infinity, respectively. Presumably, this is because all the gases are mutually soluble, and as liquid water approaches the critical point, it behaves more like a gas. Probably for the same reason, it loses its solvent power for salts. Practically, this means that leaching cannot be carried out at excessively high temperature because materials will not stay in solutions. Also, excessively high temperature would require extremely high pressure; the pressure of superheated steam at 374 °C is 218 atm (~3200 psi). The chemical reactions taking place in pressure leaching are quite complex. Following are the few simpler examples:

Pressure Leaching of Bauxite In the Bayer process, the crushed bauxite ore is ground in a fine grinding mill. Strong caustic soda solution ($130 - 350$ g Na_2O per liter) is introduced in the mill to obtain a slurry. The resultant slurry is pumped into a horizontal mild steel digester tank (autoclave) heated by steam under pressure with constant agitation. Aluminum gets dissolved as AlO_2^- anion according to the reaction (11.3) at 25 atm pressure and 200 °C. The leach liquor is separated from the insoluble residue containing oxides of iron, silicon, vanadium, and other gangue materials. The resulting sodium aluminate solution is cooled to $25 - 35$ °C to precipitate $Al(OH)_3$ and regenerate sodium hydroxide ($NaAlO_2(aq) + 2H_2O = Al(OH)_3 + NaOH$). Crystals of hydrated alumina ($Al_2O_3 \cdot 3H_2O$) are precipitated by seeding the solution with freshly precipitated aluminum hydroxide. Sodium hydroxide is recirculated after concentration and evaporation and addition of some new sodium hydroxide to compensate the loss. Pure alumina is obtained by calcination of the hydrated alumina at 1100 °C. This is dissolved in cryolite for electrolytic extraction of aluminum.

Pressure Leaching of Scheelite/Wolframite On the industrial scale, scheelite $(CaWO_4)$ concentrate ground to -150 to $+325$ mesh is leached with $10 - 18\%$ sodium carbonate solution at $190 - 225$ °C for $1.5 - 4$ h in an autoclave at a pressure of $10 - 20$ atm according to the reaction (11.4). Wolframite $[Fe(Mn)WO_4]$ containing less than 50 mol% $MnWO_4$ or mixture of scheelite and low manganese wolframite can also be leached by this method. The deleterious impurities, such as silica, arsenic, phosphorus, iron, and molybdenum present in the ore, join the leach liquor. They have to be removed by chemical treatment. Prior to the removal of the unreacted gangue by filtration, aluminum and magnesium sulfates are added at $70 - 80$ °C and the solution is stirred for about an hour at pH of $9 - 9.5$. Two-stage treatment brings down silica in the range of $0.03 - 0.06$ g l^{-2}. Removal of arsenic and phosphorus is based on the formation of poorly soluble magnesium salts of arsenic and phosphoric acids. Before precipitation, arsenite is oxidized to arsenate with sodium hypochlorite $\left(AsO_3^{3-} + NaOCl = AsO_4^{3-} + NaCl\right)$. The solution is boiled with magnesium sulfate for $3 - 4$ h and filtered to remove the residue consisting of $Mg_3(PO_4)_2$ and $Mg_3(AsO_4)_2$.

Molybdenum is removed as molybdenum trisulfide (MoS_3). Solution is first treated with sodium sulfide or sodium hydrogen sulfide at $80 - 85$ °C for 1 h at a pH of 10 to form thiomolybdate complex according to the reaction:

$$MoO_4^{2-} + 4S^{2-} + 4H_2O \rightarrow MoS_4^{2-} + 8OH^- \tag{11.7}$$

Finally, molybdenum trisulfide is precipitated by acidifying the solution (pH $= 2.5$–3.0) under stirring for $7 - 9$ h:

$$MoS_4^{2-} + 2H^+ \rightarrow MoS_3 + H_2S \tag{11.8}$$

The purified sodium tungstate solution after filtration is stored in a tank for pH and temperature adjustments. The resultant solution after removal of silica, arsenic, phosphorus, and molybdenum is further purified by the solvent extraction technique using 7% alamine-336 and 7% decanol dissolved in kerosene. Tungsten from the loaded organic phase stripped into ammonia solution produces ammonium tungstate solution that is sent to an evaporator for crystallization of ammonium paratungstate (APT). APT is subsequently decomposed in a rotary furnace at 250 °C under the flow of air to get yellow tungstic oxide (WO_3) (see Appendix A.5).

Pressure Leaching of Pentlandite Pentlandite (nickel–iron sulfide: NiS – FeS) concentrate from the Lynn Lake Mines in Manitoba (Canada) containing $10 - 14\%$ Ni, $1 - 2\%$ Cu, $0.3 - 0.4\%$ Co, $23 - 24\%$ Fe, and $28 - 34\%$ S is treated with ammonia under pressure to dissolve valuable metals and sulfur leaving iron and other impurities in the tailings. In the first step, recycled ammonia from the second-stage leaching is used for partial leaching of the Ni – Cu – Co concentrate at 8 atm and

105 °C. The leach liquor is obtained by filtering the resulting pulp, and the residue is subjected to second-stage leaching. Weaker ammonia liquor from the second stage is recirculated to the first-stage leaching. Under these conditions, nickel forms a series of soluble amines (with NH_3/Ni^{2+} ratio varying from 1 to 6), whereas sulfur is converted to several polyoxosulfates and the iron is oxidized to the insoluble hydrated oxide according to the reaction [4]:

$$2NiS + 8FeS + 14O_2 + 20NH_3 + 8H_2O$$
$$\rightarrow 2Ni(NH_3)_6SO_4 + 4Fe_2O_3.H_2O + 4(NH_4)_2S_2O_3 \qquad (11.9)$$

Insoluble hydrated oxide of iron is filtered off together with silica and other gangue residues. Sulfur is not directly converted to sulfate, instead ammonium thiosulfate $[(NH_4)_2S_2O_3]$ is formed, which subsequently gets oxidized to sulfate and trithionate:

$$2(NH_4)_2S_2O_3 + 2O_2 = (NH_4)_2SO_4 + (NH_4)_2S_3O_6 \qquad (11.10)$$

The leach liquor is boiled to decompose trithionate and precipitate copper as copper sulfide, which is filtered off and ammonia is recirculated. The filtrate is then heated and saturated with air at high temperature and pressure to ensure the oxidation of the remaining unsaturated sulfur compounds (particularly the ammonium trithionate $(NH_4)_2S_3O_6$) and to convert sulfamate into sulfate and produce amine:

$$(NH_4)_2S_3O_6 + 2O_2 + 4NH_3 + H_2O = 2(NH_4)_2SO_4 + NH_4SO_3H_2N \qquad (11.11)$$

This operation is known as oxyhydrolysis. The resulting liquor at this stage containing 40–50 g l^{-1} of nickel at NH_3/Ni^{2+} molar ratio of 2 can be treated with hydrogen under pressure to precipitate nickel powder. In Canada at the Sherritt Gordon Mines, Sudbury, the process has been in practice on a commercial scale since 1954 (see Appendix A.3b).

11.1.1.6 Bacterial Leaching

In Sect. 11.1, while listing various leaching reactions, it has been mentioned that oxide minerals of copper can be leached easily in dilute sulfuric acid, but dissolution of sulfide requires oxygen in addition to H_2SO_4 as per the reaction:

$$Cu_2S + 5/2 O_2 + H_2SO_4(aq) = 2CuSO_4(aq : 2Cu^{2+} + 2SO_4^{2-}) + H_2O \quad (11.12)$$

Bacterial action enhances the rate of dissolution. Dissolved oxygen from air acts as an oxidant and bacteria as catalyst. Industrially, it has been demonstrated that Fe^{2+} ions required for rapid dissolution are generated due to pyrite (FeS_2) present in the ore according to the reaction:

$$2FeS_2 + 7O_2 + 2H_2O \xrightarrow{\text{bacterial action}} 2FeSO_4\left(aq : 2Fe^{2+} + 2SO_4^{2-}\right)$$
$$+ 2H_2SO_4 \tag{11.13}$$

Fe^{2+} oxidizes to Fe^{3+} which enhances the rate of dissolution of Cu_2S:

$$\frac{1}{2}O_2 + 2Fe^{2+} + 2SO_4^{2-} + H_2SO_4 \xrightarrow{\text{bacterial action}} 2Fe^{3+} + 3SO_4^{2-}\left[Fe_2(SO_4)_3\right] + H_2O \tag{11.14}$$

$$Cu_2S + 10Fe^{3+} + 15SO_4^{2-} + 4H_2O \xrightarrow{\text{bacterial action}} 2CuSO_4(aq) + 10FeSO_4(aq) + 4H_2SO_4 \tag{11.15}$$

The role of ferric ions (present as $FeCl_3$ or $Fe_2(SO_4)_3$ in the leaching media) in dissolution of sulfide mineral of copper has already been stated in Sect. 11.1. The Fe^{2+} ions produced by the above reaction are reoxidized. Thus, reduction/ oxidation reactions continue in a cyclic manner, and the rate of leaching reactions (11.12–11.15) are enhanced extensively (up to million folds [5]) by the bacteria enzyme catalysts. Thiobacillus ferrooxidans, thiobacillus thiooxidans, and leptospirillum ferrooxidans [5, 6] are the most commonly known bacteria for catalytic action. Thiobacillus ferrooxidans and ferrobacillus sulfooxidans have been used successfully in bringing copper and uranium [2] from chalcopyrite and bannerite, respectively, into solution. The former bacterium prevents the dissolution of iron by oxidizing Fe^{2+} into Fe^{3+} and is thus helpful in increasing the dissolution rate of copper.

These bacteria are autotrophic, and they depend on atmospheric CO_2 for the necessary carbon for the generation of cellular tissues in the presence of many heavy metal ions. Bacterial cells basically made up of water, carbon, hydrogen, and oxygen require these nutrients as well as suitable environments for growth. In addition, little phosphorus, nitrogen, calcium, magnesium, potassium, and sodium are also needed. Most minor nutrients can be supplied by normal mineral leaching environments [6]. The growth of microorganisms is also affected by the dissolved oxygen level, temperature, pH, and oxidation/reduction potentials. In order to achieve a reasonable growth, oxygen level above 1 ppm and pH between 1.0 and 1.8 should be maintained [7]. pH increase beyond 2.0 will reduce the availability of iron due to precipitation. The bacterial action also decreases at high oxidation/reduction potentials with limited availability of iron. A temperature of about 35 °C has been found appropriate for maximum recovery in case of bacteria involved in sulfide dissolution. However, there are possibility that some bacteria are efficient at significantly higher temperatures [6].

Geological weathering due to microbiological degradation processes has been known for a long time. The commercial exploitation of bacterial leaching started for the first time in the 1950s in dump leaching at Kennecott Copper's Bingham Mines near Salt Lake City, USA [8]. It was then followed by Chile in 1980 in heap leaching [9]. The technology has been improved by the introduction of forced aeration and a variety of inoculation and is available in various trade names such as BIOX, BioHeap, GEOCOAT, BioCOP, BioNIC, BioZinc, and BIOPRO for relatively large-scale production [10]. Thermophilic microorganisms such as Acidianus, Metallosphaera, and Sulfolobus archaea capable of leaching at elevated temperatures, where traditional bacteria fail, have been employed for better recoveries [11]. After successful development of the BioCOP process for extracting copper from chalcocite concentrate, a mesophilic process with limited capabilities for recovery of copper from chalcopyrite has been reported by Clark et al. [12]. This was followed by the development of the mesophilic oxidation of refractory gold ores via the now-well-established BIOX process. In the near future, commercial exploitation of bacterial leaching is likely to be extended to possible recovery of cobalt and nickel. Next-generation technologies will be focused on pretreatment of refractory gold ores and in processing massive sulfides and silicate-locked minerals. In the more distant future, in situ leaching [8, 10] will be based on the use of thermophilic bacteria for sulfide oxidation [8]. There is also a report about commercial pilot-scale demonstration of integrated bioleaching/solvent extraction/electrowinning for zinc [6].

In view of the above facts and figures, biohydrometallurgy seems to be slowly approaching toward achieving the status of ecofriendly and economic process.

11.2   Breakdown of Refractory Minerals

Many rare and nuclear metals, for example, beryllium, zirconium, and thorium, occur in nature as silicate and phosphate. Aqueous acid and alkaline solutions are not effective in leaching of such refractory minerals. Since these minerals can be neither easily converted to oxides nor subsequently reduced to the pure metals by carbon, a different ore breakdown step [13] is incorporated in the flow sheet for the extraction of rare/refractory and nuclear metals. Strong chemical reagents, namely, concentrated acids, alkalis, and fluorides, are employed in breaking down these refractory ores at elevated temperatures [13]. As these reagents usually attack other mineral impurities, from an economic viewpoint, it is advisable to use relatively high-grade ores. Generally, extensive purification is required after the breakdown and before the metal production steps.

11.2.1   Concentrated Sulfuric Acid Breakdown

Monazite, a complex phosphate of thorium and rare earths (RE), is leached with concentrated sulfuric acid (93%). In this process, monazite sand is slowly added, while stirring, into concentrated acid heated to about 180 °C. The reaction is exothermic, and the temperature rises to 220 °C for several hours without any external heat. Thorium and rare earths are digested according to the reactions:

$$Th_3(PO_4)_4 + 6H_2SO_4 \rightleftarrows 3Th(SO_4)_2 + 4H_3PO_4 \qquad (11.16)$$

$$REPO_4 + H_2SO_4 \rightleftarrows RE(SO_4) + H_3PO_4 \qquad (11.17)$$

A total breakdown time of 4 h is required with acid/sand ratio of approximately 2:1. In order to prevent the precipitation of thorium pyrophosphate, the temperature should not be allowed to exceed beyond 220 °C.

The principal mineral of beryllium, beryl (beryllium aluminum silicate) is subjected to concentrated sulfuric acid (95%) treatment at 250–300 °C:

$$3BeO.Al_2O_3.6SiO_2 + 6H_2SO_4 = 3BeSO_4 + Al_2(SO_4)_3 + 6SiO_2$$
$$+ 6H_2O \qquad (11.18)$$

Prior to acid treatment, beryl is melted and quenched in water from 1800 °C or fused with alkaline earth carbonate at 1500 °C to destroy the original structure of the ore. This makes the ore reactive for subsequent H_2SO_4 digestion. However, concentrated sulfuric acid breakdown of monazite as well as beryl is not currently practiced.

11.2.2 Concentrated Alkali Breakdown

Sodium hydroxide either in molten state (melting point 318 °C) or as a concentrated solution is employed for the breakdown of phosphate and silicate minerals at about 200 °C. The concentrated sodium hydroxide is heated to 130 °C in a stainless steel tank fitted with stirrer. Monazite sand ground to −300 mesh is gradually added over a period of 30 min. A coarser size sand and dilute alkali require higher temperature of digestion, 220 – 230 °C, whereas in the case of fine size and concentrated alkali, the operation can be carried out at 140 – 145 °C. A very high concentration of NaOH will decrease uranium recovery; hence, an ultimate concentration of 35–40% is maintained, in any case it should not exceed 50%. For better recovery of thorium, alkali/sand ratio should be 3:1.

Since zircon ($ZrSiO_4$) is not digested to any appreciable extent with alkaline solution, a fusion breakdown method has been evolved. Zircon sand is allowed to react with fused sodium hydroxide at 500–600 °C in cast iron pots to convert silica into water-soluble sodium silicate according to the reaction:

$$ZrSiO_4 + 4NaOH = Na_2ZrO_3 + Na_2SiO_3 + 2H_2O \qquad (11.19)$$

Sodium silicate is washed off from the cooled mass leaving sodium zirconate cake for further processing by dissolving it in (55–60%) nitric acid. Many breakdown processes using fused or sintered fluorinating agents [13] such as potassium hydrogen fluoride, sodium silicofluoride, and iron cryolite have been developed for the treatment of refractory minerals. See Sect. 9.1.1 for breakdown of beryl, zircon, and niobite (columbite)-tantalite.

11.3 Physicochemical Aspects of Leaching

In this section, the thermodynamic and kinetic principles with particular reference to leaching will be discussed. As water happens to be the important component of the leaching circuit, stability of water in different conditions, for example, oxidizing, reducing, and acid and alkaline media, must be discussed before considering the details of the physicochemical aspects of leaching.

11.3.1 Thermodynamics of Aqueous Solutions

The thermodynamic laws dealing with the solutions of metals, slags, and salts are also applicable to aqueous solutions. Special features of these are summarized below:

1. Gases like oxygen and hydrogen and organic compounds like sugar dissolve molecularly in water. On increasing the dilution, the solute obeys Henry's law, whereas the Raoult's law is obeyed by the solvent. However, inorganic substances such as electrolytes (acids, bases, and salts) do not follow Henry's law.
2. At 25 °C, about 41 wt% of HCl gas dissolves in water at 1 atm pressure of HCl gas. The partial pressure of HCl falls rapidly with decrease in concentration. At low concentration, the solubility of the gas in water is proportional to the square root of the partial pressure of HCl gas in equilibrium with water. This observation is analogous to Sievert's law, which is applicable for the solubility of diatomic gases like O_2 and H_2 in metals. The proportionality relation of the solubility of HCl gas in water is based on the electrolytic dissociation reaction: $HCl(g) = H^+ + Cl^-$,

$$K = \frac{a_{H^+} \cdot a_{Cl^-}}{p_{HCl}} \tag{11.20}$$

At infinite dilution, these activities are proportional to the corresponding ionic fractions. Since the concentration of H^+ and Cl^- ions is equal (in pure water) and these ionic concentrations are separately equal to the concentration of HCl gas, the Sievert's type relation is obtained. In aqueous solutions, the concentration of electrolytes is expressed as molality, that is, the number of moles per kilogram of water. Contrary to the standard state for dilute solutions in liquid steel, where activity = weight percent at infinite dilution, the standard state for ions is defined as activity of the species, $i(a_i) = $ molality (m_i), at infinite dilution. Deviations occurring at higher concentrations are accounted by activity coefficients: $a_{H^+} = m_{H^+} \cdot f_{H^+}$ and $a_{Cl^-} = m_{Cl^-} \cdot f_{Cl^-}$, where

f denotes activity coefficient. Since activity coefficient of individual ions cannot be determined, by convention, it is agreed to use the mean activity coefficient as $f_\pm = \sqrt{f_+ \cdot f_-}$. Thus, we can determine only the activity of the neutral compound. By another convention, we can express the chemical activity of the neutral compound in solution to be equal to the product of the ionic activities:

$$a_{HCl} = a_{H^+} \cdot a_{Cl^-} = m_{H^+} \cdot m_{Cl^-} \, (f_\pm)^2 \qquad (11.21)$$

In pure water, we have $m_{H^+} = m_{Cl^-} = m_{HCl}$ and $f_\pm = 1$ as $m_{HCl} \to 0$. However, this concept is applicable in solutions having strong electrolytes undergoing almost 100% dissociation.

In the above case of dissociation of a single electrolyte, HCl, the activities of H^+ and Cl^- are equal. But in the presence of other electrolytes such as NaCl or HNO_3, the concentration of H^+ or Cl^- would be the sum of the ions generated by different electrolytes. In such cases, the activities of H^+ and Cl^- will not be equal, but the relation $a_{HCl} = a_{H^+} \cdot a_{Cl^-}$ still holds. Hence, activity of HCl will increase by addition of either NaCl or HNO_3 to HCl solution. Similarly, activity of NaCl in solution will increase by addition of HCl or $NaNO_3$. In certain cases, the increase may be large enough for the solution to be saturated with NaCl. On dissolution in water, metal salts like NaCl or $NiSO_4$ form cations, Na^+ and Ni^{2+} and anions, Cl^- and SO_4^{2-}. Some metals such as aluminum may exist as cation Al^{3+} as well as anion AlO_2^- in water according to the equilibria:

$$Al^{3+} + 2H_2O = AlO_2^- + 4H^+ \qquad (11.22)$$

In such a case, activity of the anion increases with increasing pH (decreasing H^+ activity because $pH = -\log a_{H^+}$). As the maximum pH attainable in a saturated solution is 14, many metals may not be converted to anions. Similarly on decreasing pH, some anions may undergo a change like $SO_4^{2-} + H^+ = HSO_4^-$. The concentration of HSO_4^- anion will be high at pH < 2. Different ions or ions and neutral salt may give rise to the formation of complex ions according to the reactions:

$$Ag + 2CN^- = Ag(CN)_2^- \qquad (11.23)$$

$$UO_2^{2+} + 3SO_4^{2-} = UO_2(SO_4)_3^{4-} \qquad (11.24)$$

$$Cu^{2+} + 4NH_3 = Cu(NH_3)_4^{2+} \qquad (11.25)$$

In some cases, metallic ion is converted from positive to negative charge. For example, the conversion of Ag^+ present in $AgNO_3$ as a cation to anion by forming a complex, $Ag(CN)_2^-$ on addition of NaCN or KCN. A large value of the equilibrium constant for the formation of the complex makes the complex ions very stable. This is confirmed by the fact that AgCl cannot be precipitated from $Ag(CN)_2^-$ solution by addition of HCl, whereas it is easily obtained from $AgNO_3$ solution. This principle is extremely useful in hydrometallurgy for bringing insoluble compounds in aqueous solution.

The solubility product of dissociation of compounds in solution is expressed in terms of equilibrium constant. For reactions: $NaCl = Na^+ + Cl^-$, $K = a_{Na^+}.a_{Cl^-}$. and $Al(OH)_3 = Al^{3+} + 3OH^-$, $K = a_{Al^{3+}}.a^3_{OH^-}$. When the solubility product of NaCl increases by addition of Na^+ or Cl^-, solid NaCl gets precipitated. Contrary to this, AgCl gets dissolved when activity of Ag^+ decreases by complexing with the CN^- ions. It is interesting to note that $Al(OH)_3$ precipitated at certain pH goes into solution again at higher pH by forming anion: $Al(OH)_3 + OH^- = AlO_2^- + 2H_2O$. Thus, $Al(OH)_3$ may dissolve in acidic as well as in alkaline solutions.

The solubility product varies considerably for various solids. Depending on its value, the compound may be precipitated under different conditions. For example, Ag_2S having a smaller solubility product of 5.5×10^{-51} can be precipitated from even a complex silver cyanide solution because the dissociation constant of the complex ion $Ag(CN)_2^-$ is 1.8×10^{-19}.

11.3.2 Stability Limit of Water

Leaching is one of the important steps in hydrometallurgy to bring metal value from the mineral into aqueous solution (acid or alkaline) under reducing or oxidizing conditions at normal or ambient temperature and pressure. As the entire operations of dissolution and separation are carried out in aqueous media by controlling the pH and p_{O_2} and metal values are recovered from the solution either by cementation or precipitation by blowing hydrogen gas at an appropriate pressure and pH, it is important to understand the stability limit of water under oxidizing and reducing conditions at different pH. Under reducing or oxidizing conditions, either hydrogen or oxygen will evolve due to the decomposition of water. The thermodynamic stability of water may be discussed with the aid of the following equilibria:

1. In acid as well as alkaline media, water dissociates into H^+ and OH^- ions according to the reaction:

$$H_2O\ (l) = H^+ + OH^-, \Delta G^o = 19095\ cal = -RT \ln K \qquad (11.26)$$

$$K = \frac{a_{H^+} \cdot a_{OH^-}}{a_{H_2O}}\ (\text{in pure water } a_{H_2O} = 1) \qquad (11.27)$$

$$\therefore \log K = \log a_{H^+} + \log a_{OH^-} \qquad (11.28)$$

$$\text{also } \log K = -\frac{\Delta G^o}{2.303RT} = -\frac{19095}{1363} = -14.00\ (T = 298\ K)$$

$$\therefore \log a_{H^+} + \log a_{OH^-} = -14.00 \tag{11.29}$$

Since $pH = -\log a_{H^+}$ and when $a_{H^+} = a_{OH^-}$, $pH = 7.00$, that is, water is neutral. Water is acidic at $pH < 7.00$, that is, $a_{H^+} > a_{OH^-}$, and it is alkaline at $pH > 7.00$, that is $a_{H^+} < a_{OH^-}$.

2. Under oxidizing and reducing conditions, water may be oxidized or reduced with the evolution of oxygen or hydrogen gas according to the reactions:

$$\text{Oxidation}: 2H_2O = O_2 + 4H^+ + 4e \tag{11.30}$$

$$\text{Reduction}: 2H^+ + 2e = H_2 \tag{11.31}$$

Therefore, the stability limit of water is confined to the upper and lower limits where O_2 and H_2 are, respectively, evolved. ΔG^o for reaction (11.30) at 298 K is -113.38 kcal mol^{-1}, since $\Delta G^o = -n FE^o$, $E^o = 1.23$ V. We can make use of the Nernst equation to evaluate E for reactions involving oxidation/reduction:

$$E = E^o + \frac{RT}{nF} \ln \left(\frac{a_{\text{species in oxidized state}}}{a_{\text{species in reduced state}}} \right) \tag{11.32}$$

Since $a_{H_2O} = 1$, and $n = 4$ in reaction (11.30)

$$E = E^o + \frac{RT}{4F} \ln \left(a_{H^+}^4 \cdot p_{O_2} \right) \tag{11.33}$$

at 298 K,

$$E = 1.23 + 0.059 \log a_{H^+} + 0.0147 \log p_{O_2} \tag{11.34}$$

at $p_{O_2} = 1$ atm,

$$E = 1.23 - 0.059 \text{ pH} \tag{11.35}$$

For establishing the lower limit, consider reaction (11.31). By convention, $E^o = 0$ and $n = 2$ for this reaction, at 298 K:

$$E = E^o + \frac{RT}{2F} \ln \frac{a_{H^+}^2}{p_{H_2}} = \frac{0.059}{2} \log \frac{a_{H^+}^2}{p_{H_2}} \tag{11.36}$$

$$\text{or } E = -0.059 \text{ pH} - 0.0295 \log p_{H_2} \tag{11.37}$$

at $p_{H_2} = 1$ atm,

$$E = -0.059 \text{ pH} \tag{11.38}$$

The above two equilibrium conditions are presented in Fig. 11.1 by two parallel lines of slope $= -0.059$. The vertical line at $pH = 7.00$ indicates the demarcation of acid and alkaline media.

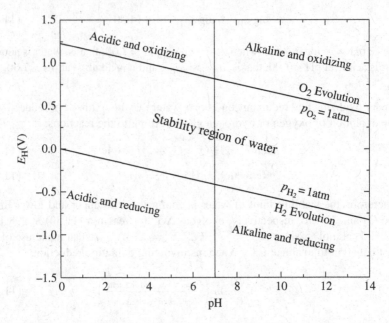

Fig. 11.1 Stability limits of water at 25 °C

The following conclusions may be drawn from the figure regarding stability of water:

1. Between the two parallel lines, the equilibrium partial pressures of both oxygen and hydrogen are less than 1 atm.
2. The region between these two lines is the area of thermodynamic stability of water under 1 atm pressure.
3. Above the line corresponding to $p_{O_2} = 1$ atm, water will tend to decompose by oxidation according to the reaction (11.30).
4. Below the line corresponding to $p_{H_2} = 1$ atm, water will tend to decompose by reduction according to the reaction (11.31).

11.3.3 Potential–pH Diagrams

On dipping a metal, M, into an aqueous solution containing its own ions, a positive charge, M^{n+}, develops on the metal at equilibrium due to the reaction $M = M^{n+} + ne$. Since the $M^{n+}(aq)$ ions are attracted to the metal, a potential difference exists between the metal and the solution. This is known as the absolute electrode potential. The value of the absolute electrode potential depends on the equilibrium state of the reaction and is affected by the metal and its purity, crystal structure, surface conditions, activity of M^{n+} ions in the solution, and temperature. The absolute electrode potential cannot be measured; however, the electrode potential of a metal can be estimated with reference to hydrogen electrode. If hydrogen gas is maintained at 1 atm pressure, the potential obtained is called standard electrode potential.

Consider a metal, M, in equilibrium with its own divalent (M^{2+}) and trivalent (M^{3+}) ions in solution. For the metal, the standard electrode potential, $E°$, can be determined by considering the following reactions:

$$M = M^{2+} + 2e \tag{11.39}$$

$$M = M^{3+} + 3e \tag{11.40}$$

$$M + 2H_2O = M(OH)_2 + 2H^+ + 2e \tag{11.41}$$

The standard electrode potential, $E°$, can be measured against the hydrogen electrode. The reactions (11.39) and (11.40) are potential dependent, whereas reaction (11.41) depends on pH, in addition to potential under standard conditions. The combined effect of electrode potential and pH is presented as the Pourbaix [14] diagram. For a given metal electrode, potentials are plotted as a function of pH in these diagrams. An arbitrary diagram [15] shown in Fig. 11.2, known as Pourbaix diagram for metal, M, is very useful in corrosion and hydrometallurgy. The boundaries in potential–pH diagrams depending on chemical equilibria set the predominance areas for various thermodynamically stable species. It is a customary to show the potential for the evolution of H_2 and O_2, both at 1 atm pressure.

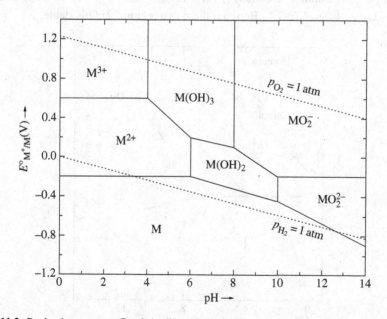

Fig. 11.2 Predominance areas (Pourbaix diagram) for metal–water system as function of pH and reduction potential (From *Principles of Extractive Metallurgy* by T. Rosenqvist [15], © 1974, p 473, McGraw-Hill Inc. Reproduced with the permission of McGraw-Hill Book Co.)

A metallic structure in contact with the aqueous solution is an electrode, and its tendency to dissolve in the solution (electrolyte) is measured by its electrode potential. In addition to the electrode potential, the pH of the solution has a profound effect on the product of anodic dissolution. Generally, metal ion concentration of 10^{-2} or 10^{-3} mol kg^{-1} of water is taken while constructing Pourbaix diagrams for hydrometallurgical applications. A brief discussion of some important metal–water systems (for a metal ion concentration of 10^{-3} mol kg^{-1} at 25 °C, that is, activity of metal ions = 10^{-3} = molality in extremely dilute solutions) has been included in this section.

The dissolution of a metal is essentially an electrochemical process. Figure 11.3 shows that the potential for the dissolution of zinc (Zn → Zn^{2+} + 2e) is more negative than the oxidation and reduction of water leading to evolution of oxygen or hydrogen (at 1 atm pressure) according to the reactions (11.30) and (11.31), respectively. Hence, either reaction can be made use of as the cathodic process to support the anodic dissolution of zinc. From the figure, it is evident that dissolution is favorable below pH 6.9 with the formation of Zn^{2+} cations and above 13.4 with ZnO_2^{2-} anions. Thus, depending on potential and pH, metallic zinc and ZnO would tend to dissolve by the following electrochemical reactions:

(i) Below pH < 6.9 with the formation of Zn^{2+} ions according to the reaction Zn = Zn^{2+} + 2e.

(ii) Above pH > 13.4 with the formation of ZnO_2^{2-} anions by the reaction (i.e., in highly alkaline solution) Zn + $2H_2O$ = ZnO_2^{2-} + $4H^+$ + 2e or ZnO + ½ O_2(g) + 2e = ZnO_2^{2-}. Between these two regions, ZnO is stable.

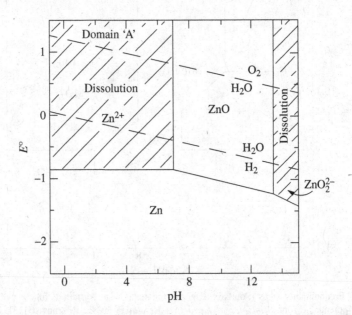

Fig. 11.3 Potential–pH diagram for zinc–water system at 25 °C (From *Hydrometallurgical Extraction and Reclamation* by E. Jackson [16], ©1986, p 30, Ellis Horwood Ltd. Reproduced with the permission of Ellis Horwood Ltd)

Fig. 11.4 Potential–pH diagram for copper–water system at 25 °C (From *Hydrometallurgical Extraction and Reclamation* by E. Jackson [16], © 1986, p 31, Ellis Horwood Ltd. Reproduced with the permission of Ellis Horwood Ltd)

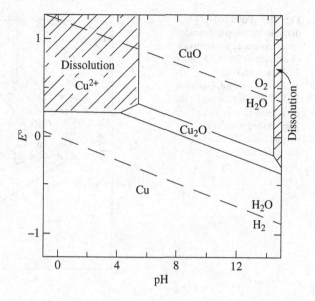

Potential–pH diagram for Cu–H$_2$O system shown in Fig. 11.4 suggests that Cu^{2+} reduction reaction Cu^{2+} + 2e = Cu has more positive potential than for H$_2$ evolution reaction and negative potential compared to oxygen reduction. Hence, hydrogen evolution reaction cannot be used as a cathodic process to dissolve copper metal anodically. However, copper being nobler, it can only be dissolved by reduction of oxygen or some other oxidizing agent. This can be achieved favorably under acid conditions at pH < 6 or by reduction of O$_2$ at high pH. The potential–pH diagrams of metal–water systems are considerably modified in the presence of complexing agents by the formation of highly stable metal complex ions. There are some major changes in the diagram on introduction of cyanide and ammonia in a Cu–H$_2$O system. Ion stable domains extending to more negative potential make copper dissolution more favorable in the entire range of pH, particularly in alkaline conditions. One can notice remarkable changes in the stability regions of different ions in Cu–H$_2$O, Cu–CN–H$_2$O, and Cu–NH$_3$–H$_2$O systems while referring to the Figs. 11.4, 11.5, and 11.6.

Gold is more noble than copper, and the potential of 1.45 V for the reaction Au^{3+} + 3e = Au at equilibrium is more than the oxygen reduction reaction. Hence, gold is highly stable in aqueous solution, and the metal remains unaffected by sulfuric acid and even by strong oxidants like nitric acid, perchloric acid, and aqua regia. Figure 11.7 clearly demonstrates that gold cannot be dissolved in noncomplexing solutions in the presence of oxygen. The potential–pH diagram of Au–H$_2$O system is drastically modified in the presence of cyanide complexing reagent due to the formation of stable Au(CN)$_2^-$ complex anions. The standard electrode potential of −0.62 V (in the presence of complexing cyanide ions) for the cathodic/anodic reaction Au(CN)$_2^-$ = Au + (CN)$^-$ is a very large shift from the highly positive value of 1.45 V in aqueous solution (Au^{3+} + 3e = Au). The modified potential of −0.62 V is negative enough to allow either reaction (11.30) or (11.31) to act as the

Fig. 11.5 Potential–pH diagram for copper-cyanide-water system at unit activity of cyanide at 25 °C (From *Hydrometallurgical Extraction and Reclamation* by E. Jackson [16], © 1986, p 32, Ellis Horwood Ltd. Reproduced with the permission of Ellis Horwood Ltd)

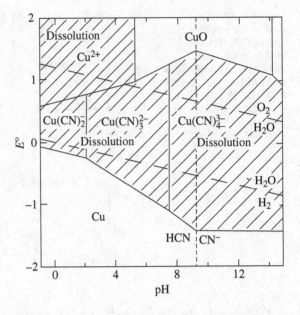

Fig. 11.6 Potential–pH diagram for copper-ammonia-water system for a combined NH_3/NH_4^+ activity of unity at 25 °C (From *Hydrometallurgical Extraction and Reclamation* by E. Jackson [16], © 1986, p 33, Ellis Horwood Ltd. Reproduced with the permission of Ellis Horwood Ltd)

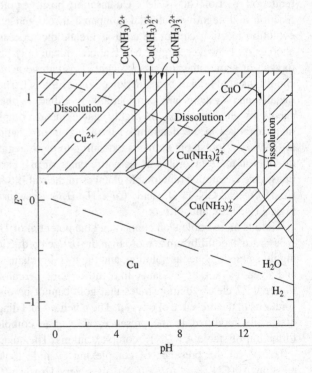

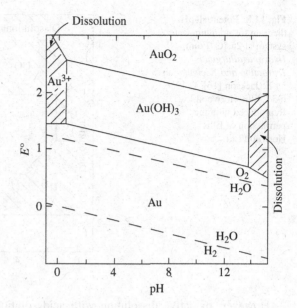

Fig. 11.7 Potential–pH diagram for gold–water system at 25 °C (From *Hydrometallurgical Extraction and Reclamation* by E. Jackson [16], © 1986, p 31, Ellis Horwood Ltd. Reproduced with the permission of Ellis Horwood Ltd)

cathodic process for gold dissolution. However, the oxygen reduction reaction is most widely practiced in the cyanide leaching of gold ores.

It may be noted that potential–pH diagrams along with complexing ions have not been compiled for many metals. Further, metal–water diagrams have limited applications because few metals only occur in native form in the earth's crust. However, these are useful in corrosion studies and can also be considered for cases of dissolution of oxides and sulfides. Leaching of an oxide may take place by either (i) chemical dissolution or (ii) oxidative dissolution or (iii) reductive dissolution.

A number of oxides get dissolved chemically without involvement of electrons. For example, potential–pH diagram of Al–H_2O (shown in Fig. 11.16, Problem 11.1) indicates that Al_2O_3 (in bauxite) undergoes chemical dissolution in acid as well as in alkaline solution according to the reactions:

$$Al_2O_3 + 6H^+ = 2Al^{3+} + 3H_2O \qquad (11.42)$$

$$Al_2O_3 + OH^- = 2AlO_2^- + H^+ \qquad (11.43)$$

Low-grade bauxite and laterites have been leached in sulfuric acid under pressure, but alkaline leaching has been practiced on a larger industrial scale. High-grade bauxite ore is digested with a strong caustic soda solution (130–350 g Na_2O g l^{-1}) at 5–10 atm pressure and 150–170 °C to achieve a high rate of dissolution.

On the other hand, the potential–pH diagram of a U–H_2O system (Fig. 11.8) demonstrates that uraninite (UO_2) can get dissolved by three types of reactions:

1. Chemical dissolution: $UO_2 + 4H^+ = U^{4+} + 2H_2O$ (11.44)
2. Oxidative dissolution: $UO_2 = UO_2^{2+} + 2e$ (11.45)
3. Reductive dissolution: $UO_2 + e + 4H^+ = U^{3+} + 2H_2O$ (11.46)

Fig. 11.8 Potential–pH diagram for uranium–water system at 25 °C (From *Hydrometallurgical Extraction and Reclamation* by E. Jackson [16], © 1986, p 34, Ellis Horwood Ltd. Reproduced with the permission of Ellis Horwood Ltd)

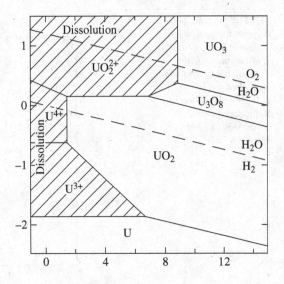

However, oxidative dissolution with acid solution in the presence of some oxidizing agent (MnO_2 – pyrolusite) is generally practiced because direct chemical dissolution in nonoxidizing acid does not take place readily. The oxidative leaching yields uranyl ions (UO_2^{2+}) or other complex uranium (VI) species that are preferred in subsequent separation and purification steps. The figure also suggests that pitchblende (U_3O_8) can only be leached by oxidative dissolution according to the reaction:

$$U_3O_8 + 4H^+ = 3UO_2^{2+} + 2H_2O + 2e \qquad (11.47)$$

Reductive dissolution of UO_2 is not practiced because of the requirement of a strong reducing agent that decomposes water. Further, the resulting U^{3+} would easily get oxidized by the atmospheric oxygen.

Since many metals occur as sulfides, potential–pH diagrams of sulfur–water and metal–sulfur–water systems are useful in explaining the leaching reactions in dissolution of sulfides in acid and alkaline solutions. The S–H_2O diagram showing the stability regions of S, H_2S, HS^-, S^{2-}, HSO_4^-, and SO_4^{2-} has to be superimposed on the metal–sulfur–water diagram to understand the dissolution reactions of metal sulfide(s). This will enable us to predict the sulfur species in equilibrium with metal species. Formation of two sulfides M_2S and MS and two oxides M_2O and MO makes such diagrams complicated. The influence of anodic potential on the dissolution of these sulfides according to the reactions can be estimated from such superimposed diagrams:

$$M_2S = MS + M^{2+} + 2e \qquad (11.48)$$

$$MS = M^{2+} + S^{2-} \tag{11.49}$$

In low pH range, the possible leaching reactions are as follows:

$$2MS + 4H_2O = M_2S + 7H^+ + HSO_4^- + 6e \tag{11.50}$$

$$M_2S + 4H_2O = 2M^{2+} + HSO_4^- + 7H^+ + 10e \tag{11.51}$$

In the presence of iron, though the system Cu–S–H$_2$O becomes even more complicated, the potential–pH diagram of Cu–Fe–S–H$_2$O is most relevant while discussing the leaching of chalcopyrite ore (CuFeS$_2$), which is the main source of copper. The oxidation reaction may be expressed as:

$$CuFeS_2 = Cu^{2+} + Fe^{2+} + 2S + 4e \tag{11.52}$$

The diagram also presents conversion of chalcopyrite into bornite (CuS·FeS$_4$), Cu$_2$S, and CuS.

11.3.4   Electrochemical Phenomenon in Leaching

From the preceding section, it may be concluded that leaching is an electrochemical phenomenon. Generally, dissolution involves an anodic reaction. Hence, oxidative leaching may be regarded as an electrochemical cell comprising the anodic reaction sustained by an appropriate cathodic reaction. The cathodic reaction may involve either the reduction of oxygen (reverse of reaction 11.30) or of the oxidized species having higher positive potential than that of the anodic reaction. A difference of at least 0.2 V [16] is necessary for this reaction to proceed. The presence of different minerals may give rise to galvanic effects. In such cases, electrically conducting mineral particles possessing different potentials may set up a galvanic cell. The mineral with more negative potential will undergo anodic dissolution, whereas the nobler mineral will act as the cathode to facilitate a reduction reaction. For example, pyrite (FeS$_2$) having high positive potential acts as cathode (Fe^{2+} + 2S + 2e = FeS$_2$, $E° = 0.757$ V) against sphalerite (ZnS) having low positive potential (Zn^{2+} + S + 2e = ZnS, $E° = 0.174$ V). Hence, a particle of ZnS that makes contact with a pyrite particle in an aerated solution will dissolve anodically.

Anodic dissolution : $ZnS = Zn^{2+} + S + 2e$ $\hspace{3cm}$ (11.53)

Cathodic reaction (on pyrite) : $1/2O_2 + 2H^+ + 2e = H_2O$ $\hspace{1cm}$ (11.30-Reverse)

Overall reaction : $ZnS + 1/2O_2 + 2H^+ = Zn^{2+} + S + H_2O$ $\hspace{1.5cm}$ (11.54)

It is important to note that the cathodic mineral does not always remain inert. Hiskey and Wardsworth [17] have shown that chalcopyrite in contact with copper in acid solution undergoes cathodic reduction with the evolution of hydrogen sulfide according to the reaction:

$$2CuFeS_2 + 6H^+ + 2e = Cu_2S + 2Fe^{2+} + 3H_2S \tag{11.55}$$

The anodic reaction may be expressed as:

$$2Cu + H_2S = Cu_2S + 2H^+ + 2e \tag{11.56}$$

The overall reaction is:

$$2CuFeS_2 + 2Cu + 4H^+ = 2Cu_2S + 2Fe^{2+} + 2H_2S \tag{11.57}$$

The electrochemical method of dissolving electrically conducting minerals or materials by direct anodic attack by applying external potential is called electroleaching [18].

The partial pressure of oxygen and pH of the solution play important roles in the dissolution and precipitation of a metal from an aqueous solution. Extensive knowledge can be derived from the thermodynamic relationship between the metal ions, hydrogen, and oxygen present in the solution. Resenquist [15] and Moore [19] have discussed this aspect by considering a hypothetical case, where metal forms cations M^{2+} and M^{3+}, anions MO_2^{2-} and MO_2^-, and solid hydroxides $M(OH)_2$ and $M(OH)_3$. In such a system, the following reactions between M, H^+, and O_2 are feasible:

$$2M + 4H^+ + O_2 \rightarrow 2M^{2+} + 2H_2O \tag{11.58}$$

$$4M^{2+} + 4H^+ + O_2 \rightarrow 4M^{3+} + 2H_2O \tag{11.59}$$

$$M^{2+} + 2H_2O \rightarrow M(OH)_2 + 2H^+ \tag{11.60}$$

$$M^{3+} + 3H_2O \rightarrow M(OH)_3 + 3H^+ \tag{11.61}$$

$$M(OH)_2 \rightarrow MO^{2-} + 2H^+ \tag{11.62}$$

$$M(OH)_3 \rightarrow MO_2^- + H^+ + H_2O \tag{11.63}$$

$$2M + 2H_2O + O_2 \rightarrow 2M(OH)_2 \tag{11.64}$$

The equilibrium constant for reaction (11.58) is given as:

$$K_{58} = \frac{a_{M^{2+}}^2 \cdot a_{H_2O}^2}{a_M^2 \cdot a_{H^+}^4 \cdot p_{O_2}} \tag{11.65}$$

Considering unit activities of H_2O and solid M, the above relationship can be expressed as:

$$a^2_{M^{2+}} = K_{58} \, a^4_{H^+} \cdot p_{O_2} \tag{11.66}$$

$$\text{or } 2 \log a_{M^{2+}} = \log K_{58} + 4 \log a_{H^+} + \log p_{O_2} \tag{11.67}$$

Since $-\log a_{H^+} = $ pH, Eq. 11.67 is modified as:

$$2 \log a_{M^{2+}} = \log K_{58} - 4 \, \text{pH} + \log p_{O_2} \tag{11.68}$$

Following a similar procedure and assuming unit activities of solids $M(OH)_2$ and $M(OH)_3$, additional relations for reactions (11.59–11.64) may be derived as under:

$$4 \log a_{M^{3+}} = \log K_{59} + 4 \log a_{M^{2+}} - 4 \, \text{pH} + \log p_{O_2} \tag{11.69}$$

$$\log a_{M^{2+}} = -\log K_{60} - 2 \, \text{pH} \tag{11.70}$$

$$\log a_{M^{3+}} = -\log K_{61} - 3 \, \text{pH} \tag{11.71}$$

$$\log a_{MO_2^{2-}} = K_{62} + 2 \, \text{pH} \tag{11.72}$$

$$\log a_{MO_2^-} = K_{63} + \text{pH} \tag{11.73}$$

$$\log p_{O_2} = -\log K_{64} \tag{11.74}$$

In order to maintain equilibrium between reactants and products, values of p_{O_2} and pH can be estimated from Eqs. 11.68, 11.69, 11.70, 11.71, 11.72, 11.73 and 11.74. This forms the basis for the representation of the thermodynamic stability of various phases in an aqueous solution. Assuming unit activity of the metal ions in solution, stability region of different phases has been shown in Fig. 11.9 as functions of pH and $\log p_{O_2}$. The figure illustrates the above relationships for a hypothetical metal with arbitrary values of different equilibrium constants. It is important to note that location of the areas get shifted by changes in the values of the equilibrium constants, but slopes of various curves (hence areas) are marked by the stoichiometry of the reactions. The lines divide the figure into predominance areas for various species. The lines will shift for different values of activities of metal ions in the solution.

Reactions (11.60, 11.61, 11.62 and 11.63) represented by vertical lines in Fig. 11.9 are independent of partial pressure of oxygen (i.e., pH dependent), reaction (11.64) represented by horizontal line is independent of pH (i.e., p_{O_2} dependent), whereas reactions (11.58 and 11.59) shown by diagonal lines are controlled by p_{O_2} and pH. The dashed horizontal lines drawn at the bottom and top of the figure correspond, respectively, to the evolution of hydrogen and oxygen separately at 1 atm pressure of each. The lower curve corresponds to $p_{O_2} = 10^{-83}$ atm. Hydrogen will evolve below this line at 1 atm pressure of hydrogen.

From Fig. 11.9, it is evident that the metallic phase, M, is stable at $p_{O_2} < 10^{-70}$ atm. On increasing p_{O_2} in low pH region, M goes into solution as per reaction (11.58). M^{2+} oxidizes to M^{3+} on further increase of p_{O_2} according to reaction (11.59). While increasing the pH of the solution at constant, p_{O_2} metal first is precipitated as $M(OH)_2$ through reaction (11.60) and subsequently gets

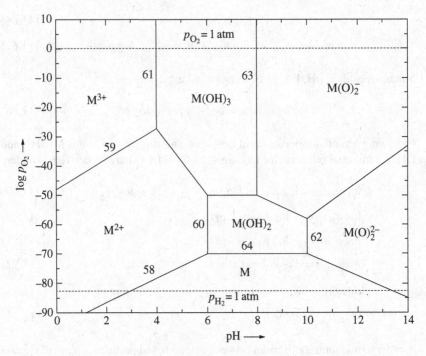

Fig. 11.9 Predominance areas for metal–water system as a function of pH and $\log p_{O_2}$ (From *Principles of Extractive Metallurgy* by T. Rosenqvist [15], © 1974, p 440, McGraw-Hill Inc. Reproduced with the permission of McGraw-Hill Book Co.)

dissolved in the form of MO_2^{2-} ions at high pH as per reaction (11.62). Since Fig. 11.9 presents a purely hypothetical case, it is not necessary that all the metals form the simple species shown in this figure. Instead, hydrated oxides, basic salts, anhydrous oxides, and complex anions like MO_n^{2-2n} and MO_n^{3-2n} and hydrated anions may be formed. The location and size of different predominance areas for different metals may differ extensively. Thus, under similar conditions, copper may exist in metallic state and iron as Fe^{2+} ions. Zinc hydroxide precipitates at a higher pH than at which ferric hydroxide precipitation takes place. Aluminum and silicon dissolve as anions under conditions where ferric hydroxide exists. The difference in location and extent of the predominance areas of various metals is one of the main principles involved in the hydrometallurgical method of extraction. Formation of more stable complex ions will affect the location and size of the predominance areas. For example, predominance area is enlarged by the formation of $Cu(NH_3)_4^{2-}$ ions (Fig. 11.6) on addition of ammonia to a solution of Cu^{2+}. This will prevent the precipitation of copper hydroxide even at high pH because activity of Cu^{2+} ions is reduced by the formation of complex ions. Any such complexing will change the position of the line due to the change in metal ion activity in reactions (11.58, 11.59, 11.60, 11.61, 11.62 and 11.63). Thus, these diagrams can be used for selective precipitation of unwanted species from solutions and for purification of solution for further recovery of metal values.

11.3.5 Kinetics of Leaching

Feasibility of a chemical reaction in a particular direction can be predicted with the aid of thermodynamic data, but it does not provide information about the rate of reaction. One has to consider kinetic aspects to understand the detailed mechanism as well as the rate of reaction. The leaching reactions between solid minerals and liquid solutions, which take place in several steps, are heterogeneous in nature. It becomes a three-phase system when a gaseous reactant such as oxygen is involved in the dissolution process. The three-phase heterogeneous system comprising solid (mineral), liquid (leachant), and gas (oxygen) has been shown schematically in Fig. 11.10.

The entire leaching operation may be divided into four steps:

Step 1: Transfer of the gaseous reactant from the gaseous phase to the liquid solution (leachant).
Step 2: Transport of the reactant through the liquid to the solid–liquid interface.
Step 3: Chemical or electrochemical reaction at the interface involving adsorption or desorption.
Step 4: Transport of the product from the interface into the bulk of the solution.

The rate of the overall process depends on one of the slowest steps. Generally, step 2 or 4, which involves diffusion, happens to be the slowest. Such processes are called diffusion controlled. Occasionally, it is observed that no single step is rate-controlling, instead more than one step may be involved in deciding the overall rate of reaction:

Fig. 11.10 Steps in a leaching process (From *Hydrometallurgical Extraction and Reclamation* by E. Jackson [16], © 1986, p 42, Ellis Horwood Ltd. Reproduced with the permission of Ellis Horwood Ltd)

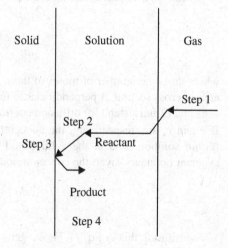

Step 1: If oxygen gas is involved in the leaching process, this step is represented by molecular dissolution of O_2 in liquid solution by the reaction:

$$O_2(g) \rightleftharpoons O_2(soln) \tag{11.75}$$

The equilibrium constant for the dissolution reaction is expressed as:

$$K = \frac{a_{O_2}(soln)}{p_{O_2}} = \frac{x_{O_2}}{p_{O_2}} \tag{11.76}$$

where x_{O_2} and p_{O_2} represent, respectively, the concentration of oxygen dissolved in the leachant and the partial pressure of oxygen gas in equilibrium with the liquid solution. Since solubility of oxygen is low, assuming validity of Henry's law, a_{O_2} has been equated to x_{O_2}. From the above expression, it is evident that solubility of oxygen in the leachant increases with increase of the partial pressure of oxygen in equilibrium with the solution. At 25 °C, the equilibrium solubility of oxygen in water is 8.26 and 40.3 mg dm^{-3} at 1 atm pressure of air ($p_{O_2} = 0.21$) and oxygen, respectively. These solubility figures decrease with increase of temperature and in the presence of other solutes in water. Under similar conditions, solubility of oxygen in one molar sulfuric acid at 1 atm pressure of oxygen is 33 mg dm^{-3}. However, it increases drastically with increase of p_{O_2}, for example, at 40 atm and 100 °C, 1 g of oxygen dissolves in water.

Step 2: Transport of a reactant from the liquid solution (leachant) to the solid–liquid interface takes place by diffusion. According to the Fick's first law of diffusion, the rate of transport of the reactant may be expressed as:

$$-\frac{dn}{dt} = -DA\frac{dc}{dx} \tag{11.77}$$

where dn is the number of moles of the reactant transferring in time, dt through an area of cross section, A perpendicular to the direction of the concentration gradient, $-\frac{dc}{dx}$ and D and x stand for diffusion coefficient and diffusion distance, respectively. If c and c_i are, respectively, the concentrations of the reactant in the bulk phase (liquid solution) and at the solid–liquid interface and δ is the thickness of the stagnant boundary layer, the concentration gradient can be expressed as:

$$-\frac{dc}{dx} = \frac{c - c_i}{\delta} \tag{11.78}$$

Substituting this in Eq. 11.77, we get:

$$\frac{dn}{dt} = \frac{DA(c - c_i)}{\delta} \tag{11.79}$$

When leachant adjacent to the solid–liquid interface is totally depleted of reactant, $c_i = 0$, the expression is modified as:

$$\frac{dn}{dt} = \frac{DAc}{\delta}$$

(11.80)

From the above expression, it can be inferred that in case of a diffusion controlled reaction, the rate is (i) directly proportional to the concentration of the reactant in the bulk of liquid solution, (ii) directly proportional to the interfacial area (A), and (iii) inversely proportional to the thickness of the stagnant boundary layer (i.e., diffusion layer δ). Thus, the rate of leaching can be maximized by increasing the interfacial area and decreasing the boundary layer thickness. For increasing the interfacial area, the ore/mineral should be ground to finer size, whereas the boundary layer thickness can be decreased by vigorous stirring. It has been reported [16] that the thickness of about 0.5 mm in the absence of stirring decreases to approximately 0.01 mm with vigorous stirring, thereby increasing the leaching rate up to 50 times.

It may not be always beneficial to increase the concentration of nongaseous reactant beyond a certain level leading to the separation of a solid from the solution on exceeding the solubility limit of the product compound. However, the situation is entirely different when oxygen is used as the gaseous reactant. As the solubility at atmospheric pressure is very low, the reaction rate is mostly directly proportional to partial pressure of oxygen over fairly long range of pressures.

Step 4: Transport of the reaction product away from the solid–liquid interface to the bulk of the solution also involves diffusion. This step may be analyzed in a manner similar to that of step 2, making use of Fick's first law. The concentration of the soluble product at the interface will be governed by the conditions prevailing in the previous step. However, the concentration in the bulk of the solution will be zero at the beginning of dissolution when a fresh solution is employed. Thus, the rate of transport of products away from the interface can be maximized by vigorous stirring and employing fresh leachant. It is important to note that step 4 is not generally rate-controlling, but it may be so when an insoluble product is formed, for example, formation of insoluble $Fe_2O_3 \cdot H_2O$ on the surface of mineral particles of chalcopyrite ($CuFeS_2$) and pentlandite (nickel–iron sulfide) in ammoniacal leaching. Under such conditions, step 4 becomes rate-controlling because the formation of insoluble product retards the diffusion of the reactants to the reaction interface and of products away from it.

Step 3: The conventional absolute reaction rate theory based on the formation of an activated complex involving an energy barrier does not provide sufficient information on the kinetics of the chemical reaction occurring at the atomic and ionic level. In majority of leaching processes, adsorption plays an important role, but all types of reactions cannot be interpreted by a single mechanism. The reactions may be classified in two groups: chemical and electrochemical. These will be discussed separately together with examples of leaching of gold and uranium ores.

At this stage, it is worth discussing briefly the effect of operating variables on leaching. Agitation or stirring is one of the important variables. In the case of diffusion controlled processes, the rate of dissolution increases by increasing the rate of stirring because the thickness of the boundary layer δ decreases. However, the rate of chemically controlled reaction is not affected by speed of stirring. The rate of dissolution does not increase indefinitely with stirring due to the formation of air pockets in the liquid at the interface. From Eqs. 11.79 and 11.80, it is clear that the rate of dissolution increases with increasing surface area (A), which can be achieved by fine grinding. The surface area increases with decrease of particle size. A diffusion controlled process is characterized by being only slightly dependent on temperature, while the chemically controlled process is strongly temperature dependent. This can be explained on the basis of linear variation of diffusion coefficient with temperature according to the Stokes–Einstein equation:

$$D = \frac{RT}{N} \frac{1}{2\pi r\eta}$$

(11.81)

where R, T, N, r, and η stand, respectively, for gas constant, temperature, Avogadro's number, radius of the mineral particle, and viscosity of the solution. The chemical rate constant (k) varies exponentially with temperature according to the Arrhenius equation:

$$k = A\, e^{-E/RT}$$

(11.82)

In other words, if temperature is doubled, D is nearly doubled but k increases about hundred times. The activation energy of the diffusion controlled process is 1–3 kcal mol^{-1}, whereas for the chemically controlled process, it is greater than 30 kcal mol^{-1}. The activation energy for intermediate- or mixed-controlled process lies between 5 and 8 kcal mol^{-1}.

In solid–solid reactions, however, the situation is different because the diffusion coefficient varies exponentially with temperature according to the equation:

$$D = D_o e^{-E/RT}$$

(11.83)

These reactions are characterized by high activation energy of the order of 200–400 kcal mol^{-1}. A chemically controlled process taking place at low temperature may change its mechanism to a diffusion controlled one at high temperature. In such cases, Arrhenius plots (log k vs $1/T$) give two straight lines with different slopes corresponding to a high activation energy at low temperature and low activation energy at high temperature. The change of mechanism is due to the fact that at low temperature, the rate of chemical reaction is much slower than the rate of diffusion, that is, the rate is chemically controlled. On the other hand, at high temperature, the chemical reaction is accelerated and hence the rate becomes much faster than the rate of diffusion, that is, it becomes diffusion controlled.

Similarly, the mechanism of solid–liquid reaction may change with increasing concentration of the reagent in the liquid phase. This is due to the fact that at a low concentration of the reactant, the reaction has low activation energy that increases with increase of temperature. At low concentration, the rate depends mainly on the speed of stirring; that is, the reaction is diffusion controlled, whereas at high concentration, the rate will be independent of stirring, that is, chemically controlled. At low concentration, the diffusion rate is very slow, and hence, the process becomes diffusion controlled. With increasing reagent concentration, the rate of diffusion increases to such an extent that it supersedes the rate of chemical reaction, and consequently, the process shifts from diffusion control to chemical control.

11.3.5.1 Kinetics of Leaching of Gold Ore in Cyanide Solution

For extraction of gold, the ore is dissolved in cyanide solution. Many theories have been proposed to explain the mechanism of dissolution of gold and silver in cyanide solutions. Habashi [2] has pointed out that the dissolution process is electrochemical in nature and the overall reaction is represented as:

$$2Au + 4NaCN + O_2 + 2H_2O = 2Na\,Au(CN)_2 + 2NaOH + H_2O_2 \quad (11.84)$$

While discussing the mechanism of gold ore dissolution according to the above reaction, the following points need to be considered:

(i). For every mole of gold, half mole of oxygen is consumed.
(ii). For every mole of gold, two moles of cyanide are consumed.
(iii). For every two moles of the metal, one mole of hydrogen peroxide is generated.
(iv). In the absence of oxygen, dissolution of gold and silver is a slow process, and hence, the reaction: $2Au + 4NaCN + H_2O_2 = 2NaAu(CN)_2 + 2NaOH$ takes place to a minor extent.

Cyanide concentration, pH, temperature, and oxygen pressure are important operating variables in dissolution of gold. Initially, the rate of dissolution of gold increases with increase of cyanide ion concentration in the solution, but after attaining the maximum, there is some retarding effect on further increase in cyanide concentration. The hydrolysis of cyanide ions at high cyanide concentration according to the reaction $CN^- + H_2O = HCN + OH^-$ increases the pH of the solution. High pH is responsible for the retarding effect, and hence, it should be maintained at approximately 11. In actual practice, protective alkalinity should be maintained to suppress the following reactions:

$$CN^- + H_2O \rightarrow HCN + OH^- \quad (11.85)$$

$$and\ H_2CO_3 + CN^- \rightarrow HCN + HCO_3^- \quad (11.86)$$

Addition of lime increases OH^- ion concentration and decreases the formation of poisonous HCN by increasing alkalinity. Hence, alkalinity of the solution should be carefully controlled.

The rate of dissolution increases with the increase of temperature up to a certain value beyond, which it decreases due to decrease in the oxygen content of the solution with increasing temperature. The maximum rate is achieved at the optimum temperature of 85 °C. However, considering the overall effects at low cyanide concentration, it can be deduced that the rate of dissolution depends only on cyanide concentration; at higher cyanide concentration, the rate becomes independent of cyanide concentration and depends only on oxygen concentration.

Mechanism In electrochemical reactions, electrons are liberated and consumed in anodic and cathodic reactions, respectively, as listed below:

$$\text{anodic reaction}: Au \rightarrow Au^+ + e$$

$$\text{or } Au^+ + 2CN^- \rightarrow Au(CN)_2^- \tag{11.87}$$

$$\text{cathodic reaction}: O_2 + 2H_2O + 2e \rightarrow H_2O_2 + 2OH^- \tag{11.88}$$

In Fig. 11.11, the reacting interface has been divided into cathodic and anodic areas A_1 and A_2, respectively. The activation energy involved in these reactions falls in the range of 2–5 kcal mol^{-1}, which is typical of diffusion controlled reactions. It is, therefore, assumed that the kinetics of dissolution of gold is basically controlled by the diffusion of both the dissolved O_2 and CN^- ions, and not by the chemical reaction at the solid–liquid interface.

Fig. 11.11 Schematic representation of the dissolution of gold in cyanide solution (From *Principles of Extractive Metallurgy*, Vol. 2, Hydrometallurgy by F. Habashi [2], © 1970, p 31, Gordon & Breach. Reproduced with the permission of the author, F. Habashi)

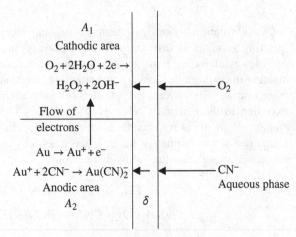

Nernst's boundary layer (δ)

In transport-controlled reactions, for diffusion of oxygen from the bulk of the solution to the cathode surface, according to Fick's first law, we can write:

$$\frac{-d[O_2]}{dt} = \frac{D_{O_2}}{\delta} A_1 \{[O_2] - [O_2]_i\} \tag{11.89}$$

where $\frac{d[O_2]}{dt}$, D_{O_2}, $[O_2]$ and $[O_2]_i$ are, respectively, the rate of diffusion and diffusion coefficient of O_2, and concentration of O_2 in the bulk of the solution and at the interface. Similarly, for the diffusion of cyanide from the bulk to the anodic surface:

$$\frac{-d[CN^-]}{dt} = \frac{D_{CN^-}}{\delta} A_2 \{[CN^-] - [CN^-]_i\} \tag{11.90}$$

where $\frac{d[CN^-]}{dt}$, D_{CN^-}, $[CN^-]$ and $[CN^-]_i$ are, respectively, the rate of diffusion and diffusion coefficient of CN^- ions, and concentration of CN^- in the bulk of the solution and at the interface.

If the chemical reactions at the metal interface are very fast as compared to the rates of diffusion of cyanide and O_2 through the stagnant boundary layer, they will be consumed immediately after reaching the solid–liquid interface. Hence, $[O_2]_i = 0$ and $[CN^-]_i = 0$, and thus, Eqs. 11.89 and 11.90 are modified as:

$$\frac{-d[O_2]}{dt} = \frac{D_{O_2}}{\delta} A_1 [O_2]$$

$$\text{and } \frac{-d[CN^-]}{dt} = \frac{D_{CN^-}}{\delta} A_2 [CN^-]$$

Since as per reaction (11.84), the rate of gold dissolution, R is twice the rate of O_2 consumption and half the rate of cyanide consumption, we can write:

$$\text{Rate of gold dissolution, } R = \frac{2d[O_2]}{dt} = \frac{2D_{O_2}}{\delta} A_1 [O_2] \tag{11.91}$$

$$\text{and also } R = \frac{1}{2} \frac{d[CN^-]}{dt} = \frac{1}{2} \frac{D_{CN^-}}{\delta} A_2 [CN^-] \tag{11.92}$$

At steady state, the rates expressed by Eqs. 11.91 and 11.92 are equal; hence,

$$R = \frac{2D_{O_2}}{\delta} A_1 [O_2] = \frac{1}{2} \frac{D_{CN^-}}{\delta} A_2 [CN^-]$$

from Eq. 11.91, $A_1 = \frac{R\delta}{2D_{O_2}[O_2]}$ and from (11.92), $A_2 = \frac{2R\delta}{D_{CN^-}[CN^-]}$.

The total interface area, $A = A_1 + A_2$

$$= \frac{R\delta}{2D_{O_2}[O_2]} + \frac{2R\delta}{D_{CN^-}[CN^-]}$$

$$= \frac{R\delta\{D_{CN^-}[CN^-] + 4D_{O_2}[O_2]\}}{2D_{O_2}[O_2].D_{CN^-}[CN^-]}$$

$$\therefore R = \frac{2AD_{O_2}D_{CN^-}[O_2][CN^-]}{\delta\{D_{CN^-}[CN^-] + 4D_{O_2}[O_2]\}} \tag{11.93}$$

From Eq. 11.93, it is clear that at low cyanide concentration, the first term in the denominator is negligible as compared to the second term:

$$\therefore \text{Rate} = \frac{1}{2}A\frac{D_{CN^-}}{\delta}[CN^-] = k[CN^-] \propto [CN^-]$$

This is in agreement with the experimental observation that at low cyanide concentration, the rate of dissolution depends only on cyanide ion concentration. On the other hand, at high cyanide concentration, the second term can be neglected, and hence, we can write:

$$\text{Rate} = 2\frac{AD_{O_2}}{\delta}[O_2] = k_2[O_2] \propto [O_2] \text{ or } p_{O_2}$$

This is also in accordance with the experimental fact that at high cyanide concentration, the rate of gold dissolution depends only on oxygen concentration. From Eqs. 11.91 and 11.92, we can also express the rate of dissolution "R" as:

$$R = \frac{2D_{O_2}}{\delta}A_1[O_2] = \frac{1}{2}\frac{D_{CN^-}}{\delta}A_2[CN^-] \tag{11.94}$$

when $A_1 = A_2$, $D_{CN^-}[CN^-] = 4 D_{O_2}[O_2]$ that is, when $\frac{[CN^-]}{[O_2]} = 4\frac{D_{O_2}}{D_{CN^-}}$, the rate of dissolution reaches its limiting value.

From the known values of $D_{O_2} = 2.76 \times 10^{-5}$ cm^2 s^{-1} and $D_{CN^-} = 1.83 \times 10^{-5}$ cm^2 s^{-1} we get $D_{O_2}/D_{CN^-} = 1.5$, and hence, the limiting rate of dissolution is achieved when the concentration ratio of cyanide ion to oxygen is 6 { i.e., $[CN^-]/[O_2] = 4 \times 1.5 = 6$}. Thus, the experimentally observed value ranging between 4.6 and 7.4 is in good agreement with the theoretical value of 6.

The rate of dissolution of gold in cyanide solution is affected by the elements and ions present in the leach liquor. For example, the presence of small amounts of lead, bismuth, thallium, and mercury enhances the dissolution process. As the dissolution of gold is a diffusion controlled process, the increase may be due to reduction in the thickness of the stagnant boundary layer by the presence of these elements. On the other hand, the following factors may retard the rate of dissolution:

1. *Consumption of oxygen by side reactions*: Pyrrhotite present in gold ore may form ferrous hydroxide and sodium sulfide:

$$FeS + 2OH^- \rightarrow Fe(OH)_2 + S^{2-} \tag{11.95}$$

Ferrous hydroxide gets easily oxidized to ferric hydroxide in the presence of oxygen, and sulfide ion is partially oxidized to thiosulfate:

$$2Fe(OH)_2 + \frac{1}{2}O_2 + H_2O \rightarrow 2\,Fe(OH)_3 \qquad (11.96)$$

Thus, the rate of dissolution of gold is affected by consumption of oxygen in side reactions.

2. *Consumption of cyanide by side reactions*: Minerals of copper, zinc, and lead associated with gold ore dissolve in cyanide and, thus, reduce the cyanide concentration:

$$ZnS + 4CN^- = \left[Zn(CN)_4\right]^{2-} + S^{2-} \qquad (11.97)$$

In the presence of oxygen, sulfide minerals react with cyanide to form thiocyanate:

$$S^{2-} + CN^- + \frac{1}{2}O_2 + H_2O \rightarrow CNS^- + 2OH^- \qquad (11.98)$$

The rate also decreases by adsorption of sodium cyanide on ferric hydroxide formed due to the presence of iron sulfides. Similarly, adsorption also occurs on quartz, aluminosilicates, or other silicates.

Gold is recovered from the cyanide leach liquor after solid–liquid separation by the well-established Merrill–Crowe process [20] based on zinc cementation. The process has been in practice throughout the world for more than a century due to its easier operation. But the pollution caused by the toxic sodium/potassium cyanide has forced the industries and chemical/metallurgical engineers to find an alternative lixiviant for gold leaching. In recent years, extensive research has been directed in testing the utility of different chemicals in order to develop a non/less toxic commercial leachant for gold dissolution.

Although less toxic thiourea [21–24] can dissolve gold at reasonably fast rate, its use is restricted because it decomposes into complex compounds, which retard the leaching process by passivating the gold surface. Jeffrey et al. [25] and Feng and van Deventer [26] carried out extensive studies on thiosulfate leaching, but the process suffers from the disadvantage of high thiosulfate consumption due to its oxidation into polythionates. Less toxic and stable thiocyanate has been another possibility but was discouraged on account of slow rate of leaching [27–29]. However, this problem has been sorted out by a combination of acid thiocyanate and ferric sulfate as the oxidant [30–32]. The mixture has proved to be a potent lixiviant with better stability and lower thiocyanate consumption in bringing gold into solution at a faster rate.

Li et al. [32] have employed the rotating disk method to study the effects of time, thiocyanate concentration, addition of thiourea, and concentrations of Fe^{2+}, Fe^{3+}, Cu^+(added as CuCl), Cu^{2+}(as CuSO$_4$), and Ag$^+$ (as AgCl) ions on the rate of leaching of gold. Gold is brought in the leach liquor by the complexes [32] formed in the solution mixture by the ferric sulfate and acidic thiocyanate according to the reactions:

$$3Fe(SCN)_4^- + Au = 3Fe^{2+} + Au(SCN)_4^- + 8SCN^- \tag{11.99}$$

$$Fe(SCN)_4^- + Au = Fe^{2+} + Au(SCN)_2^- + 2SCN^- \tag{11.100}$$

The rate of dissolution in the acid thiocyanate containing ferric sulfate as oxidant is not affected to any appreciable extent by higher molar ratios of Fe (III)/SCN [32]. Ferrous and cupric ions have no effect on the leaching kinetics, whereas the rate of leaching of gold is retarded significantly in the presence of silver and cuprous ions. Based on their electrochemical investigations, Li et al. [32] have proposed that the anodic reaction in dissolution of gold in acid thiocyanate solution happens to be the rate-limiting step in the overall leaching process.

11.3.5.2   Kinetics of Leaching of Copper in Cyanide Solution

In a similar manner, copper complex cyanide ions: $Cu(CN)_2^-$, $Cu(CN)_3^{2-}$, and $Cu(CN)_4^{3-}$ are formed on dissolution of copper in cyanide solution. The reaction proceeds electrochemically [33]. The $Cu(CN)_3^{2-}$ formed by consuming $3CN^-$ ions (instead of $2CN^-$ as in cyanidation gold) is the most stable phase. The rate of dissolution of copper in cyanide solution has been derived as:

$$\text{Rate} = \frac{2AD_{O_2}D_{CN^-}\,[O_2][CN^-]}{\delta\{D_{CN^-}\,[CN^-] + 6\,D_{O_2}[O_2]\}} \tag{11.101}$$

In this case, the limiting rate of dissolution corresponds to the ratio $[CN^-]/[O_2] = 9$.

11.3.5.3   Kinetics of Leaching of Pitchblende in Sodium Carbonate
Solution

The Canadian high-grade pitchblende concentrate [34] containing 62.4% U_3O_8, 19.4% SiO_2, 11.4% Pb, 0.62% S, and 3.4% CaO is leached with Na_2CO_3 and $NaHCO_3$ solution under pressure in an autoclave. In the presence of oxygen, uranium is oxidized and goes into solution according to the reaction:

$$U_3O_8(s) + \frac{1}{2}O_2(g) + 9CO_3^{2-} + 3H_2O(l) \rightarrow 3UO_2(CO_3)_3^{4-} + 6OH^- \tag{11.102}$$

In the presence of bicarbonate, the dissolution reaction is:

$$U_3O_8(s) + \frac{1}{2}O_2(g) + 3CO_3^{2-} + 6HCO_3^-$$

$$\rightarrow 3UO_2(CO_3)_3^{4-}(aq) + 3H_2O(l) \tag{11.103}$$

The process of dissolution is retarded due to the formation of insoluble sodium urinate (Na_2UO_4) on the surface of ore particles as per the reaction:

$$UO_2(CO_3)_3^{4-} + 4OH^- + 2Na^+ \rightarrow Na_2UO_4 + 3CO_3^{2-} + 2H_2O \tag{11.104}$$

Calcium bicarbonate is used to stop the formation of sodium uranate.

The effects of operating variables such as concentration of carbonate and bicarbonate ions, partial pressure of oxygen, temperature, and particle size of the concentrate have been studied to establish the kinetics and mechanism of the dissolution process. As expected, the rate of dissolution increases with increase of temperature and decrease of particle size, that is, increase in surface area. Peters and Halpern [34] have established that both Na_2CO_3 and $NaHCO_3$ are essential for leaching of uranium. The rate of leaching is proportional to the square root of the partial pressure of oxygen gas in equilibrium with the leach liquor. This means oxygen gets dissociated prior to dissolution. Based on the experimental investigations of Peters and Halpern [34], the following conclusions may be drawn about the rate-determining step:

(i) The heterogeneous reaction occurring at the pitchblende surface and not the homogeneous reaction in the bulk of the solution is the rate-determining step.

(ii) Oxygen takes part in reaction after dissociation. This is evident from the fact that the rate of dissolution varies linearly with the square root of the partial pressure of oxygen in equilibrium with the leach liquor.

(iii) Although the presence of both CO_3^{2-} and HCO_3^- in the leach liquor is essential for the dissolution of U_3O_8, these ions do not seem to participate directly in the rate-determining step. This is based on the fact that increasing the concentration of either Na_2CO_3 or $NaHCO_3$ beyond 80 g l^{-1} does not affect the rate of dissolution. However, these salts help in preventing hydrolysis and precipitation.

(iv) Transport of reactants (i.e., O_2, CO_3^{2-}, and HCO_3^-) to the pitchblende surface or transport of products away from the surface does not seem to be a rate-controlling step because the activation energy of 12.3 kcal mol^{-1} is much higher than that expected for a diffusion controlled process. The rate of dissolution of pitchblende has been expressed as:

$$\text{Rate} = \frac{d[U_3O_8]}{dt} = KAp_{O_2}^{1/2}e^{(12300/RT)} \tag{11.105}$$

where K, A, p_{O_2}, and T are, respectively, a constant, the surface area of the mineral pitchblende, partial pressure of oxygen gas in equilibrium with the solution, and absolute temperature. The experimental rate of dissolution at 100 °C and 4 atm O_2 pressure was found to be 0.112 mg U_3O_8 cm^{-2} min^{-1} in solutions containing an excess of Na_2CO_3 and $NaHCO_3$ (more than 80 g l^{-1} of each).

Mechanism A mechanism for the dissolution of pitchblende in solutions containing sodium carbonate and sodium bicarbonate may be analyzed in the following steps:

Step I: Dissolution of gaseous oxygen in the aqueous solution: Dissolution of oxygen (reaction 11.75) is reversible, and equilibrium is rapidly established:

$$\frac{1}{2}O_2(g) \overset{fast}{\rightleftharpoons} \frac{1}{2}O_2(g-soln)$$

Step II: Adsorption and dissociation of oxygen on the pitchblende surface along with the oxidation of U_3O_8 to UO_3: This is a slow step, hence rate determining.

$$U_3O_8(s) + \tfrac{1}{2}O_2(g) \overset{slow}{\rightarrow} [\text{activated complex}] \rightarrow 3UO_3(s) \qquad (11.106)$$

UO_3 formed in step II reacts rapidly with carbonate and bicarbonate and dissolves in the aqueous solution.

Step III:Reaction of UO_3 with CO_3^{2-} and HCO_3^- and dissolution as a carbonate:

$$3UO_3\ (s) + 3CO_3^{2-}\ (aq) + 6HCO_3^- \overset{fast}{\rightarrow} 3UO_2(CO_3)_3^{4-}\ (aq) + 3H_2O(l)\ \ (11.107)$$

The overall reaction can be expressed as:

$$U_3O_8(s) + \frac{1}{2}O_2 + 3CO_3^{2-}(aq) + 6HCO_3^-(aq) \rightarrow 3UO_2(CO_3)^{4-}(aq) + 3H_2O(l)$$
$$(11.108)$$

According to the absolute reaction rate theory, the rate of a chemical reaction is given as:

$$v = \frac{KT}{h}c_1c_2\,e^{-\Delta G^{\ddagger}/RT} \qquad (11.109)$$

where K, h, (c_1 and c_2), and $\Delta G^{\ddagger}$ stand for Boltzmann constant, Planck constant, concentrations of reactants, and free energy of activation, respectively. Application of Eq. 11.109 to the rate-controlling step II yields the following expression for the specific reaction rate per unit area at the pitchblende surface:

$$\text{Specific rate} = \frac{d[U_3O_8]}{dt} = \frac{KT}{h}\,c_{U_3O_8} \cdot c_{\frac{1}{2}O_2(aq)} \cdot e^{-\Delta G^{\ddagger}/RT} \qquad (11.110)$$

where $[U_3O_8]$ is the number of molecules of U_3O_8 dissolved from a surface area A in time dt, $c_{U_3O_8}$ is the concentration of U_3O_8 surface sites, and $c_{O_2(aq)}$ is the concentration of the dissolved O_2 molecules. The free energy of activation is obtained as:

$$\Delta G^{\ddagger} = G^{\ddagger} - G_{U_3O_8} - \frac{1}{2}G_{O_2(aq)} \tag{11.111}$$

Thermodynamic consideration of the equilibrium in step I yields the following expression:

$$c_{O_2(aq)} = c_{O_2(g)}e^{-\Delta G_1/RT} \tag{11.112}$$

where $c_{O_2(g)}$ is the concentration of gaseous oxygen. ΔG_1, the standard molar free energy of dissolution of oxygen in water in step I is given as:

$$\Delta G_1 = \frac{1}{2}G_{O_2(aq)} - \frac{1}{2}G_{O_2(g)} \tag{11.113}$$

From Eqs. 11.110 and 11.112, we get the following expression for the specific rate:

$$\begin{aligned}
\frac{d[U_3O_8]}{dt} &= \frac{KT}{h}c_{U_3O_8}c_{\frac{1}{2}O_2(g)}.e^{-\left(\Delta G^{\ddagger}+\Delta G_1\right)/RT} \\
&= \frac{KT}{h}c_{U_3O_8}c_{\frac{1}{2}O_2(g)}e^{-\left(G^{\ddagger}-G_{U_3O_8}-\frac{1}{2}G_{O_2(g)}\right)/RT} \\
&= \frac{KT}{h}c_{U_3O_8}c_{\frac{1}{2}O_2(g)}.e^{-\left(H^{\ddagger}-H_{U_3O_8}-\frac{1}{2}H_{O_2(g)}\right)/RT}.e^{\left(S^{\ddagger}-S_{U_3O_8}-\frac{1}{2}S_{O_2(g)}\right)/R} \\
&= \frac{KT}{h}c_{U_3O_8}c_{\frac{1}{2}O_2(g)}.e^{-E/RT}.e^{\left(S^{\ddagger}-S_{U_3O_8}-\frac{1}{2}S_{O_2(g)}\right)/R} \tag{11.114}
\end{aligned}$$

$G = H - TS$, where H and S are, respectively, the standard enthalpies and entropies of the concerned species. The entropies of the solid activated surface state, $S^{\ddagger}$ and $S_{U_3O_8}$, can be neglected because they have only one degree of vibrational freedom. Hence, Eq. 11.114 can be expressed as:

$$v = \frac{KT}{h}c_{U_3O_8}c_{\frac{1}{2}O_2(g)}.e^{-E/RT}.e^{-\frac{1}{2}S_{O_2(g)}/R} \tag{11.115}$$

The theoretical rate of 0.37 mg U_3O_8 cm^{-2} min^{-1}, estimated by this equation at 100 °C and 4 atm pressure of oxygen, may be compared with the experimentally determined rate, 0.112 mg cm^{-2} min^{-1}. The agreement further confirms that dissolution of U_3O_8 in carbonate leach liquor in the presence of oxygen is a chemically controlled process.

In view of the above discussion, it may be concluded that formation of UO_3 in step 2 by the aqueous oxidation of UO_2 is doubtful. Since the rate of dissolution of UO_2 in aqueous solution of Na_2CO_3 and $NaHCO_3$ under oxygen gas pressure follows a trend similar to that of dissolution of gold in cyanide solution, an electrochemical mechanism based on the following two reactions may be considered:

(i). The reduction of oxygen in the cathodic area:

$$O_2 + 2H_2O + 4e \rightarrow 4OH^-$$

(ii). The oxidation of UO_2 in the anodic area:

$$UO_2 \rightarrow UO_2^{2+} + 2e$$

The uranyl ion forms a soluble carbonate complex in the presence of carbonate ions. This mechanism is similar to the corrosion of metals as discussed in the preceding section. Hiskey [35] has studied the kinetics of dissolution of UO_2 in ammoniacal carbonate solutions as a function of total carbonate concentration, oxygen pressure, temperature, and pH. According to him, the nature of the dissolution is electrochemical involving the anodic oxidation of UO_2 coupled with reduction of oxygen.

11.3.5.4 Kinetics of Leaching of Gibbsite in Sodium Hydroxide Solution

Glastonbury [36] studied the effect of surface area, sodium hydroxide concentration, and temperature on the kinetics of dissolution of gibbsite in sodium hydroxide solution according to the reaction:

$$Al(OH)_3 + NaOH = NaAlO_2 + 2H_2O \qquad (11.116)$$

Based on his investigations, Glastonbury [36] proposed the following mathematical model to deduce the rate of extraction in the temperature range, 50–100 °C:

$$Rate = 4.60 \times 10^5 . A (c_{NaOH})^{1.73} \left(e^{23850/RT}\right) \qquad (11.117)$$

where A and c represent surface area (cm^2) and concentration of sodium hydroxide solution $(mol\ l^{-1})$, respectively. The kinetics of the reaction was chemically controlled. The large activation energy of 23,850 cal mol^{-1} causes large temperature effects on the kinetics, which is reflected by the fact that the reaction rate at 100 °C is approximately 150 times faster than the reaction rate at 50 °C. From the fractional disorder of 1.73, it seems reasonable to assume that reaction is second order with respect to $[OH^-]$ ion concentration.

11.3.5.5 Kinetics of Leaching of Cuprite in Sulfuric Acid

The rate of dissolution of oxide minerals [33] in acids is dependent on the activity of hydrogen ions without involving oxidation–reduction couples. As an exception, dissolution of cuprite (Cu_2O) is affected by surface area and geometric factors. In the absence of oxygen for the reaction:

$$Cu_2O + 2H^+ \rightarrow Cu^{2+} + Cu + H_2O \tag{11.118}$$

the oxidation–reduction couple is completed by the disproportionation of cuprous copper. In the presence of oxygen, cuprite forms cupric ions in the solution according to the reaction:

$$Cu_2O + 4H^+ + \frac{1}{2}O_2 \rightarrow 2Cu^{2+} + 2H_2O \tag{11.119}$$

If the same rate-controlling process accounts for both the reactions, the reaction rate [33] is given by:

$$\text{Rate} = \frac{d[Cu_2O]}{dt} = \frac{d[Cu^{2+}]}{dt} = \frac{1}{2}d[Cu^{2+}]_{O_2}dt \tag{11.120}$$

where subscript O_2 refers to the rate of appearance of cupric ions in the presence of oxygen.

11.3.5.6 Kinetics of Leaching of Scheelite in Sodium Carbonate Solution

Queneau and Cooke [37] investigated the kinetics of dissolution of scheelite in basic solutions of carbonate, fluoride, phosphate, and mixed fluoride and phosphate. Based on their studies, they have concluded that sodium carbonate solution is a more selective leachant for industrial scheelite concentrates associated with excessive amount of calcium gangue. The dissolution of scheelite in sodium carbonate solution along with the formation of calcite crystals (according to the reaction $CaWO_4(s) + Na_2CO_3(aq) = CaCO_3(s) + Na_2WO_4(aq)$) on the attacked surface follows a parabolic rate law with an activation energy of 14.1 kcal mol^{-1}. In the temperature range 115–135 °C, transport of CO_3^{2-} ions through the calcite to the scheelite surface happens to be the rate-controlling factor, not the diffusion of tungstate, WO_4^{2-} ions. This has been confirmed by the fact that the rate of dissolution of scheelite depends on the pH of the carbonate solution. However, a positive deviation from parabolic behavior in the temperature range 135–155 °C points out toward calcite grain growth and formation of noncoherent calcite layer. In the light of these observations and findings, one important conclusion has been drawn that to get better leaching efficiency, a procedure must be incorporated for removal of calcite film from the calcite particles. Tungsten recovery up to 99.5% can be achieved by employing ball-mill autoclave at 200 °C using double the stoichiometric requirement of sodium carbonate solution. This amount is much less compared to that required in the absence of attrition. A plant based on these principles is under operation at Bishop, California.

11.3.5.7   Kinetics of Leaching of Sulfides

Habashi [2] has reported that dissolution of ZnS in dilute sulfuric acid at 100 °C follows a trend similar to that of rate of dissolution of gold in cyanide solution. At constant oxygen pressure, the rate of dissolution increases with increasing H_2SO_4 concentration and finally approaches a plateau when the rate is proportional to the partial pressure of oxygen. Kinetics of dissolution of sulfide minerals such as chalcopyrite, sphalerite, galena, and so on can be analyzed by the electrochemical process involving the following reactions:

$$\text{anodic dissolution} : MS \rightarrow M^{2+} + S + 2e \text{ in anodic area}$$

$$\text{cathodic reaction} : M^{2+} + S + 4e \rightarrow M + S^{2-} \text{ in cathodic area}$$

$$\text{and in the presence of oxygen} : O_2 + H_2O + 4e \rightarrow 4OH^-$$

Thus, sulfides produce elemental sulfur or sulfide ions. Wadsworth [33, 38] has reviewed the kinetics of leaching of sulfide and oxide minerals in different media and analyzed the dissolution kinetics in the light of electrochemical reactions. It has long been recognized that dissolution of chalcocite [38] (Cu_2S) in ferric sulfate/ferric chloride solutions takes place in two steps according to the reactions:

$$Cu_2S + 2Fe^{3+} \rightarrow CuS + Cu^{2+} + 2Fe^{2+} \tag{11.121}$$

$$CuS + 2Fe^{3+} \rightarrow Cu^{2+} + S + 2Fe^{2+} \tag{11.122}$$

At low temperature, reaction (11.121) is completed before the formation of elemental sulfur. In Sect. 11.3.4, the electrochemical nature of dissolution of chalcopyrite ($CuFeS_2$) has been mentioned according to reactions (11.55), (11.56) and (11.57) listed below:

$$\text{Cathodic reaction} : 2CuFeS_2 + 6H^+ + 2e = Cu_2S + 2Fe^{2+} + 3H_2S \tag{11.55}$$

$$\text{Anodic reaction} : 2Cu + H_2S = Cu_2S + 2 H^+ + 2e \tag{11.56}$$

$$\text{Overall reaction} : 2CuFeS_2 + 2Cu + 4H^+ = 2Cu_2S + 2Fe^{2+} + 2H_2S \tag{11.57}$$

Since the rest potential of chalcopyrite is approximately 0.46–0.55 V, it will dissolve cathodically according to the reaction (11.55), if contacted with a metal (Cu, Fe, Pb, Zn, etc.) that has lesser potentials. In this case, anode is not provided cathodic protection but is induced to react anodically as per reaction (11.56). Hiskey and Wadsworth [17] have measured the kinetics of the galvanic conversion of $CuFeS_2$ using copper according to reactions (11.55) and (11.56). Both cathodic and anodic reactions produce Cu_2S, which gets welded and provides good electrical contact. At lower temperature, stirring disrupts the bond and retards the reaction. At 90 °C, stirring has less effect on the overall kinetics. Initially, the rate was found to be proportional to the anodic surface area (Cu), but as the reaction proceeded, discharge at the cathodic surface became rate limiting due to the dissolution of $CuFeS_2$ cathodically and the diminishing surface area. The rate of dissolution of chalcopyrite has been expressed by the following general equation:

$$\text{Rate} = \frac{a_{H^+}^{1/2} A_1 A_2 k_1}{(A_1 + k_2 A_2)^{1/2} (k_3 A_1 + k_4 A_2)^{1/2}} \tag{11.123}$$

where A_1 and A_2 represent cathodic and anodic surface areas in the electrochemical system, respectively, and k_1, k_2, k_3, and k_4 are empirical rate constants.

On the other hand, the anodic dissolution of chalcopyrite in the presence of ferric ions [$Fe_2(SO_4)_3$ or $FeCl_3$] has been explained on the basis of diffusion of ferric ions through the deposited sulfur film resulting from the dissolution:

$$CuFeS_2 \rightarrow Cu^{2+} + Fe^{2+} + 2S + 4e \tag{11.124}$$

The depletion of ferric ions in solution results in the buildup of a sulfur layer of a thickness assumed to be directly proportional to the amount of copper entering in the solution. Beckstead and Miller [39] studied the kinetics of dissolution of chalcopyrite in oxygenated ammoniacal solution according to the reaction:

$$CuFeS_2 + 4NH_3 + \frac{17}{4}O_2 + 2OH^- = Cu(NH_3)_4^{2+} + \frac{1}{2}Fe_2O_3 + 2SO_4^{2-} + H_2O \tag{11.125}$$

They have reported that ammonia oxidation leaching of chalcopyrite is controlled by a catalytic electrochemical surface reaction. The kinetics of the electrochemical reaction was analyzed using Butler–Volmer equation. Beckstead and Miller [39] have expressed dissolution kinetics of the reaction by the under mentioned equation valid over a wide range of temperature, pressure, and cupric ion concentrations:

Low pressure, $p_{O_2} < 1$ atm:

$$\text{Rate} = \frac{k_l}{d_o} [OH^-]^{1/2} [O_2]^{1/2} (k_1 + k_2 [Cu^{2+}])^{1/2} (1 - f)^{2/3} \tag{11.126}$$

High pressure, $p_{O_2} > 1$ atm:

$$\text{Rate} = \frac{k_h}{d_o}[\text{OH}^-]^{1/2}\left(k_1 + k_2[\text{Cu}^{2+}]\right)^{1/2}(1-f)^{2/3} \qquad (11.127)$$

where k_1, k_2, k_l, and k_h are empirical rate constants, d_o is the initial particle diameter, and f is the fraction of CuFeS_2 reacted. At high p_{O_2}, the rate becomes zero order with respect to oxygen; otherwise, the rate is approximately proportional to half order with respect to hydroxyl and cupric ions and p_{O_2} (at low pressure). The half-cell reaction proposed includes the anodic dissolution of CuFeS_2 as per reaction:

$$\text{CuFeS}_2 + 19\text{OH}^- = \text{Cu}^{2+} + \frac{1}{2}\text{Fe}_2\text{O}_3 + 2\text{SO}_4^{2-} + \frac{19}{2}\text{H}_2\text{O} + 17e \qquad (11.128)$$

The rate of reaction was independent of ammonia concentration but was significantly dependent on temperature, hydroxyl, and cupric ion concentration. The activation energy of 10 kcal mol^{-1} suggests that the electrochemical reaction was limited by a surface reaction mechanism.

Ngoc, Shamsuddin, and Prasad [40] conducted studies on chloride leaching for simultaneous recovery of copper, zinc, and lead from an off-grade and complex chalcopyrite concentrate associated with appreciable amount of sphalerite, galena, and pyrite. They investigated the effects of temperature, concentration and quantity of ferric chloride, stirring speed, and leaching time on metal dissolution. Leaching tests were also conducted with individual (HCl, NaCl, CuCl_2, FeCl_3) and mixed chlorides (two-, three-, and four-component mixtures). Results showed the possibility of recovery of not only 99% copper and 89% zinc but also 82% lead and 58% elemental sulfur by the treatment of the concentrate with 4 M FeCl_3 at 110 °C for 2 h employing 25% excess FeCl_3 at a stirring speed of 700 rev min^{-1}. Kinetic studies indicated that the chalcopyrite, sphalerite, and galena of the concentrate dissolved simultaneously in the FeCl_3 lixiviant as if each mineral was separately leached and the copper and zinc dissolution reactions were under chemical control (linear kinetics). The dissolution rate of all the metal values (Cu, Zn, and Pb) followed the Arrhenius relationship (log k vs. $1/T$). The calculated activation energy values of 37.4, 40.3, and 15.8 kJ mol^{-1}, respectively, for dissolution of copper, zinc, and lead were in agreement with those reported in literature. From the agreement in the characteristic activation energy values, Ngoc et al. [40] have suggested that the constituent minerals, chalcopyrite, sphalerite, and galena of the concentrate dissolved simultaneously in ferric chloride as if each individual mineral was leached separately. The dissolution of copper and zinc appeared to be under chemical control. As regard to the dissolution of lead, no definite conclusion on its mechanism could be drawn due to the contradicting results, namely, linear rate law (indicative of chemical control) and very low activation energy value (similar to diffusion control).

11.4 Treatment of Leach Liquor

The resultant liquor obtained after leaching the mineral with dilute acidic or alkaline solution or after dilution of the pasty mass obtained by chemical break-down of refractory minerals by fusion with concentrated acid, alkali, or fluorides contains impurities, undissolved species, and suspended solid particles of the gangue minerals. The leach liquor has to be separated from suspended particles, concentrated, and purified prior to the recovery of the metal value. However, the extent of concentration and purification required varies from metal to metal depending on the level and nature of impurities and the subsequent method of recovery. Suspended solids are removed by settling, classification, or filtration. For handling large quantities of solution, the Dorr thickener is generally used. It consists of a large-diameter flat-bottomed tank of relatively less height fitted with a central shaft carrying slow revolving arms with scrappers. The leach liquor from agitators or leaching tanks is directly fed to a central well at the top of the shaft. After allowing sufficient time for settling and thickening, clear solution containing the metal value is taken into a channel around the thickener. The residues left at the bottom of the tank are gradually collected by scrappers to the center and finally withdrawn. Settling of fine particles is facilitated by addition of organic coagulating agents such as glue.

The resultant clear solution after solid–liquid separation is purified by suitable techniques, such as chemical precipitation, fractional crystallization, carbon-in-pulp, ion exchange, or solvent extraction. In rare metal extraction, ion exchange or/and solvent extraction is/are being used extensively for purification and concentration of dilute solutions and separation of chemically similar elements. In this section, the principle and application of these techniques will be briefly discussed.

11.4.1 Chemical Precipitation

Generally, the chemical precipitation technique is adopted for removal of impurities and recovery of certain metals from the leach liquor under conditions of controlled pH, temperature, and concentration in the presence of oxidizing or reducing agent. Traces of radium present in sulfuric acid leach liquor of uranium do not precipitate as $RaSO_4$. The precipitation of $RaSO_4$ is energized by addition of barium chloride to the leach liquor to form $BaSO_4$ precipitate. The mixture containing the precipitates of $RaSO_4$, $BaSO_4$, and $PbSO_4$ is filtered to obtain clear uranyl sulfate solution for further operation. Arsenic and phosphorus are removed as poorly soluble magnesium arsenate and magnesium phosphate, respectively, from the leach liquor obtained by pressure leaching of scheelite/wolframite concentrate in sodium carbonate solution. The filtrate is treated with sodium sulfide to precipitate molybdenum as molybdenum trisulfide, MoS_3. Thus, some metals may be removed and also recovered as sulfides. Recovery of metal sulfides will be discussed in Sect. 11.5.1.

Many other impurities present in minor quantities may be removed by precipitating them as hydroxides at appropriate pH. In treatment of ores of uranium and other rare metals, often iron is precipitated as ferric hydroxide by adjusting the pH to about 3.2 with lime or soda ash. In almost all the leach liquors, iron (ferrous as well as ferric ions) is present as a common impurity. The two hydroxides of iron, namely, $Fe(OH)_2$ and $Fe(OH)_3$, are precipitated at pH > 7 and pH ~ 3.2, respectively. Since many metals get precipitated in this pH range, Fe^{2+} present in the leach liquor is first oxidized to Fe^{3+} by adding MnO_2 in the leaching circuit or blowing oxygen. The precipitate of $Fe(OH)_3$ is voluminous; absorbs precipitates of arsenic, antimony, tin, and germanium; and poses problems in settling and washing. In order to avoid these problems, iron is precipitated as a dense and crystalline basic sulfate having higher settling rate and better filtering characteristics. Toward this end, there has been an important development known as the "jarosite process," which is extremely useful in the extraction of zinc via the hydrometallurgical route from sphalerite containing relatively higher percentage of iron.

It is interesting to note that the jarosite process has derived its name from jarosite, a naturally occurring mineral with chemical formula, $K_2Fe_6(OH)_{12}(SO_4)_4$, named after the place of its first location, Jarosite in Spain. Other jarosite-type compounds may be formed by substituting K^+ with NH_4^+, Na^+, or Rb^+. The process has solved the problem of zinc ferrite ($ZnO.Fe_2O_3$) formation during roasting of sphalerite (particularly in fluidized bed roasters). Zinc ferrite from the calcine can be dissolved in more concentrated acid at relatively higher temperature (90–95 °C) because excess iron can be precipitated as jarosite that can be easily removed from the leach liquor. In the jarosite process, calcine (ZnO) is dissolved in a neutral leach tank and Fe^{2+} is oxidized to Fe^{3+} in the presence of MnO_2. $Fe(OH)_3$ is precipitated at a pH of 5. The overflow is purified and subjected to electrowinning, and the underflow slurry is leached with spent electrolyte at 90–95 °C with addition of fresh H_2SO_4. Ferric hydroxide precipitate is redissolved, and zinc ferrite is attacked. The liquor is subjected to thickening to remove insoluble constituents as underflow. Yellow jarosite is precipitated by the addition of $NH_4OH/NaOH$ to the overflow taken in the precipitation tank (see Appendix A.4b) according to the reaction:

$$3Fe_2(SO_4)_3 + 10H_2O + 2NH_4OH = (NH_4)_2Fe_6(SO_4)_4(OH)_{12}$$
$$+ 5H_2SO_4 \tag{11.129}$$

In order to counterbalance the effect of H_2SO_4 released and to maintain the pH ~ 5, some more calcine (ZnO) is added. The slurry is taken to a thickener to collect jarosite and recycle the solution to the neutral leach.

Prior to the development of the jarosite process, in order to produce leach liquor with minimum iron, only iron-free zinc oxide was readily dissolved in sulfuric acid at

60 °C leaving zinc ferrite as insoluble residue. Hence, recovery of zinc rarely exceeded 85% due to the loss of zinc in ferrite. With the advent of this process, recovery has increased up to 96%. Most of the zinc ferrite can be dissolved in hot acid leach solution, and even higher amount of iron can be easily precipitated and filtered.

11.4.2   Fractional Crystallization

This method of purification/separation makes use of the advantage of the difference in the solubilities of salt in water. This principle was employed first for separation of niobium (columbium) from tantalum due to the difference in solubility of K_2TaF_7 and K_2NbF_7. It is one of the oldest methods for the separation of compounds of chemically similar metals. It depends on small differences that are magnified in repeated operations. However, in recent years, it has been replaced by faster and less cumbersome methods of ion exchange and solvent extraction. In view of its historical importance, a brief account of the technique used earlier in the separation of chemically similar zirconium and hafnium, and niobium and tantalum is given here.

Zircon sand broken by potassium silicofluoride according to the reaction Zr (Hf) $SiO_4 + K_2SiF_6 = K_2Zr(Hf)F_6 + 2SiO_2$ contains double fluorides of zirconium and hafnium. The ground sintered mass after leaching with water retains the entire hafnium (~1.6%). Since hafnium has a high neutron absorption cross section, it has to be reduced to less than 200 ppm for nuclear applications of zirconium. For separation of Zr and Hf, a 16- to 18-stage crystallization in stainless steel vessels operating between temperatures of 100 and 19 °C has been found necessary to reduce the hafnium content to the prescribed level. At these temperatures, the respective solubilities [13] of zirconium salt are 250 and 16.3 g l^{-1}. Solubility of K_2HfF_6 is somewhat higher. The mother liquor from each stage is passed back to a previous stage for a latter batch because the overall yield using fresh water for each crystallization stage is about 10%. In this way, the overall yield of zirconium is 80%. For rapid salt dissolution, all the tanks are equipped with immersion heater and paddle stirrer. The tanks are arranged to facilitate gravity flow of mother liquors down to cascade. During separation of K_2ZrF_6 and K_2HfF_6, K_2ZrF_6 crystallizes, whereas K_2HfF_6 is left in the mother liquor. During the treatment of the niobite-tantalite mineral, K_2TaF_7 crystallizes and K_2NbF_7 is collected in the mother liquor.

11.4.3   Activated Carbon

The leach liquor resulting from the sodium cyanide leaching of the gold ore is generally poor in gold concentration. It needs to be concentrated for recovery of gold in the next processing step. The most common method to increase the dissolved gold concentration involves treatment of the leach liquor with activated carbon [1]. The carbonaceous materials such as coal, wood, and plants like coconut, sized between

1 and 4 mm diameter, are thermally activated to get internal surface area of about 1000 m^2 per kg of carbon. Gold cyanide from the leach liquor is adsorbed (loaded) on the carbon when the former is brought in contact with the latter for about 1.5 h in a mechanically agitated tank. The gold adsorption efficiency increases with the number of stages (60 – 70% adsorption per stage). In general, six adsorption stages have worked out to be optimum for adsorption of the order of 99.5%. The gold adsorption is affected by temperature, concentration of gold, and cyanide in the leach liquor. It decreases with rise in temperature, increases with increase of gold concentration, and decreases with increasing cyanide concentration. Gold from the loaded carbon is eluted (stripped) in concentrated cyanide solution at elevated temperature of about 95 – 150 °C. Activated carbon is regenerated by the activation process in a rotary kiln at 650 °C in steam to remove organic waste. This is followed by HCl treatment to remove inorganic precipitate like carbonates.

11.4.4   Ion Exchange

The principle of ion exchange seems to be derived from the special characteristic power of soil to absorb ammonia from ammonium sulfate by releasing calcium as per the reaction, calcium soil + ammonium sulfate = ammonium soil + calcium sulfate. Ion exchange technique was first employed [13] on commercial scale by synthesizing inorganic zeolite: $Na_2Al_2Si_3O_{16}$, capable of exchanging sodium ions in water softening and sugar treatment. Since natural and synthetic zeolites having limited exchange capacity are stable only in a neutral environment, efforts were made to develop rugged polymeric materials. In 1935, Adams and Holmes [41] synthesized the first resinous type of exchange product from phenol and formaldehyde and anilines and formaldehyde. But these resins had limited applications in industrial processes because of poor stability and low absorption capacity. Since 1944, styrene polymers with high physical and chemical stability have been used in the manufacture of ion exchange resins.

11.4.4.1   Ion Exchange Resins

The most important type of resin is manufactured by suspension polymerization of styrene plus variable amounts of a cross-linked agent divinyl benzene. Cation exchange resins are obtained by sulfonation or carboxylation of the resulting polymer, whereas the anion exchangers are obtained by amination after chloromethylation of chloromethyl ether and a Friedel–Crafts type catalyst. Based on these principles, a three-dimensional polystyrene network is prepared by copolymerization of styrene (phenylethene) with divinyl benzene (diethenyl benzene). The amount of divinyl benzene in the mixture controls the degree of cross-linkage, which in turn determines the physical properties of the resin. Generally, 8% divinyl benzene is used. Sulfonation of the cross-linked polystyrene results in a strong acid cation exchange resin. The strong acid–SO_3H functional group occupying the meta or para position on the phenyl ring [16] is shown below:

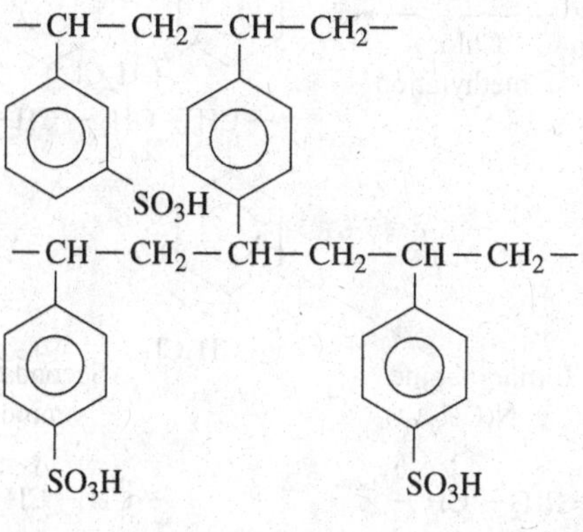

$$-CH-CH_2-CH-CH_2-$$

For preparation of weak acid cation exchanger, a weak acid functional group such as a phosphoric acid ($-PO_3H_2$) or carboxylic acid ($-COOH$) is introduced into a polystyrene network. The carboxylic acid resin may also be obtained by copolymerization of methacrylic acid and variable amount of divinyl benzene depending on the requirement of cross-linkage as per the reaction [16]:

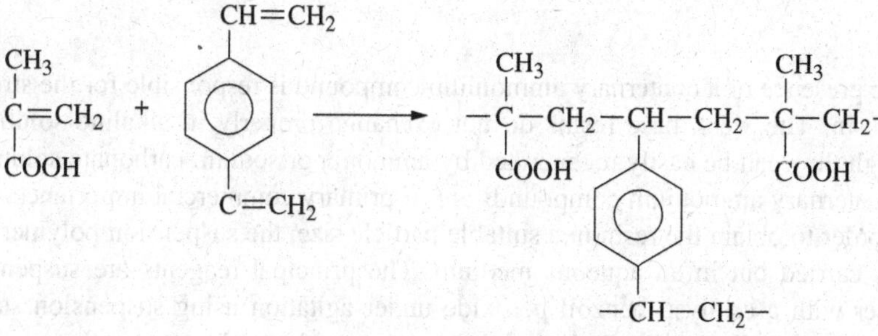

Polyacrylate weak acid
cation exchange resin

Anion exchange resins are prepared by introducing an amine group into the cross-linked polystyrene network by chloromethylation reaction. For preparing a strong base functional group, a tertiary amine [16] is used, and for weak base, a secondary amine is used:

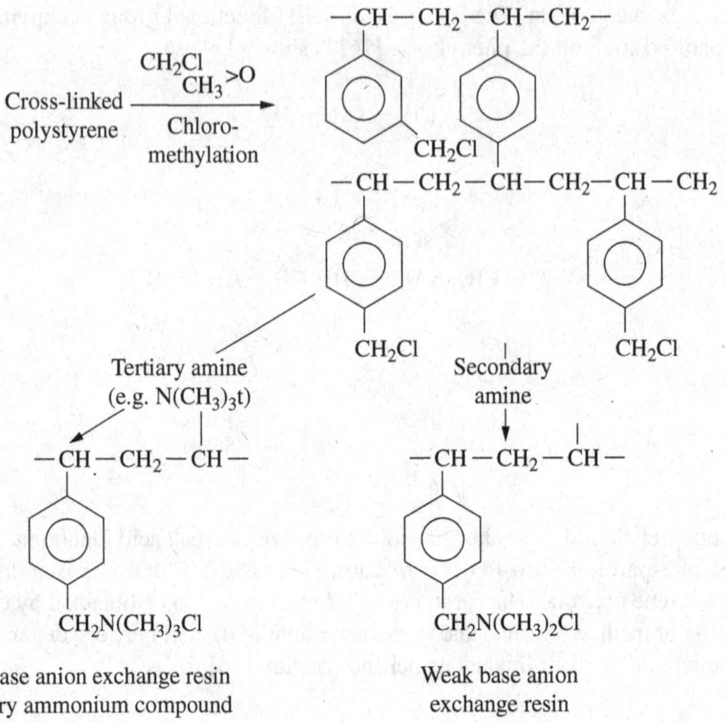

$$-CH-CH_2-CH-CH_2-$$

Cross-linked polystyrene $\xrightarrow[\text{Chloro-methylation}]{\overset{CH_2Cl}{\underset{CH_3}{}}>O}$

CH_2Cl

$$-CH-CH_2-CH-CH_2-CH-CH_2$$

CH_2Cl

Tertiary amine
(e.g. $N(CH_3)_3t$)

CH_2Cl Secondary amine CH_2Cl

$$-CH-CH_2-CH-$$

$CH_2N(CH_3)_3Cl$

Strong base anion exchange resin
quaternary ammonium compound

$$-CH-CH_2-CH-$$

$CH_2N(CH_3)_2Cl$

Weak base anion
exchange resin

The presence of a quaternary ammonium compound is responsible for the strong base resin. The weak base resins do not exchange properly in alkaline solutions although they can be easily regenerated by ammonia or sodium carbonate solution. The quaternary ammonium compounds are of primary commercial importance.

In order to obtain the resin in a suitable particle size, the suspension polymerization is carried out in an aqueous medium. The principal reagents are suspended together with a catalyst, benzoil peroxide under agitation using suspension stabilizers, for example, polyvinyl alcohol, bentonite, starch, or alginates in the aqueous medium. The copolymer can have one — SO_3H group introduced into each aromatic nucleus by treatment with sulfuric acid, chlorosulfonic acid, or sulfur trioxide in nitrobenzene. The cation exchange capacity of the resulting resin is 5 mg equivalent per gram of dry resin. The resinous products exceptionally stable up to 120 °C and resistant to acids, alkalis, and oxidizing and reducing agents are commercially available in spherical shape with size ranging from 20 to 50 mesh as well as in the form of membranes of usually 1–2 mm thickness.

Thus, ion exchange resins may be considered as insoluble [42] high molecular acids or bases, which can form insoluble salts. They have loosely cross-linked polymerized organic immobile structure, attached to a large number of mobile, active, or functional groups. The functional groups may be acidic or basic in character. Strongly acid resins with cation exchanger contain a sulfonic acid

$(-SO_3H)$ group, whereas weakly acid resins have carboxylic $(-COOH)$ or phenolic $(-OH)$ active groups. Such resins capable of exchanging H^+ or Na^+ for ions of similar charge are marketed in hydrogen or sodium form. The active group in anion exchange resins of the strong base type is a quaternary ammonium group $(-NX_3)$, and the weak base resins contain an amino $(-NX_2)$ or imino $(=NX)$ group, where X may be hydrogen or an organic radical. Anion exchange resins are generally prepared in the chloride or hydroxyl form and can exchange the active ion for ions of similar charge. The porosity, stability, and swelling characteristics of a resin are governed by the degree of cross-linkage, which in turn increases with the proportion of divinyl benzene used in synthesis.

11.4.4.2   Principle

There is an exchange reaction between the ions in the solution and those of like charge in the resin when the two are brought into contact. For example, the cation exchange process is represented by the following reaction:

$$n\,HR(resin) + M^{n+}(solution) \rightleftarrows MR_n(resin) + n\,H^+(solution) \qquad (11.130)$$

where R is the insoluble organic part of the resin and H is the ionizable part of the functional group of the resin. In a similar way, an anion exchange reaction may be expressed as:

$$n\,RCl\,(resin) + A^{n-}(solution) \rightleftarrows R_nA(resin) + n\,Cl^-(solution) \qquad (11.131)$$

In the above exchange processes, equivalent amounts of H^+ and Cl^- are released into the solution. The extent of exchange is limited by the capacity of the resin which is of the order of 2–10 g equivalent per kilogram [43] of the dry resin. The ion exchange equilibria representing the ion transfer between a solid resin and liquid solution are valid only for dilute solutions. The equilibrium constant or the distribution coefficient for transfer of any ion, I, taking part in reaction of the type (11.130) or (11.131) may be expressed as:

$$K_I = \frac{\text{concentration of I in the resin}}{\text{concentration of I in the solution}} \qquad (11.132)$$

The extent to which an ion is retained or sorbed in the resin depends on the nature of the ion, the resin, and the solution under consideration [43]. Ions of greater charge and lesser hydrated ionic radii are more easily sorbed in a resin than those of lesser charge and greater hydration. For example, the order of retention of ions of thorium; lanthanum, magnesium, and sodium by a resin from equivalent solutions would be $Th^{4+} > La^{3+} > Mg^{2+} > Na^+$. The order of retention in case of ions of the same valence like those of an alkali metal group and the alkaline earths are $Cs^+ > Rb^+ > K^+ > Na^+ > Li^+$ and $Ba^{2+} > Sr^{2+} > Ca^{2+} > Mg^{2+}$, respectively. However, the degree of hydration plays an important role in the sorption of tripositive elements of lanthanide and actinide series. Since the radii of the hydrated tripositive ions increase with atomic number, the order of retention by a resin decreases with increase of the atomic number. The order of retention for the actinide series is as follows:

$$U^{3+} > Np^{3+} > Pu^{3+} > Am^{3+} > Cm^{3+}$$

Similarly, the order of retention of anions by a strong base resin follows the trend:

$$SO_4^{2-} > CrO_4^{2-} > NO_3^- > HPO_4^{2-} > I^- = Br^- > Cl^-$$

The apparently incorrect position of phosphate ions may be due to their large size. Weak acid and weak basic resins exhibit preferences toward certain ions, but strong resins do not exhibit such a selectivity toward ions. For example, carboxylic resins are selective toward Ca^{2+} and Cu^+ among a mixture of different ions. Degree of cross-linkage affects sorption. A stable resin with considerable cross-linkage possesses smaller pores and preferentially absorbs ions with a smaller hydrated volume. As ion exchange equilibria are valid in dilute solutions, K values are affected to a great extent by the total concentration of ions. A resin absorbs Ca^{2+} in preference to Na^+ from a dilute solution containing calcium and sodium ions, whereas the order is reversed in the concentrated solution.

11.4.4.3   Kinetics of Exchange Reaction

An ion exchange resin may be regarded as a three-dimensional network of polymer chains bonded to each other at regular intervals by cross-linking molecules. The attached functional groups undergo ionization to generate cations (or anions). These ions are capable of exchanging other ions present in the solution after coming into contact. The ion exchange process is heterogenous presenting a case of solid (resin)–liquid (aqueous solution) reaction, but it complies with homogeneous second-order reaction kinetics. The exchange reaction between a strong acid cation exchange resin in the hydrogen form (RH) and a dilute solution containing I^+ ions may be represented as:

$$RH + I^+ = RI + H^+ \tag{11.133}$$

The reaction may be considered to occur in the following steps [16]:

(i)   Transport of I^+ ions from the bulk of the solution across the stagnant boundary layer surrounding the resin bed

(ii)   Diffusion of I^+ ions into the interior of the resin

(iii)   Exchange reaction between H^+ ions and I^+ ions

(iv)   Diffusion of H^+ ions outward to the surface of the bed

(v)   Diffusion of H^+ ions across the stagnant boundary layer into the bulk of the solution

The diffusion occurring in the boundary layer (steps i and v) may be termed as "film diffusion," whereas steps (ii) and (iv) are generally referred as "particle diffusion". The diffusion in the solid resin is complicated due to the restriction of polymer network, charge distribution associated with the fixed ions of the functional groups, porosity of the solid bed, and effective diffusivity of metal ions. Higher cross-linkage may favor particle diffusion as the rate-controlling process. As a large range of metal ion concentrations are encountered in hydrometallurgical processing, the rate-controlling step in the initial stages of the ion exchange reaction is likely to be film diffusion. The diffusion path lengthens progressively as the actual reaction zone moves toward the resin bead surface. This gradually leads to a change from a film diffusion controlled process to a particle diffusion-controlled process.

11.4.4.4 Ion Exchange Process

Ion exchange is carried out in two steps [13, 43], namely, "sorption" and "elution." In the first step, the solution containing ions of interest is allowed to pass through a column filled with grains of a resin in an appropriate form. During the downward movement in the column, the ions of the same charge as those on the resin are taken up (sorbed) by the resin and the depleted solution flows out of the column. In the second step, the ions retained by the resin are recovered by percolating a suitable solvent (water, appropriate acidic, or alkaline solution) through the bed. The liquid eluted from the column contains the desired ions. Thus, elution is just the reverse of sorption. A resin can be employed for thousands of cycles. Often, the resin bed is regenerated during elution. When the bed becomes impervious due to pick up of dust and insoluble matter, a "back-wash" with water is carried out to "loosen" the bed and to remove mechanically entrapped matter.

If one is interested in recovering two ions, I_1 and I_2 of similar charge present in the solution, advantage is taken of the separation factor [43]. The separation factor (α) of the resin is defined as the ratio of two distribution coefficients, K_{I_1} and K_{I_2}:

$$
\begin{aligned}
\alpha &= \frac{K_{I_1}}{K_{I_2}} \\
&= \frac{\text{concentration of } I_1 \text{ in the resin}}{\text{concentration of } I_2 \text{ in the resin}} \times \frac{\text{concentration of } I_2 \text{ in the solution}}{\text{concentration of } I_1 \text{ in the solution}}
\end{aligned}
\tag{11.134}
$$

$$
\text{or} \quad \frac{\text{concentration of } I_1 \text{ in the resin}}{\text{concentration of } I_2 \text{ in the resin}} = \alpha \frac{\text{concentration of } I_1 \text{ in the solution}}{\text{concentration of } I_2 \text{ in the solution}}
$$

As both I_1 and I_2 may be sorbed on passing the solution through the resin bed, it is necessary that the left-hand side of the above equation should be large to achieve a good separation. The species I_1 will then tend to be retained by the resin, while I_2 is carried away by the solution, that is, I_2 ions will move ahead of I_1 in the exchange column during the elution and will leave the column first. This can be achieved by having either a large value of α or by having a large ratio of I_1 to I_2 in the solution.

On the other hand, when the affinities of the ions for the resin are similar, that is, α is small, the second condition may be beneficial. For this purpose, eluting solution with anions that may form complexes with I_1 and I_2 (cations) may be employed. If the complex formed with I_2 is stronger, the ratio of free I_1 ions to free I_2 ions in solution will be large and this will favor good separation. For example, citric acid partly neutralized with ammonia to a pH of 3–3.5 has been used in the separation of tripositive ions of the lanthanide and actinide series. In this case $(H_2 \text{ cit})^-$, anions present in the solution form neutral complexes of the type $M(H_2 \text{ cit})$ with the tri-positive ions. The preferential elution of the component metal is facilitated because the stability of the complex increases with decreasing atomic number. During the separation of ferric ions from potassium ions, excess chloride is added to the solution to form negatively charged complex of iron, which can be removed by passing the solution over an anion exchange resin. Potassium is allowed to move first through the column. Subsequently, iron is eluted with dilute hydrochloric acid. Alternatively, the mixed solution with excess of chloride can be passed over a cation exchange resin that allows the complex ferric ions to pass through and retain potassium. Potassium can be subsequently recovered by elution with an acid solution.

· Ion exchange has been a batch process for several years in the past. After extensive research investigations, the "Resin-in Pulp" (RIP) technique [13, 42] and ion exchange membranes were introduced, but these developments could not fulfill the prime objective of making ion exchange a continuous process. In the RIP technique, unclarified slurries or pulps obtained by leaching are brought into contact with resin and eluted with a suitable solvent. The technique employs tanks, moving baskets of resins in solution tanks, pulse columns, and jerked beds. In uranium processing, a semicontinuous operation has been achieved by the use of a few columns in series, some for sorption and some for elution. Continuous ion exchange equipments have also been developed. Ion exchange membranes are being used increasingly in electrochemical industries to minimize energy losses due to mixing of anolyte and catholyte. These membranes are used as a partition between anolyte and catholyte.

11.4.4.5 Applications in Metallurgy

Ion exchange technique has been successfully employed in four distinct classes of applications, namely, concentration, separation, purification, and metathesis. However, it has been most extensively used in concentration of solutions. When other methods like precipitation, evaporation, or concentration fail, ion exchange is very well suited for economic treatment of extremely dilute solutions. Concentration of the order of thousand fold can easily be achieved. Many metals in dilute solutions constitute harmful wastes requiring careful attention in disposal from the viewpoint of health hazards. Ion exchange is very well suited for the disposal of such wastes because it can quite effectively remove and recover even traces of ions present in any solution. Separation of ions is widely used for analytical purposes. Ions with very small difference in the strength of adsorption can be easily separated by this technique. Purification of a very high degree can be achieved. Salts can be converted into acids and bases and other salts by metathesis, although it has a limited application in metallurgy. Some of the unique examples where ion exchange plays an important role in extraction, concentration, separation, or purification of metals are briefly discussed here.

1. *Extraction of uranium*: Pitchblende, U_3O_8, is leached with dilute sulfuric acid in the presence of pyrolusite at a pH of 1.8 as per reaction (11.1) in rubber-lined mild steel pachucas [44, 45]. The resultant leach liquor containing about 0.4 g U_3O_8 per liter solution together with appreciable amounts of V, Al, Fe, PO_4^{3-}, Cl^- and a large amount of sulfate ions after filtration is passed through an anionic exchange resin, Amberlite IRA-400. Uranyl sulfate of the lean leach liquor complexes with the excess sulfate ions present in the solution to form negatively charged complexes, such as $[UO_2(SO_4)_3]^{4-}$ and $[UO_2(SO_4)_2]^{2-}$. These sulfato complexes are selectively adsorbed on the resin according to the reaction:

$$4RCl + [UO_2(SO_4)_3]^{4-} \rightleftarrows R_4UO_2(SO_4)_3 + 4Cl^- \qquad (11.135)$$

where RCl is the resin in chloride form. Elution with 1 M NH_4NO_3–0.1 M HNO_3 or M NaCl–0.1 M H_2SO_4 yields effluent containing an almost pure uranium of concentration: 5 g U_3O_8 per liter. This concentrated solution is followed by diuranate precipitation (see Appendix A.7). The uranium recovery in the ion exchange process is over 99%.

2. *Treatment of gold ore*: In the treatment of complex Canadian gold ore, cyanide solutions were passed through a richly basic anion exchange resin for the adsorption of cyanide complexes of gold, silver, copper, iron, cobalt, and nickel. By following a selective elution procedure, separation of different metals from the mixture was possible. Nickel and zinc were eluted with hydrochloric acid, iron and copper with sodium cyanide solution, gold and silver with acetone acidified with hydrochloric acid, and finally cobalticyanide complex with strong potassium thiocyanate solution. This demonstrates the importance of ion exchange technique in concentration, separation, and purification.

3. *Separation of zirconium and hafnium*: Zirconium and hafnium that occur together have very similar chemical characteristics. Zirconium metal produced from its ores contains small quantities of hafnium. With low absorption cross section for thermal neutrons, zirconium is suitable for structural purposes in nuclear reactor construction. Hafnium having high neutron absorption cross section has to be reduced to less than 200 ppm in zirconium. In order to produce hafnium-free zirconium, a separation process step has to be added after the ore breakdown step. Prior to the development of the ion exchange process, fractional crystallization of K_2ZrF_6 was carried out in 16 – 18 stages for zirconium–hafnium separation. As fractional crystallization is a very tedious and cumbersome process, ion exchange has proved to be a boon in Zr–Hf separation. Both cation and anion exchange resins have been used for this purpose.

From a nitric acid solution, both the metals are adsorbed in a thin layer at the top of the resin Zeokarb-225. On elution with 0.5 M H_2SO_4, zirconium moves down the column ahead of hafnium. About 98% Zr containing less than 200 ppm Hf can be recovered by employing a suitable length of the column. Hafnium is then eluted rapidly with 1.5 M H_2SO_4. The anion exchanger method is based on the fact that Zr and Hf form stable anions such as fluorozirconate and fluorohafnate. These complexes are adsorbed on a suitable anion resin and subsequently selectively eluted with a mixture of HF–HCl acid solution to yield hafnium-free zirconium.

Similarly, spent fuel elements are treated for purification, separation, and recovery of unburnt uranium, plutonium, and radioactive fission products useful for industrial and research purposes. However, in recent years, solvent extraction has superseded the ion exchange technique in separation and purification of chemically similar elements like Zr and Hf, Nb and Ta, U and Pu, V and Th, and fission products elements. Currently, ion exchange happens to be an important tool for concentration of dilute solutions obtained by leaching, plating rinse waste, and disposal of industrial wastes.

4. *Treatment of plating solutions*: It is necessary to purify the contaminated plating bath for reuse and recovery of valuable constituents. For example, chromic acid is used extensively in hard chrome or decorative coating baths or for formation of a protective oxide coating. In these operations besides the desired plating or oxidizing reactions, dissolution of some of the base metals in chromic acid takes place. On the industrial scale, chromic acid is recovered by cation exchanger and chromates by anion exchanger resins. Another example is that of tin recovery from hot alkaline tin plating wastes. Hot dilute sodium stannate is passed through a hydrogen exchanger bed to form insoluble stannic acid that separates as a flocculants material of 2% solids in a decantation tank. The stannic acid is subsequently treated for the recovery of tin.

5. *Disposal of industrial wastes*: Waste solutions from many industries contain relatively valuable materials in very low concentrations whose recovery by conventional methods is not economical. Several industries are employing ion exchange columns for waste treatment, not only for prevention of loss of valuable metals but also from the viewpoint of public health. Purification of industrial wastes before disposal into rivers is definitely more beneficial compared to the recovery of metal values. Ion exchange technique has been successfully employed in the treatment of waste solutions from brass, textiles and rayon mills, anodizing, and plating units. Waste liquids from atomic energy installations are radioactive in character and contain many fission product ions. These would cause great harm to public health as well as vegetation when they enter water supplies. Ion exchange has proved to be a valuable tool in decontamination of radioactive wastes. The technique has also been useful in recovery of some radioactive isotopes for research and medicinal purposes.

11.4.5   Solvent Extraction

Ion exchange is a batch process. Despite extensive and serious efforts, it has not been possible to develop it on a continuous scale due to the limited capacity of the resins and slower adsorption/desorption kinetics of the solid–liquid exchange reactions. Thus, the technique has been confined to a few specific applications. Contrary to this, solvent extraction has emerged as a continuous process on account of a much faster rate of transfer of ionic species in the liquid – liquid exchange process (from the aqueous phase to the organic phase and vice versa). In recent years, the solvent extraction technique has been exploited on a tonnage scale in a number of metallurgical and chemical industries. It is generally considered to have come into vogue about eight decades ago in the pioneering observations of Peligot [43] on the transfer of uranyl nitrate from an aqueous solution to an organic solvent, diethyl ether. Since then, it has been employed for a variety of purposes including separation of chemically similar elements, concentration of dilute solutions, extraction of desired species from a solution, purification of salts, recovery of valuable constituents from waste solutions, and so on in the field of chemistry, chemical engineering, and extractive metallurgy. The first metallurgical application may be said to have begun with the commercial exploitation of the process (in 1942) for the recovery and purification of uranium by the US Atomic Energy Commission. The technique is being used on an industrial scale not only in rare metal extraction but also in the winning and purification of common metals.

11.4.5.1   Basic Principles and Process

Solvent extraction process consists of bringing into contact the aqueous solution containing the metal value with a suitable organic phase (an organic compound, i.e., extractant, dissolved in a diluent). The metal value is distributed between the two phases and under equilibrium [45] the distribution coefficient, D is defined as:

$$D = \frac{c_o}{c_a} \tag{11.136}$$

where c_o and c_a are, respectively, the concentration of metal ion in the organic phase and in the aqueous phase. The relation holds good provided there is no association or dissociation in either phase. If α_o and α_a are the association or dissociation coefficient in the organic and aqueous phases, respectively, the relationship is modified as $D = \frac{c_o(1-\alpha_o)}{c_a(1-\alpha_a)}$. This coefficient helps in assessing the extent of extraction in a given number of stages as total extraction does not occur in only one step. Further, the aqueous solution may contain a number of impurity metals ($i_1, i_2, \ldots, i_n$). along with the metal (m) of interest. In such a case, the ratios of distribution coefficients $D_m/D_{i_1}, D_m/D_{i_2}$, and so on known as the separation factors [13], are indicative of the degree of separation achieved in one equilibrium stage. For good separation that is required in the extraction as well as purification operations, these factors should obviously be much larger than or much smaller than unity [2].

After extraction, the metal value is recovered from the loaded organic phase in a subsequent step known as "stripping" where the loaded organic phase is brought into contact with a suitable aqueous stripping solution. The ease with which metal ions are transferred from one phase to the other may be controlled by adjusting the compositions and operating conditions properly (discussed in a subsequent section). Provision is made to strip only one metal, if more metal ions are extracted in the first stage, by adding one more step known as "scrubbing" to each of the extraction and stripping columns. Generally, the process requires several steps to attain the maximum amount of extraction. It is necessary to optimize [2, 13] the number of extraction stages and volume of the solvent in each stage for a given quantity of solvent. For maximum extraction, the solvent is used in a number of volume fractions rather than using all the solvent at a time. After n extractions, the concentration of the solute left behind in the aqueous solution, $(c_a)_n$, is given by

$$(c_a)_n = (c_a)_i \left[\frac{1}{1+D\left(v_o / v_a \right)} \right]^n \; ; \text{where } v_o \text{ and } v_a \text{ are the volumes of organic and aqueous}$$

phases, respectively, and $(c_a)_i$ is the initial concentration of metal ions in the aqueous phase.

Batch and continuous counter current extraction methods have been employed for solvent extraction. Generally, feed preparation is carried out in stainless steel or concrete tanks provided with a coil for heating or cooling, a stirrer, and a manhole. For extraction, vertical or horizontal vessels [2, 13] known as columns and mixer-settlers, respectively, are in use. Pumper decanters and centrifugal extractors have been employed to a lesser extent. In every type of extraction unit, the main object is to obtain rapid and efficient transfer of solute from one phase to the other.

11.4.5.2 Mechanism

Some of the important mechanisms of solvent extraction are outlined below:

1. *Charged complex formation followed by ion exchange*: The solvent acts either as an anion or a cation exchanger [2, 13, 45]. The metal ions are readily transferred to the organic phase by the formation of more soluble charged complexes between the metal species and the solvent:

   a. Anion exchangers include the basic organic extractants like amines. The above principle is illustrated by the following equations in the extraction of uranyl ions in sulfuric acid medium:

$$UO_2^{2+} + 2SO_4^{2-} = UO_2(SO_4)_2^{2-}$$

$$2R_3N + H_2SO_4 = (R_3NH)_2SO_4$$

$$\text{and} \quad (R_3NH)_2SO_4 + UO_2(SO_4)_2^{2-} = (R_3NH)_2UO_2(SO_4)_2 + SO_4^{2-} \quad (11.137)$$

b. Cation exchanger solvents include acid organic extractants like dialkyl phosphoric acid and carboxylic acid. The extraction of a divalent metal ion by dialkyl phosphoric acid may be represented according to the equation:

$$2HR_2PO_4 + M^{2+} \rightarrow M(R_2PO_4)_2 + 2H^+ \quad (11.138)$$

General Mills Incorporated was the first to demonstrate that an aliphatic α-hydroxy-oxime (LIX 63) could be used to extract copper from dilute sulfate leach liquor. LIX has been classified as liquid cation exchanger [16] on account of the following exchange reaction in extraction of copper:

$$2(RH)_{org} + (Cu^{2+})_{aq} = (R_2Cu)_{org} + (2H^+)_{aq} \quad (11.139)$$

For which, the equilibrium constant,

$$K = \frac{[R_2Cu]_{org}[H^+]_{aq}^2}{[RH]_{org}^{2+}[Cu^{2+}]_{aq}} \quad (11.140)$$

since $D = \frac{[R_2Cu]_{org}}{[Cu^{2+}]_{aq}}$,

we can express the distribution coefficient as:

$$D = \frac{K[RH]_{org}^2}{[H^+]_{aq}^2}$$

$$\therefore \log D = \log K + 2\log[RH]_{org} + 2pH_{aq} \quad (11.141)$$

Thus, the distribution coefficient depends on the pH of the solution. The pH value corresponding to 50% extraction by LIX63 is 4.8. This is too high for extracting copper from heap leach liquor together with poor selectivity over iron. The rate of extraction was enhanced by mixing aromatic β-hydroxyoxime with α-hydroxyoxime (LIX-63). The mixture called LIX 64 has better selectivity over iron at pH = 3.3. LIX64 has been used commercially at the Bluebird Mine Miami [16]. On further modification, LIX65N was developed to obtain satisfactory rates of extraction at elevated temperatures. A small proportion of LIX63 with LIX65N designated as LIX64N works better in terms of greater selectivity of copper over iron [46] at lower pH of 2.9. Kelex reagents based on 8-hydroxyquinoline were developed by Ashland

Chemical Company. These reagents extract acid along with copper and show poor selectivity against iron; however; they possess excellent stripping characteristics.

In recent years, a mixture of aldoximes and ketoximes [5] dissolved in petroleum distillate (to the extent of 10–20% by volume) is being used in most of the copper-producing plants based on the hydrometallurgical route. The appropriate mixture can be easily prepared to achieve optimum conditions for efficient and effective extraction, stripping, and separation. Low acid concentration (0.5–5 g H_2SO_4 per liter of solution) favors extraction, whereas stripping of copper from the loaded organic solvent is favored at high acid concentration (150–200 g H_2SO_4 per liter). Alternatively, aldoximes modified with ester can also be used. The formation of the copper complex with two oxime molecules is shown in Fig. 11.12. In the exchange reaction (11.139), RH is the extractant, aldoxime, or ketoxime. In aldoxime, R stands for C_9H_{19} or $C_{12}H_{25}$ and A stands for H, whereas in ketoxime, R is C_9H_{19} and A is CH_3 (Fig. 11.12).

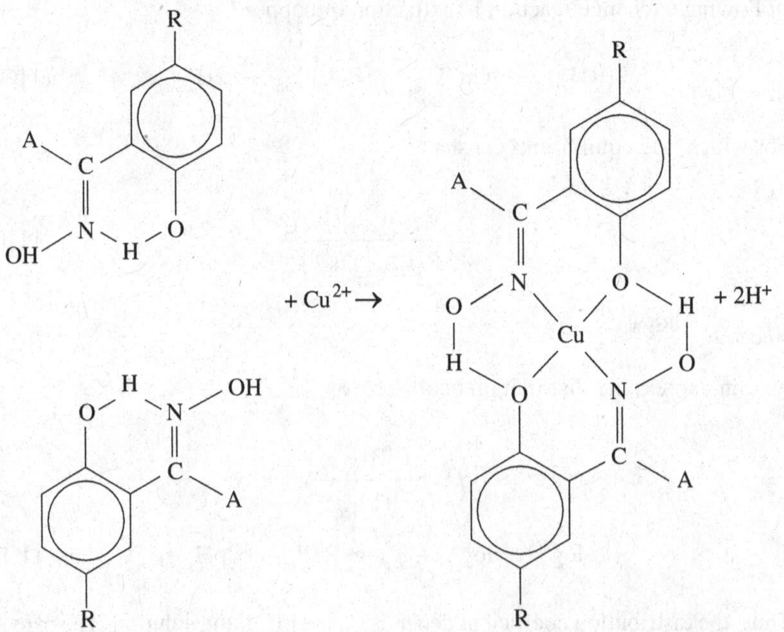

Fig. 11.12 Formation of copper complex with two oxime molecules (From *Extractive Metallurgy of Copper* by W. G. Davenport et al. [5], © 2002, p 310, Elsevier Science Ltd. Reproduced with the permission of Elsevier Science Ltd)

2. *Additive product formation:* Readily extractable additive complexes are formed by the replacement of water molecules attached to the metal chelates when a hydrophobic extractant such as tri-butylphosphate (TBP) or tri-*n* butylphosphine oxide is used as the organic solvent. These solvents are characterized by electron donor [45] oxygen atoms, which are helpful in the removal of water molecules during extraction. For example, uranyl ions formed by ionization of uranyl nitrate

during further hydration gets associated with a ring of six oxygen atoms contributed by water molecules. Removal of these water oxygen atoms can be affected by a donor group of extractant to form a neutral molecule. Thus, distribution of metal ions to the organic phase depends on the donor capacity of the extractant; for example, TBP is a strong donor and can replace all the water of hydration, whereas diethyl ether replaces only a part of water bound to ions. The transfer [45] may be represented by the reaction:

$$2RO\cdots \underset{RO}{\overset{RO}{P}} = O + UO_2(NO_3)_2$$

$$\rightarrow \left[RO\cdots \underset{RO}{\overset{RO}{P}} = O \right]_2 UO_2(NO_3)_2 \text{ where R is } C_4H_9$$

3. *Salt formation*: While using chelating (synergistic) agents, the solute metal ions form complexes that are more soluble in the organic phase. Thenoyltrifluoroacetone (TTA) added to TBP has been found to promote extraction of many metals. If both acid and basic functional groups of an organic solvent molecule combine with a metallic ion, a chelate is formed. Acetyl acetone in the enol form on reaction with beryllium ion gives rise to the formation of the following chelate by replacing enolic hydrogen [2].

When sodium tungstate solution is allowed to come into contact with the extractant [47] (7% alamine-336 and 7% decanol dissolved in kerosene), tungstate ions are transferred from the aqueous phase to the organic phase by forming a salt according to the reaction:

$$\left[\frac{n}{2}\left(R_3NH \cdot HSO_4\right)_2\right]_{org} + \left[\left(W_xO_yH_z\right)^{n-}\right]_{aq} \rightarrow \left[\left(R_3NH\right)_n W_xO_yH_z\right]_{org} + \left[n\left(HSO_4^-\right)\right]_{aq}$$

11.4.5.3 Desirable Characteristics of a Solvent and Various Solvents

In order to select a proper solvent, a number of characteristic features [2, 43] are to be kept in mind. A solvent should show a high selectivity, which may be expressed by the separation factor as discussed earlier. The solvent should be immiscible with the solution to be extracted, chemically stable against various chemical reagents (including strong acids/bases, oxidizing/reducing reagents, and α, β, and γ radiations for use with radioactive materials), less viscous to reduce the costs of pumping, and noninflammable and nontoxic from the point of view of in-plant safety. There must be enough difference in densities and proper interfacial tension between the two phases so as to have good contact as well as separation. The solvent should also be readily available in pure state at a reasonable cost and should be capable of easy recovery, for example, by distillation. The solvents of interest are widely known by abbreviated and trade names. They can be divided into four main groups [2, 48]:

1. *Organophosphorus compounds*: This group provides a number of compounds for extraction and purification of metals, such as tri-butyl-phosphate (TBP), tri-octyl phosphine oxide (TOPO), tri-phenyl phosphite (TPP), tetrabutyl ethylene diphosphonate (TBEDP), and di2ethylhexyl phosphoric acid (D2EHPA). Each solvent is effective in a specific medium. For example, TBP and D2EHPA are powerful extractants in HNO_3 and H_2SO_4 media, respectively. The former is less effective in HCl and H_3PO_4 solutions and shows poor extraction through H_2SO_4 media. TBP has been used for the extraction of a number of metals U, Th, Fe, Zr, Hf, Ta, and La. TBP-D2EHPA and TOPO-TBP systems are good examples of synergistic [49, 50] action.

2. *Nitrogen containing organic compounds*: This group consists of amines [51] and ammonium compounds [52] (mol wt > 200) such as tricapryl amine (Alamine-336), methyl trioctylamine (Aliquat 336), α hydroxyoxime (LIX-63), triiso-octylamine (TIOA), methyldioctylamine (MDOA), amberlite, ethylenediamine tetra acetic acid (EDTA), and diethylene triamine penta acetic acid (DTPA). Tertiary amines are more powerful than primary and secondary amines. A number of common metals such as Co, Ni, and Zn have been successfully extracted through HCl media by the use of either quaternary ammonium compounds (trialkyl monomethyl ammonium chlorides) or tertiary amines. Hydroxyquinoline has been reported to be a powerful synergistic agent in this

group. Hydroxyoximes [53] have been used for the extraction of copper from leach liquor.

3. *Carboxylic acids, soaps, and surface active agents*: These solvents are less effective as compared to organophosphorus compounds, but a number of high-molecular-weight carboxylic acids (mol wt 165–300) have been used for the extraction of common metals [54, 55]. pH plays an important role during extraction with carboxylic acid. Fatty acid soaps [48] diluted in kerosene have been successfully used for the extraction of heavy metals. The use of surface active agents such as polypropylene glycol and high molecular sulfonic acids such as dinonyl naphthalene sulfonic acid [56] has been reported for the extraction of common as well as base metals through sulfate leach liquors.

4. *Miscellaneous group*: Apart from the above three major groups, many other organic solvents [48] are in use, for example, ethers (diethyl ether and isopropyl ether), alcohols (1-octanol. 2-octanol, amyl- and octyl-alcohol), ketones (methyl isobutyl ketone), benzene, toluene, and so on. Dithizone, EDTA, DTPA, TTA, and so on show synergistic action with this group of solvents. Ethyl and isopropyl ether [57] have been used for separation of rare earths.

11.4.5.4 Factors Influencing Extraction

The extraction of an inorganic compound from an aqueous solution by means of an organic solvent is affected by a number of factors listed below:

1. *Salting agent*: An agent that increases the extraction by increasing the distribution coefficient is known as a salting agent. It may be a salt or an acid possessing the same anion as the solute. For example, addition of either HNO_3, $NaNO_3$, or $Mg(NO_3)_2$ favors the extraction of uranyl nitrate by diethyl ether [45].

2. *Nature of anion*: Different solvents may require different media for the extraction of the same metal ion, that is, the extractability is dependent not only on the metal ions but also on the anion to which it is associated in the solution. It is known that in the case of uranyl ion extraction, TBP or diethyl ether functions satisfactorily in a nitric acid medium, whereas D2EHPA requires strictly a sulfuric acid medium (i.e., uranyl sulfate solution).

3. *Oxidizing and reducing agent*: For better extraction, the metal should be in its most soluble state [43]. In some cases, this may be affected by its valence, as in the case of Pu, where Pu^{6+} is more easily extracted as compared to Pu^{3+} by TBP. Thus, extraction may be favored by oxidizing agents while stripping may be favored by reducing agents.

4. *Chelating compounds*: Additives to the organic phase, which enhance extraction by increasing c_o/c_a ratio, are known as synergistic [2] or chelating agents. They form more soluble complexes with metal species. Generally, neutral organophosphorus compounds show synergistic effects when added to acidic organophosphorus compounds, for example, TBP in D2EHPA.

5. *Modifiers*: Additives that increase the solubility of the extractant are known as modifiers [2]. They are of particular use where the complex formed by the

chelating agent is less soluble in the extractant resulting in a three-phase system. High-molecular-weight alcohols (e.g., capryl alcohol, isododecanol) are often used as modifiers.

6. *Sols and cruds formation*: Solvent extraction has been described as a colloidal phenomenon [58]. Salting out agents promote extraction by the formation of "sols," provided they are easily dispersed in the organic phase; however, if these "sols" segregate to form nonmiscible "cruds" or "emulsion," the extraction is hindered.

7. *pH*: When there are two metals of interest, selective extraction and stripping may be achieved by variation of pH. For example, in the case of uranium and vanadium extraction [13], the former is extracted at a pH of 1.0, whereas the latter at 1.7. Better stripping may be achieved at a particular pH.

8. *Phase ratio*: Phase ratio [2, 13] is defined as the ratio of volume of organic phase to the volume of aqueous phase. This ratio is adjusted according to the solvent loading capacity.

11.4.5.5 Kinetics of Solvent Extraction

Theories regarding the kinetics of solvent extraction are still controversial. In the case of chelating extractants, one group considers a completely interfacial reaction process, while another assumes the reaction to occur in the aqueous phase adjacent to the interface. In fact, no single theory can satisfactorily explain all the transfer mechanisms. Despite these drawbacks, the classical two-film theory has been successful in explaining the factors affecting the rate of extraction.

When two immiscible phases (aqueous phase containing metal ions and organic phase containing the extractant) are allowed to come into contact (not in equilibrium), the rate of transport of metal ions from the aqueous phase (a) to organic phase (o) is proportional to the concentration difference and the area of contact, A, of the two phases and inversely proportional to the thickness of the boundary layer or film thickness, δ. Transfer of metal from phase (a) to phase (o) via two films is shown in Fig. 11.13. If c, D, and δ stand, respectively, for concentration, diffusion coefficient, and film thickness with appropriate subscripts as indicated in the figure, the rate of mass transport in the aqueous phase may be expressed according to the Fick's law as:

$$\left(\frac{dm}{dt}\right)_a = D.A. \frac{(c_a - c_{ai})}{\delta_a} \tag{11.142}$$

Denoting $D/\delta_a = k_a$ as the film mass-transfer coefficient in the aqueous phase, we can write:

$$\left(\frac{dm}{dt}\right)_a = k_a.A(c_a - c_{ai}) \tag{11.143}$$

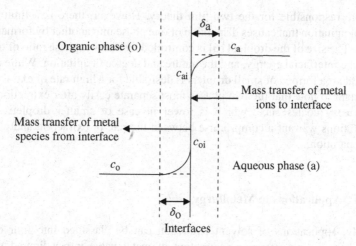

Fig. 11.13 Transfer of metal ions via two films (From *Hydrometallurgical Extraction and Reclamation* by E. Jackson [16], © 1986, p 117, Ellis Horwood Ltd. Reproduced with the permission of Ellis Horwood Ltd)

The metal ion produces a new species in the organic phase (o) by undergoing reaction at the interface. The new species is then transferred in phase (o) away from the interface at a rate given by

$$\left(\frac{dm}{dt}\right)_o = D.A. \frac{(c_{oi} - c_o)}{\delta_o} = k_o A (c_{oi} - c_o) \qquad (11.144)$$

The two Eqs. 11.142 and 11.144 can be equated at steady state:

$$\left(\frac{dm}{dt}\right)_a = \frac{D_a A}{\delta_a}(c_a - c_{ai}) = \left(\frac{dm}{dt}\right)_o = \frac{D_o A}{\delta_o}(c_{oi} - c_o) \qquad (11.145)$$

The above equations clearly demonstrate that the rate of extraction can be enhanced by increasing the interfacial area and decreasing the film thickness. Agitation of the fluids will increase the interfacial area by creating a more intimate dispersion of the two phases as well as decrease the film thickness on both sides of the interface. The phase ratio and temperature are among the other factors that affect extraction. The former controls the interfacial area, and the latter changes the diffusion coefficient.

The two-film theory is based on molecular diffusion under laminar flow conditions. On the other hand, introduction of agitation generates turbulent flow conditions where eddy diffusion dominates. As eddies move rapidly from one position to another, interface will frequently receive solutes carrying eddies, which will rapidly increase the mass transfer. In brief, the agitation by way of introducing turbulent flow conditions will disturb the interface and the concentration gradient, which are responsible for the two-film theory. However, there is a limit to the degree of agitation that causes dispersion of one phase into another by formation of droplets. The size of the droplet [16] is controlled by density, phase ratio of the two phases, the interfacial energy, and the nature and degree of agitation. While a large interfacial area formed of small droplets is desirable for a high rate of extraction, it is important to note that the two phases must separate easily after extraction. This requires easy coalescence, which is lower in case of smaller droplets. These considerations warrant a compromise between the rate of extraction and efficient phase separation.

11.4.5.6 Applications in Metallurgy

Generally, applications of solvent extraction can be classified into four distinct groups, namely, concentration (extraction of metal value into a limited volume from a large volume of aqueous solution), separation (of elements chemically similar in nature), extraction (recovery of metal value from solutions obtained by leaching of ores), and purification (to obtain a purer starting material for further metallurgical operation). Many a time, extraction, concentration, separation, and purification may be all achieved simultaneously. Some of the typical applications of solvent extraction in the field of extractive metallurgy are discussed briefly in this section.

1. *Extraction of uranium*: Solvent extraction is most widely used in the uranium industry [13, 59, 60]. Uranium occurs essentially as pitchblende ($x\mathrm{UO}_2 \cdot y\mathrm{UO}_3$), carnotite ($\mathrm{K}_2\mathrm{O} \cdot 2\mathrm{UO}_3 \cdot \mathrm{V}_2\mathrm{O}_5 \cdot 3\mathrm{H}_2\mathrm{O}$), autunite ($\mathrm{CaO} \cdot 2\mathrm{UO}_3 \cdot \mathrm{P}_2\mathrm{O}_5 \cdot 8\mathrm{H}_2\mathrm{O}$), and turbernite ($\mathrm{CuO} \cdot 2\mathrm{UO}_3 \cdot \mathrm{P}_2\mathrm{O}_5 \cdot 8\mathrm{H}_2\mathrm{O}$). Two flowsheets are generally adopted for extraction: one involving sulfuric acid leaching, ion exchange purification, precipitation as diuranate, and purification of the compound by solvent extraction (see Appendix A.7); and the other comprising pressure leaching with sodium carbonate solution, precipitation as diuranate, and further purification of this compound by solvent extraction. Solvent extraction may be employed along with ion exchange as indicated above, or alternatively, the solvent can be directly contacted with the pulp obtained by acid leaching. The compound (diuranate) obtained by either method contains a number of impurities such as vanadium, aluminum, molybdenum, titanium, boron, cerium, iron, and magnesium. Removal of these impurities is essential in order to get nuclear-grade uranium, and the same is possible by the use of solvent extraction.

   Various extractant used for this purpose are TBP, diethyl ether, D2EHPA, and amines. TBP is most widely used because of its cheapness, stability against concentrated nitric acid, and safety considerations. A 20–35% TBP dissolved in kerosene is the commercial solvent. HNO_3, NaNO_3, or $\mathrm{Mg(NO}_3)_2$ may be used as salting agent. From the loaded solvent, uranium ions are stripped into dilute nitric acid in a separate column at 60 °C. This elevated temperature stripping decreases the distribution coefficient and allows a lower aqueous to organic phase ratio, thereby giving a more concentrated solution requiring less subsequent evaporation. For the treatment of ores containing vanadium (e.g., carnotite ore of Colorado [2] plateau), use of D2EHPA has been found to be suitable. A 10% D2EHPA and 2.5% TBP in kerosene extracts uranium in the first cycle. In the second cycle, vanadium is extracted from the raffinate by the use of a different proportion of the solvent. Sodium carbonate solution and 10% dilute sulfuric acid are used for stripping of uranium and vanadium, respectively.'

2. *Extraction of thorium*: The most important mineral of thorium is monazite (complex phosphate of thorium and rare earths). Important steps in thorium extraction are sulfuric acid or sodium hydroxide breakdown of monazite, digestion, selective precipitation, and purification to separate thorium from uranium and rare earths [13, 61]. The feed is obtained by dissolving thorium oxalate or hydroxide (resulting from the previous operations) in nitric acid. Alternatively, feed solution is obtained by direct nitric acid leaching of thorite ($ThSiO_4$) ore. TBP dissolved in xylene is the suitable extractant for this purpose. All thorium deposits contain a small amount of uranium, and it is, therefore, essential and useful to remove and recover uranium in the first stage of extraction. Advantage is taken of the fact that uranium has a much higher [13] distribution coefficient than thorium (e.g., 20 and 0.5 for uranium and thorium, respectively, for 40% TBP in xylene; and 6 and 0.04, respectively, for 5% TBP in xylene). Thus, the more dilute solution having a higher separation factor is suitable for uranium extraction, whereas thorium is extracted by 40% TBP in xylene in the second cycle (see Appendix A.8). Both uranium and thorium are stripped by 0.02 M nitric acid. Cerium is kept in an unextractable state by adding hydrogen peroxide to the feed.

3. *Extraction of cesium*: A solvent extraction process for the recovery of cesium from pollucite ore ($Cs_2O \cdot Al_2O_3 \cdot 4SiO_2$) has been developed [2, 57, 62]. The ore is leached in some mineral acid. BAMBP [4-sec-butyl-2(α-methylbenzyl) phenol] dissolved in kerosene extracts cesium ions from the leach liquor by the exchange of alkaline cesium with the acidic hydrogen of phenolic group as represented by the general equation:

$$aCs_{aq}^+ + bRH_{org} \rightleftarrows [Cs_aR_bH_{b-a}]_{org} + aH_{aq}^+ \tag{11.146}$$

where RH is a substituted phenol and subscripts "aq" and "org" refer to the aqueous and organic phases, respectively. The complex Cs-BAMBP formed by such an exchange is more soluble in kerosene than in the aqueous phase. Subsequently, cesium is stripped into strong acid and recovered as cesium salt by evaporation and crystallization.

4. *Extraction of boron*: Boric acid can be obtained by the solvent extraction technique from certain low-grade brines containing approximately 1% of $Na_2B_4O_7$ using aliphatic or aromatic polyols dissolved in kerosene as the extractant [62]. A complex of borate-didiol chelate anion with an alkali cation is formed when the alkaline brine is brought into contact with this solvent. The borate is recovered by stripping the loaded solvent with dilute sulfuric acid.

5. *Extraction of beryllium*: Beryllium is usually extracted from sulfuric acid leach liquor of beryl ($3BeO\cdot Al_2O_3\cdot 6SiO_2$) [62]. A number of organic solvents TBP, D2EHPA, PDA, MIBK, and Methyl-p-*tert*-octyl-phenylphosphate have been found to be suitable. Isodecanol may be used as a synergistic agent while extracting with D2EHPA. Sodium hydroxide is a suitable stripping agent. BeO obtained by heating the final liquor (obtained by this technique) is about 99% pure and cheaper than that produced by the conventional method, which gives a purity of only 92–95%.

6. *Extraction of copper*: About 80% of primary copper is produced via the pyrometallurgical route incorporating steps like concentration, smelting, and electrorefining and the remaining 20% is obtained through the hydrometallurgical route based on a combination of leaching, solvent extraction, and electrowinning [63–65]. The leach liquor (containing 1–3 g copper and 1–10 g iron per liter of solution) obtained from the leaching circuit is very dilute and impure. In order to get smooth, dense, and high-purity cathode copper for industrial use, the electrolyte must contain more than 30 g copper per liter of the solution. Solvent extraction offers attractive possibilities in this respect since it has the ability to concentrate very dilute leach liquor to a level needed for electrolysis and at the same time yields very pure electrolyte for electrowinning.

Currently, most copper hydrometallurgical plants use the recently developed mixture of aldoximes and ketoximes [5]. This combination fulfills all the desirable characteristics of organic solvents (as mentioned in Sect. 11.4.5.3) as it is capable of faster and efficient transfer of copper ions from the leach liquor to the organic phase (extraction step) and from the loaded organic phase to the aqueous electrolyte (stripping). Aldoximes–ketoximes solvent combination not only extracts copper in preference to iron and manganese but also prevents the transfer of harmful species from the leach liquor to the electrolyte. Copper is stripped into a concentrated sulfate–sulfuric acid solution.

7. *Uranium, plutonium, and fission product separation*: As U-235 burns in a natural uranium reactor, there is an accumulation of fission product elements that absorb thermal neutrons and thereby hinder the chain reaction. Also, there is valuable plutonium formed and a large proportion of uranium remains unused. Hence, spent fuel elements are treated for the purification and recovery of plutonium and also some of the radioactive fission products (like Sr-90, Ce-144, Pm-147, and Cs-137) that are useful for agricultural, industrial, and space research applications. In breeder reactors, thorium blankets are treated similarly for the separation and recovery of valuable U-233, some of the fission product elements formed, and unused thorium. In all these cases of spent fuel treatment, solvent extraction [57, 66, 67] has played an important role.

Spent fuel from a nuclear reactor is dissolved in nitric acid in the presence of some oxidizing agent in order to bring plutonium into a more soluble state (Pu^{6+}). The feed is subjected to TBP extraction when uranium and plutonium are extracted, and any fission product going to the organic phase is scrubbed into the aqueous phase by means of an aqueous solution of sodium nitrate. In the next stage, plutonium is stripped into an aqueous phase containing a reducing and a salting agent. Any uranium coming to the aqueous phase is re-extracted with a small quantity of the organic solvent in the scrub section of the plutonium column. Lastly, pure uranium from the organic phase is stripped into distilled water.

Recovery, separation, and purification of other fission products have been achieved by pH control as well as by the use of different solvents and strip solutions. D2EHPA has also been found useful for the recovery of both Sr and Ce, whereas Ce is recovered from the first cycle raffinate after oxidation. Pm and Cs can be separated by the use of TBP and a substituted phenol, respectively.

8. *Zirconium–hafnium separation*: Zirconium and hafnium are chemically similar elements and occur together in nature. Since the reactor-grade zirconium must contain less than 200 ppm of hafnium, they are separated by the solvent extraction technique. The feed is prepared by dissolving the product (obtained by fusion of zircon with concentrated sodium hydroxide) in nitric acid so as to have 126 g per liter of Zr-Hf in solution [13] with the acidity of 13.5 N. Most commonly used extractant for this purpose is 50% TBP in xylene. Both Zr and Hf are extracted into the organic phase. Hf is stripped into 5 N nitric acid. Later on, the solvent containing only zirconium is treated with water to bring zirconium back into aqueous phase. Separation can also be effected by the use of MIBK through a hydrochloric acid medium containing ammonium thiocyanate.

9. *Niobium–Tantalum separation*: Niobium and tantalum have similar chemical properties and their separation by fractional crystallization is a tedious process [13]. This has been greatly simplified by solvent extraction technique. MIBK and TBP are the principal extractants for this purpose. Methyl di-octyl amine and di-isopropyl ketone can also be used. Extraction is carried out through a hydrofluoric acid medium containing sulfuric acid. Tantalum is extracted into MIBK in four extractions and two stripping stages (2 N HF stripping agent). Niobium remaining in the raffinate is precipitated as the double fluoride K_2NbF_7 by adding a solution of K_2CO_3 and HF. Tantalum is recovered from the loaded solvent as sodium tantalate by using a solution of 5% Na_2CO_3.

10. *Cobalt, iron, nickel, and copper separation*: Nickel and copper occur together in nature along with iron and cobalt. Since their separation is difficult by conventional pyrometallurgical and hydrometallurgical methods, attention has been paid to solvent extraction techniques [68]. These elements can be separated in stages by using different extractants. Copper is removed from nickel by the use of naphthenic acid, whereas cobalt and nickel are separated by TBP or D2EHPA. Carboxylic acid and tertiary amine have been used to affect the separation of iron and cobalt. Further, iron and cobalt can be separated from nickel in hydrochloric acid solution by means of tri-octyl amine.

11. *Extraction of tungsten*: In order to produce pure yellow tungstic oxide, WO_3 (for the production of pure tungsten metal), the sodium tungstate solution after removal of silica, phosphorus, arsenic, iron, and molybdenum is further purified by the solvent extraction process [47]. The solution containing 5 g WO_3 per liter of solution is contacted with the extractant (comprising 7% alamine-336, 7% decanol, and 86% kerosene, by volume). During countercurrent flow of the two

phases, tungstate ions are transferred from the aqueous phase to the organic phase. The extraction is pH dependent and sufficiently enhanced in the rage of 1–3. The organic phase (loaded with 40 g WO_3 per liter) leaving the extraction column is scrubbed with de-ionized water to remove contaminated sodium ions. Tungsten from the loaded organic phase is stripped into ammonia solution (5% NH_3). The presence of free ammonia in the aqueous phase at 60 °C prevents the precipitation of ammonium paratungstate (APT) in the stripping circuit. The organic phase after stripping is recirculated for extraction, and the ammonium paratungstate solution is sent to the evaporator for crystallization of APT. APT is decomposed to yellow oxide (WO_3) at above 250 °C in a furnace under the flow of air (see Appendix A.5).

12. *Extraction of gold*: Solvent extraction technique has not been successfully employed on industrial scale in the recovery of gold from the cyanide leach liquor. However, in recent years, there have been serious attempts to find a suitable organic solvent for extraction of gold from solutions. As gold forms anionic complexes with thiocyanate solutions [69], alamine 336 [70] was considered as a suitable extractant for the extraction of the anionic auric complex, $Au(SCN)_2^-$. The kinetics of extraction was fast and the two phases could be easily separated. Li et al. [70] have reported that the extraction of gold with about 98% recovery was independent of alamine 336 concentration at a molar ratio of alamine 336/gold above 10. However, it was dependent on alamine 336 concentration at lower ratios. Acidic thiourea was found to be most effective solution for stripping gold from the loaded amine extractant [70].

13. *Extraction of zinc*: In order to produce special high-grade zinc from the sulfate, leach liquor obtained after jarosite precipitation is further purified by solvent extraction. Zinc is first extracted with di-2-ethyl hexyl phosphoric acid and then stripped into the spent sulfate electrolyte (from the electrolytic plant) from the loaded organic phase [71] to procure a strong electrolyte containing 90 g Zn per liter of the solution.

14. *Miscellaneous uses*: Waste solutions from some industries contain relatively valuable materials in very dilute concentrations. Recovery by conventional methods is not economical. Solvent extraction can be used for the treatment of such wastes. In this context, recovery of isotopes formed during nuclear fission and concentration of mine wastes, particularly for copper recovery, are note-worthy. Solvent extraction can also be used for purification of electrolytic solutions used for electroplating, electrowinning, and electrorefining.

11.5 Recovery of Metals from Leach Liquor

Chemical precipitation, cementation, gaseous reduction, and electrowinning are four important methods for recovery of metals from solutions. Electrowinning will be discussed along with electrorefining in Chap. 12 on electrometallurgy.

11.5.1   Precipitation of Metal Sulfides

Chemical precipitation as a means of purification has been discussed in Sect. 11.4.1. In extraction via the hydrometallurgical route, precipitation plays an important role in the recovery of metals from leach liquors. A number of metals can be precipitated as sulfides by introducing hydrogen sulfide gas in the solution. Some sulfides can be selectively precipitated by adjustment of pH. Solubility of H_2S in water depends on the temperature and partial pressure of H_2S gas in equilibrium with water. After dissolution, H_2S undergoes dissociation in water according to the following two reactions:

$$H_2S \text{ (g)} = H^+ + HS^- \tag{11.147}$$

$$K_{147} = \frac{a_{H^+} . a_{HS^-}}{p_{H_2S}} = 1.02 \times 10^{-7} \tag{11.148}$$

and

$$HS^- = H^+ + S^{2-} \tag{11.149}$$

$$K_{149} = \frac{a_{H^+} . a_{S^{2-}}}{a_{HS^-}} = 1.66 \times 10^{-14} \tag{11.150}$$

K_{147} increases with temperature and attains a maximum value [16] ($= 2.8 \times 10^{-7}$) at about 90 °C and decreases thereafter. Solubility of H_2S in water increases with partial pressure of the gas in equilibrium with water which may be represented as:

$$H_2S(g) \rightleftharpoons H_2S(aq) \tag{11.151}$$

$$K = \frac{a_{H_2S \text{ (aq)}}}{p_{H_2S}} = \frac{[H_2S]_{aq} . [f_{H_2S}]_{aq}}{p_{H_2S}} \tag{11.152}$$

where $[H_2S]_{aq}$ and $[f_{H_2S}]_{aq}$ represent concentration and activity coefficient of H_2S, respectively, in the aqueous phase. At 1 atm pressure and at 25 °C, water gets saturated [72] with H_2S by dissolving approximately 0.09 mol dm^{-3} (i.e., 0.09 M).

In dilute solution, activity may be assumed to be equal to concentration; hence, from Eqs. 11.148 and 11.150, we get:

$$K_{147} . K_{149} = \frac{[H^+]^2 . [S^{2-}]}{p_{H_2S}} = 1.69 \times 10^{-21} \tag{11.153}$$

Under standard conditions, $p_{H_2S} = 1$ atm

$$\therefore \left[S^{2-}\right] = \frac{1.69 \times 10^{-21}}{[H^+]^2} \qquad (11.154)$$

$$\text{or } \log\left[S^{2-}\right] = \log\left(1.69 \times 10^{-21}\right) + 2pH \qquad (11.155)$$

Since both the ionization constants K_{147} and K_{149} increase with temperature up to about 90 °C and attain respective values of 2.8×10^{-7} and 1×10^{-20}, and solubility of H_2S in water decreases from 0.09 M at 25 °C to 0.043 M at about 90 °C, Eq. 11.155 is modified [16] as:

$$\log\left[S^{2-}\right] = \log\left(1.2 \times 10^{-20}\right) + 2pH \qquad (11.156)$$

Thus, the sulfide ion concentration depends on the pH of the solution in addition to p_{H_2S} and temperature. But pH has a far greater effect as compared to partial pressure and temperature. Blowing H_2S at higher pressure in the leach liquor increases the working temperature by about 100 °C, which increases the rate of sulfide precipitation. The sulfide ion concentration is affected by the presence of other solutes in solution. Hence, the combined effect of variables, namely, pH, p_{H_2S}, and temperature may be optimized to achieve selective precipitation of sulfides of the metal ions present in the leach liquor. From Eq. 11.155, it is possible to calculate the pH at which a particular metal sulfide may be precipitated or else the thermodynamic feasibility of precipitation at a given pH may be analyzed (see Problem 11.3).

11.5.1.1 Kinetics of Precipitation of Sulfides

Formation of a metal sulfide, MS, by reaction between metal ions, M^{2+}, present in the aqueous solution and H_2S gas dissolved takes place in the following three steps:

Step 1: Dissolution of H_2S gas in aqueous solution (reaction 11.151) after transfer from the gaseous phase:

$$H_2S(g) \rightleftharpoons H_2S(aq)$$

Step 2: Generation of S^{2-} ions by dissociation of the dissolved H_2S according to reactions (11.147 and 11.149):

$$H_2S(g) = H^+ + HS^-$$

and

$$HS^- = H^+ + S^{2-}$$

Step 3: Formation of a metal sulfide:

$$M^{2+} + S^{2-} = MS \qquad (11.157)$$

Hence, the overall reaction in sulfide formation is:

$$M^{2+} + H_2S = MS + 2H^+ \qquad (11.158)$$

The reaction between the divalent metal ions, M^{2+}, and the sulfide ions, S^{2-}, seems to be homogeneous and expected to obey the second-order reaction kinetics. But when H_2S gas is quickly supplied and transferred at a constant partial pressure in the leach liquor at equilibrium under steady-state conditions, there is a likelihood of attaining a constant S^{2-} ion concentration in the gaseous phase. Under such a condition of constant S^{2-} ion concentration, reaction (11.157) becomes unimolecular and follows the first-order reaction kinetics.

The precipitation of a metal sulfide can be optimized by simultaneously controlling the operating variables, namely, pH, temperature, and pressure of the H_2S gas. NiS can be precipitated at 120 °C from a solution at a pH of 3 by blowing H_2S gas at 3 atm pressure. However, there is no precipitation at this pH on introducing the gas at 1 atm pressure and 25 °C. NiS can be precipitated from such a solution at 25 °C by blowing H_2S at 7 atm in the presence of catalytic nickel powder. Precipitation can be achieved without any catalyst by increasing the temperature to 120 °C. It has been reported [72] that initially rapid precipitation of sulfide takes place homogeneously, but it slows down due to the increase in acidity (generation of H^+ ions on formation of the sulfide, reaction 11.158), which in turn reduces sulfide ion concentration. At this stage, catalytic action is required for precipitation to continue further. The rate of sulfide precipitation decreases as it becomes diffusion controlled.

11.5.1.2 Sulfide Precipitation Practice

Sulfide precipitation has been employed for purification and recovery, particularly in the large-scale production of nickel and cobalt. From the ammoniacal leach liquor obtained in the Sherrit Gordon pressure leaching process (Sect. 11.1.1.5, Appendix A.3b), copper is completely removed at an early stage of the flow sheet. The leach liquor containing metal amines and ammonium sulfate and sulfamate is boiled to decompose trithionate to precipitate copper as sulfide. Evolved ammonia is recycled to the leaching circuit and copper sulfide is filtered off. In the next stage of the process (Appendix A.3b), the resultant solution is treated at 170 °C by blowing hydrogen gas at 30 atm to precipitate nickel powder. After filtration of the nickel powder, the filtrate contains 95% Co. Cobalt along with the remaining nickel is precipitated with H_2S. The mixture containing precipitates of sulfides of cobalt and nickel is leached with ammonia in the presence of air and separated into cobalt and nickel salts. Nickel salt is recycled to the nickel precipitation circuit, and the cobalt compound is finally precipitated as cobalt metal by hydrogen under pressure.

The cobalt concentrate after sulfatizing roast is leached with water in the Kokkila cobalt plant of the Outokumpu Company of Finland. The leach liquor obtained at a pH of 1.5 is first treated directly with H_2S at atmospheric pressure to precipitate CuS. After removal of CuS, zinc and nickel are selectively precipitated as sulfides. Finally, cobalt is recovered by hydrogen reduction under pressure from the ammoniacal solution.

11.5.2   Cementation

Precipitation of a metal from an aqueous solution of its salts by another metal is known as cementation because the precipitated metal is usually cemented on the metal introduced in the solution. It is based on the principle of electrochemical reduction of a metal ion from solution by adding a less noble metal in the solution or by passing the solution over less noble (more anodic) metal. For example, copper is precipitated from solution containing Cu^{2+} ions by adding zinc powder which dissolves anodically due to the difference in their standard electrode potentials:

$$Cu^{2+} + 2e = Cu; E^o = +0.337 \text{ V}$$

$$Zn^{2+} + 2e = Zn; E^o = -0.763 \text{ V}$$

Since Zn^{2+}/Zn system has more negative potential, zinc will dissolve anodically, while copper ions will get reduced by the cathodic reaction as shown below:

Anodic dissolution : $Zn = Zn^{2+} + 2e$ (11.159)

Cathodic precipitation : $Cu^{2+} + 2e = Cu$ (11.160)

Overall cell reaction : $Zn + Cu^{2+} = Zn^{2+} + Cu$ (11.161)

The potential for the cell: Zn^{2+}/Zn : Cu^{2+}/Cu is obtained as $E^o_{cell} = 0.337 - (-0.763) = 1.10 \text{ V}$.

The standard free energy change for the overall cell reaction is calculated as $\Delta G^o = -zFE^o_{cell} = -212.3 \text{ kJ mol}^{-1}_{Zn}$ ($-50.7 \text{ kcal mol}^{-1}_{Zn}$). The high negative value of ΔG^o confirms the thermodynamic feasibility of precipitation of copper by zinc from the solution.

But the cell potential rapidly falls as reaction proceeds due to the increase of zinc concentration with precipitation of the metallic copper. After achieving equilibrium, the potentials corresponding to the two half cells (Zn^{2+}/Zn and Cu^{2+}/Cu) become equal, and hence, at 25 °C, we have:

$$E_{Cu}^o + \frac{0.0591}{2} \log a_{Cu^{2+}} = E_{Zn}^o + \frac{0.0591}{2} \log a_{Zn^{2+}}$$

$$\text{and, thus, we get } \frac{a_{Cu^{2+}}}{a_{Zn^{2+}}} = 7 \times 10^{-38} = \frac{[Cu^{2+}]}{[Zn^{2+}]}$$

The extremely low ratio shows that the copper ion concentration left in the solution is going to be extremely low. Hence, copper can be recovered very effectively by cementation with zinc.

For the cell with systems having closer E_M^o values, for example, Cu^{2+}/Cu (0.337 V) and Ni^{2+}/Ni (−0.25 V), equilibrium is established at higher ratio of activities/concentrations. Hence, the Cu–Ni system with $\frac{a_{Cu^{2+}}}{a_{Ni^{2+}}} = 10^{-20} = \frac{[Cu^{2+}]}{[Ni^{2+}]}$ will offer lesser selectivity. The ratio further indicates that if $a_{Cu^{2+}} < 10^{-20}.a_{Ni^{2+}}$ the electron flow will be in reverse direction that is from Cu to Ni. Under these conditions, copper will act as anode and nickel as the cathode.

The above cases refer to simple cationic systems, and the electrode potential may change where complex ions are involved. The recovery of gold from cyanide leach liquor by zinc powder according to the reaction:

$$2Au(CN)_2^- + Zn = 2Au + Zn(CN)_4^{2-} \tag{11.162}$$

involves the following two electrode systems:

$$Au(CN)_2^- + e = Au + 2CN^-; E^\circ = -0.67 \text{ V}$$

$$\text{and } Zn(CN)_4^{2-} + 2e = Zn + 4CN^-; E^\circ = -1.26 \text{ V}$$

The resulting cell potential for the reaction (11.162) would be 0.59 V. From the standard electrode potentials of some metals listed in Table 11.1, it is concluded that chances of precipitation of impurity elements increase with increase of the cell potential. Thus, lead, tin, nickel, and cobalt present in the solution are likely to contaminate the precipitated copper. In the recovery of platinum and palladium [16] by zinc, the likelihood of precipitation of impurity is even more. Thus, cementation happens to be a less selective process compared to precipitation because reduction of all the more noble metals present in the solution will continue until all the less noble metals are dissolved or covered with the precipitated metals. However, despite these problems, selective precipitation has been practiced. Silver has been selectively recovered from spent copper contaminated silver electrolytes by reduction with copper: $2Ag^+ + Cu = 2Ag + Cu^{2+}$. Similarly, copper can be cemented with nickel powder in electro-winning of nickel: $Cu^{2+} + Ni = Cu + Ni^{2+}$. Copper is recovered on a commercial scale from leach liquor obtained by dump and heap leaching by cementation with iron scrap.

Table 11.1 Standard electrode potential of some metal ion/metal systems

Electrode system	$E°$ (V)
$Au^{3+} + 3e = Au$	+1.45
$Pt^{2+} + 2e = Pt$	+1.20
$Pd^{2+} + 2e = Pd$	+0.92
$Ag^+ + e = Ag$	+0.80
$Cu^{2+} + 2e = Cu$	+0.34
$2H^+ + 2e = H_2$	0.0
$Pb^+ + 2e = Pb$	−0.13
$Sn^{2+} + 2e = Sn$	−0.14
$Ni^{2+} + 2e = Ni$	−0.25
$Co^{2+} + 2e = Co$	−0.28
$Cd^{2+} + 2e = Cd$	−0.40
$Fe^{2+} + 2e = Fe$	−0.44
$Fe^{3+} + e = Fe^{2+}$	+0.77
$Zn^{2+} + 2e = Zn$	−0.76
$Zr^{4+} + 4e = Zr$	−1.53
$Al^{3+} + 3e = Al$	−1.66
$Be^{2+} + 2e = Be$	−1.85
$Th^{4+} + 4e = Th$	−1.90
$Mg^{2+} + 2e = Mg$	−2.37
$Na^+ + e = Na$	−2.71
$Ca^{2+} + 2e = Ca$	−2.87

11.5.2.1 Kinetics of Cementation

Cementation being the oldest process among all the methods of recovery of metals from solutions, it has been studied most extensively. Habashi [2] has given a very simple approach to the kinetics of cementation. The rate of change of concentration of ions of the more noble metal due to its precipitation by introduction of a less noble metal is given by:

$$-\frac{dc}{dt} = k\frac{A}{v}c \tag{11.163}$$

where c, k, A, and v stand for concentration of the more noble ions, rate constant, surface area of the metal in contact with the solution, and volume of the reacting solution, respectively. On integration, we get:

$$k = -\frac{v}{A} \cdot \frac{1}{t} \ln \frac{c_2}{c_1} \tag{11.164}$$

where c_1 is the initial concentration of the more noble metal and c_2 is the concentration at time t. The process may be diffusion controlled or chemically controlled. For the cementation of M_1 by M_2, we have:

$$M_1^{2+} + M_2 = M_1 + M_2^{2+} \tag{11.165}$$

$$\text{Rate of cementation reaction} = kA\left[M_1^{2+}\right]_i$$

According to the Fick's law, the rate of diffusion

$$= \frac{D}{\delta}A\left\{\left[M_1^{2+}\right] - \left[M_1^{2+}\right]_i\right\} \tag{11.166}$$

where $\left[M_1^{2+}\right]_i$ is the concentration of the metal ions, M_1^{2+} at the interface, shown in Fig. 11.14. $\left[M_1^{2+}\right]$, D, and δ refer, respectively, to the initial metal ion concentration, diffusion coefficient, and thickness of the Nernst boundary layer. Under steady state:
Rate of diffusion = Rate of chemical reaction (cementation) at the interface

$$\therefore \frac{D}{\delta}A\left\{\left[M_1^{2+}\right] - \left[M_1^{2+}\right]_i\right\} = kA\left[M_1^{2+}\right]_i \tag{11.167}$$

$$\text{or } \frac{D}{\delta}\left[M_1^{2+}\right] = \left[M_1^{2+}\right]_i\left(k + \frac{D}{\delta}\right)$$

$$\text{or } \left[M_1^{2+}\right]_i = \frac{D/\delta}{k + D/\delta}\left[M_1^{2+}\right] \tag{11.168}$$

$$\therefore \text{Rate of cementation at the surface area}, A = \frac{k.D/\delta}{k + D/\delta}A\left[M_1^{2+}\right] \tag{11.169}$$

Figure 11.14 refers to a case where the metal precipitated does not adhere to the cementing metal. When $k << \frac{D}{\delta}$, rate $= k.A\left[M_1^{2+}\right]$, that is, process is chemically controlled; when $k >> \frac{D}{\delta}$, rate $= \frac{D}{\delta}A\left[M_1^{2+}\right]$, that is, the process is diffusion controlled. For example, the precipitation of copper by iron in chloride solution (Cu^{2+} + Fe = Cu + Fe^{2+}) is a diffusion controlled process with the activation energy of 3 kcal mol^{-1}, whereas cementation of lead by iron in chloride medium is chemically controlled with an activation energy of 12 kcal mol^{-1}. The rate of cementation also depends on the metal employed, for example, precipitation of lead by zinc is much faster than precipitation by iron.

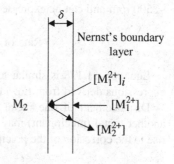

Fig. 11.14 Schematic representation of cementation process (From *Principles of Extractive Metallurgy*, Vol. 2, Hydrometallurgy by F. Habashi [2], © 1970, p 230, Gordon & Breach. Reproduced with the permission of the author, F. Habashi)

Cementation of copper by iron has been studied most extensively. However, the precipitation is complicated due the two oxidation states of iron. Involvement of side reactions increases iron consumption as per the following reactions:

$$Cu^{2+} + Fe = Cu + Fe^{2+} \tag{11.170}$$

$$Fe + \tfrac{1}{2}O_2 + 2H^+ = Fe^{2+} + H_2O \tag{11.171}$$

$$Cu + \tfrac{1}{2}O_2 + 2H^+ = Cu^{2+} + H_2O \tag{11.172}$$

$$2Fe^{2+} + \tfrac{1}{2}O_2 + 2H^+ = 2Fe^{3+} + H_2O \tag{11.173}$$

$$2Fe^{3+} + Fe = 3Fe^{2+} \tag{11.174}$$

$$Fe + 2H^+ = Fe^{2+} + H_2 \tag{11.175}$$

$$Cu + 2Fe^{3+} = 2Fe^{2+} + Cu^{2+} \tag{11.176}$$

From the above equations, it is clear that the presence of oxygen in solution is detrimental. It helps in dissolution of both iron and copper and also oxidizes iron (II) to iron (III). The formation of iron (III) results in wastage of iron. The elimination of oxygen decreases iron consumption. Cementation is more effective in hydrogen atmosphere. pH has virtually no effect on copper precipitation, but iron dissolution increases at lower pH in the later stages and in the post-period of precipitation [16]. Schalch and Nicol [73] have reported that the initial copper ion concentration has a significant effect on the cementation kinetics as well as on the physical properties of the deposit.

Nadkarni and Wadsworth [74] have observed that cementation of copper by iron is a first-order reaction with an activation of energy of 21.3 kJ mol^{-1} (5.09 kcal mol^{-1}). Based on detailed studies on the effect of rate of stirring (n) and concentration [Cu^{2+}], they arrived at the following rate equation:

$$\text{Rate of cementation} = k_o A [Cu^{2+}] n^{1/2} \tag{11.177}$$

where A and k_o are the surface area and modified rate constant, respectively. The rate becomes independent of stirring speed after achieving the limiting speed of ~2500 rpm and can be expressed as:

$$\text{Rate of cementation} = k_o A [Cu^{2+}] \tag{11.178}$$

Equation 11.178 is similar to one of the special cases of chemically controlled expressions derived from Eq. 11.169.

Displacement of one metal (cemented M_c) from the solution by addition of another metal (precipitant) may be considered as a corrosion process, which gives rise to the corrosion of the precipitant metal. The main cathodic process responsible for anodic dissolution is reduction of the more noble metal ion from the solution. Only a small potential difference develops at the beginning along with the surface of the precipitant metal. Areas with more negative potential behave anodically and tend to dissolve, whereas rest areas become cathodic. The following cathodic reactions are feasible:

$$O_2 + 4H^+ + 4e = 2H_2O \tag{11.179}$$

$$2H^+ + 2e = H_2 \tag{11.180}$$

$$M_c^{z+} + ze = M_c \tag{11.181}$$

The oxygen reduction reaction, which has a high positive potential in acid solutions, occurs preferentially as compared to the reduction of most of the metal ions on dissolution of oxygen. Hence, oxygen should be eliminated as far as possible for efficient cementation.

The extent of competition between the discharge of hydrogen ions (reaction 11.180) and the reduction of metal ions (11.181) depends on the potential of the metal system, pH of the solution, and the hydrogen over voltage on the cemented metal. Noble metals like silver and copper are reduced in preference to hydrogen. Alternatively, for efficient reduction of the metal, the hydrogen over voltage should be large enough to effectively shift the hydrogen ion discharge potential to a more negative value than that of the cemented metal. Increase of pH may be helpful, but hydrolysis must be avoided to maintain the desired rate of cementation. This will also lead to wasteful consumption of the precipitant metal. It is observed that the rate of reduction remains low as cementation starts. However, with reduction in dissolved oxygen together with deposition of some metal, formation of a galvanic cell between the precipitant metal and the cemented metal deposited enhances the rate of cementation. The galvanic cell formed will have a greater cell potential [16] compared to the value listed in Table 11.1 for the particular system. With progress of cementation, the cathodic areas continuously increase in size due to the accumulation of deposited/cemented metal. This increases the overall rate of precipitation. The cementation process is most likely to proceed in the following steps:

(i) Transfer of the cementing metal ions from the solution to the cathode surface/ solution interface
(ii) Transfer across the boundary layer of the metal ions to the cathodic reduction sites
(iii) Chemical reaction involving anodic dissolution of the precipitant metal along with simultaneous cathodic reduction of the cementing metal ions and incorporation of the metal atoms in the lattice or accumulation as a powder (cemented metal)
(iv) Transport of the precipitant metal ions to the cemented metal/solution interface
(v) Transport of the precipitant metal ions from the cemented metal/solution interface into the bulk of the solution

The slowest of the above steps controls the overall rate of cementation. However, in some cases, there may be two equally slow steps categorizing the process under mixed control. Generally, the reaction kinetics of the cementation process is governed by mass transfer in step (i) involving three types of transport, namely, diffusion due to concentration gradient, migration of charge due to cell potential, and convection current. Hydrodynamic conditions prevailing in the system will dictate the contribution of each to the overall transport. It is important to note that step (iv) may not exist in the early stage of the process. Step (iv) may become rate-controlling by the formation of fine, coherent, and adherent deposit as cementation progresses. It may be chemically controlled (step iii) in some cases, such as cementation of cobalt [75] from zinc sulfate solution.

11.5.2.2 Cementation Practice

Cementation occupies a prominent place in recovery of copper from leach liquors obtained from dump and heap leaching. For cementation of copper, inclined open launders containing iron/steel scrap have been extensively used. The leach liquor flowing under gravity makes contact with the scrap. The process is labor intensive because cement copper has to be removed from the scrap resting on the wooden frame placed in the launder. As launders are directly exposed to atmosphere, oxygen pick up by solution increases iron consumption three times the theoretical requirements. The process variables and purity of the iron/steel scrap affect the purity of the cement copper. There have been efforts to improve the efficiency of launder plants by developing reactor vessels to reduce oxygen pick up by solutions and improve mass transfer in the solution. The cone precipitator developed by Kennecott [76] is more popular due to its high volumetric capacity, provision of automation, and ability to treat a variety of solutions. It provides a high degree of turbulence and reduces retention time as compared to launders.

Reid and Biswas [77] were successful in producing high-purity copper (>99.8% pure) for direct metallurgical usage by carrying out cementation at elevated temperatures (95 °C) on 0.25% carbon steel strip under nitrogen atmosphere.

Cementation also plays an important role in the purification of leach liquor generated by leaching of the roasted sphalerite concentrate. After removal of iron by the jarosite process (Sect. 11.4.1), the filtrate is subjected to a series of cementation reactions with zinc dust to remove copper, cadmium, cobalt, and nickel before electrowinning of zinc metal. The operation varies from plant to plant in terms of number of stages, temperature of cementation, and particularly in the use of reagents in removal of cobalt. In some plants, cadmium is cemented by passing the filtrate through a series of five fluidized beds of zinc dust.

11.5.3 Gaseous Reduction

Electrochemical oxidation and reduction have been used for selective precipitation of metals from aqueous solutions. Gaseous reduction using carbon monoxide and sulfur dioxide has been carried out, but hydrogen has been used most widely. The technique employed at Sherritt Gordon Mines demonstrates a remarkable and important development in pressure leaching technology. In this section, the principle of this technique has been discussed briefly.

Metal ions can be selectively precipitated by passing hydrogen gas through the purified leach liquor at controlled pH and activity of the metal ions. Although a number of metal ions may be present in the solution, for simplicity, we consider precipitation of only two metal ions, M^{2+} and Me^{2+}, by a reduction reaction of the type:

$$M^{2+} + 2e \rightarrow M$$

$$\text{and } Me^{2+} + 2e \rightarrow Me$$

According to the Nernst equation, the respective electrode potentials can be expressed as:

$$E_M = E_M^o + \frac{RT}{2F} \ln a_{M^{2+}} \tag{11.182}$$

$$\text{and } E_{Me} = E_{Me}^o + \frac{RT}{2F} \ln a_{Me^{2+}} \tag{11.183}$$

Simultaneously, the reducing gas blown in the solution is oxidized as:

$$\frac{1}{2}H_2 = H^+ + e \tag{11.184}$$

for which

$$E_H = E_H^o + \frac{RT}{F} \ln \left(\frac{a_{H^+}}{p_{H_2}^{1/2}} \right)$$

$$= E_H^o + \frac{RT}{F} \ln a_{H^+} - \frac{RT}{F} \ln \left[p_{H_2} \right]^{1/2} \tag{11.185}$$

Since $E_H^o = 0$, and $-\log a_{H^+} = $ pH, at 298 K, on substituting the values of R and F, we get:

$$E_H = -0.059\text{pH} - \frac{0.059}{2} \log \left(p_{H_2} \right) \tag{11.186}$$

A critical analysis of Eqs. 11.182 and 11.183 demonstrates that at constant temperature, E_M and E_{Me} depend on the activity of metal ions present in the solution, whereas Eq. 11.186 indicates that at constant temperature, E_H depends on pH of the solution as well as on the partial pressure of hydrogen gas blown in the solution. Hence, plot of E versus pH for each reaction at constant metal ion activity (unimolar concentration of M/Me) will give straight lines of zero slope for E_M and E_{Me} and a straight line with negative slope for E_H. The lower the electrode potential, the greater the tendency of the metal to dissolve in the electrolyte. The variation of E_H and E_M with pH of the solution is shown in Fig. 11.15. E_M plotted for various metals, namely, Cu, Ni, Co, Cd, Fe, and Zn, is independent of pH, and E_H has been plotted for $p_{H_2} = 10$ and 100 atm. In these presentations, the temperature and the activity of metal ions have been assumed constant for the entire pH range. When $E_H < E_M$ (or E_{Me}), reduction will occur according to reactions:

$$M^{2+} + H_2 = M + 2H^+ \tag{11.187}$$

$$Me^{2+} + H_2 = Me + 2H^+ \tag{11.188}$$

Fig. 11.15 E_M and E_H as a function of pH for metal and hydrogen electrodes at unimolar concentration of M and 25 °C

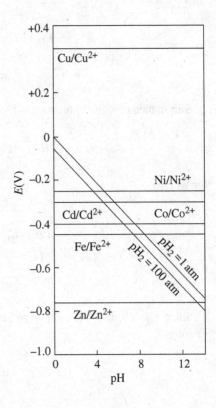

From the figure, it is clear that the point of intersection of E_{Cu} and E_H lines lies well into the negative pH region; hence, copper can be displaced by hydrogen from a molar solution of copper at any pH. Similarly, other noble metals can be displaced, theoretically, at all practical pH values. Contrary to this, zinc reduction (displacement) from aqueous solution requires a high pH of 14. This is too high, and as such, zinc cannot be precipitated from the solution at any pH. But a pH of 5 and 9 would be sufficient for displacement of Ni and Fe, respectively. Co and Cd can be precipitated at pH between 6 and 8. The discussion is based on ignoring over potentials of hydrogen as well as those of metals. Displacement or precipitation reaction can be carried out at lower pH by increasing the partial pressure of hydrogen blown into the solution. Thus, a method of selective reduction (precipitation) from a solution containing a number of metal ions can be worked out. For example, copper powder can be precipitated at pH of 2 from a solution containing Cu^{2+}, Ni^{2+}, and Co^{2+}. By increasing the pH, nickel and cobalt can be precipitated. A pH > 7 would precipitate $Ni(OH)_2$ due to hydrolysis. This is prevented by addition of ammonia [78], which increases the pH and reduces $a_{Ni^{2+}}$ by formation of nickel complex $[Ni(NH_3)_6]^{2+}$. As a result, the position of nickel line in Fig. 11.15 will be shifted to provide further selectivity. At elevated temperature and pressure, the rate of reaction is enhanced not only due to the increase of temperature but also due to the increased solubility of hydrogen in water at high pressure. In Sherritt Gordon Process, nickel is precipitated by blowing hydrogen at 30 atm pressure and 170 °C according to the reaction:

$$Ni(NH_3)_2SO_4 + H_2 = Ni + (NH_4)_2SO_4 \qquad (11.189)$$

11.5.3.1 Kinetics of Precipitation by Hydrogen

Although precipitation of nickel and cobalt from aqueous solutions by hydrogen is thermodynamically feasible at room temperature, on the commercial scale, it is carried out at elevated temperature and pressure in order to achieve a reasonable rate of production. Reduction of metal ions is not effective with molecular hydrogen. Its dissociation into atomic hydrogen makes it reactive. Since dissociation energy involved is high (about 430 kJ mol^{-1}), the reaction is facilitated by means of catalysts. As catalysts, many metal ions promote homogeneous reaction and many solids provide catalytic active surfaces to facilitate heterogeneous reaction. Ag^+, Hg_2^{2+}, and Cu^{2+} ions are well-known homogeneous hydrogenation catalysts. These ions can successfully split molecular hydrogen by the formation of intermediate hydride complexes [79]. Metals precipitated by homogeneous reaction from the solution are in the form of finely divided powder. On the other hand, when the reaction is heterogeneous, a catalytic active surface is provided by seeding with finely divided solid, as practiced in the reduction of nickel and cobalt from ammoniacal solution. In the absence of seeding, deposition takes place on the walls and agitators of the autoclave. Under certain conditions, a metal can be reduced by both homogenous and heterogeneous reactions. First, nuclei are formed homogeneously in the absence of seeding. These nuclei then act as seeds for the heterogeneous reaction. The rate of reaction increases with increase of the surface area of the precipitated metal. Such a process, called autocatalytic, takes place during the precipitation of copper from ammoniacal solution.

In both categories of homogeneous and heterogeneous reactions, the rate of precipitation increases with temperature and partial pressure of hydrogen. However, the two types of reactions differ by the fact that the former depends on the initial metal ion concentration, whereas the latter is highly dependent on the area of the catalytic surface. The maximum temperature of reduction depends on the solubility of salts. Solubility of many metal sulfates increases with temperature and attains a maximum value in the temperature range of 80–150 °C and falls rapidly with further increase of temperature. In addition, stability of metal complexes decreases with increase of temperature. Decomposition of complexes generally results in the precipitation of basic salts and hydroxides. Owing to these limitations, precipitation is always carried out below 200 °C. Hydrolysis can be minimized by using solutions of high concentration, which increases the ionic strength and decreases the activity coefficient. The solubility of hydrogen in aqueous solution decreases up to approximately 80 °C. It increases thereafter, and the increase is rapid above 140 °C. At 200 °C and 20 atm pressure of hydrogen, the solubility is approximately 0.2 mol dm^{-3} (0.2 M). As a result, optimum conditions for reduction is within 140–200 °C. The solubility and boiling point of the solution can be simultaneously increased by increasing hydrogen gas pressure.

Meddings and Mackiw [80] have reported that the rate of reduction of nickel from ammoniacal sulfate solution containing diamine nickel ion is independent of nickel concentration up to 85% precipitation. The rate has been found to be directly proportional to the surface area of the nickel powder (A) and the partial pressure of hydrogen gas (p_{H_2}) at equilibrium:

$$\text{Rate} = -\frac{d\left[Ni^{2+}\right]}{dt} = kAp_{H_2} \tag{11.190}$$

According to Sircar and Wiles [81], the rate is proportional to the surface area, nickel ion concentration, and square root of the partial pressure of hydrogen gas:

$$\text{Rate} = -\frac{d\left[Ni^{2+}\right]}{dt} = k_1 A\left[Ni^{2+}\right]p_{H_2}^{\frac{1}{2}} \tag{11.191}$$

The kinetics of reduction of cobalt has not been very extensively studied as compared to nickel. From ammoniacal ammonium sulfate solution of high cobalt concentration, the rate of reduction that occurs on cobalt metal catalyst has been reported to be directly proportional to surface area and partial pressure of hydrogen:

$$\text{Rate} = -\frac{d[Co^{2+}]}{dt} = kAp_{H_2} \tag{11.192}$$

The highest rate of precipitation was achieved at the $[NH_3]:[Co^{2+}]$ ratio of 2.5:1 with activation energy [82] of 37.6 kJ mol^{-1}. This is comparable with the activation energy of 42.6 kJ mol^{-1} for nickel reduction [80] under identical conditions.

Reduction of copper takes place homogeneously in noncomplexing solutions. The rate of precipitation has been found to be independent of the surface area of copper powder but increases with increase of temperature and pressure [83]. However, the reduction of copper from ammoniacal sulfate solutions is highly complex due to the contamination of triamino copper (II) ion solution by suspension of basic copper sulfate [84] up to a $[NH_3]/[Cu^2+]$ ratio of 3:1. Several insoluble compounds including CuO and Cu_2O are formed during the course of reduction with hydrogen. Copper (I) amines are formed by homogeneous and heterogeneous reactions. Copper powder is precipitated by disproportionate reaction of the reduced amines according to the reaction:

$$2Cu(NH_3)_n^+ = Cu + Cu(NH_3)_{2n}^{2+} \tag{11.193}$$

11.5.3.2   Hydrogen Reduction Practice

Sherritt–Gordon Mines Limited adopted the hydrogen precipitation technique for recovery of nickel and cobalt as a part of the development of ammonia pressure leaching of copper–nickel–cobalt sulfide concentrates from Lynn Lake Mines. In this process, two-stage ammonia leaching is carried out in autoclaves at 80 °C under 8 atm (Sect. 11.1.1.5). The clarified solution is boiled to promote decomposition of thiosulfate and thionate and to precipitate Cu_2S simultaneously. The solution is treated with H_2S to precipitate the remaining copper. The copper-free solution is then heated and allowed to saturate with air at high temperature and pressure to generate conditions for the oxidation of the remaining unsaturated sulfur compounds and to convert sulfamate to sulfate. Nickel is recovered from the solution by blowing hydrogen gas at a pressure of 30 atm in an autoclave at 200 °C (reaction 11.189).

The precipitated nickel powder is filtered off (see Appendix A.3b). Filtrate containing 95% cobalt and the residual nickel is treated with H_2S. The solution after filtration of sulfides of nickel and cobalt is evaporated to obtain ammonium sulfate crystals. The Sherritt Gordon plant at Fort Saskatchewan, Alberta, Canada, is based on operating cycles comprising nucleation, densification, and leaching of nickel deposited on the autoclave walls.

The autoclave containing ammonia leach liquor is purged with nitrogen to remove air before blowing hydrogen. Ferrous sulfate catalyst in the form of slurry is added, and the liquor is heated to 118 °C by steam-heated coils. Under agitation, pressure of hydrogen is raised to 23 atm. Fine particles of nickel get precipitated on ferrous oxide nuclei, and precipitation is completed within 30–40 min. The resultant powder is allowed to settle in the autoclave, and solution is discharged. During the densification stage, batches of nickel solution after oxyhydrolysis are reduced on the catalytic nickel seed which becomes more dense as growth progresses. Finally, nickel deposited on the walls and internal equipments of the autoclave is dissolved in ammoniacal ammonium sulfate solution in the presence of air.

11.6 Problems

Problem 11.1

(a) With the help of the following data construct a Pourbaix diagram for $Al–H_2O$ system assuming Al^{3+} ion concentration of 10^{-3} mol kg^{-1} water at 298 K.

$$Al^{3+} + 3e = Al, \quad E^\circ = -1.633 \text{ V} \tag{1}$$

$$2Al^{3+} + 3H_2O = Al_2O_3 + 6H^+, \log K = -11.4 \tag{2}$$

$$Al_2O_3 + H_2O = 2AlO_2^- + 2H^+, \log K = -29.4 \tag{3}$$

$$Al_2O_3 + 6H^+ + 6e = 2Al + 3H_2O, E^\circ = -1.55 \text{ V} \tag{4}$$

$$AlO_2^- + 4H^+ + 3e = Al + 2H_2O, E^\circ = -1.262 \text{ V} \tag{5}$$

(b) Given for the reactions at 298 K: (1) $Al(OH)_3 = Al^{3+} + 3OH^-, K = 5 \times 10^{-33}$ (2) $H_2O = H^+ + OH^-, K = 10^{-14}$ and (3) $Al(OH)_3 + OH^- = AlO_2^- + 2H_2O$, $K = 40$, calculate the pH range in which $Al(OH)_3$ exists at 298 K in $Al - H_2O$ system at unimolar concentration of Al^{3+} ions.

(c) Show the two predominance areas of dissolution of Al_2O_3 in the diagram. Write the corresponding chemical reactions in acid and alkaline solutions.

Solution

(a) Since equilibrium (1): $Al^{3+} + 3e = Al$ involves three electrons, hence according to Nernst expression:

$$E_{Al} = E^\circ_{Al} + \frac{RT}{3F} \ln a_{Al^{3+}} \left(a_{Al^{3+}} = 10^{-3}\right)$$

$$= -1.633 + \frac{0.059}{3} \log\left(10^{-3}\right) = -1.633 - \frac{0.059}{3} \times 3 = -1.692 \text{ V}$$

Equilibrium (2): $2Al^{3+} + 3H_2O = Al_2O_3 + 6H^+$ involves H^+ ions, not electrons,

$$K = \frac{a_{H^+}^6}{a_{Al^{3+}}^2}, \left(\log K = -11.4, a_{Al^{3+}} = 10^{-3}\right)$$

$$\therefore \log K = 6 \log a_{H^+} - 2 \log a_{Al^{3+}}$$

$$\text{or} -11.4 = -6 \text{ pH} - 2 \log\left(10^{-3}\right)$$

$$\text{or} -11.4 = -6pH + 6$$

$$\therefore pH = 2.9$$

Equilibria (1) and (2) are, respectively, independent of pH and potential.
Equilibrium (3): $Al_2O_3 + H_2O = 2AlO_2^- + 2H^+$ involves H^+ ions but not electrons

$$K = a_{AlO_2^-}^2 \cdot a_{H^+}^2 \left(\text{as } a_{Al^{3+}} = 10^{-3}, a_{AlO_2^-} = 10^{-3}\right)$$

$$\therefore \log K = 2 \log a_{AlO_2^-} + 2 \log a_{H^+}$$

$$-29.4 = 2 \log\left(10^{-3}\right) - 2 \text{ pH}$$

$$\therefore pH = 11.7$$

Equilibrium (4): $Al_2O_3 + 6H^+ + 6e = 2Al + 3H_2O$ involving H^+ ions and electrons depends on pH as well as E. According to Nernst equation, we can write:

$$E = E^o + \frac{0.059}{6} \log a_{H^+}^6$$

$$= E^o - \frac{0.059}{6}.6.pH = -1.55 - 0.059.pH = -1.55 - 0.059 \times 11.7 = -2.24 \text{ V}$$

Thus, equilibrium (3) is terminated at -2.24 V.
Equilibrium (5): $AlO_2^- + 4H^+ + 3e = Al + 2H_2O$ involves H^+ ions and electrons, hence,

$$E = E^o + \frac{0.059}{3} \log\left(a_{AlO_2^-} \cdot a_{H^+}^4\right)$$

$$= -1.262 + \frac{0.059}{3} \log a_{AlO_2^-} - \frac{0.059}{3} 4 \text{ pH}$$

$$= -1.262 + \frac{0.059}{3} \log\left(10^{-3}\right) - \frac{0.059}{3} 4 \text{ pH}$$

$$= -1.262 - 0.059 - 0.079 \text{ pH}$$

$$= -1.321 - 0.079 \text{ pH}$$

In order to complete the boundary of equilibrium (5), substitute pH = 14,

$$E = -1.321 - 0.079 \times 14 = -2.427 \text{ V}$$

It is customary to superimpose the thermodynamic stability limit of water in the Pourbaix diagram (at $p_{H_2} = 1$ atm, $E = -0.059.pH$, Eq. 11.38, hence $E = 0$ at pH $= 0$ and $E = -0.826$ V at pH $= 14$). Such a diagram for Al-H$_2$O system displaying different stability regions is shown in Fig. 11.16.

(b) For the reaction: $Al(OH)_3 = Al^{3+} + 3OH^-$, $K = 5 \times 10^{-33}$

$$K = \frac{a_{Al^{3+}} \cdot a_{OH^-}^3}{a_{Al(OH)_3}} = a_{OH^-}^3 = 5 \times 10^{-33} \left[a_{Al^{3+}} = 1 = a_{Al(OH)_3} \right]$$

$$\therefore a_{OH^-} = 1.7 \times 10^{-11}$$

For the reaction: $H_2O = H^+ + OH^-$, $K = 10^{-14} = a_{H^+} \cdot a_{OH^-}$

$$\therefore a_{H^+} = \frac{10^{-14}}{a_{OH^-}} = \frac{10^{-14}}{1.7 \times 10^{-11}} = 5.88 \times 10^{-4} \quad (pH = -\log a_{H^+})$$

$$\therefore pH = 3.23$$

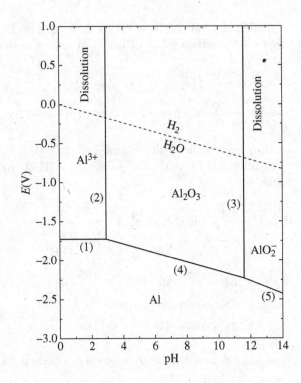

Fig. 11.16 Pourbaix diagram for Al–H$_2$O system at 298 K

For the reaction: $Al(OH)_3 + OH^- = AlO_2^- + 2H_2O$, $K = 40$.

$$\therefore K = \frac{a_{AlO_2^-}}{a_{OH^-}} = 40$$

$$a_{OH^-} = \frac{1}{40} \left(a_{AlO_2^-} = 1 \right)$$

$$\therefore a_{H^+} = \frac{10^{-14}}{a_{OH^-}} = \frac{10^{-14}}{1/40} = 4 \times 10^{-13}$$

$$pH = -\log a_{H^+} = 12.394$$

Thus, at the unimolar concentration of Al^{3+} ions, $Al(OH)_3$ in $Al - H_2O$ system exists in the pH range of 3.23–12.4 at 298 K.

(c) The two predominance areas of dissolution of Al_2O_3 in acid and alkaline solutions have been shown in Fig. 11.16. The chemical reactions are given below:

In acid solutions (i.e., at low pH), $Al_2O_3 + 6H^+ = 2Al^{3+} + 3H_2O$.

In basic solutions (i.e., at high pH), $Al_2O_3 + OH^- = 2AlO_2^- + H^+$.

Problem 11.2

Show that after n number of extractions, the concentration of the solute left in the aqueous phase, $(c_a)_n$ is given by $(c_a)_n = (c_a)_i \left[\frac{1}{1+D\left(v_o/v_a \right)} \right]^n$, where v_a and v_o are the volumes of the aqueous and organic phases, respectively, and $(c_a)_i$ is the initial concentration of the metal in the aqueous phase.

Solution

Derivation of expression for calculation of optimum number of stages in multiple extraction: If v_a is the volume of the aqueous phase containing w g of the metal ions and v_o is the volume of the organic phase and after the first extraction, w_1 g of the solute remains in the aqueous phase.

Average concentration in the aqueous phase after the first stage of extraction $= \frac{w_1}{v_a} = c_a$.

Average concentration of the metal in the organic phase $= \frac{w-w_1}{v_o} = c_o$.

Therefore, by definition distribution coefficient,

$$D = \frac{c_o}{c_a} = \frac{\frac{w-w_1}{v_o}}{\frac{w_1}{v_a}}$$

$$\text{or } D \cdot \frac{w_1}{v_a} = \frac{w - w_1}{v_o}$$

$$\text{or } D\, w_1 v_o = v_a w - v_a w_1$$

$$\text{or } (v_a + D.v_o)w_1 = v_a.w$$

$$\therefore w_1 = \frac{v_a.w}{v_a + D.v_o} = w\left[\frac{1}{1 + D\left(v_o/v_a\right)}\right]$$

After the second extraction, weight of the solute remaining in the aqueous phase $= w_2$.

The equilibrium concentration of the aqueous phase $= \frac{w_2}{v_a}$

and the equilibrium concentration of the organic phase $= \frac{w_1 - w_2}{v_o}$

$$\therefore D = \frac{\dfrac{w_1 - w_2}{v_o}}{\dfrac{w_2}{v_a}}$$

$$\text{or } w_2 = w_1\left[\frac{1}{1 + D\left(v_o/v_a\right)}\right] = w\left[\frac{1}{1 + D\left(v_o/v_a\right)}\right]^2$$

Similarly, after n number of extractions, amount of metal remaining in the aqueous

$$\text{phase, } w_n = w\left[\frac{1}{1 + D\left(v_o/v_a\right)}\right]^n$$

$$\text{or } \frac{w_n}{v_a} = \frac{w}{v_a}\left[\frac{1}{1 + D\left(v_o/v_a\right)}\right]^n$$

$$\text{or } (c_a)_n = (c_a)_i\left[\frac{1}{1 + D\left(v_o/v_a\right)}\right]^n$$

where $(c_a)_i$ is the initial concentration of the metal in the aqueous phase, which is reduced to $(c_a)_n$ after n number of extractions.

Problem 11.3

Estimate the optimum number of stages in extracting the metal value almost completely from a solution containing 100 g l^{-1} of metal ions from three different organic extractants having distribution coefficients of 9, 4, and 3.

Solution
When $D = 9$

		Concentration	
Initial concentration (g l^{-1})	Stage	c_o	c_a
100	1	90	10
10	2	9	1
1	3	0.9	0.1

Since the total amount of the metal in the organic extractant after three stages of extraction $= 99.9$ g 1^{-1} and left in the raffinate $= 0.1$ g 1^{-1}, the optimum number of extraction stages would be 3.

When $D = 4$

Initial concentration (g 1^{-1})	Stage	Concentration	
		c_o	c_a
100	1	80	20
20	2	16	4
4	3	3.2	0.8
1	4	0.64	0.16

Since the total amount of the metal in the organic extractant after four stages of extraction $= 99.84$ g 1^{-1} and left in the raffinate $= 0.16$ g 1^{-1}, the optimum number of extraction stages would be 4.

When $D = 3$

Initial concentration (g 1^{-1})	Stage	Concentration	
		c_o	c_a
100	1	75	25
25	2	18.75	6.25
6.25	3	4.65	1.6
1.6	4	1.2	0.4
0.4	5	0.3	0.1

Since the total amount of the metal in the organic extractant after five stages of extraction $= 99.9$ g 1^{-1} and left in the raffinate $= 0.1$ g 1^{-1}, the optimum number of extraction stages would be 5.

Problem 11.4

A leach liquor (at pH $= 1.5$) contains Fe^{3+} and Fe^{2+} ions (respectively at the concentration of 10^{-5} and 10^{-4} M) along with some other impurities at 25 °C. If pH of the leach liquor is gradually increased by addition of some alkali at what pH will $Fe(OH)_3$ and $Fe(OH)_2$ get precipitated?

Given that:

$$Fe\,(OH)_2 = Fe^{2+} + 2OH^-, \quad K_1 = 7.94 \times 10^{-16}$$
$$Fe\,(OH)_3 = Fe^{3+} + 3OH^-, \quad K_2 = 1.58 \times 10^{-39}$$
$$H_2O\,(l) = H^+ + OH^-, \quad \Delta G^o = 19095 \text{ cal}$$

Solution

$$H_2O\,(l) = H^+ + OH^-, K = \frac{a_{H^+} . a_{OH^-}}{a_{H_2O}} = a_{H^+} . a_{OH^-}$$

$$\therefore \log K = \log a_{H^+} + \log a_{OH^-}$$

$$\Delta G^o = 19095 = -RT \ln K$$

$$\log K = -\frac{\Delta G^o}{2.303.RT} = -\frac{19095}{1363} = -14.00$$

Hence,

$$\log a_{H^+} + \log a_{OH^-} = -14.00 \tag{1}$$

$$Fe(OH)_3 = Fe^{3+} + 3OH^-,$$

$$K_2 = \frac{a_{Fe^{3+}} \cdot a_{OH^-}^3}{a_{Fe(OH)_3}} = a_{Fe^{3+}} \cdot a_{OH^-}^3 = 1.58 \times 10^{-39}$$

$$a_{OH^-}^3 = \frac{1.58 \times 10^{-39}}{a_{Fe^{3+}}} = \frac{1.58 \times 10^{-39}}{10^{-5}} = 15.8 \times 10^{-33}$$

$$\therefore a_{OH^-} = 2.5 \times 10^{-11}, \log a_{OH^-} = -10.6 \tag{2}$$

From Eqs. 1 and 2, we get: $\log a_{H^+} = -14.00 - (-10.6) = -3.4$
As $pH = -\log a_{H^+}$
Hence, $Fe(OH)_3$ can be precipitated at a pH of 3.4.

$$Fe(OH)_2 = Fe^{2+} + 2OH^-,$$

$$K_1 = \frac{a_{Fe^{2+}} \cdot a_{OH^-}^2}{a_{Fe(OH)_2}} = a_{Fe^{2+}} \cdot a_{OH^-}^2 = 7.94 \times 10^{-16}$$

$$a_{OH^-}^2 = \frac{7.94 \times 10^{-16}}{a_{Fe^{2+}}} = \frac{7.94 \times 10^{-16}}{10^{-4}} = 7.94 \times 10^{-12}$$

$$a_{OH^-} = 2.82 \times 10^{-6}, \log a_{OH^-} = -5.55 \tag{3}$$

From Eqs. 1 and 3, we get: $\log a_{H^+} = -14.00 - (-5.55) = -8.45$
Hence, $Fe(OH)_2$ can be precipitated at a pH of 8.45. Ans.

Problem 11.5

A leach liquor at a pH of 2 containing metal ions: Cu^{2+}, Ni^{2+}, and Mn^{2+}, each at a concentration of 0.01 M, is treated with H_2S gas at 25 °C. In the light of the given data, what is your opinion about selective precipitation of CuS at this pH? What is the concentration of Cu^{2+} ions left in the leach liquor after equilibrium is attained? Calculate the pH required for precipitation of MnS.

Given that:

$$CuS = Cu^{2+} + S^{2-}, \quad K_{CuS} = 7.94 \times 10^{-37}$$
$$MnS = Mn^{2+} + S^{2-}, \quad K_{MnS} = 3.16 \times 10^{-10}$$
$$NiS = Ni^{2+} + S^{2-}, \quad K_{NiS} = 3.98 \times 10^{-20}$$

Solution

At 25 °C, $[S^{2-}]$ ion concentration can be calculated by Eq. 11.155, pH = 2

$$\log \left[S^{2-}\right] = \log \left(1.69 \times 10^{-21}\right) + 2pH$$
$$= \log \left(1.69 \times 10^{-21}\right) + 2 \times 2 = -16.77$$
$$\therefore \left[S^{2-}\right] = 1.69 \times 10^{-17} \text{ and } \left[M^{2+}\right] = 0.01 \text{ (given)}$$

The ionic concentration product, K, for each metal sulfide would be:

$$\left[M^{2+}\right] \times \left[S^{2-}\right] = 0.01 \times 1.69 \times 10^{-17} = 1.69 \times 10^{-19} = K$$

M^{2+} stands for Cu^{2+}, Ni^{2+}, and Mn^{2+}. Since this ionic concentration product, K (1.69×10^{-19}), is much higher than $K_{CuS} = 7.94 \times 10^{-37}$ but less than $K_{MnS} = 3.16 \times 10^{-10}$ and very close to $K_{NiS} = 3.98 \times 10^{-20}$, CuS cannot be very selectively precipitated at the pH of 2. There is no possibility of precipitation of MnS, but precipitation of NiS cannot be ruled out.

$$\text{Concentration of } \left[Cu^{2+}\right] \text{ left in the solution at equilibrium} = \frac{K_{CuS}}{\left[S^{2-}\right]}$$
$$= \frac{7.94 \times 10^{-37}}{1.69 \times 10^{-17}}$$
$$= 4.7 \times 10^{-20} \text{ Ans.}$$

For precipitation of MnS:

$$\left[S^{2-}\right] = \frac{K_{MnS}}{\left[Mn^{2+}\right]} = \frac{3.16 \times 10^{-10}}{10^{-2}} = 3.16 \times 10^{-8}$$

Use Eq. 11.155 to calculate pH at the above $[S^{2-}]$ ion concentration:

$$\log \left[S^{2-}\right] = \log \left(1.69 \times 10^{-21}\right) + 2pH$$

$$pH = \frac{1}{2} \left[\log \left(3.16 \times 10^{-8}\right) - \log \left(1.69 \times 10^{-21}\right)\right] = 6.64 \text{ Ans.}$$

But at this pH, all the three sulfides, CuS, MnS, and NiS will get precipitated.

Problem 11.6

From the standard electrode potentials of the following half cells listed below suggest two pairs of metal ionic systems equilibrating at the highest and lowest ratio of concentrations at 25 °C. Calculate the ratio of concentration in each case.

$$Fe^{2+} + 2e = Fe, E^o = -0.440 \text{ V}$$

$$Zn^{2+} + 2e = Zn, E^o = -0.763 \text{ V}$$

$$Ni^{2+} + 2e = Ni, E^o = -0.250 \text{ V}$$

$$Cd^{2+} + 2e = Cd, E^o = -0.403 \text{ V}$$

$$Cu^{2+} + 2e = Cu, E^o = 0.337 \text{ V}$$

Can iron effectively displace cadmium from the aqueous solution? Which element can most effectively cement out cadmium from the solution? Give a quantitative estimate.

Solution

The potential of the two half cells in any electrochemical cell comprising any two metals after achieving equilibrium becomes equal. For the cell in which the two metals have the largest difference in E^o, for example, Cu^{2+}/Cu and Zn^{2+}/Zn will equilibrate at the least ratio of activities $\frac{a_{Cu^{2+}}}{a_{Zn^{2+}}}$ (hence, at the least metal ion concentration ratio). Therefore, at 25 °C, we can write:

$$E^o = E^o_{Cu} + \frac{RT}{nF} \log a_{Cu^{2+}} = E^o_{Zn} + \frac{RT}{nF} \log a_{Zn^{2+}}$$

$$0.337 + \frac{4.575 \times 298}{2 \times 23066} \log a_{Cu^{2+}} = -0.763 + \frac{4.575 \times 298}{2 \times 23066} \log a_{Zn^{2+}}$$

and thus we get : $\log \left[\dfrac{a_{Cu^{2+}}}{a_{Zn^{2+}}}\right] = -\dfrac{1.10}{0.0296} = -37.162$

$$\therefore \left[\frac{a_{Cu^{2+}}}{a_{Zn^{2+}}}\right] = 6.88 \times 10^{-38} = \frac{[Cu^{2+}]}{[Zn^{2+}]} \text{ the least ratio Ans.}$$

The system with lowest difference in E^o, for example, Cd^{2+}/Cd and Fe^{2+}/Fe, will equilibrate at larger concentration ratio of $\frac{[Cd^{2+}]}{[Fe^{2+}]}$. Following the above procedure, we get:

$$\log \left[\frac{a_{Cd^{2+}}}{a_{Fe^{2+}}}\right] = \frac{-0.440 - (-0.403)}{0.0296} = -\frac{0.037}{0.0296} = -1.25$$

$$\therefore \left[\frac{a_{Cd^{2+}}}{a_{Fe^{2+}}}\right] = 5.6 \times 10^{-2} = \frac{[Cu^{2+}]}{[Fe^{2+}]} = \text{ the largest ratio Ans.}$$

Since the $\left[\frac{a_{Cd^{2+}}}{a_{Fe^{2+}}}\right]$ ratio is rather close to unity, iron cannot effectively displace cadmium from the aqueous solution. As the electrode potentials of Cd^{2+}/Cd and Zn^{2+}/Zn are widely apart, zinc can most effectively cement out cadmium from the solution.

$$\log \left[\frac{a_{Cd^{2+}}}{a_{Zn^{2+}}}\right] = \frac{-0.763 - (-0.403)}{0.0296} = -\frac{0.360}{0.0296} = -12.16$$

$$\therefore \left[\frac{a_{Cd^{2+}}}{a_{Zn^{2+}}}\right] = 6.88 \times 10^{-13} \frac{[Cd^{2+}]}{[Zn^{2+}]} \text{ Ans.}$$

Since $\frac{[Cd^{2+}]}{[Zn^{2+}]} = 6.88 \times 10^{-13}$ is relatively very small, zinc can most effectively cement out cadmium from the solution.

Problem 11.7

Concentration of copper in the leach liquor can be reduced to a very low level by cementing out copper with iron. The liquor leaving the cementation launder at 27 °C contains 0.5 g l^{-1} of iron. What is the residual equilibrium copper content of the solution if the activity coefficient of the ferrous ions in dilute sulfuric acid solution is 0.2?

Solution

The cementation reaction: $Cu^{2+} + Fe = Fe^{2+} + Cu$

The half-cell potential for iron is:

$$E = E^0_{Fe} + \frac{RT}{nF} \log a_{Fe^{2+}} = -0.44 + \frac{4.575 \times 300}{2 \times 23066} \log \left(\gamma_{Fe^{2+}} \cdot [Fe^{2+}]\right)$$

$$= -0.44 + 0.0296 \cdot \log \left(0.2 \cdot \left[\frac{0.5}{55.85}\right]\right)$$

$$= -0.44 - 0.0813 = -0.5213 \text{ V}$$

At equilibrium, the copper half-cell potential will have the same potential:

$$E = E^0_{Cu} + \frac{RT}{nF} \log a_{Cu^{2+}} = 0.337 + \frac{4.575 \times 300}{2 \times 23066} \log a_{Cu^{2+}}$$

$$\therefore -0.5213 = 0.337 + 0.0296 \log \left(a_{Cu^{2+}}\right)$$

$$\text{or} \log \left(a_{Cu^{2+}}\right) = -\frac{0.5213 + 0.337}{0.0296} = -28.9966 \simeq -29$$

Assuming activity coefficient of Cu^{2+} ions to be unity in the dilute solution and substituting concentration for activity, we get:

$$\log [Cu^{2+}] \simeq -29$$

$$\therefore [Cu^{2+}] = 10^{-29} M = 63.54 \times 10^{-29} = 6.4 \times 10^{-28} \text{ g l}^{-1}$$

Hence, the residual equilibrium copper content of the solution $= 6.4 \times 10^{-28}$ g l^{-1} Ans.

Problem 11.8

Hydrogen gas is bubbled at 1 atm pressure through aqueous solutions, each containing either Zn^{2+}, Fe^{2+}, Cd^{2+}, Cu^{2+}, or Ni^{2+} ion. If concentration of the individual ion in the solution is 10^{-2} M, calculate the minimum pH required for the precipitation of Zn, Fe, Cd, Cu, and Ni from the aqueous solution at 25 °C.

Given that: $E^o_{Zn^{2+}/Zn} = -0.762$, $E^o_{Fe^{2+}/Fe} = -0.44$, $E^o_{Cd^{2+}/Cd} = -0.403$,

$E^o_{Cu^{2+}/Cu} = 0.337$ V, and $E^o_{Ni^{2+}/Ni} = -0.241$ V.

Solution

For the precipitation of the metal from an aqueous solution, the necessary condition is $E_M = E_H$

at 25 °C we have $E_H = -0.059$ pH (when $p_{H_2} = 1$ atm)

and

$$E_M = E^o_M + \frac{0.059}{2} \log a_{M^{2+}} = E^o_M + \frac{0.059}{2} \log 10^{-2} = E^o_M - 0.059$$

$\therefore$ when $E_H = E_M$, we have $- 0.059$ pH $= E^o_M - 0.059$,

for Zn precipitation: -0.059 pH $= -0.762 - 0.059$

$$pH = 0.821/0.059 = 13.9$$

for Fe, -0.059 pH $= -0.44 - 0.059$

$$pH = 0.499/0.059 = 8.5$$

for Cd, -0.059 pH $= -0.403 - 0.059$

$$pH = 0.462/0.059 = 7.8$$

for Cu, -0.059 pH $= 0.337 - 0.059$

pH $= - 0.278/0.059 = - 4.7$ (since pH is negative, Cu can be precipitated at any pH)

for Ni, -0.059 pH $= -0.241 - 0.059$

$$pH = 0.30/0.059 = 5.1 \ \text{Ans.}$$

Problem 11.9

Calculate the minimum pressure of hydrogen gas required to precipitate Zn^{2+}, Fe^{2+}, and Cd^{2+} at 25 °C from an aqueous solution at pH = 6. The potential for the precipitation of Zn^{2+}, Fe^{2+}, and Cd^{2+} can be taken as -0.83, -0.50, and -0.47 V, respectively.

Solution

For the precipitation of a metal from solution: the potential for metal precipitation = potential for hydrogen evolution, that is, $E_M = E_H$
at 25 °C,

$$E_H = E_H^o - 0.059 \, pH - \frac{0.059}{2} \log p_{H_2}$$
$$E_H = 0 - 0.059 \times 6 - 0.0295 \log p_{H_2}$$
$$= -0.354 - 0.0295 \log p_{H_2}$$

now since $E_M = E_H$
for Zn^{2+}, $-0.83 = -0.354 - 0.0295 \log p_{H_2}$

$$or - 0.476 = -0.0295 \log p_{H_2}$$

$$\log p_{H_2} = \frac{0.476}{0.0295} = 16.14$$

$$that \, is, p_{H_2} = 1.37 \times 10^{16} atm$$

for Fe^{2+}, $-0.50 = -0.354 - 0.0295 \log p_{H_2}$

$$or \log p_{H_2} = \frac{0.146}{0.0295} = 4.95$$

$$\therefore p_{H_2} = 8.9 \times 10^4 atm$$

for Cd^{2+}, $-0.47 = -0.354 - 0.0295 \log p_{H_2}$

$$\log p_{H_2} = \frac{0.116}{0.0295} = 3.93$$

$$p_{H_2} = 8.6 \times 10^3 atm \, Ans.$$

References

1. Free, M. L., & Moats, M. S. (2014). Hydrometallurgy. In S. Seetharaman (Ed.), *Treatise on process metallurgy. Vol. 3, Industrial process, Part A* (pp. 949–982). Oxford: Elsevier. (Chapter 2.7).
2. Habashi, F. (1970). *Principles of extractive metallurgy, Vol. 2, Hydrometallurgy.* New York: Gordon & Breach. (Chapter 2).
3. Newton, J. (1967). *Extractive metallurgy.* London: Wiley Eastern. (Chapter 7).
4. Dennis, W. H. (1954). *Metallurgy of non-ferrous metals.* London: Pitman. (Chapter 8).
5. Davenport, W. G., King, M., Schlesinger, M., & Biswas, A. K. (2002). *Extractive metallurgy of copper* (4th ed.). Oxford: Elsevier Science Ltd. (Chapters 17 and 18).
6. Free, M. L. (2014). Biohydrometallurgy. In S. Seetharaman (Ed.), *Treatise on process metallurgy. Vol. 3, Industrial process, Part A* (pp. 983–994). Oxford: Elsevier. (Chapter 2.8).

7. Pooley, F. D., Shrestha, G. N., Errington, M. T., & Gibbs, H. E. (1987). In G. A. Davies (Ed.), *Separation processes in hydrometallurgy* (pp. 58–67). New York: Ellis Horwood Ltd. (Chapter 2).

8. Brierley, J. A., & Brierley, C. L. (2001). Present and future commercial applications of biohydrometallurgy. *Hydrometallurgy, 59* (2–3), 233–239.

9. Montealegre, R., & Bustos, S. (1990). In R. Badilla, T. Vargas, & L. Herrera (Eds.), *From molecular biology to industrial applications* (pp. 95–121). Santiogo de Chile: Universitaria. (Chapter 5).

10. Brierley, C. L. (2009). Biohydrometallurgy: What is its future? *Advances in Materials Research, 71-73,* 3–10.

11. Brierley, J. A. (1997). In D. E. Rawlings (Ed.), *Biomining, theory, microbes, and industrial processes* (pp. 103–110). Springer-Verlag. (Chapter 4).

12. Clark, M. E., Batty, J. D., van Buuren, C. B., Dew, D. W., & Eamon, M. A. (2006). Biotechnology in minerals processing: Technological breakthroughs creating value. *Hydrometallurgy, 83*(1–4), 3–9.

13. Jamrack, W. D. (1963). *Rare metal extraction by chemical engineering techniques.* New York: Pergamon. (Chapter 2, 3 and 4).

14. Pourbaix, M. (1966). *Atlas of electrochemical equilibria in aqueous solutions.* London: Pergamon.

15. Rosenqvist, T. (1974). *Principles of extractive metallurgy.* New York: McGraw–Hill. (Chapter 15).

16. Jackson, E. (1986). *Hydrometallurgical extraction and reclamation.* New York: Ellis Horwood Ltd. (Chapters 2, 3, 4 and 5).

17. Hiskey, J. B., & Wadsworth, M. E. (1975). Galvanic conversion of chalcopyrite. *Metallurgical Transactions B, 6,* 183–190.

18. Boustista, R. G., & Flett, D. S. (1976). *Electrodissolution in hydrometallurgy* (Report no. LR 226 (NE)). Warren Spring Laboratory.

19. Moore, J. J. (1990). *Chemical metallurgy* (2nd ed.). London: Butterworth Heinemann Ltd. (Chapter 7).

20. Gold Metallurgy 2011(a) Merrill-Crowe process chemistry [online]. Accessed 25 Febr 2011. Available from http://goldmetallurgy.com/merrill-crowe-process-chemistry. Gold Metallurgy. 2011(b). Merrill-Crowe process operation [online]. Accessed 18 Mar 2011. Available from http://goldmetallurgy.com/merrill-crowe-process-operation.

21. Çelik, H. (2004). Extraction of gold and silver from a Turkish gold ore through thiourea leaching. *Minerals and Metallurgical Processing, 21,* 144–148.

22. Li, J., & Miller, J. D. (2002). Reaction kinetics for gold dissolution in acid thiourea solution using formamidine disulfide as oxidant. *Hydrometallurgy, 63,* 215–223.

23. Li, J., & Miller, J. D. (2006). A review of gold leaching in acid thiourea solutions. *Mineral Processing and Extractive Metallurgy Review, 27,* 177–214.

24. Li, J., & Miller, J. D. (2007). Reaction kinetics of gold dissolution in acid thiourea solution using ferric sulfate as oxidant. *Hydrometallurgy, 89,* 279–288.

25. Jeffrey, M., Heath, J., Hewitt, D., Brunt, S., & Dai, X. (2008). A thiosulfate process for recovering gold from refractory ores which encompasses pressure oxidation, leaching, resin adsorption, elution and electrowinning. In C. A. Young, P. R. Taylor, C. G. Anderson, & Y. Choi (Eds.), *Hydrometallurgy 2008, Proceedings of the sixth international symposium* (pp. 824–836). Phoenix.

26. Feng, D., & van Deventer, J. S. J. (2011). Thiosulphate leaching of gold in the presence of carboxymethyl cellulose (CMC). *Minerals Engineering, 24,* 115–121.

27. Broadhurst, J. L., & Du Preez, J. G. H. (1993). A thermodynamic study of the dissolution of gold in an acidic aqueous thiocyanate medium using iron (III) sulfate as an oxidant. *Hydrometallurgy, 32,* 317–344.

28. Barbosa-Filho, O., & Monhemius, A. J. (1994). Leaching of gold in thiocyanate solutions—Part 1: Chemistry and thermodynamics. *Transactions of the Institution of Mining and Metallurgy Section C, 103*, C105–C110.
29. Barbosa-Filho, O., & Monhemius, A. J. (1994). Leaching of gold in thiocyanate solutions—Part 3: Rates and mechanism of gold dissolution. *Transactions of the Institution of Mining and Metallurgy Section C, 103*, C117–C125.
30. Yang, X., Moats, M. S., & Miller, J. D. (2010). Gold dissolution in acidic thiourea and thiocyanate solutions. *Electrochimica Acta, 55*, 3643–3649.
31. Yang, X., Moats, M. S., Miller, J. D., Wang, X., Shi, X., & Xu, H. (2011). Thiourea–thiocyanate leaching system for gold. *Hydrometallurgy, 106*, 58–63.
32. Li, J., Safarzadeh, M. S., Moats, M. S., Miller, J. D., LeVier, K. M., Dietrich, M., & Wan, R. U. (2012). Thiocyanate hydrometallurgy for the recovery of gold, part II: The leaching kinetics. *Hydrometallurgy, 113–114*, 10–18.
33. Wadsworth, M. E. (1979). Hydrometallurgical processes. In H. Y. Sohn & M. E. Wadsworth (Eds.), *Rate processes in extractive metallurgy*. New York: Plenum. (Chapter 3, Section 3.1).
34. Peters, E., & Halpern, J. (1953). Kinetics of the dissolution of pitchblende. *Transactions of the Canadian Mining and Metallurgy Bulletin, 56*, 344–364.
35. Hiskey, J. B. (1979). Kinetics of uranium oxide dissolution in ammonium carbonate. *Transactions of the Institution of Mining and Metallurgy, Section C, 88*, C145–C152.
36. Glastonbury, J. R. (1968). Kinetics of gibbsite extraction. In *Advances in extractive metallurgy* (pp. 908–917). London: Institution of Mining and Metallurgy.
37. Queneau, P. B., & Cooke, S. R. B. (1969). Kinetics of dissolution of scheelite in alkaline aqueous solutions. *Transactions of the Metallurgical Society, AIME, 245*, 2451–2459.
38. Wadsworth, M. E. (1981). Electrochemical reactions in hydrometallurgy. In J. K. Tein & J. F. Elliott (Eds.), *Metallurgical treatises* (pp. 1–22). Warrendale: Metallurgical Society, AIME.
39. Beckstead, L. W., & Miller, J. D. (1977). Ammonia oxidation leaching of chalcopyrite-reaction kinetics. *Metallurgical Transactions B, 8*, 19–38.
40. Ngoc, N. V., Shamsuddin, M., & Prasad, P. M. (1990). Oxidative leaching of an off-grade/complex copper concentrate in chloride lixiviants. *Metallurgical Transactions B, 21B*, 611–620.
41. Adams, B. A., & Homes, E. L. (1935). Adsorptive properties of synthetic resins. *Journal of the Society of Chemical Industry, 54*, 1–5.
42. Kitchener, J. A. (1957). *Ion Exchange Resins*. New York: Wiley. (Chapter 1).
43. Glasstone, S. (1956). *Principles of nuclear reactor engineering*. London: D. Van Nostrand. (Chapter 5).
44. Street, M., & Quassim, R. Y. (1973). Recovery of uranium from unclarified liquors by ion exchange. In D. J. I. Evans & R. S. Shoemaker (Eds.), *International symposium on hydrometallurgy, Chicago* (pp. 425–437). New York: AIME.
45. Harrington, C. D., & Ruehle, A. E. (1959). *Uranium production technology*. New York: D. Van Nostrand. (Chapter 4).
46. Flett, D. S., & Melling, J. (1979). Selectivity of LIX64N and LIX65N for copper (II) over iron (III). *Transactions of the Institution of Mining and Metallurgy, Section C, 88*, C256–C259.
47. Yih, S. W. H., & Wang, C. T. (1979). *Tungsten sources, metallurgy, properties and applications*. New York: Plenum. (Chapter 2).
48. Zakarias, M. J., & Cahalan, M. J. (1966). Solvent extraction for metal recovery. *Transactions of the Institution of Mining and Metallurgy (London), Section C, 75*, C245–C259.
49. Blake, C. A., Horner, D. E., & Schmitt, J. M. (1959). *Synergistic uranium extractants: combination of neutral organophosphorus compounds with dialkyl phosphoric acids* (US Atomic Energy Commission Report, Oak Ridge National Laboratory. Report No. 2259).
50. Deane, A. M., Kennedy, J., & Sammes, P. G. (1960). Synergism in the extraction of uranium (U^{vi}) from aqueous solution by combination of acidic and non-ionic phosphorylated reagents. *Chemistry & Industry*, 443–451.
51. Scribner, G. W., & Kotecki, A. (1965). Solvent extraction of metal ions with trifluoroacetyl acetone iso-butyl amine. *Analytical Chemistry, 37*, 1304–1306.

52. Haffenden, W. J., & Lawson, G. J. (1968). Effect of ammonium salts on the solvent extraction of metal ions with cation exchanger. In *Advances in extractive metallurgy* (pp. 678–685). London: Institution of Mining and Metallurgy.

53. Flett, D. S., Okuhara, D. N., & Spink, D. R. (1973). Solvent extraction of copper by hydroxyoximes. *Journal of Inorganic and Nuclear Chemistry, 35*, 2471–2475.

54. Fletcher, A. W., & Hester, K. D. (1964). A new approach to copper-nickel ore processing. *Transactions of the Metallurgical Society, AIME, 229*, 282–291.

55. Fletcher, A. W., Flett, D. S., Pegler, J. L., Pepworth, D. P., & Wilson, J. C. (1968). Separation of zinc and cadmium by solvent extraction. In *Advances in extractive metallurgy* (pp. 686–711). London, Institution of Mining and Metallurgy.

56. Spitzer, E. L. T. M., & Radder, J. (1968). Metal extraction by means of trialkylsulfonium compounds. In *Advances in extractive metallurgy* (pp. 669–677). London: Institution of Mining and Metallurgy.

57. Lever, M., & Payne, J. B. (1968). Separation of rare earths and production of metals. In *Advances in extractive metallurgy* (pp. 789–817). London: Institution of Mining and Metallurgy.

58. Marcus, Y. (1963). Solvent extraction of inorganic species. *Chemical Reviews, 63*, 139–170.

59. Ritcey, G. M., & Ashbrook, A. W. (1979). *Solvent extraction: Principles and applications to process metallurgy*. London: Elsevier. (Chapter 1).

60. Hurst, F. J., Arnold, W. D., & Ryan, A. D. (1979). Recovering uranium from wet process phosphoric acid. *Chemical Engineer, 84*(1), 56–63.

61. Cuthbert, F. L. (1979). *Thorium production technology*. Reading: Addition Wesley. (Chapter 5).

62. Klopfenstein, R. K., & Arnold, D. S. (1966). Recent developments in solvent extraction technology. *JOM, 18*(11), 1195–1202.

63. Warshawsky, A., & Berkovitz, H. (1979). Hydrxyoxime solvent impregnated resins for selective copper extraction. *Transactions of the Institution of Mining and Metallurgy, Section C, 88*, C36–C43.

64. Rawling, K. R. (1969). Commercial solvent extraction plant recovers copper from leach liquor. *World Mining, 22*, 34–37.

65. Flett, D. S. (1977). Solvent extraction in hydrometallurgy. *Chemistry & Industry*, 706–714.

66. Horner, D. E., Crouse, D. I., Brown, K. B., & Weaver, B. (1964). The recovery of fission products from aqueous solutions by solvent extraction. In M. E. Wadsworth & F. T. Davis (Eds.), *Unit processes in hydrometallurgy* (pp. 545–568). New York: Gordon & Breach.

67. Shamsuddin, M., & Prasad, P. M. (1971). *Solvent extraction and its metallurgical applications, Eastern Metals Review, Mid Year Number* (pp. 85–92).

68. Smithson, G. R., Shea, J. F., & Tewksbury, T. L. (1966). Liquid-liquid extraction of nonferrous metals-review of the art. *JOM, 18*(9), 1037–1046.

69. Li, J., Wan, R. Y., LeVier, K. M., & Miller, J. D. (2008). Thiocyanate process chemistry for gold recovery. In C. A. Young, P. R. Taylor, C. G. Anderson, & Y. Choi (Eds.), *Hydrometallurgy 2008, Proceedings of the sixth international symposium* (pp. 824–836). Phoenix.

70. Li, J., Safarzadeh, M. S., Moats, M. S., Miller, J. D., LeVier, K. M., Dietrich, M., & Wan, R. U. (2012). Thiocyanate hydrometallurgy for the recovery of gold, part IV: Solvent extraction of gold with alamine 336. *Hydrometallurgy, 113–114*, 25–30.

71. Wells, J. A., Hopkins, W. R., & Stein, R. B. (1992). Chemical and electrolytic processing. In H. L. Hartman (Ed.), *SME mining engineering handbook* (pp. 2250–2258). New York: Society of Mining Engineers of AIME. (Chapter 25.4.3).

72. Simons, C. S. (1964). Hydrogen sulfide as a hydrometallurgical reagent. In M. E. Wadsworth & F. T. Davis (Eds.), *Unit processes in hydrometallurgy* (pp. 592–615). New York: Gordon & Breach.

73. Schalch, E., & Nicol, M. J. (1978). Electrochemical investigation of copper cementation on iron. *Transactions of the Institution of Mining and Metallurgy, Section C, 87*, C16–C22.

74. Nadkarni, R. M., & Wadsworth, M. E. (1968). Kinetic studies of the cementation of copper with iron. In *Advances in extractive metallurgy* (pp. 918–941). London: Institution of Mining and Metallurgy.

75. Lawson, F., & Nhan, L. (1981) Kinetics of removal of cobalt from zinc sulfate electrolyte by cementation. In *Hydrometallurgy 81, Proceedings of symposium on hydrometallurgy, UMIST, July 1981*. Society of Chemical Industry, G4.

76. Schlitt, W. J., & Richards, K. J. (1974). Chemical aspects of copper cementation. In *Proceedings of solution mining symposium, Dallas, AIME* (pp. 401–422). New York: Gordon & Breach.

77. Reid, J. G., & Biswas, A. K. (1973). Continuous flow production of pure copper by high temperature cementation on iron. *Transactions of the Institution of Mining and Metallurgy, Section C, 82*, C221–C225.

78. Parker, R. H. (1978). *An introduction to chemical metallurgy* (2nd ed.). Oxford: Pergamon. (Chapter 7).

79. Halpern, J. (1959). Homogeneous catalytic activation of molecular hydrogen by metal ions and complexes. *The Journal of Physical Chemistry, 63*, 398–403.

80. Meddings, B., & Mackiw, V. N. (1964). The gaseous reduction of metals from aqueous solutions. In M. E. Wadsworth & F. T. Davis (Eds.), *Unit processes in hydrometallurgy* (pp. 345–385). New York: Gordon & Breach.

81. Sircar, S. C., & Wiles, D. R. (1960). Hydrogen precipitation of nickel from buffered acid solutions. *Transactions of the Metallurgical Society, AIME, 218*, 891–893.

82. Kunda, W., Warner, J. P., & Machiw, V. N. (1962). The hydrometallurgical production of cobalt. *Transactions Canadian Institute of Mining, Metallurgy and Petroleum, 65*, 21–27.

83. Schaufelberger, F. A. (1956). Precipitation of metal from salt solution. *JOM, 8*(5), 659–664.

84. Evans, D. J. I. (1968). Production of metals by gaseous reduction from solutions—Processes and chemistry. In *Advances in extractive metallurgy* (pp. 831–907). London: Institution of Mining and Metallurgy.

Chapter 12
Electrometallurgy

Electrometallurgy is an important branch of metallurgy which uses electrical energy in winning of metals from leach liquors and refining of crude metals obtained via the pyrometallurgical route (e.g., blister copper and lead bullion) as well as by hydro-metallurgical and electrometallurgical routes (e.g., aluminum). This branch of met-allurgy also includes the extraction of metals and alloys by electrothermic smelting. In this chapter, discussion is confined to electrowinning and electrorefining. Before discussing the principle of electrometallurgy, it would be interesting to note the difference between electrowinning and electrorefining. Electrowinning is an elec-trolytic extraction method in which cathodic reduction is made use of to deposit the desired metal from the electrolyte that is obtained from a leaching step in the flow sheet. The process uses insoluble or inert anodes and the anodic process supports a gaseous evolution reaction. Contrary to this, electrorefining uses anodes of impure metal to be refined and the cathode is of pure metal starting sheet or a blank of another metal. In electrorefining, the anodic reaction is a metal dissolution reaction, which is reverse of the cathodic reduction. In electrowinning, the anodic reaction is usually oxygen evolution from aqueous solutions:

$$2H_2O = 4H^+ + O_2 + 4e \qquad (12.1)$$

However, in the electrowinning of a metal from a sulfide or a matte, anode dissolution takes place, for example, electrowinning of nickel from nickel matte (Ni_3S_2). Dissolution of Ni_3S_2 generates Ni^{2+} ions and elemental sulfur. Such pro-cesses are categorized as electrowinning with soluble anode. On the other hand, in fused salt electrowinning processes, the electrolyte and anode compositions dictate the anodic reaction. In aluminum electrowinning using fused cryolite electrolyte containing dissolved Al_2O_3, CO, and CO_2 are generated, and, in production of magnesium from fused chloride electrolyte, chlorine is evolved at the anode. Depending on the operating conditions, pure metal may be occasionally deposited on a pool of molten metal covering the conducting hearth of the cell and protected by

M. Shamsuddin, *Physical Chemistry of Metallurgical Processes, Second Edition*,
The Minerals, Metals & Materials Series, https://doi.org/10.1007/978-3-030-58069-8_12

the electrolyte (e.g., aluminum production). On the other hand, magnesium floats on the electrolyte that is in contact with the cathode and a specially designed cell is employed to prevent the metal coming into contact with the anodic chlorine.

During the past four decades, a number of plants based on hydrometallurgical route incorporating leaching–solvent extraction–electrowinning and leaching–chemical purification (Jarosite process)–electrowinning have come up in the production of relatively high-purity copper and zinc, respectively. The production of high-purity cathode has been achieved by successive improvement in the tankhouse automation, development of superior anodes and cathodes, and efficient electrolyte handling [1–3]. In recent years, there have been serious attempts to produce rare metals/alloys by fused salt electrolytic process at lower cost [3–9].

Both electrowinning and electrorefining are carried out in electrolytic cells fabricated out of reinforced concrete tank lined with either lead or lead alloy or PVC or high-density polyethylene. In zinc electrowinning, a new technique utilizes upside-down concrete "T" sections [10] that fit close together to form the cells lined with plastic liners. The size of the cells varies according to the number of electrodes and the interelectrode spacing. In copper and zinc electrowinning, the number of electrodes varies significantly from plant to plant, but the center-to-center distance between cathodes is normally 90–100 mm. In each electrolytic cell, cathodes are arranged alternately with anodes. In order to ensure even and uniform deposits on the end cathodes, one additional anode over the number of cathodes is used. The cathodes and anodes are connected separately, in parallel. The cells are connected in series. There is a large variation in the number of cells in a row from plant to plant.

In normal aqueous electrowinning processes, the anodes should be completely insoluble. Any dissolution will contaminate the electrolyte (leach liquor) by additional metal ions that may reduce on the cathode and deposit along with the desired metal. Traditionally, lead alloys containing 5–10% antimony have been used as anodes. There is a slight contamination of cathode deposit by lead due to some anodic corrosion in dilute acid electrolytes, which increases with an increase in acid concentration. However, this can be minimized by the addition of Co^{2+} ions in the electrolytes, which decreases the anodic corrosion over-potential of oxygen evolution reaction. In recent years, the use of more stable anode of Pb–Ca–Sn [2] has increased in solvent extraction copper electrowinning circuit. Pb–0.5/0.75% Ag alloys have been developed for zinc electrowinning. In order to reduce the energy consumption of copper electrowinning, dimensionally stable anodes of titanium coated with iridium oxide and ruthenium oxide [11–14] have been developed. In addition to lowering energy consumption, they protect the environment by avoiding lead sludge formation and its contamination with the product, and cobalt sulfate addition.

Carbon anodes [15, 16] are exclusively used in the electrolysis of fused cryolite with dissolved Al_2O_3 for aluminum production. These may be either prebaked or self-baking. Prebaked anodes made of individual carbon blocks are replaced with new ones after the one has been nearly consumed. The self-baking anode made of carbon paste (Soderberg) is contained in a steel frame. While descending toward the hot zone, the paste becomes hard and conducting. On consumption from below new carbon paste is added. Steel studs embedded in the anode carry electric current. They are pulled out and moved up as the anode descends. The anodes undergo oxidation to form CO and CO_2 during usage. In recent years, the use of the Soderberg anodes has a decreasing trend due to their lower production efficiency.

In electrowinning from aqueous solutions, the cathodes are usually in the form of starting thin sheets of the same metal, which has to be deposited. These are prepared by depositing on cathode blanks in special separate electrolytic cells. The thin electrodeposited sheet can be easily stripped. Titanium and stainless steel blanks [10] are used for preparation of copper and nickel starting sheets, respectively. The blank surface is wetted with a parting agent to promote stripping which may be done manually or by using automatic stripping machines. In zinc electrowinning, pure aluminum cathode blanks are used directly. After carrying out deposition operation for 24 h or longer, the cathodes are stripped. In aluminum production, an iron pot lined with carbon, which acts as cathode, is used. In electrolytic production of magnesium, steel cathodes [17] are used because magnesium does not alloy with iron.

12.1 Principle

Electrolysis is basically the decomposition of a liquid (aqueous solution of a salt or mixture of fused salts or ionic compounds) by the passage of an electric current. Thus, the electrolytic process is essentially based on chemical reactions involving ions (anions and/or cations) and electrons which may be represented by electrochemical equations. Potential–pH diagram for the M-H_2O system discussed in Sect. 11.3.3 is useful in electrometallurgy. The diagram may be used as a guide for selective dissolution of metals from anode and deposition of metal ions from the solution (electrolyte) at the cathode. During electrolysis, basically a liquid (aqueous solution or fused salts) ionic compound decomposes by passage of current. In the electrolytic cell, anode is the electrode at which electrons are generated whereas electrons are consumed at the cathode. Thus, the electrolytic cell is similar to the electrochemical cell discussed in Sect. 11.5.2 under the heading "cementation," but the electrode polarities are reversed. The following cathodic reactions may be considered during electrolysis:

1(a). In acid solution, hydrogen evolution is a common cathodic reaction:

$$2H^+ + 2e \rightarrow H_2 \tag{12.2}$$

1(b). In neutral and alkaline solutions, hydrogen evolution takes as:

$$H_2O + e \rightarrow \frac{1}{2}H_2 + OH^- \tag{12.3}$$

2(a). In acid solution, oxygen may be reduced as:

$$\frac{1}{2}O_2 + 2H^+ + 2e \rightarrow H_2O \tag{12.4}$$

2(b). In neutral and alkaline solutions, hydroxyl ions are produced by the reduction of oxygen:

$$\frac{1}{2}O_2 + H_2O + 2e \rightarrow 2OH^- \tag{12.5}$$

3. Metal deposition takes place:

$$M^{ne} + ne \rightarrow M \tag{12.6}$$

4. Metal ions may be reduced as:

$$M^{3+} + e \rightarrow M^{2+} \tag{12.7}$$

5. Anion may get reduced as:

$$2NO_3^- + 4H^+ + 2e \rightarrow N_2O_4 + 2H_2O \tag{12.8}$$

The metal dissolution is the main anodic reaction in electrorefining:

$$M \rightarrow M^{ne} + ne \tag{12.9}$$

Other possible reactions at the anode include reverse of reactions (12.1, 12.2, and 12.3: gas adsorption), and (12.7: oxidation of metal ion: $M^{2+} \rightarrow M^{3+} + e$).

During the electrolysis of molten sodium chloride at above 600 °C, the only two ionic species, Na^+ and Cl^-, produced by the ionization of NaCl (NaCl $\rightarrow Na^+ + Cl^-$) are, respectively, reduced at the cathode, and oxidized at the anode according to the reactions:

$$\text{Reduction at the cathode}: Na^+ + e \rightarrow Na \tag{12.10}$$

$$\text{Oxidation at the anode}: 2Cl^- \rightarrow Cl_2 + 2e \tag{12.11}$$

The chlorine gas produced at the anode is collected as a by-product. On the other hand, in the electrolysis of dilute sulfuric acid, OH^- ions are also generated from water in addition to H^+ and SO_4^{2-}. SO_4^{2-} and OH^- ions move toward the anode and

H^+ ions are converted to hydrogen gas at the cathode (reaction: 12.2). On account of a lower discharge potential, OH^- ions lose electrons easily compared to SO_4^{2-} ions and undergo the anodic reaction to generate oxygen:

$$4OH^- \rightarrow 2H_2O + O_2 + 4e \qquad (12.12)$$

Thus, the electrolysis of dilute sulfuric acid is equivalent to the decomposition of water [18]:

$$2H_2O \rightarrow 2H_2 + O_2 \qquad (12.13)$$

However, in the electrolysis of aqueous copper sulfate solution, there are two important ionization reactions:

$$CuSO_4 \rightarrow Cu^{2+} + SO_4^{2-} \qquad (12.14)$$

$$H_2O \rightarrow H^+ + OH^- \qquad (12.15)$$

In the solution electrolyte, copper ions migrate to the negatively charged electrode (the cathode) under the influence of electrical potential between the two electrodes and combine with the electrons released at the anode to form copper atoms that are discharged at the cathode according to the half-cell reaction:

$$Cu^{2+} + 2e \rightarrow Cu \quad \text{(at the cathode)} \qquad (12.16)$$

The reaction at the anode depends on the nature of the electrode material. For example, the discharge of hydroxyl ions at the nonconsumable anode (inert, e.g., carbon, platinum, etc.) generates oxygen gas (reaction 12.12) whereas in the case of consumable crude copper anode, Cu^{2+} ions are produced as per reaction:

$$Cu \rightarrow Cu^{2+} + 2e \quad \text{(at the anode)} \qquad (12.17)$$

Combining the above two reactions at anode and cathode, the net reaction can be represented as:

$$Cu \text{ (at anode)} \rightarrow Cu \quad \text{(at cathode)} \qquad (12.18)$$

The above cell reaction summarizes the basis of electrorefining.

12.1.1   Cell Potential

The minimum theoretical cell voltage, E_{cell}, required for electrolysis significantly varies for aqueous and fused electrolytes. During the electrolysis of aqueous solutions, the theoretical voltage depends on the reversible electrode potentials of anode and cathode. E_{cell} can be estimated with the aid of Nernst's equation and the electrochemical series, as discussed in Chap. 11. In this chapter, it has been demonstrated that water decomposes with the evolution of oxygen and hydrogen gases, respectively, at the anode and cathode on application of 1.23 V (i.e., $E_{cell} = -1.23$ V) at 298 K. E_{cell} can also be estimated from the knowledge of free energy ($\Delta G^o = -nFE_{cell}$). For example,

$$H_2O \rightarrow H_2 + \frac{1}{2}O_2, \Delta G^o_{298} = +237 \text{ k J} \tag{12.19}$$

In the above reaction, $n = 2$ because two electrons are transferred in reactions:

$$2H^+ + 2e \rightarrow H_2 \tag{12.20}$$

$$\text{and } 2OH^- \rightarrow H_2O + \frac{1}{2}O_2 + 2 \text{ e} \tag{12.21}$$

$$\therefore E_{cell} = -\frac{237000}{2 \times 96500} = -1.23 \text{ V}$$

The positive value of ΔG^o (and the negative value of E_{cell}) indicates that the decomposition of water does not proceed spontaneously. Hence, to produce hydrogen and oxygen, a voltage slightly more than 1.23 V must be applied. Equation, $\Delta G^o = -nFE_{cell}$, is more useful in the calculation of E_{cell} in fused salt electrolysis due to lack of availability of E^o data in fused salt mixture.

The voltage required to operate an electrolytic cell is the sum of (i) emf for the chemical reaction (E_{cell}), (ii) emf due to polarization at the anode (η_a) and cathode (η_c), (iii) emf to overcome the ohmic resistance of the electrolyte and the contact potentials between electrodes and bus bars (E_{res}). The contribution of each may be accounted as:

1. **Chemical reaction:** The deposition of copper from an aqueous solution of $CuSO_4$ may be represented as:

$$CuSO_4 + H_2O = Cu + H_2SO_4 + \frac{1}{2} O_2 \tag{12.22}$$

$$\text{or } Cu^{2+} + H_2O = Cu + 2H^+ + \frac{1}{2} O_2 \tag{12.23}$$

The standard free energy change for reaction (12.23) is the sum of the free energies of the following two half-cell reactions:

at the cathode : $Cu^{2+} + 2e = Cu$, $\Delta G_{298}^o = -65620$ J (12.16)

and at the anode : $H_2O = H_2 + \frac{1}{2} O_2$, $\Delta G_{298}^o = 237000$ J (12.19)

and $H_2 = 2H^+ + 2e$, $\Delta G_{298}^o = 0$ (12.24)

The standard free energy of ionization of hydrogen is zero because E^o for the standard hydrogen electrode has been arbitrarily chosen as zero. Hence, the free energy, ΔG_{298}^o of the overall/net reaction (12.23) is 171380 J:

$$\therefore E^o = -\frac{171380}{2 \times 96500} = -0.89 \text{ V}$$

The positive value of the free energy (hence the negative value of emf) suggests that reaction (12.16) will not occur spontaneously from left to right. The negative value of -0.89 V is a measure of the emf, which has to be applied to allow reaction (12.23) to occur. Thus, application of 0.89 V would bring the cell into equilibrium without any net change. On the other hand, application of a higher voltage will lead to copper deposition according to reaction (12.23). The emf of the net chemical reaction (E_{cell}) may be calculated from Nernst's equation: $\left[E = E^o + \frac{RT}{nF} \ln \frac{\text{(activity of the species in oxidized state)}}{\text{(activity of the species in reduced state)}}\right]$. Hence, for the reaction at the cathode: $Cu^{2+} + 2e = Cu$ (12.16), we can write:

$$E_{cell} = E^o + \frac{RT}{nF} \ln \frac{a_{Cu^{2+}}}{a_{Cu}}$$ (12.25)

As $n = 2$ for reaction (12.16), at 298 K Eq. 12.25 may be simplified as:

$$E_{cell} = E^o + 0.0295 \cdot \log \left[\frac{a_{Cu^{2+}}}{a_{Cu}}\right]$$ (12.26)

The second term ($0.0295 \cdot \Sigma \log a_i$) on the right-hand side acts as the correction factor for activities. A change by a factor of 10 in the activity of ions in the electrolyte will bring a change in the cell emf by only 0.059 V for $n = 1$, 0.0259 V if $n = 2$, and so on. Thus, E_{cell} is not much affected due to activity changes and hence it may be assumed to be constant and has a value equal to the standard value (E^o). For example, a tenfold decrease in the copper activity together with a tenfold increase in the activity of acid increases E_{cell} by only 0.06 V (about equal contribution of 0.0295 V by each). Contrary to this, during electrorefining, $\Delta G = 0$ when the thermodynamic state of the metal being dissolved is almost identical to that of the metal being deposited; hence, $E_{cell} = 0$. Although the cell voltage is zero, about 0.3 V is normally required to overcome ohmic resistance and over-voltages.

2. **Anode (η_a) and cathode (η_c) polarization**: The anode polarization is due to oxygen over-voltage caused by local surplus in sulfuric acid concentration and because of the difficulty of nucleating oxygen bubbles. η_a can be reduced by agitation or by using a catalyst or a reducing agent to pick up oxygen without forming gas bubbles. On the other hand, the cathode polarization (η_c) occurs near the cathode surface due to the concentration gradient in the metal ions caused by the depletion in the solution.

3. **Ohmic resistance of the electrolyte:** The voltage, E_{res} required to overcome the ohmic resistance of the electrolyte and the connection between electrodes and bus bars is called ohmic drop. It can be lowered by optimizing the composition of the electrolyte to have a higher proportion of conducting ions and by cell design to have wide and short electrical paths leaving space for the growth of cathodes. Since resistance R of a solid conductor is directly proportional to length l and inversely proportional to cross-sectional area A, we can write:

$$R = \rho \frac{l}{A} \tag{12.27}$$

where ρ is the specific resistivity of the electrolyte in $\Omega \mathrm{cm}^{-3}$. Ionic species present affect the value of ρ. Hydrogen and hydroxyl ions make least contributions in the value of ρ. Thus, low resistances are obtained in fairly strong acid or alkaline solutions. Hence, the total operating voltage required for a cell is given as:

$$E = E_{cell} + \eta_a + \eta_c + E_{res} \tag{12.28}$$

In general, plant operation data show that E is normally two or three times larger than E_{cell}. For example, in deposition of copper, a potential drop of 2.3 V per cell occurs although E_{cell} is only 0.89 V. Since resistivity of the electrolyte decreases with an increase in temperature, electrolytic tanks are operated at about $40 - 50\,°C$. However, in case of fused salts or oxides, higher resistance is desired to generate larger heat (I^2R) in order to maintain the cell at a higher temperature. As a result, a higher cell voltage is required. This is exemplified in aluminum extraction where E_{cell} is only 1.2 V for the net chemical reaction, but the total emf required for deposition of aluminum from Al_2O_3 dissolved in cryolite melt is > 4.2 V ($E_a \simeq 0.50$, $E_c \simeq 0.15$, $E_{res} \simeq 2.45$ V).

The actual cell potential employed depends on the magnitude of current required to attain the desired rate of production. E_{cell} is independent of the value of current, but polarization at the electrodes and voltage drops across the electrolyte, the bus bars, and the contacts increase with the use of higher current.

12.1.2 Discharge Potential

The discharge potential refers to the minimum potential between the electrode and electrolyte required for continuous discharge of an ion. Depending on the electrode reaction, it is the sum of the reversible electrode potential (E_r) and over-potential (η), that is, $E_{discharge} = (E_r + \eta)$. For example, the discharge potential of Zn ions on zinc electrode is -0.96 V because $E_r = -0.76$ V $\left(E_{Zn}^o\right)$ and $\eta = -0.2$ V. Similarly, the discharge potential of hydrogen [18, 19] is -1.13 V because, for hydrogen, $E_r = 0$ $\left(E_H^o\right)$ and $\eta = -1.13$ V. It is important to note that hydrogen over-potential (more commonly known as hydrogen over-voltage) plays a significant role in the deposition/discharge of metal ions on a cathode. Zinc will be deposited without hydrogen evolution provided the applied voltage is such that the electrode potential lies between the discharge potentials of zinc and hydrogen. The absolute value of over-potential is controlled by current density, pH, and nature of the bath. On the other hand, only hydrogen will evolve at the cathode without any metal deposition/discharge if the electrode potential of the metal is less than -1.13 V. Thus, zinc, despite being relatively negative to hydrogen in the electrochemical series, is deposited without hydrogen evolution due to hydrogen over-voltage. In a similar manner, Mn, Cr, Fe, Cd, Co, Ni, Sn, and Pb are deposited from aqueous electrolytes on account of higher hydrogen over-voltage. Noble metals like Ag and Cu (electropositive to hydrogen) can be deposited from aqueous solutions without hydrogen evolution whereas reactive base metals with more negative discharge potentials would evolve hydrogen instead of metal discharge at the cathode (e.g., electrolysis of NaCl solution). Hence, fused salt electrolysis becomes essential in electrowinning/electrorefining of calcium, sodium, magnesium, beryllium, aluminum, titanium, thorium, and niobium. In contrast to this statement, sodium is deposited from aqueous solutions on a mercury cathode due to its high hydrogen over-voltage on the extremely smooth surface of the liquid metal and low activity of sodium in dilute solution of mercury. Table 12.1 classifies metals under three categories according to their chances of deposition from aqueous electrolytes without hydrogen evolution, under hydrogen over-voltage, and only from fused salt electrolytes.

The above knowledge can be made use of in selective removal of impurities during electrolysis. For example, since the discharge potentials of nickel $(-0.45$ V$)$ $[= E_r (-0.25) + \eta (-0.2 \text{ V})]$, cobalt $(-0.40$ V$)$, and iron $(-0.48$ V$)$ are comparable, some Co and Fe will deposit along with nickel in the electrowinning of nickel from soluble Ni_3S_2 cast anodes. The problem of the deposition of Co and Fe has been sorted out by separating the electrolytic cell into anode and cathode compartments with the aid of suitable diaphragms together with the provision of anolyte purification.

The generation and deposition/discharge of ions, respectively, at the anode and cathode can be estimated with the help of Nernst's equation using activity values of

Table 12.1 Normal electrode potentials ($E°$ V) of some metals at 25 °C (with reference to hydrogen, E_H^o)

	Electronegative metals with reference to hydrogen	
Electropositive metals that can be deposited from aqueous solutions	Metals that can be deposited by electrolysis of aqueous solutions under hydrogen over voltage	Metals that can be deposited by electrolysis of fused salts only
Au^+, Au +1.68	Pb^{2+}, Pb −0.126	Nb^{3+}, Nb −1.10
Au^{3+}, Au +1.42	Sn^{2+}, Sn −0.140	V^{3+}, V −1.51
Pt^{2+}, Pt +1.215	Mo^{2+}, Mo −0.20	Al^{3+}, Al −1.667
Pd^{2+}, Pd +0.83	Ni^{2+}, Ni −0.250	Be^{2+}, Be −1.72
Ag^+, Ag +0.799	Co^{2+}, Co −0.277	Ti^{2+}, Ti −1.75
Hg^+, Hg +0.799	Cd^{2+}, Cd −0.402	U^{3+}, U −1.80
Cu^{2+}, Cu +0.344	Fe^{2+}, Fe −0.441	Th^{4+}, Th −1.90
Bi^{3+}, Bi +0.23	Cr^{3+}, Cr −0.744	Mg^{2+}, Mg −2.376
Sb^{3+}, Sb +0.10	Zn^{2+}, Zn −0.761	Na^+, Na −2.714
W^{3+}, W +0.05	Mn^{2+}, Mn −1.05	Ca^{2+}, Ca −2.87
		K^+, K −2.924
		Li^+, Li −3.024

ions in the electrolyte and the metals in the solid electrode. If $a_{M^{2+}}$ and a_M, respectively, refer to the activity of M^{2+} ions in the electrolyte and the activity of M in the solid electrode, a high value of the ratio $a_{M^{2+}}/a_M$ (i.e., $a_{M^{2+}} \gg a_M$) means that the dissolution of M is highly favorable at the anode. On the other hand, if $a_{M^{2+}} \ll a_M$, the deposition of M^{2+} ions will take place at the cathode (see Problems 12.5 and 12.6).

12.1.3 Current and Energy Efficiency

According to Faraday's Law of electrolysis, the amount of metal deposited, m, is calculated as $m = zct$, where z, c, and t are the electrochemical equivalent, current, and time, respectively. Generally, the current used exceeds the theoretically required value in depositing the metal from the leach liquor electrolyte. This introduces a term "current efficiency" which is defined as the ratio of the weight of the metal actually deposited to that which is expected to be deposited, theoretically, in accordance with the Faraday Law. It is usually expressed in percentage. This anomaly is mainly due to the following unavoidable side reactions that occur during electrolysis:

1. Some hydrogen may be occluded in the deposited metal due to hydrogen evolution. Discharge of hydrogen at the cathode should be controlled by limiting the cell voltage.
2. A fraction of iron present in the electrolyte oxidizes ($Fe^{2+} \rightarrow Fe^{3+}$) at the free surface of the anode with air or oxygen. Fe^{3+} is then reduced at the cathode. Thus, the current efficiency decreases due to the loss of current in oxidation/reduction

reactions. The current efficiency goes below 80% when iron content in copper leach liquor is very high compared to a more normal value of 95% at low iron concentrations. Hence, iron must be reduced to the required level prior to leaching by converting it to an insoluble product. Alternatively, the leach liquor is purified for removal of iron prior to electrolysis, for example, removal of iron by jarosite precipitation in zinc extraction (see Sect. 11.4.1).
3. The re-dissolution of cathode material may also reduce current efficiency. This may occur when acidity reaches a very high value.

In addition to the above side reactions, current may be lost through leakages to earth and between bus bars. Due to the humid conditions prevailing in tank houses, these losses cannot be eliminated, but they can be minimized by maintenance of a high standard of cleanliness in the plant.

Some heat ($\propto I^2 R$) is generated due to the current and the resistance of the electrolyte. In order to pass high current without any appreciable rise in temperature, the resistance of aqueous solution electrolyte should be kept low. The resistivity of acid solutions is lower than that of alkaline solutions. Neutral solutions have the highest resistivity and hence must be avoided. As resistivity of electrolytes decreases with an increase in temperature, electrolytic tanks are commonly operated at about 40–50 °C. While using fused salts/oxides as electrolytes, a higher resistance is necessary to maintain a high cell temperature at any suitable current. This will require a corresponding high voltage. Occasionally, supplementary heating may be required.

In order to improve the process efficiency and purity of the product, leach liquor may need purification to eliminate certain ions prior to electrolysis. The quality of the cathode deposit is affected by temperature, pH, and current density. Normally, current density varies from 100 to 300 A m^{-2}. The highest value is used to achieve the maximum production rate. However, irregular, loose, and spongy deposits are produced at high current densities. Occasionally, metal may even fall in the cell, resulting in a considerable decrease in its current efficiency. Additives to the electrolyte, such as colloids, glues, and gum may improve the physical properties of the deposit at even high current densities.

Energy efficiency is generally reported as electric power required to produce 1 kg of the metal. It is expressed as kWh kg^{-1}. Energy efficiency is calculated as:

$$\text{Energy efficiency } (\%) = \text{current efficiency } (\%) \times \frac{\text{theoretical voltage}}{\text{applied voltage}} \quad (12.29)$$

At a constant applied voltage, the energy efficiency increases with an increase in the current efficiency, that is, more useful utilization of current with the decreasing overall power. The situation becomes more complicated when the applied voltage changes at constant current efficiency. Power input increases with an increase in voltage, but useful work does not vary proportionately because the applied voltage is lost in overcoming polarization and resistance effect in the cell. A reduction in the applied voltage decreases power input, but a certain minimum voltage is required to achieve an appropriate rate of electrolysis. Therefore, optimum conditions are adjusted to maintain a balance between current efficiency and energy efficiency. The energy efficiency in electrowinning/electrorefining of aluminum is poor because energy is consumed in keeping the electrolyte in the molten state that is required for carrying out electrolysis. There is no alternative route to get higher energy efficiency, except that larger pot cells are employed for electrolytic reduction of alumina. The economics of the electrolytic process may be expressed in terms of energy consumption per kg of metal deposited by the relation [10]:

$$\text{Energy consumption } \left(\text{kWh kg}^{-1}\right) = \frac{E.F}{\left(M/_z\right) \times 3600 \times 0.01(\text{CE})} \qquad (12.30)$$

where E is the applied voltage in volts, F the Faraday constant, M is the relative atomic mass, and CE is the percentage current efficiency.

12.2   Applications

12.2.1   Electrowinning

Table 12.1 classifies metals into three groups which can be electrowon/electrorefined either from aqueous solutions or fused salt electrolytes. There are two categories of metals which can be deposited from aqueous electrolytes. Metals electropositive to hydrogen can be deposited without hydrogen evolution whereas metals electronegative to hydrogen are deposited under hydrogen over-voltage. Highly reactive metals with negative discharge potentials can only be deposited from fused salt electrolytes.

12.2.1.1   Electrowinning from Aqueous Electrolytes

From the preceding section and Table 12.1, it is clear that gold, silver, and copper can be electrowon from aqueous solutions whereas zinc, chromium, cadmium, cobalt, iron, nickel, and manganese can be discharged on the cathode from aqueous solutions under hydrogen over-voltage. During extraction of zinc and manganese, simultaneous requirements of avoiding hydrogen evolution as well as preventing co-deposition of copper, cobalt, antimony, and so on pose severe problems. This warrants removal of impurities from leach liquors prior to electrolysis by precipitation, cementation, and/or use of diaphragms. In the electrowinning of zinc from leach liquors obtained by sulfuric acid leaching of roasted sphalerite concentrate, iron is eliminated by jarosite precipitation whereas cadmium and copper are removed by cementation. Diaphragms separating the electrolytic cell into anode and cathode compartments are used to prevent deposition of iron and cobalt along with nickel in electrowinning nickel from Ni_3S_2 anodes. If iron present in the leach liquor is not removed, Fe^{2+} and Fe^{3+} ions will alternatively get oxidized and reduced at anode and cathode, respectively. This back-and-forth oxidation/reduction would increase the power consumption of the electrolytic process and hence the current and energy efficiency will decrease. Generally, inert or insoluble anodes of lead or lead alloys are employed in electrowinning. However, occasionally, there may be some exception, for example, use of cast Ni_3S_2 anodes in extraction of nickel. Hydrogen over-voltage is controlled by pH and use of cathodes of different metals and alloys (other than the metal under extraction). During extraction of a metal, the more noble metals must be eliminated from the electrolyte or else they will get preferentially deposited along with the metal. Owing to this problem, more negative metals, namely, zinc and manganese, require a higher degree of purification. The electrowinning of these metals requires thorough elimination of impurities which reduce hydrogen over-voltage.

Electrowinning of Copper About 20% of primary copper amounting to 2.5 million tons per year is obtained through the hydrometallurgical route involving leaching, solvent extraction, and electrowinning from aqueous solutions [20, 21]. The leach liquor is generated by heap and dump leaching of oxide ores in dilute sulfuric acid. Chalcocite, bornite, and covellite are easily leached in sulfuric acid whereas leaching of chalcopyrite requires an oxidant. It enhances in the presence of bacteria (Sect. 11. 1.6).

The leach liquor so obtained is not suitable for direct electrodeposition of high-purity copper because it is not only extremely dilute in copper content (1 – 6 kg Cu^{2+} m^{-3} of solution) but also highly impure due to high iron content (1 – 10 kg m^{-3}). For production of smooth, dense, and high-purity cathode copper on an industrial scale, the copper content of the electrolyte should be more than 35 kg Cu^{2+} m^{-3} of solution. The leach liquor is therefore concentrated by the solvent extraction technique using the oxime group of organic extractant, namely, aldoximes and ketoximes [22]. The extractant, 5 – 10% (by volume) dissolved in purified kerosene when contacted with the leach liquor, extracts most of the copper, leaving very little in the raffinate (0.3 kg Cu^{2+} m^{-3}) according to the reaction: $2RH$ (org) + Cu^{2+}(aq) = R_2Cu (org) + $2H^+$(aq). In the next step, copper is stripped into Cu^{2+}-depleted aqueous phase in order to enrich the electrolyte to the desired high Cu^{2+} concentration (45 kg m^{-3}) for efficient electrowinning. The electrolyte almost free from zinc and iron is electrolyzed in a PVC-lined concrete tank having a rolled Pb-Sn-Ca alloy anode and a 316 L stainless steel cathode [20]. The electrolytic reactions may be represented as:

$$\text{At the cathode}: Cu^{2+} + 2e \rightarrow Cu, E^o = 0.34 \text{ V}$$
$$\text{At the anode}: H_2O \rightarrow \frac{1}{2}O_2 + 2H^+ + 2e, E^o = -1.23 \text{ V}$$

Instead of theoretically required 0.89 V for the electrolysis of one molal $CuSO_4$ solution, about 2 V are actually used in order to overcome (i) oxygen over-potential (0.5 V), (ii) copper deposition over-voltage (0.05 V) and (iii) ohmic resistance at a cathode current density of 300 A m^{-2} (0.5 V). Well-managed electrolytic plants operate at current efficiencies of more than 90% [3, 20]. Copper stripped from the cathode is washed and stored, the spent electrolyte is recycled to the solvent extraction units [22], and oxygen is released into the atmosphere.

The depleted solution (raffinate), containing approximately 0.4 kg Cu and 5 kg H_2SO_4 m^{-3} of solution from the solvent extraction unit, is used as lixiviant for industrial-scale leaching. Before recycling, the concentration of H_2SO_4 in the raffinate is increased to 10 kg m^{-3}. In all electrowinning plants, guar [23] gum is dissolved in electrolytes (about 250 g per ton of cathode copper) to produce smooth and dense copper deposits. About 150 ppm of Co^{2+} [3, 20] in the electrolyte (added as $CoSO_4$ solution) facilitates O_2 evolution at the anode. This is helpful in retarding oxidation of lead anodes. In turn, anode life increases and contamination in copper deposit decreases.

Electrowinning of Zinc   About 80% of primary zinc is produced via the hydromet-allurgical route involving leaching and electrowinning from aqueous solutions. As zinc is electronegative to hydrogen, the electrowinning of zinc is made possible by hydrogen over-voltage. From the discussion in Sect. 12.2.1.1, it is clear that, for efficient zinc deposition, sulfate solution obtained from the leaching circuit needs thorough purification because the purified electrolyte renders high hydrogen over-voltage suitable for electrodeposition of zinc. Hydrogen over-voltage decreases significantly in the presence of even traces of arsenic, antimony, germanium, cobalt, and nickel to a level to not only stop zinc deposition but also cause dissolution of the deposited zinc. Cobalt and nickel decrease cathode current efficiency [24].

After removal of iron by jarosite process (see Sect. 11.4.1), a series of cementa-tion reactions are carried out to remove copper, cadmium, cobalt, and nickel. The procedure varies from plant to plant in terms of number of steps, temperature, and reagents used for cobalt removal. In this section, the steps adopted in the Outokumpu process [10] are briefly outlined:

1. In the first step, a calculated amount of zinc dust is added to cement out copper, leaving about 50–80 gm^{-3} of copper in the solution:

$$Cu^{2+} + Zn = Cu + Zn^{2+} \tag{12.31}$$

$$\text{and } Cu^{2+} + Cu + H_2O = Cu_2O + 2H^+ \tag{12.32}$$

   The cemented copper is of high purity, and a higher proportion of Cu_2O may be produced for reducing zinc consumption.

2. Cobalt, nickel, and remaining copper can be precipitated by treating the solution at higher temperatures with As (III) oxide and zinc:

$$2HAsO_2 + 2Co^{2+} + 6H^+ + 5Zn = 2CoAs + 5Zn^{2+} + 4H_2O \tag{12.33}$$

$$2HAsO_2 + 2Ni^{2+} + 6H^+ + 5Zn = 2NiAs + 5Zn^{2+} + 4H_2O \tag{12.34}$$

$$Cu^{2+} + Zn = Cu + Zn^{2+} \tag{12.31}$$

$$2HAsO_2 + 6Cu^{2+} + 6H^+ + 9Zn = 2Cu_3As + 9Zn^{2+} + 4H_2O \tag{12.35}$$

3. The filtrate after separation of the residue is passed through a series of fluidized beds of zinc dust to cement out cadmium:

$$Cd^{2+} + Zn = Cd + Zn^{2+} \tag{12.36}$$

The large surface area of zinc powder eliminates formation of long cadmium dendrites and gives a coarse product, easily separable in a hydro-cyclone. The resulting high-purity solution can now be fed to the electrolytic tank for zinc deposition. Manganese can only be tolerated as an impurity because it is more electronegative as compared to zinc and does not affect the cathode reaction. However, it partly oxidizes at the anode to form MnO_2, which settles at the bottom of the cell. This MnO_2 may be recycled to the neutral leach to oxidize Fe^{2+} to Fe^{3+}.

The purified leach liquor containing 55–90 kg Zn^{2+} and 125–175 kg H_2SO_4 m^{-3} of solution is electrolyzed at about 30–35 °C. Hard rolled aluminum cathodes are employed to attain a higher over-voltage and Pb–0.5–1.0% Ag alloy anodes [25] to lower oxygen over-voltage for oxygen evolution, which is advantageous in retarding the corrosion of the anode. At a current density of 300 A m^{-2}, the cathode current efficiency of zinc deposition is 90% and energy consumption works out to be 3.2 kWh kg^{-1}. The spent electrolyte containing 50–65 kg Zn^{2+} and 150–200 kg H_2SO_4 m^{-3} of solution is recycled to the leaching plant. At the Cominco zinc plant, cathode current efficiency of 94% has been achieved by careful control of glue addition.

In addition to the purification of the electrolyte, attention has to be paid to the operating variables that increase hydrogen over-voltage and produce better deposits with high current efficiency: (i) smooth zinc cathode surface maximizes hydrogen over-voltage; (ii) cathode material, that is, aluminum sheet is preferred to get higher hydrogen over-voltage; (iii) addition of glue in the electrolyte increases hydrogen over-voltage as well as current efficiency; (iv) proper circulation of electrolyte avoids concentration polarization (e.g., use of grid anode provides better circulation); and (v) temperature rise reduces hydrogen over-voltage, hence bath is maintained at 30–35 °C.

Electrowinning of Nickel In the production of nickel, the normal extraction procedure of partial roasting, smelting, and converting renders pentlandite [(NiFe)$_4$S$_8$] concentrate into a product, which cannot be easily separated. The resulting matte is very slowly cooled from 1100 °C to 400 °C (in 3 days), ground, and magnetically separated to obtain Cu–Ni alloy and nickel–copper matte [26]. Ni$_3$S$_2$ and Cu$_2$S can be separated by froth flotation technique after fine grinding.

In electrowinning with insoluble anodes, electrolyte is purified leach liquor obtained by leaching nickel–copper matte (Ni_3S_2–Cu_2S) in sulfuric acid. Both nickel and copper may be recovered by adjusting the process variables. Purification of electrolyte is an important step in the electrowinning of nickel. Cobalt is precipitated as stable Co (III) hydroxide by Ni (III) hydroxide, prepared electrolytically. The filtrate is further purified by solvent extraction using LIX 64N extractant. Copper is first extracted at a pH of 2. Nickel is then extracted from the raffinate at a pH of 9 by pH adjustment with ammonia. Leach liquor obtained by acid pressure leaching of laterite ores of nickel is also purified by the solvent extraction method. Nickel from the purified liquor is electrodeposited on nickel starting cathode using a Pb–Sb alloy anode. The spent electrolyte that is rich in acid is recycled to the leaching circuit for additional dissolution of the matte. In the presence of cobalt, chloride media work better over sulfate electrolyte. Anionic cobalt complex, $CoCl_4^{2-}$, formed in the chloride solution is more effectively separated from the uncomplexed nickel. Higher conductivity of the chloride electrolyte and lower over-potential associated with chlorine evolution permit electrodeposition at higher current densities with lower energy consumption.

Nickel can also be electrowon with soluble anodes of cast Ni_3S_2. For this purpose, Ni_3S_2 obtained by froth flotation containing 26.7% S is melted and cast into anodes. The solidified anodes kept in an insulated annealing box are very slowly cooled so that βNi_3S_2 transforms completely into $\beta' Ni_3S_2$ at 505 °C. The latter form of Ni_3S_2 is stronger and does not disintegrate or crack during usage. By mass, the anode analyzes 70% Ni, 3% Cu, 0.5% Co, 0.5% Fe, and about 20% S together with small quantities of Pb, Zn, Se, and precious metals. Ni_3S_2 anode is electrolyzed using mainly sulfate electrolyte and pure nickel starting cathode. The anodic reaction is:

$$Ni_3S_2 \rightarrow 3Ni^{2+} + 2S + 6e \tag{12.37}$$

Nickel ions in the purified catholyte compartment are deposited on the cathode ($3Ni^{2+} + 6e \rightarrow 3Ni$). Thus, the overall cell reaction is:

$$Ni_3S_2 \rightarrow 3Ni + 2S, \Delta G^o = +203 \text{ kJ} \tag{12.38}$$

$$\text{For this reaction, } n = 6, E_{cell} = -\frac{\Delta G^o}{nF} = -0.35 \text{ V}$$

However, in practice, about 2.85 V is applied. The components of the applied voltage are $E_{cell} = -0.35$ V, $\eta_a = 1.1$ V, $\eta_c = 0.2$ V, and $E_{res} = 1.2$ V. The negative sign is ignored because all components have to be overcome by the applied voltage. Initial dissolution of the anode at a potential of about 2.85 V produces the anode slime containing more than 90% elemental sulfur. The anode slime occupying twice the volume of the starting anode is quite adherent. Hence, during electrolysis, cell resistance increases due to the increase in thickness. As a result, the applied voltage has to be increased from about 2.85 to 4.5 V. At this high positive potential [10], most of the impure metals present in the anode dissolve along with nickel sulfide:

$$FeS = Fe^{2+} + S + 2e \tag{12.39}$$

$$Cu_2S = Cu^{2+} + CuS + 2e \tag{12.40}$$

$$CuS = Cu^{2+} + S + 2e \tag{12.41}$$

$$CoS = Co^{2+} + S + 2e \tag{12.42}$$

In addition to the above dissolution reactions at the anode, some oxygen is evolved: $2H_2O = O_2 + 4H^+ + 4e$ because the anode potential is large enough to decompose some water. As a result, current efficiency decreases. Thus, there is an imbalance between the nickel entering in the electrolyte and that depositing on the cathode. Precious metals and selenium remain undissolved and join the anode slime. The resulting anolyte at a pH of 1.9 has high amounts of impurities such as Cu 0.4, Fe 0.12, and Co 0.08 kg m^{-3} of solution. Such an electrolyte is not suitable for nickel electrowinning. The deficiency of nickel and excess of acid have to be adjusted during anolyte purification. By adjusting pH to 4–5, purification maximizes the current efficiency. For purification of the electrolyte, diaphragm cells are employed to separate anode and cathode compartments. The diaphragm is a canvas screen which does not obstruct the flow of electric current. In addition, anodes are enclosed in loosely woven bags to facilitate removal of anode slimes. The anolyte is pumped into the chemical treatment plant and the resulting purified nickel solution is fed to the cathode compartment to deposit nickel. The process is continuous. Diffusion of impure anolyte to the cathode compartment is prevented by maintaining a slightly higher hydrostatic head of the purified electrolyte in the cathode compartment. In this way, catholyte seeps to the anode compartment by gravity but does not get contaminated by the untreated solution.

In the first step of electrolyte purification, pH of the anolyte is increased to 3.7 by the addition of nickel hydroxide slurry:

$$2H^+ + Ni(OH)_2 = Ni^{2+} + 2H_2O \tag{12.43}$$

Chlorine is injected to oxidize Fe (II) to Fe (III) and precipitate iron as Fe(OH)$_3$. The clarified solution is then allowed to react with nickel carbonate to oxidize Co (II) and precipitate cobalt as Co(OH)$_3$:

$$2Co^{2+} + Cl_2 + 3H_2O + 3NiCO_3 = 2Co(OH)_3 + 3Ni^{2+} + 3CO_2 + 2Cl^- \tag{12.44}$$

The above precipitation reactions also energize coprecipitation of arsenic and lead that are removed along with $Co(OH)_3$ precipitation and filtration. Finally, copper is removed as CuS by stirring the solution with nickel powder and elemental sulfur:

$$Cu^{2+} + Ni + S = CuS + Ni^{2+} \tag{12.45}$$

The sequence of removal of impurities may vary from plant to plant. Copper may be first precipitated as Cu_2S by introducing H_2S gas along with the coprecipitation of arsenic and lead. Iron and cobalt may be precipitated simultaneously with chlorine and nickel carbonate. Before feeding the electrolyte to the cathode compartments the chemical composition of the purified solution is adjusted to have approximately [10], Ni^{2+} 60, SO_4^{2-} 100, NaCl 100 and H_3BO_3 20 kg m^{-3} of the solution.

Electrowinning of Manganese It is more difficult to deposit manganese electrolytically from aqueous solutions because its standard electrode potential (-1.05 V) is more electronegative than zinc (-0.76 V). Since hydrogen over-voltage decreases with an increase in pH, manganese deposition is facilitated at higher pH values. As the electrolytic process of cathodic deposition of manganese requires a relatively high applied voltage, there would be some decomposition of water: $2H_2O = O_2 + 4H^+ + 4e$. Thus, hydrogen ion concentration at the anode would increase and pH will decrease. Under this condition, manganese can only be deposited if anolyte and catholyte compartments are separated from each other by a diaphragm. This requires very careful control of temperature and composition of the electrolyte to achieve a reasonable deposition. In order to improve the current efficiency, the electrolyte must be made free from impure metals nobler than manganese.

The leach liquor for manganese electrolysis is obtained by leaching manganese ores in sulfuric acid. Ores containing higher oxides of manganese are given reduction roast treatment to enhance the kinetics of dissolution. Leaching of the calcine is carried out with the spent electrolyte at a pH of 2.5. Subsequently, pH is raised to 6.5 by introducing ammonia and additional calcine to precipitate hydroxides of iron and aluminum, which also carry away molybdenum, arsenic, and silica. In order to prevent the formation of $Mn(OH)_2$ in the catholyte compartment, SO_2 is injected. The resultant solution [10] containing manganous sulfate 30–40 and ammonium sulfate 125 kg m^{-3} of solution at a pH of 8.3 is electrolyzed using stainless steel cathode and insoluble Pb–1% Ag alloy anode at 5.1 V and a current density of 450–650 A m^{-2}. The current efficiency of manganese deposition is low, of the order of 50%. Since high cell voltages are used, the energy consumption is large, about 8.5 kWh kg^{-1}. The acid anolyte still containing some manganese is recycled to the leaching circuit.

12.2.1.2 Electrowinning from Fused Salt Electrolytes

Since the base metals with higher negative discharge potentials evolve hydrogen instead of metal deposition at the cathode, calcium, sodium, magnesium, beryllium, aluminum, titanium, thorium, and niobium are deposited from fused salt electrolytes at relatively high temperatures. Although the basic principles of electrolytic operations involving fused salts and aqueous solutions are the same, processes differ from each other in several aspects. The technical problems involved in the purification of electrolytes and the recovery of metals are similar in both cases. In the production of reactive as well as highly electronegative metals, production and purification of a suitable salt and electrolysis of the salt/compound dissolved in a suitable salt mixture for metal recovery are the two major steps. The fused salt electrolyte bath is prepared by dissolving a pure salt or a compound of the desired metal in an inert salt mixture of reactive metals. The decomposition of the salt/compound is carried out by passing DC electric power through the fused electrolyte contained in a pot with the provision of collecting the produced metal.

As the fused salt electrolysis is carried out at relatively high temperatures, metal deposited at cathodes requires protection from the atmosphere. For example, metallic aluminum is collected on the cathode at the bottom of the electrolytic cell and is protected by the electrolyte above it whereas magnesium being lighter floats on the electrolyte and a specially designed cell is needed to protect the metal from the anodic chlorine.

Electrowinning of Aluminum The entire primary aluminum is extracted by fused salt electrolytic process developed independently in 1886 by Charles Martin Hall in the United States and Paul L.T. Heroult in France. The pure Al_2O_3 produced from bauxite ($Al_2O_3.H_2O$) by the Bayer process is first exposed to the cell exhaust gas in a scrubbing chamber to capture gaseous and particulate fluorides. The resulting alumina with adsorbed moisture and fluoride, on charging into the molten cryolite (Na_3AlF_6) bath contained in the electrolytic cells, reacts to form gaseous species [$(4Na_3AlF_6 + Al_2O_3(dissolved) + 3C\,(s) = 4Al\,(l) + 6NaF\,(dissolved) + 3CF_4(g)$], that causes stirring and thus assists in the dissolution of alumina in the electrolyte. Alumina dissolves in cryolite by forming complex oxyfluoride aluminate anions [27]. The cryolite (melting point: 1012 °C) fuses at lower temperatures by the addition of AlF_3 and CaF_2 so as to conduct the electrolytic operation at about 960 °C. A flow sheet showing various steps in the extraction of aluminum from bauxite is presented in Appendix A.11. The solubility of alumina varies largely with the concentration of AlF_3 in the electrolyte and temperature. Although saturation ranges between 6% at 960 °C and 13% at 1000 °C [27], modern electrolytic cells are generally operated with 2–4% Al_2O_3 in the electrolyte. As per recent practice, the typical electrolyte [27] contains 9–11% AlF_3, 4–6% CaF_2 and 1.5–4% Al_2O_3. The electrolyte having a liquidus temperature in the range of 945–950 °C is subjected to electrolysis in the Hall–Heroult cells in the temperature range of 955–965 °C. The cells require 5–10 °C of superheat (difference between the operating temperature and the liquidus temperature) to dissolve alumina powder in the bath. The pool of molten aluminum in the cathodic area is nearly 300 °C above its melting point, 660 °C.

Since these fluorides are extremely corrosive, iron pots lined with carbon acts as the cathode whereas carbon anodes are dipped into the electrolyte and are consumed by oxidation as CO and CO_2. The anodes consumed during electrolysis are continuously renewed. During electrolysis, aluminum is deposited at the cathode and oxygen is evolved at the anode and reacts with carbon to produce CO_2 by the exothermic reaction. This helps in controlling the electrolyte temperature but consumes carbon anode. Fresh Al_2O_3 has to be added as it becomes depleted in the electrolyte with the progress of electrolysis. More noble impurities (as compared to aluminum) such as iron and silicon present in the raw materials or in the anode will deposit along with aluminum on the cathode. Hence, aluminum oxide and anode should be free of impurities as far as possible. The purity of aluminum produced may be achieved up to 99.5% or better by careful control of the raw materials. The less noble sodium (compared to aluminum) will not deposit on the cathode, but vapors of sodium may diffuse into the cooler parts of the cathode to dissolve in the carbon. The dissolution of sodium will reduce service life of the cathode.

Although the Hall–Heroult process has been in use for more than a century for primary production of aluminum on commercial scale, the electrolytic cell reactions are not yet well understood. According to Tabereaux and Peterson [27], alumina dissolves in the electrolyte by forming oxyfluoroaluminate anions ($Al_2OF_6^{2-}$ and $Al_2O_2F_6^{4-}$), but, for calculation of specific energy consumption, they have considered a simplified approach. The cell reactions are briefly summarized below:

Dissociation of alumina in the electrolyte:

$$2Al_2O_3 \rightarrow 2AlO_2^- + Al^{3+} \tag{12.46}$$

and the cathodic reaction:

$$Al^{3+} + 3e \rightarrow Al \tag{12.47}$$

and the anodic reaction:

$$3AlO_2^- \rightarrow 3Al + 3O_2 + 3e \tag{12.48}$$

also at the anode:

$$3C(\text{electrode}) + 3O_2(g) = 3CO_2(g) \tag{12.49}$$

On addition of the anodic and cathodic reactions, the overall cell reaction is:

$$Al^{3+} + 3AlO_2^- + 3C(s) \rightarrow 4Al(l) + 3CO_2(g)$$

In a simplified manner, the above reaction can be presented as:

$$2Al_2O_3(s) + 3C(s) \rightarrow 4Al(l) + 3CO_2(g), \quad \Delta G_{1273}^0 = +332 \quad \text{kcal} \quad (12.50)$$

The required theoretical standard decomposition potential at 1000 °C is given as

$$E_{1273}^o = -\frac{\Delta G^o}{nF} = -\frac{332000}{3 \times 23066} = -4.8 \text{ volts per 4 mol of Al}$$
$$= -1.2 \text{ V per mol of Al}$$

Assuming, $a_C = a_{Al} = 1$ and $p_{CO_2} = 1$ atm, the theoretical decomposition depends on the activity of Al_2O_3 in the electrolyte. The saturation limit of Al_2O_3 in the electrolyte at the working temperature reaches 6% Al_2O_3. At this concentration, $a_{Al_2O_3} = 1$, but, since during electrolysis concentration of Al_2O_3 decreases, $a_{Al_2O_3}$ will change and accordingly decomposition voltage will vary. For example, if $a_{Al_2O_3} = 0.1$,

$$E_{1273} = E_{1273}^o + \frac{RT}{nF} \ln a_{Al_2O_3}^2$$
$$= -1.2 + \frac{1.987 \times 1273}{3 \times 23066} \ln (0.1)^2 = -1.37 \text{ V}$$

Thus, we see that cell voltage increases with a decrease in the activity of $a_{Al_2O_3}$ in the cryolite bath. Although the theoretical decomposition voltage is 1.2 V, there is over-voltage for CO_2 evolution on the anode. In actual practice, in order to achieve a favorable production rate, a minimum voltage of 4.2 V is applied to overcome the IR drop of the electrolyte, concentration polarization, charge transfer, and bubble nucleation. This amounts to a high power consumption of about 13.2 kWh kg^{-1} of aluminum for a cell operating at 350 k A with 95% current efficiency. In addition to this power consumption, 1920 kg of Al_2O_3, 16 kg of AlF_3, and 420 kg of carbon are required to produce 1 ton of aluminum. Although the theoretical requirement of carbon is only 333 kg [27], the excess consumption is due to the oxidation of the hot anodes exposed to the atmosphere.

The current efficiency of modern electrolytic aluminum plants based on new cell technology using prebaked anodes of high-quality carbon has improved from 82% in 1990 to 96% in 2000. The maximum current efficiency of the cells using Soderberg anodes is only 92% due to its inferior quality carbon, and magnetic design [27]. The lower efficiency may be due to the dissolution of a fraction of aluminum in the electrolyte, followed by its transfer and oxidation at the anode. Although the possibility of electrochemical oxidation cannot be ruled out, there is likelihood of oxidation of aluminum by the CO_2 gas formed at the anode:

$$3CO_2(g) + 2[Al]_{dissolved} = Al_2O_3 + 3CO(g) \qquad (12.51)$$

A correlation [15, 16] has been observed between current efficiency and CO content in the gas leaving the anode. The higher the CO content, the lower is the current efficiency. The current efficiency can be increased by decreasing the transport of aluminum to the anode compartment, which can be achieved by increasing the distance between the anode and the cathode. Current efficiency also increases with increasing current density at a constant bath composition. Currently, highly automated aluminum electrolytic cells are operated with the aid of sophisticated computers.

An important feature of the Hall–Heroult cell is the abrupt increase in voltage across the cell from 4.2 V to > 30 V and a corresponding decrease in amperage. This is known as the "anode effect" and occurs when the alumina (Al_2O_3) concentration in cryolite goes below one-third of its normal value. At this concentration, the arrival of oxygen-containing ions at the electrode surface gets reduced to about two-thirds of the normal rate. At a lower concentration of Al_2O_3, the electrolyte does not wet the anode and hence a gas-filled gap is created between the anode and the electrolyte. The formation and discharge of perfluorocarbon gases [16, 28–32] (mainly carbon tetrafluoride, CF_4, and minor contents of dicarbon hexafluoride, C_2F_6) further deteriorate the situation due to their strong de-wetting characteristics. The sudden rise in the voltage is localized in this region. The problem is solved by stirring and addition of fresh alumina to make up the desired composition of Al_2O_3 in the cryolite.

Electrowinning of Magnesium   About 80% of the total magnesium is produced by fused salt electrolysis (see Appendix A.12) and the remaining comes via thermal reduction processes. The electrolytic process conducted at 750 °C uses a typical electrolyte containing 25% $MgCl_2$, 15% $CaCl_2$, and 60% NaCl. Carbon anodes and steel cathodes dip in the electrolyte contained in an iron pot because magnesium does not alloy with iron [33]. Magnesium is deposited at the cathode and chlorine is evolved at the anode. Some carbon is consumed due to the presence of MgO in the electrolyte. Since the specific gravity of the molten electrolyte is greater than that of magnesium, the magnesium produced at the cathode floats on the surface of the bath. It must be protected from the anodic chlorine coming in contact by providing ceramic curtains. The curtains dipping into the electrolyte separate the anode and cathode compartments. The molten magnesium recovered in the subsequent operation is protected from atmospheric oxidation either by molten sulfur or chloride flux. Chlorine evolved at the anode is collected for hydrochloric acid manufacture. The theoretical decomposition voltage of pure $MgCl_2$ is above 2.5 V, but, in order to overcome polarization and to achieve a reasonable production rate, about 6–7 V is required. With a current efficiency of 75–90% about 18 kWh of power is required to produce one kilogram of magnesium.

Electrowinning of Sodium Sodium can be obtained by electrolysis [33] of either sodium hydroxide in Castner cell or NaCl dissolved in $CaCl_2$ in Downs cell. In the latter cell operated at 600 °C, the electrolyte containing approximately 40% NaCl and 60% $CaCl_2$ is maintained in molten condition by thermal energy furnished by the cell current. Sodium and chlorine are, respectively, collected at wire gauze diaphragms and in a dome, located at the top of the cell. The chloride-based electrolytic process works out to be cheaper coupled with the advantage of getting chlorine gas as a by-product as compared to the electrolysis of NaOH. With a current efficiency of 75–80%, the Downs cell consumes about 10 kWh kg^{-1} of sodium produced.

Electrowinning of Calcium Calcium is extracted by electrolysis of fused salt mixture of $CaCl_2$–10% CaF_2 in a graphite container using iron cathode and graphite anode. About 50 kWh of power is consumed to produce 1 kg of 98% pure calcium at a current efficiency of 65–82% [34].

Electrowinning of Lithium Lithium is obtained by electrolysis of LiCl–48% KCl at 400–420 °C, using cast iron cathode and carbon anode. The cell is operated at 8–9 V, using a current of 0.4–2 A cm^{-2} with a current efficiency of 85–90% and power consumption of 34–40 kWh kg^{-1} Li [34].

Electrowinning of Rare Metals In the production of rare metals by electrolysis advantage is taken of the fact that the decomposition potentials of chlorides or fluorides of rare metals are lower than the inert salts of alkali and alkaline earth metals. The electrolytic process produces purer metals and may work out to be cheaper compared to the metallothermic reduction using calcium, magnesium, or sodium. The electrolyte is prepared by dissolving chlorides or fluorides of rare metals in the fused mixture of inert salts of alkali or alkaline earth metals. The decomposition potential of the inert salt should be higher than the decomposition potential of the salt being electrolyzed at unit activity. This is essential because a larger difference in the two potentials at unit activity gives sufficient margin for safe operation when activity of the salt decreases during electrolysis. A low melting inert salt mixture reduces the cost of maintenance, but the higher melting product metal will deposit in solid form as a powder or dendritic crystals. In order to produce the electrodeposited metal in liquid state, salt baths with higher melting points should be employed. The small electrolytic cell may be prepared from a single block [35] of graphite, alumina, or silica whereas, for large-scale production, it may be constructed from refractory bricks bonded with appropriate cement to obtain the cell walls that are impervious and resistant to the molten salt.

Difficulties are encountered in the extraction of rare metals by fused salt electrolysis due to (i) the low solubility of their halides in salt mixtures, (ii) the deposition of the high melting metals in solid form, and (iii) the formation of ions of variable valence (in the case of metals possessing multiple valencies, e.g., Ti). Despite these difficulties, there have been efforts to produce some rare metals by fused salt electrolysis. For example, the electrolysis of $BeCl_2$ dissolved in a eutectic melt of LiCl–KCl–NaCl, in the proportion of 40.4%, 53.2%, and 6.4% (melting point 362 °C), is carried out at a fairly low temperature to yield beryllium metal flakes [35]. Thorium metal powder can be obtained by the electrolysis of $ThCl_4$ dissolved in a eutectic melt of NaCl–KCl, carried out at 700–800 °C [35] (see Appendix A.8). Loosely adhering dendritic zirconium metal powder is deposited at the cathode by the electrolysis of K_2ZrF_6 dissolved in NaCl melt at 850 °C [35]. The metal powder/flake collected from the cathode compartment is crushed, leached with water, filtered, and dried in air after separating the metal from the electrolyte.

In order to overcome the problem of low solubility of the rare earth oxides in molten fluoride salts, Abbasalizadeh et al. [9] conducted a series of experiments to convert rare earth oxides into rare earth fluorides using AlF_3, ZnF_2, FeF_3, and Na_3AlF_6 as fluorinating agents in a $LiF–Nd_2O_3$ system. The resulting neodymium fluoride can be subsequently electrolyzed in the same reactor to deposit neodymium on the cathode. This concept has been successfully utilized in the production of other rare earths.

There have been constant efforts during the past several decades to develop new processes based on electrochemical principles to lower down the cost of production of titanium. The extraction of titanium possessing multiple valencies has been difficult due to its strong chemical affinity for oxygen and high melting point (1670 °C). Presently, Kroll process is the only well-established commercial process. Being a multistep batch process, it suffers from several disadvantages: (i) high capital cost involving large labor force, (ii) the titanium sponge produced requires vacuum melting, and (iii) the process has an undesirable impact on the environment, more than aluminum and steel [36]. As the process is energy incentive and time consuming, the cost of titanium produced is on the higher side, which limits its applications for special purposes only, despite its attractive properties of light weight, high strength, corrosion resistance, and bio-compatibility [37]. Electrochemical processes for the extraction of titanium under development may be classified into three categories depending upon the form in which Ti is present in the electrolytic cells: (i) titanium compounds dissolved in the electrolyte, (ii) used as an anode, and (iii) as a cathode.

Electrolytic Methods Based on Dissolved Titanium Compounds   During the 1950s, titanium was produced by the electrolysis of a eutectic melt of LiCl–KCl at 355 °C, or a eutectic melt of NaCl–KCl at 650 °C, containing $TiCl_3$, which was obtained by the reduction of $TiCl_4$ with hydrogen. At a current density of 250 A dm^{-2} [35] for the electrolyte containing 15% $TiCl_3$, the cell achieved a current efficiency of 30–40%. To obtain a higher current efficiency of 50–60%, a sillimanite diaphragm was used to separate the anolyte and catholyte compartments [35]. The lower current efficiency is due to the loss of current in repeated back-and-forth oxidation and reduction of Ti^{2+}, Ti^{3+}, and Ti^{4+} ions at the anode and cathode, respectively. Furthermore, electrolysis at lower temperatures, 355–650 °C, produces solid titanium with large surface area and low bulk density, that has to be removed intermittently for further deposition of titanium on the cathode.

In order to overcome the problem of the presence of multivalent Ti ions in the electrolyte, Ginata [4, 5] operated the cell at 1700 °C with $TiCl_4$ dissolved in the mixture of fused fluoride and chloride to extract liquid titanium. The dissolution of $TiCl_4$ in the fused salt mixture may be represented according to the reaction:

$$3CaF_2(l) + KCl(l) + TiCl_4(\text{in soln}) = TiKCaF_6(l) + 2CaCl_2(l) + \tfrac{1}{2}Cl_2(g)$$
$$(12.52)$$

The deposition of titanium on the cathode is assumed to take place according to the reaction [36]:

$$TiKCaF_6(l) + 2CaCl_2(l) = Ti(l) + 3CaF_2(l) + KCl(l) + 3/2Cl_2(g) \qquad (12.53)$$

The presence of a monovalent alkali metal together with a divalent alkaline earth metal maintains Ti in the trivalent state only [4].

Since the above electrolytic operation is based on the use of expensive $TiCl_4$, the cost of titanium produced will not be lower than what is achieved by the Kroll process, and hence, unlikely to achieve the economic status. In addition, the process needs special refractory material to withstand the high temperature in contact with the corrosive molten titanium.

As oxide is easier to handle than halides, the reduction of TiO_2 seems to be a simpler approach from economic viewpoint by avoiding the expensive and time-consuming carbochlorination step. The cost of chlorination is slightly more than half the total cost of production of titanium metal from rutile by the Kroll process [36]. However, the objective of developing an electrolytic cell for the extraction of titanium similar to the Hall–Heroult cell for the production of aluminum has not been fulfilled because no salt that can dissolve an appropriate amount of TiO_2 to carry out efficient electrolysis has yet been found. The Avanti process [36] developed at MIT produces titanium metal by conducting electrolysis of oxide melt containing TiO_2. The cell is operated at 1700 °C using an inert anode of an expensive metal (iridium) instead of carbon to avoid carbon pick-up in the melt. Nevertheless, it seems difficult to obtain titanium metal with low acceptable oxygen content (< 0.2%) when the liquid titanium deposited at the bottom of the cell (made cathode) is in contact with the electrolyte containing TiO_2.

In the QIT process [36], molten CaF_2 electrolyte floats on a slag containing TiO_2. During electrolysis, the less electropositive elements, iron, chromium, manganese, and vanadium, settle at the bottom of the electrolytic cell after reduction and titanium is deposited on the cathode. Again, in this process, it is doubtful whether a titanium metal with acceptable oxygen content can be obtained when molten metal remains in contact with the slag containing oxygen. Furthermore, there is a strong chance of attack of the refractory lining by molten titanium.

Electrolytic Methods Based on Anodes Containing Ti-Bearing Materials From the above discussion, it is clear that several attempts to obtain titanium at lower cost by the electrolysis of the chloride melts as well as TiO_2 dissolved in oxide melt have not succeeded. The electrolysis of the chloride melt was uneconomical due to the high cost of $TiCl_4$ and poor current efficiency on account of multivalent titanium. On the other hand, electrolysis of the oxide melt does not produce titanium metal with acceptable oxygen content (lower than 0.2%). These problems led to the development of a few electrolytic techniques in which a Ti-bearing material (containing low-valence Ti) is used as the anode. Under this category, the noteworthy processes, namely, USTB, MER, and Chinuka, employ consumable anodes of electronically conducting titanium oxycarbide (TiO–TiC solid solutions). Oxycarbide anodes can be prepared by heating a mixture of TiO_2 and carbon at an elevated temperature in an inert atmosphere. The resulting mixture, after pressing and sintering, is used as the anode in a high-temperature electrolytic cell having either iron, molybdenum, or stainless steel cathode, and molten salt electrolyte consisting of a mixture of fluorides and chlorides of alkali and alkaline earth metals [38–46].

Withers et al. [40] have produced titanium powder at an estimated cost of about half that of the Kroll process by conducting electrolysis using a suboxide–carbide anode. The process has been demonstrated on a continuous basis to the scale of 250 kg per day of titanium containing less than 500 ppm oxygen. Jiao and Zhu [42] conducted a series of electrolyses on the possibility of extracting titanium, using a Ti_2OC solid solution as an anode, in a NaCl–KCl melt. Carbon monoxide (CO) was monitored at the anode during electrolysis together with the cathode product analysis. Kjos et al. [44] recovered titanium by an electrochemical process using titanium oxycarbide anode, prepared by heating a mixture of TiC and TiO. A liquid metal cathode was employed to carry out electrolysis in the molten electrolyte of NaCl–Na_3AlF_6 as well as NaCl–KCl.

A single-phase consumable titanium oxycarbonitride $TiC_{0.25}O_{0.25}N_{0.5}$ anode [45] was prepared out of a homogenous mixture of TiO, TiC, and TiN with a molar ratio of 1:1:2 by spark plasma sintering at 1873 K for the USTB titanium process. The anode dissolution process showed that $TiO_{0.25}C_{0.25}N_{0.5}$ was electrochemically dissolved to yield Ti^{2+} in the NaCl–KCl melts with the simultaneous evolution of CO and N_2 at the anode. The deposition of pure titanium powders on the cathode demonstrated that $TiO_{0.25}C_{0.25}N_{0.5}$ is a suitable consumable anode

In the MER process [6], a mixture of TiO, Ti_2O_3, Ti_3O_5, and Ti_4O_7 is prepared by carbothermic reduction of TiO_2 or Ti slag. Of these, TiO is the most appropriate suboxide. The oxide is mixed with carbon and pressed and sintered at 1100–1300 °C to obtain a well-bonded composite anode.

In the Chinuka process [6], consumable anode of titanium oxycarbide Ti_2OC is prepared by a thermal treatment of impure TiO_2 with TiC or carbon. During electrolysis conducted at 570–900 °C, the impurities (Al, Ca, Cr, Fe, and Si) present in the anode dissolve in the electrolyte. Thus, it is advisable to purify the electrolyte during the operation in order to get a reasonably pure Ti deposit.

A combination of electro-reduction and refining has been proposed by Jewell et al. [46] to obtain pure titanium metal crystallite sponge or powder of particle size 1–30 μm from an impure raw material. The technique is based on the anodic dissolution of titanium oxycarbide together with cathodic deposition of Ti in a eutectic mixture of NaCl–KCl at 973–1173 K.

Electrolytic Methods Based on Cathodes Containing Ti-Bearing Materials Although reduction of TiO_2 with calcium as well as magnesium is thermodynamically feasible, the rate of reaction is slow because the reactant TiO_2 is covered by the product, CaO (and MgO). In addition, an appreciable amount of oxygen is picked up by the reduced titanium when the system attains equilibrium with CaO/Ca (MgO/Mg). The dissolution of oxygen in the titanium metal deteriorates its mechanical properties. In order to carry out the reaction without hindrance, TiO_2 is reduced in the $CaCl_2$ melt, which dissolves the by-product, CaO. The process of reduction can be made continuous by restricting the concentration of CaO in the $CaCl_2$ melt below the saturation limit. In this way, the activity of CaO can be lowered down to <1, which is helpful in reducing the oxygen pick-up by the metal. As $MgCl_2$ dissolves very little MgO, magnesiothermic reduction is not practiced, despite being cheaper and readily available as compared to calcium.

Okabe et al. [47] adopted the above principle to produce nearly oxygen-free titanium using titanium and carbon as the cathode and anode, respectively, immersed in molten $CaCl_2$. $CaCl_2$ produces the deoxidant calcium and decreases the activity of the by-product, CaO. The titanium sample at the cathode is deoxidized by the electrolytically produced deoxidant Ca and the resulting O^{2-} ions react at the carbon anode to evolve CO (CO_2) gas. Okabe et al. [47] could reduce the oxygen content in the titanium wire sample from 1400 ppm to <100 ppm at the cost of carbon pick-up of about 50 ppm. This principle led to the development of many electrochemical methods [48–55] where TiO_2 has been made cathode to recover the metallic titanium. OS, EMR, FFC-Cambridge, and SOM processes are based on this principle.

The concept of calciothermic direct reduction of titanium dioxide in molten $CaCl_2$ as proposed above has been experimentally tested in the OS process [50]. The process produces a metallic titanium sponge containing 2000 ppm oxygen at 1173 K in the $CaCl_2$ bath employing a carbon anode and a Ti basket-type cathode filled with TiO_2 powder. The operation is conducted in a single cell, where the calciothermic reduction and the electrochemical reaction for recovery of the reducing agent, Ca, are carried out simultaneously in the same molten $CaCl_2$ bath. The melt dissolves a few molar percent of Ca, which gives the media a strong reducing power. To shorten the operating time and achieve a lower oxygen content in the titanium metal, a concentration of 0.5–1 mol% CaO in the molten $CaCl_2$ bath was found to be optimum. The cathodic reduction of titanium dioxide occurs through deposited calcium metal.

Park et al. [52] developed the EMR process to produce titanium powder by the direct reduction of TiO_2 in molten $CaCl_2$ through an electronically mediated reaction (EMR). The electrolytic cell is divided into two parts. In one part, $CaCl_2$ melt is electrolyzed to generate Ca at an auxiliary nickel cathode ($Ca^{2+} + 2e = Ca$) . The resulting Ca is dissolved into the nickel to form a Ni–Ca alloy rich in Ca. Depending on the composition of the melt, CO/CO_2 and Cl_2 gases are evolved at the carbon anode. In the second part of the electrolytic cell, TiO_2 is electronically connected to the alloy. The following two reactions occur in this part: (i) generation of Ca^{2+} ions by ionization of Ca present in the alloy ($Ca = Ca^{2+} + 2e$) and (ii) reduction of TiO_2 ($TiO_2 + 4e = Ti + 2O^{2-}$). Thus, metallic titanium is produced by the electronically mediated reduction process.

The electrolytic process developed at the University of Cambridge has been popularized as the FFC-Cambridge process after the names of researchers, Fray, Farthing, and Chen, and the place of research [8, 48, 53]. In this process, a metal basket containing the sintered pellets, prepared by pressing the TiO_2 powder, serves as the cathode in the electrolytic cell with $CaCl_2$ melt and consumable graphite anode. Ca^{2+} ions generated by the ionization of $CaCl_2$ transform into Ca. During electrolysis the applied potential is controlled in such a way that the activity of Ca in the electrolyte melt remains below unity at the cathode without any deposition of Ca metal. TiO_2 is reduced at the cathode by the ionization of oxygen from the cathode itself. This suggests that the overall reaction in the FFC cell at the cathode is the removal of oxygen from TiO_2 by the displacement of oxygen anions with electrons according to the reaction:

$$TiO_2 + 4e = Ti + 2O^{2-} \tag{12.54}$$

The cathodic reduction of titanium dioxide takes place through oxygen ionization. It has been reported that the cathode reactions taking place in a number of steps, are very complex [8, 36]. The oxygen anions get discharged at the graphite anode with the evolution of CO_2 gas. It seems that the optimized FFC-Cambridge process can produce titanium metal directly from high-purity TiO_2 at lower cost with energy consumption of 12 kWh kg^{-1} at a current efficiency of 50% at 900 °C [8]. The Metalysis Ltd. [8] has adopted this process on commercial scale.

The solid oxide membrane (SOM) process, initially developed for the production of magnesium [54] from magnesium oxide dissolved in fluoride-based fluxes in the temperature range of 1150–1300 °C, has been successfully extended to titanium extraction [55] with the primary objective of reducing the number of processing steps involved in the conventional Kroll process. In the SOM process, a cell made of steel contains the electrolyte of TiO_2 dissolved in a eutectic mixture of CaF_2–MgF_2. The cell is equipped with a carbon cathode and an oxygen ion conducting yttria-stabilized zirconia (YSZ) membrane in the form of a one-end closed tube, which contains either a liquid metal that acts as the anode or a liquid MgF_2–CaF_2 ionic flux that connects the YSZ membrane to an anode. During electrolysis, titanium ions are reduced at the cathode while the oxygen anions pass through the YSZ membrane and are discharged at the anode to evolve molecular oxygen. The SOM process has been commercialized [8].

Electrowinning of Rare Earth Alloys Han et al. [56] have been successful in preparing Mg–Li–Al–La alloys by the electrolysis of the LiCl–KCl–$MgCl_2$–KF melt containing La_2O_3. Aluminum and tungsten were used as the anode and cathode, respectively. They have confirmed the presence of Al_2La, $La_{12}Mg_{17}$, $LiMg/Li_3Mg_7$

phases by X-ray diffraction analysis and homogeneous distribution of Mg, Al, and La in the alloys by scanning electron microscopy (SEM) and energy dispersive spectrometry (EDS). Ye et al. [57] have prepared Ti–Fe-based quaternary alloys for hydrogen storage by the SOM method from oxides dissolved in molten $CaCl_2$. SOM-assisted electro-deoxidation has been used for extraction of the high-temperature structural material, Ti_5Si_3 silicide [58], directly from complex multicomponent Ti/Si containing metal oxides dissolved in molten salts. Abbasalizadeh et al. [59] have carried out electrolytic deposition of neodymium alloy from the neodymium fluoride salt containing neodymium oxide using a consumable iron anode. The dissolution of iron does not only prevent the evolution of perfluoro carbon gases (CF_4 and C_2F_6) at the anode but also form iron fluoride that serves as the fluorinating agent. The constant generation of in situ FeF_3 enables the continuous conversion of neodymium oxide into neodymium fluoride. The presence of NdF_3 as well as FeF_3 in the fused electrolyte produces Nd–Fe alloy at the cathode, which can be used as a master alloy for the production of Nd–Fe–B magnets.

12.2.2 Electrorefining

The principle of electrowinning and electrorefining using aqueous or fused salt electrolytes is the same. In electrorefining, the crude metal is electrolytically dissolved at the anode and deposited as pure metal at the cathode according to the reactions:

$$M(\text{impure}) \rightarrow M^{n+} + ne \quad (\text{at the anode})$$
$$M^{n+} + ne \rightarrow M \text{ (pure)} \quad (\text{at the cathode})$$

The overall reaction is:

$$M(\text{impure}) \rightarrow M(\text{pure}) \tag{12.55}$$

Adoption of this process requires fulfilling the following criteria:

1. Ionic species in the electrolyte should support an anodic reaction.
2. Ionic species in the electrolyte should get reduced at the cathode.
3. The electrolyte should be capable of ionic transport.
4. A metallic conductor is required for transporting the electrons through the external circuit from the anodic to cathodic site.

In the preceding paragraphs, it has been stated that during electrolysis the impure metals of the anode having higher negative potentials will dissolve along with the metal to be refined (anodic dissolution process) whereas the impurities with higher positive electrode potentials will form a residue (known as the anode slime). In the cathodic reduction process, the metals with higher positive electrode potential will be deposited preferentially leaving the metals having higher negative potential in the electrolyte solution. Thus, electrorefining is a two-stage purification process where the metals more noble than the metal to be refined remain insoluble at the anode and the less noble metals dissolve in the electrolyte but do not deposit at the cathode. As the solubility of oxide, nitride, and carbide in the fused halides, particularly in a chloride bath, is very low, the interstitial impurities, namely, oxygen, nitrogen, and carbon, present in the anode, do not take part in the cell reaction, and hence they remain in the anode feed.

The cell operates at a very low voltage at the unit activity of the metal at the anode (to be refined) containing little impurities. Under these circumstances, sufficient energy is released by the dissolution of metal at the anode for the deposition of the metal at the cathode. This enables electrorefining a highly efficient process from the energy viewpoint.

In order to make electrorefining an effective process, standard electrode potential of the metal to be refined should be more positive (e.g., the refining of copper and silver) than that of the impure metals present in the crude anode. The process of refining becomes less effective when the metal to be refined and impurities have closer electrode potentials. In such cases, the co-deposition of impurities is prevented by the use of diaphragm cells together with the facility of anolyte purification before feeding the electrolyte to the cathode compartment. For example, the co-deposition of iron and cobalt is suppressed in diaphragm cell in the electrorefining of nickel. Common metals (e.g., Cu, Ag, Au, Ni, Pb, and Sn) as well as rare metals (e.g., Al, Nb, and Be) have been successfully electrorefined using, respectively, aqueous solutions and fused salt electrolytes.

12.2.2.1   Electrorefining from Aqueous Electrolytes

Copper is refined to remove arsenic and antimony and for the recovery of silver and gold, which presents the most important application of electrorefining through aqueous solution. The fired-refined copper and blister copper mixed with anode scrap and copper scrap are subjected to melting and casting to obtain suitable anodes. Pure thin sheets of electrodeposited copper serving as cathodes end up as the final product. The electrolysis is conducted through the acid solution of copper sulfate containing 40–45 kg m^{-3} Cu and 180–200 kg m^{-3} H$_2$SO$_4$ [22]. In order to obtain smooth and bright deposits, glue and thiourea are added in small proportions to the electrolyte under circulation. To achieve an appropriate resistivity of the electrolyte, the solution is heated by steam to maintain the required working temperature of 50–60 °C. During the process of electrodeposition, gold, silver, and platinum group metals (more noble than copper) and also sulfur, selenium, and tellurium (impurities present in the starting anode as copper compounds: Cu$_2$S, Cu$_2$Se, and Cu$_2$Te) remain undissolved and form the anode slime. Copper may also form compounds with arsenic and antimony whereas lead forms insoluble PbSO$_4$ with the electrolyte. All these compounds join the anode slime along with the noble metals.

The base metals, like bismuth, iron, nickel and cobalt undergo complete dissolution along with copper but only copper is deposited on the cathode leaving the less noble metals in the solution. Both the electrolyte and the anode slime get contaminated due to the partial dissolution of arsenic and antimony. In order to avoid the problem of co-deposition of impurities along with copper at their optimum concentration in the electrolyte, part of the electrolyte is withdrawn, purified, and fed back to the circulating electrolyte.

Although the theoretical potential for copper is zero, about 0.3 V is required to overcome ohmic resistance in the electrolyte and contacts and possible overvoltages. At a current density of 200–300 A m^{-2}, the current efficiency is about 95%, consuming on average approximately 0.25 kWh of energy to produce 1 kg of the refined copper. About 1 kg of steam is required to maintain the desired temperature of the solution during electrolysis.

The anode slime is treated for the recovery of silver and gold. In the first step, copper and other impurities are dissolved in concentrated sulfuric acid. The residue containing mainly noble metals is cast into anodes and electrolyzed. Silver is first deposited on the cathode leaving less noble copper in the solution and gold into the anode mud. This mud cast into anodes is electrolyzed to deposit gold.

Electrorefining of nickel is more troublesome because impurities like cobalt and iron are close to nickel in the electrochemical series. In order to avoid co-deposition of these metals with nickel, the anode and cathode compartments are separated by a diaphragm. In the anode compartment, nickel along with copper, cobalt, and iron gets dissolved and the more noble platinum group metals join the anode slime. From time to time, anolyte is pumped from the electrolytic cell and purified. Iron, cobalt, and copper are, respectively, precipitated as hydroxide by air oxidation, chlorine oxidation, and cementation on nickel powder. The purified solution containing about 60 kg Ni m^{-3} of solution is fed to the cathode compartment after adjusting the chloride concentration to approximately 55 kg m^{-3}, boric acid to approximately 16 kg m^{-3}, and sulfate to approximately 95 kg m^{-3}. Nickel is deposited on a stainless steel blank cathode. Chloride ions prevent passivation of nickel anodes and boric acid controls the pH of the solution to about 5. The bath maintained at 60 °C is electrolyzed at an applied voltage of 1.9 V and current density of 200 A m^{-2} with current efficiency of about 95%. A slightly higher hydrostatic head of the electrolyte is maintained in the cathode compartment to prevent flow of anolyte into the cathode compartment. The use of diaphragm increases the applied voltage resulting in higher energy consumption. The energy requirement of 1.9 kWh is significantly higher for refining 1 kg of nickel as compared to that of copper electrorefining (~ 0.25 kWh kg^{-1} of copper).

Lead is electrorefined to remove bismuth. As lead forms insoluble compound with H$_2$SO$_4$ and HCl, a fluorosilicic acid bath containing approximately 9% lead as PbSiF$_6$ and 8% free fluorosilicic acid (H$_2$SiF$_6$) is used as electrolyte. During electrorefining, lead, zinc, and tin enter into the solution from the impure lead anode. Impurities nobler than lead sink to the bottom of the electrolytic tank forming anode slime. Since copper causes anode polarization and hinders normal electrode reaction, it must be removed prior to electrolysis. The cell is operated at 35 – 40 °C with an initial voltage of about 0.5 V. Cathode lead is purer than that produced by the conventional furnace refining, but it is hardly preferred due to the toxic nature of the electrolyte.

12.2.2.2   Electrorefining from Fused Salt Electrolytes

The fused salt electrolytic process essentially, carried out at relatively high temperatures, is expensive and troublesome, and hence employed only when there are chances of deriving adequate benefits or in the absence of alternative method of refining. Under these circumstances, the commercial exploitation of the fused salt electrorefining technology is limited only to the refining of aluminum although the processes have been developed for a number of metals [60].

Aluminum produced by the Hall–Heroult process is refined by Hoopes' three-layer process using fused salt electrolyte to obtain high-purity aluminum. In the original process developed in 1901, the electrolyte (specific gravity 2.6) containing 30 – 37% AlF_3, 25 – 30% NaF, 30 – 38% BaF_2, and 7% Al_2O_3 was sandwiched between Al–Cu anode of specific gravity 2.8 (at the bottom) containing 25% Cu with some silicon to lower the melting point and cathode of high-purity aluminum (specific gravity 2.3) floating at the top. The electrolysis was carried out in a graphite-lined steel pot with a coating of alumina for better and adequate insulation. The liquid aluminum cathode was electrically connected to suspended carbon electrodes and the anode was fed power via the graphite lining at the bottom. During electrolysis, metals nobler than aluminum, for example, copper and iron, remained undissolved whereas the less noble metals dissolved and joined the electrolyte. In the presence of Al_2O_3 in fluoride - based electrolyte (AlF_3 – NaF), the cathodic reactions are similar to those reported in the electrowinning of aluminum from cryolite (3NaF–AlF_3) bath, discussed in the previous section. The electrolytic cell was operated at 950–1000 °C at an applied voltage of 5–7 V. Adequate thickness of the electrolyte layer had to be maintained to prevent contact between anode and cathode layers. Proper arrangement was made to feed aluminum into the anode layer. The refined aluminum of 99.98% purity accumulated at the cathode was tapped simultaneously.

Another three-layer process developed by the Pechiney Company of France is known as the Gadeau process. Though it is similar to the Hoopes' process, it employs a relatively low melting chloride–fluoride mixed electrolyte bath containing 23% AlF_3, 17% NaF, and 60% $BaCl_2$ (specific gravity 2.7), which can be operated at about 750 °C. Heavier electrolyte demands heavier Al – Cu alloy anode of specific gravity 3.0. This can be obtained by increasing the copper content in the alloy up to 30%. The mixed electrolyte is less stable as compared to the fluoride electrolyte, and hence needs periodic addition of constituents to maintain the desired composition. As it is less corrosive, the electrolytic cell can be operated safely with magnesite lining on the walls. The process produces aluminum of 99.99% purity at a current efficiency of approximately 100% at an operating voltage of 7 V.

A low-melting, highly stable electrolyte containing 48% AlF_3, 18% NaF, 18% BaF_2, and 16% CaF_2 was developed in Switzerland. The electrolytic cell can be operated at 740 °C, but the electrolyte has lower conductivity compared to that of the above two electrolytes.

In the past, electrorefining of aluminum was practiced at a number of places. However, the process has not been in much use in recent years due to significant improvement in the purity of aluminum (of the order of 99.9%) produced by the Hall–Heroult process. This purity is sufficient for most purposes. In order to meet the increasing demand of high-purity aluminum for superconductors, semiconductors, and other electronic industries, fractional solidification is preferred over electrorefining. The former technique can lower down copper, iron, and silicon below 10 ppm with energy saving of almost 50% as compared to the latter.

Magnesium can be electrorefined in a similar manner employing $MgCl_2$ electrolyte and anodes of either Mg – Cu, Mg – Zn, or Mg – Zn – Al alloy. However, this technique does not provide any benefit over other refining techniques.

Titanium, zirconium, and molybdenum have been electrorefined on laboratory scale using molten halides dissolved in alkali metal chlorides. In the case of titanium refining, titanium carbide formed by carbothermic reduction of titanium dioxide may be used as anode. During electrolysis, carbon will drop into the anode slime. Economically, this technique may compete with the commonly employed process of metallothermic reduction of titanium tetrachloride.

12.3 Problems [18, 61]

Problem 12.1
At 900 K, sodium is deposited by electrolysis of the fused salt electrolyte containing 35 mol% NaCl and 65 mol% $CaCl_2$. Assuming that the fused salt mixture behaves ideally, calculate the reversible cell potential of the electrolytic cell producing sodium and chlorine gas at 1 atm pressure. The standard free energy of formation of pure liquid NaCl at 900 K is -346 kJ mol^{-1}. Neglect the effect of heat of fusion of NaCl (melting point 800 °C).

Solution
During the electrolysis of the NaCl–$CaCl_2$ fused salt mixture, sodium and chlorine are generated according to the reaction:

$$NaCl(l) = Na(l) + \frac{1}{2}Cl_2(g), \Delta G^o = 346000 \text{ J}$$

$$K = \frac{a_{Na} \cdot p_{Cl_2}^{\frac{1}{2}}}{a_{NaCl}} = \frac{1}{a_{NaCl}}, \left(a_{Na} = 1 \text{ and } p_{Cl_2} = 1 \text{ atm}\right)$$

$$\Delta G^o = -n\,FE^o., \qquad E^o = -\frac{346000}{1 \times 96500} = -3.59 \text{ V}$$

$$E_{cell} = E^o + \frac{RT}{nF} \ln\left(a_{NaCl}\right)$$

Since NaCl–CaCl$_2$ melt behaves ideally, $x_{NaCl} = a_{NaCl} = 0.35$

$$E_{cell} = -3.59 + \frac{8.314 \times 900}{1 \times 96500} \ln\left(0.35\right)$$

$$= -3.59 - 0.08 = -3.67 \text{ V Ans}$$

Problem 12.2

The usual decomposition potential of alumina by the reaction:

$$Al_2O_3(s) + C(s) = 2\,Al(l) + CO_2(g)$$

is -1.15 V in the Hall–Heroult cell at 920 °C. In actual practice, the applied potential to the industrial cell is between 4.5 and 6.0 V. It is known that the reduction reactions involve CO as well as CO$_2$, and the electrolyte may not be saturated with Al$_2$O$_3$(s). Calculate what the reversible potential would be for a cell where the gas released from the anode consists of a p_{CO}/p_{CO_2} ratio of 1/1 and the electrolyte contains 3% Al$_2$O$_3$ rather than being saturated with solid Al$_2$O$_3$, the value of the latter condition being 8% Al$_2$O$_3$. Account for the difference in the voltage.

Given that $C(s) + O_2(g) = CO_2(g)$, $\Delta G_1^o = -94200 - 0.2\,T$ cal

$$C(s) + \tfrac{1}{2}\,O_2(g) = CO(g), \Delta G_2^o = -26700 - 20.95\,T \text{ cal}$$

$$2\,Al(l) + 3/2\,O_2(g) = Al_2O_3(s), \Delta G_3^o = -405760 - 3.75\,T\,\log\,T + 92.22\,T \text{ cal}$$

Solution

Hall Cell : $p_{CO}/p_{CO_2} = 1$, $T = 1193$ K, $a_{Al_2O_3} = \frac{0.03}{0.08} = 0.375$, $a_{Al} = 1$

For the desired reaction: Al$_2$O$_3$(s) + 3 CO(g) = 2 Al(l) + 3CO$_2$(g)

$$\Delta G^o = 3\Delta G_1^o - 3\Delta G_2^o - \Delta G_3^o = 203260 + 3.75\,T\,\log\,T - 29.97\,T \text{ cal}$$

At 920 °C: $\Delta G_{1193}^o = 181270$ cal
Cathodic reaction: $2Al^{3+} + 6e = 2Al$ (l)
Anodic reaction: $3CO + 3O^{2-} = 3CO_2$
$\Delta G^o = -nFE^o, n = 6$

$$\therefore E^o = -\frac{181270}{6 \times 23066} = -1.31 \ V$$

$$E = E^o - \frac{RT}{nF} \ln \left[\left\{ \frac{p_{CO_2}}{p_{CO}} \right\}^3 \left(\frac{a_{Al}^2}{a_{Al_2O_3}} \right) \right] \quad (a_{Al} = 1, a_{Al_2O_3} = 0.375)$$

$$= -1.31 - \frac{1.987 \times 1193}{6 \times 23066} \ln \left(\frac{1^3 \times 1^2}{0.375} \right)$$

$$= -1.31 - 0.0168 = -1.33 \ V$$

From the above calculations, the theoretical decomposition voltage of alumina in the Hall–Heroult cell is -1.33 V whereas in practice it is much larger (4.5–6.0 V). The reason for a large increase is accounted as (i) IR drop of the electrolyte/electrode, (ii) concentration polarization, (iii) charge transfer, and (iv) bubble nucleation.

Problem 12.3

From the following data, explain why is it essential to keep the cast anodes in canvas bags and use diaphragm cells in electrorefining of nickel from Ni_3S_2?

Given that

$$Ni_3S_2 = 3Ni + 2S, \Delta G^o = 48500 \ cal \ mol^{-1}, T = 298 \ K$$

$$Fe^{2+} + 2e = Fe, \quad E_{Fe}^o = -0.44 \ V$$

$$S^{2-} = S + 2e, E_S^o = 0.48 \ V$$

Solution

In the electrorefining of nickel from Ni_3S_2,

$$Ni_3S_2 = 3Ni + 2S, \quad \Delta G^o = 48500 \ cal \ mol^{-1} = -nFE^o, n = 6$$

$$E_{cell}^o = -\frac{48500}{6 \times 23066} = -0.35 \ V$$

$$Fe^{2+} + 2e = Fe, n = 2 \text{ and at the anode}, E_{Fe}^o = -0.44 \ V,$$

$$E_{cell}^o = E_{Fe}^o + \frac{RT}{nF} \ln \left[\frac{a_{Fe^{2+}}}{a_{Fe}} \right]$$

$$\therefore \ln \left[\frac{a_{Fe^{2+}}}{a_{Fe}} \right] = \frac{(E_{cell}^o - E_{Fe}^o) \times nF}{RT}$$

$$= \frac{\{-0.35 - (-0.44)\} \times 2 \times 23066}{1.987 \times 298} = 7.01$$

$$\left[\frac{a_{Fe^{2+}}}{a_{Fe}} \right] = 1.1 \times 10^3$$

At the cathode $E_{Fe}^o = +0.44 \ V, n = 2$

$$\therefore \ln \left[\frac{a_{Fe^{2+}}}{a_{Fe}} \right] = \frac{\left(E^o_{cell} - E^o_{Fe} \right) \times nF}{RT}$$

$$= \frac{\{-0.35 - (0.44)\} \times 2 \times 23066}{1.987 \times 298} = -61.5$$

$$\left[\frac{a_{Fe^{2+}}}{a_{Fe}} \right] = 1.8 \times 10^{-27}$$

Since at the anode, $\left[\frac{a_{Fe^{2+}}}{a_{Fe}} \right] = 1.1 \times 10^3$, that is, $a_{Fe^{2+}} \gg a_{Fe}$, iron will get dissolved at the anode. On the other hand, at the cathode, $\left[\frac{a_{Fe^{2+}}}{a_{Fe}} \right] = 1.8 \times 10^{-27}$, that is, $a_{Fe} \gg a_{Fe^{2+}}$, Fe^{2+} ions will discharge at the cathode. In order to prevent discharge of Fe^{2+} ions at the cathode, diaphragm cells together with the provision of anolyte purification is employed in the electrorefining of nickel.

Sulfur problem: $S^{2-} = S + 2e$, $E^o_S = 0.48$ V, $n = 2, T = 298$ K

$$E^o_{cell} = E^o_S + \frac{RT}{nF} \ln \left[\frac{a_{S^{2-}}}{a_S} \right] = E^o_S - \frac{RT}{nF} \ln \left[\frac{a_S}{a_{S^{2-}}} \right]$$

$$\ln \left[\frac{a_S}{a_{S^{2-}}} \right] = \frac{\left(E^o_S - E^o_{cell} \right) \times nF}{RT}$$

$$= \frac{\{0.48 - (-0.35)\} \times 2 \times 23066}{1.987 \times 298} = 64.66$$

$$\therefore \left[\frac{a_S}{a_{S^{2-}}} \right] = 1.2 \times 10^{28}$$

The extremely high activity ratio of $\left[\frac{a_S}{a_{S^{2-}}} \right]$ at the anode means sulfur does not dissolve; hence there is no need to consider discharge of sulfur at the cathode. In order to facilitate collection of sulfur, Ni_3S_2 anodes are kept in canvas bags.

Problem 12.4

In the electrolytic refining of copper at 50 °C, the voltage drop from anode to cathode is 0.035 V. If electrolyte contains 0.7 kg Cu^{2+} ions per m^3 of the solution electrolyte, calculate the minimum silver ion concentration in the electrolyte for silver deposition.

Given that $Ag^+ + e = Ag, E^o_{Ag} = 0.799$ V

$$Cu^{2+} + 2e = Cu, E^o_{Cu} = 0.344 \text{ V}$$

Solution

For oxidation of copper at the anode, the half-cell potential can be calculated from Nernst's equation:

$$E = E_{Cu}^o + \frac{RT}{nF} \ln \left[\frac{a_{Cu^{2+}}}{a_{Cu}}\right] \quad (0.7 \text{ kg m}^{-3} = 0.7 \text{ gl}^{-1}, \text{hence } a_{Cu^{2+}} = 0.7)$$

$$= 0.344 + \frac{8.314 \times 323}{2 \times 96500} \ln(0.7) \quad (a_{Cu} = 1)$$

$$= 0.344 - 0.005 = 0.339 \text{ V} = E_{anode}$$

For silver reduction at the cathode:

$$E = E_{Ag}^o + \frac{RT}{nF} \ln \left[\frac{a_{Ag^+}}{a_{Ag}}\right]$$

$$= 0.799 + \frac{8.314 \times 323}{1 \times 96500} \ln(a_{Ag^+}) \quad (a_{Ag} = 1)$$

$$= 0.799 + 0.0278 \ln(a_{Ag^+}) = E_{cathode}$$

$$E_{cell} = E_{anode} - E_{cathode}$$

$$\therefore 0.035 = 0.399 - [0.799 + 0.0278 \ln(a_{Ag^+})]$$

$$0.0278 \ln(a_{Ag^+}) = 0.399 - 0.799 - 0.035 = -0.435$$

$$\ln(a_{Ag^+}) = -\frac{0.435}{0.0278} = -15.6316$$

$$a_{Ag^+} = 1.6 \times 10^{-7}$$

Activity = concentration.

$\therefore$ concentration of silver ions in the electrolyte for deposition

$$= 1.6 \times 10^{-7} \text{g l}^{-1} = 1.6 \times 10^{-7} \text{ kg m}^{-3} \quad \text{Ans.}$$

Problem 12.5

In the electrorefining of copper at 25 °C, explain the following with the aid of the given data:

1. Iron gets dissolved at the anode but not discharged at the cathode.
2. Silver and gold do not dissolve at the anode but join the anode slime.

$$Fe^{2+} + 2e = Fe, E_{Fe}^o = -0.44 \text{ V}$$

$$Ag^+ + e = Ag, E_{Ag}^o = 0.799 \text{ V}$$

$$Au^+ + e = Au, E_{Au}^o = 1.551 \text{ V}$$

$$E_a \approx E_c = 0.28 \text{ V}$$

Solution

The half-cell potential at the anode: $E_{cell} = 0.28$ V

According to Nernst's equation:

$$E^o_{cell} = E^o_{Fe} + \frac{RT}{nF} \ln \left[\frac{a_{Fe^{2+}}}{a_{Fe}} \right] \quad n = 2 \text{ and at the anode, } E^o_{Fe} = -0.44 \text{ V}$$

$$\therefore \ln \left[\frac{a_{Fe^{2+}}}{a_{Fe}} \right] = \frac{\left(E^o_{cell} - E^o_{Fe} \right) \times nF}{RT}$$

$$= \frac{\{0.28 - (-0.44)\} \times 2 \times 96500}{8.314 \times 298} = 56.087$$

$$\left[\frac{a_{Fe^{2+}}}{a_{Fe}} \right] = 2.3 \times 10^{24}$$

The extremely large value of $\left[\frac{a_{Fe^{2+}}}{a_{Fe}} \right]$ at the anode indicates that $a_{Fe^{2+}} \gg a_{Fe}$, hence dissolution of iron is highly favorable at the anode.

The half-cell potential at the cathode: $E_{cell} = 0.28$ V, $E^o_{Fe} = 0.44$ V, $n = 2$
From Nernst's equation, we can write:

$$\ln \left[\frac{a_{Fe^{2+}}}{a_{Fe}} \right] = \frac{\left(E^o_{cell} - E^o_{Fe} \right) \times nF}{RT}$$

$$= \frac{\{0.28 - (0.44)\} \times 2 \times 96500}{8.314 \times 298} = -12.464$$

$$\therefore \left[\frac{a_{Fe^{2+}}}{a_{Fe}} \right] = 3.8 \times 10^{-6}$$

The low value of $\left[\frac{a_{Fe^{2+}}}{a_{Fe}} \right]$, that is, $a_{Fe^{2+}} \ll a_{Fe}$, at the cathode indicates that iron is not discharged at the cathode.

Similarly, $\quad Ag^+ + e = Ag, \quad E^o_{Ag} = 0.799$ V $\quad$ and $\quad Au^+ + e = Au, E^o_{Au} = 1.551$ V, $n = 1$, $E_{cell} = 0.28$ V, $T = 298$ K

Following the above procedure at the anode, we can write:

$$\ln \left[\frac{a_{Ag^+}}{a_{Ag}} \right] = \frac{\left(E^o_{cell} - E^o_{Ag} \right) \times n \times F}{RT}$$

$$= \frac{(0.28 - 0.799) \times 1 \times 96500}{8.314 \times 298} = -20.215$$

$$\therefore \left[\frac{a_{Ag^+}}{a_{Ag}} \right] = 2 \times 10^{-9}$$

$$\text{and} \quad \left[\frac{a_{Au^+}}{a_{Au}} \right] = 3.2 \times 10^{-22}$$

Since $\left[\frac{a_{Ag^+}}{a_{Ag}}\right]$ and $\left[\frac{a_{Au^+}}{a_{Au}}\right]$ are very low at the anode, silver and gold will not dissolve but join the anode slime.

Problem 12.6

During electrolysis, cobalt easily dissolves at the anode and also deposits at the cathode. Explain this with the aid of the following data and suitable calculations at 25 °C. Given that

(i) $Co^{2+} + 2e = Co, E^o_{Co} = -0.28$ V
(ii) $E_{cell} = 1.2$ V and -0.45 V, depending upon the anode and cathode reactions, respectively.

Solution

For oxidation of cobalt at the anode, the half-cell potential:

$$E^o_{cell} = E^o_{Co} + \frac{RT}{nF} \ln \left[\frac{a_{Co^{2+}}}{a_{Co}}\right] \quad n = 2$$

and at the anode:

$$E^o_{cell} = 1.2 \text{ V}, E^o_{Co} = -0.28 \text{ V}$$

$$\therefore \ln \left[\frac{a_{Co^{2+}}}{a_{Co}}\right] = \frac{(E^o_{cell} - E^o_{Co}) \times n \times F}{RT}$$

$$= \frac{\{1.2 - (-0.28)\} \times 2 \times 96500}{8.314 \times 298} = 115.29$$

$$\left[\frac{a_{Co^{2+}}}{a_{Co}}\right] = 1.2 \times 10^{50}$$

The extremely large value of $\left[\frac{a_{Co^{2+}}}{a_{Co}}\right]$ at the anode means $a_{Co^{2+}} \gg a_{Co}$, hence dissolution of cobalt is highly favorable at the anode.

Similarly, for discharge of cobalt at the cathode: $E^o_{cell} = -0.45$ V, $E^o_{Co} = 0.28$ V

$$\therefore \ln \left[\frac{a_{Co^{2+}}}{a_{Co}}\right] = \frac{(E^o_{cell} - E^o_{Co}) \times n \times F}{RT}$$

$$= \frac{\{-0.45 - (0.28)\} \times 2 \times 96500}{8.314 \times 298} = -56.87$$

hence $\left[\frac{a_{Co^{2+}}}{a_{Co}}\right] = 2 \times 10^{-25}$

The extremely low value of $\left[\frac{a_{Co^{2+}}}{a_{Co}}\right]$ at the cathode indicates that $a_{Co} \gg a_{Co^{2+}}$, hence cobalt will deposit at the cathode.

Problem 12.7

If 0.5 g of chromium is deposited in 1 h on passing 10 A current through a chromium electrolytic plating bath containing Cr^{6+} ions, what is the current efficiency of the cell? Atomic weight of chromium is 52.

Solution

According to Faraday's law of electrolysis, the theoretical mass deposited can be calculated as:

$$m = zct = \frac{52}{6 \times 96500} \times 10 \times (60 \times 60) = 3.23 \text{ g}$$

$$\text{current efficiency} = \frac{\text{actual weight deposited}}{\text{theoretical weight to be deposited}}$$

$$= \frac{0.5}{3.23} \times 100 = 15.5\% \text{ Ans.}$$

Problem 12.8

An electrolytic tank for refining copper has 41 anodes and 40 cathodes, each of 1 m^2 cross-sectional area. If they are arranged in parallel and the cells are operated at a current density of 200 A m^{-2} with a current efficiency of 90%, calculate the amount of refined copper deposited in 24 h. Atomic weight of copper is 63.54.

Solution

The amount of copper deposited in 24 h on the surface of 1 m^2

$$= \text{current efficiency} \times z.c.t$$

$$= \frac{90}{100} \times \left(\frac{63.54}{2 \times 96500} \right)(200 \times 1)(24 \times 60 \times 60)$$

$$= 5120 \text{ g m}^{-2}$$

$$\text{Total cathode area} = 40 \times 2 \times 1 = 80 \text{ m}^2$$

$$\therefore \text{Total amount of copper deposited} = \frac{5120 \times 80}{1000} = 409.6 \text{ kg} \quad \text{Ans.}$$

Problem 12.9

Copper is electrowon from $CuSO_4$ solution at 40 °C with a current efficiency of 86% at an applied voltage of 2.0 V. Calculate the energy efficiency and the energy consumption of the electrolytic cell that produces copper.

Given that $Cu^{2+} + H_2O = Cu + 2H^+ + \frac{1}{2} O_2$, $\Delta G^o_{313} = 170 \text{ kJ}$

Solution

$$E^o = -\frac{\Delta G^o_{313}}{RT} = -\frac{170000}{2 \times 96500} = -0.88 \text{ V}$$

The positive value of the free energy (hence the negative value of emf) suggests that the electrodeposition of copper ($Cu^{2+} + 2e = Cu$) will not take place spontaneously. The negative value of -0.88 V is a measure of the emf, which has to be applied to allow the given reaction ($Cu^{2+} + H_2O = Cu + 2H^+ + \frac{1}{2} O_2$) to occur. Hence, from Eq. 12.29,

$$\text{Energy efficiency } (\%) = \text{current efficiency } (\%) \times \frac{\text{theoretical voltage}}{\text{applied voltage}}$$

$$= 86 \times \frac{0.88}{2.0} = 37.8\% \text{ Ans.}$$

From Eq. 12.30,

$$\text{Energy consumption } \left(kWh\ kg^{-1}\right) = \frac{E.F}{\left(M_{/z}\right) \times 3600 \times 0.01(CE)}$$

$$= \frac{2 \times 96500}{\left(\dfrac{64}{2}\right) \times 3600 \times 0.01 \times 86} = 1.95 \text{ Ans.}$$

Problem 12.10
Assuming complete dissociation of $CuSO_4$ and H_2SO_4 in the electrolyte containing 0.6 m $CuSO_4$ and 1.5 m H_2SO_4 in Problem 12.9, calculate the percent error involved in setting activity = molality.

Solution
On complete dissociation of $CuSO_4$ and H_2SO_4 in the electrolyte,

$$a_{Cu^{2+}} = m_{Cu^{2+}} \text{ and } a_{H^+} = m_{H^+}$$
$$\text{and } m_{Cu^{2+}} = m_{CuSO_4}, m_{H^+} = 2m_{H_2SO_4}$$

For the reaction $Cu^{2+} + H_2O = Cu + 2H^+ + \frac{1}{2} O_2$ (refer the relevant equation on page 537)

$$E = E^o_{Cu} + \frac{RT}{nF} \ln \left[\frac{a_{Cu^{2+}} \cdot a^2_{H^+} \cdot p^{\frac{1}{2}}_{O_2}}{a_{Cu} \cdot a_{H_2O}} \right],$$

$a_{H_2O} = 1 = a_{Cu}, p_{O_2} = 1$ atm, $E^o_{Cu} = 0.88$ (from the Problem 12.9), $a_{H^+} = m_{H^+} = 2 \times 1.5, a_{Cu^{2+}} = m_{Cu^{2+}} = 0.6.$

$$\therefore E = 0.88 + \frac{8.314 \times 313}{2 \times 96500} \ln \left[0.6(2 \times 1.5)^2 \right] = 0.903 \text{ V}$$

According to Faraday's law, $(64/2 = 32)$ g Cu will be deposited on the cathode, hence the current required for the deposition of 1 kg Cu $= \frac{96500}{60 \times 60} \times \frac{1}{32} = 0.838\,A$
The theoretical energy consumption $= 0.838 \times 0.903 = 0.756$ kWh kg^{-1}.
Since the cell operates at 2.0 V with a current efficiency of 86%,
The energy consumption would be $= 0.756 \times \frac{100}{86} \times 2 = 1.759$ kWh kg^{-1}.

$$\text{The energy efficiency} = \frac{0.756}{1.759} \times 100 = 43\%$$

$$\%\text{error} = \frac{43 - 37.8}{43} \times 100 = 12.1\% \text{ Ans.}$$

References

1. Wells, J. A., Hopkins, W. R., & Stein, R. B. (1992). Chemical and electrolytic processing. In H. L. Hartman (Ed.), *SME mining engineering handbook* (pp. 2250–2258). New York: Society of Mining Engineers of AIME. (Chapter 25.4.3).
2. Moats, M., & Free, M. (2007). A bright future of copper electrometallurgy. *JOM, 60*(10), 34–36.
3. Robinson, T., Moats, M., Davenport, W., Karcas, G., & Demetrio, S. (2007). Electrolytic copper electrowinning – 2007 world tankhouse operating data. In G. E. Houlachi, J. D. Edwards, & T. G. Robinson (Eds.), *Copper – Cobre 2007 international conference. Vol. V* (pp. 375–424). Montreal: The Metallurgical Society of CIM.
4. Ginata, M. V. (2003, June 27–28). Titanium electrowinning. In *International symposium on ionic liquids*, Cary ie Rouet, France, pp. 1–15.
5. Ginata, M. V. (2000). Why produce titanium by EW? *JOM, 52*(5), 18–20.
6. Zhang, Y., Fang, Z. Z., Sun, P., Zheng, S., Xia, Y., & Free, M. (2017). A perspective on thermochemical and electrochemical processes for titanium metal production. *JOM, 69*(10), 1861–1868.
7. Stefanidaki, E., Hasiotis, C., & Kontoyannis, C. (2001). Electrodeposition of neodymium from LiF-NdF$_3$-Nd$_2$O$_3$ melts. *Electrochimica Acta, 46*(17), 2665–2670.
8. Fray, D., & Schwandt, C. (2017). Aspects of the application of electrochemistry to the extraction of titanium and its applications. *Metallurgical and Materials Transactions B: Process Metallurgy and Materials Processing Science, 58*(3), 306–312.
9. Abbasalizadeh, A., Malfliet, A., Seetharaman, S., Sietsma, J., & Yang, Y. (2017). Electrochemical extraction of rare earth metals in molten fluorides; conversion of rare earth oxides into rare earth fluorides using fluoride additives. *Journal of Sustainable Metallurgy, 3*(4), 627–637.
10. Jackson, E. (1986). *Hydrometallurgical extraction and reclamation*. New York: Ellis Horwood Limited/Wiley. (Chapter 5).
11. Bewer, G., Debrodt, H., & Herbst, H. (1982). Titanium for electrochemical processes. *JOM, 34*(1), 37–41.
12. Moats, M., Hardee, K., & Brown, C., Jr. (2003). Copper electrowinning in cobalt-free electrolyte using mesh on lead anodes. In J. E. Dutrizac & C. Clement (Eds.), *Proceedings of the copper–cobre 2003 international conference. Vol. V* (pp. 543–553). Warrendale: TMS.
13. Moats, M., Hardee, K., & Brown, C., Jr. (2003). Mesh-on-lead anodes for copper electrowinning. *JOM, 55*(7), 46–48.

14. Hardee, K., & Moats, M. (2003). Application of titanium mesh-on-lead technology to metal electrowinning systems. In R. Woods & F. M. Doyle (Eds.), *Electrochemistry in mineral and metal processing* (pp. 294–302). Pennington: The Electrochemical Society.

15. Grjotheim, K., & Welch, B. J. (1987). *Aluminium smelter technology* (2nd ed.). Dusseldorf: Aluminium-Verlag. (Chapter 4).

16. Thonstad, J., Fellner, P., Haarberg, G. M., Hives, J., Kvande, H., & Sterten, A. (2001). *Aluminium electrolysis* (3rd ed.). Aluminium Verlag, Marketing & Kommunikation GmbH. (Chapter 7).

17. Sevryukov, N., Kuzmin, B., & Chelishchev, Y. (translated: Kuznetsov, B.) (1960). *General metallurgy*. Moscow: Peace Publishers. (Chapter 17).

18. Moore, J. J. (1990). *Chemical metallurgy* (2nd ed.). Oxford: Butterworth–Heinemann. (Chapter 6).

19. Parker, R. H. (1978). *An introduction to chemical metallurgy* (2nd ed.). Oxford: Pergamon. (Chapter 7).

20. Free, M. L., & Moats, M. S. (2014). Hydrometallurgy. In S. Seetharaman (Ed.), *Treatise of process metallurgy. Vol. 3, Industrial process, Part A* (pp. 949–982). Oxford: Elsevier. (Chapter 2.7).

21. Jenkins, J., et al. (1999). Electrolytic copper – Leach, solvent extraction and electrowinning world operating. In S. K. Young et al. (Eds.), *Copper 99–Cobre 99 – Fourth international conference, Vol. 4, Hydrometallurgy of copper* (pp. 493–566). Warrendale: TMS.

22. Davenport, D. W., King, M., Schlesinger, M., & Biswas, A. K. (2002). *Extractive metallurgy of copper* (4th ed.). Oxford: Pergamon. (Chapter 16).

23. Gilchrist, J. D. (1980). *Extraction metallurgy* (2nd ed.). New York: Pergamon. (Chapter 12).

24. Kerby, R. C., & Krauss, C. J. (1980). Continuous monitoring of zinc electrolyte quality at Cominco cathodic by overpotential measurements. In J. M. Cigan, T. S. Mackay, & T. J. O'Keefe (Eds.), *Lead-Zinc-Tin '80* (pp. 187–193). Las Vegas: TMS-AIME.

25. Moats, M., Guerra, E., & Gonzalez, J. A. (2008). Zinc electrowinning – Operating data. In L. Centomo, M. Collins, J. Harlamovs, & J. Liu (Eds.), *Zinc and lead metallurgy* (pp. 307–314). Montréal: METSOC.

26. Shamsuddin, M., & Sohn, H. Y. (2019). Constitutive topics in physical chemistry of high-temperature nonferrous metallurgy – A review: Part 1. Sulfide roasting and smelting. *JOM, 71* (9), 2353–2365.

27. Tabereaux, A. T., & Peterson, R. D. (2014). Aluminum production. In S. Seetharaman (Ed.), *Treatise of process metallurgy, non ferrous production technology. Vol. 3, Industrial process, Part A* (pp. 839–917). Oxford: Elsevier (Chapter 2.5).

28. Marks, J. (1998). PFC emission measurements from Alcoa aluminum smelters. In *Light metals* (pp. 287–291). Warrendale: The Minerals, Metals & Materials Society.

29. Leber, B. P., Jr., Tabereaux, A. T., Marks, J., Lamb, H., Howard, T., Kantamaneni, R., Gibbs, M., Bakshi, V., & Dolin, E. J. (1998). Perfluorocarbon (PFC) generation at primary aluminium smelters. In *Light metals* (pp. 277–285). Warrendale: The Minerals, Metals & Materials Society.

30. Bouzat, G., Carraz, J. C., & Meyer, M. (1996). Measurement of CF_4 and C_2F_6 emissions from prebake pots. In W. Hale (Ed.), *Light metals* (pp. 413–4417). Warrendale: The Minerals, Metals & Materials Society.

31. Tabereaux, A. T., Richards, N. E., & Satchel, C. E. (1995). Composition of reduction cell anode gas during normal conditions and anode effects. In U. Mannweiler (Ed.), *Light metals* (pp. 325–333). Warrendale: The Minerals, Metals & Materials Society.

32. Roberts, R. A., & Ramsey, P. J. (1994). Evaluation of fluorocarbon emissions from the aluminum smelting process. In U. Mannweiler (Ed.), *Light metals* (pp. 31–388). Warrendale: The Minerals, Metals & Materials Society.

33. Mantell, C. L. (1960). *Electrochemical engineering*. New York: McGraw-Hill. (Chapter 5).

34. Takeda, O., Uda, T., & Okabe, T. H. (2014). Rare earth, titanium group metal, and reactive metal production. In S. Seetharaman (Ed.), *Treatise on process metallurgy, non ferrous*

production technology. Vol. 3, Industrial process, Part A (pp. 995–1069). Oxford: Elsevier (Chapter 2.9).

35. Jamrack, W. D. (1963). *Rare metal extraction by chemical engineering techniques.* New York: Macmillan. (Chapter 7).

36. Fray, D. J. (2008). Novel methods for the production of titanium. *International Materials Review, 53,* 317–325.

37. Fang, Z. Z., Paramore, J. D., Sun, P., Ravichandran, K. S., Zhang, Y., Xia, Y., Cao, F., Koopman, M., & Free, M. (2018). Powder metallurgy of titanium-past, present and future. *International Materials Review, 63,* 407–459.

38. Fray 17. Wainer, E. (1959). *Cell feed material for the production of titanium.* U. S. Patent 2,868,703.

39. Withers, J. C., Loutfy, R. O., & Laughtin, P. (2007). Electrolytic process to produce titanium from TiO_2 feed. *Materials and Technologies, 22,* 66–70.

40. Withers, J., Laughlin, J., Elkadi, Y., DeSilva, J., & Loutfy, R. (2010). The electrolytic production of Ti from a TiO_2 feed. *Key Engineering Materials, 436,* 61–74.

41. Jiao, S. Q., & Zhu, H. M. (2006). Novel metallurgical process for titanium production. *Journal of Materials Research, 21*(9), 2172–2175.

42. Jiao, S. Q., & Zhu, H. M. (2006). Electrolysis of Ti_2CO solid solution prepared by TiC and TiO_2. *The Journal of Alloys and Compounds, 438,* 243–246.

43. Tang, D., Xiao, W., Tian, L., & Wang, D. (2013). Electrosynthesis of Ti_2CO_n from TiO_2/C composite in molten $CaCl_2$: Effect of electrolysis voltage and duration. *Journal of the Electrochemical Society, 160,* 259–2866.

44. Kjos, O. S., Haarberg, G. M., & Martinez, A. M. (2010). Electrochemical production of titanium from oxycarbide anodes. *Key Engineering Materials, 436,* 93–101.

45. Wang, Q. Y., Song, J., Wu, J., Jiao, S. Q., Hou, J., & Zhu, H. M. (2014). A new consumable anode material of titanium oxycarbonitride for the USTB titanium process. *Physical Chemistry Chemical Physics, 16,* 8086–8091.

46. Jewell, D., Jiao, S. Q., Kurtanjek, M., & Fray, D. J. (2012). *Titanium metal production via oxycarbide electrorefining.* International Titanium Association.

47. Okabe, T. H., Nakamura, M., Oishi, T., & Ono, K. (1993). Electrochemical deoxidation of titanium. *Metallurgical Transactions B, 24,* 449–455.

48. Fray, D.J., Farthing, T.W., & Chen, Z. (1999). *Removal of oxygen from metal oxides.* Patent, International Publication No: WO9964638.

49. Chen, G. Z., Fray, D. J., & Farthing, T. W. (2001). Cathodic deoxygenation of the alpha case on titanium and alloys in molten calcium chloride. *Metallurgical and Materials Transactions B: Process Metallurgy and Materials Processing Science, 32B,* 1041–1052.

50. Suzuki, R. O., Ono, K., & Teranuma, K. (2003). Calciothermic reduction of titanium oxide and *in-situ* electrolysis in molten $CaCl_2$. *Metallurgical and Materials Transactions B: Process Metallurgy and Materials Processing Science, 34,* 287–295.

51. Schwandt, C., Alexander, D. T. L., & Fray, D. J. (2009). The electro-deoxidation of porous titanium. *Electrochimica Acta, 54,* 3819–3829.

52. Park, I., Abiko, T., & Okabe, T. H. (2005). Production of titanium powder directly from TiO_2 in $CaCl_2$ through an electronically mediated reaction (EMR). *Journal of Physics and Chemistry of Solids, 66*(2–40), 410–413.

53. Chen, G. Z., Fray, D. J., & Farthing, T. W. (2000). Direct lectrochemical reduction of titanium dioxide to titanium in molten calcium chloride. *Nature, 407,* 361–364. https://doi.org/10.1038/35030069.

54. Krishnan, A., Pal, U. B., & Lu, X. G. (2005). Solid oxide membrane process for magnesium production directly from magnesium oxide. *Metallurgical and Materials Transactions B: Process Metallurgy and Materials Processing Science, 36B,* 463–473.

55. Suput, M., Delucas, R., Pati, S., Ye, G., Pal, A., & Powell, A. C. (2008). Solid oxide membrane technology for environmentally sound production of titanium. *Transactions of the Institution of*

Mining and Metallurgy, Section C, Mineral Processing and Extractive Metallurgy Review, 117, C118–C122.

56. Han, W., Chen, Q., Sun, Y., & Jiang, T. (2011). Electrodeposition of Mg-Li-Al-La alloys on inert cathode in molten LiCl-KCl eutectic salt. *Metallurgical and Materials Transactions B: Process Metallurgy and Materials Processing Science, 42B,* 1367–1375.

57. Ye, X. S., GangLu, X., Wei, C. H. L., Xing, Z. D., Yong, L. Z., Qing, H. G., & Zhong, D. (2011). Preparation of Ti–Fe based hydrogen storage alloy by SOM method. *International Journal of Hydrogen Energy, 36,* 4573–4579.

58. Zou, X., Lu, X., Zhou, Z., Xiao, W., Zhong, Q., Li, C., & Ding, W. (2014). Electrochemical extraction of Ti_5Si_3 silicide from multicomponent Ti/Si – Containing metal oxide compounds in molten salt. *Journal of Materials Chemistry A, 2,* 7421–7431.

59. Abbasalizadeh, A., Seetharaman, S., Venkatesan, P., Sietsma, J., & Yang, Y. (2019). Novel reactive anode for electrochemical extraction of rare earth metals from rare earth oxides. *Electrochemica Acta, 310,* 146–152. https://doi.org/10.1016/j.electacta.2019.03.161.

60. Hart, P. F., & Hill, A. W. D. (1971). Electrorefining in molten salts. In A. T. Kuhn (Ed.), *Industrial electrochemical processes* (pp. 245–254). Oxford: Elsevier.

61. Kubaschewski, O., & Alcock, C. B. (1979). *Metallurgical thermochemistry* (5th ed.). Oxford: Pergamon.

Appendixes

In this section flow sheets for extraction of copper, lead, nickel, zinc, tungsten, beryllium, uranium, thorium, titanium, zirconium, aluminum and magnesium are respectively, presented in Appendixes A.1–A.12.

Appendix B Recommended physical constants

Appendix C Conversion factors.

A.1: Flow Sheet for Extraction of Copper from Chalcopyrite

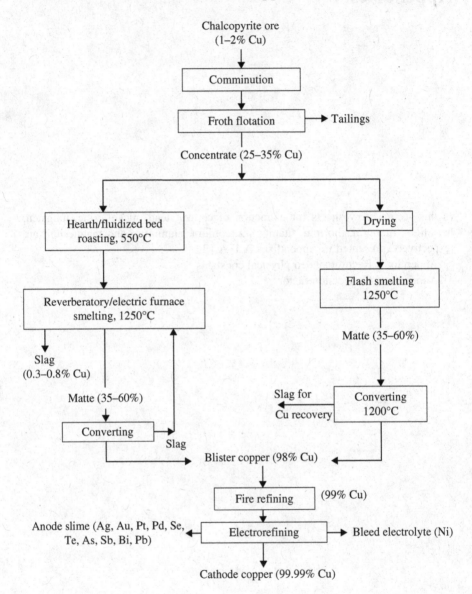

A.2: Flow Sheet for Extraction of Lead from Galena

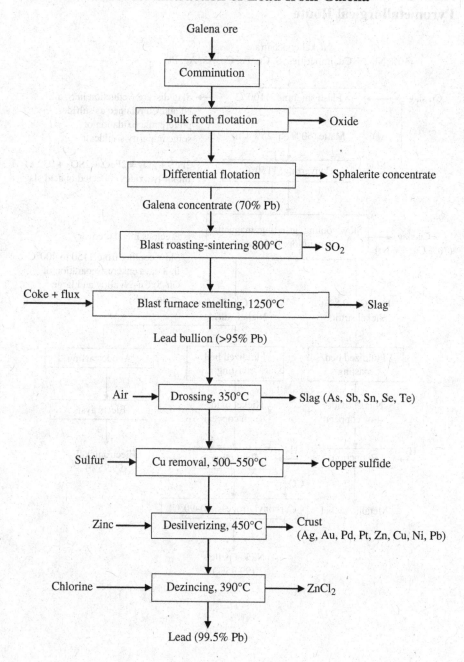

Galena ore

Comminution

Bulk froth flotation → Oxide

Differential flotation → Sphalerite concentrate

Galena concentrate (70% Pb)

Blast roasting-sintering 800°C → SO₂

Coke + flux → Blast furnace smelting, 1250°C → Slag

Lead bullion (>95% Pb)

Air → Drossing, 350°C → Slag (As, Sb, Sn, Se, Te)

Sulfur → Cu removal, 500–550°C → Copper sulfide

Zinc → Desilverizing, 450°C → Crust (Ag, Au, Pd, Pt, Zn, Cu, Ni, Pb)

Chlorine → Dezincing, 390°C → ZnCl₂

Lead (99.5% Pb)

A.3a: Flow Sheet for Extraction of Nickel from Pentlandite Via Pyrometallurgical Route

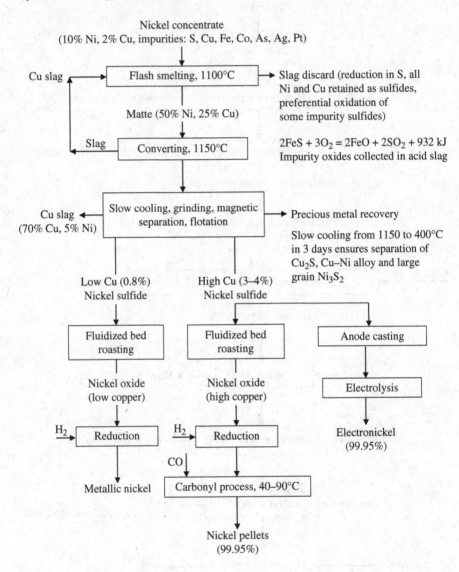

Nickel concentrate
(10% Ni, 2% Cu, impurities: S, Cu, Fe, Co, As, Ag, Pt)

Flash smelting, 1100°C → Slag discard (reduction in S, all Ni and Cu retained as sulfides, preferential oxidation of some impurity sulfides)

Cu slag

Matte (50% Ni, 25% Cu)

Slag

Converting, 1150°C

$2FeS + 3O_2 = 2FeO + 2SO_2 + 932$ kJ
Impurity oxides collected in acid slag

Slow cooling, grinding, magnetic separation, flotation → Precious metal recovery

Cu slag (70% Cu, 5% Ni)

Slow cooling from 1150 to 400°C in 3 days ensures separation of Cu_2S, Cu–Ni alloy and large grain Ni_3S_2

Low Cu (0.8%) Nickel sulfide

High Cu (3–4%) Nickel sulfide

Fluidized bed roasting

Fluidized bed roasting

Anode casting

Nickel oxide (low copper)

Nickel oxide (high copper)

Electrolysis

H_2 → Reduction

H_2 → Reduction

CO

Electronickel (99.95%)

Metallic nickel

Carbonyl process, 40–90°C

Nickel pellets (99.95%)

A.3b: Flow Sheet for Extraction of Nickel from Pentlandite Via Hydrometallurgical Route

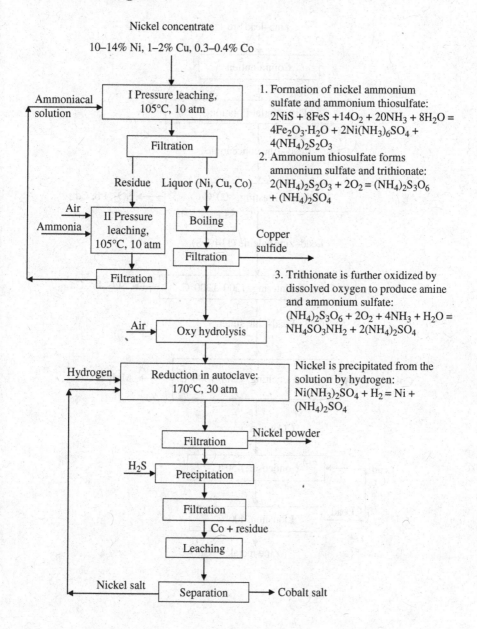

Nickel concentrate

10–14% Ni, 1–2% Cu, 0.3–0.4% Co

I Pressure leaching, 105°C, 10 atm

Ammoniacal solution

Filtration

Residue   Liquor (Ni, Cu, Co)

Air

Ammonia

II Pressure leaching, 105°C, 10 atm

Boiling

Filtration

Copper sulfide

Filtration

Air

Oxy hydrolysis

Hydrogen

Reduction in autoclave: 170°C, 30 atm

Filtration   Nickel powder

H₂S

Precipitation

Filtration

Co + residue

Leaching

Nickel salt

Separation   → Cobalt salt

1. Formation of nickel ammonium sulfate and ammonium thiosulfate:
$$2NiS + 8FeS + 14O_2 + 20NH_3 + 8H_2O = 4Fe_2O_3 \cdot H_2O + 2Ni(NH_3)_6SO_4 + 4(NH_4)_2S_2O_3$$

2. Ammonium thiosulfate forms ammonium sulfate and trithionate:
$$2(NH_4)_2S_2O_3 + 2O_2 = (NH_4)_2S_3O_6 + (NH_4)_2SO_4$$

3. Trithionate is further oxidized by dissolved oxygen to produce amine and ammonium sulfate:
$$(NH_4)_2S_3O_6 + 2O_2 + 4NH_3 + H_2O = NH_4SO_3NH_2 + 2(NH_4)_2SO_4$$

Nickel is precipitated from the solution by hydrogen:
$$Ni(NH_3)_2SO_4 + H_2 = Ni + (NH_4)_2SO_4$$

A.4a: Flow Sheet for Extraction of Zinc from Sphalerite Via Pyrometallurgical Route

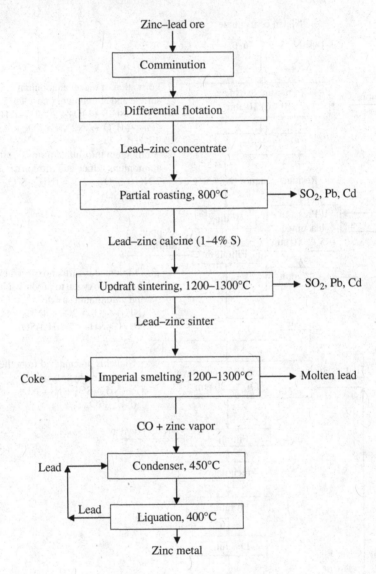

A.4b: Flow Sheet for Extraction of Zinc from Sphalerite Via Hydrometallurgical Route

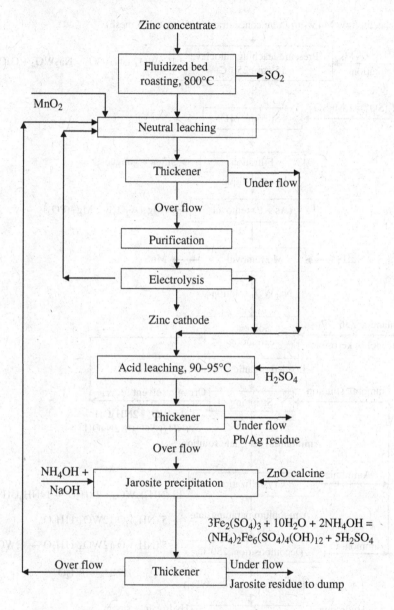

A.5: Flow Sheet for Extraction of Tungsten from Scheelite/Wulframite

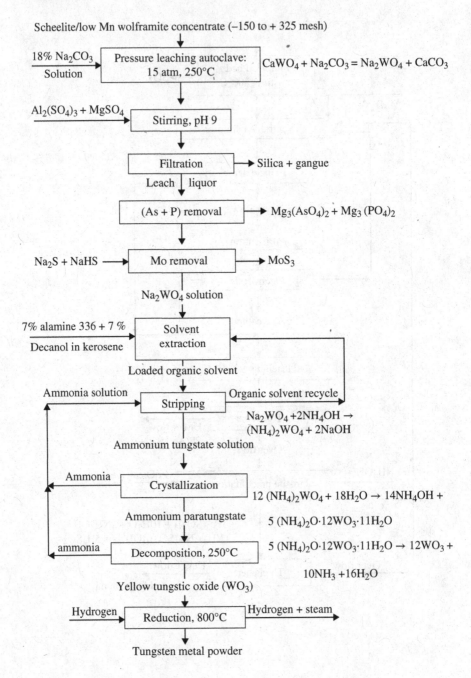

Scheelite/low Mn wolframite concentrate (−150 to + 325 mesh)

$\dfrac{18\% \, Na_2CO_3}{Solution}$ → Pressure leaching autoclave: 15 atm, 250°C → $CaWO_4 + Na_2CO_3 = Na_2WO_4 + CaCO_3$

$Al_2(SO_4)_3 + MgSO_4$ → Stirring, pH 9

Filtration → Silica + gangue

Leach liquor

(As + P) removal → $Mg_3(AsO_4)_2 + Mg_3(PO_4)_2$

$Na_2S + NaHS$ → Mo removal → MoS_3

Na_2WO_4 solution

$\dfrac{7\% \, alamine \, 336 + 7\%}{Decanol \, in \, kerosene}$ → Solvent extraction

Loaded organic solvent

Ammonia solution → Stripping → Organic solvent recycle

$Na_2WO_4 + 2NH_4OH \rightarrow (NH_4)_2WO_4 + 2NaOH$

Ammonium tungstate solution

Ammonia ← Crystallization

$12 \, (NH_4)_2WO_4 + 18H_2O \rightarrow 14NH_4OH +$

$5 \, (NH_4)_2O \cdot 12WO_3 \cdot 11H_2O$

Ammonium paratungstate

ammonia ← Decomposition, 250°C

$5 \, (NH_4)_2O \cdot 12WO_3 \cdot 11H_2O \rightarrow 12WO_3 +$

$10NH_3 + 16H_2O$

Yellow tungstic oxide (WO_3)

Hydrogen → Reduction, 800°C → Hydrogen + steam

Tungsten metal powder

A.6: Flow Sheet for Extraction of Beryllium from Beryl

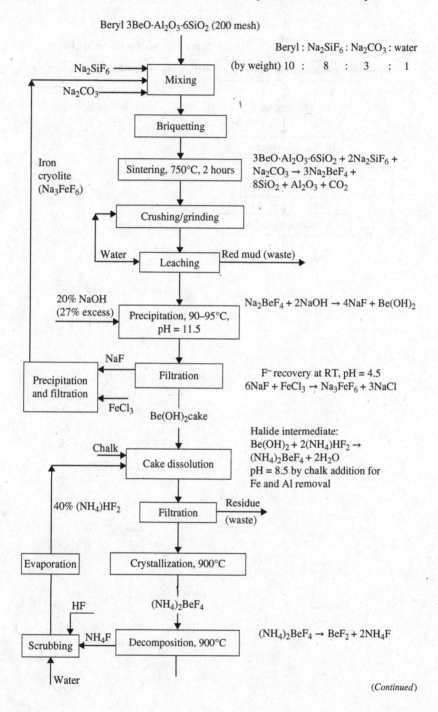

Beryl $3BeO \cdot Al_2O_3 \cdot 6SiO_2$ (200 mesh)

Beryl : Na_2SiF_6 : Na_2CO_3 : water
(by weight) 10 : 8 : 3 : 1

Na_2SiF_6 →

Na_2CO_3 →

Mixing

Briquetting

Iron cryolite (Na_3FeF_6)

Sintering, 750°C, 2 hours

$3BeO \cdot Al_2O_3 \cdot 6SiO_2 + 2Na_2SiF_6 + Na_2CO_3 \rightarrow 3Na_2BeF_4 + 8SiO_2 + Al_2O_3 + CO_2$

Crushing/grinding

Water → Leaching → Red mud (waste)

20% NaOH (27% excess) → Precipitation, 90–95°C, pH = 11.5

$Na_2BeF_4 + 2NaOH \rightarrow 4NaF + Be(OH)_2$

NaF

Precipitation and filtration

Filtration

F^- recovery at RT, pH = 4.5
$6NaF + FeCl_3 \rightarrow Na_3FeF_6 + 3NaCl$

$FeCl_3$

$Be(OH)_2$ cake

Chalk → Cake dissolution

Halide intermediate:
$Be(OH)_2 + 2(NH_4)HF_2 \rightarrow (NH_4)_2BeF_4 + 2H_2O$
pH = 8.5 by chalk addition for Fe and Al removal

40% $(NH_4)HF_2$ → Filtration → Residue (waste)

Evaporation

Crystallization, 900°C

HF

$(NH_4)_2BeF_4$

Scrubbing ← NH_4F ← Decomposition, 900°C

$(NH_4)_2BeF_4 \rightarrow BeF_2 + 2NH_4F$

Water

(Continued)

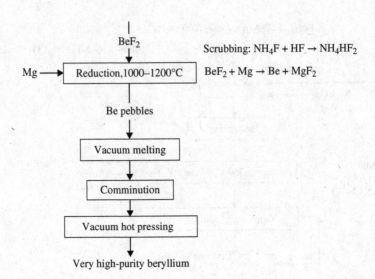

Very high-purity beryllium

A.7: Flow Sheet for Extraction of Uranium from Pitchblende

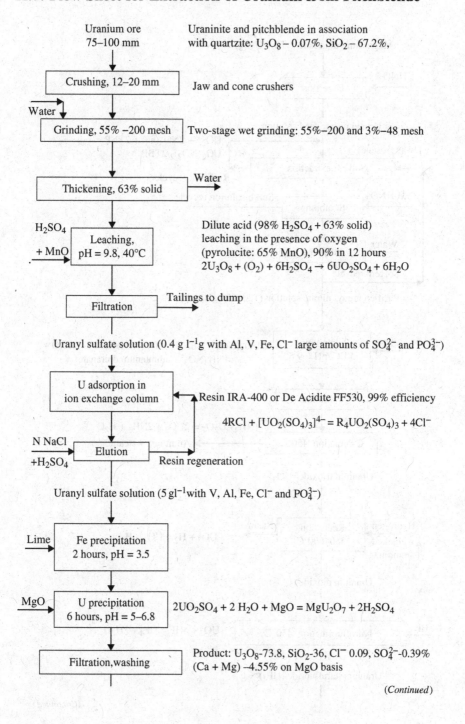

Uranium ore 75–100 mm — Uraninite and pitchblende in association with quartzite: $U_3O_8 - 0.07\%$, $SiO_2 - 67.2\%$,

Crushing, 12–20 mm — Jaw and cone crushers

Water

Grinding, 55% –200 mesh — Two-stage wet grinding: 55%–200 and 3%–48 mesh

Thickening, 63% solid — Water

H_2SO_4 + MnO

Leaching, pH = 9.8, 40°C — Dilute acid (98% H_2SO_4 + 63% solid) leaching in the presence of oxygen (pyrolucite: 65% MnO), 90% in 12 hours
$$2U_3O_8 + (O_2) + 6H_2SO_4 \rightarrow 6UO_2SO_4 + 6H_2O$$

Filtration — Tailings to dump

Uranyl sulfate solution (0.4 g l^{-1}g with Al, V, Fe, Cl^- large amounts of SO_4^{2-} and PO_4^{3-})

U adsorption in ion exchange column — Resin IRA-400 or De Acidite FF530, 99% efficiency
$$4RCl + [UO_2(SO_4)_3]^{4-} = R_4UO_2(SO_4)_3 + 4Cl^-$$

N NaCl +H_2SO_4

Elution — Resin regeneration

Uranyl sulfate solution (5 g$l^{-1}$with V, Al, Fe, Cl^- and PO_4^{3-})

Lime

Fe precipitation 2 hours, pH = 3.5

MgO

U precipitation 6 hours, pH = 5–6.8 — $2UO_2SO_4 + 2 H_2O + MgO = MgU_2O_7 + 2H_2SO_4$

Filtration, washing — Product: U_3O_8-73.8, SiO_2-36, Cl^- 0.09, SO_4^{2-}-0.39% (Ca + Mg) –4.55% on MgO basis

(*Continued*)

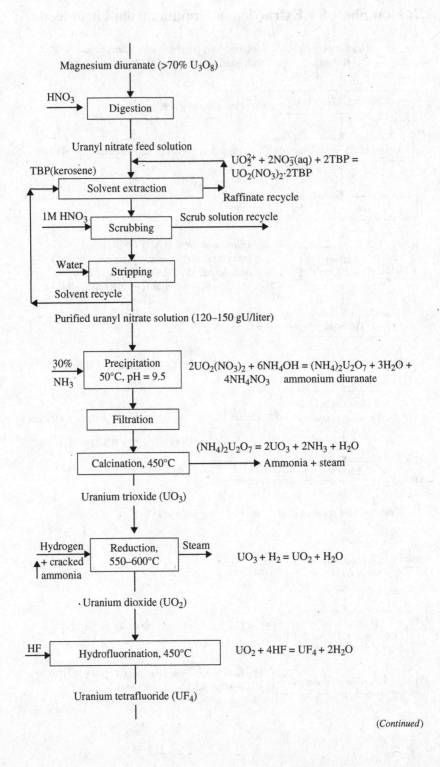

Magnesium diuranate (>70% U_3O_8)

HNO$_3$ → **Digestion**

Uranyl nitrate feed solution

$UO_2^{2+} + 2NO_3^-(aq) + 2TBP =$
$UO_2(NO_3)_2 \cdot 2TBP$

TBP(kerosene)

Solvent extraction

Raffinate recycle

1M HNO$_3$ → **Scrubbing** → Scrub solution recycle

Water → **Stripping**

Solvent recycle

Purified uranyl nitrate solution (120–150 gU/liter)

$\dfrac{30\%}{NH_3}$ → **Precipitation 50°C, pH = 9.5**

$2UO_2(NO_3)_2 + 6NH_4OH = (NH_4)_2U_2O_7 + 3H_2O +$
$4NH_4NO_3$ ammonium diuranate

Filtration

$(NH_4)_2U_2O_7 = 2UO_3 + 2NH_3 + H_2O$

Calcination, 450°C → Ammonia + steam

Uranium trioxide (UO_3)

Hydrogen + cracked ammonia → **Reduction, 550–600°C** → Steam

$UO_3 + H_2 = UO_2 + H_2O$

Uranium dioxide (UO_2)

HF → **Hydrofluorination, 450°C**

$UO_2 + 4HF = UF_4 + 2H_2O$

Uranium tetrafluoride (UF_4)

(Continued)

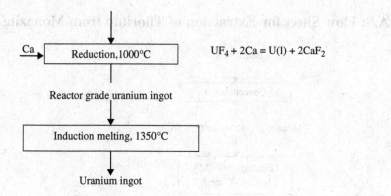

$$UF_4 + 2Ca = U(l) + 2CaF_2$$

A.8: Flow Sheet for Extraction of Thorium from Monazite

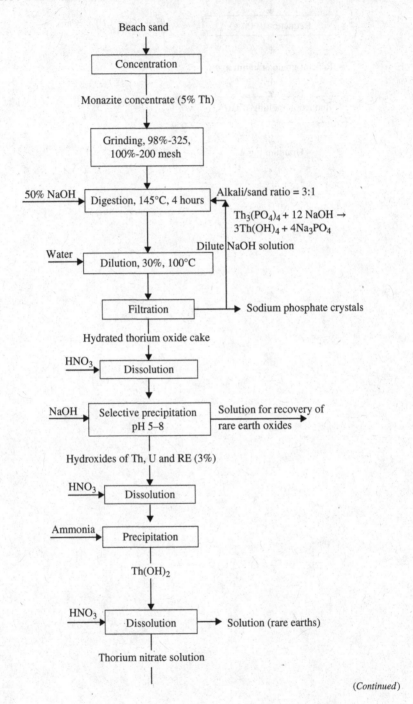

Beach sand

Concentration

Monazite concentrate (5% Th)

Grinding, 98%-325, 100%-200 mesh

50% NaOH → Digestion, 145°C, 4 hours ← Alkali/sand ratio = 3:1

$$Th_3(PO_4)_4 + 12 \, NaOH \rightarrow 3Th(OH)_4 + 4Na_3PO_4$$

Dilute NaOH solution

Water → Dilution, 30%, 100°C

Filtration → Sodium phosphate crystals

Hydrated thorium oxide cake

HNO₃ → Dissolution

NaOH → Selective precipitation pH 5–8 → Solution for recovery of rare earth oxides

Hydroxides of Th, U and RE (3%)

HNO₃ → Dissolution

Ammonia → Precipitation

$Th(OH)_2$

HNO₃ → Dissolution → Solution (rare earths)

Thorium nitrate solution

(Continued)

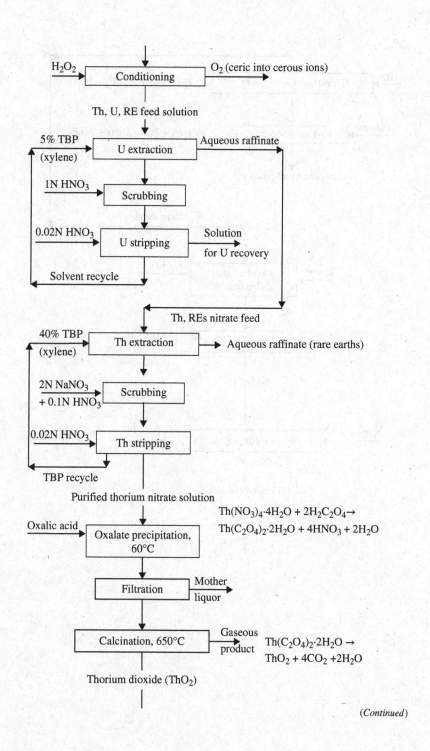

$$Th(NO_3)_4 \cdot 4H_2O + 2H_2C_2O_4 \rightarrow$$
$$Th(C_2O_4)_2 \cdot 2H_2O + 4HNO_3 + 2H_2O$$

$$Th(C_2O_4)_2 \cdot 2H_2O \rightarrow$$
$$ThO_2 + 4CO_2 + 2H_2O$$

(*Continued*)

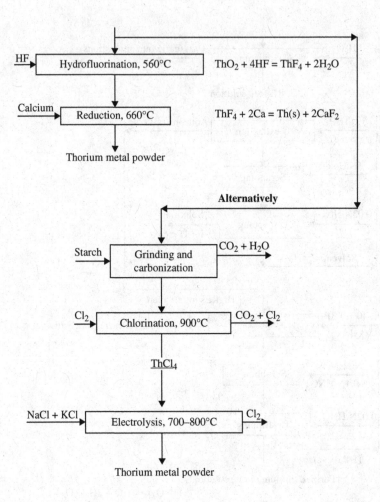

A.9: Flow Sheet for Extraction of Titanium from Rutile

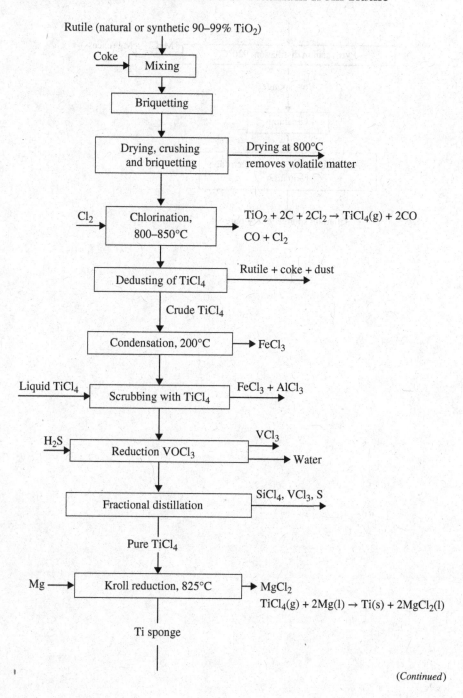

Rutile (natural or synthetic 90–99% TiO_2)

Coke → Mixing

Briquetting

Drying, crushing and briquetting → Drying at 800°C removes volatile matter

Cl_2 → Chlorination, 800–850°C → $TiO_2 + 2C + 2Cl_2 \rightarrow TiCl_4(g) + 2CO$
$CO + Cl_2$

Dedusting of $TiCl_4$ → Rutile + coke + dust

Crude $TiCl_4$

Condensation, 200°C → $FeCl_3$

Liquid $TiCl_4$ → Scrubbing with $TiCl_4$ → $FeCl_3 + AlCl_3$

H_2S → Reduction $VOCl_3$ → VCl_3 → Water

Fractional distillation → $SiCl_4$, VCl_3, S

Pure $TiCl_4$

Mg → Kroll reduction, 825°C → $MgCl_2$
$TiCl_4(g) + 2Mg(l) \rightarrow Ti(s) + 2MgCl_2(l)$

Ti sponge

(Continued)

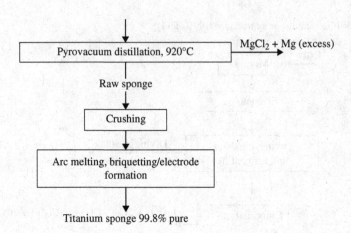

A.10: Flow Sheet for Extraction of Zirconium from Zircon Sand

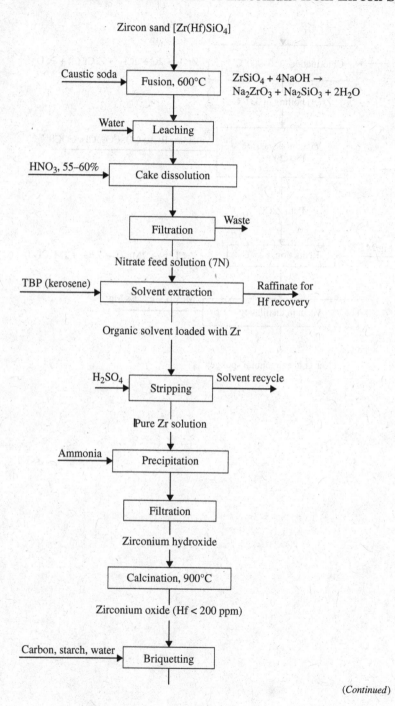

Zircon sand [Zr(Hf)SiO$_4$]

Caustic soda → Fusion, 600°C

$$ZrSiO_4 + 4NaOH \rightarrow Na_2ZrO_3 + Na_2SiO_3 + 2H_2O$$

Water → Leaching

HNO$_3$, 55–60% → Cake dissolution

Filtration → Waste

Nitrate feed solution (7N)

TBP (kerosene) → Solvent extraction → Raffinate for Hf recovery

Organic solvent loaded with Zr

H$_2$SO$_4$ → Stripping → Solvent recycle

Pure Zr solution

Ammonia → Precipitation

Filtration

Zirconium hydroxide

Calcination, 900°C

Zirconium oxide (Hf < 200 ppm)

Carbon, starch, water → Briquetting

(*Continued*)

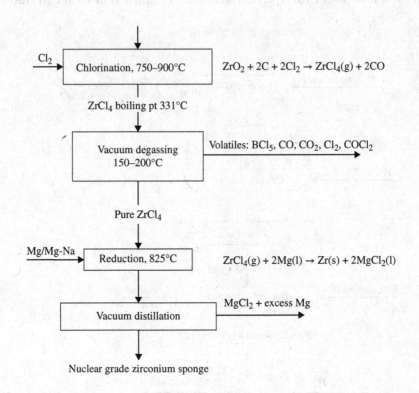

Cl$_2$

Chlorination, 750–900°C $ZrO_2 + 2C + 2Cl_2 \rightarrow ZrCl_4(g) + 2CO$

ZrCl$_4$ boiling pt 331°C

Vacuum degassing   Volatiles: BCl$_5$, CO, CO$_2$, Cl$_2$, COCl$_2$
150–200°C

Pure ZrCl$_4$

Mg/Mg-Na

Reduction, 825°C $ZrCl_4(g) + 2Mg(l) \rightarrow Zr(s) + 2MgCl_2(l)$

MgCl$_2$ + excess Mg

Vacuum distillation

Nuclear grade zirconium sponge

A.11: Flow Sheet for Extraction of Aluminum from Bauxite

Bauxite
(40–50% Al_2O_3, <5% SiO_2, 9.30% TiO_2, 4.30% Fe_2O_3)

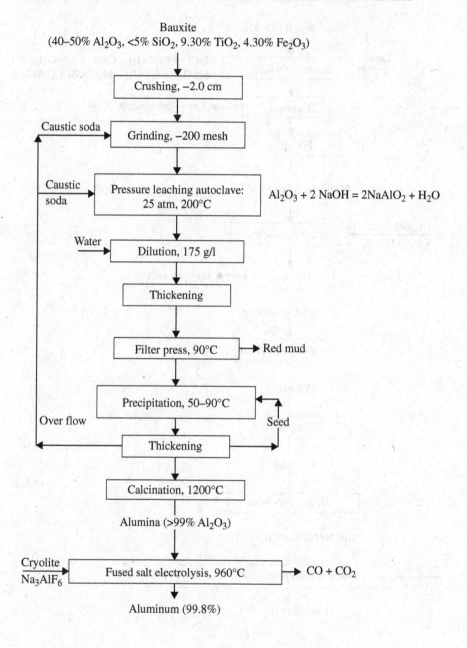

Crushing, −2.0 cm

Caustic soda

Grinding, −200 mesh

Caustic soda

Pressure leaching autoclave:
25 atm, 200°C

$$Al_2O_3 + 2\,NaOH = 2NaAlO_2 + H_2O$$

Water

Dilution, 175 g/l

Thickening

Filter press, 90°C → Red mud

Precipitation, 50–90°C

Over flow   Seed

Thickening

Calcination, 1200°C

Alumina (>99% Al_2O_3)

Cryolite
Na_3AlF_6 → Fused salt electrolysis, 960°C → $CO + CO_2$

Aluminum (99.8%)

A.12: Flow Sheet for Extraction of Magnesium from Sea Water

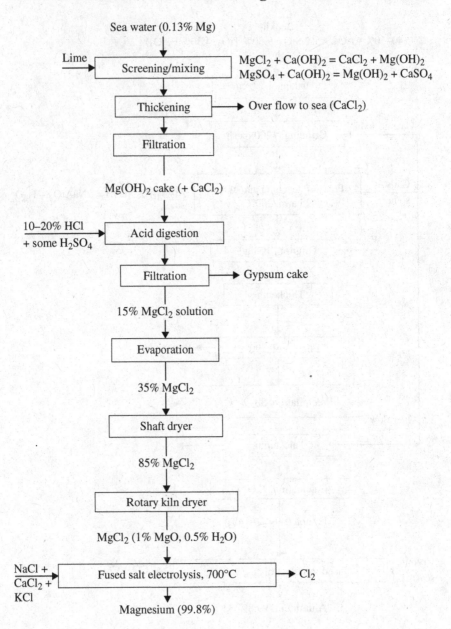

Sea water (0.13% Mg)

Lime → Screening/mixing

$MgCl_2 + Ca(OH)_2 = CaCl_2 + Mg(OH)_2$
$MgSO_4 + Ca(OH)_2 = Mg(OH)_2 + CaSO_4$

Thickening → Over flow to sea ($CaCl_2$)

Filtration

$Mg(OH)_2$ cake (+ $CaCl_2$)

10–20% HCl + some H_2SO_4 → Acid digestion

Filtration → Gypsum cake

15% $MgCl_2$ solution

Evaporation

35% $MgCl_2$

Shaft dryer

85% $MgCl_2$

Rotary kiln dryer

$MgCl_2$ (1% MgO, 0.5% H_2O)

NaCl + CaCl_2 + KCl → Fused salt electrolysis, 700°C → Cl_2

Magnesium (99.8%)

Appendix: B – Recommended Values of Physical Constants

Physical constant	Symbol	Value
Acceleration due to gravity	g	9.81 m s^{-2}
Avogadro number	N	6.02252×10^{23} molecules mol^{-1}
Boltzmann constant	k	1.38054×10^{-23} J K^{-1}
Faraday constant	F	96495 Coulombs mol^{-1} $= 23066$ cal V^{-1} mol^{-1}
Gas constant	R	1.987 cal K^{-1} mol^{-1} $= 8.3143$ J K^{-1} mol^{-1} $= 0.08206$ liter atm K^{-1} mol^{-1}
Planck constant	h	6.6256×10^{-34} J s

Appendix: C – SI Units and Conversion Factors

Physical Quantity	SI unit (a)	cgs unit (b)	conversion factor (from b to a)
Length	m	cm	10^{-2}
Mass	kg	g	10^{-3}
Time	s	s	1
Volume	m^3	cm^3	10^{-6}
Density	kg m^{-3}	g cm^{-3}	10^{3}
Force	$N = $ kg m s$^{-1} = $ J m^{-1}	dyne	10^{-5}
Pressure	N m$^{-2} = $ kg m^{-1}s$^{-2} = $ J m^{-3}	dyne cm^{-2}	10^{-1} (1 atm $= 760$ mm Hg $= 760$ Torr $= 1.013$ bar $= 1.033 \times 10^4$ kg. m^{-2} $= 1.013 \times 10^5$ N. m^{-1} $= 1.013 \times 10^5$ Pascal
Surface tension	N m$^{-1} = $ J m^{-2}	dyne cm^{-1}	10^{-3}
Energy, work	$J = $ kg m^{2}s^{-2}	cal erg	4.184 10^{-7} (1 J $= 0.102$ kg m)
Molar free energy, enthalpy	J mol^{-1}	cal mol^{-1}	4.184
Molar entropy, heat capacity	J K^{-1}mol^{-1}	cal K^{-1}mol^{-1}	4.184
Concentration	mol m^{-3}	mol l^{-1}	10^{3}
	mol dm^{-3}	mol l^{-1}	1
Molality	mol kg^{-1}	mol kg^{-1}	1
Ionic strength	mol kg^{-1} mol m^{-3} mol dm^{-3}	mol kg^{-1} mol l^{-1} mol l^{-1}	1 10^{3} 1

Index

Note: Page numbers in normal refer to subject contents; those in *italics* to Figures; those in **bold** to Tables.

© The Minerals, Metals & Materials Society 2021
M. Shamsuddin, *Physical Chemistry of Metallurgical Processes*, The Minerals,
Metals & Materials Series, https://doi.org/10.1007/978-3-030-58069-8

Printed in the United States
by Baker & Taylor Publisher Services